ION IMPLANTATION
IN SEMICONDUCTORS
1976

ION IMPLANTATION IN SEMICONDUCTORS 1976

Edited by
Fred Chernow
University of Colorado
Boulder, Colorado

and
James A. Borders and David K. Brice
Sandia Laboratories
Albuquerque, New Mexico

Plenum Press · New York and London

Library of Congress Cataloging in Publication Data

International Conference on Ion Implantation in Semiconductors and Other Materials, 5th, Boulder, Colo., 1976.
Ion implantation in semiconductors, 1976.

Includes bibliographical references and index.
1. Ion implantation—Congresses. 2. Semiconductors, Effect of radiation on—Congresses. I. Chernow, Fred, II. Borders, James A. III. Brice, David K. IV. Title.
TK7871.85.I5762 1976   621.3815'2   77-2980
ISBN-13: 978-1-4613-4198-7   e-ISBN-13: 978-1-4613-4196-3
DOI: 10.1007/ 978-1-4613-4196-3

Proceedings of the Fifth International Conference on Ion Implantation in Semiconductors and Other Materials, held in Boulder, Colorado, August 8-13, 1976

Conference Chairman: Fred Chernow

Program Committee: Fred H. Eisen, Wie Kan Chu, James F. Gibbons, David K. Brice, and James A. Borders

Local Arrangements: Graeme Eldridge

© 1977 Plenum Press, New York
Softcover reprint of the hardcover 1st edition 1977
A Division of Plenum Publishing Corporation
227 West 17th Street, New York, N.Y. 10011

All rights reserved

No part of this book may be reproduced, stored in a retrieval system, or transmitted, in any form or by any means, electronic, mechanical, photocopying, microfilming, recording, or otherwise, without written permission from the Publisher

PREFACE

The Fifth International Conference on Ion Implantation took place in Boulder, Colorado between the 9th and 13th of August 1976. Papers were delivered by scientists and engineers from 15 countries, and the attendees represented 19 countries. As has become the custom at these conferences, the sessions were intense with the coffee breaks and evenings given to informal meetings among the participants. It was a time to renew old friendships, begin new ones, exchange ideas, personally question authors of papers that appeared in the literature since the last conference and find out what was generally happening in Ion Implantation.

In recent years it has beome more difficult to get funding to travel to such meetings. To assist the participating authors financial aid was solicited from industry and the Office of Naval Research. We are most grateful for their positive response to our requests. The success of the conference was in part due to their generous contributions.

The Program Committee had the unhappy task of the reviewing of more than 170 abstracts. The result of their labors was well worth their effort. Much thanks goes to them for molding the conference into an accurate representation of activities in the field.

Behind the scenes in Boulder, local arrangements were handled ably by Graeme Eldridge. The difficulty of this task cannot be overemphasized. Our thanks to him for a job well done.

The Bureau of Conferences at the University of Colorado outdid themselves in their effort to make the conference run smoothly. We were all impressed by their efficient handling of the many details that would have been otherwise overlooked.

Last but not least, our hats go off to the attendees who made the conference the success that it was.

December 1976
Boulder, Colorado

Fred Chernow
James A. Borders
David K. Brice
The Editors

CONTENTS

SILICON IMPLANTATION

Annealing Behaviors of Phosphorus Implanted in Silicon. . .   1
    T. Hirao, K. Inoue, S. Takayanagi and Y. Yaegashi

Sheet Resistivity and Hall Effect Measurements of Aluminium
Implanted Silicon and Silicon Films on Sapphire   11
    G. Holmén and S. Peterström

Annealing of Room Temperature Implants of Indium in
Silicon   21
    P. Blood, W. L. Brown and G. L. Miller

Electrical and Electron Microscope Studies of Boron
Molecular Ion Implants Into Silicon   31
    D. G. Beanland

Properties of Amorphous Silicon Layers Formed by Ion
Implantation and Vapor Deposition   39
    I. Ohdomari, M. Ikeda, H. Yoshimoto, N. Onoda,
    Y. Tanabe and T. Itoh

Enhanced and Inhibited Oxidation of Implanted Silicon . . .   49
    G. Mezey, T. Nagy, J. Gyulai, E. Kotái, A. Manuaba
    T. Lohner and J. W. Mayer

High Dose Implantation of Au and Cu into Si Studied by
Auger Electron and Backscattering Spectroscopies.   57
    A. Hiraki, M. Iwami, K. Shuto, T. Saegusa,
    T. Narusawa, K. Gamo and S. Namba

Defects Introduced Into Silicon by Boron Implantation
in the MeV Energy Range   65
    J. C. Muller, R. Stuck, P. Siffert and S. Kalbitzer

GaAs

Vacancy-Impurity Complexes in Implanted and High
Temperature Annealed n-GaAs   77
    E.V.K. Rao, N. Duhamel, P. N. Favennec and H. L'Haridon

Sulfur Ion Implantation in Gallium Arsenide   89
    M. Fujimoto, H. Yamazaki and T. Honda

Radiotracer Profiles in Sulfur Implanted GaAs   97
    F. H. Eisen and B. M. Welch

Electrical Properties of Cd- and Te-Implanted GaAs. . . .   107
    B. K. Shin, Y. S. Park and J. E. Ehret

Characteristics of Implanted n-Type Profiles in GaAs
Annealed in a Controlled Atmosphere   115
    D. H. Lee, R. M. Malbon and J. M. Whelan

Photoluminescence of Cd-Ion Implanted GaAs.   123
    K. Aoki, K. Gamo, K. Masuda and S. Namba

Some Structural and Electrical Characteristics of GaAs
Annealed After Implantation With Be, Mg, Zn and Cd. . . .   131
    R. B. Benson, M. A. Littlejohn, K. Lee and R. E. Ricker

Impurity Distribution of Ion-Implanted Be in GaAs by
SIMS, Photoluminescence, and Electrical Profiling   141
    J. Comas, L. Plew, P. K. Chatterjee, W. V. McLevige
    K. V. Vaidyanathan and B. G. Streetman

Dual Species Ion Implantation Into GaAs   149
    C. A. Stolte

Lateral Spread of the Proton Isolation Layer in GaAs. . .   159
    H. Matsumura, S. Gecim and K. G. Stephens

METALS

Metallurgical Applications of Ion Implantation.   167
    S. M. Myers

CONTENTS

Solubility Enhancement in Ion Implanted Cu Alloys. 179
   J. M. Poate, J. A. Borders, A. G. Cullis and
   J. K. Hirvonen

Dose Rate Effects in a Precipitation Hardened Nickel-
Aluminium Alloy. 181
   J. E. Westmoreland. P. R. Malmberg, J. A. Sprague,
   F. A. Smidt and L. G. Kirchner

AℓSb Preciptate Evolution During Sb Implantation in Aℓ:
Experiment and Theory. 191
   R. A. Kant, S. M. Myers and S. T. Picraux

Formation of Corrosion-Resistant Surface Alloys by
Implantation of Low-Energy Chromium and Nickel Ions
Into Polycrystalline Iron. 201
   B. D. Sartwell, A. B. Campbell and P. B. Needham

Ion Implantation Induced Disorder in Ni Studied by
Rutherford Backscattering and Electron Microscopy. 213
   J. S. Williams, R. Andrew, C. E. Christodoulides,
   W. A. Grant, P. J. Grundy and G. A. Stephens

Nitrogen Implantation into Molybdenum: Superconducting
Properties and Compound Formation. 223
   G. Linker and O. Meyer

Ion Trapping, Sputtering and Structural Changes in O_2^+
and N_2^+ Bombardment of Polycrystalline Aluminium Films . . . 231
   O. Auciello, R. A. Baragiola, E. R. Salvatelli and
   J. L. Spino

Implantation of Co into Aluminum: Damage and Lattice
Location Studies 239
   R. Kalish and L. C. Feldman

Dechanneling by Dislocations in Zn-Implanted Al. 247
   G. Foti, S. T. Picraux, S. U. Campisano, E. Rimini
   and R. A. Kant

INSULATORS

Ion-Implantation in Piezoelectric Substrates 257
   P. Hartemann and M. Morizot

Association of the 6-eV Optical Band in Sapphire
with Oxygen Vacancies 265
    B. D. Evans, H. D. Hendricks, F. D. Bazzarre
    and J. M. Bunch

Thermoluminescence of Ion-Implanted SiO_2 275
    G. W. Arnold

Studies of Radiation Damage Produced by Ion
Implantation in Sapphire 285
    T. F. Luera, J. A. Borders and G. W. Arnold

The Structure Damage, Phase Formation and Si Depth
Distribution in the Implanted Natural Diamond 295
    V. V. Krasnopevtsev, Ju. V. Milyutin, V. S. Vavilov,
    P. N. Lebedev, A. E. Gorodetsky, A. N. Khodan and
    A. P. Zakharov

Expansion of Thermally Grown SiO_2 Thin Films Upon
Irradiation With Energetic Ions 305
    D. W. Ormond, E. A. Irene, J.E.E. Baglin and
    B. L. Crowder

RECOIL IMPLANTATION

Recoil Implantation 319
    R. A. Moline

Application of the Boltzmann Transport Equation to the
Calculation of Range Profiles and Recoil Implantation
in Multilayered Media 333
    D. H. Smith and J. F. Gibbons

Preferential Sputtering and Recoil Implantation
During Depth Profiling 347
    D. K. Murti and R. Kelly

The Influence of Recoil Implantation of Absorbed Oxygen
on the Entrapment of Xenon in Aluminum and Silicon 363
    K. Wittmaack and P. Blank

CONTENTS xi

Formation of Highly-Doped Thin Layers by Using
Knock-on Effect 375
   H. Ishiwara and S. Furukawa

Damage Production and Annealing in Ion Implanted $Si-SiO_2$
Structure as Studied by EPR 383
   T. Izumi, T. Taku and T. Matsumori

Anomalous Residual Defects in Silicon After Annealing
of Through-Oxide Phosphorus Implantations 391
   M. Tamura, N. Natsuaki, M. Miyao and T. Tokuyama

SILICON DAMAGE

Divacancy Formation by Polyatomic Ion Implantation 401
   H. J. Stein

Investigation of Ion Implantation Damage with
X-Ray Double Reflection 409
   D.P. Lecrosnier, G. P. Pelous and J. Burgeat

EPR Study of Oxygen-Implanted Silicon 417
   P. R. Brosius

EPR of the Lattice Damage From Energetic Si in
Silicon at 4 K. 427
   K. L. Brower

Internal Friction Study of Vacancy-Oxygen Centers in
Ion-Implanted Silicon 435
   B. S. Berry and W. C. Pritchet

Annealing of Defects in Ion Implanted Layers by
Pulsed Laser Radiation 445
   G. A. Kachurin. V. A. Bogatyriov, S. I. Romanov
   and L. S. Smirnov

Annealing Behavior of Proton Bombardment Damage in
P-Type Silicon 453
   A. Jain, B. J. Smith and J. Stephen

Residual Damage in Silicon Implanted and Post-Annealed
Silicon 461
   L. D. Glowinski, P. S. Ho and K. N. Tu

Recovery of Radiation Damage Produced by Phosphorus
Implantation in Silicon: T.E.M. and Proton
Back-Scattering Analysis   471
    F. Cembali, R. Galloni, M. Servidori and
    F. Zignani

Radiation Damage of 50-250 keV Hydrogen Ions in
Silicon.   483
    W. K. Chu, R. H. Kastl, R. F. Lever, S. Mader
    and B. J. Masters

Anomalous Annealing Behavior of Secondary Defects
in Si Implanted with As Ions Through Dielectric Layer. . . .   493
    G. Nakamura, Y. Yukimoto, Y. Akasaka
    and K. Horie

Analysis of Defect Structures in Recrystallized Amorphous
Layers of Self-Ion Irradiated Silicon by Channeling
and Transmission Electron Microscopy Measurements.   503
    P. P. Pronko, M. D. Rechtin, G. Foti, L. Csepregi,
    E. F. Kennedy and J. W. Mayer

Dependence of Residual Damage in "Through-Oxide" Implants
on Substrate Orientation and Anneal Sequence   511
    E. F. Kennedy, L. Csepregi, J. W. Mayer
    and T. W. Sigmon

DEVICES

Use of Ion Implantation in Device Fabrication at
Hitachi CRL.   519
    T. Tokuyama

Sb^+-Implanted Buried Layer Beneath Thick Oxide
Applied for Vertical FET   535
    Y. Akasaka, K. Horie, G. Mitarai, Y. Hirose
    K. Nomura and H. Nishiumi

Ion Implanted Solar Cells.   543
    J. B. Neilson, T. M. Vanderwel, J. Shewchun
    and D. A. Thompson

The Electrical Effects of Radiation Damage Near the
Interface of Schottky Barrier Contacts   555
    D. V. Morgan and P. D. Taylor

CONTENTS

Novel Microfabrication Process Without Lithography
Using an Ion-Projection System 563
   R. Sacher, G. Stengl, P. Wolf and R. Kaitna

COMPOUND SEMICONDUCTORS

Formation of New Radiative Recombination Centers in
$Al_xGa_{1-x}As$ by Nitrogen-Ion Implantation 575
   Y. Makita, S. Gonda, H. Tanoue and T. Tsurushima

Effects of Dual Implantations and Annealing Atmosphere
on Lattice Locations and Atom Profiles of Sn and Sb
Implanted in GaP 585
   M. Takai, K. Gamo, T. Ishida, K. Masuda,
   S. Namba and A. Mizobuchi

Implantation of Be, Cd, Mg and Zn in GaAs and $GaAs_{1-x}P_x$ . . . 593
   R. Zülch, H. Ryssel, H. Kranz, H. Reichl and
   I. Ruge

Effects of Electrically and Optically Inactive Ion
Implantation in $N-GaAs_{1-x}P_x$ ($x \approx 0.37$) on Photoluminescent
Properties 603
   H. Okabayashi

Electrical and Photoluminescence Properties of Be-
Implanted GaAs and $GaAs_{0.62}P_{0.38}$ 611
   P. K. Chatterjee, W. V. McLevige, B. G. Streetman
   and K. V. Vaidyanathan

Ag-Ion Implantation into ZnSe. 621
   D. Haberland, H. Nelkowski and W. Schlaak

Resistance Control of SnO_2 Films by Ion Implantation 629
   O. Tabata, S. Kimura, and Y. Sato

Ion-Bombardment of Amorphous Semiconductors and Related
Evolution of Structural and Electrical Properties. 637
   M. Benmalek, J. P. Thomas and J. M. Mackowski

MIS Structure in As^+ Implanted CdS 649
   J. A. Hutchby

Long Range Migration of Defects During Low Temperature
Boron Implantation in ZnTe 663
   P. F. Engel, J. C. Pfister, J. Marine
   and D. Thomas

PROFILES

Structural Rearrangement in Dielectric Films
Under Ion Bombardment 671
  N. N. Gerasimenko

About the Determination of Lattice Defects in
Backscattering Experiments 687
  Y. Quéré

Heavy Ion Ranges in Silicon and Aluminium 693
  W. A. Grant, D. Dodds, J. S. Williams,
  C. E. Christodoulides, R. A. Baragiola
  and D. Chivers

Range Distributions and Electronic Stopping Powers
of Energetic $^{14}N^+$ Ions. 703
  D. G. Simons, D. J. Land, J. G. Brennan
  and M. D. Brown

A Theoretical Approach to the Calculation of Impurity
Profiles for Annealed, Ion-Implanted B in Si. 711
  A. Chu and J. F. Gibbons

Boron Profiles and Diffusion Behavior in SiO_2-Si
Structures. 727
  H. Ryssel, H. Kranz, J. Biersack, K. Müller
  and R. A. Henkelmann

Anomalous Redistribution of Ion-Implanted Dopants 735
  H. B. Dietrich and J. Comas

Index 743

Note: "Secondary Photon Emission From Ion Bombarded Oxides,"
by R. Kelly, C. J. Good and M. T. Shehata, was a paper
presented at the conference but not submitted for
publication in this volume.

ANNEALING BEHAVIORS OF PHOSPHORUS IMPLANTED IN SILICON

Takashi Hirao, Kaoru Inoue and Shigetoshi Takayanagi
Central Research Laboratories, Matsushita Electric
Industrial Co., Ltd.
Moriguchi, Osaka, Japan

Yuki Yaegashi
Research Laboratory, Matsushita Electronics
Corporation
Takatsuki, Osaka, Japan

ABSTRACT

Annealing behavior of phosphorus implanted in silicon has been studied using secondary ion mass spectrometry.

Room temperature phosphorus implants with doses from 1×10^{15} to 3×10^{16} ions/cm^2 at 70 keV into (111) Si were annealed over the range of 700°C to 1150°C.

Anomalous diffusion in the tail region was observed below 800°C and the slope of the tail was independent of dose, annealing temperature and time. Interstitial diffusion is considered to be the probable mechanism for the tail formation. The conventional thermal diffusion predominantly contributes to the broadening of the concentration profiles above 1000°C.

Diffusion coefficients have been determined as a function of dose and annealing temperature above 1000°C.

At 1000°C the diffusion coefficients show dose-dependency in the dose range 10^{13} to 10^{16} ions/cm^2. A higher diffusion coefficient of the initial enhanced diffusion is observed at this temperature. The enhanced diffusion is not so influenced by preannealing at lower temperatures.

Above 1100°C, it was found that the diffusion coefficients are nearly dose-independent in the dose range from 5×10^{12} to 1×10^{14} ions/cm^2. Above a dose of ~10^{14} ions/cm^2, the diffusion coefficients increase with dose. For a dose of 10^{15} ions/cm^2, the diffusion coefficients are not so much influenced by the radiation damage produced by phosphorus implantation itself and are in good agreement with those obtained by Fuller and Ditzenberger.

For doses higher than ~10^{15} ions/cm^2, the concentration profiles could not be estimated without taking concentration dependent diffusion coefficients into account.

INTRODUCTION

The characteristics of implanted phosphorus layers in silicon have been investigated by a number of authors using radio-tracer or electrical measurements. Many of these investigations explain the origin of the exponential tail and range distribution of phosphorus ions either channeled along the major axis or implanted away from any major axis in silicon. (1)(2)

Recently, Cembali et al compared the shape of carrier profiles of phosphorus implanted into silicon crystals along (110) axis with the damage distribution due to the implantation itself as a function of annealing temperatures between 100 and 900°C and correlation between phosphorus electrical activation and damage was shown. (3)

Although several features of ion-implanted phosphorus layers are well established, little is known about the redistribution of ion-implanted total phosphorus atoms due to heat treatments.

Very few reports in which secondary ion mass spectrometry has been used to determine the in-depth total phosphorus atom profiles are available. (4)

This paper describes the annealing behavior of total phosphorus atoms and diffusion coefficients determined as a function of ion dose and diffusion temperatures, using secondary ion mass spectrometry.

Effects of preannealing at relatively low temperatures on the diffusion coefficients are described.

EXPERIMENTAL PROCEDURE

Silicon slices of (111) orientation, phosphorus-doped n-type, 0.8–1.2 Ω-cm and boron-doped p-type, 0.5–20 Ω-cm were used.

Phosphorus implantations were performed at 70 keV in the dose range of 5×10^{12} to 3×10^{16} ions/cm^2.

The implantations were directed in a random direction (slices ~8° toward the ion beam) to minimize channeling. After the implantation, the implanted slices were annealed in an N$_2$ atmosphere in the temperature range 700 to 1150°C. For annealing above a temperature of 1000°C, the silicon dioxide was deposited on the implanted slices by the thermal decomposition of SiH$_4$ at 450°C to a thickness of 5000 Å. The in-depth total impurity profiles above a dose of 1×10^{15} ions/cm^2 were determined by mass analysis of secondary ions ejected from the samples by oxygen ion bombardment using a Cameca IMS 300 Ion Analyzer. PO⁻ions were selected for mass analysis of phosphorus. The analyzer data was converted to phosphorus concentration by knowing the total implanted dose in the silicon.

The primary beam was overscanned in two directions perpendicular to each other and secondary ions from the central part of the area were detected in order to avoid the influence of the secondary ions from the edge of the crater.

The concentration profiles of phosphorus in the dose range 5×10^{12} to 1×10^{14} ions/cm^2 were determined by C-V measurements together with junction location technique or by combining Hall effect measurements with a successive layer removal technique by anodic oxidation.

THEORY USED IN THE PRESENT STUDY

The one dimensional theory for the diffusion of impurities in a nonoxidizing atmosphere when ion implantations are conducted at room temperature into bare Si wafers at a nonchanneling direction and then an oxide layer is deposited at low temperatures has been developed by Douglas et al. (5)

The as-implanted profiles are assumed to be gaussians in the theory though they are not in fact gaussians.

The solution to a neutral ambient diffusion problem for the SiO$_2$ - Si system is presented by the following formula.

$$N(x,t) = \frac{N_\square}{2\Delta R_p (2\pi D_{si} t)^{1/2}} [\Omega(x,t) + \beta_1 \Omega(-x,t)]$$

$$\beta_1 = \{(D_{si})^{1/2} - K_s(D_{ox})^{1/2}\}/\{(D_{si})^{1/2} + K_s(D_{ox})^{1/2}\}$$

$$\Omega(x,t) = C^{1/2} \exp[-(A - \frac{B^2}{4C})] \frac{\pi^{1/2}}{2} \{1 + \mathrm{erf}(\frac{x - B/2}{C^{1/2}})\}$$

$$A = (x - R_p)^2/(2\Delta R_p^2), \quad B = 4D_{si}t(x - R_p)/(\Delta R_p^2 + 2D_{si}t)$$

$$C = 4D_{si}t\Delta R_p^2/(\Delta R_p^2 + 2D_{si}t)$$

where $N_\square$, R_p and ΔR_p are the dose, the projected range and the range straggling, respectively. D_{si} and D_{ox} are the diffusion coefficients in silicon and in silicon dioxide. K_s is the segregation constant at the interface.

RESULTS AND DISCUSSIONS

The Influence of Annealing on the Concentration Profiles of Phosphorus

Fig. 1 exhibits the concentration profiles of phosphorus implanted in silicon as a function of the annealing temperatures between 700 °C and 1100 °C for 30 minutes for a dose of 3 x 10^{16} ions/cm^2. Anomalous diffusion of phosphorus in the tail region is observed and the tail is stable at temperatures between 700 °C and 800 °C. In addition to the anomalous diffusion in the tail region broadening of the concentration profiles around the gaussian region due to thermal diffusion is observed at 900°C.

Fig. 2 exhibits the concentration profiles of phosphorus after annealings at 550°C and 750°C for 30 minutes and 72 hours for doses of 10^{15} and 10^{16} ions/cm^2, respectively.

The concentration profiles of phosphorus after annealing at 550°C are nearly the same as non-annealed profiles.

At 750°C, in addition to the pronounced exponential tails observed after annealing for 30 minutes, the decrease of peak concentration and broadening of the concentration profiles around the gaussian regions are observed after 72 hours.

Also shown is the concentration profile of phosphorus after annealing at 800°C for 30 minutes for a dose of 10^{15} ions/cm^2. As can be seen from the figure, the shape of those exponential tails are independent of dose, the annealing times and temperatures.

This suggests that interstitial diffusion is the most probable mechanism for the tail formation. (6)

At temperatures above 1000°C, the conventional thermal diffusion predominantly contributes to the broadening of the concentration profiles of phosphorus.

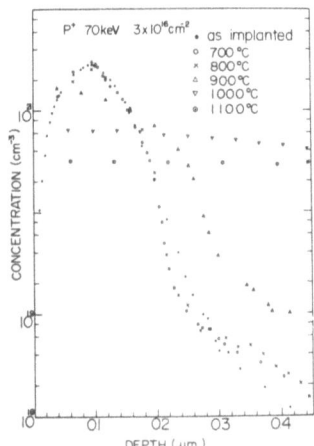

Fig. 1 Profiles of P implantations after annealing at different temperatures.

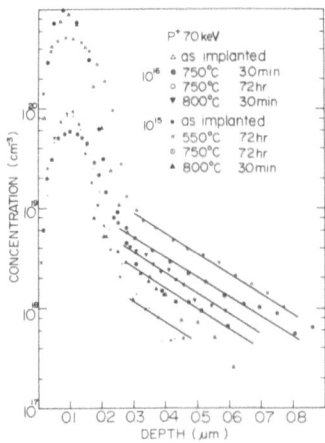

Fig. 2 Dependence of the concentration profiles of P on doses, annealing temperatures and times.

Diffusion Behaviors of Phosphorus above 1000°C

1) Estimation of Diffusion Coefficients

Sheet resistances after thermal diffusion were theoretically calculated by numerical integrations over the carrier concentration profiles, using successively assumed values for diffusion coefficients and mobility values for electrons in bulk silicon appropriate to the concentration. (7)

The calculated curves of sheet resistances for three kinds of phosphorus doses are shown in Fig. 3 as a function of diffusion coefficients assumed for the calculation.

Sheet resistances after the diffusion were measured by the four point probe method.

The point at which the calculated curve coincides with a measured sheet resistance gives a diffusion coefficient of phosphorus for a certain implantation and diffusion condition.

For an example, it can be predicted that the diffusion coefficient of phosphorus is approximately 8×10^{-14} cm^2/sec for a dose of 1×10^{13} ions/cm^2.

The best fitting to a measured profile was made by iterating computer calculation around the predicted values in the dose range 5×10^{12} to 1×10^{16} ions/cm^2.

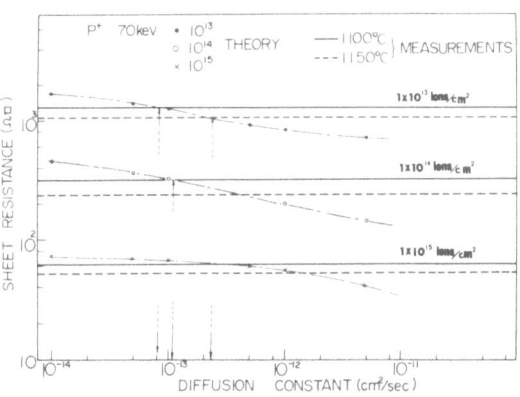

Fig. 3 Method to estimate the diffusion coefficients is shown. Theoretically calculated sheet resistances are plotted as a function of assumed diffusion coefficients for three doses of 10^{13}, 10^{14} and 10^{15} ions/cm^2.

1) Diffusion Coefficients as a Function of Dose and Diffusion Temperature

Fig. 4 shows the concentration profiles of phosphorus after diffusion at 1000, 1100 and 1150°C for 30 minutes for a dose of 10^{15} ions/cm^2.

Also shown in the figure is the as-implanted profile. Solid lines in the figure represent the calculated profiles of phosphorus which are considered to be optimum fittings to the measured profiles. The agreement between experimental and calculated results is excellent.

Fig. 4 Dependence of the concentration profiles of phosphorus on the diffusion temperatures.
Implantion energy and dose are 70 keV and 1 x 10^{15} ions/cm², respectively.

   no anneal (Δ)
   1000°C, 30 mins. (•)
   1100°C, 30 mins. (x)
   1150°C, 30 mins. (o)

Optimum fittings to the measured profiles are shown by solid lines.

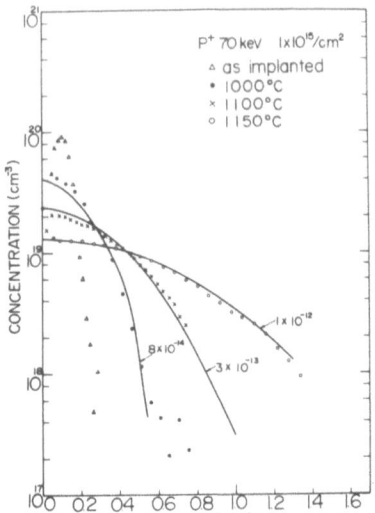

For a dose of 1 x 10^{16} ions/cm² the profiles tend to step distributions with steeply decaying distributions which are very similar to those of conventional thermal diffusion with high concentrations, as is shown in Fig. 5. We define diffusion coefficient as the values for which optimum fittings between calculated and actual profiles are obtained around the high impurity concentration.

Fig. 6 shows a comparison of experimental and predicted diffusion coefficients as a function of phosphorus dose at diffusion temperatures of 1000, 1100 and 1150°C. The broken lines show predicted diffusion coefficients. From a practical point of view, sheet resistance method appears to be valid to estimate diffusion coefficients although some deviations are observed.

Above a diffusion temperature of 1100°C, the diffusion coefficients are nearly dose-independent in the dose range 5 x 10^{12} to 1 x 10^{14} ions/cm² while they increase with dose above a dose of 10^{14} ions/cm².

Diffusion coefficients of phosphorus at 1000°C show dose-dependency in the dose range 10^{13} to 10^{16} ions/cm².

The phosphorus diffusion coefficients are shown in Fig. 7 as a function of inverse temperature for four doses of 10^{13}, 10^{14}, 10^{15} and 10^{16} ions/cm². Also shown in the figure by a solid line is those values obtained by Fuller and Ditzenberger for conventional thermal diffusion. (8) As can be seen from the figure, diffusion coefficients of phosphorus implanted with a dose of 10^{15} ions/cm² are in good agreement with those values above 1100°C while at 1000°C its value shows about three times larger than that of Fuller et al. The fact suggests that the diffusion coefficients are not so much influenced by the radiation damage produced by phosphorus implantation itself above 1100 °C while for the diffusion at 1000°C and for doses of 10^{14} and 10^{15} ions/cm², it seems that enhanced diffusion is observed.

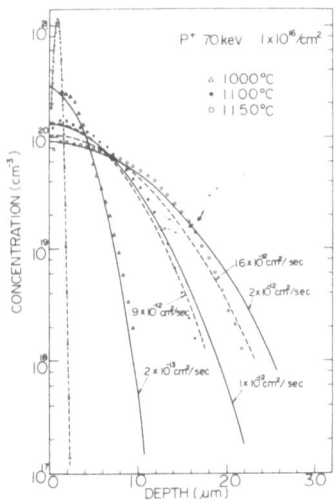 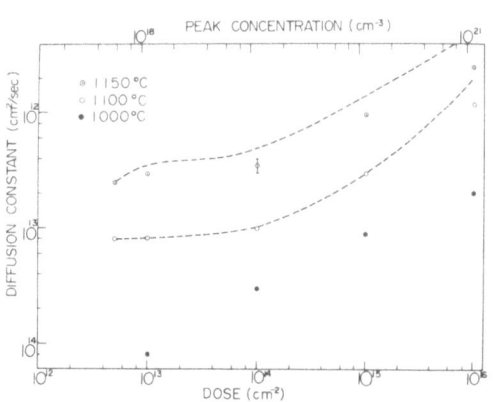

Fig. 5 Dependence of the concentration profiles of phosphorus on the diffusion temperatures. Dose : 1 x 10^{16} ions/cm²

Fig. 6 Dependence of diffusion coefficients on doses. Diffusion temperatures: 1000°C(●), 1100 °C(○), 1150 °C(●)

Fig. 7 Diffusion coefficients of phosphorus in Si as a function of diffusion temperatures. Doses: 1 x 10^{13}(●), 1 x 10^{14}(x) 1 x 10^{15}(○), 1 x 10^{16}(△) (ions/cm²)

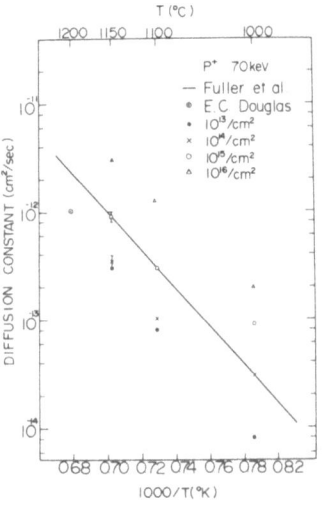

However, for a dose of 10^{15}ions/cm^2, initial fast diffusion was followed by much slower diffusion and this diffusion coefficient was calculated to be 2.8×10^{-14} cm^2/sec using the formula developed by Seidel et al. (9) This value agrees well with that obtained by Fuller et al. The fact suggests that after the initial fast diffusion enhanced due to defects produced during ion implantation, normal type of diffusion process becomes predominant.

In order to get some information about the correlation between radiation damage produced by phosphorus implantation itself and diffusion coefficients, the samples implanted with a dose of 1×10^{15}ions/cm^2 was annealed in the temperature range 400 to 800°C for 30 minutes without the silicon dioxide on the silicon, prior to the diffusion at 1000°C. After the annealing, those samples were additionally diffused at 1000°C for 30 minutes in dry N_2. As can be seen from Fig. 8, the redistributions of phosphorus in high concentration region are not so much influenced by the preannealing. Further investigation is needed to better understand the correlation between secondary defects and diffusion coefficients.

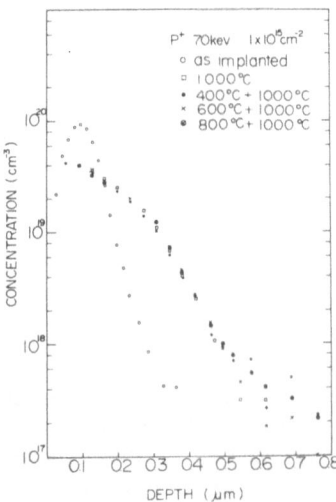

Fig. 8 Effect of preannealing at lower temperatures on the redistribution of total phosphorus atoms.

CONCLUSION

1) Anomalous diffusion of phosphorus in the tail region was observed at 700-800°C and the slope of the tail was independent of dose, annealing temperature and time. Interstitial diffusion is considered to be the most probable mechanism for the tail formation.

2) At 1000°C the diffusion coefficients show dose-dependency in the dose range 10^{13} to 10^{16}ions/cm^2. A higher diffusion coef-

ficient of the initial enhanced diffusion is observed at this temperatures.

3) Above 1100°C, the diffusion coefficients are nearly dose-independent in the dose range 5×10^{12} to 1×10^{14} ions/cm^2. Above a dose of $\sim 10^{14}$ ions/cm^2, they increase with dose. For a dose of 10^{15} ions/cm^2, the diffusion coefficients are not so much influenced by the radiation damage produced by phosphorus implantation itself and are in good agreement with the values obtained by Fuller et al.

4) For doses higher than $\sim 10^{15}$ ions/cm^2, concentration dependent diffusion coefficients must be taken into consideration.

ACKNOWLEDGEMENT

The authors wish to thank Dr. S. Horiuchi and E. Ichinoe for valuable discussions and Mrs. T. Hirao for help in preparing the manuscript.

REFERENCES

1) G. Dearnaley et al, Proc. of the 2nd Int. Conf. on Ion Inplant. in Semicond., Garmish, Germany, 439 (1971)
2) P. Blood et al, J. Appl. Phys., 45, 5123 (1974)
3) F. Cembali et al, Rad. Effects, 26, 161 (1975)
4) J. M. Morabito et al, Surface Science, 33, 422 (1972)
5) E. C. Douglas et al, IEEE. Trans. on Electron Devices, ED-21, No. 6, 324 (1974)
6) J. A. Davies et al, Can. J. Phys., 44, 1631 (1966)
7) J. C. Irvin, Bell System Tech. J., 41, 387 (1962)
8) C. S. Fuller et al, J. Appl. Phys., 27, 544 (1955)
9) T. E. Seidel et al, Trans. Met. Soc. AIME 245, 491 (1969)

SHEET RESISTIVITY AND HALL EFFECT MEASUREMENTS OF ALUMINIUM IMPLANTED SILICON AND SILICON FILMS ON SAPPHIRE

G. Holmén and S. Peterström

Department of Physics, Chalmers University of Technology

Fack, S-402 20 Gothenburg 5, Sweden

ABSTRACT

Bulk silicon and silicon films on sapphire have been implanted with 40 keV aluminium ions. The doses used were between 2.5×10^{12} and 1.0×10^{16} ions/cm^2. The annealing behaviour of the implanted layers was studied by sheet resistivity and Hall effect measurements. The annealing was performed in the temperature range 300 to 800°C. As a rule the sheet resistivity was found to be higher for aluminium implantations in silicon on sapphire than corresponding implantations in bulk silicon. It was also found that the recrystallization of a continuous amorphous layer occurs at 500°C in silicon (100) on sapphire, which is about 50°C lower than in bulk silicon (111).

1. INTRODUCTION

Ion implantation has by this time become an established technique for the manufacture of semiconductor devices in silicon. This production has been facilitated by a great number of investigations concerning the properties of different kinds of dopants implanted in bulk silicon [1,2]. However, the field of ion implantation in silicon films on insulating substrates has not been so carefully researched. The annealing behaviour of boron and phosphorus ions implanted in silicon on sapphire (SOS) has been investigated by Eklund, Holmén and Peterström [3], who found that the sheet resistivity of implanted layers in 1-2 Ωcm SOS and bulk silicon is almost the same. This indicates that it could be possible to use the ion implantation technique to produce semiconductor devices in SOS, superior to corresponding components in bulk silicon with respect to speed, packing density and insulation. The aim of the present work has been to expand the investigation of boron and phosphorus implantations [3] to include also aluminium implantations in bulk silicon and SOS.

2. EXPERIMENTAL PROCEDURE

Cleaned and etched 175 Ωcm n-type silicon slices, orientation (111), and SOS wafers with a silicon film of thickness 1 μ, orientation (100), 1 Ωcm n-type have been used. 40 keV aluminium ions were implanted at room temperature. In bulk silicon, implantations were also performed at 400°C. Six implantations of 5 mm diameter were made in the same wafer. The ion doses used, between 2.5×10^{12} and 1.0×10^{16} ions/cm^2, were implanted at a dose rate of 8.0×10^{11} ions/cm^2 s. In order to reduce channeling the silicon crystal was tilted 10° off the direction of the ion beam.

After implantation the wafers were isochronically annealed for 30 minutes in a vacuum better than 10^{-5} torr. The annealing was per-

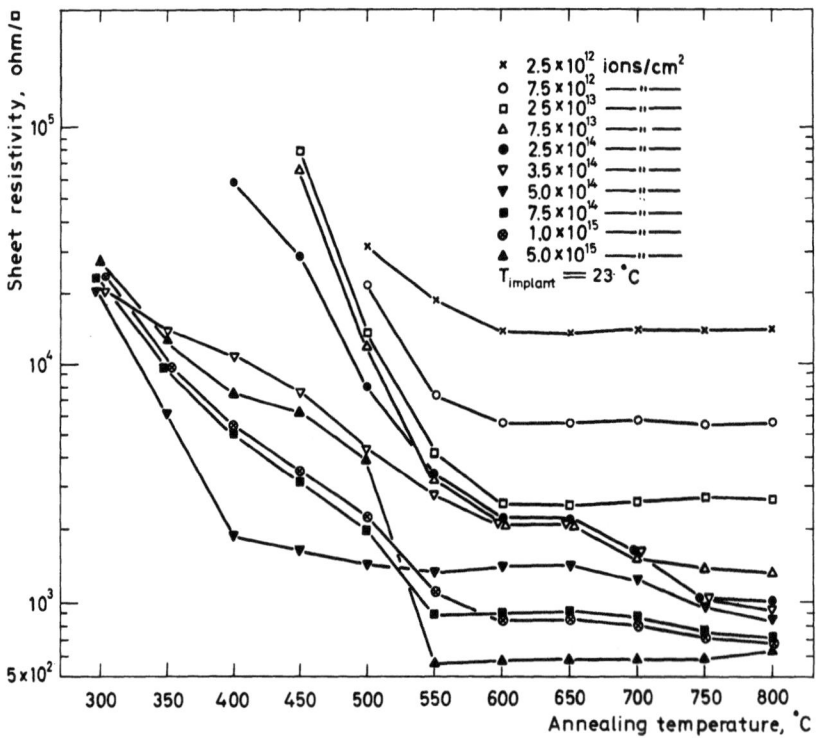

Fig. 1. The annealing behaviour of bulk silicon for different doses of 40 keV aluminium ions implanted at room temperature.

formed in the temperature range 300 to 800°C with a temperature interval of 50°C between different annealing treatments. By means of a four point probe the sheet resistivity was measured at room temperature. For the doses 10^{13} and 10^{14} ions/cm^2 the effective number of carriers $(N_s)_{eff}$, and the effective mobility, μ_{eff}, were calculated from Hall effect measurements performed at room temperature and at 77 K, using the van de Pauw method [1,2].

3. RESULTS AND DISCUSSION

Figure 1 shows the sheet resistivity as a function of the annealing temperature for bulk silicon implanted with 40 keV aluminium ions at room temperature. The low dose implantations, 2.5×10^{12} to 2.5×10^{13} ions/cm^2, have reached their lowest resistivity value after annealing at 600°C. Hall effect measurements (see Fig. 2 and ref. 4) show that the decrease in sheet resistivity for the annealing temperatures 300 to 600°C is caused by an increase in the effective number of carriers during the reordering of the silicon lattice. This is proposed to be due to an increase in the number of aluminium atoms in substitutional positions combined with a decrease in the number of compensating electrically active defect centers, possibly interstitial aluminium atoms [4].

As can be seen in Figure 2 the effective number of carriers is of the order of 10^2 times lower at 77 K than at room temperature. This great decrease in the number of carriers indicates that the effective number of carriers which occurs at room temperature consists of both substitutional aluminium atoms and electrically active defect centers with ionization energies greater than that of aluminium. Similar results have been received for boron implanted silicon annealed at 300°C [5]. Figure 2 also shows that the mobility is higher at 77 K than at room temperature depending on a decrease in lattice scattering. The increase in mobility at 77 K is less pronounced for SOS than for bulk silicon probably due to a greater number of lattice defects in the former material.

In Figure 3 the sheet resistivity of bulk silicon as a function of the implanted dose is plotted for different annealing temperatures. In the annealing range 550 to 700°C an increase in sheet resistivity occurs for doses between 10^{14} and 2.5×10^{14} ions/cm^2 (see also Fig. 1). One possible explanation for this is that silicon interstitials created during the implantation process displace substitutional electrically active aluminium atoms. Such a process has also been proposed to occur for phosphorus [6] and boron [7] implantations. In Figure 1 an annealing step can be seen at about 700°C for doses around 10^{14} ions/cm^2. Above the annealing temperature 600°C the mobility is essentially constant for the dose 10^{14} ions/cm^2 and the decrease in sheet resistivity is proposed to depend on a transi-

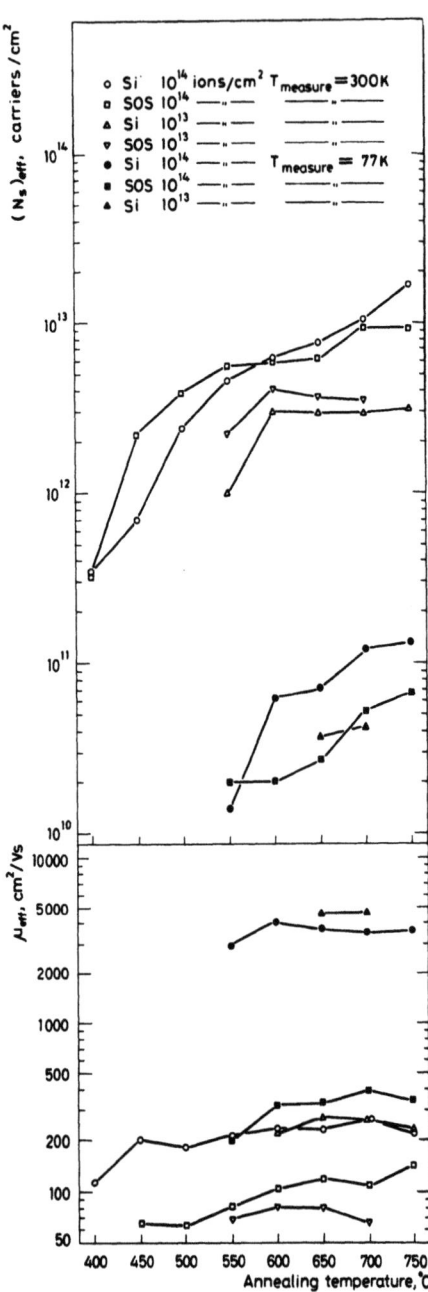

Fig. 2. The effective carrier concentration and the effective mobility, measured at 77 K and 300 K, as a function of annealing temperature for bulk silicon and silicon on sapphire implanted with 10^{13} and 10^{14} aluminium ions/cm^2.

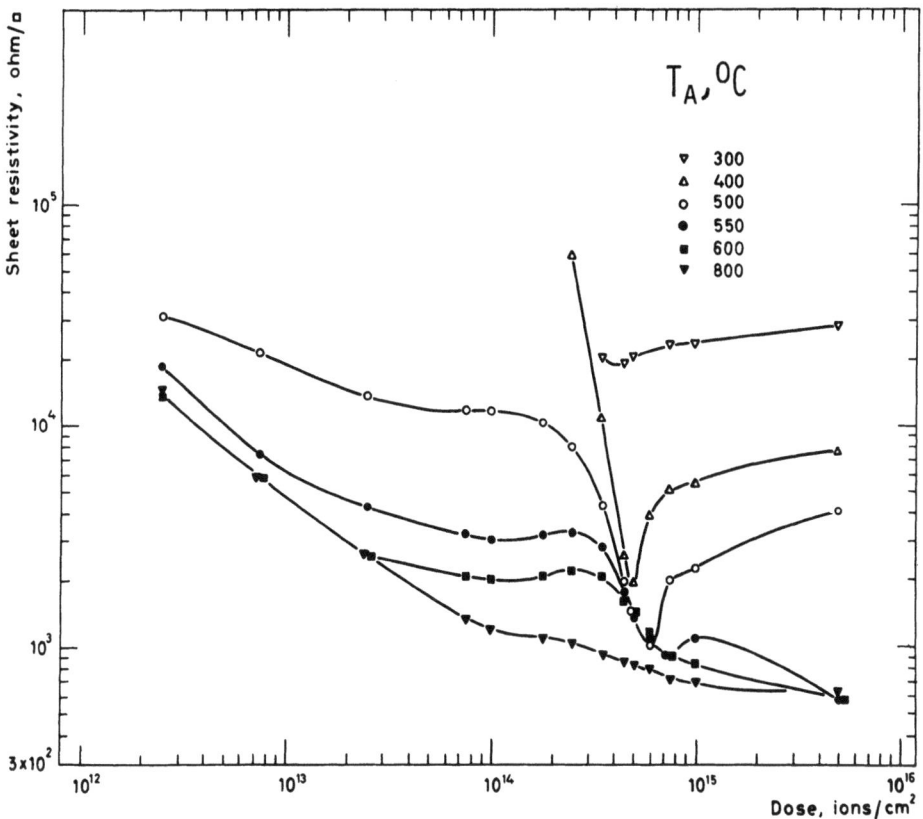

Fig. 3. The sheet resistivity of bulk silicon as a function of aluminium ion dose after annealing at different temperatures.

tion of aluminium atoms from inactive to substitutional, electrically active positions.

For a dose between 4×10^{14} and 6×10^{14} ions/cm^2, depending on the temperature, Figure 3 shows that a minimum in sheet resistivity occurs after annealing at temperatures between 300 and 500°C. These minima depend on the degree of disorder in the crystal. According to sheet resistivity measurements of phosphorus implantations in silicon performed by Eklund and Andersson [8] the critical dose for amorphization is defined as the dose were the sheet resistivity is a minimum after annealing at 500°C. Figure 3 shows that the corresponding minimum in the present case occurs at about 7×10^{14} ions/cm^2. In Figure 1 the curves for implantations with the ion doses 7.5×10^{14}, 1.0×10^{15} and 5.0×10^{15} ions/cm^2 have an annealing step between 500 and 550°C. This step is consequently attributed to the recrystallization of a continuous amorphous layer formed by a high implantation dose.

The annealing curves for bulk silicon implanted with 40 keV

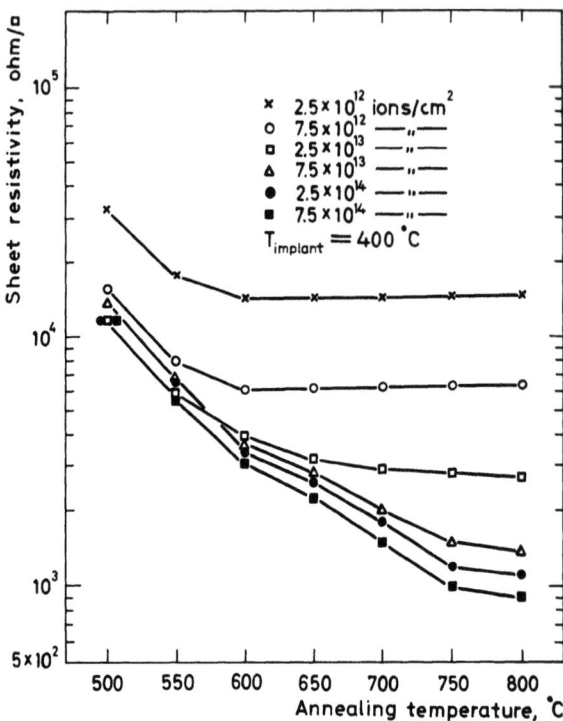

Fig. 4. The annealing behaviour of bulk silicon for different doses of 40 keV aluminium ions implanted at 400°C.

aluminium ions at 400°C are shown in Figure 4. The sheet resistivity for implantations with the doses 2.5×10^{12} and 7.5×10^{12} ions/cm^2 is almost the same as for corresponding room temperature implantations. However, the sheet resistivity for implantations with higher doses decreases monotonically with annealing temperature. During implantation of aluminium ions in silicon at 400°C no amorphous layer is formed even at the highest dose, 7.5×10^{14} ions/cm^2, used in this work [9]. On annealing of an amorphous layer the dopant atoms are supposed to occupy electrically active sites in the silicon lattice with great preference [10]. This may explain the lower sheet resistivity for high dose implantations performed at room temperature compared to corresponding implantations at 400°C, as no amorphous layer will be created during implantation at this high temperature.

Figure 5 shows the annealing behaviour of SOS implanted with 40 keV aluminium ions at room temperature. There is no marked divergence in sheet resistivity between SOS and bulk silicon implantations during annealing at temperatures lower than 450°C. At higher annealing temperatures and with doses lower than the amorphization dose the sheet resistivity is higher for SOS than

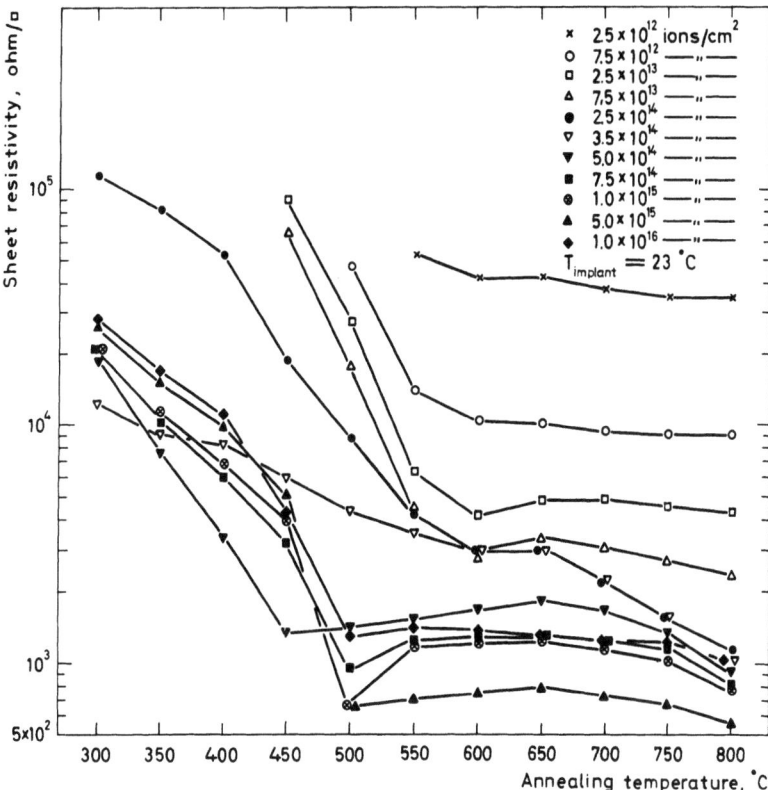

Fig. 5. The annealing behaviour of silicon on sapphire for different doses of 40 keV aluminium ions implanted at room temperatures.

for bulk silicon because of a lower mobility in the former material (see Fig. 2).

Aluminium implantations in SOS (100) with doses above the amorphization dose give rise to a decrease in sheet resistivity at the annealing temperatures 450 to 500°C, while in bulk silicon (111) this annealing step occurs at about 50°C higher temperature. Hall effect measurements show that the decrease in sheet resistivity depends on an increase in the effective number of carriers at these annealing temperatures. The same annealing behaviour has been observed for a 10^{15} ions/cm² phosphorus implantation in SOS [3]. The lower annealing temperature for recrystallization of a continuous amorphous layer may be explained by a higher regrowth rate for Si (100) than for Si (111) [11]. However, it can not be excluded that the lower recrystallization temperature in SOS may depend on a larger number of vacancies in this material, compared with bulk silicon. Holmén et al. [12] have proposed an annealing model in

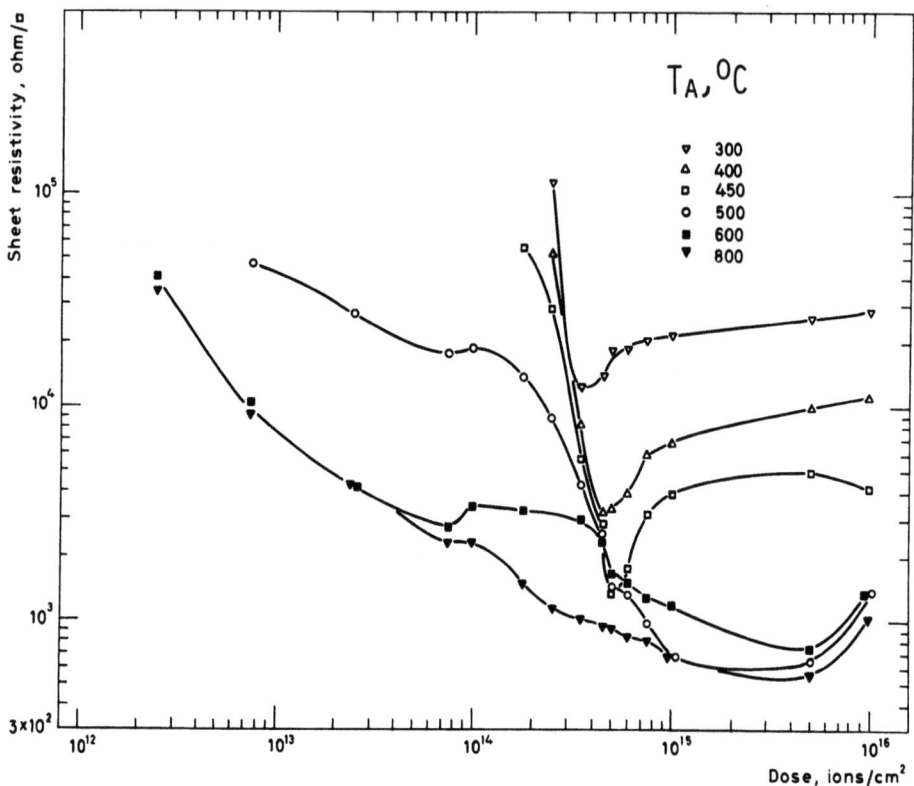

Fig. 6. The sheet resistivity of silicon on sapphire as a function of aluminium ion dose after annealing at different temperatures.

which interstitials and vacancies below the continuously disordered layer participate in the recrystallization process. This model describes the enhanced annealing of germanium during bombardment with germanium ions at temperatures well above room temperature, but it may also be valid for thermal annealing of silicon.

The sheet resistivity of SOS as a function of the implantation dose at different annealing temperatures is plotted in Figure 6. The overall annealing behaviour is the same as in bulk silicon. However the lower annealing temperature for recrystallization of an amorphous layer in SOS (100) can be seen by comparing Figures 3 and 6.

In SOS an aluminium implantation with the dose 1.0×10^{16} ions/cm^2 was performed. Figures 5 and 6 show that the sheet resistivity for this dose is considerably higher than for the dose 5×10^{15} ions/cm^2, after annealing at temperatures above 450°C. This is proposed to depend on the fact that a maximum number of carriers, 1.5×10^{19} carriers/cm^3 [1], has been reached with the implantation dose 5.0×10^{15} ions/cm^2 and the excess number of aluminium atoms act as compensating centers.

Figure 7 shows the sheet resistivity after annealing at 800°C for different doses of aluminium implanted in bulk silicon and SOS. For comparison with boron and phosphorus implantations, data published by Eklund, Holmén, and Peterström [3] are used. The higher sheet resistivity for the aluminium case depends on a lower concentration of electrically active dopants. The electrical activity is about 80 to 100% for boron and phosphorus implantations after annealing at 800°C but only about 50% or less for aluminium (ref. [1,2] and figure 2). Although the sheet resistivity is somewhat higher for aluminium implantations in SOS compared with bulk material it should be possible to use aluminium as dopant when producing semiconductor devices both in bulk silicon and SOS. As in the phosphorus case [3] the lower annealing temperature 500°C for recrystallization of a continuously disordered layer in SOS (100) may be valuable in MOS technology using an aluminium gate since the annealing temperature in this case should be below about 550°C [1].

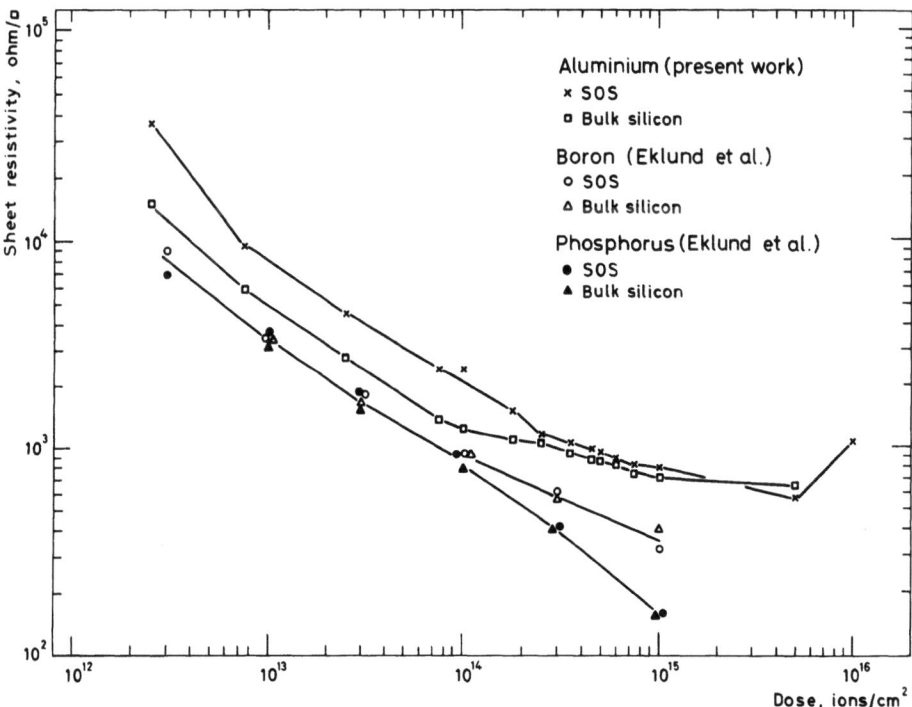

Fig. 7. The sheet resistivity after annealing at 800°C for different doses of aluminium, boron and phosphorous ions implanted at room temperature in bulk silicon and silicon on sapphire.

REFERENCES

1. J.W. Mayer, L. Eriksson, and J.A. Davies, Ion Implantation in Semiconductors, Academic Press, New York, 1970.

2. G. Dearnaley, J.H. Freeman, R.S. Nelson, and J.H. Stephen, Ion Implantation. North Holland, Amsterdam, 1973.

3. K.H. Eklund, G. Holmén, and S. Peterström, Appl. Phys. Lett. 24, 283 (1974).

4. R. Baron, G.A. Shifrin, D.J. Marsh, and J.W. Mayer, J. Appl. Phys. 40, 3702 (1969).

5. T.E. Seidel and A.U. MacRae, Trans. AIME 245, 491 (1969).

6. Å. Andersson and G. Swenson, Rad. Effects 15, 231 (1972).

7. S. Peterström and G. Holmén, Physica Scripta 10, 142 (1974).

8. K.H. Eklund and Å. Andersson, Proc. II Int. Conf. on Ion Implantation in Semiconductors (ed. I. Ruge and J. Graul), Springer Verlag, Berlin, 1971, p. 103.

9. F.L. Vook, Proc. Int. Conf. on Radiation Damage and Defects in Semiconductors, Reading, 1972, p. 60.

10. N.G. Blamires, M.D. Matthews, and R.S. Nelson, Phys. Lett. 28A, no 3, 178 (1968).

11. L. Csepregi, J.W. Mayer, and T.W. Sigmon, Appl. Phys. Lett. 29, 92 (1976).

12. G. Holmén, S. Peterström, A. Burén, and E. Bøgh, Rad. Effects 24, 45 (1975).

ANNEALING OF ROOM TEMPERATURE IMPLANTS OF INDIUM IN SILICON

P. Blood[*]

Mullard Research Laboratories, Redhill, Surrey, England

W. L. Brown and G. L. Miller

Bell Laboratories, Murray Hill, New Jersey 07974 USA

1. INTRODUCTION

The implantation and annealing of the heavy group III p-type dopants in silicon (Ga, In, Tl) has received much less attention than boron. However these dopants have a number of specialized applications, and the recent use of indium implantation for high value resistors (1) was part of the motivation for this study.

The lattice location of hot indium implants has been studied by Eriksson et al. (2). Doses of $\sim 2 \times 10^{14}$ indium ions cm^{-2} were implanted at 40 keV into $\langle 111 \rangle$ silicon substrates heated to temperatures between 350°C and 600°C to avoid the build-up of lattice damage and subsequently annealed for 10-minute periods up to 800°C in an argon atmosphere.

For practical applications (ref. 1 for example) implantation into hot substrates is an undesirable complication, and we have studied the annealing of room temperature implantations. The sensitivity limitations of backscattering demand doses in the region of 3×10^{14} cm^{-2}. Such doses will create an amorphous silicon layer and give volume concentrations which exceed the solid solubility of indium (3). We will therefore be interested in both the recovery of the silicon lattice and the behavior of the indium atoms.

The specimens were Syton polished n-type crystals ($\sim 10\ \Omega$cm) implanted in a misaligned direction and annealed in flowing nitrogen.

---
[*]Resident visitor at Bell Laboratories, Murray Hill, N. J.

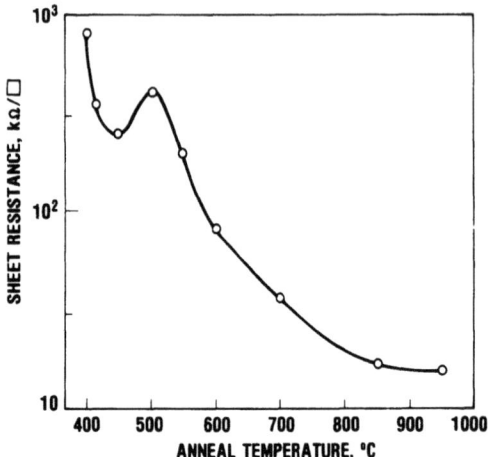

Figure 1. The sheet resistance of (111) silicon implanted at 95 keV with 1×10^{14} indium ions cm^{-2} for annealing temperatures between 400°C and 950°C (from ref. 1).

Backscattering spectra were obtained with 3 µC of 1.9 MeV α-particles incident on the target at a current of ~3 nA in a beam 1.5 mm diameter. An annular surface barrier detector was used with a scattering angle of 180°.

Figure 1 shows the sheet resistance of a 1×10^{14} cm^{-2} 95 keV In implanted specimen for sequential annealing up to 950°C (1). Two regions of interest can be identified: the reverse annealing stage around 500°C, similar to other group III impurities, and the high temperature region where the electrical conductivity reaches a maximum saturation value corresponding to a maximum number of active indium impurities and a minimum concentration of compensating defect centres. In this paper we concentrate on this latter region: annealing at 940°C.

2. RESULTS AND DISCUSSION

Figure 2 shows backscattering spectra for a ~5 x 10^{14} cm^{-2} 100 keV indium implant into (111) silicon, before and after annealing (R_p = 460 Å). The backscattering yield from silicon in the aligned spectrum shows that initially the material was amorphous with a χ_{min}, measured on the $\langle 110 \rangle$ axis behind the damaged layer, of 11%. After annealing χ_{min} recovered to only 7.4% compared with ~3% for good quality silicon. These data are listed in Table I. Misaligned spectra for the indium peak (also in Fig. 2) give the total concentration of indium as 4×10^{14} cm^{-2} after implantation but only 0.7×10^{14} cm^{-2} after annealing. This loss of indium was not expected.

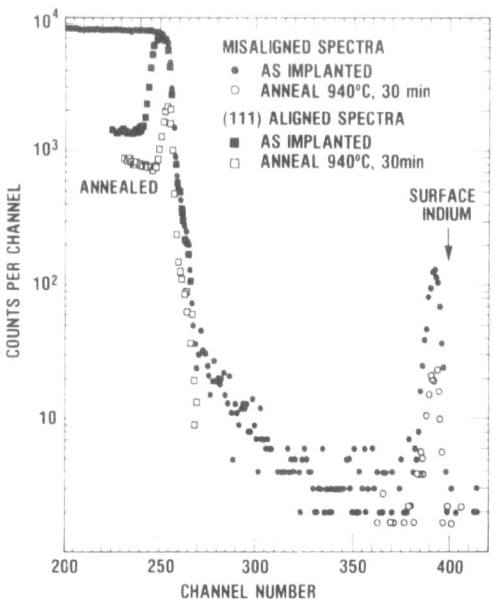

Figure 2. Backscattering spectra from (111) silicon implanted at 100 keV with 5×10^{14} indium ions cm^{-2}. The figure shows aligned spectra from the silicon and misaligned spectra from the indium, both before and after annealing at 940°C for 30 mins.

However we found that it was much reduced by implanting through a nitride surface cap, showing that the loss probably occurs by diffusion to the surface and evaporation to the ambient.

Similar implantation and annealing of (100) and (110) material however gave different results: low values of χ_{min} were obtained indicative of good crystal quality, and the majority of indium was retained. The indium loss has been observed for implantation into crystals cut with the $\langle 111 \rangle$ axis parallel and at 4° to the surface normal. Gallium implanted at a dose of 5×10^{14} cm^{-2} and energy of 100 keV into (111) and (100) crystals gave similar qualitative results (Table I).

The loss of indium cannot be due to conventional diffusion since at 950°C $\sqrt{D} \sim 200$ Å $hr^{-1/2}$(4). More striking is the marked difference in annealing behavior of (111) and (100) material-- both in terms of impurity retention and χ_{min}.

Table I. High temperature annealing of 100 keV, 5×10^{14} cm^{-2} indium and gallium implantations. All anneals are at 940°C for 30 minutes in flowing nitrogen.

impurity	specimen	fraction of impurity remaining	$\langle 110 \rangle$ impurity yield	silicon $\langle 110 \rangle$* χ_{min}
indium	(111)	~0.2	0.6	7.4%
indium	(110)	0.6	0.8	2.6%
indium	(100)	0.8	0.8	2.8%
indium	(111)+ nitride	0.76	0.9	12.5%
gallium	(111)	0.25	0.46	11.0%
gallium	(100)	0.50	0.55	2.0%

*Measured behind the regrown layer

We believe that these features can be explained by an extension of the recent work of Csepregi et al. (5). At 550°C epitaxial regrowth of amorphous layers on (111) material is a factor 25 slower than for the (100) orientation (6). This slow epitaxial regrowth means that for annealing temperatures above about 600°C it is possible for polycrystallites to nucleate and grow; these crystallites are supposed to have orientations that may be misaligned several degrees from the $\langle 111 \rangle$ axis of the underlying crystal (5), so they give an increased aligned backscattering yield. However, the fast epitaxial regrowth on (100) substrates produces a single crystal layer giving a low aligned yield. We observe these high and low values of χ_{min} for annealed (111) and (100) implanted specimens and we further suggest that indium is lost from (111) material by migration along polycrystallite grain boundaries. The remainder of this paper is concerned with demonstrating the validity of this suggestion.

For these experiments 3×10^{14} cm^{-2} doubly charged In ions were accelerated through 170 keV into (111) silicon. The greater penetration of these higher energy indium ions provided a thicker amorphous layer for easier observation of regrowth behavior and facilitated the resolution of surface indium from buried indium.

Figure 3 shows $\langle 111 \rangle$ aligned spectra for an unannealed specimen, and specimens annealed for periods of up to 10 hours at 550°C (some intermediate annealing times have been omitted for clarity). The indium concentration derived from random spectra is 2.5×10^{14} cm^{-2} and the peak is 1420 Å below the surface, in good agreement with the anticipated mean projected range of 1320 Å. The figure clearly shows that the silicon has regrown epitaxially on the underlying crystal and random spectra show no redistribution of the indium. For anneal times longer than about 6 hours a peak appeared in the silicon yield behind the remaining amorphous layer. This peak remained at the same position for longer annealing times and is at the same depth as the indium peak. It may be due to scattering from silicon atoms displaced a small distance from their regular sites by the high indium concentration.

Figure 4 shows the thickness of the regrown layer as a function of annealing time at 550°C. This confirms that for In implanted Si regrowth rate is much slower than in (100) material and the rate of 190 Å hr^{-1} agrees well with the data of Csepregi et al. (6) for times shorter 4.5 hours. For longer anneals Csepregi et al. (6) find an increase in the regrowth rate, whereas our data

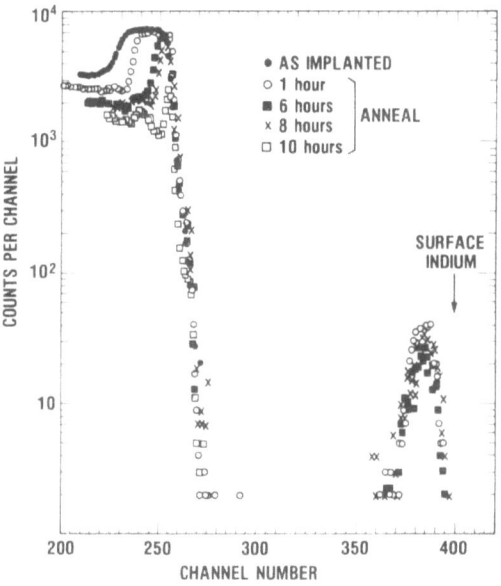

Figure 3. Backscattering spectra for (111) silicon implanted at 170 keV with 3×10^{14} In^{++} ions cm^{-2}. Aligned $\langle 111 \rangle$ spectra are shown for specimens before and after annealing at 550°C.

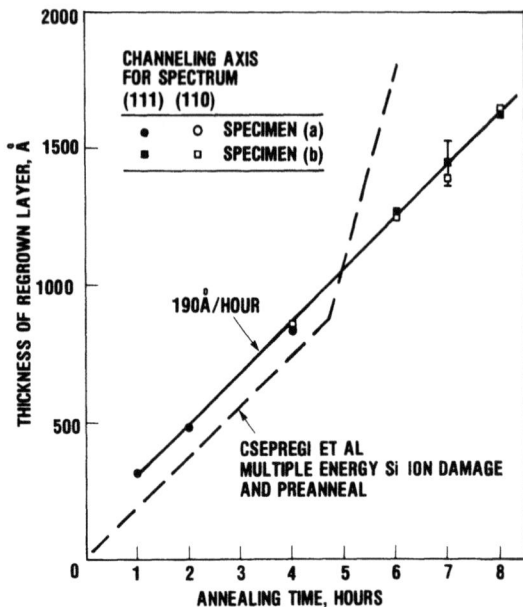

Figure 4. Thickness of the regrown layer for annealing of amorphous layers on (111) silicon at 550°C. Measurements were taken from ⟨111⟩ and ⟨110⟩ aligned spectra.

remains linear. The reason for this difference is not known. The intercept on the regrowth axis of Fig. 4 occurs because the initial thickness of the amorphous layer is not well defined due to an extended partially damaged interface region. This was removed by Csepregi et al. (6) by a low temperature heat treatment.

Spectra for various high temperature anneals are shown in Fig. 5 and data for minimum silicon yields and indium retention are summarized in Table II.

Direct annealing at 940°C for 30 mins. left a region behind the silicon surface where the yield was a factor 5 higher than for good single crystal silicon, but not as high as the yield from amorphous material. Detailed examination showed a narrow surface peak followed by a region of gradually increasing yield extending to about 1000 Å below the surface. We interpret this as due to scattering by defects and misoriented polycrystallites. After the initial 30 min. anneal 0.67 of the initial indium remained, and after a further 30 mins. this fell to 0.36 with no further change in the silicon spectrum, showing that the regrown layer is stable.

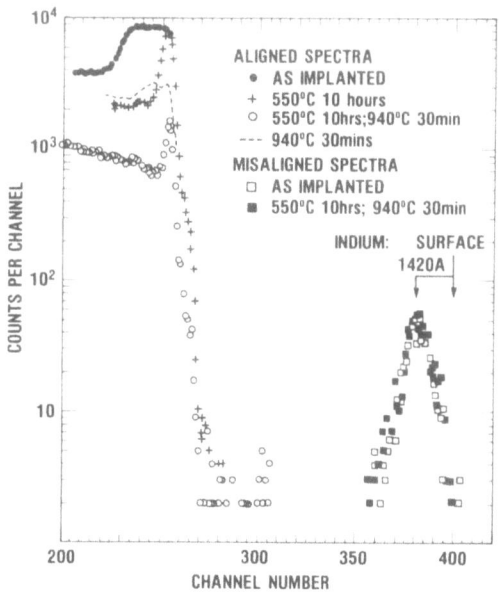

Figure 5. High temperature heat treatments of 170 keV implants of 3×10^{14} In^{++} ions cm^{-2} in (111) silicon. Aligned $\langle 111 \rangle$ spectra are shown for the silicon and misaligned spectra for the indium peak.

Spectra obtained with 1 MeV α-particles, giving better depth resolution, showed about 1×10^{13} cm^{-2} indium atoms at the surface after the 940°C anneal (this feature is present but not so clear on the 1.9 MeV spectra). This concentration represents the imbalance between the rate of migration to the surface and the rate of evaporation to the ambient.

Csepregi et al. (5) have shown that the formation of the polycrystalline region can be prevented by a prolonged anneal at 550°C, where epitaxial regrowth occurs, followed by a short anneal at 940°C to remove the residual disorder in the layer. A spectrum for an indium implanted sample annealed in this way is shown in Fig. 5, with the spectrum after the 550°C anneal shown for comparison. After this two step annealing χ_{min} was low (4.9%) and all the indium was retained (Table II). Furthermore, even when using 1 MeV α-particles there was no evidence of indium at the surface. We therefore conclude that when the formation of a polycrystalline structure is prevented by a suitable two step anneal indium is not lost from the specimen. This suggests that indium migrates to the surface along grain boundaries. (The modest loss of indium in (110) and (100)

Table II. Annealing of 3×10^{14} cm^{-2} 170 keV In^{++} implants in (111) silicon.

anneal treatment	specimen	fraction of indium remaining	fractional $\langle 110 \rangle$ In yield	silicon $\langle 110 \rangle$* X_{min}
not annealed	c	--	1.0	39.4%
550°C 10 hr	f	1.0±0.1	0.7	22%
550°C 10 hr +940°C 30 min	f	0.9$_5$	0.7	5.8%
550°C 12 hr +940°C 30 min	b	1.0	0.7	4.9%
940°C 30 min	d	0.6$_7$	0.8	31%
940°C 30+30 min	d	0.3$_6$	0.9	29%

*Measured behind the regrown layer

specimens implanted at only 100 keV, shown in Table I, is probably due to the close proximity of the surface; the important feature is the marked difference in indium retention between these samples and the (111).)

The $\langle 110 \rangle$ aligned backscattering yield from indium (see Table II) shows that only a small fraction of the impurity is substitutional, as would be expected for volume concentrations above the solid solubility.

3. SUMMARY

For doses of implanted indium or gallium sufficient to make silicon amorphous we have observed that annealing (111) material at 940°C leaves considerable residual silicon disorder and causes a large loss of impurity atoms, in contrast to (100) and (110) slices where the disorder is low and most of the implanted atoms are retained. Both effects are due to the slow epitaxial regrowth rate on (111) silicon. This permits the formation of a polycrystalline region during regrowth so the impurity can then migrate to the surface along grain boundaries. The formation of the polycrystalline region can be prevented by a two-step anneal at

550°C and 940°C. This produces a single crystal layer which retains most of indium.

The regrowth rate for an amorphous layer produced by indium implantation of (111) silicon is 190 Å hr^{-1} at 550°C. This is in good agreement with the results of Csepregi et al. (6) for short anneal times, but we do not find an increase in rate for times longer than 4.5 hrs. However we do confirm the general features of regrowth on (111) silicon which have been observed by Csepregi et al. (5).

4. ACKNOWLEDGMENTS

We thank A. Hartman and R. S. D'Angelo for providing the silicon used in these experiments, and R. A. Boie for his help with instrumentation, particularly the pulse pile-up rejection system.

REFERENCES

(1) K. R. Whight, P. Blood and K. H. Nicholas, to be published in Solid State Electronics, 1976.

(2) L. Eriksson, J. A. Davies, N. G. E. Johansson and J. W. Mayer, Journ. Appl. Phys. 40, 842-854 (1969).

(3) S. Fischler, Journ. Appl. Phys. 33, 1615 (1962).

(4) A. S. Grove, Physics and Technology of Semiconductor Devices, Wiley, 1967, p. 38.

(5) L. Csepregi, W. K. Chu, H. Muller, J. W. Mayer and T. W. Sigmon, Rad. Effects, 1976 to be published.

(6) L. Csepregi, J. W. Mayer and T. W. Sigmon, Appl. Phys. Letts. 29, 92-93 (1976).

ELECTRICAL AND ELECTRON MICROSCOPE STUDIES OF BORON

MOLECULAR ION IMPLANTS INTO SILICON

D.G. Beanland

Chemistry Division, Harwell, on attachment from Royal
Melbourne Institute of Technology, Australia

ABSTRACT

The implantation of molecular ions permits a variation in the damage produced by a particular ion species under conditions of fixed dose, dose rate and energy. In this study boron has been implanted from ion beams of B, BCl, Cl + B, BCl_2, BF, B + F and BF_2 at doses of 10^{14} and 10^{15} cm^{-2} with the energy of the boron atoms constant at 25 keV. The sheet resistance has been measured during isochronal annealing and significant differences are shown to exist between B implanted alone and as a molecule with Cl or F. The annealing characteristics are discussed with reference to the different damage structures observed by a transmission electron microscopy study of the damaged regions.

INTRODUCTION

Boron is the most important p-type dopant used for controlling the electrical properties of silicon. When it is introduced into the semiconductor by implantation the resulting electrical properties are determined by the ion dose, the damage produced by the implant and the subsequent annealing treatment[1]. While the behaviour of boron implants into silicon has been studied as a function of dose and annealing temperature using a variety of analysis techniques[2-5], few investigations have been made of the changes which result when the damage associated with the implant is varied. Blamires[3] used neon implants to pre-damage the silicon before implanting boron and showed that by causing the silicon to become amorphous annealing occurs at lower temperatures. Doses of 10^{15} cm^{-2} were used; at this dose neon implants generate an amorphous region in the silicon while boron implants do not. The degree of damage associated with an implant can also be modified by implanting molecular ions. Muller et al[6] studied the effect of implanting BF_2^+ molecules by measuring the sheet resistance and mobility using Hall measurements with van der Pauw patterns. Variation of the implantation temperature[7] can also be used to modify the amount of damage produced by a particular implant. Unfortunately this requires wafers to be thermally bonded to a heated stage during implantation, a technique which is presently inconvenient for routine implantations.

This paper considers the implantation of molecular boron ions into silicon as a method of modifying the radiation damage which occurs in association with the implantation of the boron ions. Separate implantations of the boron ion and the other constituent ion of the molecule also have been made to permit comparisons to be made with the molecular ion case. It has been shown[8-10] that the damage produced by molecular ions is greater than that produced by the implantation of the equivalent atomic ions. This is because the energy density of the individual damage cascades is greater for molecular ion implantation. The molecules are considered to split into their component atoms upon impact with the substrate surface[11] with energy in proportion to their atomic masses. The atoms have nearly identical LSS predicted range and range straggling. Since B_2^+ is not generated in the ion source the molecular ions selected for investigation were BF^+, BF_2^+, BCl^+ and BCl_2^+.

The BF_2^+ implants reported by Muller et al[6] at 10^{15} cm^{-2} and 150 keV gave a sheet resistance, after annealing at 650°C, ten times lower than an equivalent energy boron implant, while little difference was seen at 10^{14} cm^{-2}. They also reported no electrical influence of the fluorine atoms and a reduction of enhanced diffusion in the case of 10^{15} cm^{-2} BF_2^+ implants to negligible proportions. Prussin[12] has studied the ternary defects arising after various annealing treatments of 10^{14} and 10^{15} cm^{-2} BF_2^+ implants and found their behaviour to be similar to that of B^+ implants. In addition he noted that diodes made with the BF_2^+ implants showed reduced reverse leakage currents.

EXPERIMENTAL DETAILS

Implantations were undertaken into 5 cm diameter 5-20 Ωcm <111> n-type silicon wafers at room temperature in the Harwell Mk 4 Ion Implanter with the specimens tilted 7° and using potentials which gave an effective 25 keV for the boron atoms. The 10^{15} and 10^{14} cm^{-2} implants used beam currents of 100 µA and 10 µA respectively. Two series of implants were conducted. The chlorine series was BCl_2^+, BCl^+ and $Cl^+ + B^+$ and the fluorine series was BF_2^+, BF^+ and $B^+ + F^+$. The B^+ implant was used as a control for each series.

ELECTRICAL AND ELECTRON MICROSCOPE STUDIES   33

The sheet resistance measurements were made using a 4 point in-line probe using two wafers for each implant condition. Five readings were made on each wafer at points randomly chosen near its centre. The results given are the averages of the ten readings. The isochronal anneals were of 30 minutes duration. They were undertaken each 50°C in the range 400 to 700°C using a dry nitrogen atmosphere and each 100°C between 700 and 1100°C when the specimens were protected by an R.F. sputtered SiO_2 film of approximately 1000Å thickness which was removed in buffered HF to permit measurements. The 3 mm transmission electron microscopy specimens were ultrasonically cut from 5 cm wafers implanted with the sheet resistance wafers. After vacuum annealing for 30 minutes at 800°C and jet polishing in 1:7 HF/HNO_3 solution, they were examined in a Siemens 102 electron microscope at 100 keV and subsequently annealed at 900°C and 1000°C for further examination.

RESULTS AND DISCUSSION

The isochronal anneal sheet resistance data for the 10^{15} cm^{-2} chlorine and fluorine implants is plotted in Figures 1 and 2 respectively. Using the milkiness criterion all the 10^{15} cm^{-2} wafers were amorphous following implantation, with the exception of those implanted with B^+. Figure 1 shows that the wafers with implants in the chlorine series have very high sheet resistances below 600°C, indicating that intense damage has occurred. At 700°C, following the epitaxial recrystallisation phase, they exhibit a lower sheet resistance than the boron implant, but at higher temperatures they have a significantly higher sheet resistance. The fluorine series of implants, however, behave quite differently. From Figure 2 it can be seen that, although the initial damage is high, the epitaxial recrystallisation of the amorphous region is efficient, and by 650°C the sheet resistance is approximately 6 times lower than that of the boron implant. This value represents an electrical activity of approximately 50% of that attained after an 1100°C anneal, and could be utilised if a low temperature anneal process was required. At 900°C and above, the sheet resistances of the fluorine series implants follow that of the boron implant. During the annealing cycles and within the temperature range 450-600°C, the wafers implanted with BF^+, $F^+ + B^+$ and BF_2^+ were observed to become distinctly blue. The nature of these colours, which disappeared after annealing to 650°C, has been investigated and will be separately reported.

At doses of 10^{14} cm^{-2} the chlorine series implants exhibit sheet resistances higher than B^+ throughout the entire range of annealing temperatures, although the values are approximately equal at 1000°C and above. This result is depicted in Figure 3, while Figure 4 gives the fluorine series result for 10^{14} cm^{-2}. The fluorine series

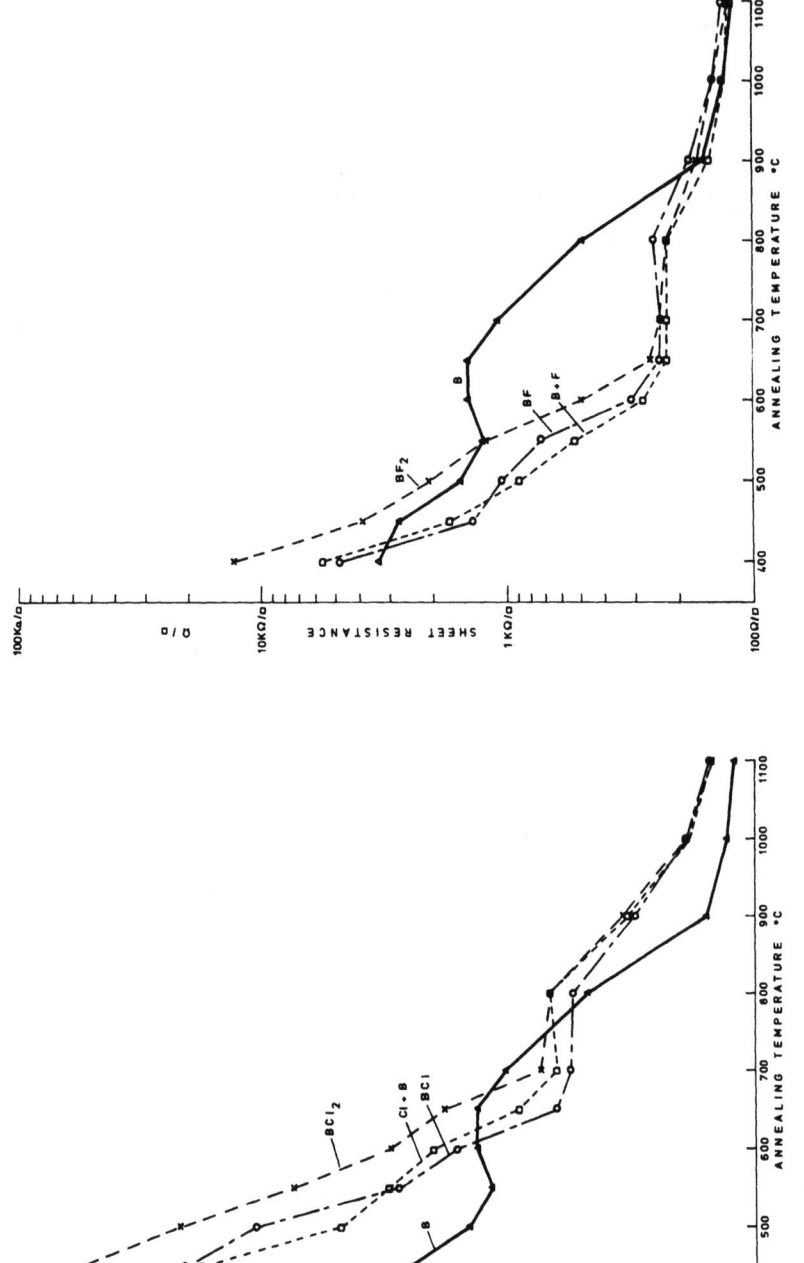

Figure 1. Isochronal anneal characteristic of sheet resistance for 10¹⁵ cm⁻² implanted silicon: B^+ 25 keV, Cl^+ 80 keV + B^+ 25 keV, BCl^+ 105 keV, BCl_2^+ 184 keV

Figure 2. Isochronal anneal characteristic of sheet resistance for 10¹⁵ cm⁻² implanted silicon: B^+ 25 keV, F^+ 43 keV + B^+ 25 keV, BF^+ 68 keV, BF_2^+ 111 keV

ELECTRICAL AND ELECTRON MICROSCOPE STUDIES 35

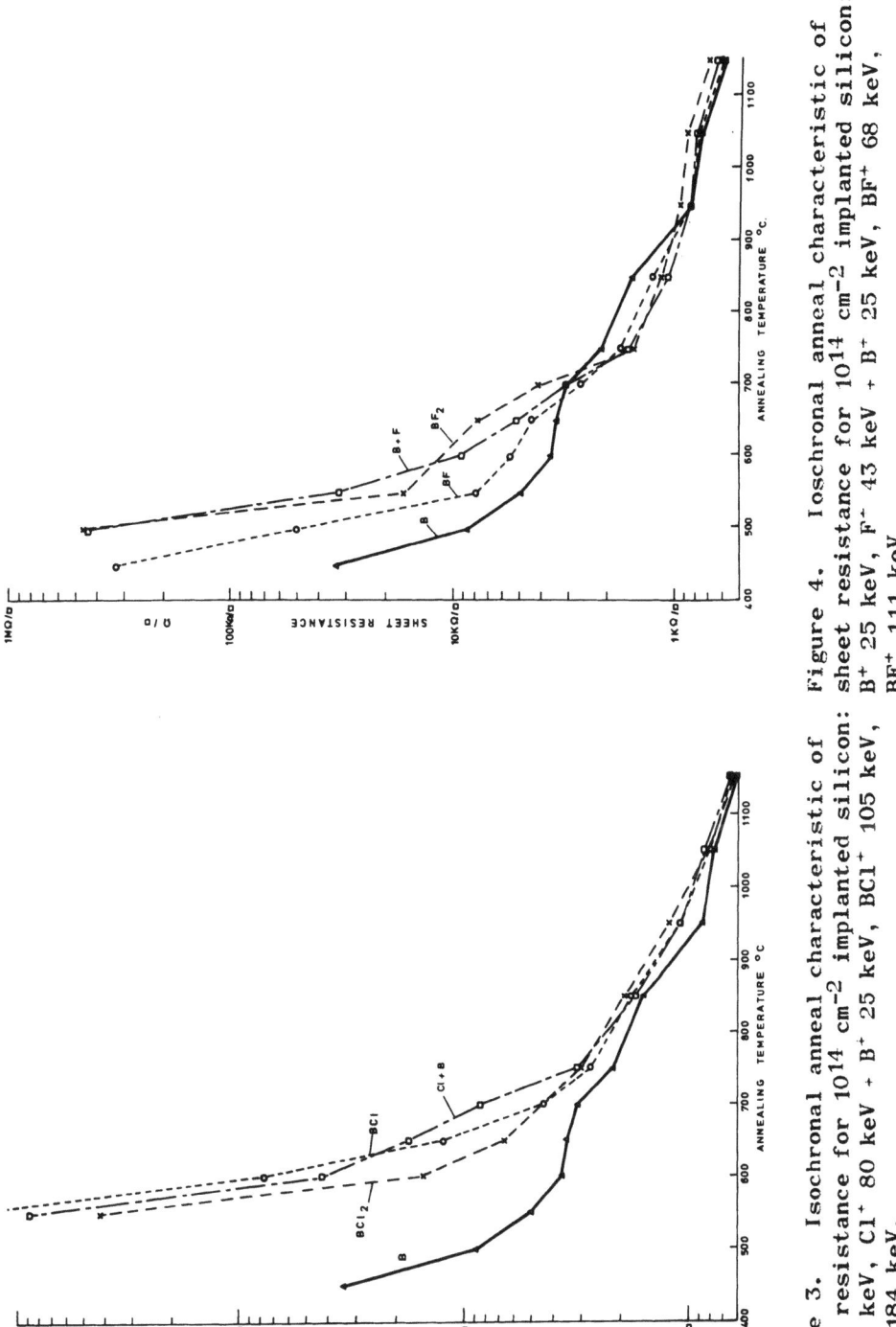

Figure 3. Isochronal anneal characteristic of sheet resistance for 10^{14} cm^{-2} implanted silicon: B^+ 25 keV, Cl^+ 80 keV + B^+ 25 keV, BCl^+ 105 keV, BCl_2^+ 184 keV.

Figure 4. Isochronal anneal characteristic of sheet resistance for 10^{14} cm^{-2} implanted silicon: B^+ 25 keV, F^+ 43 keV + B^+ 25 keV, BF^+ 68 keV, BF_2^+ 111 keV.

implants have sheet resistances lower than boron at 700°C and 800°C, while above this temperature they are approximately equal. Following the implantation at 10^{14} cm^{-2} the only wafers exhibiting amorphousness were those implanted with BCl_2^+. The behaviour of the molecular implants can be seen to be similar to that of the equivalent separate ion implants.

Transmission electron microscope observations can be used to provide useful supportive information about the residual damage in the silicon lattice after the annealing process. The defect structure observed for the 10^{15} cm^{-2} implants annealed at 900°C is shown in Figure 5. In the boron implant the widely reported rod and loop dislocation structure is observed, while for the other implants, which all caused amorphous regions to be generated within the silicon, far more complex dislocation networks are observed. Their structure exhibited microtwinning which is typical for implants beyond the amorphous threshold dose[4]. The fluorine series implants, however, exhibit a more open type of dislocation structure than the finely structured dislocation networks seen for BCl^+ and BCl_2^+. The observation of less damage in fluorine implanted specimens correlates with the lower sheet resistances which they exhibited in comparison to the chlorine implanted specimens. The exception, in Figure 5, is the $Cl^+ + B^+$ implant, which shows very few dislocations and a heavy concentration of irregular gas bubbles and surface pits, presumably arising from the chlorine.

The observed defect structures for the 10^{14} cm^{-2} implants, following annealing at 900°C, are shown in Figure 6. At this much lower dose simple dislocation loops are evident in all specimens. Rods were observed for B^+, BCl^+, $Cl^+ + B^+$, BCl_2^+ and BF_2^+ implants at 800°C, but they have all annealed by 900°C. The observed density of loops is least for B^+, BF^+ and $B^+ + F^+$ and progressively greater for BF_2^+, BCl^+, $Cl^+ + B^+$ and BCl_2^+. While the defect density observed approximately follows the damage density of the various implants within each series, the fluorine series implants have somewhat less damage than anticipated. The BF^+ and $B^+ + F^+$ implants at 10^{14} cm^{-2} show no more damage than the B^+ implant.

The $^{19}F(p,\alpha)^{16}O$ nuclear reaction[13] was used to compare the fluorine content of BF_2^+ implanted specimens following anneals at 700, 900 and 1100°C with that of an "as implanted" specimen. Protons of 2.65 MeV were directed onto the specimens and the scattered alphas detected. The yield after 700°C annealing showed no decrease on the "as implanted" value, but after 900°C the fluorine content had dropped to approximately 1/3 and after 1100°C it had almost completely disappeared. Since the proton probe detects fluorine at considerable depth it can be concluded that the gas has escaped from the surface and not diffused into the silicon. This result is consistent with observations[14] made for Xe when approximately 2/3 was found to escape by 700°C.

ELECTRICAL AND ELECTRON MICROSCOPE STUDIES

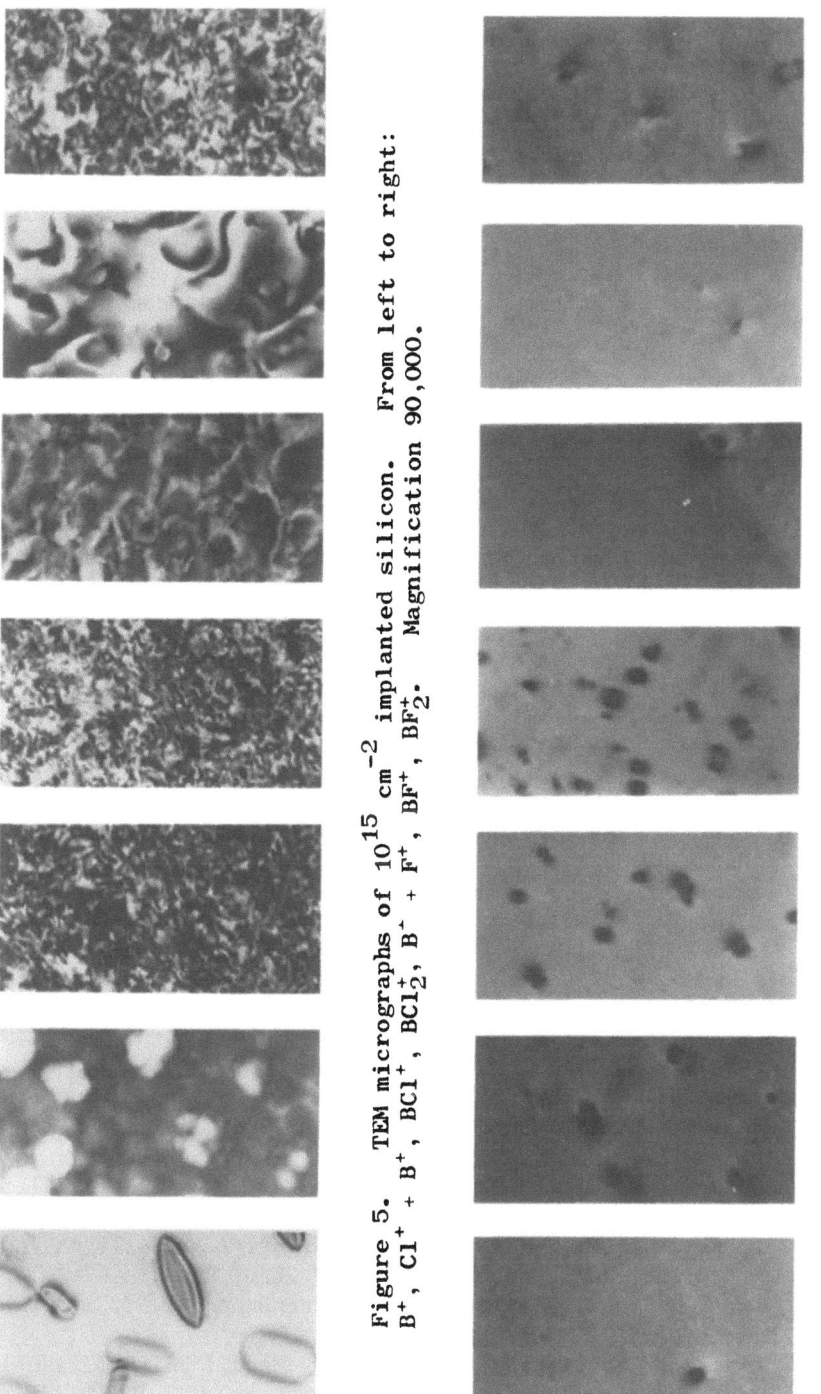

Figure 5. TEM micrographs of 10^{15} cm^{-2} implanted silicon. From left to right: B^+, $Cl^+ + B^+$, BCl^+, BCl_2^+, $B^+ + F^+$, BF^+, BF_2^+. Magnification 90,000.

Figure 6. TEM micrographs of 10^{14} cm^{-2} implanted silicon. From left to right: B^+, $Cl^+ + B^+$, BCl^+, BCl_2^+, $B^+ + F^+$, BF^+, BF_2^+. Magnification 90,000.

CONCLUSIONS

Molecular ion implants can be used to vary the damage created when a specific ion of a particular energy is implanted. Since many device parameters are sensitive to the residual damage after annealing it is possible that molecular ions can give improved performance. Further investigations of this aspect are being undertaken. It is apparent that fluorine molecular implants rather than chlorine molecular implants, with boron, are likely to be more successful. Fluorine does little to impede and may enhance the achievement of active locations for the boron atoms. Chlorine does not behave quite as favourably, as the implants studied were disproportionately inferior. This effect cannot be ascribed only to the additional damage.

ACKNOWLEDGEMENT

The assistance of Mr. D. Chivers in undertaking the implants, Dr. G. Dearnaley with the fluorine nuclear reaction experiment and Mr. E. Wittam with the diode fabrication is thankfully acknowledged. Dr. J. Stephen and Dr. B. Smith generously allowed their equipment to be used for the measurements and provided many useful discussions.

REFERENCES

1. G. Dearnaley, J.H. Freeman, R.S. Nelson and J. Stephen, "Ion Implantation", (North Holland) (1973).
2. D.E. Davies, App. Phy. Lett., 14, 227 (1969).
3. N.G. Blamires, Proc. Eur. Conf. on Ion Implantation (Peregrinus), 52 (1970).
4. S.M. Davidson and G.R. Booker, Rad. Effects 6, 33 (1970).
5. D.J. Mazey, R.S. Nelson and R.S. Barnes, Phil. Mag. 17, 1145, (1968).
6. H. Muller, H. Ryssel and I. Ruge, "Ion Implantation in Semiconductors", Ed. I. Ruge and J. Graul (Springer-Verlag), 85, (1971)
7. L. Eriksson, J.A. Davies, N.G.E. Johansson and J.W. Mayer, J. Appl. Phys., 40, 842 (1969).
8. J.A. Moore, G. Carter and A.W. Tinsley, Rad. Effects, 25, 49, (1975).
9. J.B. Mitchell, J.A. Davies, L.M. Howe, G. Foti and J.A. Moore, "Ion Implantation in Semiconductors", Ed. S. Namba (Plenum) 493 (1975).
10. H.H. Andersen and H.L. Bay, J. Appl. Phys., 45, 953 (1974).
11. D.G. Beanland, J.H. Freeman and C.A. English, "Applications of Ion Beams to Materials", Ed. G. Carter, J.S. Colligon and W.A. Grant (Inst. of Phys. London) 262 (1975).
12. S. Prussin, "Ion Implantation in Semiconductors, Ed. S. Namba (Plenum) 449 (1975).
13. E. Moller and N. Starfelt, Nuc. Instr. Meth., 50, 225 (1967).
14. S. Mader and K.N. Tu, J. Vac. Sci. Technol., 12, 501 (1975).

PROPERTIES OF AMORPHOUS SILICON LAYERS FORMED BY ION IMPLANTATION
AND VAPOR DEPOSITION

I. Ohdomari, M. Ikeda, H. Yoshimoto, N. Onoda,

Y. Tanabe and T. Itoh

School of Science and Engineering, Waseda University,

Tokyo, Japan

ABSTRACT

Amorphous silicon layers have been formed on single crystal silicon by ion implantation, vapor deposition and additional ion implantation into deposited films. Properties and crystallization processes of amorphous silicon have been measured by using TEM, TED, ESR and backscattering. A close relationship has been observed between the changes in surface area of voids which have been observed in vapor-deposited films and in the density of amorphous ESR center. Thus void network has been identified to be a main source of amorphous ESR center. Ranking of the "degrees of amorphousness" in amorphous silicon has been tried in accordance with the crystallinity after crystallization by annealing. TED patterns of twin structure which were observed in common in post-crystallized structures of deposited films and argon-implanted layers have been analyzed and explained by a model which takes into account the secondary twins joined on the "downwards-pointing" faces of primary twins.

INTRODUCTION

Since the pioneering work by Mazey et al[1], the properties and the annealing behaviors of implantation-induced damage have been extensively studied by TEM, TED, backscattering, electrical measurements and so on. If once the amorphous region is formed, however, the result is expressed in a few words; "continuous amorphous layer has been formed", and the various structures which might be in the amorphous state have scarcely been studied. While it is well known that the recrystallization of the continuous amorphous region occurs epitaxially on the underlying substrate by annealing[2], it has been occasionally reported also that the implanted layers with the doses far beyond the critical one do not recrystallize to single crystal. This fact is indicative that there could be considerable variation in amorphous state.

On the other hand, there have been reported many studies about the structure and the crystallization of amorphous germanium and silicon prepared by vacuum deposition or sputtering. Most of them are concerned with phenomena which occur on foreign substrates[3], and there are few studies with respect to the crystallization of amorphous silicon on crystalline silicon.

The object of this work is to make clear the relation among the "degrees of amorphousness" in as-prepared samples, crystallization processes and post-crystallized structures, by examining each item precisely using transmission electron microscopy, electron diffraction, ESR and backscattering.

EXPERIMENTAL

Amorphous silicon layers were formed on single crystal silicon substrates by the following three ways.
(1) Single crystal silicon substrates were implanted with between 5×10^{13} and 5×10^{15} argon ions/cm^2 at room temperature at energies of 50, 70, 100 and 130 keV.
(2) Nominally undoped silicon was evaporated onto (111) silicon wafers to the thickness of 1000-7000 Å. Deposition rates were in the range between 80 and 120 Å/min. The operating pressure did not exceed 2×10^{-7} Torr. during deposition.
(3) The samples of group (2) (1000 Å thick) were implanted with argon under the same condition as group (1).

Annealing of these samples were performed for 30 min. at various temperatures up to 900 °C in vacuum. The ESR measurements were done using an X-band microwave spectrometer with 100 kHz modulation at room temperature. The samples for electron microscopy were thinned in a dilute HF+HNO$_3$ solution. TED photographs of annealed samples were taken under an accelerating voltage of 100 keV. Backscattering measurements were performed with He$^+$ ions under an accelerating voltage of 350 keV.

RESULTS AND DISCUSSION

Initial Amorphous States

All the evaporated films and implanted layers except a case of the lowest dose showed halo patterns characteristic of an amorphous state. The observed intensity profiles in diffraction photographs have 4 maxima at about 2.0, 3.6, 5.8 and 8.3 Å^{-1} in s-value.

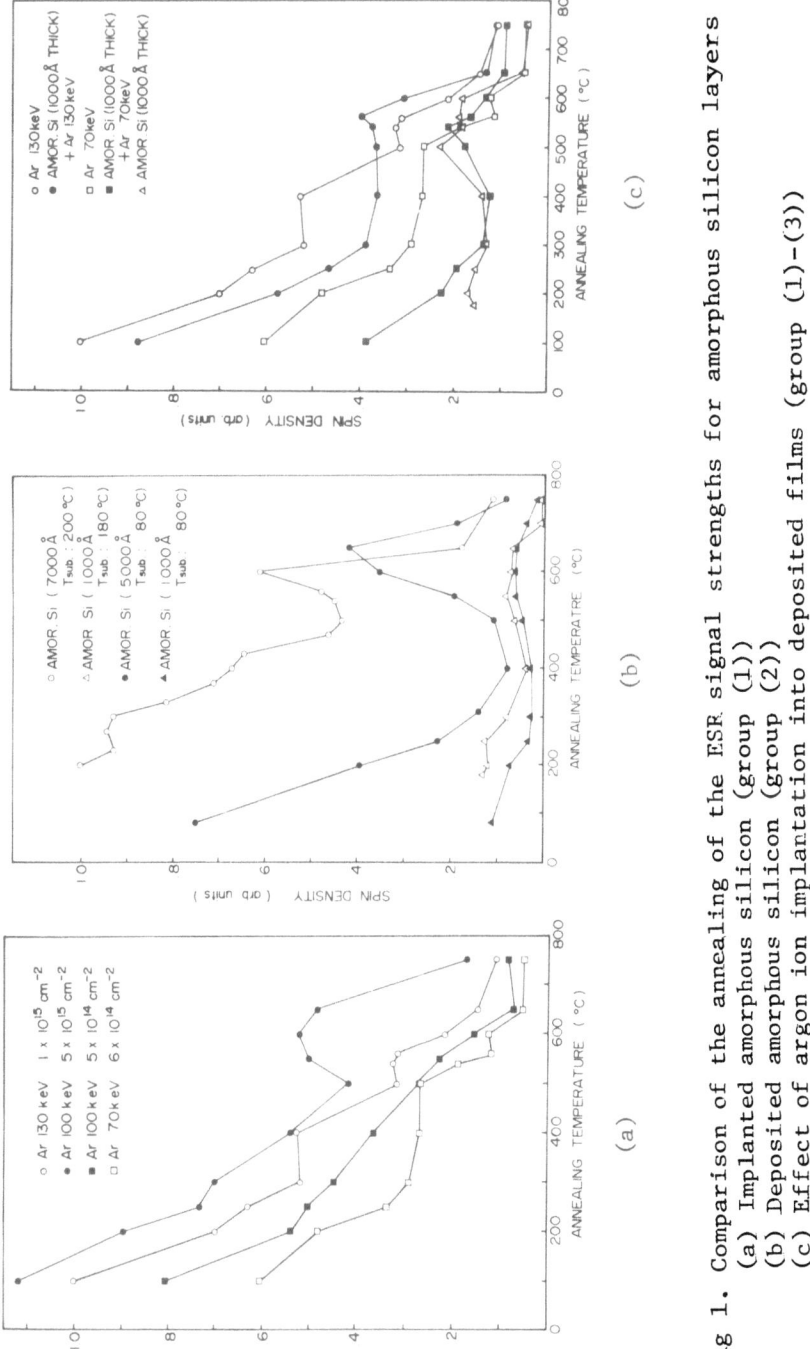

Fig 1. Comparison of the annealing of the ESR signal strengths for amorphous silicon layers
(a) Implanted amorphous silicon (group (1))
(b) Deposited amorphous silicon (group (2))
(c) Effect of argon ion implantation into deposited films (group (1)-(3))

According to our preliminary result of RDF analysis, rms deviation in bond length of nearest neighbours and second nearest neighbours are smaller in lightly implanted samples and thin deposited films than in heavily implanted samples. However, ranking of amorphous silicon layers prepared by various ways according to their "degrees of amorphousness" is rather difficult at present because of very slight differences in photometer traces of the diffraction plates among those samples. More accurate results could be obtained by using a secter, sufficient re-estimation of background intensity and taking photographs at larger s-values.

In the as-deposited silicon films (from 2000 to 7000 Å in thickness), void network regions of appreciable density deffiency within the films were observed. Regions surrounded by void network are amorphous and 200-300 Å in diameter.

Argon implantation diminish the void network in the as-deposited films. It is not observed in purely implanted samples either.

Crystallization Processes

Annealing behaviour (up to 750 °C) of amorphous silicon prepared by three different ways have been studied using ESR measurements. Fig. 1 (a) shows the annealing processes of amorphous center in

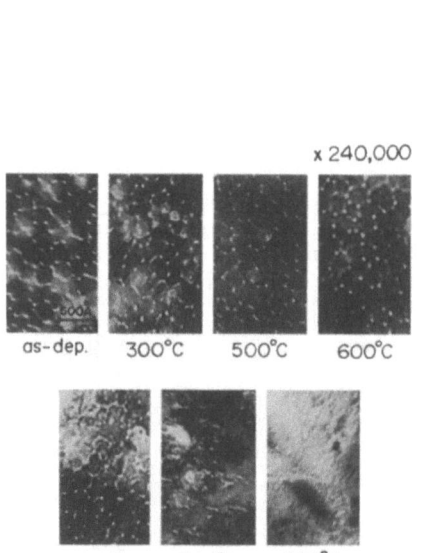

Fig. 2 Transmission electron micrographs of deposited silicon films (5800 Å in thickness)

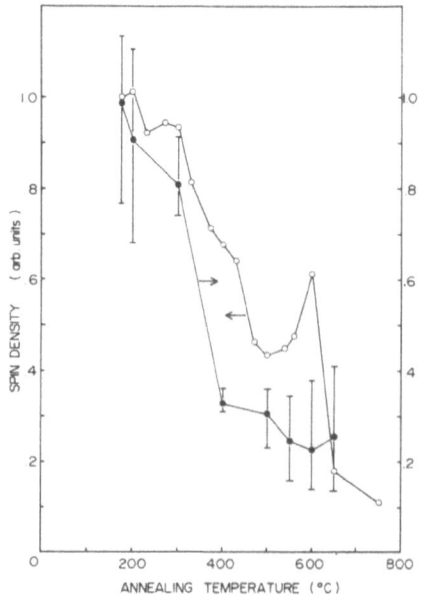

Fig. 3 Comparison of the annealing between the surface area of voids and the ESR signal strength

argon-implanted samples. Each annealing curve decreases with the rise of annealing temperatures, but in the case of the sample implanted with 5×10^{15} ions/cm^2 at 100keV, it increases temporarily at a temperature around 600 °C. Fig. 1 (b) shows the case of deposited amorphous silicon. The decrease in spin density up to about 450 °C and the subsequent rapid increase around 650 °C are observed in these cases also. Since, as shown later, each sample which shows the rapid increase in spin density crystallizes to polycrystalites, this increase may have some influence on crystallization of amorphous silicon.

In Fig. 1 (c), we compare the spin densities of three kinds of samples (group (1)-(3)). Argon implantations bring about an increase in dangling bonds in deposited amorphous films. Crystallinity after 650 °C annealing of amorphous silicon deposited to a thickness of 1000 Å at 180 °C is twin, but in the case of the film additionally implanted with argon, polycrystalites are obtained after some annealing. Thus the additional argon implantation increases the degree of amorphousness in the initial as-deposited films.

In Fig. 2, we show the transmission electron micrographs of amorphous silicon films deposited at 200 °C and annealed at various temperatures up to 900 °C. The total length of white lines in these micrographs decreases gradually as the annealing temperature is raised, while the diameter of amorphous grains remains constant. No voids were observed after annealing at 900 °C.

In Fig. 3, the surface area of voids (i.e. square sum of the length of each void) is plotted as a function of annealing temperature. The surface area and spin density decrease with a close relationship as annealing temperature is raised, in spite of ambiguity caused by focus of the photographs. Accordingly it may be tentatively concluded that surfaces of amorphous grains are predominant sources of dangling bonds. As mentioned earlier, argon implantation into deposited films diminishes the voids, but doesn't decrease the number of dangling bonds. Presumablly it relocates the atoms in the deposited films, and another kind of amorphous state results.

Fig. 4 shows the backscattering yields of deposited films annealed at various temperatures. The height of the plateau of the as-deposited sample

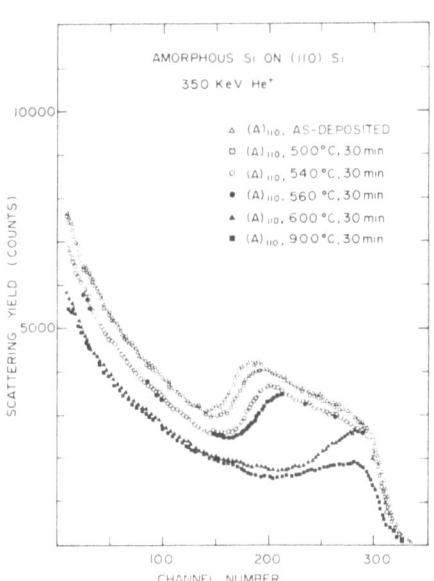

Fig. 4 Backscattering spectra for deposited films

coincides with the random yield. The amorphous-crystalline interface shifts toward the surface by annealing at 500°C, and lowering of the scattering yield occurs at 540°C. The former shows the epitaxial crystallization of a deposited film onto single crystal silicon, and the latter shows the crystallization inside the deposited film itself. Crystallization temperatures in both cases are considerably low compared with the case of deposited films on amorphous substrates.

Crystallinity after Thermal Treatment
    Results of TED are summarized in Table 1. Crystalline states after 900 °C annealing distribute between single-crystalline and polycrystalline according to the "degrees of amorphousness" and thickness of amorphous layer. The ranking of amorphous silicon layers formed under various experimental conditions are summarized in Table 2. As far as the diffraction patterns after crystallization are concerned, the degrees of amorphousness in argon-implanted samples (100 keV-5×10^{14}) are comparable to that of thin deposited films. Additional argon implantation increases the degrees of amorphousness in thin deposited films. The diffraction patterns obtained from heavily implanted and deposited films after annealing at 900 °C are the same twin pattern as shown in Table 2. It is clear that these two amorphous layers are of the same structure after crystallization, the matrix as which has a {111} plane parallel to the substrate. Holt [4] applied the analysis of Pashley and Stowell [5] to the TED pattern from zinc-blende structure epitaxial films. His analysis can be applied to our case of diamond structure, but there are 24 anomalous {111} spots which can not be explained by his analysis.

Table 1. TED patterns of initial states and crystalline states

group	initial state (existence of void network)	TED patterns after 900 °C anneal
(1) impl.		
5×10^{13} cm^{-2}, 100 keV	spot (no)	spot
5×10^{14} cm^{-2}, 50 keV	spot + diffuse Debye-Scherrer ring (no)	spot
5×10^{14} cm^{-2}, 100 keV	spot + diffuse Debye-Scherrer ring (no)	twin spot
5×10^{15} cm^{-2}, 50 keV	diffuse Debye-Scherrer ring (no)	twin spot
5×10^{15} cm^{-2}, 100 keV	diffuse Debye-Scherrer ring (no)	thin broken ring
(2) dep.		
1000 Å	diffuse Debye-Scherrer ring (no)	twin spot
2000-7000 Å	diffuse Debye-Scherrer ring (yes)	thin broken ring
(3) dep. + impl.		
1000 Å, 5×10^{14} cm^{-2} (100 keV)	diffuse Debye-Scherrer ring (no)	thin broken ring

Table 2. Three different crystalline states

TED PATTERNS AFTER CRYSTALLIZATION	EXPERIMENTAL CONDITIONS		
	IMPL.	DEPO.	DEPO + IMPL
(image)	Ar 100 keV 5 x 10^{13} cm^{-2} Ar 50 keV 5 x 10^{14} cm^{-2}		
(image)	Ar 50 keV 5 x 10^{15} cm^{-2} Ar 100 keV 5 x 10^{14} cm^{-2} Ar 130 keV 1 x 10^{15} cm^{-2}	1000 Å (180°C) 1200 Å (180°C)	
(image)	Ar 100 keV 5 x 10^{15} cm^{-2}	2000 — 7000 Å (200°C) 5000 Å (80°C)	1000 Å + Ar 100 keV 5 x 10^{14} cm^{-2}

These anomalous {111} spots can be separated into two groups as follows; (1) six pairs of adjacent spots in <2$\bar{2}$0> directions and (2) six pairs of spots near the 1/3 {4$\bar{2}\bar{2}$} spots due to double positioning. The former can be explained by taking into account tipl twinned particles[6], but the latter can not.

To explain the spots of group (2), we propose a new model shown in Fig. 5, which includes secondary twins, such as a tetrahedron ABPF joined on the "downwards-pointing" face AFB of primary twin OAFB. Then the resultant {111} faces of this secondary twin are nearly parallel to the beam, resulting in the formation of {111} diffraction spots belonging to group (2). The diffraction pattern calculated for the new model of multiply twinned particles is shown in Fig. 6. This pattern is in very good agreement with experiment.

It is well known that ion species affect the recrystallization processes of implanted amorphous layers. In the case of argon implantation, formation of gas bubbles can inhibit the epitaxial recrystalization of the amorphous layer[7]. Therefore such concluding remarks as that an ion-implanted layer with some dose and a vapor deposited film are of the dame degree of amorphousness might not be reasonable. Strictly speaking, the comparison should be made under the condition free of precipitation and gas bubbles. Almost all models of amorphous silicon and germanium, are consistent with the point that an amorphous structure consists of three kinds of building units, i.e. a chair-like ring, a boat-like ring and a five-fold ring, so the authors would like to think that the degrees of amorphousness differ in the mixing ratio of three kinds of building units. Since single crystal consists of chair-like rings only, epitaxial crystallization of an implanted layer to single crystal might be a reasonable result of crystallization in an amorphous layer which originally

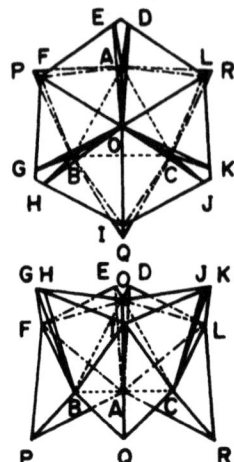

Fig. 5   Modified Ino's MTPs model which includes secondary twins on the "downwards-pointing" {111} faces of primary twins.

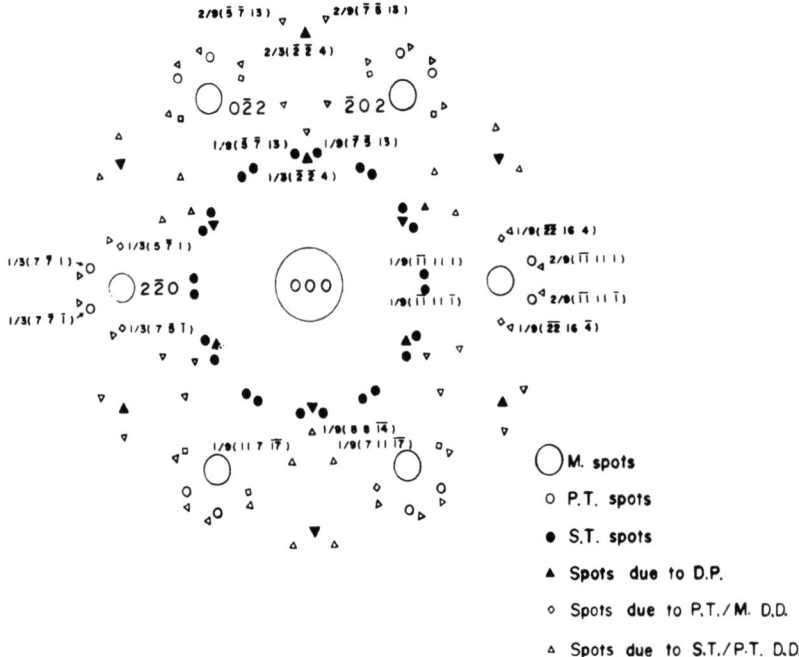

Fig. 6   Expected diffraction pattern for the modified MTPs model in Fig. 5 where the effect of the double positioning is taken into account.

had contained chair-like rings only and a low degree of amorphousness.

CONCLUSION

Our results are summarized as follows.
1) Void network in vacuum-deposited films is a main source of the amorphous ESR center.
2) Existence of single crystal substrates trigger the crystallization of deposited films. Crystallization takes place at the amorphous--single interface and inside the film itself.
3) The thicker the amorphous layer is, or the higher the degree of amorphousness is, the worse the crystallinity after crystallization.
4) Additional argon ion implantation into deposited films increases the degree of amorphousness.
5) Twin patterns obtained from post-crystallized implanted layers and deposited films can be explained by taking into account the secondary twins joined on the "downwards-pointing" faces of primary twins.

REFERENCES

1) D.J. Mazey, R.S. Nelson and R.S. Barnes, 1968, Phil. Mag., 17, 1145.

2) L. Csepregi, J.W. Mayer and T.W. Sigmon, 1976, Appl. Phys. Letters, 29, No.2, 92.

3) S.C. Moss and J.F. Graczyk, 1969, Phys. Rev. Letters, 23, 1167.

4) D.B. Holt, 1974, Thin Solid Films, 24, 1.

5) D.W. Pashley and M.J. Stowell, 1963, Phil. Mag., 8, 1605.

6) S. Ino, 1966, J. Phys. Soc. Japan, 21, 346.

7) B.L. Crowder, private communication.

ENHANCED AND INHIBITED OXIDATION OF IMPLANTED SILICON*

G. Mezey, T. Nagy, J. Gyulai,** E. Kotai, A. Manuaba and T. Lohner

Central Research Institute for Physics, 1125 Budapest, Hungary

and

J.W. Mayer

California Institute of Technology, Pasadena, California, 91125

ABSTRACT

Silicon samples of different orientation were implanted with ions of group IV, III and V and subsequently oxidized (sometimes after a preanneal). Among species investigated Ge, Si and Ga showed strong passivation. Passivation proved to be partly connected with presence of disordered layers, therefore, oxidation rates correlate well with lattice reordering. As in case of lattice reordering, correlation was found between oxide growth rate and substrate orientation.

INTRODUCTION

The influence of ion-implantation on oxidation properties of silicon was an early finding (1). The problem, however, is of great interest, especially in cases where implantation is used as a predeposition step and heat treatment is supposed to drive atoms to proper depth and this heat treatment is performed under an oxidizing ambient.

On the other hand, if ion implantation can cause passivation of the silicon surface without deteriorating its characteristics, new ideas in device fabrication may also arise.

The present experiments represent first results of a systematic study of oxidation phenomena. Stimulus was also taken from the work of the Harwell group on titanium (2), who found correlation between oxide growth and difference in electronegativity of matrix and dopant atoms.

Some of the investigated species were chosen to explore whether similar relationships would hold in a silicon matrix. Most emphasis was, however, put on experiments to separate the effect of dopants and of the disorder on oxide growth rates. When talking of reordering the lattice, the influence of substrate orientation was also investigated.

The measuring technique for experiments in this paper is partially based on a very sensitive oxygen determination worked out in our laboratory (3,4). This technique uses an $(\alpha\alpha)$-type nuclear reaction, i.e. a "resonance" at 3.05 MeV for He^+ ions scattered on oxygen atoms. As we showed earlier, with such conditions the sensitivity of detecting oxygen is enhanced by a factor of 30 over regular RBS.

As an addition to show sensitivity of the technique, we present determination of amount of oxygen on a thin aluminum foil. (A collimator system was applied in front of the detector to reduce background.) Thicknesses of thinnest oxide layers were determined by this technique, Fig. 1, to be about 10^{15} O/cm^2.

To understand more about the role of the implantation disorder in oxide growth, we first separated dopant effects from radiation damage. Therefore, Si, Ge, and O were implanted in Si of <111> and <100> orientation. Dopant effects were also established using B, Al, P, Ga, As and Sb. Some basic phenomena were clearly established, but reproducibility was not good enough to make quantitative statements. We will briefly discuss also these findings, though further work is necessary.

EXPERIMENTAL

Silicon wafers used in these experiments were 1-5 ohm cm, <111> and <100> oriented crystals. Implantations for lower energy (40-80 keV) were performed at room temperature. Variable energy implantation of Si and O was made at LN_2 temperature to get a fully amorphous surface. For lower energy implants the investigated dose ranged from 1×10^{14} to 6×10^{15} cm^{-2}. For the variable energy implantation 50 keV, 4×10^{16}; 100 keV, 8×10^{14}; 150 keV, 1×10^{15}; 200 keV, 4×10^{15}; 250 keV, 4×10^{15} cm^{-2} were used.

The oxygen implantation was made into a pre-amorphized surface using O_2^+ beam: 50, 4.4 x 10^{14}; 80, 1.3 x 10^{15}; 140, 2 x 10^{15}; 200 keV, 2.54 x 10^{15} cm^{-2}. Most of the oxidation was made in a dry O_2 atmosphere at 900°C. Special care was taken to the reference samples: all samples were partly masked against ions and the implanted and non-implanted parts were not separated throughout. To detect oxide thicknesses two closely adjacent points were chosen. To cross-check homogeneity, some extra reference samples were also oxidized at the same time. If preanneals were made, usually 550°C, 3 hr was chosen in dry N_2 to achieve full regrowth (5). In some cases, 900°C 30 minutes preanneal in dry N_2 gas was applied.

To analyze oxide layers, mostly nuclear methods were applied. The number of oxygen atoms on the surface was compared on implanted and non-implanted portions. For the short oxidation times the 3.05 MeV reaction was used to get accurate results. For thicker layers 2 MeV backscattering cross-checked with ellipsometer measurements were used to get information also on composition. To get more reliable results, in most cases alignment was made on the silicon substrate and channeling spectra were taken to reduce the yield from the substrate.

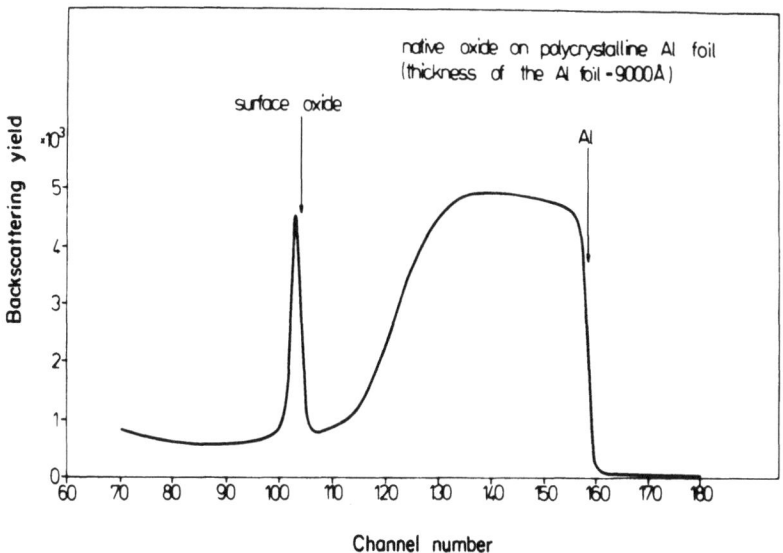

Fig. 1. Example for analyzing thin oxide layers by 3.05 MeV ($\alpha\alpha$) reaction.

RESULTS

Two sets of experiments were performed on <111> oriented silicon. In one case the implantation was done with 40 or 80 keV ions of silicon and germanium at room temperature, while in the other, amorphous layer was formed at LN_2. Two spectra taken on an implanted and a non-implanted area at resonance energy are shown in Fig. 2. Though it is a thin oxide, the ($\alpha\alpha$) reaction makes it clearly visible even riding in this case on a random spectrum. Special care must be taken, however, that the energy-loss corresponding to the oxide thickness should not exceed half-width of the resonance to keep conditions of a "thin layer approximation" (4). RBS and ellipsometry or mechanical step-height analyzer are proper methods to use for analysis of thicker layers.

Curves in Fig. 3 show relative oxide growth rate data for Si and Ge implants in terms of passivation, defined as $(N_i - N_o)/N_o$ (%), where N_i is the number of oxygen atoms (per unit area) on the implanted portion, N_o is the same quantity on the reference surface. Note that for the higher dose Ge implant, the passivation for a 20-40 min. oxidation may reach more than 90%, i.e. slowing down the oxidation rate by more than an order of magnitude.

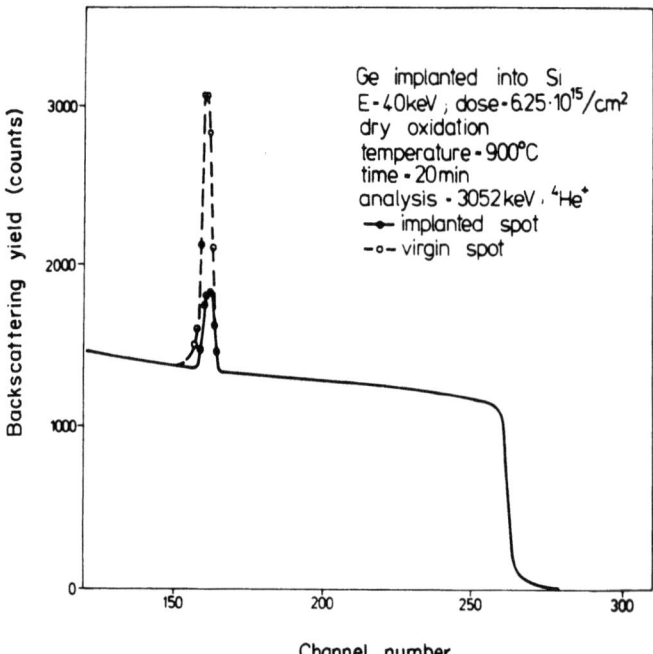

Fig. 2. Effect of Ge implantation on oxide growth rate of <111> Si.

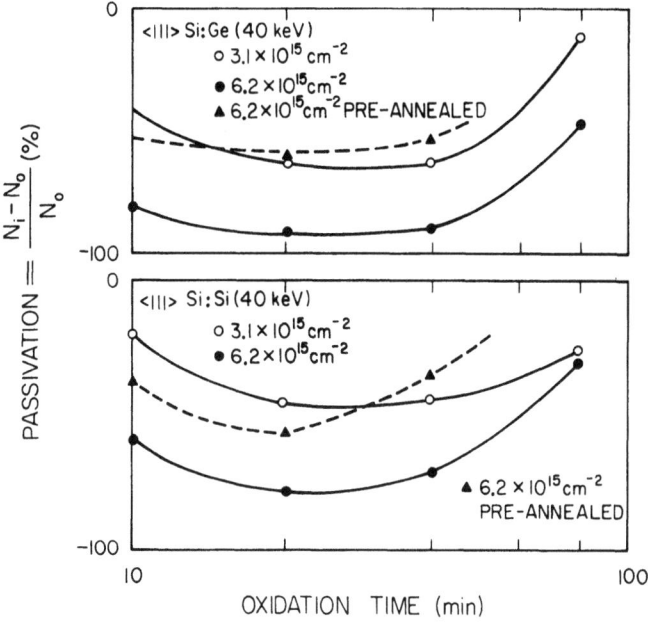

Fig. 3. Passivation of <111> Si by implantation of Ge and Si against a 900°C dry oxidation (N_i = number of O atoms/cm^2 for implanted, N_o the same for non-implanted case).

Silicon is also active, but to a smaller extent. Preanneals at 900°C, 30 minutes in dry N_2, where the <111>-Si is not perfectly annealed, also produce an effect.

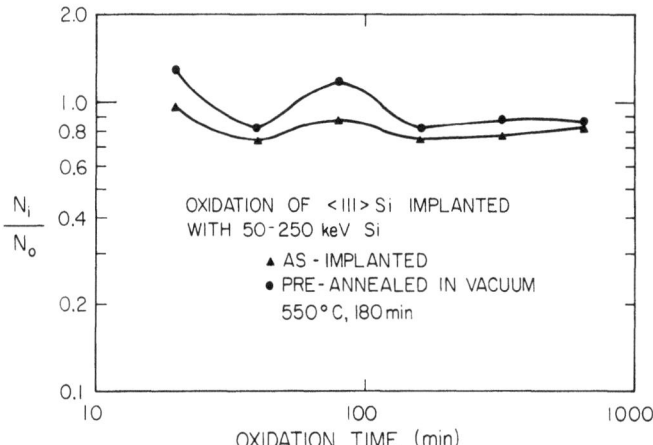

Fig. 4. Effect of implantation on oxidation properties of amorphized <111> silicon. Oxidation at 900°C, in dry oxygen.

Figure 4 shows similar behavior for amorphized Si. In this case longer oxidation was also made to look at oxide thicknesses where the effect of implantation becomes negligible. The fact that this leveling off is not at the $N_i/N_o = 1$ line, arises from differences of oxide thickness grown during the first period of oxidation. Absolute oxide thickness for 640 min. oxidation was 1475 Å on the implanted portion.

Oxidation of <100> oriented samples implanted with variable dose implantation is presented in Fig. 5, both for unannealed and pre-annealed cases. No enhanced oxidation was found for samples pre-annealed to reorder the amorphous layer. The large N_i/N_o ratio for Si implanted samples shows the influence of disorder. The ratio is smaller for samples implanted with Si and O. These curves are different than for <111> Si; we attribute the difference to the faster initial oxidation rate of <111> Si.

Implantation of dopant species in <111> Si resulted generally in an enhancement of the oxide growth rate. The most outstanding effect was found for antimony and arsenic. The only species that caused passivation was the gallium. Reproducibility of the measurement was not fully satisfactory, therefore, only a table is given on the results showing the overall picture (see Table 1). Table 2 shows the dose dependence of N_i/N_o for 40 keV Sb implantation. Results for 80 keV implantation were of the same character; the minimum value, $N_i/N_o = 0.61$ (in Table 2) for 1×10^{15} cm^{-2} was shifted, however, to a dose of 6×10^{14} cm^{-2} ($N_i/N_o = 0.63$).

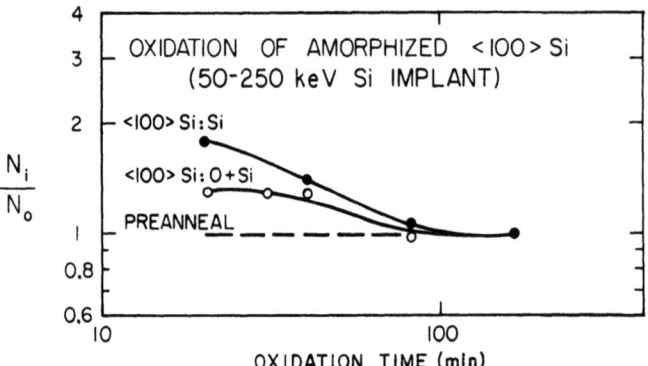

Fig. 5. Effect of implantation on oxidation properties of amorphized <100> Si. Case for successive variable energy oxygen implantation is also shown. Pre-anneal at 550°C, 3 hrs. in vacuum.

Table 1. Effect of 40 keV implantation on dry oxidation of <111> Si (relative oxide thickness for 900°C, 20 min.)

Dose (cm^{-2})	N_i/N_o					
	B	Al	Ga	P	As	Sb
1×10^{14}	-	-	-	-	1.51	2.5
1×10^{15}	1.71	1.25	0.74	1.52	1.65	1.2
3×10^{15}	-	1.91	-	-	2.3	3.8
6×10^{15}	1.25	-	0.37	1.61	2.3	3.1

Table 2. Effect of 40 keV antimony implantation on dry oxidation of <100> Si (relative oxide thickness for 900°C, 20 min.)

Dose (cm^{-2})	N_i/N_o
1×10^{14}	0.97
3×10^{14}	1.1
6×10^{14}	.92
1×10^{15}	.61
2×10^{15}	.83
3.9×10^{15}	1.4
6.2×10^{15}	3.1

DISCUSSION

The results are part of a detailed investigation to find correlation between implantation (doping and radiation damage) and oxidation of silicon surfaces.

It is clear by this stage that:
i. Implantation of group IV elements have a great influence on oxidation (passivation).
ii. Most other implants cause enhancement of oxidation except gallium.
iii. Substrate orientation plays a substantial role.

From (i) and (iii) it may be concluded that oxidation, when no dopant effects are present, is mainly governed by lattice perfection. Strict correlation was found between oxide growth and lattice reordering. That is the reason that implantation has a slight effect only on oxidation of <100> Si, which reorders faster than the <111> oriented crystal (6).

The results presented here suggest in connection with (ii) that a simple correlation between oxide growth rate and electronegativity cannot be found, as is the case for titanium (2). The lattice disorder effects dominate or at least compensate simple chemistry rules. To date, we have not found the final factors governing oxidation of silicon. However, from a practical point of view, the above results might find use to adjust oxidation rates on specific surface areas during, say, device fabrication.

ACKNOWLEDGEMENTS

Samples with variable energy implantation are acknowledged to Dr. T.W. Sigmon (Hewlett-Packard). Thanks are due to Ing. E. Pasztor for the other implantations and to Ing. J. Gyimesi (both at CRIP, Budapest) for the careful oxidations.

REFERENCES

1. O. Meyer and J.W. Mayer, Rad. Effects $\underline{3}$, 139 (1970).
2. e.g. J.O. Benjamin and G. Dearnaley in Application of Ion Beams to Materials, 1975, Eds. G. Carter, J.S. Colligon and W.A. Grant, Inst. of Physics Series #28, London and Bristol, 1976) p. 141.
3. L. Keszthelyi, I. Demeter, G. Mezey, Z. Szokefalvi-Nagy and L. Varga, Proc. Intl. Meeting on Ion Implantation in Semiconductors, Rossendorf, GDR, ZfK-236, p. 111 (1972).
4. G. Mezey, J. Gyulai, T. Nagy, E. Kotai and A. Manuaba, in Ion Beam Surface Layer Analysis, Eds. O. Meyer, G. Linker and F. Kappeler, Plenum Press, New York-London, 1976) p. 303.
5. L. Csepregi, J.W. Mayer and T.W. Sigmon, Appl. Phys. Lett. $\underline{29}$, 92 (1976).
6. H. Muller, W.K. Chu, J. Gyulai, J.W. Mayer, T.W. Sigmon and T.R. Cass, Appl. Phys. Lett. $\underline{26}$, 292 (1975).

* Supported in part by the Institute of Cultural Relations, Budapest, and NSF (R. Hull).

** Also at California Institute of Technology, Summer 1976.

HIGH DOSE IMPLANTATION OF AU AND CU INTO SI STUDIED

BY AUGER ELECTRON AND BACKSCATTERING SPECTROSCOPIES

A. HIRAKI, M. IWAMI, K. SHUTO, T. SAEGUSA
and T. NARUSAWA

Faculty of Engineering, Osaka University
Suita, Osaka, JAPAN

K. GAMO and S. NAMBA

Faculty of Engineering Science, Osaka
University, Toyonaka, Osaka, JAPAN

Metastable metallic Si-Au and -Cu alloys were formed at room temperature by high dose implantation (60KeV: $\sim 10^{17}$ ions/cm^2) of Au and Cu into Si single crystals. The alloy phases were identified by Auger Electron Spectroscopy(AES).

1. INTRODUCTION

Ion implantation can be a powerful tool of producing directly, without any metallurgical process, stable and metastable(or new) alloys. This possibility has been tried, in several laboratories, to fabricate superconducting alloy materials with high transition temperature.

Hiraki and his coworkers have studied Si-Au metastable alloy in connection with the low temperature(100~200°C) SiO$_2$ formation over evaporated thin films of Au on Si single crystal substrates[1]. This metastable alloy was found to be present at the Si(substrate)-Au(film) interface[2], although no stable(or equilibrium) alloy phase exists in the Si-Au system.

An interesting point about this finding is that at the interface, the bonding nature of Si crystal changes from

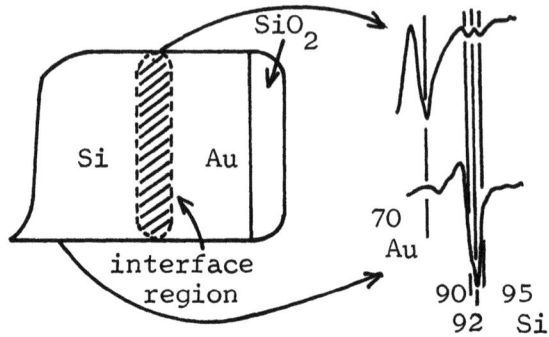

Fig.1

Si(LVV) Auger spectra from Si substrate and Si-Au interface.

covalent to metallic to form the metastable alloyed interface and, in addition, this change is induced by just the room temperature deposition of Au onto Si[3].

It is known that Si becomes metal in a liquid state and this metallic Si can be frozen in a rapidly quenched near eutectic($Si_{30} Au_{70}$) liquid. And this quenched Si-Au liquid has been shown to be very similar to the metastable phase found at the interface[4].

The present paper reports that the transition of Si from covalent to metallic can be induced by implantation of Au and also Cu into covalent Si crystals. Both Auger Electron Spectroscopy(AES), with aid of Ar^+ ion sputtering technique, and He^+ ion Rutherford Back Scattering (RBS) methods have been employed, respectively, to identify the Si metallic state and check distribution of implanted Au and Cu in the substrates.

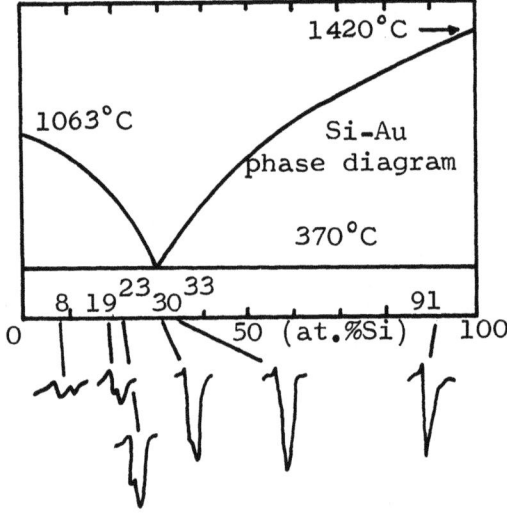

Fig.2

Si(LVV) Auger peaks of Si-Au vapor quenched alloys as a function of Si concentration (in at.%).

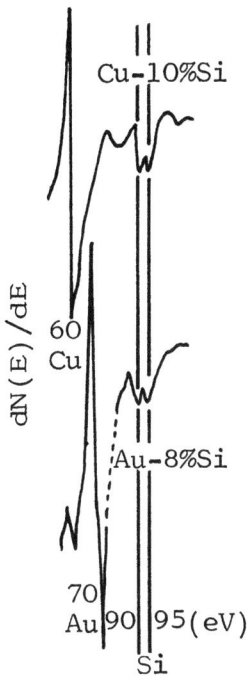

Fig.3

Si(LVV) Auger spectra from vapor quenched alloys(Si-Au and Si-Cu). In both alloys the Auger spectrum has the double peak.

2. AES OBSERVATION OF METALLIC SI IN METASTABLE SI-AU ALLOYS

Hiraki et al studied Si-Au interface using AES. In this study, onto a clean surface of a Si substrate a Au film was deposited at temperature lower than 50°C in high vacuum($\sim 10^{-9}$ Torr) and then Si(LVV) or Si(LMM) Auger spectra, where V represents a valence energy state, were taken from both the substrate and the interface[2,3]. As is shown schematically in Fig.1, the Si spectrum of the interface region has two peaks at 90 and 95eV and clearly differs from that of pure(covalent) Si at 92eV.

With regard to the double-peaked Si spectrum, it was proposed, through AES study of thin films of Si-Au alloys(with various composition ratios) obtained by very rapid vapor quenching , that the spectrum correspoded to the valence electronic state of Si atoms forming metallic bonds with Au atoms from the following.

Figure 2 shows AES spectra of several vapor quenched Si-Au films together with the phase diagram. The double peak was observed only for alloy films containing more than 70at.% Au; in films with much more Si only the single peak(92eV) could be seen. This result is explained as follows: each Si atom can form stable

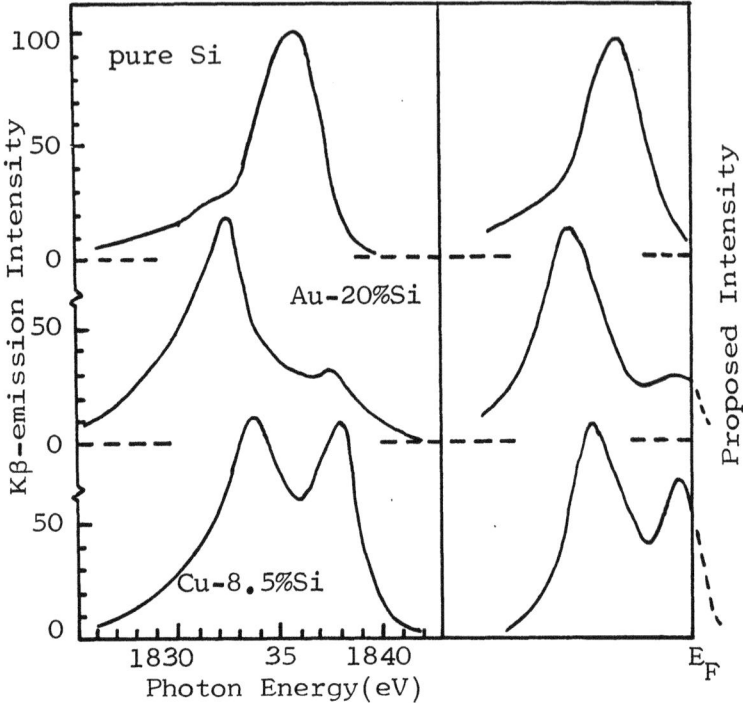

Fig. 4 Si Kβ-emission(soft X-ray) from pure Si, Au-20%Si and Cu-8.5%Si alloys. Estimated density of state of Si 3p-electrons are also shown.

metallic bonds with surrounding metal atoms only if sufficient Au atoms are present; in Au-deficient alloys only Si-Si covalent bonds arise. And the critical amount of Au for the stabilization of Au-Si metallic bonds is estimated from Fig.2 to be around 70at.%.

In addition, essentially the same double peaked spectrum was observed from the vapor quenched Si-Cu film which contained ~90at.% of Cu atoms as shown in Fig.3. The origin of this similarity between Si-Au and Si-Cu alloys was understood by the following Soft X-ray Spectroscopy(SXS)[5]. Figure 4 shows the results from the measurements of Si-Kβ X-ray emission, which yields information about the local density of 3p-electrons (because of the restriction due to the selection rule). The Si 3p-band has only a single peak in pure(or covalent) Si but has two peaks in both metal rich(~90at. %) Si-Au and Si-Cu alloys owing to the interaction of Si 3p-electrons with Au 5d- and Cu 3d-electrons[6] increasing appreciably the density of states of the 3p-

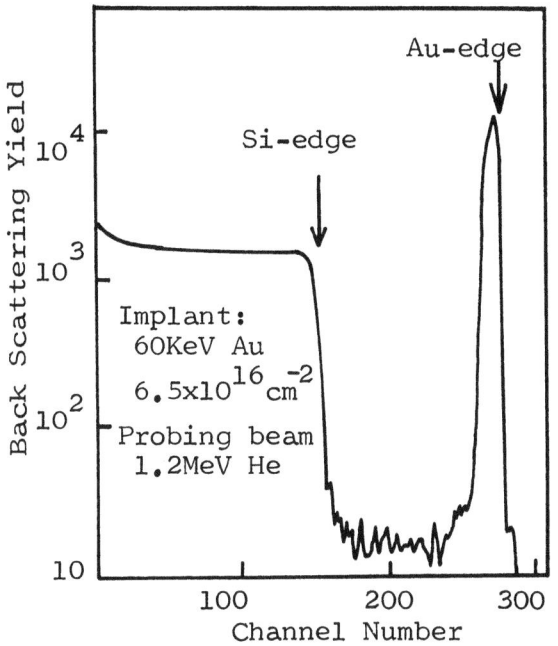

Fig. 5

Back scattering spectra of Au-implanted Si.

electrons at the Fermi energy(E_F) and therefore Si is metallic here. Also similar density of states of Si 3p-electrons in both alloys is consistent with the result of AES shown in Fig.3.

From above mentioned facts, the identification of metallic Si in Si-Au and Si-Cu systems by the double-peaked Si(LVV) Auger spectrum may be approved.

3. HIGH DOSE IMPLANTATION OF AU AND CU INTO SI CRYSTAL

To provide a situation for Si to transform from covalent to metallic, since the stabilization of metallic Si requires a fair amount(more than 50at.%) of metal atoms(as recognized from Fig.2), it may be necessary to implant metal(Au or Cu) ions to very high doses. Therefore, high dose implantation were performed at 60KeV to a dose around $\sim 10^{17}$ ions/cm² into chemically polished Si (111) wafers, 8° off the [111] axis to avoid channeling, at room temperature.

After the implantation, at first 1.2MeV He^+ RBS spectra of the specimens were taken to get the distribution of the implanted metal ions and then AES profiles were obtained by thin layer removal using Ar^+ ion

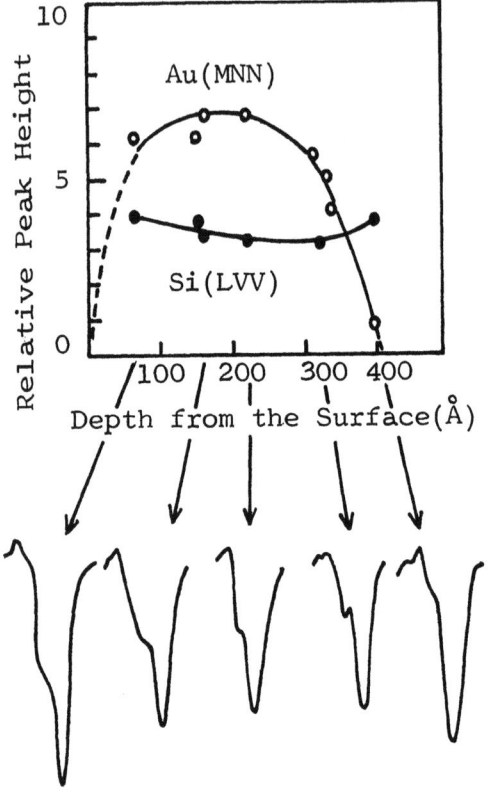

Fig.6

In-depth Auger profiles of Au(MNN) and Si(LVV) of Au-implanted Si.

sputtering at 1KeV with current density of ~10μA/cm^2.

Figure 5 shows a RBS spectra of Au implanted specimen from which the dose (6.5×10^{16} ions/cm^2) and depth profile can be obtained.

The amount of the implanted Au in the Si substrate (or the Au peak height) did not exceed 40at.% even if the implantation was undertaken for long time indicating that the sputtering due to the implantation determined the actual concentration of implanted Au.

Figure 6 is an AES profile of the same specimen (of Fig.5): AES spectra from Si(LVV) and Au(MNN) transition are plotted against depth from the surface (which is converted from the Ar$^+$ ion sputtering time and RBS data). From Si(LVV) spectra, it can be said that although there is some symptom of covalent to metallic transformation which can be detected by the shape of the spectra about to split from the single peak to the double peak, the complete transformation did not take place presumably due to the deficiency of implanted Au owing to the above mentioned sputtering effect.

Then as an alternative to Au which can not be

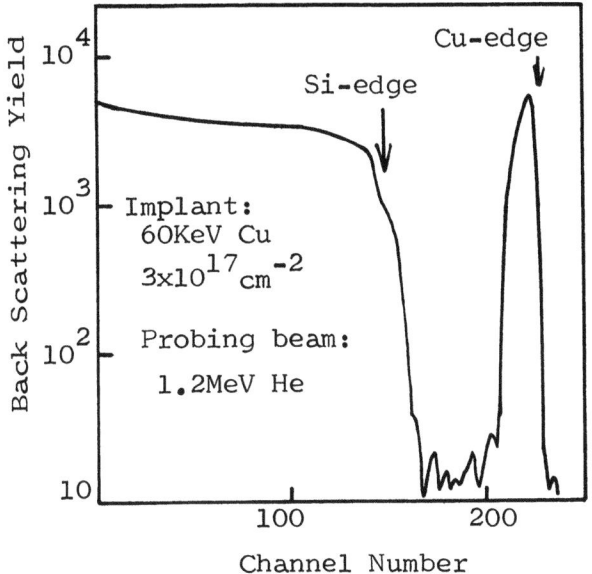

Fig. 7

Back scattering spectra of Cu-implanted Si.

implanted more than 40at.%, Cu implantation was tried in hope that enough Cu can be implanted to induce the transformation which can be checked also by AES as was described in section 2.

As seen from Fig. 7, it was possible to implant more Cu(3×10^{17}/cm^2) than Au(6.5×10^{16}/cm^2: Fig. 5) —— this value was also limited by the sputtering effect as in the case of Au implantation. Due to lighter mass than Au, the implanted region was more than twice as thick ($\sim$850Å) and the highest concentration of the implanted Cu was estimated to be $\sim$80at.%. And in this case, as expected, Si(LVV) spectra in AES profile of Fig. 8 indicate that the covalent to metallic transformation has taken place in the implanted region where the double peaked Si spectra are clearly observed. In other words, by the implantation it is possible to induce directly the formation of Si-Cu metallic alloy in which Si atoms form metallic bonds with the implanted Cu atoms.

4. CONCLUDING REMARKS

In the present study, it has been shown that high dose implantation of Au and Cu into single crystal of Si can induce formation of metastable metallic alloys (Si-Au and Si-Cu) through the transformation of Si from covalent to metallic state.

There have been several reports claiming the formation of alloy phase by implantation based upon the

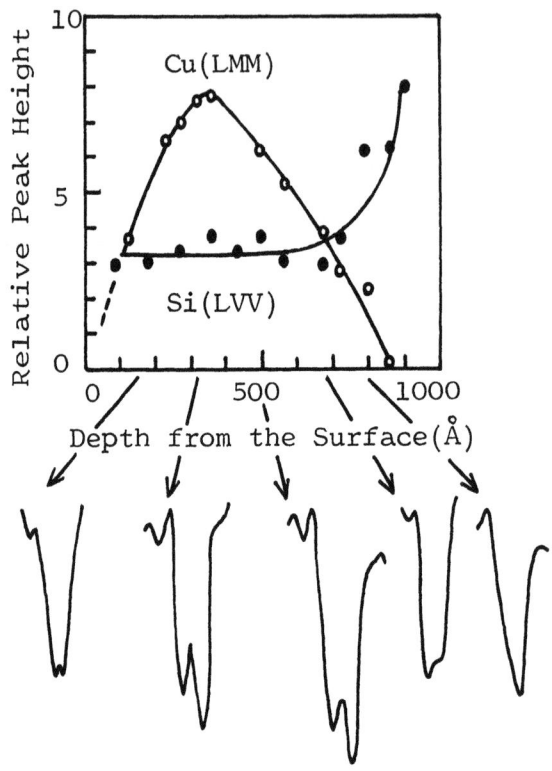

Fig. 8

In-depth Auger profiles of Cu(LMM) and Si(LVV) of Cu-implanted Si.

macroscopic measurements, such as critical temperature (Tc) measurements of the implantation induced (or modified) superconducting alloys. The present work also supports more directly the possibility of formation of alloy phase by implantation through the microscopic observation (AES) of changes in the valence electronic states of constituent atoms, i.e. covalent to metallic transformation of Si atoms.

REFERENCES

[1] A. Hiraki, E. Lugujjo and J. W. Mayer, J. appl. Phys. 43 (1972) 3643.
[2] T. Narusawa, S. Komiya and A. Hiraki, Appl. Phys. Letters, 21 (1972) 272.
[3] T. Narusawa, S. Komiya and A. Hiraki, Appl. Phys. Letters, 22 (1973) 389.
[4] A. Hiraki, A. Shimizu, M. Iwami, T. Narusawa and S. Komiya Appl. Phys. Letters, 26 (1975) 57.
[5] K. Tanaka, M. Matsumoto, S. Maruno and A. Hiraki, Appl. Phys. Letters, 27 (1975) 529.
[6] K. Terakura, J. Phys. Soc. Japan, 42 (1976) 450.

DEFECTS INTRODUCED INTO SILICON BY BORON

IMPLANTATION IN THE MeV ENERGY RANGE

J.C. Muller, R. Stuck, P. Siffert

Centre de Recherches Nucléaires et Université
Louis Pasteur Laboratoire de Physique des Rayonnements
et d'Electronique Nucléaire
F 67037 Strasbourg Cedex (France)

S. Kalbitzer

Max Planck Institut für Kernphysik
69 Heidelberg (Germany)

ABSTRACT

The nature of the defects introduced into n-type silicon by implantation of 8 MeV boron ions has been investigated by thermally stimulated current measurements performed directly on the produced p-n junctions. These curves, analysed by the "delaying heating method", are compared to those obtained for low energy implants. Levels located at $E_c - 0.18$, $E_c - 0.42$, $E_v + 0.26$, $E_v + 0.32$ eV are identified. They are quite the same as those observed for low energy implants achieved in the same materials, their annealing however occurs at higher temperature.

INTRODUCTION

During the last years, strong efforts have been put into the study of ion implantation into semiconductors, mainly silicon. Detailed results concerning the macroscopic properties of the layers are available for low energy implantation especially relative to electrical conductivity, damage distribution and annealing. Much less has been undertaken to identify the damage levels created in the forbidden bandgap by the implantation process. Results have been obtained only by a few authors by capacitance [1], photoconductivity [2], junction current [3] or thermally stimulated current (TSC) measurements [4,5].

For higher implant energies (>1 MeV) of heavy ions the problem is still open. There are only data of Davies and Roosild [6,7] obtained by TSC measurements on 2 MeV $^{16}O^+$ and $^{12}C^+$ implanted silicon layers. Here, the results on microscopic defects introduced into silicon by high energy implantation of boron ions are compared with our previous studies of the same material at low energies. Mainly TSC measurements have been used, since a.c capacitance techniques as performed by Chan and Sah [1] are not so useful for multi-level systems.

The principle of the TSC method is well known. It consists in filling the traps after cooling down ($\simeq 77$ K) the implanted diode and then heating the sample at a constant rate in order to observe the thermally stimulated current as the traps become emptied. The energy level within the bandgap of a particular trapping center can be deduced from the TSC curve, since it is related to the temperature at which the level is emptied.

The analysis of the deep levels from the frequency dependence of the capacitance has been employed by several authors [8,11]. When a small a.c signal is applied to the reverse biased junction, the defects at the space charge region boundary are alternatively charged and discharged. When the frequency of the a.c signal is increased, the charge state of a deep level no longer can follow at high frequencies. From the change of capacitance with applied voltage, frequency and temperature, the main parameters of a deep level can be evaluated.

EXPERIMENTAL

After conventional etching, 150 Ω cm, n-type silicon samples have been implanted with 8 MeV boron ions at the Heidelberg Tandem Van de Graaff accelerator at doses ranging between 10^{13} and 10^{14} cm^{-2}. The beam intensity was 20 nA of $^{11}B^+$. In order to obtain homogeneous implantations the wafers were rotated through the beam [12]. The projected range of these ions is approximately 10 μ. Annealing was performed for 30 min. under vacuum at temperatures up to 800°C.

Carrier density and mobility in the layer were measured by the van der Pauw technique. The capacitance measurements were performed in the frequency range of 5-500 kHz. For TSC measurement, the device was mounted in a cryostat and heated at a constant rate $\beta = 0.75$ K/sec. The details of the experimental procedure have been described in a previous paper [13].

RESULTS

*Carrier density: Effective carrier density measurements as a function of annealing temperature (Fig. 1) show that 600°C is sufficient to bring about full electrical activity for an 8 MeV - 10^{13} B/cm^2 implant. At a dose of about 10^{14} cm^{-2} the pattern is more complex. It is interesting to notice that there is a tendency toward higher annealing temperatures with decreasing implantation energies for obtaining 100% yield.

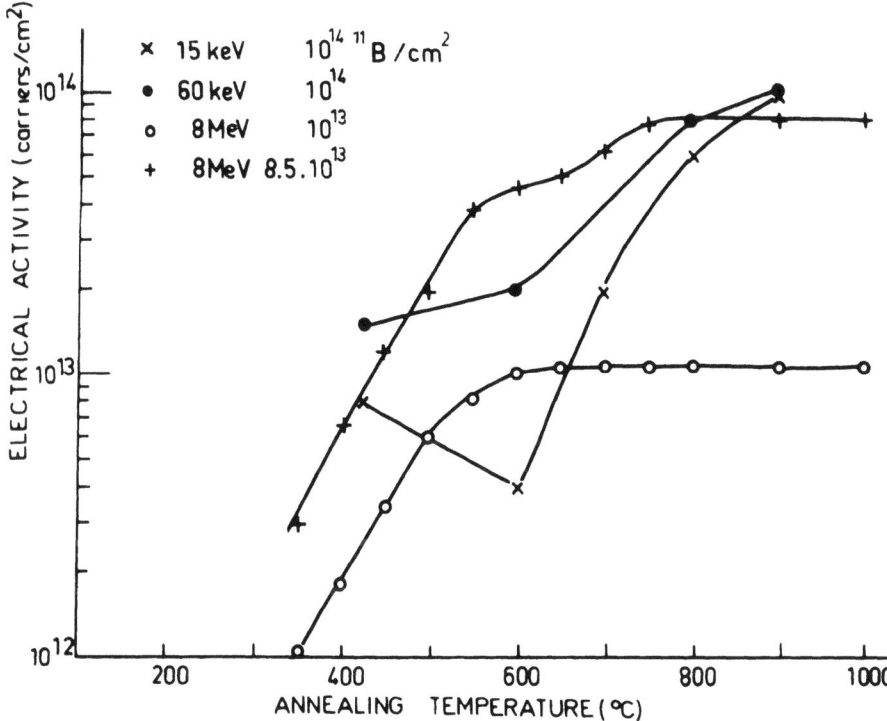

Fig. 1. Electrical activity vs anneal temperature of boron implanted layers at various energies and doses.

*Capacitance measurement: Fig. 2 shows the results of C-V measurements performed on the 8 MeV implanted diode as a function of frequency and annealing temperature of the device. Contrary to the low energy case, even after 400°C annealing a strong frequency dependence is observed, which decreases at 600°C. Following the model of Sah and Reddi [8] and Schibili and Milnes [10] the defect can be characterized from this variation of capacitance. Especially, the slope of the temperature dependence of the frequency dependent part of the functions

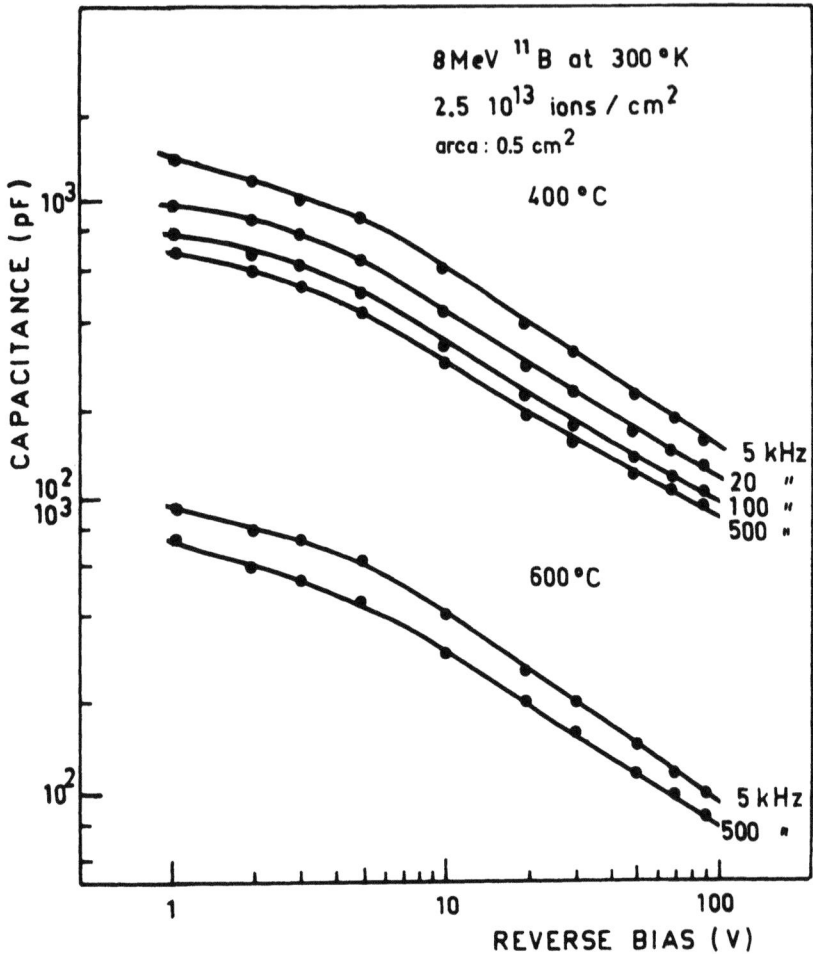

Fig 2. Capacitance-voltage characteristics of an 8 MeV ^{11}B implanted diode at various isochronal (30') anneal temperatures and different frequencies (5 kHg to 500 kHg) of the measuring bridge.

capacitance directly gives the activation energy of the trapping center. We used this procedure with success for low energy boron implants (Fig. 3) where only one level has been observed [13]. The application of this method to high energy implants is not possible due to the presence of several levels; the capacity frequency dependence no longer would follow the predicted law. For qualitative purposes, however, the differential capacitance ΔC between low and high frequency versus annealing (Fig. 4) is still useful.

DEFECTS INTRODUCED INTO SILICON

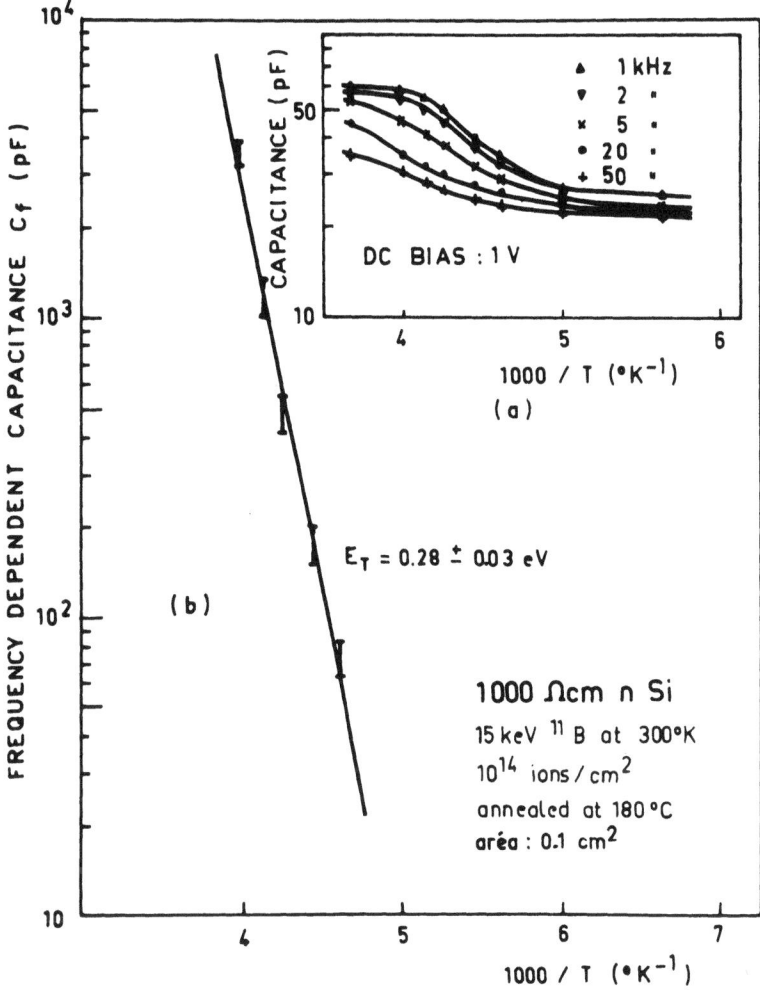

Fig. 3 a) Capacitance-temperature curves of 15 keV ^{11}B implanted at various frequencies. b) Plot of the temperature dependent part of the junction capacitance.

Temperatures of 600–700°C are required for the high energy implants to get a substantial reduction of the frequency dependence. For low energies, 300°C is sufficient to achieve the same result.

*TSC measurements: Fig. 5a shows typical TSC curves obtained with diodes implanted at 8 MeV after two different excitation processes, namely: 1) forward bias before switching to 10 V reverse voltage at 77 K and 2) light excitation by 0.37 μ wavelength source, which is absorbed to 99% within 0.5 μ of silicon. For comparison, the results observed after low energy

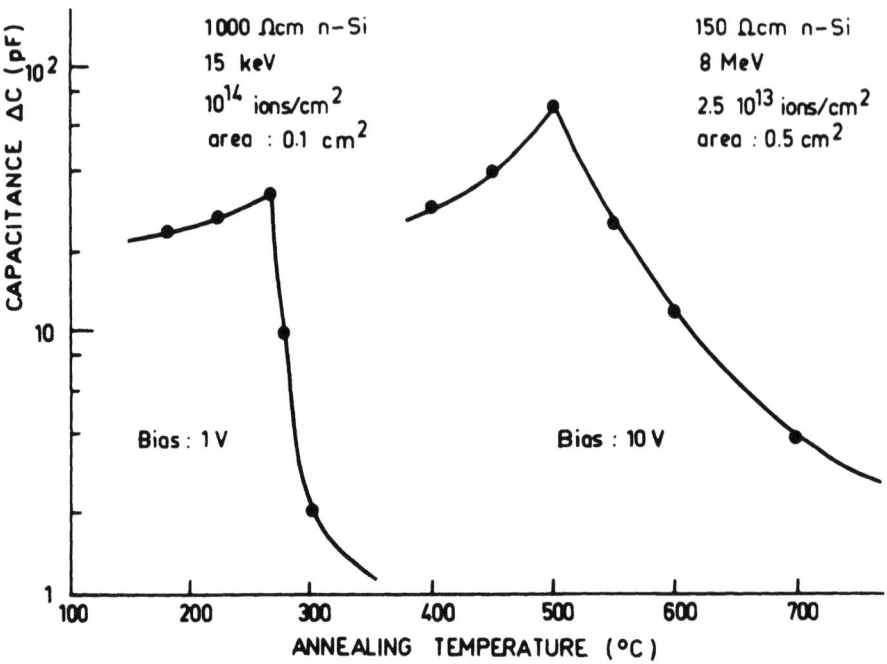

Fig. 4 Differential capacitance ΔC between 5kHz and 500 kHz vs annealing temperature for 15 keV and 8 MeV boron implants.

implants on 1000 Ω cm n-type silicon are shown on the same figure 5b. The analysis of the peaks has been performed by making use of the delayed heating method we developed in order to determine the activation energy (with respect to E_v or E_c) and capture cross section [13]. The results for both low and high energy implants are shown in Table 1. The nature of the corresponding defects will be discussed later. Furthermore, the annealing of the various levels has been followed. Generally speaking, they anneal completely around 600-650°C (Fig. 6a). A comparison with the behavior of low energy implanted layers is quite interesting since the defects in this latter case disappear at 300°C already (Fig. 6b).

DISCUSSION

The results reported on Table 1 indicate the presence of six different energy levels. The three major peaks are identical to those observed by the TSC technique on the same material after irradiation with a strong ^{60}Co source (7 x 10^4 rad/h)[14], except for that at E_c - 0.18 which has not been seen. We have also compared our results with those of Davies [7] for $^{12}C^+$ implanted layers; the temperature of the peaks we recorded (123, 147 and

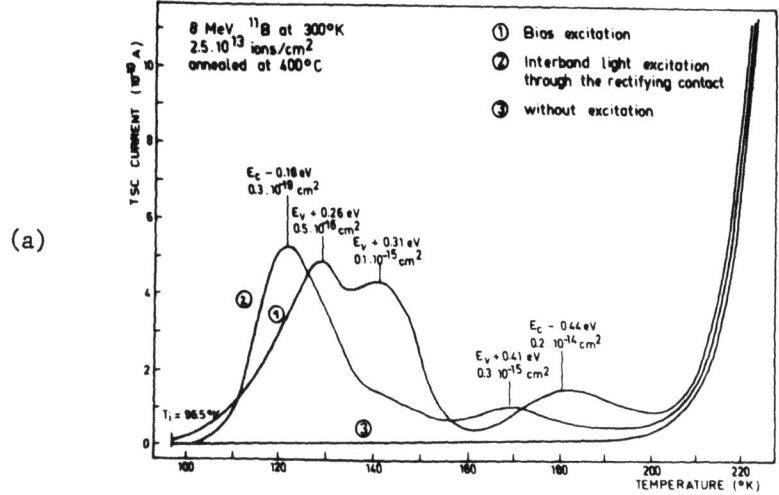

(a)

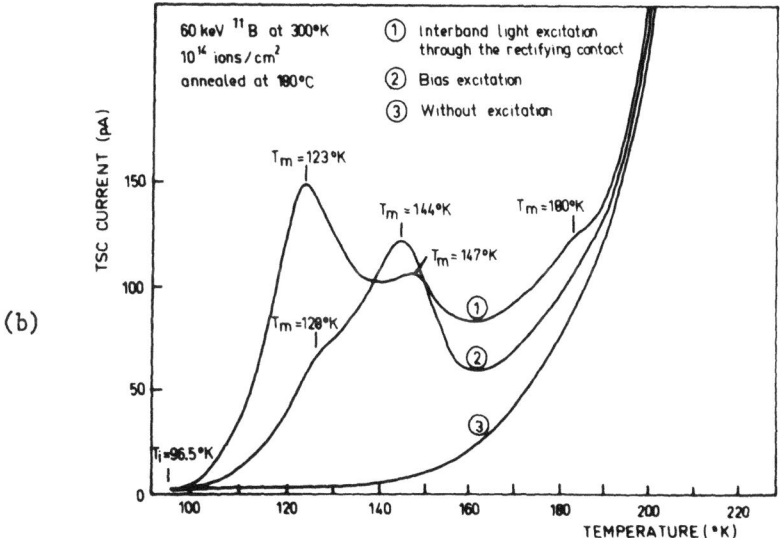

(b)

Fig. 5 a) TSC curves obtained on a 8 MeV ^{11}B implanted p-n junction (area 0.5 cm^2) realized on 150 Ωcm n-type silicon, for a starting temperature T_i = 96.5 K, a constant heating rate of β = 0.75 deg/sec, an applied bias of -10 v and different excitation modes. b) TSC curves recorded in the same conditions for a diode (area 0.12 cm^2) realized by implantation of boron ions at low energy (60 keV) on 1000 Ωcm n-type silicon.

TABLE I

Peak Temperature (°K)	Trapping level E_T (eV)	Cross section σ (cm^2)	Boron Ions RT 60 keV	Boron Ions RT 8 MeV	Light Particle ^{60}Co
123	$E_C - 0.18 \pm 0.01$	$0.3 \cdot 10^{-19}$	X	X	
128	$E_V + 0.26 \pm 0.02$	$0.5 \cdot 10^{-16}$	X	X	X
144	$E_V + 0.31 \pm 0.02$	$0.1 \cdot 10^{-15}$	X	X	
147	$E_C - 0.34 \pm 0.02$	$0.3 \cdot 10^{-15}$	X	(X)	X
170	$E_V + 0.41 \pm 0.02$	$0.3 \cdot 10^{-15}$	(X)	X	
180	$E_C - 0.44 \pm 0.02$	$0.2 \cdot 10^{-14}$	X	X	X

TABLE II

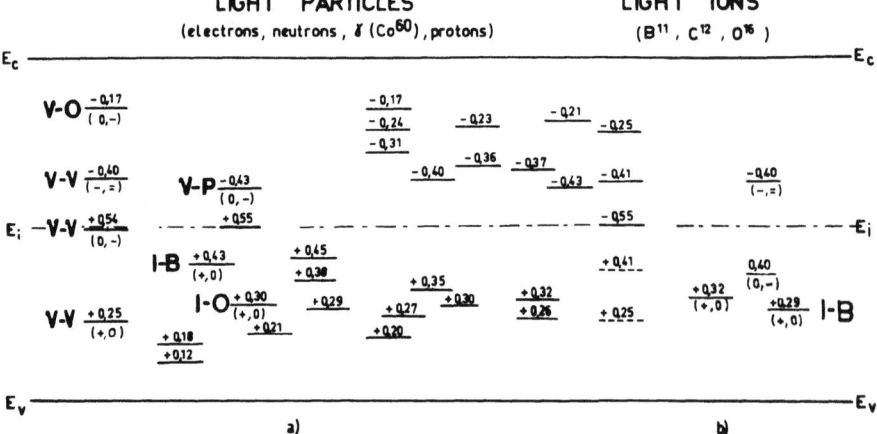

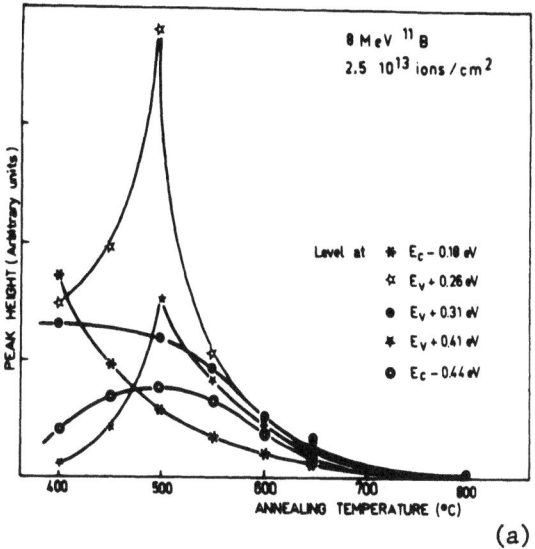

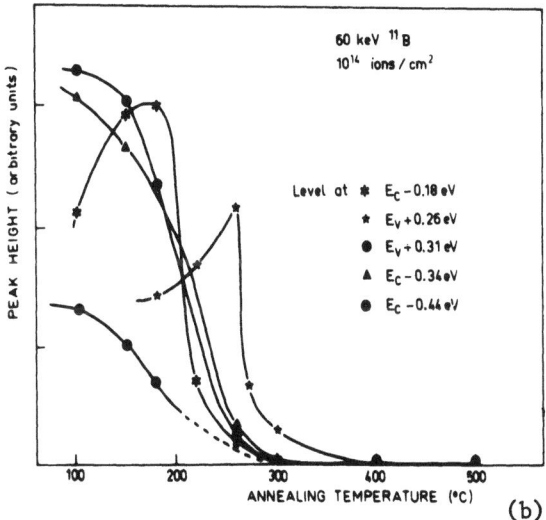

Fig 6 Evolution of the TSC peak heights as a function of annealing temperature for 8 MeV (a) and 60 keV (b) implants.

and 180 K) is very close to that of this author (121, 146, and 177 K) for the same excitation mode. But, the simplified analysis of the TSC curve he used leads to different values of the defects energy. A first conclusion we draw is that the nature of the defects is rather independent of the heavy ion species. For light particles this has been known quite a time [15]. This is plausible if the material is either pure or at least of identical impurity content. By following the concentration of the different levels (peak area on TSC curves) as a function of implant energy

(60 keV - 8 MeV), it appeared that the levels of $E_v + 0.26$ and $E_c - 0.44$ eV have increased by the same proportion; a similar correlation exists between $E_c - 0.18$ and $E_v + 0.31$ eV. To identify the nature of the defects, we compared our results with the generally accepted levels of certain well identified atomic structures in silicon. (Table II). The most probable associations are the V-V, V-O, I-O complexes, in view of the good agreement in the energies. This is compatible with the oxygen content usually found in FZ material. We cannot decide presently whether I-B (at $E_v + 0.29$ eV) contributes to the $E_v + 0.31$ eV level or not. (However, Watkins [16] has recently shown that the assignment of the level at $E_v + 0.29$ eV to the interstial boron cannot be correct.)

However, a major difference appeared in the annealing behavior between low energy (or light particles) and high energy bombarded layers. Both TSC and differential capacitance measurements have shown that the primary defects anneal at 300 and 600°C, respectively. This result is rather unexpected in the light of the preceding explanations. We would expect the defects to anneal at the same temperature independent of the bombarding energy. Maybe the distance from the heavily damaged region toward the surface plays a greater role than expected.

CONCLUSION

It appears that the TSC technique is well adapted to investigate ion implanted layers. By a proper choice of the conductivity type of the silicon, the diode realized by the ion bombardment can be used directly for the measurements. The "delayed heating" method we developed to analyze the experimental curves strongly improves the precision of the energy determination. Since the process gives access directly to the microscopic nature defects, it is complementary to the usually employed backscattering technique for characterizing the damaged layers. However, the simplicity of the TSC procedure, the high sensitivity (10^{11} cm^{-3} compared to 10^{16} cm^{-3}), and the selectivity to the different microscopic damages constitute strong advantages. By a proper choice of the experimental conditions, it will be possible to determine the concentration of the individual levels identified.

The increase of the annealing temperature of the defects when the implant energy becomes higher cannot be explained yet and further experiments are required.

ACKNOWLEDGEMENTS

We want to thank Mrs. Weymann and Mrs. Klotz for their assistance in the preparation of wafers and diagrams.

REFERENCES

1. W. W. Chan and C. T. Sah, J. Appl. Phys. 42, 4768(1971).
2. B. Netange, M. Cherke and P. Baruch, Appl. Phys. Lett. 20, 349(1972).
3. A. P. Karatsyuba, V. I. Kurinny, S. V. Rytchkova, T. P. Timashova and V. V. Yudin, Radiation Damage and Defects in Semiconductors, p. 81 (Institue of Physics, London, 1972).
4. P. Ashburn, D. V. Morgan, Solic State Electr. 17, 689(1974).
5. J. C. Muller, R. Stuck and P. Siffert, Int. Conf. on Lattice Defects in Semiconductors, Freiburg (1974), Inst. Phys. Conf. Sev. No. 230, p. 513(1975).
6. D. E. Davies and S. Roosild, Appl. Phys. Lett. 18, 548(1971).
7. D. E. Davies and S. Roosild, Int. Conf. of Ion Implantation in Semiconductors (Garmisch), ed. I. Ruge and J. Graul (Berlin, Springer-Verlag) p. 23(1971).
8. C. T. Sah and V. G. K. Reddi, IEEE Elect. Devices 11, 345(1964).
9. R. R. Senechal and J. Basinski, J. Appl. Phys. 39, 3723(1968).
10. E. Schibli and A. G. Milnes, Solid State Elect. 11, 323(1968).
11. W. G. Oldham and S. S. Naik, Solid State Elect. 15, 1085(1972).
12. A. Kostka and S. Kalbitzer, Rad. Effects 19, 77(1973).
13. J. C. Muller R. Stuck, R. Berger and P. Siffert, Solid State Electr. 17, 1293(1974).
14. J. C. Muller, P. Siffert and H. M. Heijne, to be presented at the Int. Conf. on Lattice Defects in Semiconductors, Dubrovnik (1976).
15. L. J. Cheng, J. C. Corelli, J. W. Corbett and G. D. Watkins, Phys. Rev. 152, 761(1966).
16. G. D. Watkins, Phys. Rev. B13, 2511(1976).

VACANCY-IMPURITY COMPLEXES IN IMPLANTED AND HIGH TEMPERATURE

ANNEALED n-GaAs

E.V.K. Rao, N. Duhamel, P.N. Favennec[†] and H. L'haridon[†]

Centre National d'Etudes des Télécommunications

92220 BAGNEUX - [†]22301 LANNION - FRANCE

ABSTRACT

In this paper we present different experimental results that show the existence of compensation in n-GaAs when implanted and subsequently annealed at high temperatures. This is realized by studying the layers implanted with nondopant or electrically inactive impurities like, H, B and As. We also present the results of photoluminescence measurements performed to investigate the nature of the defects responsible for compensation.

INTRODUCTION

The successful application of ion implantation for making devices in GaAs warrants a better understanding of the defects in implanted layers. Recently several papers are published on the study of native vacancy associated complexes in GaAs substrates (1,2,3). However, to our knowledge only few people have worked on the problem of defect compensation in implanted GaAs (4). Davies et al (5) have implanted non dopant B^+, N^+ and F^+ ions at low dose levels ($\leq 10^{11}$ ions/cm^2) in n-GaAs to investigate the defect compensation. Their results show that after a 600°C anneal the original electrical activity of the substrate is not entirely recovered. At the same time, few other authors have proposed the technique of coimplantation to inhibit the formation of intrinsic layers (6), to improve the electrical activity of the dopant impurities (7), and also to control the impurity diffusion (8).

In the present work we make an attempt to determine the nature of the defect compensation in layers implanted and subsequently annealed at high temperatures. This is accomplished by working with layers in which nondopant impurities, like H, B and As, are implanted in n-GaAs substrates of varied initial doping levels. These impurities being electrically inactive should help to detect the compensation due to defects when introduced in a substrate of known initial doping level.

EXPERIMENTAL PROCEDURES

The substrates used in this study are n-GaAs samples of <100> orientation. They are either undoped or doped with Si or Te with free carrier concentrations ranging from 2×10^{16} to 5×10^{18} cm^{-3}. The implants of H, B and As ions are carried out in a Van de Graaff accelerator. The incident ion beam is 7° away from the <100> axis. All the implants are conducted at room temperature. The ion doses are varied in the range of 10^{12} to 10^{15} ions/cm^2 and the dose rate is kept below 300 nA/cm^2. All the implanted surfaces are coated either with silicafilm deposited SiO_2 or with rf sputtered Si_3N_4 protective layers before anneal treatment. Isochronal anneals of 15 min to a maximum temperature of 850°C are conducted in an open quartz tube furance under pure argon flow.

Capacitance-voltage (C-V) measurements with the Al Schottky barrier diodes are conducted for determining the free carrier concentration profiles. In a few cases, Scanning Electron Microscope (SEM) and 1.15 μm HeNe Laser beam waveguiding measurements are performed to detect the defect compensation in the implanted layers realized in heavily n doped substrates (n ≥ 1×10^{18} cm^{-3}). The photoluminescence (PL) spectra are taken at 77°K with the sample immersed in Liquid Nitrogen. A laboratory made 30 mw - 6328 Å HeNe laser beam is used as the excitation source. The power of the focussed laser beam at the sample level is about 15 mw which corresponds to an excitation level of 5×10^{18} photons cm^{-2} sec^{-1}. The emitted radiation is detected with a S1 response photomultiplier mounted on the exit slit of the spectrometer. All the spectra are taken with the photomultiplier at room temperature whose output signal is detected by a lock-in amplifier. No corrections for the detector and spectrometer response are made for the spectra presented in this work.

RESULTS AND DISCUSSION

In the following we present and discuss different experimental results that show the existence of compensation in the layers implanted respectively with H, B and As ions and subsequently annealed at high temperatures.

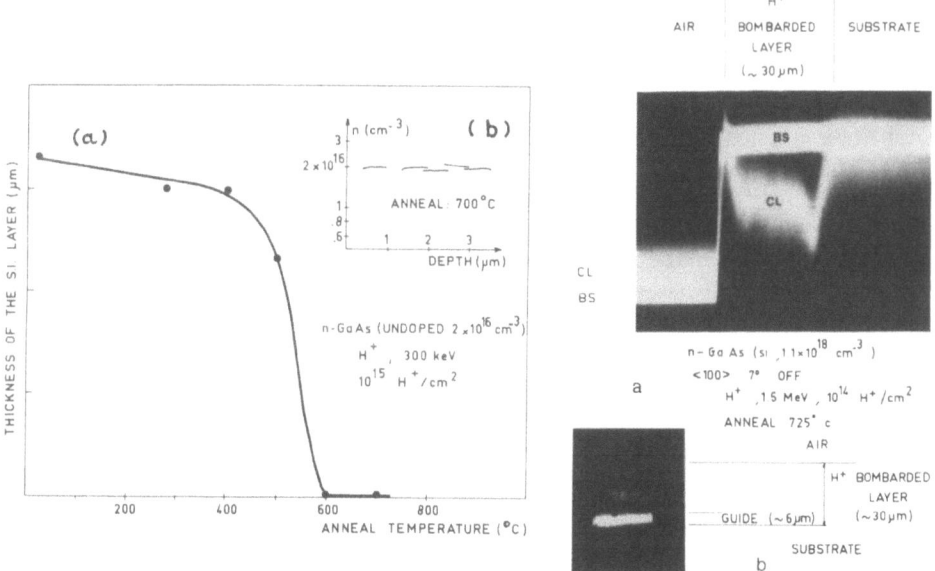

Fig. 1a- Thickness of semi-insulated or compensated layer as a function of 15 min isochronal annealing temperature.

Fig. 1b- Carrier concentration profile measured in a layer annealed at 700°C for 15 min with SiO_2 protection.

Fig. 2a- Scanning Electron Microscope oscillograph of a cleaved face perpendicular to the bombarded layer.

Fig. 2b- Photograph showing buried guiding of 1.15 μm HeNe laser beam in the sample of (a).

H^+ or Proton Bombarded Layers

Figure 1 represents the typical results of the electrical (C-V) measurements of the proton bombarded layers in an undoped n-GaAs substrate of n = 2×10^{16} cm^{-3}. The energy of protons is 300 Kev and the dose level is 10^{15} H^+/cm^2. Figure 1a shows the 15 min isochronal annealing behavior of the above layers where the measured thickness of the semi-insulated or compensated layer is plotted against the anneal temperature. We find here that the compensated layer due to the proton bombardment gets annealed above 600°C.
In the figure 1b we have shown the measured carrier concentration profile in the layer annealed at 700°C. These measurements are made after successively etching the layer up to a depth of 3 μm which is the projected range of the 300 Kev protons. The compensation due to the proton bombardment is annealed out at 700°C. The original carrier concentration of the substrate, 2×10^{16} cm^{-3}, is observed.

However, when similar proton bombardments are conducted in heavily n doped substrates, we are able to demonstrate the existence of defect associated compensation in similarlly annealed layers. The typical results of these experiments are shown in fig. 2. This figure shows the results obtained in the case of a heavily Si doped n-GaAs substrate of n = 1×10^{18} cm^{-3} bombarded with 10^{14} protons/cm^2 at 1.5 Mev and annealed at 725°C. A high energy for proton bombardment is chosen to realize a defect region which will not be influenced by the in-diffusion of the vacancies from the surface during anneal (3). Figure 2a shows the SEM oscillograph of a cleaved face (a 110 plane) perpendicular to the bombarded layer. B.S and C.L designate the oscilloscope traces obtained in the back-scattered primaries mode and the cathodoluminescent mode. Each of these traces is obtained by scanning the primary electron beam across the cleaved face successively in the two modes. B.S trace permits the location of the cleaved edge while the C.L trace allows the estimation of the depth of the bombarded layer-substrate interface. The form of the C.L trace in fig. 2a suggests that the luminescence properties of the bombarded layer after a 725°C anneal are recovered to a level equal to that of the substrate at all depths except in a region close to the bombarded layer-substrate interface where the unannealed defects are localized. After cleaving the above sample to obtain two parallelly cleaved faces, we have conducted the 1.15 µm He-Ne laser beam waveguiding measurements. These are performed by focussing the laser beam on one cleaved face and examining the other cleaved face through an IR image converter to observe the guiding layer. Fig. 2b shows the photography of the magnified image of the output face taken on the visible screen of an IR image converter. The air-sample interface is made visible by flashing with white light, while the photograph is taken. In this figure, one can clearly note the presence of a guiding layer of about 6 µm width and buried at a depth of about 24 µm. Note the close agreement in the measured depths of the guide-substrate interface in fig. 2b and the bombarded layer-substrate interface in fig. 2a. These results show that the confinement of the laser beam is taking place in the region where unannealed defects are localized. It is known from elsewhere (9) that the refractive index increase needed for waveguiding in the proton bombarded layers is achieved through the bombardment induced defects. The defects create a depression in the concentration of the free carriers originally present in the substrate (10). Consequently, the results of the figs. 2a and 2b together, prove the existence of defect **compensation even though they** do not give a measure of the compensation.

Since H in GaAs is a nondopant impurity, the results of figures 1 and 2 can only be explained if we suppose the participation of substrate impurity atoms (dopant or residual impurity atoms) to form compensation centers. Also we know that both the single vacancies and

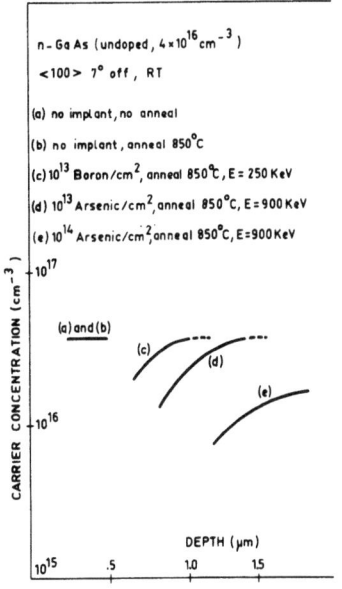
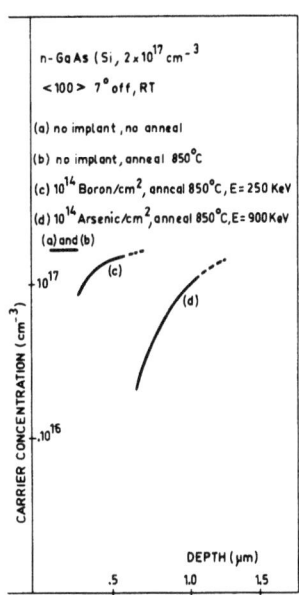

Fig. 3 - Typical results of electrical measurements on B and As implanted layers in undoped n-GaAs substrates (n = 4x10^{16} cm^{-3}). The samples are annealed at 850°C for 15 min with Si$_3$N$_4$ protection.

Fig. 4 - Typical results of electrical measurements on B and As implanted layers in Si doped n-GaAs substrates (n = 2x10^{17} cm^{-3}). The samples are annealed at 850°C for 15 min with Si$_3$N$_4$ protection.

the vacancy associated complexes can be active electrically. We therefore conclude that the compensation observed in the proton bombarded layers is probably due to a vacancy-substrate impurity atom associated defects. Even though we do not know the nature of the vacancy participating in the formation of the compensation centers, our results confirm that it is generated during the bombardment but not during the annealing process. This is because the compensated layer is situated quite far away from the surface to be influenced by the indiffusion of vacancies (3).

B and As Implanted Layers

Figures 3 and 4 represent the typical results of the C-V measurements on B and As implanted and 850°C annealed layers. They show the results obtained respectively in undoped n-GaAs (n = 4x10^{16}cm^{-3}) and Si doped n-GaAs (n = 2x10^{17} cm^{-3}) substrates. The implantation energies for B and As are 250 Kev and 900 Kev. The corresponding projected ranges are 0.5 and 0.6 µm. The dose for each implantation are given in the figures. The striking feature of the above figures

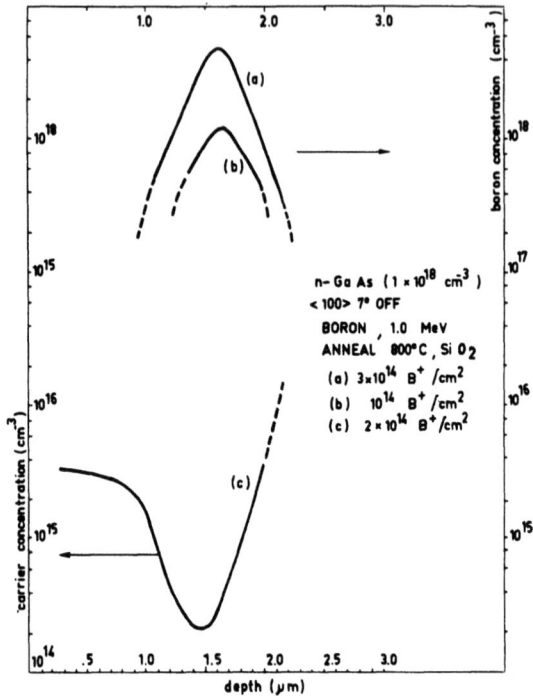

Fig. 5 - Chemical (a) and (b) and free carrier concentration (c) profiles of 1 Mev B implanted layers in heavily Si doped n-GaAs substrates (n = 1×10^{18} cm^{-3}). Dose : (a) = 3×10^{14} B$^+$/cm^2 ; (b) = 1×10^{14} B$^+$/cm^2 ; (c) = 2×10^{14} B$^+$/cm^2. Vertical axis : right. Boron concentration ; left. Carrier concentration.

is the existence of the compensation in both B and As implanted and 850°C annealed layers. Although, we find variations in the widths of the compensated layers with respect to the implant dose and the original doping level of the substrate, we do not consider them to be important for further discussion. On the contrary, we remark that unlike the proton bombarded layers, we observe compensation when B and As implanted layers are made in undoped substrates. In fact compensation is observed even when small doses ($\sim 10^{12}$ ions/cm^2) implantations for B or As are carried out in undoped n-GaAs substrates.

Similar results are obtained when B implants are carried out in heavily doped substrates. An example of which is shown in figure 5. In this figure curve (c) shows the results of C-V measurements on a sample implanted with 1 Mev Borons of dose 2×10^{14} B$^+$/cm^2 in a heavily Si doped n-GaAs (n = 1×10^{18} cm^{-3}) substrate and subsequently annealed at 800°C for 15 min with SiO$_2$ protection. Curves (a) and (b) represent the chemical profiles of Boron as obtained from Secondary Ion Microprobe analyses on two samples realized under the same conditions as above but with Boron doses of 1×10^{14} B$^+$/cm^2 and 3×10^{14} B$^+$/cm^2

respectively. One can make the following observations from a comparative study of the curves (a), (b), and (c) : i) existence of compensation with a maximum in the free carrier depression situated at 1.5 µm from the surface (see curve c). ii) peak of the Boron concentration in the chemical profiles (curves a and b) is situated at a depth of about 1.6 µm. This value is in close agreement with the calculated R_p value of 1 Mev Borons in GaAs reported by Davies et al (5), 1.55 µm. It shows that the diffusion of B in GaAs during a 15 min 800°C anneal treatment is negligible and consequently rules out the possibility of interstial Boron diffusion.

Since the depth of the maximum in the Boron concentration coincides with the depth of the minimum in the uncompensated free carrier concentration, one is naturally tempted to attribute the observed compensation to the presence of Boron. That is an acceptor-like behavior for Boron in GaAs. However, to our knowledge no electrical activity for Boron in GaAs is reported in the literature. Comparing the results of proton bombarded and B implanted layers shown in figures 1 to 5, we make the following remarks : i) unlike the proton bombarded layers, the B or As implanted layers exhibit compensation in the lightly doped substrates. This suggests a contribution to compensation from centers other than the simple implantation induced defects as assumed in the case of proton bombarded layers. ii) B and As implanted layers behave in a similar manner, i.e., they exhibit compensation eventhough both of these impurities are supposed to be electrically inactive in GaAs. iii) B implanted layers exhibit a compensation that seems to be related to its presence in GaAs. All the above tend to prove that the compensation observed in the B implanted layers is perhaps due to a Boron related defect. That is a defect generated due to the presence of the Boron or a defect complex associated with the Boron atom is responsible for the compensation. We suppose that a similar explanation holds good for compensation observed in the As implanted layers.

PL Measurements

Photoluminescence (PL) analysis helps to detect the nature of the vacancy associated defect centers since some of these are known to be radiative. We have conducted PL measurements on different samples implanted with B or As in heavily n doped (Si or Te) substrates and subsequently annealed at high temperatures.

Figure 6 shows the PL spectra at 77°K taken on three samples : (a) unimplanted unannealed, (b) unimplanted and annealed, and (c) Boron implanted and annealed samples. The anneal treatment is carried out at 800°C for 15 min with SiO_2 protection. The implanted layer is realized by performing multi implantations of Boron in a Si-doped n-GaAs substrate of n = $3.8 \times 10^{18} cm^{-3}$. Boron implant doses and energies are so adjusted as to obtain a uniform Boron concentra-

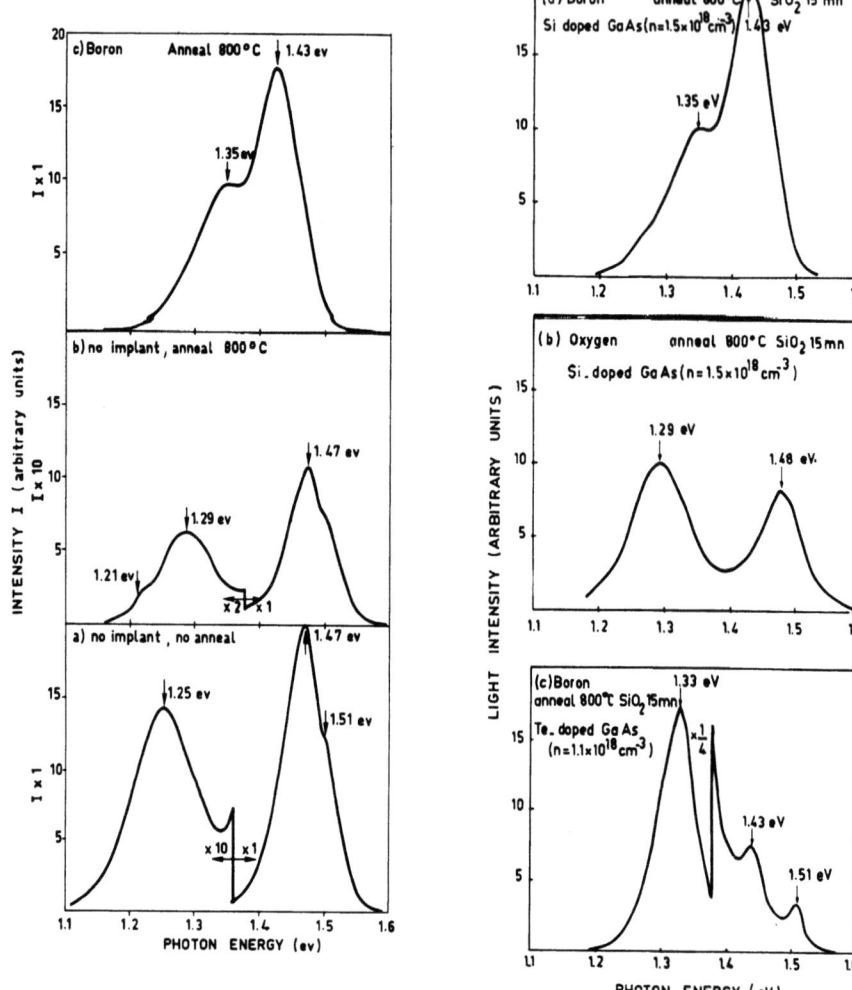

Fig. 6 - Photoluminescence spectra at 77°K showing the influence of B implantation in heavily n doped GaAs. See text for the details of the samples.

Fig. 7 - Comparison of the photoluminescence spectra at 77°K of 3 samples. See text for the details of the samples.

tion of 3×10^{18} B/cm3 over a depth of about 3 µm from the surface. Note that the concentration of Boron is almost equal to that of the free carriers in the substrate. In the PL spectrum of the original substrate (fig. 6a), we observe three emission peaks situated at 1.51, 1.47 and 1.25 eV respectively. The peak at 1.51 eV is due to the band-edge-emission while the peak at 1.47 eV involves the recombination at a hydrogenic acceptor level, which in this case is a Si-acceptor (Si on As site, Si_{As}) (11). The origin of the broad emission band with its peak at 1.25 eV is not known. We presume that this does not correspond to the well known V_{Ga} - Donor (Si_{Ga}) complex centers reported by Williams (12). In the PL spectrum of (b) we observe that the heat treatment for anneal has decreased the intensity of 1.51 and 1.47 eV emission bands. This indicates the formation of nonradiative defect centers during the anneal. Also we observe a broad band with a peak at 1.29 eV and a shoulder at 1.21 eV. Although the origin of the 1.29 eV band is not clear we suppose that it is related to the 1.25 eV band observed in figure 6a. Finally, in the PL spectrum of Boron implanted and annealed layer, figure 6c, we observe the following important changes : i) an increase in the total PL intensity to a level almost equal to that of the original substrate indicating a decrease in the concentration of nonradiative defects formed during anneal, ii) disappearance of the emission bands at 1.51 and 1.47 eV, and iii) emergence of a strong new emission band with a peak at 1.43 eV with a shoulder at 1.35 eV. From the results of figure 6 we relate the emergence of 1.43 eV band to the presence of B in GaAs. As regards the peak at 1.35 eV (obtained after decomposition) we found it to be related to the substrate and so it is not discussed further.

In figure 7 we have compared the PL spectra at 77°K of three different samples. a) Boron multi implant in a Si doped n-GaAs substrate of $n = 1.5 \times 10^{18}$ cm^{-3}, b) oxygen multi implant in a substrate similar to that in (a), and c) Boron multi implant in a Te doped n-GaAs substrate of $n = 1.1 \times 10^{18}$ cm^{-3}. Boron or oxygen ion implant doses and energies are so adjusted to obtain a uniform concentration of 1×10^{18} cm^{-3} over a depth of about 3 µm. All the above layers are annealed at 800°C for 15 min with SiO_2 protection. A comparison of the spectra of (a) and (b) shows that the 1.43 eV band with a strong emission intensity exists only in the Boron but not in the Oxygen implanted layer. Comparing the spectra of (a) and (c) we find that the 1.43 eV band also emerges when B is implanted in Te doped substrates. The 1.33 eV emission band observed in the spectrum (c) is seen also in the unimplanted and annealed samples (not presented) and so it is not important for further discussion.

From the results of the PL spectra shown in figures 6 and 7, we conclude that the strong emission peak observed at 1.43 eV is related to the presence of Boron atoms in GaAs. In the following we show that this peak is only indirectly related to the presence of Boron atoms. Queisser (13) observed an intense emission band at 1.42 eV (at 77°K) in melt-grown Si-doped GaAs following a high temperature treatment in hydrogen atmosphere. He also observed that the peak position of the 1.42 eV band depends on Si concentration. Later on Kressel et al (11) reported a similar band in liquid phase epitaxially grown layers doped with Si. They attributed it to the deep acceptor level of Si formed of a complex of Si on an As site with either vacancies or donors. Also, a similar band is observed (14) in Ge doped GaAs which is attributed to the complexes formed of the association of an Arsenic vacancy (V_{As}) with a Ge atom on an As site (Ge_{As}). Jeong et al (15) in their investigation of electron irradiated GaAs samples by PL analyses have found that the 1.42 eV emission band is related to a complex formed of the association of Arsenic vacancy with Se on an As site ($V_{As}-Si_{As}$). We thus suppose that the 1.43 eV emission observed in the Boron implanted layers (fig. 6c, 7a and 7c) is due to an Arsenic vacancy associated complex. This is most probably a complex formed of an As vacancy (V_{As}) associated with a Si atom on an As site (Si_{As}). The small differences in the emission energies, 1.43 eV instead of 1.42 eV could be due to the differences in the concentration of Si as pointed out by Queisser (13). The presence of the 1.43 eV band in a Te doped substrate implanted with Boron could be explained as due to the contamination of Si in the original substrate.

As a consequence of the above discussion, we tentatively attribute the compensation observed in the B implanted layers to an acceptor defect center formed by the association of V_{As} with Si_{As}. As mentionned earlier Kressel et al (11) have reported a deep acceptor level for this center. Since the compensation is found to be related to Boron (fig. 5) we hypothesize that the presence of B in the lattice has favored an increase in the concentration of the arsenic vacancies and the arsenic vacancy associated centers in the annealed layers. The later includes the newly formed Si-acceptors (Si_{As}) other than those originally present in the substrate. These acceptors may result from the migration of Si atoms from Ga sites to As sites. Here we suppose that B is replacing Ga substitutionally, ie., B is incorporated on Ga sites after an anneal at 800°C. This supposition is valid since the localized mode vibration studies of Thompson and Newman (16) on GaAs containing Boron, revealed that a great fraction of B takes Ga sites. According to the above hypothesis one would expect a compensation due to the excess arsenic vacancy associated defect centers in addition to that due to unannealed implantation induced defects. The later should have an origin similar to that observed in the case of proton bombarded layers. In both types of compensation, we think that the impurity atoms originally present in the substrate are participating.

The PL measurements conducted on As implanted samples did not permit us to investigate the compensation observed in these layers. We have observed a complete quenching of the photoluminescence intensity on the samples implanted with As at doses above 10^{13} As/cm^2 in heavily n-doped substrates and subsequently annealed at 850°C. We however wish to remark on the following: i) Introduction of As, unlike B, creates a large concentration of luminescence killing centers. ii) The compensation in these layers can perhaps be explained interms of a model similar to that discussed for B. In this model in addition to the unannealed defect compensation, we expect an excess Ga vacancy associated defect compensation due to the presence of As in GaAs.

The tentative interpretation of compensation in B (or As) implanted layers is based on certain important assumptions that need to be clarified. For example, before one could arrive at a satisfactory explanation, the following questions should be answered : i) Is B electrically inactive ? ii) What is its location ?. In a recent paper, Laithwaite et al (17) have suggested an acceptor behavior for B when it is incorporated on As sites, ie., antistructure defects. Their suggestion is based on the localized mode vibration measurements made on GaAs substrates containing Boron. The above is in contradiction with our results of PL measurements (fig. 6 and 7) since we observed a strong emission band related to a known arsenic vacancy associated complex. We therefore believe that more work is needed to understand the compensating behavior of B (or As) in implanted layers.

CONCLUSIONS

We have shown that the implantation of nondopant impurities, like H, B and As in n-GaAs exhibit compensation even when these layers are annealed at high temperatures. In the case of proton bombarded layers we have suggested that the compensation is due to unannealed defects in which a substrate impurity atom (dopant or residual) is participating. In the case of B implanted layers, the compensation observed is tentatively attributed to excess arsenic vacancy associated complexes formed in association with an impurity atom originally present in the substrate. This is, in addition to the compensation due to unannealed defects. Unfortunately our present experimental results did not permit us to estimate the extent of the compensation due to unannealed defects. We predict it to be smaller than that due to the excess Arsenic vacancy associated complexes. An explaination similar to that proposed for B is advanced to interpret the compensation observed in the As implanted layers. In that case we expect excess Gallium vacancy associated complexes to contribute to the compensation. Presently more work is in progress to estimate the compensation in B and As implanted layers.

REFERENCES

1. P.K. Chatterjee, K.V. Vaidyanathan, M.S. Durschland and B.G. Streetman, Solid State Commun. 17, 1421 (1975).

2 C.W. Wolfe and G.E. Stillman, Appl. Phys. Lett. 27, 564 (1975).

3 S.Y. Chiang and G.L. Pearson, J. Appl. Phys. 46, 2986 (1975) ; J. Luminescence 10, 313 (1975).

4. T. Itoh and J. Kasahara, J. Appl. Phys. 45, 4915 (1974).

5. D.E. Davies, J.K. Kennedy and A.C. Yang, Appl. Phys. Lett. 23, 615 (1973).

6. T. Itoh and Y. Kushido, J. Appl. Phys. 42, 5120 (1971).

7. J.M. Woodcock, Appl. Phys. Lett. 28, 226 (1976).

8. E.B. Stoneham and J.F. Gibbons, in Ion Implantation in Semiconductors, edited by S. Namba (Plenum, Newyork, 1975) p. 57.

9. E. Garmire, H. Stoll, A. Yariv and R.G. Hunsperger, Appl. Phys. Lett. 21, 87 (1972).

10. J.C. Dyment, J.C. North and L.A. D'Asaro, J. Appl. Phys. 44, 207 (1973).

11. H. Kressel, J.U. Dunse, H. Nelson and F.Z. Hawoylo, J. Appl. Phys. 39, 2006 (1968).

12. E.W. Williams, Phys. Rev. 168, 922 (1968).

13. H.J. Queisser, J. Appl. Phys. 37, 2909 (1967).

14. E.W. Williams and C.T. Elliott, Brit. J. Appl. Phys (J. Phys. D) 2, 1657 (1969).

15. M. Jeong, J. Shirafuju and Y. Inuishi, J. J. Appl. Phys. 12, 109 (1973).

16. R.C. Newman, F. Thompson, M. Hyliands and R.F. Peart, Solid State Commun. 10, 505 (1972).

17. K. Laithwaite, R.C. Newman and P.D. Green, J. Phys. C : Solid State Phys. 8, L77 (1975).

SULFUR ION IMPLANTATION IN GALLIUM ARSENIDE

M. Fujimoto, H. Yamazaki and T. Honda

Musashino Electrical Communication Laboratory

Musashino-shi, Tokyo 180   JAPAN

ABSTRACT

   The electrical properties of a sulfur ion implanted layer have been investigated to produce a thin n-type layer. The implantations were performed in doses ranging from 2×10^{12} cm^{-2} to 10^{16} cm^{-2} at energies of 100, 200 and 350 keV through 500 Å thick sputtered SiO_2 film on the sample surface. The dependence of sheet carrier density, carrier profile and mobility on ion dose, implantation energy and annealing conditions was determined. Carrier concentration profiles in the implanted layer were closely approximated by curves calculated with a diffusion coefficient of sulfur.
   Gallium arsenide field effect transistors and Gunn effect digital devices, fabricated from sulfur ion implanted layers, showed good performances.

INTRODUCTION

   Several investigations have been reported on the electrical properties of sulfur ion implanted layer[1]-[3]. In those reports, some important facts, which affect the amounts of electrically activated implanted sulfur, have been clear. Doping efficiency of implanted sulfur after annealing depends on the quality of semi-insulating substrates, substrate temperature during implantations and the encapsulation film.
   This paper describes doping efficiency, carrier and mobility profiles in the sulfur ion implanted layers as a function of dose, implantation energy and annealing conditions. Gallium arsenide FET's and Schottky barrier gate Gunn effect devices were fabricated by using sulfur ion implanted layers. These devices exhibited good performances.

EXPERIMENTAL

Gallium arsenide samples used in this experiment were chromium doped semi-insulating wafers having an etch pit density of less than 2000 cm^{-2} and a low chromium content of less than 0.2 ppm. The samples were coated with a 500 Å thick sputtered SiO_2 film prior to the ion implantations. The SiO_2 film was effective in preventing oil contamination during the ion implantations. The calculated sulfur ions in a 500 Å SiO_2 film are 2.3 % of the total implanted 10^{13} cm^{-2} sulfur ions at 100 keV. At 200 keV, 1.6 % is contained in the film and, at 350 keV, the amount in the film is almost negligible. Thus, the implanted sulfur ions were neglected in the film.

Ion implantations were carried out at room temperature. Following the implantations, the SiO_2 film was removed and a new 5000 Å sputtered SiO_2 was deposited on both the implanted and the rear surfaces of the sample to prevent gallium arsenide decomposition during annealing. Sample annealing was carried out in flowing hydrogen gas.

Carrier concentration and Hall mobility were measured by using the van der Pauw method. The carrier concentration profiles were determined by combining Hall effect measurements and layer removal techniques.

RESULTS AND DISCUSSIONS

a. Dose and Anneal Temperature Dependence

Figures 1 and 2 show the normalized sheet carrier densities as a function of annealing temperatures for samples doses ranging from 2×10^{13} to 10^{16} cm^{-2} at 200 keV and 350 keV, respectively. Samples implanted with a dose of 2×10^{13} cm^{-2} at 350 keV have about 70 % doping efficiency above 750°C annealing. However, samples with doses of more than 10^{14} cm^{-2} require annealing at higher temperatures before the implanted ions become electrically active. The activation energy for the annealing process is about 2.2 eV.

It is known that the surface conductive layer is produced by the diffusion of gallium and chromium during annealing of the chromium doped gallium arsenide covered with thermally deposited SiO_2 film. However, for sputtered SiO_2 film, no surface conductive layer was observed after annealing for 20 minutes at 900°C.

The annealing temperature relation to Hall mobility for samples implanted at 200 keV is shown in Figure 3. Annealing time is 20 minutes. There is a marked difference in the annealing temperature effect on Hall mobility between samples with low and high ion doses. Hall mobility of samples with a dose of 2×10^{13} cm^{-2} ions increases proportionally to annealing temperature above 750°C. These results indicate a reduction in the damage of the layer. For samples with a dose of more than 10^{14} cm^{-2}, Hall mobility decreases with the

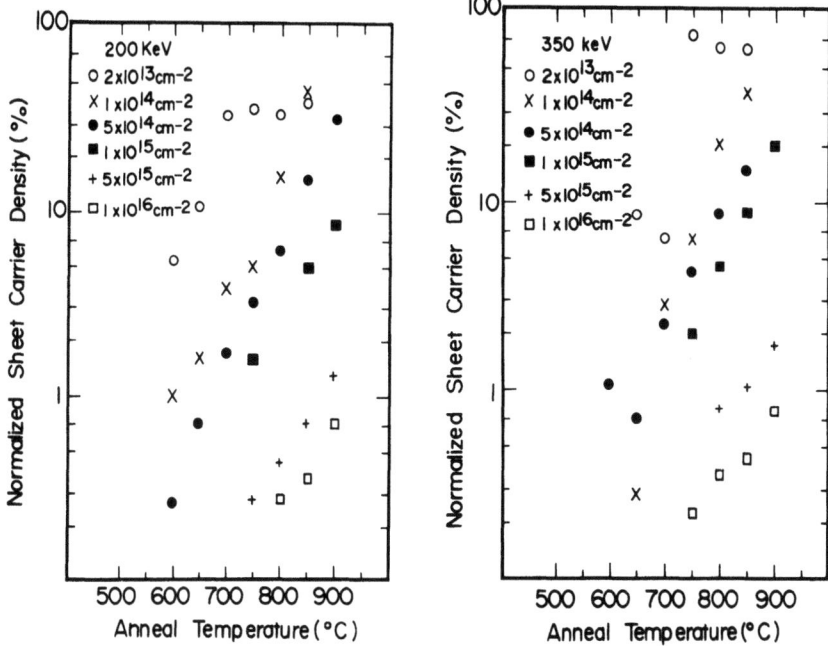

Fig. 1 Normalized sheet carrier density for samples implanted at 200 keV.

Fig. 2 Normalized sheet carrier density for samples implanted at 350 keV.

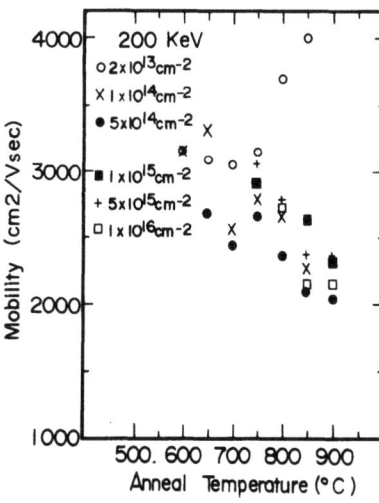

Fig. 3 Hall mobility as a function of annealing temperature. Annealing time is 20 minutes.

increase in annealing temperature. This results from the increase in the carrier concentrations. 750°C seems to be a transition temperature for the annealing of sulfur ion implanted samples.
Sheet carrier density dependence on sulfur ion doses at 100, 200 and 350 keV implantations is shown in Figure 4. Samples were annealed for 20 minutes at 850°C. For samples with a dose of 2×10^{12} cm^{-2} at 350 keV, nearly 100% doping efficiency was obtained.

b. Annealing Time Dependence

Figure 5 shows sheet carrier concentrations as a function of annealing time. Ion dose is 2×10^{13} cm^{-2} and annealing temperature is 850°C. Sheet carrier concentrations almost completely saturate after one hour annealing. It is interesting to note that doping efficiency depends on ion implantation energy. Higher doping efficiency was obtained at higher implantation energies. For comparison, sulfur ions were implanted into cadmium doped p-type gallium arsenide. In this case, higher implantation energies show a lower doping efficiency, apparently due to the relatively high compensation centers in the high implantation energy layer. At present, the reason for the implantation energy dependence of doping efficiency for chromium doped substrates is not clear.

The mobility of the sulfur ion implanted layer as a function of annealing time is shown in Figure 6. Mobilities are almost the same at each implantation energy for annealing time of more than one hour. Mobilities of 5000 cm^2/volt-sec were obtained after two to three hours annealing. These values are comparable to those obtained in n-type epitaxial layers grown on chromium doped semi-insulating gallium arsenide.

c. Profiles

Carrier concentration profiles in the sulfur implanted layers were determined by combining Hall effect measurements and layer removal techniques. In Figure 7, the carrier concentration profile for a sample with a dose of 2×10^{13} cm^{-2} at 350 keV is shown. The sample was annealed at 850°C for 20 minutes. The solid line in Figure 7 was calculated from an analytical solution of Fick's law, assuming Gaussian initial distribution, 100% doping efficiency and no sulfur out diffusion. The dotted line is a calculated profile, assuming 51% doping efficiency. The implanted sulfur distribution after annealing is well approximated by the sulfur profile. $D_0=10.9$ cm^2/sec and activation energy E=2.95 eV were used as the diffusion coefficient of sulfur in gallium arsenide. These values were obtained by Matino for sulfur diffusion into gallium arsenide from gallium sulfide sources (4). The carrier concentration profile for a sample implanted at 200 keV after two hours annealing is shown in Figure 8.

SULFUR ION IMPLANTATION IN GALLIUM ARSENIDE 93

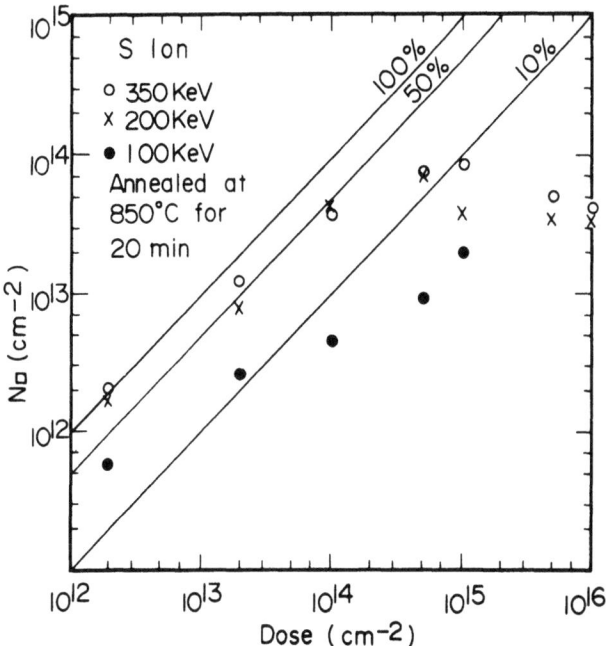

Fig. 4 Sheet carrier density as a function of sulfur ion doses.

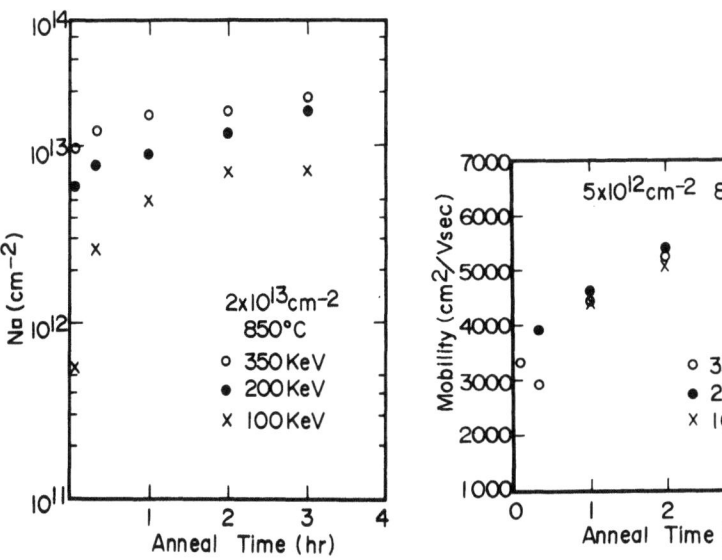

Fig. 5 Sheet carrier density as a function of annealing time.

Fig. 6 Mobility as a function of annealing time.

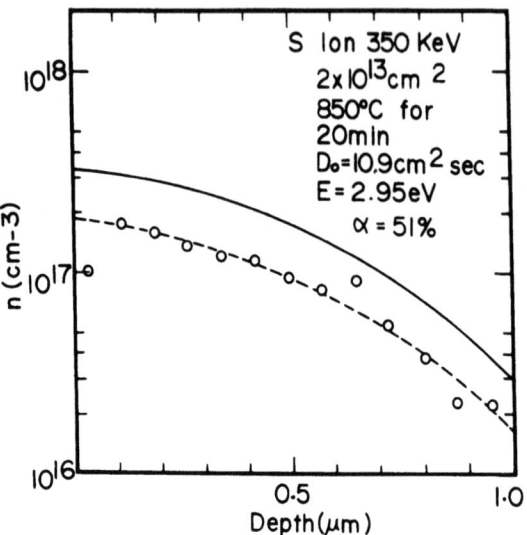

Fig. 7 Carrier concentration profile in sulfur ion implanted layer

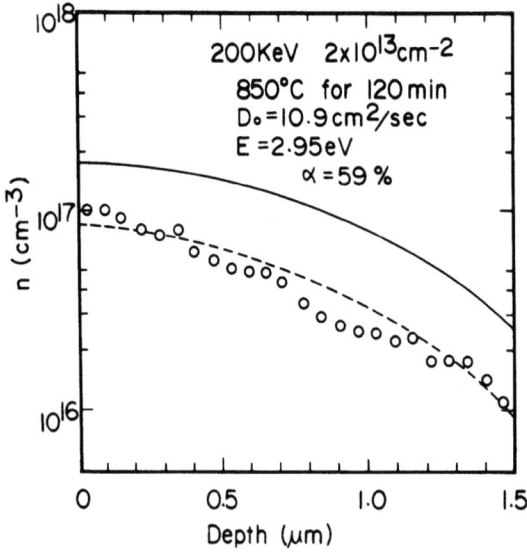

Fig. 8 Carrier concentration profile in sulfur ion implanted layer.

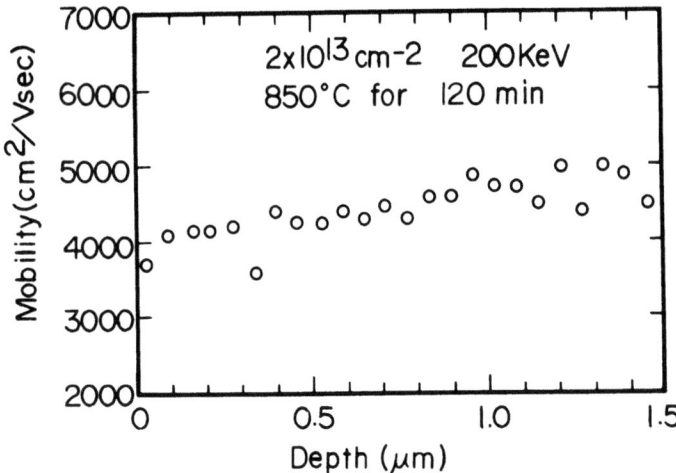

Fig. 9 Mobility profile in sulfur ion implanted layer

Mobility profiles in the implanted layer after annealing at 850°C for two hours are shown in Figure 9. Samples were implanted at 200 keV with a dose of 2×10^{13} cm^{-2}. It is well known that, in the n-type epitaxial layer grown on chromium doped semi-insulating gallium arsenide substrates, mobility decreases near the substrate interface. However, in sulfur ion implanted layers annealed for long time, high mobility was obtained at the substrate interface. In this case, the mobility increases from 4000 cm^2/volt-sec near the surface to 5000 cm^2/volt-sec at the substrate interface.

d. Devices

Recently, gallium arsenide field effect transistors, fabricated using an ion implanted layer, have been reported (5)-(8). Gallium arsenide FET's and Schottky barrier gate Gunn effect devices were fabricated by sulfur ion implantation into chromium doped semi-insulating substrates. Detailed fabrication techniques and results will be submitted eleswhere (9),(10).

For fabricating gallium arsenide FET's, sulfur ions were selectively implanted in the device area at 200 keV with a dose of 5×10^{12} cm^{-2}. Aluminum film was used as the implantation mask. Samples were annealed at 800°C for 210 minutes. Aluminum gate with 1.4 micron length and 400 microns width were made on the implanted layer. The transconductance of 16 mhos and the projected f_{max} = 30 GHz were obtained from S parameter measurements.

For Schottky barrier gate Gunn effect devices, 5×10^{12} cm^{-2} were implanted at 200 keV. Samples were annealed at 800°C for three and a half hours. Spacing between cathode and anode is 100 microns.

Minimum gate trigger voltage obtained ranged from 100-350 mV. These values were comparable to those obtained by using an n-type epitaxial layer grown on the semi-insulating substrates.

SUMMARY

1) The choice of semi-insulating substrates with a low chromium density is important for obtaining high doping efficiency and reproducibility.
2) Sputtered SiO_2 layers are suitable for encapsulating gallium arsenide substrates.
3) Doping efficiency depends on both the ion dose and the implantation energy.
4) Carrier concentration profiles are closely approximated by the implanted sulfur diffusion.
5) Mobility of 5000 cm^2/volt-sec at the substrate interface was obtained by annealing at 850°C for two hours.
6) FET's and Gunn devices fabricated from sulfur ion implanted layer have fairly good performances.

ACKNOWLEDGEMENTS

The authors would like to thank Mr.T.Mizutani and Mr.S.Ishida for fabricating FET's and Gunn devices. It is pleasure to thank Drs.H.Harada, Y.Sato, M.Watanabe and H.Toyoda for many valuable suggestions and encouragement.

REFERENCES

1) Y.Kato, T.Shimada, Y.Shiraki and K.F.Komatsubara, J.A.P. 45,1044 (1974)
2) F.H.Eisen, Proc. 4th Conf. Ion Implantation p.3 (1974)
3) J.M.Woodcock, J.M.Shannon and D.J.Clark, Solid State Electronics 18,267 (1975)
4) H.Matino, Solid State Electronics, 17,35 (1974)
5) R.G.Hunsperger and N.Hirsch, Electron. Lett. 9, 577 (1973)
6) B.M.Welch, F.H.Eisen and J.A.Higgins, J.A.P. 45,3685 (1974)
7) R.G.Hunsperger and N.Hirsch, Solid State Electronics 18,349 (1975)
8) J.A.Higgins, B.M.Welch, F.H.Eisen and G.D.Robinson, Electron. Lett. 12,17 (1976)
9) T.Mizutani and S.Ishida, submitted to Electron. Lett.
10) T.Mizutani, submitted to Solid State Electronics.

RADIOTRACER PROFILES IN SULFUR IMPLANTED GaAs

F. H. Eisen and B. M. Welch

Science Center, Rockwell International

Thousand Oaks, California 91360

ABSTRACT

We have measured profiles of S^{35} implanted in GaAs at 100 keV, using an anodization and stripping technique, in an effort to better understand the characteristics of sulfur implanted layers in GaAs. Profile data from unannealed samples implanted at either room temperature or 350°C show that the S^{35} distribution is close to the LSS prediction. Measurement of the S^{35} counting rate before and after annealing at 850°C for 30 min using a silicon nitride layer as a protective cap shows that 50 and 75% of the implanted tracer diffuses to the surface of the GaAs sample, and is removed during dissolution of the nitride layer and removal of the layer of GaAs (about 130Å) by anodization and stripping. This amount of out-diffusion of the implanted sulfur may largely account for the apparent low doping efficiencies achieved by sulfur implantation.

INTRODUCTION

The n-type doping of GaAs by sulfur implantation is useful in the production of microwave devices in GaAs. Work has been reported in which sulfur implantation was used to produce the active regions for Schottky barrier FETs [1-3], to make improved contacts to Gunn Diodes [4], to make Gunn type digital devices [5] and in the fabrication of Read type IMPATT diodes [6]. The doping efficiencies (ratio of implanted dose to sheet electron concentration) observed for low dose sulfur implants are usually appreciably less than 100%. For example, for doses near 1×10^{13} S ions/cm^2 at implantation energies of about 100 keV, doping efficiencies in the range of about

10 to 40% have been reported [2,7,8]. Also, the doping profiles resulting from sulfur implantation in GaAs are often much deeper than the implanted sulfur distribution calculated from LSS range parameters [9]. In contrast, the doping efficiency observed for selenium implanted samples approaches 100% at low doses and the doping profile lies much closer to the expected distribution of dopant atoms than is the case for sulfur [10].

In an attempt to attain a better understanding of the behavior of implanted sulfur in GaAs, we have measured the distribution of radioactive S^{35} implanted into GaAs samples. Profiles have been determined for unannealed samples implanted at room temperature or at 350°C and for samples annealed at 850°C using a reactively sputtered silicon nitride cap. The most important finding is that there is appreciable outdiffusion of the sulfur during annealing. This outdiffusion may account for the apparent low doping efficiencies achieved by sulfur implantation in GaAs.

EXPERIMENTAL

All samples were <100> oriented and were either Te doped n^+ material or Cr-doped semi-insulating material. Implantations were performed at an energy of 100 keV. In most cases, stable S^{34} was implanted in addition to the radioactive S^{35} so that the total S dose would be comparable to that employed in S doping experiments (about 10^{13} to 10^{14} ions/cm^2). The tracer profile was measured by removing GaAs layers of known thickness using an anodization technique described elsewhere [11], and counting the activity remaining in the sample following removal of the anodized layer. An end window Geiger counter housed in a lead safe was used for this purpose. The use of n^+ substrate material permitted the anodization of samples which had not been annealed following implantation. Profiles for annealed samples were measured on both n^+ substrates and for Cr-doped semi-insulating substrates. Reactively sputtered silicon nitride was employed as a protective layer in order to retard dissociation of the GaAs during annealing. The properties of this reactively sputtered nitride have been described in some detail elsewhere [10].

RESULTS AND DISCUSSION

Fig. 1 shows a profile measured for an unannealed sample implanted at room temperature. Also shown is the distribution of 100 keV S predicted from LSS range parameters [12], matched to the initial S^{35} counting rate. The absence of data points near the surface is due to a difficulty encountered in measuring the profile for this sample. The reduction of activity after the first few anodization and dissolution steps was appreciably lower than expected.

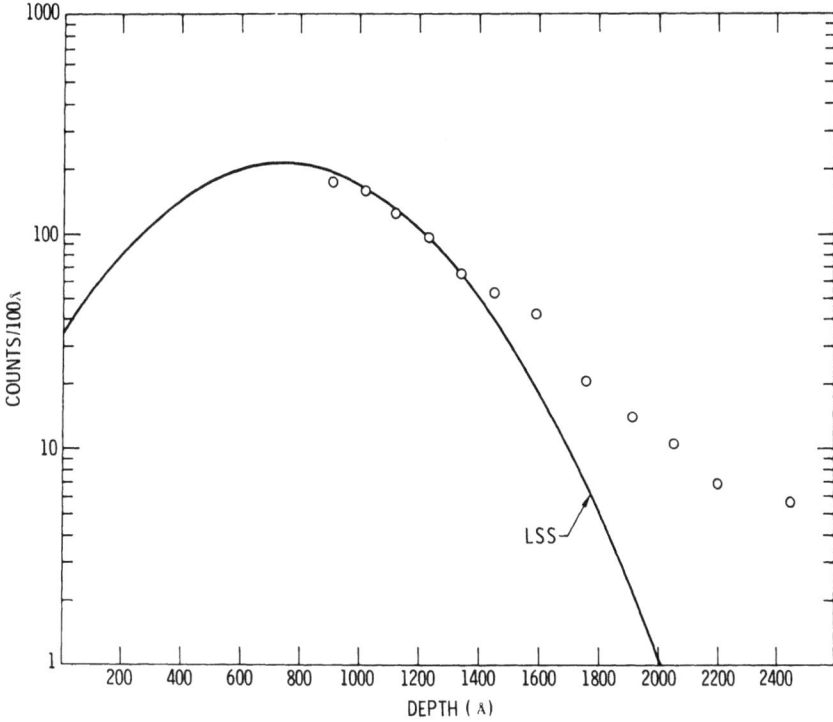

Fig. 1. S^{35} profile for an unannealed sample implanted at room temperature.

Experiments carried out after the removal of these first layers of GaAs indicated that there was an appreciable S^{35} concentration at the surface of the sample. It is believed that this occurred because sulfur moved through the anodized layer toward the GaAs surface during the anodization process, and it was not possible to completely remove this sulfur from the surface of the GaAs by the process of dissolving the anodized layer. It was possible to overcome the problems caused by the pileup of S^{35} on the GaAs surface by placing the sample in the anoidization cell with the polarity reversed, after dissolving the anodized layer, and then immersing it in boiling HCl for 1 min. The points in Fig. 1 are those obtained after adopting this procedure. They lie close to the S distribution calculated from LSS range parameters to a depth of about 1300Å and then fall somewhat higher than this calculated profile. This tail on the S^{35} distribution seems similar to the tails observed for phosphorus implantation in silicon, which have been shown to be due to channeling [13]. It was felt necessary to confirm that this tail was a real effect and not due to the inability to remove a layer of S^{35} lying on the surface of the sample.

This was accomplished by removing approximately 50Å of GaAs by sputtering. No significant decrease in the S^{35} activity remaining in the sample was observed following the sputtering of the GaAs which was carried out at two different times during the measurement of the profile in the tail region. This indicates that there was not a large concentration of S^{35} on the surface of the sample so that the tail is a real effect.

The S^{35} profile measured for an unannealed sample which was implanted at 350°C is shown in Fig. 2. This profile lies quite close to the prediction from LSS range parameters except for the presence of a tail at depths greater than about 1500Å. The tail, in this case, appears to fall off more rapidly than it did for the sample implanted at room temperature. This suggests that the tail may be due to channeling effects since, if it were produced by diffusion during implantation, it might be expected to be larger in the case of the hot implant. The close agreement between the profile measured after implantation at 350°C and the LSS range curve for 100 keV sulfur in GaAs indicates that there is no significant diffusion of the implanted sulfur during implantation at temperatures as high as 350°C.

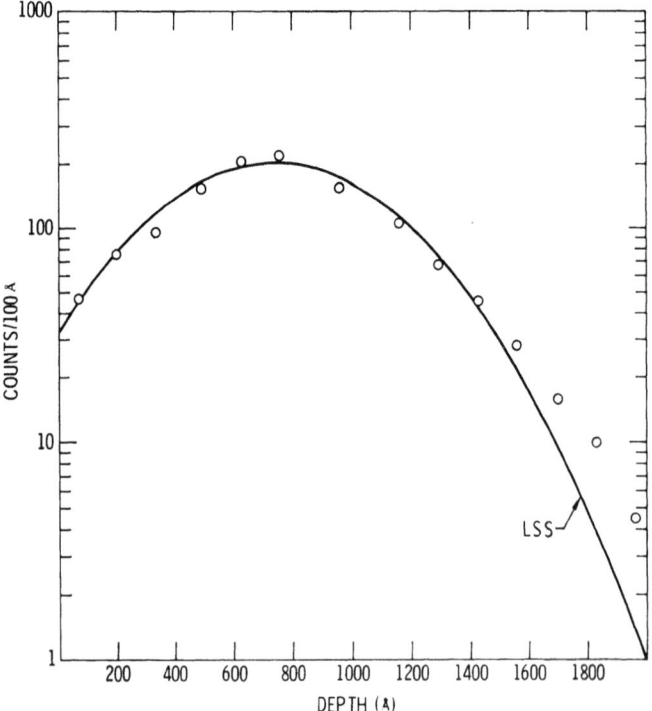

Fig. 2. S^{35} profile for an unannealed sample implanted at 350°C.

Several samples were annealed at a temperature of 850°C for 30 min using reactively sputtered silicon nitride as a protective layer (cap) to retard dissociation of the GaAs during annealing. The S^{35} activity remaining in the sample after various treatments is summarized in Table I. When the silicon nitride was removed from the surface of these annealed samples using concentrated hydrofluoric acid, a decrease in the S^{35} activity was observed. A further decrease in activity occurred after the samples had been placed in hot HCl for a period of about 1 min. S^{35} profiles were measured for 3 of these samples. A very large decrease in the activity in the sample took place when the first anodized layer was removed. The thickness of GaAs removed in this depth was about 130Å, however, it seems likely that the S^{35} removed by anodization was trapped near the surface of the GaAs.

S^{35} profiles measured for 3 annealed samples are shown in Fig. 3. The sulfur is seen to have diffused several thousand Å into the GaAs. The profile obtained for implantation into a Cr-doped semi-insulating substrate may be slightly different from those measured for implantation into n^+ substrate material. However, there are not sufficient points for the semi-insulating substrate to establish that the difference is significant. The depth of the S^{35} distributions in these samples is comparable to the depths of the electrical profiles observed for sulfur implantation doping of GaAs. In addition, the fraction of the initial S^{35} activity remaining in the GaAs samples after removal of the first layer by anodization and dissolution of the oxide layer is in agreement with the doping efficiency observed in sulfur implantation doping using 100 keV sulfur ions and a reactively sputtered silicon nitride annealing cap. This suggests that the apparent low doping efficiencies observed in sulfur implantation doping of GaAs are a result of the substantial outdiffusion of the sulfur during annealing.

We have observed that the doping efficiency obtained by sulfur implantation increases as the implantation energy is increased from 100 keV to 200 or 400 keV [14]. These results are summarized in Table II for samples annealed with a reactively sputtered silicon nitride cap. Similar results have been seen by other workers using a different annealing cap [15]. It is likely that this apparent dependence of doping efficiency on implantation energy can be explained by the outdiffusion results discussed above. Implantations carried out at higher energies would result in a sulfur distribution which was initially deeper than for lower energy implants and therefore, one would expect to lose a smaller fraction of the implanted sulfur by outdiffusion.

It is tempting to conclude that outdiffusion of sulfur will account for all cases in which a doping efficiency appreciably lower

Table I

S^{35} activity remaining in annealed samples (850°C-30 min)

SAMPLE MATERIAL	IMPLANT	S^{34} DOSE (ions/cm^2)	ACTIVITY AFTER CAP REMOVAL	ACTIVITY AFTER HOT HCl	ACTIVITY AFTER FIRST LAYER REMOVAL
Cr-doped	RT	1×10^{13}	87%	62%	
N^+	350°	1×10^{13}	55%	46%	18%
Cr-doped	350°	3×10^{13}	83%	60%	24%
Cr-doped	350°	1×10^{13}	84%	65%	
N^+	350°	1×10^{14}	62%	43%	16%

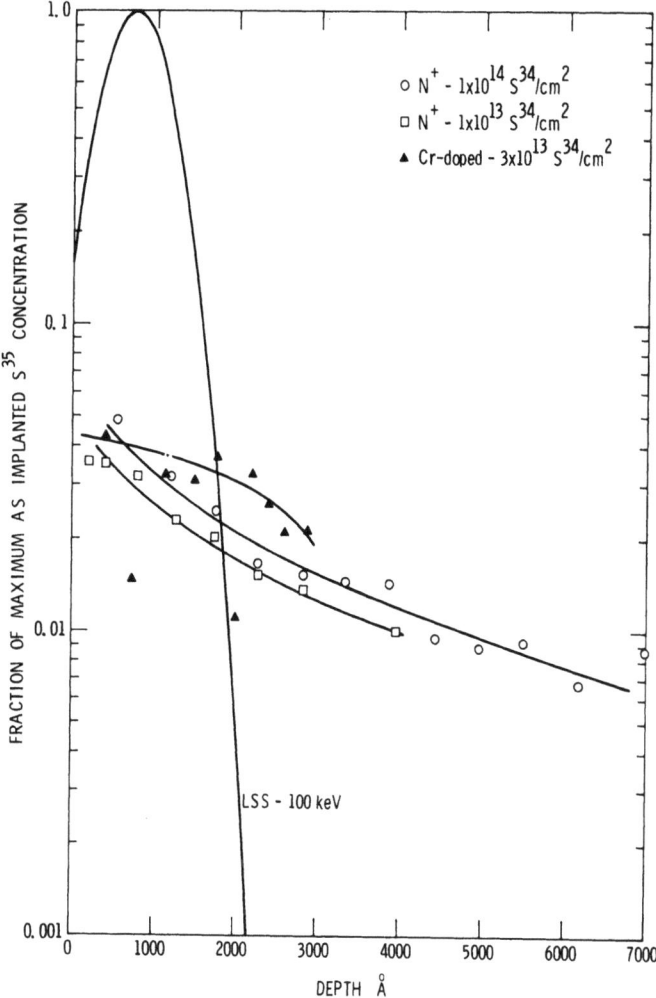

Fig. 3. S^{35} profiles for samples implanted at 350°C and annealed at 850°C for 30 min with a silicon nitride cap.

than 100% is observed for low doses of implanted sulfur. However, we have found large differences in the electron concentration profile resulting from sulfur implantation, which depend on the material used for the annealing cap. Some typical results are shown in Fig. 4. It is possible that these results can be accounted for by different amounts of outdiffusion for the different annealing caps. This is not an unreasonable idea since the trapping of the sulfur in the region near the surface of the GaAs and outdiffusion into the cap may well depend on the local strain and stoichiometry at the surface of the GaAs and therefore, on the nature of the material used for the annealing cap. However, it is also possible that with

Table II

Doping efficiency for sulfur implanted samples annealed at 850°C
for 30 min with a reactively sputtered silicon nitride cap

Ions/cm^2	100 keV RT	100 keV 350	200 keV RT	200 keV 350	400 keV RT	400 keV 350
1E13		~ 20%	35%	48%	53%	
3E13			35%		45%	44%
1E14	8.3%	13%		26%	9.6%	38%

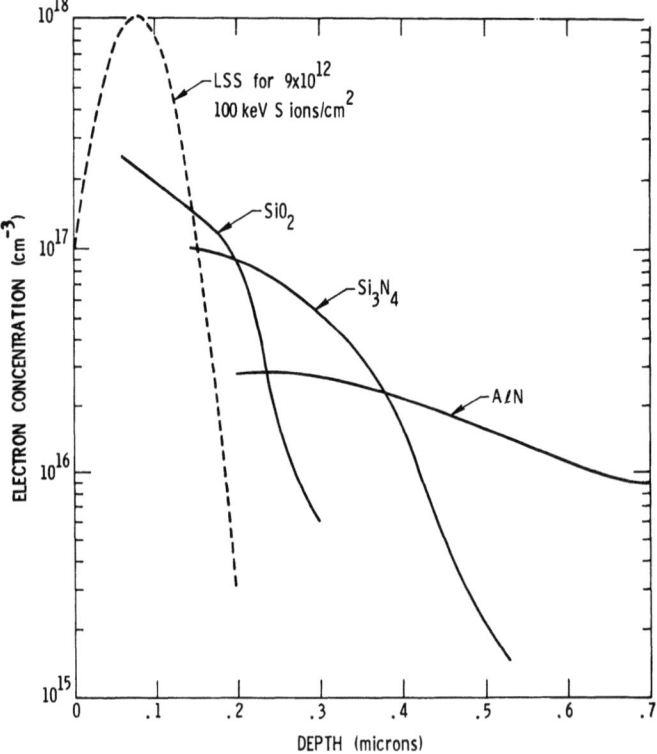

Fig. 4. Electron concentration profiles obtained from Schottky
barrier C-V measurements for semi-insulating GaAs samples implanted
at 350°C with 9x10^{12} 100 keV S ions/cm^2 and annealed at 850°C
for 30 min with the indicated caps.

some annealing caps, defect centers may be introduced into the GaAs which would compensate the sulfur remaining in the sample after the outdiffusion process has taken place. Further work is required before it is possible to generalize our results and so decide to what extent outdiffusion processes will account for low doping efficiencies observed in sulfur implantation doping of GaAs.

SUMMARY

The data presented above show that there is no significant diffusion during hot implantations of sulfur into GaAs at implantation temperatures as high as 350°C. A substantial outdiffusion of S^{35} implanted at an energy of 100 keV has been observed for annealing carried out with a silicon nitride cap. This outdiffusion seems to be largely responsible for the low doping efficiencies observed in our work on sulfur implantation doping of GaAs using a reactively sputtered silicon nitride cap. Outdiffusion may also account for low doping efficiencies which occur when annealing is carried out using other cap materials.

ACKNOWLEDGEMENT

We wish to thank J. A. Davies and O. M. Westcott of Chalk River Nuclear Laboratories for assistance with implantation and profiling problems.

REFERENCES

1. B. M. Welch, F. H. Eisen and J. A. Higgins, J. Appl. Phys. $\underline{45}$, 3685 (1974).
2. R. G. Hunsperger and N. Hirsch, Solid State Electron. $\underline{18}$, 349 (1975).
3. J. A. Higgins, B. M. Welch, F. H. Eisen and G. D. Robinson, Electron. Lett. $\underline{12}$, 17 (1976).
4. D. H. Lee, J. J. Berenz and R. L. Bernick, Electron. Lett. $\underline{11}$, 191 (1975).
5. T. Mizutani and K. Kurumada, Electron. Lett. $\underline{11}$, 638 (1975).
6. J. J. Berenz, R. S. Ying and D. H. Lee, Electron. Lett. $\underline{10}$, 157 (1974).
7. F. H. Eisen, B. Welch, K. Gamo, T. Inada, H. Muller, M-A. Nicolet and J. W. Mayer, Application of Ion Beams to Materials 1975, ed. by G. Carter, J. S. Colligon, and W. A. Grant (The Institute of Physics, London, 1976) p. 64.
8. H. Muller, J. Gyulai, J. W. Mayer, F. H. Eisen and B. M. Welch, Ion Implantation in Semiconductors, ed. by S. Namba (Plenum Press, New York, 1975) p. 19.

9. F. H. Eisen, Ibid, p. 3.
10. K. Gamo, T. Inada, S. Krekler, J. W. Mayer, F. H. Eisen and B. M. Welch, Solid State Electron. (to be published).
11. H. Muller, F. H. Eisen and J. W. Mayer, J. Electrochem Soc. 122, 651 (1975).
12. J. F. Gibbons, W. S. Johnson and S. W. Mylroie, Projected Range Statistics (Dowden, Hutchinson and Ross, Inc. Stroudsburg, Pennsylvania, 1975).
13. P. Blood, G. Dearnaley and M. A. Wilkins, J. Appl. Phys. 45, 5123 (1974).
14. F. H. Eisen, J. A. Higgins, A. A. Immorlica, Jr., R. L. Kuvas, B. W. Ludington, B. M. Welch, C. P. Wen and R. R. Zucca, "Investigation of Technological Problems in GaAs, " Semi-annual Technical Report No. 1, AFCRL-TR-75-0435, Contract No. F19628-75-C-0113, and F. H. Eisen, B. M. Welch, K. Gamo and T. Inada (unpublished work).
15. M. Fujimoto, T. Honda and H. Yamazaki, this conference.

ELECTRICAL PROPERTIES OF Cd- AND Te-IMPLANTED GaAs*

B. K. Shin
Systems Research Laboratories, Inc.
Dayton, Ohio 45440

Y. S. Park and J. E. Ehret
Air Force Avionics Laboratory
Wright-Patterson Air Force Base, Ohio 45433

ABSTRACT

Sheet-resistivity and Hall-effect measurements were made on Cd- or Te-implanted GaAs samples. During annealing the sample surfaces were protected with pyrolytically deposited Si_3N_4. In the Cd-implanted samples, nearly complete electrical activity was obtained after an 800-900°C anneal for doses below 10^{14} Cd/cm^2. For Te implants the doping efficiencies were lower even after the 900°C anneal, and the Te-donor-induced self-compensation of residual defects (of the $V_{Ga}Te$ complex type) associated with the implantation process is thought to be responsible for the results obtained.

INTRODUCTION

It is well known that when GaAs is implanted by n-type dopants,[1,2] high doping efficiency cannot be obtained (unlike p-type dopant implantation[3]). Several methods are under consideration for increasing doping efficiency in n-type implanted GaAs. For example, recent studies of n-type implanted dopants have revealed that increased doping efficiency can be obtained by implantation at elevated substrate temperatures[4,5] and through the use of improved encapsulants during annealing.[4,6] Inasmuch as attaining efficient doping is important in ion implantation, electrical studies are essential for characterizing implanted layers. A careful analysis of sheet-resistivity and Hall-effect results yielded detailed electrical properties of the implanted Cd and Te; these results are discussed and experimental conditions are outlined briefly.

EXPERIMENTAL

Careful selection of implantation parameters is important in achieving the desired implantation doping. Parameters which influence doping characteristics in GaAs are (i) implanted dopant ion (mass, species, energy, dose, and dose rate), (ii) substrate (temperature, crystal orientation, and impurity content), (iii) annealing conditions (temperature, time, and ambient), and (iv) encapsulation (method, material, and temperature).

Sample Preparation

The substrate material was (100)-oriented single-crystal wafers obtained from Laser Diode Laboratories. The substrates were either undoped ($\sim 10^{16}$ cm^{-3}) n-type GaAs (for Cd implantation) or high-resistivity ($\sim 10^9$ Ω-cm) Cr-doped GaAs (for Te implantation). Prior to implantation, samples were carefully cleaned with repeated applications of solvents such as TCE, acetone, and isopropyl alcohol to remove organic substances present on the polished surface of the wafer. Etching of GaAs was accomplished using freshly prepared $H_2SO_4 : 30\%H_2O_2 : H_2O$ solution in a 5 : 1 : 1 ratio by volume for a short period of time. After the cleaning samples were encapsulated with a 300-500 Å uniform layer of pyrolytic Si_3N_4 at 720°C.[7] With this encapsulation the sample can withstand temperatures up to 950°C and the stoichiometric integrity of the GaAs surface is preserved. Such encapsulation when deposited prior to implantation can protect the sample surface, prevent beam-enhanced loss of Ga or As during implantation, and control the implantation profile. If the implanted samples had been encapsulated after implantation, the 720°C encapsulation temperature would have been sufficiently high for some annealing to have occurred, which is not desirable when a lower temperature annealing is to be used for damage-effect studies.

Implantation Machine

The ion machine employed--a modified Implanter-I manufactured by Accelerators, Inc.--is a pre-accelerated machine with a usable energy range of 35 to 150 keV. The machine is divided into four basic sections: the ion source and accelerating column, mass analyzer, beam shaper and scanner, and target chamber. A hot-cathode source was used to ionize both Cd and Te. The accelerating columns are composed of a series of cylindrical lenses which give equal potential planes to the ground. For the mass analysis a magnet of $\sim$ 12 KG is used which is sufficient for separating a mass ratio between 1 and 200 amu into a 30-deg. port. The beam-defining electronics consist of mechanical slits and electrostatic quadrupoles for beam shaping, a 5-deg. neutral particle separator for beam purifying, and an X-Y scanner for uniform implantation

ELECTRICAL PROPERTIES OF Cd- AND Te-IMPLANTED GaAs

(2×2-cm area). The target chamber is equipped with a carrousel arrangement of eight-position rotary wheels. The heater is capable of heating the sample during implantation and holding the temperature stable up to 600°C.

Electrical Measurements

The sheet-resistivity and Hall-effect measurements were made by means of an apparatus which makes use of electrometers operated in a unity-gain mode to isolate the sample and effectively reduce cable capacitance. The measured quantities were sheet resistivity (ρ_s) and sheet-Hall coefficient (R_s). From these the effective sheet-carrier concentration n_s (or p_s) $\equiv (eR_s)^{-1}$ (assuming a Hall factor of unity) and the effective Hall mobility $\mu_{eff} \equiv R_s/\rho_s$ were calculated. The bulk (volume) carrier concentration n (or p) and mobility μ were calculated as a function of depth by measuring R_s and ρ_s as successive layers of the implanted section were removed using a diluted solution of $H_2SO_4 : 30\%H_2O_2 : H_2O$ in a $1:1:50$ ratio by volume at 0°C. Such an etch produces uniform, damage-free GaAs surfaces, with a thin layer of the material being removed at a rate of about 50 Å/min as determined by a Sloan Dektak microtopographer.

RESULTS AND DISCUSSION

Cd Implantation

Cadmium was implanted into undoped n-type GaAs at room temperature to a dose (ϕ) of 10^{13}, 10^{14}, 10^{15}, and 10^{16} Cd/cm² at 135 keV. The dependence of the sheet-hole concentration (p_s) and the effective mobility (μ_{eff}) upon annealing temperature (T_A) (700, 800, and 900°C) is shown in Fig. 1 for various doses. The general annealing characteristics are quite similar to those reported previously by Hunsperger and Marsh.[3] The effective mobility recovers rapidly at annealing temperatures between 700 and 800°C, although it never reaches the values (~ 450 cm²/Vsec) expected for lattice scattering alone. The sheet-carrier concentration is near the concentration of implanted ions for $\phi \leq 10^{14}$ cm⁻² and $T_A \simeq 800$-900°C.

The concentration profiles of the 10^{14}, 10^{15}, and 10^{16} cm⁻² samples are displayed in Fig. 2 for $T_A = 900°C$. The LSS distribution[8] for $\phi = 10^{14}$ cm⁻² is also shown, and it can be seen that the actual electrical profiles are much flatter than the LSS profile. Also displayed is the diffusion profile expected when a source of 10^{14} cm⁻² Cd ions is placed on a GaAs surface and then heated for 10 min at 900°C.[9] This diffusion profile is very similar to the actual profile. Such a simplified diffusion model is sufficiently accurate since the diffusion range at 900°C is much greater than

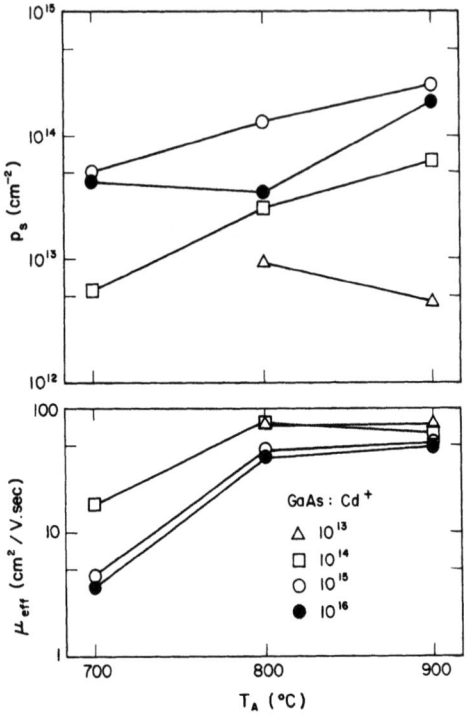

Figure 1. Dependence of p_s and μ_{eff} upon T_A for Cd-implanted GaAs

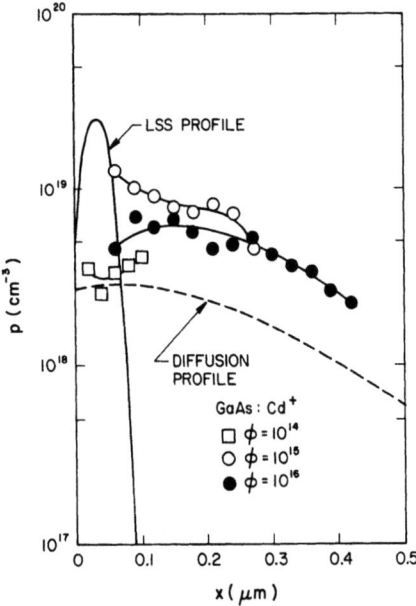

Figure 2. Electrical Profiles of Cd-implanted GaAs Annealed at 900°C

the implantation (LSS) range for 135-keV Cd ions. (In fact, even at 700°C the diffusion range is somewhat greater than the LSS range.) It should be noted that the p(x) curves for various doses do not differ markedly and that the $\phi = 10^{16}$ cm^{-2} curve actually lies below the $\phi = 10^{15}$ cm^{-2} curve. One reason for the $\phi = 10^{16}$ cm^{-2} curve not being higher is the solid solubility limit of Cd in GaAs, which is reported to be $\sim 2 \times 10^{19}$ cm^{-3} for a vapor source at 900°C.[10] This value is being approached for $\phi = 10^{15}$ cm^{-2}.

Te Implantation

Tellurium was implanted into Cr-doped GaAs at 250°C to a dose of 10^{13} or 10^{14} Te/cm^2 at 120 keV. Figures 3 and 4 give annealing characteristics of n_S and μ_{eff} for anneals at 600, 700, 800, and 900°C. Annealing at temperatures of $\gtrsim 600$°C produced electrically active n-type layers for both doses. The effective depth of the implanted layers was determined by means of the chemical-etch thin-layer-removal technique and found to be $\lesssim 0.1$ μm in all the annealed samples.

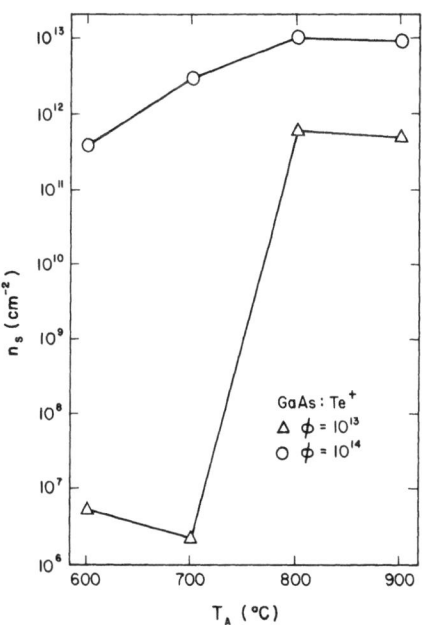

Figure 3. Dependence of n_s Upon T_A for Te-implanted GaAs

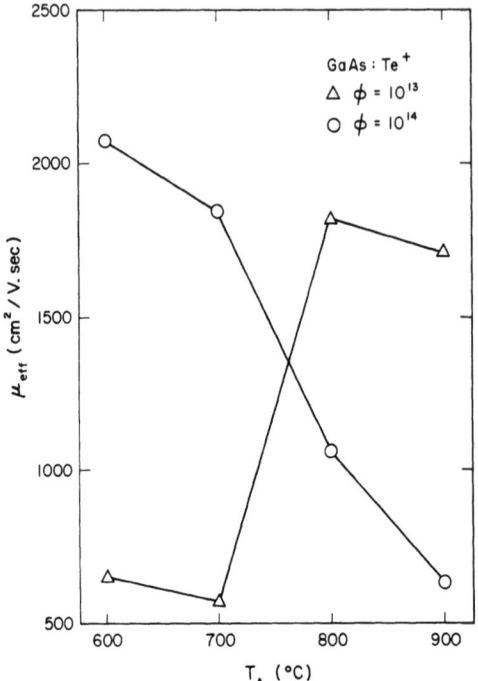

Figure 4. Dependence of μ_{eff} upon T_A for Te-implanted GaAs

The annealing mechanism in the 600-900°C region was carefully examined through a separate determination of the concentration of donors (N_D) and acceptors (N_A) from the known values of μ and n (with $\mu = \mu_{eff}$ and n being the effective bulk concentration in cm^{-3} determined from the assumed effective implantation depth of 0.1 µm) for the 10^{14} sample. A summary of the results for N_D, N_A, and N_I for given n and T_A is given in Fig. 5.

The behavior of μ_{eff} with T_A shown in Fig. 4 is consistent with that of N_I (= N_D + N_A); an increased concentration of ionized centers with increasing T_A leads to a decrease in mobility. The value of N_D increases with T_A for T_A between 600° and 800°C, implying that the amount of substitutional Te increases with increasing annealing temperature. Further increase in T_A does not produce an indefinite increase in N_D, but N_D approaches a maximum at 800-900°C, since the total amount of the Te source is limited by the implanted dose. The value of N_A is approximately proportional to N_D, while the value of K ($\equiv N_A/N_D$) remains nearly constant at a value of ~ 0.6 and relatively independent of T_A. This fact suggests that as the amount of substitutional Te increases with increasing T_A, a portion of the activated Te readily becomes "self-compensated," forming a V_{Ga}Te-type complex (Ga-vacancy-Te-donor

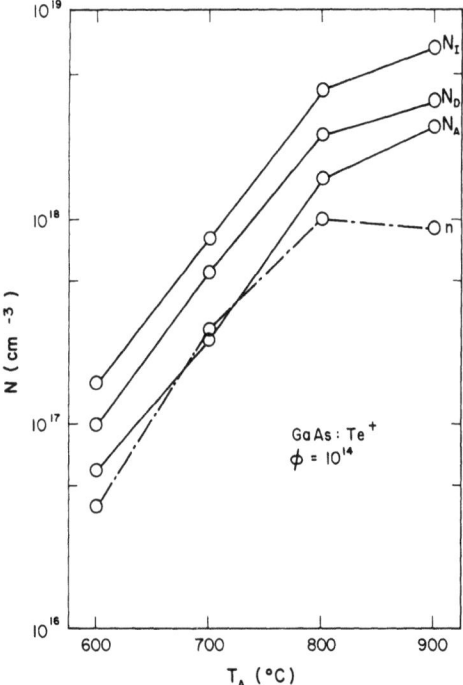

Figure 5. Dependence of N_D, N_A, and N_I Upon Given n and T_A for 10^{14} Te/cm^2 Implanted GaAs

complex). This confirms the results of a previous photoluminescence study on Te-implanted GaAs.[4] The formation of the V_{Ga}Te-complex center has also been observed in Te-doped GaAs when annealed above 600°C.[11] The total amount of N_I at T_A = 900°C is approximately equal to the Te dose, indicating that most of the implanted Te ions became substitutional after annealing at 900°C.

SUMMARY

Results of sheet-resistivity and Hall-effect measurements on Cd-implanted GaAs showed that almost 100% doping efficiency was obtained in the samples with $\phi \lesssim 10^{14}$ Cd/cm^2 after annealing at 800-900°C. Electrical profiles in these samples follow the diffusion profile and $\phi = 10^{15}$ Cd/cm^2 yields a solid solubility concentration. On the other hand, much lower efficiency was obtained in the Te-implanted samples even with "hot-implantation" and 900°C annealing. A detailed analysis of the annealing characteristics showed that electrical compensation due to the formation of a V_{Ga}Te-complex center is responsible for the low electrical activity obtained.

ACKNOWLEDGMENTS

The authors wish to thank M. Whitaker for editorial assistance.

REFERENCES

1. A. G. Foyt, J. P. Donnelly, and W. T. Lindley, Appl. Phys. Lett. 14, 372 (1969).

2. J. D. Sansbury and J. F. Gibbons, Rad. Eff. 6, 269 (1970).

3. R. G. Hunsperger and O. J. March, J. Electrochem. Soc. 116, 488 (1969); Rad. Eff. 6, 236 (1970).

4. J. S. Harris, F. H. Eisen, B. Welch, J. D. Haskell, R. D. Pashley, and J. W. Mayer, Appl. Phys. Lett. 21, 601 (1972).

5. D. E. Davies, S. Roosild, and L. Lowe, Solid State Electron. 18, 733 (1975).

6. J. P. Donnelly, W. T. Lindley, and C. E. Hurwitz, Appl. Phys. Lett. 27, 41 (1975).

7. The pyrolytic encapsulations were performed at MIT Lincoln Laboratory, Lexington, MA.

8. J. Lindhard, M. Scharff, and H. E. Schiott, Mat. Fys. Medd. Dan. Vid. Selsk. 33, No. 14 (1963); see also, J. F. Gibbons, W. S. Johnson, and S. W. Mylroie, Projected Range Statistics (Halsted Press, New York, 1975).

9. M. Fujimoto, K. Kudo, and N. Hishinuma, J. Appl. Phys. Japan 8, 725 (1969).

10. B. Goldstein, Phys. Rev. 118, 1024 (1960).

11. C. J. Hwang, J. Appl. Phys. 40, 4584 (1969).

*Work supported in part under Contract F33615-72-C-1099.

CHARACTERISTICS OF IMPLANTED N-TYPE PROFILES IN GaAs ANNEALED

IN A CONTROLLED ATMOSPHERE

D.H. Lee and R.M. Malbon

Torrance Research Center
Hughes Aircraft Company
Torrance, California 90509

J.M. Whelan

Materials Science Department
University of Southern California
Los Angeles, California 90007

ABSTRACT

A controlled atmosphere technique (CAT) has been developed as an alternative to dielectric encapsulation for the anneal of implanted GaAs. In the CAT, a flowing hydrogen-arsenic atmosphere is adjusted to be near equilibrium with the surface of the implanted GaAs during the heat treatment. The CAT anneal was used to study the characteristics of n-type impurities (Si and S), random implanted at room temperature into GaAs epitaxial layers. In situ implanted silicon distributions were measured before anneal by secondary ion mass spectroscopy and the projected range data show agreement to within 10% with tabulated values for energies 50 to 400 keV. Silicon and sulfur implantations of $>2 \times 10^{13}$ cm^{-2} annealed at 800°C for 20 min resulted in apparent electrical conversion efficiencies >75%. The CAT annealed profiles are approximately Gaussian near the peaks with exponential tail behavior at lower concentrations.

INTRODUCTION

Ion implanted compound semiconductors are usually encapsulated with a dielectric material to prevent dissociation of the surface during high temperature anneals. High temperature anneals are required after implantation to remove radiation damage and to establish high electrical activity in the implanted layer. For example, sputtered or pyrolytically deposited thin films of SiO_2, Si_3N_4, Al_2O_3, and AlN have been used as encapsulants for the anneal of implanted GaAs.[1-4] However, when n-type impurities (e.g., Si and S) are implanted into GaAs and annealed, the resultant doping profiles generally exhibit low electrical conversion efficiencies ($\leq 50\%$) and/or anomalous diffusional broadening.[5] Loss of profile integrity in annealed GaAs has been attributed to interdiffusion effects between the constituents of the dielectric encapsulant and the implanted layer or, in the case of Cr-doped semi-insulating substrates, to the influence of high background levels of compensation.[6]

A controlled atmosphere was therefore investigated as a method to replace dielectric encapsulation for the anneal of implanted GaAs.[7] Moreover, vapor-phase-epitaxial (VPE) layers were employed in an effort to minimize high levels of compensation in the starting GaAs. Characteristics of CAT annealed $^{28}Si^+$ and $^{32}S^+$ implantations in VPE GaAs layers are discussed.

EXPERIMENTAL PROCEDURE

All implantations were performed at room temperature under random alignment conditions into 3 to 5 μm thick epitaxial layers grown on <100> GaAs substrates. To facilitate the electrical measurements, n-type epitaxial layers with constant background doping levels that ranged from 8×10^{14} to 1×10^{17} cm^{-3} were grown on heavily doped n^+ substrates. At the onset of the silicon implantation studies, it was recognized that an accurate assessment of the CAT anneal could only be made if a measure of the implanted silicon atom distribution was obtained before heat treatment. Therefore, in situ silicon distributions, implanted at energies 50 to 400 keV, were measured by secondary ion mass spectroscopy (SIMS).

A schematic illustration of the apparatus used for the controlled atmosphere high temperature anneals is given in Fig. 1. Surface erosion effects are minimized by means of a flowing hydrogen-arsenic atmosphere that is adjusted to be in near equilibrium with the surface of the implanted GaAs during the anneal. To supply an arsenic overpressure in the vicinity of the implanted GaAs sample, preheated hydrogen is passed over a quartz boat which contains crushed GaAs/Ga. During the heat treatment, the temperature zone over the arsenic source is maintained at a temperature 20 to 40°C higher than the desired anneal temperature maintained over the implanted sample. After the CAT anneal, electrical profiles were extracted from reverse-biased Schottky-barrier

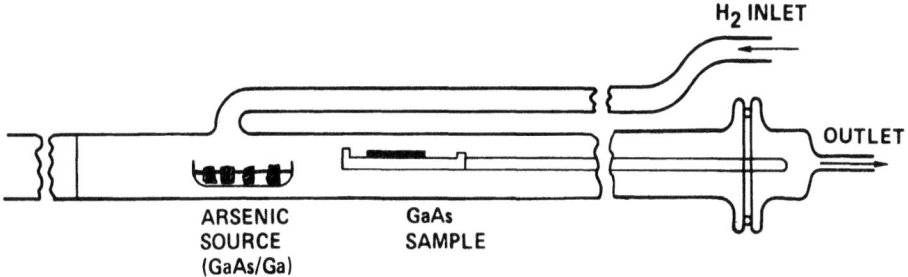

Fig. 1. Cross-section of the quartz assembly used for the controlled atmosphere anneal of implanted GaAs.

capacitance-voltage (C-V) measurements taken at 1 MHz. Where necessary, step etching techniques were used to construct the complete doping profile.

RESULTS AND DISCUSSION

An initial evaluation of the controlled atmosphere anneal process was made with unimplanted VPE layers. Several epitaxial layers with doping levels as low as 1×10^{15} cm^{-3} were annealed in the temperature range 600 to 800°C for times as long as 30 min. In all cases, there was no measurable change in the uniform background doping of the epitaxial layers. Moreover, for these anneal conditions, no surface erosion was observed with 900X optical interference microscopy. However, when the 800°C anneals were extended to times ≥60 min, the samples showed signs of degradation in surface morphology.

Silicon Profiles

Silicon (^{28}Si) ions were implanted at energies 50 to 400 keV with doses between 6×10^{13} and 4×10^{14} cm^{-2}. In Fig. 2, the solid points are the silicon atom distributions for the 150 and 400 keV implantations as measured by the SIMS technique. The normalization factor for a unity concentration at the peak is 1×10^{19} cm^{-3}. To first order, all of the as-implanted silicon atom distributions were Gaussian. Table 1 compares the measured values for the projected range R_p and standard deviation ΔR_p with the theoretical estimates tabulated by Gibbons et al.[8] for 50, 150, 300, and 400 keV. Over this energy range, the agreement between theory and experiment is better than 10% for R_p and within 20% for ΔR_p.

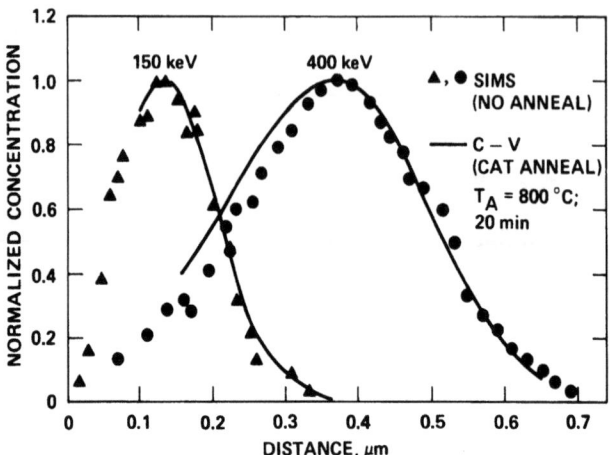

Fig. 2. Normalized depth distributions of silicon atoms implanted into GaAs at 150 keV (triangles) and 400 keV (dots). Solid lines are normalized doping profiles after a 800°C, 20 min CAT anneal.

Table I. Projected Range Data for ^{28}Si in GaAs

^{28}Si → GaAs	Experiment (SIMS)		Theory (Ref. 8)	
Energy (keV)	R_p (μm)	ΔR_p (μm)	R_p (μm)	ΔR_p (μm)
50	0.040	0.026	0.042	0.025
150	0.140	0.068	0.129	0.061
300	0.280	0.118	0.263	0.100
400	0.375	0.133	0.351	0.121

Also plotted in Fig. 2 are normalized curves of the measured doping profiles for 150 and 400 keV, 5 x 10^{12} cm^{-2} silicon implantations after a 800°C, 20 min CAT anneal. Comparison between the profiles before and after heat treatment clearly illustrates that the CAT anneal does not produce significant profile broadening. Good agreement between the shallow 150 keV implanted profiles also confirms the visual observation that a minimum of surface erosion occurs during the anneal.

Profiles for 120 keV silicon ions implanted at various doses and CAT annealed are given in Fig. 3. The broken curve is a Gaussian distribution with $R_p = 0.103$ µm, $\Delta R_p = 0.051$ µm and a peak ion concentration of 1.8×10^{18} cm^{-3}. After an 800°C, 20 min anneal, the resultant doping profiles show: (i) **reasonable agreement with theory** in the location of the maximum (R_p), (ii) Gaussian characteristics around the peaks with some diffusional broadening, and (iii) penetrating tail behavior in the leading edges at low doping levels. For a dose of 2.3×10^{13} cm^{-2} the apparent electrical conversion efficiency η is ~80% and the maximum carrier concentration is ~1.1×10^{18} cm^{-3}. However, when the dose is decreased to ~5×10^{12} cm^{-2}, η decreases to ~30% accompanied by a loss of profile integrity. Below 2×10^{12} cm^{-2}, an intrinsic "i" layer is formed, which extends throughout the predicted width of the implanted ion distribution. Additional measurements are needed to clarify the mechanism for this acceptor-like compensation.

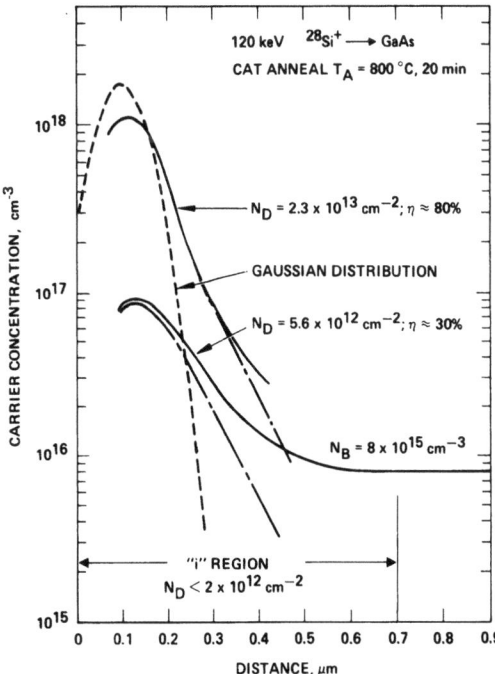

Fig. 3. Doping profiles for 120 keV silicon ions implanted into epitaxial GaAs at various doses.

Subtraction of the background doping level N_B of the epitaxial layer from the measured profiles in Fig. 3 shows that the tail behavior at low concentrations can be represented by an exponential $\exp(-x/\lambda)$ with a characteristic length $\lambda = 0.083$ µm. At the intermediate dose of 5.3×10^{12} cm^{-2}, the implanted profile is dominated by the tail; however, at the higher dose of 2.3×10^{13} cm^{-2} the tail has less influence on the doping profile around the peak. This suggests a saturation effect for the exponential tail behavior as the dose is increased.

Figure 4 shows the measured doping profile after a 120 keV, 2.3×10^{13} cm^{-2} silicon implantation and 800°C, 20 min CAT anneal. In this case, the profile is nearly identical to the one discussed in Fig. 3 and attests to the repeatability of the controlled atmosphere anneal. After background subtraction, an exponential tail is again observed in the leading edge (dashed-dot curve in Fig. 4); however, for this case $\lambda = 0.048$ µm. This is almost a factor of two smaller than the previous λ value of 0.083 µm (Fig. 3). A series of silicon implantations into different resistivity VPE layers revealed that λ was a function of the doping level N_B in the epitaxial layer. A summary of these results is presented in Fig. 5 where the solid line represents the relationship $\lambda = N_B^{-0.62}$. Also included are estimates of λ obtained from profiles reported by Woodcock et al.[9] on annealed 200 and 400 keV silicon implantations in high-resistivity n-type GaAs epitaxial layers. The data of Fig. 5 indicate that the differential C-V measurement technique may be complicated by problems associated with Debye screening lengths (dashed line in Fig. 5). Further investigations are required to ascertain the exact nature of the tail characteristics.

SULFUR PROFILES

A parallel set of experiments similar to the silicon implantation and CAT anneal studies was repeated for sulfur ions ($^{32}S^+$). After a 800°C, 20 min anneal, the following features were observed: (i) reasonable agreement ($\leq 10\%$) between the peak location of the doping profiles and the projected range values predicted from theory, (ii) good profile integrity for doses $> 2 \times 10^{13}$ cm^{-2} with apparent electrical conversion efficiencies as high as 86%,[7] and (iii) "i" layer formation at doses $< 2 \times 10^{12}$ cm^{-2}. In general, the CAT annealed sulfur profiles had more diffusional broadening and deeper penetrating tails than did similar silicon implanted profiles annealed under the same conditions.

CONCLUSION

Monoenergetic silicon ($^{28}Si^+$) and sulfur ($^{32}S^+$) impurities implanted at doses of $>2 \times 10^{13}$ cm^{-2} into GaAs epitaxial layers exhibited apparent electrical conversion efficiencies >75% when

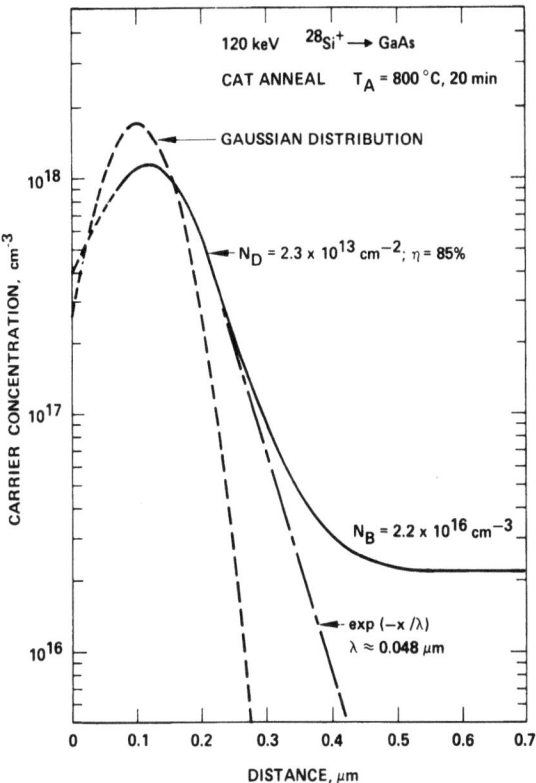

Fig. 4.
Doping profile for 120 keV silicon ions implanted into an epitaxial GaAs layer with $N_B = 2.2 \times 10^{16}$ cm^{-3}.

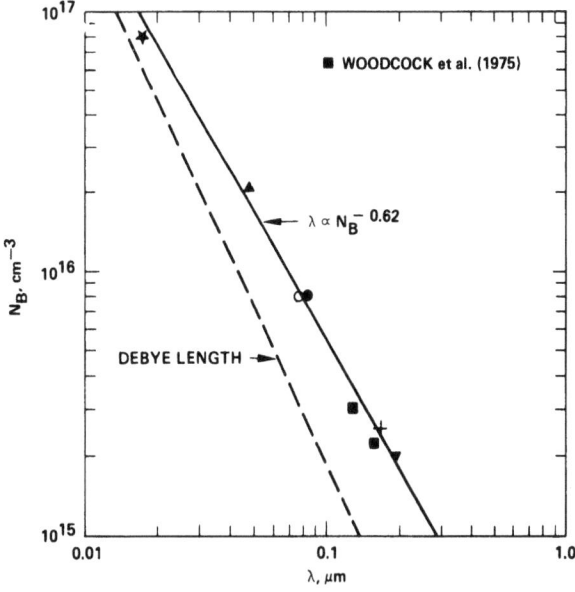

Fig. 5.
Plot of epitaxial layer doping N_B versus characteristic length λ of exponential tail behavior for annealed silicon profiles in GaAs.

annealed at 800°C for 20 min in a controlled atmosphere of hydrogen and arsenic. To a first approximation, the electrical conversion efficiency was found to be independent of incident ion energy. The annealed profiles are characterized by a Gaussian near the peaks with exponential tail behavior in the leading edges at lower concentrations. Depth locations of the annealed silicon and sulfur profiles were within 10% of the theoretical projected ranges tabulated by Gibbons et al.[8] for energies 50 to 400 keV. Although a decrease in apparent electrical conversion efficiency and loss of profile integrity was observed for ion doses $<5 \times 10^{12}$ cm^{-2}, the encouraging results for the higher dose implantation support the conclusion that the controlled atmosphere technique may be an attractive alternative to dielectric encapsulation for the high temperature anneal of implanted GaAs layers.

ACKNOWLEDGMENTS

The authors wish to thank W. B. Henderson for the expitaxial material, C. A. Reamer for the sample preparation and T. Whatley and D. Comaford of ARL for the SIMS.

REFERENCES

1. A.G. Foyt, J.P. Donnelly, and W.T. Lindley, Appl. Phys. Letters 14, 372 (1969).

2. E.C. Bell, A.E. Glaccum, P.L.F.H. Hemet, and B.J. Sealy, Rad. Eff. 22, 253 (1974).

3. B.J. Sealy and A.D.E. D'Cruz, Electronics Letters 11, 323 (1975).

4. D. Eirug Davies, J.K. Kennedy, and C.E. Ludington, J. Electrochem. Soc. 122, 1374 (1975).

5. J.D. Sansbury and J.F. Gibbons, Rad. Eff. 6, 269 (1970).

6. F.H. Eisen, in *Ion Implantation in Semiconductors: Science and Technology*, ed. by S. Namba (Plenum, New York, 1975), p. 3.

7. R.M. Malbon, D.H. Lee, and J.M. Whelan, J. Electrochem. Soc. (to be published).

8. J.F. Gibbons, W.S. Johnson, and S.W. Mylroie, *Projected Range Statistics* 2nd Ed, Distributed by Halstead Press/A Division of John Wiley & Sons, Inc. (New York, 1975).

9. J.M. Woodcock, J.M. Shannon, and D.J. Clark, Solid-State Electronics 18, 267 (1975).

PHOTOLUMINESCENCE OF Cd-ION IMPLANTED GaAs

K. Aoki, K. Gamo, K. Masuda and S. Namba

Faculty of Engineering Science, Osaka University.

Toyonaka, Osaka, Japan

Photoluminescence of Cd implanted GaAs with and without As preimplantation has been measured to study the effect of As dual implantation. Three emission bands at 1.479, 1.454 and 1.392 eV at 90 K were mainly observed after the implantation and annealing at 800°C, these were ascribed to Cd acceptors, defects which were observed only for As implantation, and As and Cd dual implantation, and defects which seem to be associated with As vacancy, respectively. Temperature dependence of emission intensity, linewidth and peak energy was measured to characterize the emission centers. It was found from the profiling of the 1.392 eV emission that the diffusion of As vacancy is suppressed by As implantation.

INTRODUCTION

Recent studies suggest that stoichiometry or vacancy concentration affects the characteristics of implanted layer in compound semiconductors such as GaAs. Hunsperger et al. [1] have found that insulating layers(i-layer) can be formed by Cd or Zn implantation in GaAs and suggested that a complex center associated with As vacancy (level of 0.29 eV above valence band) is responsible for the formation of i-layers. Although preimplantation of As has been considered to be effective to prevent the formation of i-layers[2], characteristics of the defect and the role of preimplanted As are not still obvious. It is also important to reveal these problems from standpoint of stoichiometry control by dual implantation.

Many defects show characteristic emission. Vacancies for example, show emmission at ~1.39 eV [3,4] and defect complexes associated with As vacancy show emission at around 1.37~1.41 eV [5,6]. Ga vacancy-donor complexes show emission at around 1.2 eV [7].

Therefore photoluminescence (PL) is a powerful technique to investigate the effects of vacancies and As preimplantation on the resulting characteristics of implanted GaAs. We implanted Cd with or without As preimplantation and measured PL of implanted GaAs in order to study these effects.

EXPERIMENTAL

Samples used were undoped, <111> oriented, n-type GaAs (n= $4.6 \times 10^{16}/cm^2$). After mechanical polishing, the sample surfaces were etched chemically by an aquasolution of 1 mol NaOH and 0.7 mol H_2O_2. 40 keV As and 50 keV Cd were implanted at room temperature (R.T.) or 300°C to a dose of $2 \times 10^{13}/cm^2$ to $1 \times 10^{15}/cm^2$, in a direction of 8° off <111> axis to reduce channeling effects. 40 keV As has an almost same projected range as 50 keV Cd ion (R=180 Å). After ion implantation, the sample surfaces were coated with SiO_2 and annealing was done at 800°C in flowing hydrogen gas for 15 minutes.

A 1 kW Xe-lamp filtered by a $CuSO_4$-saturated solution or an argon-ion laser were used for PL excitation. PL measurements were done at 4.2 K 273 K by using a grating monochromator, a cooled RCA-7102 photomultiplier and a lock-in amplifier. The depth profiling of the PL intensities was done by successive layer removal of the surface by the solution of 1 mol NaOH and 0.7 mol H_2O_2.

RESULTS AND DISCUSSION

Figure 1 shows PL spectra of Cd implanted GaAs measured at 90 K. Solid and dot-dashed lines are the PL spectra for implantation at room temperature and 300°C, respectively. For comparison, the spectrum of an unimplanted, 800°C annealed sample is also shown by dotted line. The emission at 1.490 eV (A) of the unimplanted sample is likely due to Si acceptor. The 1.479 eV emmission (B) observed for implanted samples is due to Cd acceptors, identified from the temperature dependence of emmission spectrum [8]. It is clear that the intensity of the Cd emission is 10 times more intense for hot implantations than for R.T. implantations. The peak energy for the emission D (1.392 eV) observed for hot implantation is the same as that often proposed to be associated with As vacancy by several authors [3,4]. In case of R.T. implantation, the emission E (1.351 eV) associated with Cu acceptors [9] was clearly observed.

Figure 2 shows the depth distribution of the intensities of the emission B (Cd acceptor) for implantation at R.T. and 300°C.

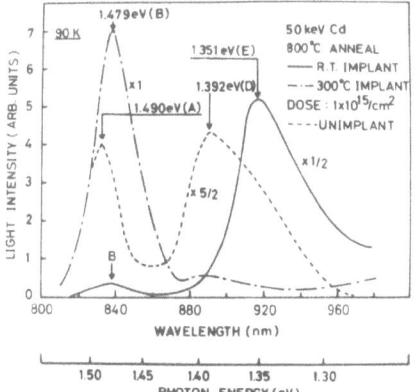

Fig. 1 P.L spectra at 90 K of Cd implanted GaAs.

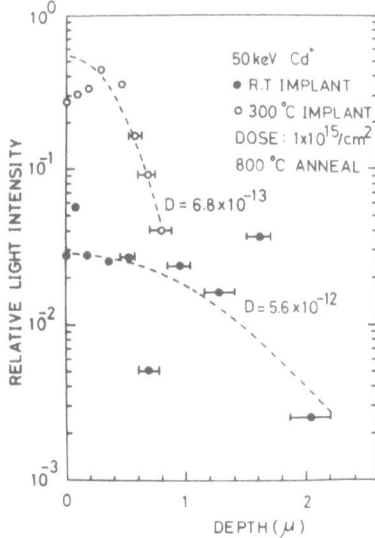

Fig. 2 Depth distribution of Cd acceptor emission measured at 90 K in Cd implanted GaAs.

The emission was observed up to a depth of 2.0 μm for R.T. implantation and 0.8 μm for 300°C implantation, a depth which is far deeper than the projected range (~180 Å). These results suggest that implanted Cd atoms diffuse into bulk during 800°C annealing. The dashed lines show the intensity profile calculated from the equation,

$$I_{PL}(\ell) = K \int_0^\infty n(x+\ell) I_0 e^{-\alpha x} dx$$

where $n(x + \ell)$ is the concentration of Cd acceptors and is given by $N/2\sqrt{\pi Dt} \cdot \exp(-(x+\ell)^2/4Dt)$ and $I_0 e^{-\alpha x}$ is the excitation intensity at a depth x from the surface, and ℓ is the depth removed. A diffusion coefficient D of 5.6×10^{-12} cm²/sec for R.T. implantation and 6.8×10^{-13} cm²/sec for 300°C implantation was used. These values are larger than the diffusion coefficient reported by Goldstein[10]. The larger value for R.T. implantation suggest that more defects which enhance the diffusion are produced by R.T. implantation. In the calculation, we neglected surface recombination and diffusion of excess carriers and the effect of the residual defects.

Figure 3 shows depth-distribution of the intensity of the emission D observed for Cd and As implantation at 300°C, normalized to the light intensity of 1.503 eV emission which is due to Si donors. This emission is distributed uniformly before implantation. We can exclude the effects of nonradiative centers by the normalization, because the emission due to Si donors is also quenched

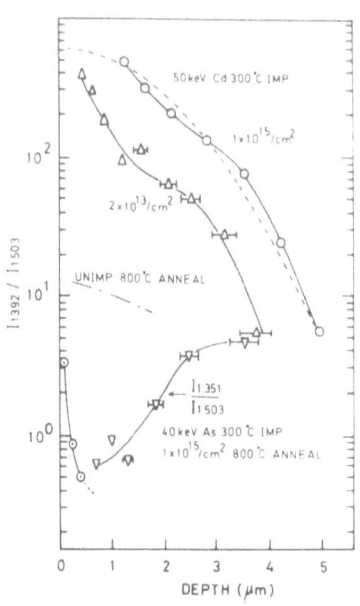

Fig. 3 Depth distribution of normalized emission intensity $I_{1.392}/I_{1.502}$ at 90 K of Cd implanted and As implanted GaAs. The distribution of an unimplanted, 800°C annealed GaAs is also shown.

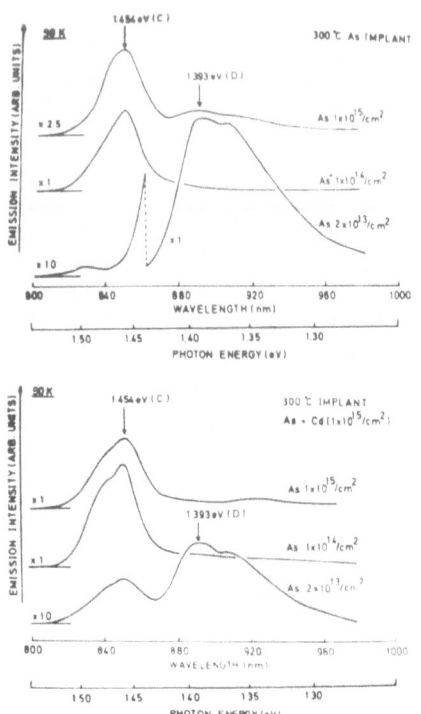

Fig. 4 PL spectra of As implanted sample and dual implanted sample. Both As and Cd ions were implanted at 300°C.

by the nonradiative centers. The 1.503 eV emission was not observed at the surface region shallower than a depth of 0.4 μm and 1.3 μm for Cd implantation with doses of 2×10^{13}/cm^2 and 1×10^{15}/cm^2 respectively, and a depth of 0.1 μm for As implantation with a dose of 1×10^{15}/cm^2. As can be seen from Fig. 3, in cases of Cd implantation, the emission D was distributed up to a depth of ~5 μm. On the other hand, in case of 300°C As implantation, the emission D was distributed only near the surface. At depths beyond ~0.4 μm, the emission E was observed. The similar depth-distribution of the emission D was also observed for As and Cd dual implantation. (c.f. Fig. 6) Intensity profiles of the emission D for As implantations suggest that the diffusion of the defects responsible for the emission D can be suppressed by As preimplantation. From the peak energy of the emission, the defects seem to be As vacancy [3,4].

Figure 4 shows the PL spectra measured at 90 K for single and dual implantations. For both cases, a new emission at 1.454 eV (C) was observed. The line is superposed by the emission A in cases of 300°C As implantation and the emission B in cases of 300°C dual

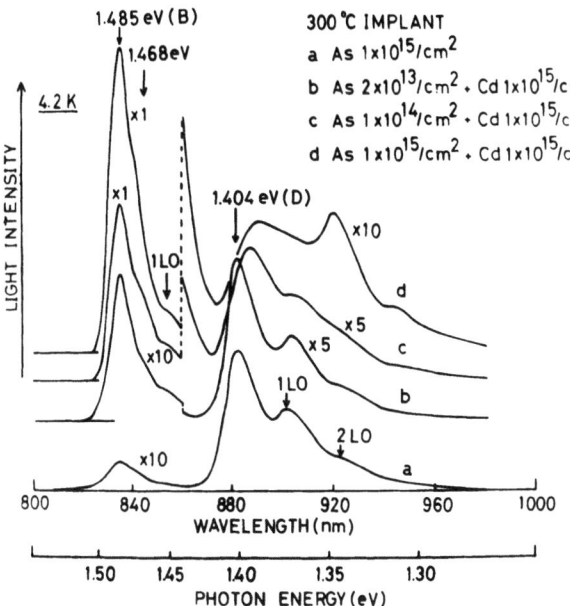

Fig. 5 PL spectra at 4.2 K of dual implanted GaAs. Both ions of As and Cd were implanted at 300°C

implantation. The emission D was also observed. In cases of 300°C As implantation, the emission D was intense for the low dose of $2 \times 10^{13}/cm^2$ and weak for the higher doses of $1 \times 10^{14}/cm^2$ and $1 \times 10^{15}/cm^2$. It seems likely that more defects which act as nonradiative centers still remain for high dose implantation. In case of dual implantation, the intensity of the emission D depends on total dose (c.f. Fig. 5).

Figure 5 shows PL spectra measured at 4.2 K for dual implanted GaAs. Cd ion dose is $1 \times 10^{15}/cm^2$. At 4.2 K, the emission B (Cd acceptor) appeared at 1.485 eV and the emission C appeared as a shoulder near 1.468 eV. The 1-LO replica of the emission B was also seen at 1.452 eV. It was found that the intensity of emission B increases with increasing As ion dose. For example, for As preimplantation with a dose of $1 \times 10^{15}/cm^2$, the intensity of the emission B is 20 times more intense than that for As preimplantation with a dose of $2 \times 10^{13}/cm^2$. The emission C appeared at ~1.404 eV showed phonon structure of 1 LO and 2 LO, which suggests that the radiative center responsible to this emission is not a localized center such as As vacancy-Ge complex.

Figure 6 shows PL spectra for dual implanted sample observed after surface removal by chemical etching. It was found that the emission C and the emission D are distributed near the implanted region. At a depth more than 1600 Å, the emission E which is due to Cu, was dominant. It is clear by comparison with Cd single implantation (c.f. Fig. 3) that As preimplantation suppresses the migration of

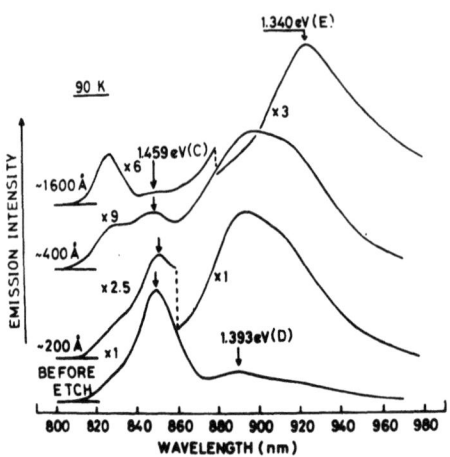

Fig. 6 Change of PL spectra at 90 K of dual implanted GaAs as a function of depth removed by chemical etching. As and Cd were implanted at R.T. and 300°C respectively.

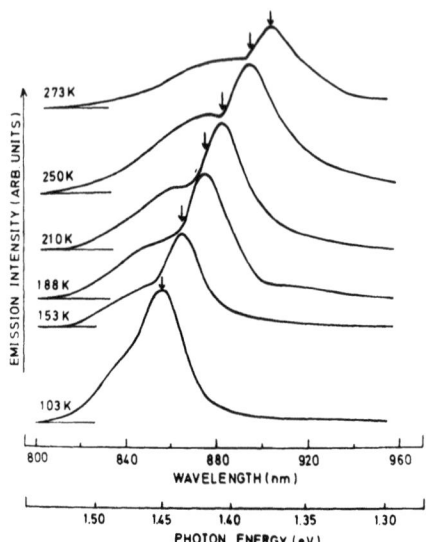

Fig. 7 PL spectra of the emission C as a function of temperature for dual implanted GaAs. Both As and Cd ions were implanted at 300°C with a dose of 1 x $10^{15}/cm^2$.

defects produced by Cd implantation.

It was found that the emission C was observed only by As implantation, and not by Cd or Ar implantation. Therefore it can be said that this emmission center is characteristic of As implantation. Although the origin is not clear, Ga vacancy associated or interstitial As associated defects are the possible centers by As implantation. We measured the temperature dependence of the line shape, the peak energy and the linewidth in order to characterize this emmission center. Figure 7 shows the temperature dependence of PL spectra of the emission C and B (Cd acceptor). With increase of temperature, the peak of the emission C indicated by an arrow shifts to lower energy and the tail of the emission B at high energy side becomes pronounced. This tail reflects the Boltzman distribution of conduction electrons. On the other hand, the emission C shows almost symmetric line shape. These results indicate that the emission C can not be explained by transition between a band and an isolated level such as between conduction band and acceptor level.

Figure 8 shows peak shift of the emission C as a function of temperature. The peak shift of the emission B (Cd acceptor) is also shown [8] for comparison. The dotted line is the temperature

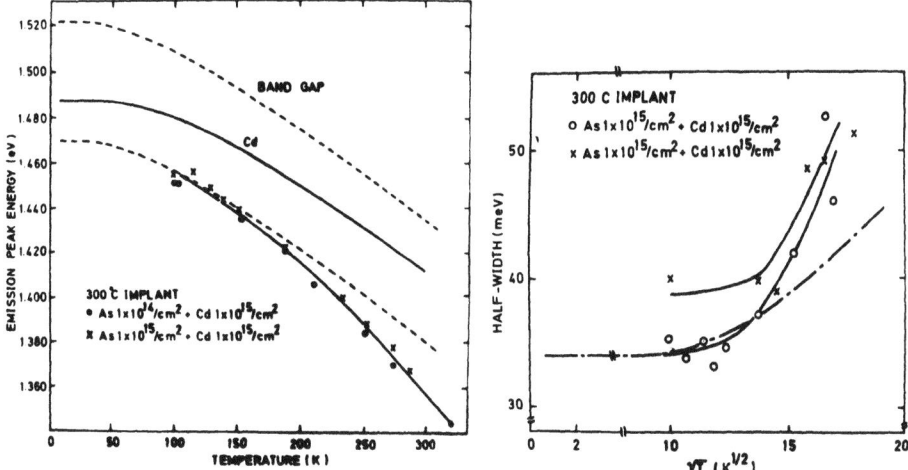

Fig. 8 Emission peak energy of the emission C as a function of temperature. Both As and Cd ions were implanted at 300°C.

Fig. 9 Temperature variation of the half-width of the emission C.

variation of band gap energy after Sturge [11]. The peak shift of the emission B is smaller than the variation of band gap energy. On the other hand, the peak shift of the emission C is larger than the shift of band gap energy. This result also suggests that the emission C is not due to a transition between a band and an isolated level.

Figure 9 shows the temperature variation of the half-width of the emission C. As is shown in this figure, the half-width is almost constant from 100 K to 170 K, and above 170 K, the half-width increases rapidly. The dot-dashed line is the half-width ΔW which is calculated by the equation,

$$\Delta W = A[\coth \hbar\omega_e/2kT]^{1/2}$$

based on a configuration coordinate (CC) model, where A is constant (34 meV), and $\hbar\omega_e$ (40 meV) is the vibration energy of the excited state. The observed variation is more rapid than the prediction by CC model.

From the temperature dependence, we can confirm that the emission C is due to a complex center. However the origin is not clear at present time, a complex center associated with Ga vacancy or interstitial As may be possible.

CONCLUSIONS

The conclusions drawn from the present work are as follows:
1) Cd acceptor emission is more intense for 300°C Cd implantation than for R.T. Cd implantation.
2) Diffusion of Cd acceptors takes place after Cd implantation and 800°C annealing. The emission center which is often proposed to be associated with As vacancies also diffuses into bulk (~5 μm).
3) As preimplantation suppresses the diffusion of the 1.392 eV emission center. Furthermore, a new complex center is formed near the implanted region by As implantation.

ACKNOWLEDGEMENT

The authors would like to thank Mr. Kawasaki for his assistance in ion implantation.

REFERENCES

1) R.G. Hunsperger and O.J. Marsh: Metal. Trans. $\underline{1}$ 603 (1970)
2) T. Itoh and Y. Kushiro: J. Appl. Phys. $\underline{42}$ 5120 (1971)
3) L.L. Chang, L. Esaki and R. Tsu: Appl. Phys. Letters $\underline{19}$ 143 (1971)
4) R. Romano-Moran and K.L. Ashley: J. Phys. Chem. Solids $\underline{34}$ 427 (1972)
5) C.J. Hwang: Phys. Rev. $\underline{180}$ 827 (1969)
6) E.W. Williams and C.T. Elliott: Brit. J. Appl. Phys. (J. Phys. D) $\underline{2}$ 1657 (1969)
7) E.W. Williams: Phys. Rev. $\underline{168}$ 922 (1968)
8) K. Aoki, K. Gamo, K. Masuda and S. Namba: Japan. J. Appl. Phys. $\underline{15}$ 145 (1976)
9) C.J. Hwang: J. Appl. Phys. $\underline{39}$ 4307 (1968)
10) B. Goldstein: Phys. Rev. $\underline{118}$ 1024 (1960)
11) M.D. Sturge: Phys. Rev. $\underline{127}$ 768 (1968)

SOME STRUCTURAL AND ELECTRICAL CHARACTERISTICS OF GaAs ANNEALED

AFTER IMPLANTATION WITH Be, Mg, Zn, AND Cd*

R. B. Benson, Jr., M. A. Littlejohn, K. Lee, and
R. E. Ricker

North Carolina State University, Raleigh, North Carolina
27607

ABSTRACT

Chromium doped semi-insulating GaAs was annealed at 800°C after implantation with Be, Mg, Zn, and Cd ions at 60 KeV and ambient temperature with a fluence of 10^{15} ions/cm^2. The defect structures in the as-implanted and as-annealed states were examined using transmission electron microscopy techniques. The differences in the electrical properties of the specimens implanted with different ions are discussed in relation to the differences in the observed defect structures for the various samples.

INTRODUCTION

Various types of crystal defects including several precipitates have been previously observed using transmission electron microscopy (TEM) techniques in GaAs annealed after implantation with Zn ions [1,2]. Reported x-ray results indicate that compounds with structures similar to the precipitates can form, with Mg or Cd replacing the Zn [3]. Crystal defects may influence to various degrees some of the electrical properties of ion implanted-annealed layers. The defect structures that were observed using TEM techniques in GaAs annealed after implantation with Be, Mg, Zn and Cd are reported here. Some electrical properties of these implanted layers were also determined and compared to the variations in observed defect structures for the different implanted ions.

---

*This work was supported by the Office of Naval Research.

EXPERIMENTAL PROCEDURES

Semi-insulating [100] Cr-doped GaAs was used in these studies. The polished wafers were examined by TEM before implantation to insure that no small defects were present, and that the as-grown dislocation density was very low. The specimens were implanted with 60 KeV Be, Mg, Cd, or Zn ions at ambient temperature in a non-channeling direction to a fluence of $10^{15}/cm^2$.

An Rf-sputtered SiO_2 passivating layer was employed, and the specimens were annealed at 800°C for 30 min. The TEM specimens were prepared by chemical polishing, and were examined using transmission electron microscopy and selected area diffraction techniques.

The method of Van der Pauw was used to determine the sheet resistivity, Hall mobility, and the sheet hole concentration of ion-implanted Cr-doped GaAs. The experimental apparatus and procedure have been discussed in a previous publication [4]. The ohmic contact material was vacuum evaporated 95% Ag-5% Mn, which has been found to yield quite good ohmic contacts to GaAs implanted with p-type dopants. Corrections were made to the measured sheet resistivity and Hall mobility in order to correct for finite contact size and for contact edge effects.

TRANSMISSION ELECTRON MICROSCOPY STRUCTURAL ANALYSIS

Electron diffraction analysis indicated that for the as-implanted specimens there was an amorphous region in the samples implanted with Zn and Cd, while those implanted with Be and Mg were completely crystalline. The differences in the as-implanted structure for the different ions for a constant fluence and implantation energy are probably due to the difference in the projected LSS range of the different ions.

The defects observed in unsectioned GaAs annealed at 800°C after implantation with a fluence of $10^{15}/cm^2$ of the various ions are illustrated in Figure 1. The micrographs for Be and Mg specimens are from thick regions of the foil with a diffraction deviation parameter of $w \gtrsim 1$ (kinematic diffraction conditions) in order to show the dislocation loops clearly. The small defects appear as dark dots in these images, whereas under the diffraction conditions $w \simeq 0$ (dynamic diffraction conditions) many of these defects would exhibit black-white (B-W) contrast, as in Figure 1d and Figure 2 [5]. The matrices of all the Be, Mg, and Cd specimens examined after annealing were always single crystalline GaAs. The specimen implanted with Be contained resolvable dislocation loops up to approximately 2000Å in size, defects exhibiting double arc

STRUCTURAL AND ELECTRICAL CHARACTERISTICS OF GaAs   133

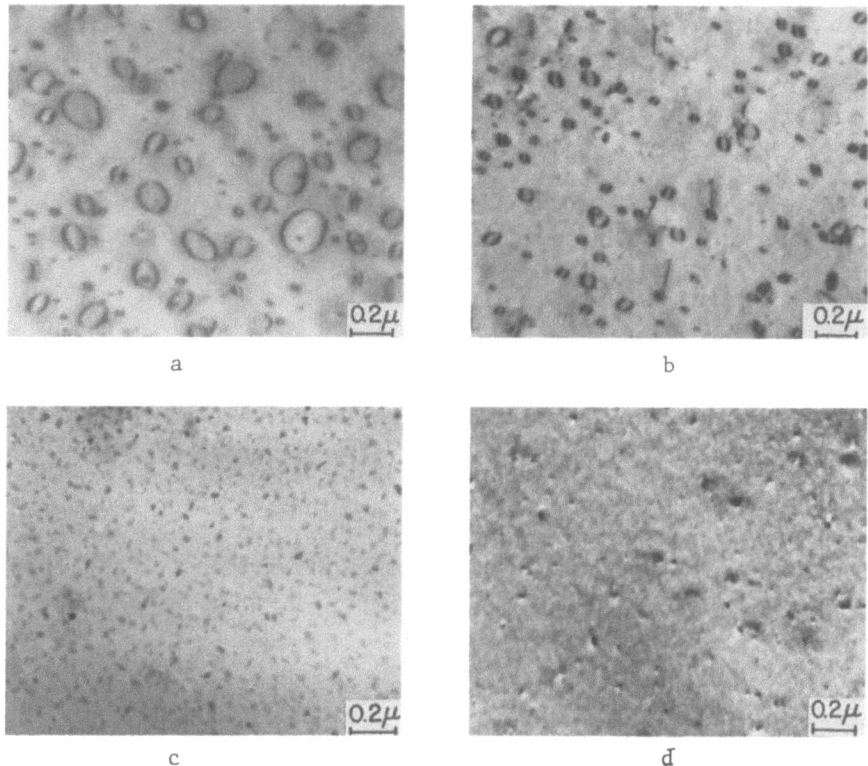

Figure 1: Electron micrographs of unsectioned GaAs annealed at 800°C after ion implantation with a fluence of 10^{15} cm^{-2} at 60KeV, a) Be, b) Mg, c) Zn, d) Cd.

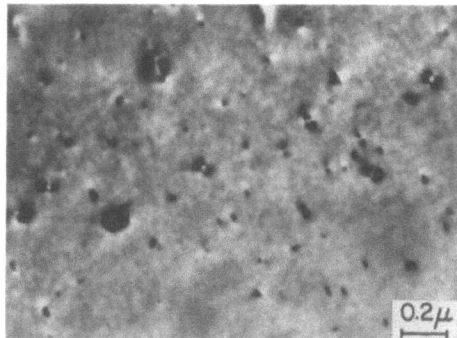

Figure 2: Electron micrograph of GaAs in the [013] orientation annealed at 800°C after implantation with a fluence of 10^{15}cm^{-2} Be.

contrast, and defects showing B-W contrast under dynamical diffraction conditions. Defects exhibiting the latter two types of contrast are shown in Figure 2. Diffraction analysis using inside-outside loop image contrast techniques with loops such as those in Figure 1a, revealed that the larger loops bounded plates of interstitials [6].

Resolvable dislocation loops up to a maximum size of approximately 800 Å are found in Mg-implanted specimens along with defects exhibiting B-W and double arc contrast, as in Be, Figure 1b.

The unsectioned-annealed specimen implanted with $10^{15}/cm^2$ of Zn ions, Figure 1c, contains a precipitate of $ZnGa_2O_4$ which was identified from the electron diffraction pattern [1]. Diffraction analysis and sectioning revealed that there was an amorphous GaAs layer adjacent to the implanted surface between 1000Å to 2000Å thick and that the $ZnGa_2O_4$ particles appeared to be contained within this layer. The formation of this type of structure may depend on the method of applying the SiO_2 mask and this is being investigated. At depths beyond this amorphous matrix layer the matrix was single crystalline GaAs and contained primarily defects less than 100Å which exhibited B-W contrast and produced no second phase diffraction pattern [2].

The Cd-implanted specimens contained primarily small defects less than approximately 200Å in size exhibiting various types of contrast (Figure 1d), of which the principal type is B-W contrast, with no resolvable dislocation loops in the specimens examined to date under these ion implantation and annealing conditions.

Preliminary measurements of the surface densities of small defects (<200Å) in Be, Mg, and Cd implanted samples were all of the same order of magnitude, with densities in the Mg- and Be-implanted samples which were greater than for the Cd by about a factor of 1.3 and 2.6, respectively. The surface density of $ZnGa_2O_4$ particles was approximately an order of magnitude greater than the surface densities of small defects in specimens implanted with the other three ions.

HALL EFFECT STUDIES

Figure 3 shows the temperature dependence of the effective Hall mobility for samples implanted at room temperature. This figure illustrates the significant differences in the temperature dependence of the mobility for these four implanted species, and at the same time shows some similarities. Here it is worth noting again that for these four implanted species at 60 KeV $10^{15}/cm^{-2}$ fluences, the Zn and Cd produce amorphous layers while Be and Mg do not.

STRUCTURAL AND ELECTRICAL CHARACTERISTICS OF GaAs

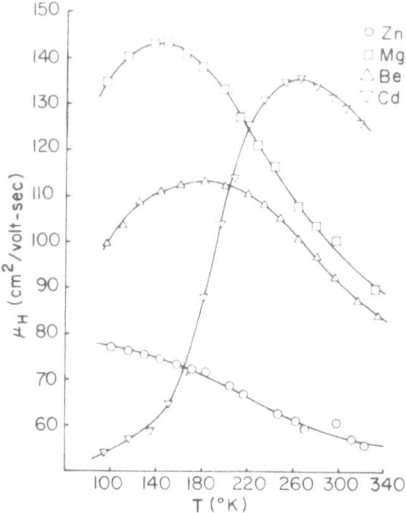

Figure 3: Effective Hall Mobility as a function of temperature for 60KeV $10^{15} cm^{-2}$ Be, Mg, Zn, and Cd implantations in GaAs.

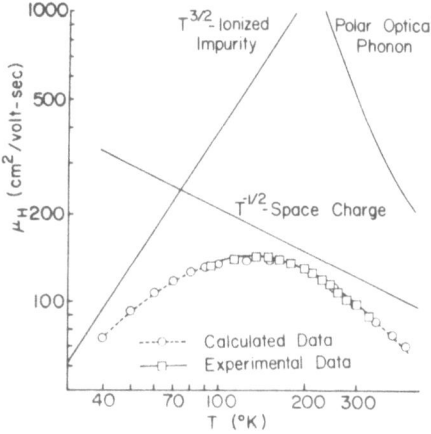

Figure 4: Hall mobility as a function of temperature for lattice scattering, ionized impurity scattering, and space charge (random potential) scattering. Also shown is Mg data from Figure 3.

Figure 4 shows the data for Mg from Figure 3, along with Hall mobilities due to specific scattering mechanisms in GaAs. The polar optical phonon curve shows the temperature dependence of lattice mobility for GaAs [7], while the ionized impurity curve shows the temperature dependence due to ionized impurity scattering in nondegenerate material. The effects of degeneracy in p-type GaAs become important for hole concentrations above approximately $10^{19} cm^{-3}$ at room temperature. In degenerate material the scattering due to ionized impurities is nearly independent of temperature [8]. Also shown is a mechanism which varies as $T^{-1/2}$, which has been called "space-charge scattering" [9], but is more appropriately a form of random potential scattering [10]. These three scattering mechanisms have been used to fit the Mg data according to Matthiesson's rule, and the calculated curve is also shown in Figure 4. The ionized impurity density for this fit is $4 \times 10^{18} cm^{-3}$. Equating this to the sheet hole concentration times the layer thickness allows a value for the layer thickness to be calculated. This value is 1100Å, which is compared to the LSS projected range of about 600Å [11]. The data for Be, Cd, and Zn can also be "fitted" in this manner. However, care must be taken to include the effects of degenerate statistics for doping levels near $10^{19} cm^{-3}$ [12].

Table 1 shows a summary of the measured room temperature electrical properties of these four implanted species, along with a brief summary of the general types of defects observed in the TEM studies. The electrical activities are generally in agreement with previous work, and seem to be in agreement with studies presented at this conference [13]. The electrical activities are quite repeatable, although for some experiments with Zn implants specimens have been examined with electrical activities of about 90% at 10^{15} cm^{-2} fluence. Electrical activities this high using Zn implants have been reported by other workers [13]. This fact could be associated with the precipitate formation, and additional investigations are in progress.

DISCUSSION

From Figure 1, Figure 3, and Table 1 there are differences in the electrical properties and in the defect structures as observed by TEM techniques for the samples implanted with the different ions. Since a major factor in the structural differences between these specimens is that those implanted with Zn and Cd have an amorphous surface layer in the as-implanted state while those implanted with Be and Mg do not, this difference will serve as a basis for discussing the differences in electrical properties between the different specimens in relation to the observed differences in structure.

STRUCTURAL AND ELECTRICAL CHARACTERISTICS OF GaAs

Table 1. Electrical Properties of Ion Implanted Unsectioned GaAs at 20°C and some Observed Microstructures. Fluence of 1×10^{15} Ions/cm^2 at 60KeV; Annealed 800°C for 30 min.

Matl.	Sheet Resis. (ohms per sq.)	Surface Carrier Concen. (10^{14} holes/cm^2)	Electr. Activ. (%)	Hall Mobility at 300°K (cm^2/v-sec)	Type of Defect	Size (Å)
Be	150	5.4	54	95	D.L. S.D.;B-W	<2000 <200
Mg	1030	0.59	6	100	D.L. S.D.;B-W	<800 <200
Zn	210	5.7	57	60	ZnGa$_2$O$_4$	<300
Cd	240	2.0	20	130	S.D.;B-W	<200

NOTE: D.L. = resolvable dislocation loops
S.D. = small defects (<200Å in size)
B-W = small defects exhibiting B-W contrast

The specimens implanted with Be and Mg contain somewhat similar types of structure, although more detailed diffraction analysis of the defects now in progress may reveal some differences in the types of defects. For both the Be- and Mg-implanted specimens the form of the mobility curves as a function of temperature are similar, the mobilities at 20°C are about the same, and the mobility in the high temperature region indicates that random potential scattering (RPS) is influencing the mobility in about the same way, all of which is consistent with the structure being similar in the two types of specimens. At low temperatures the ionized impurity scattering (IIS) is more prevalent in the specimens implanted with Be than with Mg, which is consistent with the electrical activity data in Table 1.

In the Zn-implanted specimens the surface density of the ZnGa$_2$O$_4$ precipitates, which are contained in the region with the majority of the electrical activity, is approximately an order of magnitude greater than the surface densities of small defects (<200Å) in the Be-, Cd-, and Mg-implanted specimens. The temperature dependence of mobility for the Zn-implanted specimens is predominantly influenced by RPS to a much greater extent than for samples implanted with the other three ions. This is consistent with the observed differences in surface densities of small defects and the precipitate $ZnGa_2O_4$, since both these types of defects should be a major source of RPS.

The structure of the Cd-implanted specimens consists of small defects with a lower surface density than any other of the implanted specimens and no observed resolvable dislocation loops. In samples implanted with Cd the RPS is much less important than for the Zn-doped specimens. This is consistent with the observed defect structures. However, IIS is much more important in Cd-implanted specimens in spite of the fact that the electrical activity is less than for Zn. This suggests that the formation of $ZnGa_2O_4$ particles may play a role in producing the large electrical activity in the Zn-doped specimens.

There appears to be a relationship for the four implanted ions between the mobility at 20°C and the observed defect structures. The Cd-implanted sample has the highest ambient mobility and the lowest density of small defects with no resolvable dislocation loops. The Be- and Mg-implanted samples have about the same ambient mobilities, which are lower than that of Cd-implanted samples and have very similar defect structures, with more defects than does Cd. Zn-implanted specimens with a surface density of $ZnGa_2O_4$ particles approximately an order of magnitude greater than the small defects in the Be-, Mg-, and Cd-implanted samples and with the mobility predominantly influenced by RPS has the lowest ambient temperature mobility.

The electrical properties indicate that IIS is most important for the case of Cd-implanted samples in spite of the fact of a lower electrical activity than for those implanted with Be and Zn. This anomalous behavior suggests that there are probably other types of defects which cannot be observed by the TEM techniques employed here which influence the electrical activity. This points to the advantage of using various complementary techniques such as TEM, photoluminescence, and Rutherford backscattering to investigate the effect of defects on the electrical properties of ion-implanted GaAs.

Several conclusions can be obtained from the results presented here. Different doping ions can produce considerably different defect structures under the same ion implantation and annealing conditions. Electron diffraction analysis can be used to characterize the structures. The character of the defect structures can influence the electrical properties of ion implanted GaAs. These preliminary results indicate that relationships appear to exist between some of the differences in electrical properties and some of the differences in defect structures observed by TEM techniques.

ACKNOWLEDGMENT

The authors are grateful to H. D. Hendricks of NASA Langley Research Center for the ion implantation.

REFERENCES

1. R. B. Benson, Jr., M. A. Littlejohn, P. S. Pao, and H. K. Sarin, Appl. Phys. Let., 27, 69 (1975).

2. R. B. Benson, Jr., M. A. Littlejohn, P. S. Pao, and H. K. Sarin, to be published.

3. Powder Diffraction File for Inorganic Compounds, Ed. L. G. Berry, Joint Committee on Powder Diffraction Standards, Swarthmore, Pa., 1974.

4. M. A. Littlejohn, J. R. Hauser, and L. K. Monteith, Rad. Effects, 10, 185 (1971).

5. P. W. Hutchinson and P. S. Dobson, Phil. Mag., 31, 65 (1975).

6. M. Rühle, Radiation Damage in Reactor Materials, p. 113, International Atomic Energy Agency, Vienna, STI/PUB/230, Vol. I, (1969).

7. D. Kranger, Phys. Stat. Sol. (a) 26, 11 (1974).

8. E. H. Putley, The Hall Effect and Semiconductor Physics, p. 146, Dover, New York, (1960).

9. E. M. Conwell, High Field Transport in Semiconductors, Solid State Physics, Suppl. 9, Academic Press, New York, (1967).

10. R. W. Harrison and J. R. Hauser, Phys. Rev. B 13, 5347 (1976).

11. W. S. Johnson and J. F. Gibbons, Projected Range Statistics in Semiconductors, Stanford University Bookstore, 1970.

12. B. K. Shin, D. C. Look, Y. S. Park, and J. E. Ehret, Jour. Appl. Phys 47 1574 (1976).

13. R. Zolch, H. Ryssell, H. Kranz, H. Reichl, and I. Ruge, this conference.

IMPURITY DISTRIBUTION OF ION-IMPLANTED Be IN GaAs BY

SIMS, PHOTOLUMINESCENCE, AND ELECTRICAL PROFILING*

J. Comas
Naval Research Laboratory, Washington, D.C. 20375
L. Plew
Naval Weapons Support Center, Crane, In. 47522
P.K. Chatterjee, W.V. McLevige, K.V. Vaidyanathan,
and B.G. Streetman
Coordinated Science Laboratory and Department of
Electrical Engineering
University of Illinois, Urbana, Illinois 61801

ABSTRACT

Atomic, optically active, and electrically active profiles measured for 250 keV ion-implanted Be in GaAs as a function of fluence have been correlated. Be atomic depth profiles were obtained by secondary-ion mass-spectroscopy (SIMS) techniques. Photoluminescence and Hall effect measurements made in conjunction with successive layer removal were used to obtain the optically and electrically active profiles. Atomic SIMS profiles obtained from samples implanted to a fluence of $6 \times 10^{13} cm^{-2}$ showed no major distribution changes after a 900°C, 30 min anneal treatment. Electrically active profiles obtained from duplicate samples had distributions which were similar to the SIMS results. Hall effect and resistivity measurements after annealing indicated nearly complete electrical activity of the implanted Be, with mobility values typical of p-type bulk material.

Measured Be atomic, optically active, and electrically active profiles for fluences $\geq 6 \times 10^{14} cm^{-2}$ showed significant Be redistribution after the 900°C anneal. The atomic Be profile exhibited a sharp surface peak followed by an essentially flat distribution

---

*This work supported by the Joint Services Electronics Program, Monsanto Co., Office of Naval Research, and the Naval Electronic Systems Command.

extending to approximately twice the peak position observed in unannealed samples. At approximately 1.4 μm a distinct edge was observed, at which the Be distribution dropped sharply. The general features of the optically and electrically active profiles were in qualitative agreement with the SIMS atomic profiles.

INTRODUCTION

The inherent advantages of GaAs for many device structures, such as microwave and optoelectronic devices, has placed an increasing emphasis on the controlled doping of this material. The difficulty of introducing dopants into GaAs by diffusion has resulted in the present interest in ion-implantation doping. The success of ion-implantation doping achieved in the Si technology, however, has not yet been fully realized in the compound semiconductors. Some of the difficulties include the need to distinguish effects due to the encapsulants and substrates used, as distinguished from effects due to the implantation process itself.

Considerable attention is being given to implanted Be to form p-type layers in GaAs. Be is a shallow acceptor in GaAs, and studies of its electrical, optical, and structural properties in this material have been reported elsewhere.[1-4] High electrical activation and essentially bulk mobilities have been obtained for implanted Be layers after annealing. Mobility measurements as a function of implant fluence indicate that delocalization of Be occurs after annealing for high fluence implants ($>10^{14}cm^{-2}$).[5] An ionization energy of 28.4meV has been estimated from photoluminescence measurements.[2] Atomic profile studies indicate significant concentration dependent redistribution effects after annealing.[6]

The work presented in this paper deals mainly with the depth distribution and activation of ion-implanted Be in GaAs as a function of implant fluence and annealing. Depth distributions of optically and electrically active Be are correlated with Be atomic depth profiles. In this work photoluminescence (PL) and Hall effect (HE) measurements were made in conjunction with successive layer removal to obtain the active Be distributions. The atomic depth profiles were obtained by the secondary-ion mass-spectroscopy (SIMS) technique.

EXPERIMENTAL PROCEDURE

A. Material

The material used in this work consisted mainly of undoped vapor phase epitaxial (VPE) layers grown on heavily doped n-type substrates. The net carrier concentration ($N_d \sim 8 \times 10^{13} cm^{-3}$) was obtained from Schottky barrier differential capacitance measurements. N-type, Si-doped ($N_d \sim 5 \times 10^{17} cm^{-3}$) and Te-doped substrates were used in some of the SIMS and electrical activity studies. All implants were performed at room temperature, and the samples were positioned off-axis to reduce channeling effects. Encapsulants of Si_3N_4 were deposited by the RF plasma deposition process onto substrates held at 450°C. Samples were coated on all surfaces and duplicate unimplanted samples were also annealed to check for processing effects. Isochronal anneal treatments of 30 min duration were performed in a quartz boat with a continuous flow of argon or forming gas.

B. Atomic Profiles

Atomic profiles of ion-implanted Be in GaAs were measured by the secondary-ion mass-spectroscopy (SIMS) technique using the Cameca IMS-300 ion microanalyzer. SIMS profiling consists of the continual removal of surface material by sputtering while performing a mass analysis of the sputtered charged particles.[7] The SIMS Be atomic profiles in GaAs were obtained by monitoring the $^9Be^+$ secondary-ion current as a function of sputtering time. Where changes in instrument sensitivity and normalization were of concern, the $^{71}Ga^+$ signal was also monitored. Lateral uniformity and the accumulation of Be at the encapsulant/substrate interface were investigated by use of the direct imaging mode of the ion microanalyzer.

Depth scales were established from mechanical (Dektak) and optical (Angstrometer) measurements of the sputtered craters. The peak positions of the unannealed profiles were in fair agreement with published projected ranges based on LSS calculations.[8] The Be concentration scale was established by correlating the SIMS signal at the peak position of the unannealed profile to the calculated concentration based on LSS theory.

C. Electrically Active Profile

The depth distribution of the electrically active Be was obtained using differential resistivity and Hall effect measurements made in conjunction with successive layer removal. A van der Pauw type geometry and a double ac Hall effect system[9] were used to make these measurements. Evaporated Ag:Mn contacts were sintered at 330°C in flowing H_2. A chemical etch consisting of 1 H_2SO_4 : 1 H_2O_2 : 50 H_2O at 25°C was employed to remove successive layers. Resistivity and Hall coefficient measurements were performed at intervals of approximately 500Å. To minimize uncertainty due to variations in the etch rate, the etched step height was measured every $\sim$5000Å using a Dektak mechanical stylus.

D. Optically Active Profile

Photoluminescence (PL) measurements were used to examine the optical properties of ion-implanted Be in GaAs. In the PL process electron-hole pairs are generated by the absorption of photons whose energy is larger than the GaAs bandgap energy. Recombination of these excess carriers occurs with characteristic radiative and non-radiative processes. Thus impurity incorporation as well as lattice recovery can be studied via low temperature PL techniques. By observing the relative intensity of the Be-related luminescence (6°K) as successive layers are removed by chemical etching, the optically active profile can be obtained.[9] It should be noted that due to the absorption length of the laser excitation and the diffusion length of the excited carriers, there is considerable uncertainty in the depth scale. However, this measurement is useful in providing a qualitative check of the profile of Be atoms on optically active sites.

RESULTS AND DISCUSSION

Electrically active and Be atomic profiles obtained from samples implanted to a Be fluence of $6 \times 10^{13} cm^{-2}$ are shown in Fig. 1. The anneal treatments were performed at 900°C for 30 minutes. A comparison of the unannealed and annealed atomic (SIMS) profiles shows that some broadening and diffusion toward the surface has occurred during the anneal. Hall effect data indicate that the hole mobility is comparable with bulk p-type GaAs.[9] The electrically active profile after annealing is in good agreement with the Be atomic distribution. The high % of Be activation (80-100%) is corroborated by the Be retention observed in the atomic profile. All samples implanted to fluences $\leq 6 \times 10^{13} cm^{-2}$ at 250 keV showed no major changes in the Be atomic profiles due to the anneal treatment. Samples annealed at 800°C and studied by SIMS did not show the Be build-up toward the surface at $\sim 0.2 \mu m$.

The SIMS atomic profiles obtained from samples implanted to fluences $\geq 6 \times 10^{14} cm^{-2}$ showed significant redistribution effects after annealing at temperatures above 550°C. Figure 2 illustrates the profiles obtained before annealing and after 30 min anneals at 600, 700, and 900°C. After the 600°C anneal the Be atomic distribution is skewed toward the surface with some Be loss from the initial peak. The 700°C anneal results in migration of the Be to the surface and more loss of Be at the unannealed peak position. After the 900°C anneal the atomic profile has drastically changed, exhibiting a sharp surface peak followed by a relatively flat distribution extending to a depth approximately twice the peak position of the unannealed profile. A distinct edge is observed, at

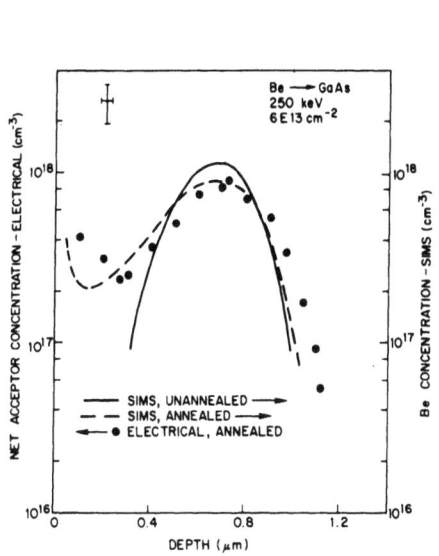 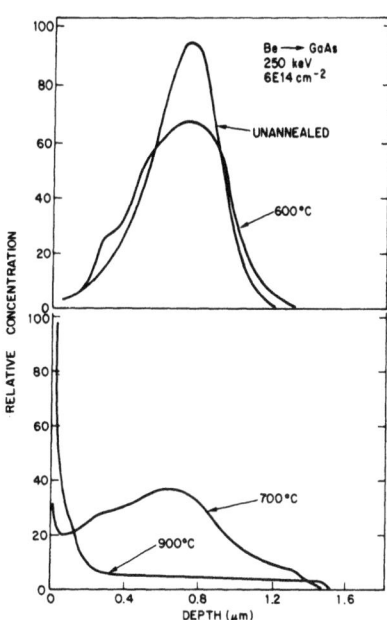

Fig. 1. Net acceptor concentration (HE) and Be atomic concentration (SIMS) versus depth for GaAs (VPE) samples implanted with Be to a fluence of $6 \times 10^{13} cm^{-2}$. The anneal was performed at 900°C for 30 min.

Fig. 2. Relative concentration versus depth for unannealed and annealed GaAs (Si-doped) samples implanted with Be to a fluence of $6 \times 10^{14} cm^{-2}$.

which the Be distribution drops sharply. Similar flat distributions, including the distinct edge at approximately twice the depth of the unannealed peak position, have been observed for samples implanted to corresponding concentrations at 40, 100, and 150 keV. The redistribution effects observed were similar for the VPE, Te-doped, and Si-doped substrates studied.

The Be concentration in the region of the as-implanted peak position (0.7-0.8μm) was measured as a function of implant fluence after annealing at 900°C for 30 minutes. Shown in Fig. 3 is the Be concentration in n-type, Si-doped GaAs substrates as a function of implant fluence ranging from 1×10^{13} to $5 \times 10^{15} cm^{-2}$. The dashed line represents the manner in which the concentration would increase if there were no Be losses. For Be fluences $< 10^{14} cm^{-2}$ there are relatively small Be losses, whereas for fluences $> 10^{14} cm^{-2}$ the Be losses are significant. The relatively constant concentration for fluences $> 10^{14} cm^{-2}$ suggests that solubility

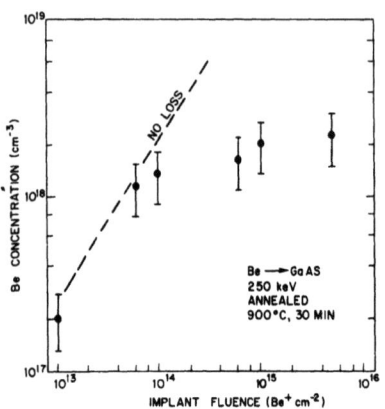
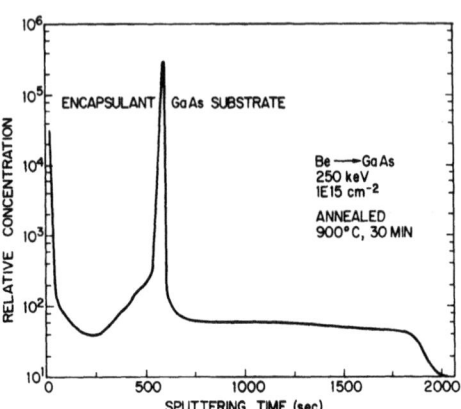

Fig. 3. Be concentration in the 0.7-0.8μm region versus implant fluence in GaAs (Si-doped) after annealing at 900°C for 30 min.

Fig. 4. Relative concentration of Be in the Si_3N_4 encapsulant and the GaAs (Te-doped) substrate after annealing.

associated effects are occurring. The maximum concentration after annealing at 900°C for 30 minutes is approximately $1-3 \times 10^{18}$ Be atoms cm^{-3}.

The profiles shown in Fig. 2 indicate that there is significant Be migration toward the surface, with Be accumulation at the encapsulant/substrate interface. This is substantiated by the Be atomic profile in the Si_3N_4 encapsulant and GaAs substrate shown in Fig. 4. A GaAs substrate was implanted at 250 keV to a fluence of $1 \times 10^{15} cm^{-2}$, encapsulated, and then annealed at 900°C for 30 min. The Be distribution shown in Fig. 4 has not been corrected for changes in the SIMS sensitivity for Be in the encapsulant and the GaAs substrate. The results indicate there is an accumulation of Be at the encapsulant/GaAs interface which acts as a source for diffusion into the encapsulant. Lateral uniformity of the Be in the encapsulant/substrate region was also studied using the direct imaging mode of the Cameca ion microanalyzer. The lateral distribution of the Be was observed to be uniform in the substrate, interface, and encapsulant.

Profiles of electrically active Be from samples implanted to higher fluences ($> 10^{14} cm^{-2}$) also showed that major redistribution effects had occurred during annealing. Atomic and electrically

IMPURITY DISTRIBUTION OF ION-IMPLANTED Be IN GaAs

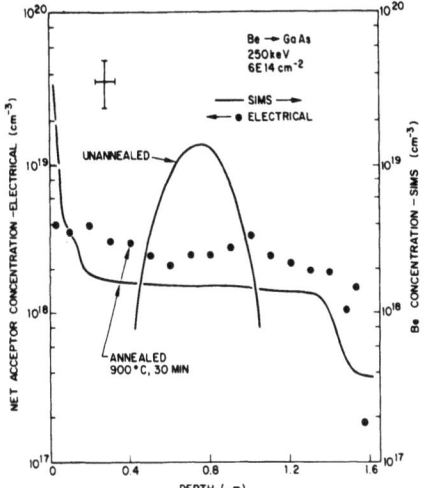
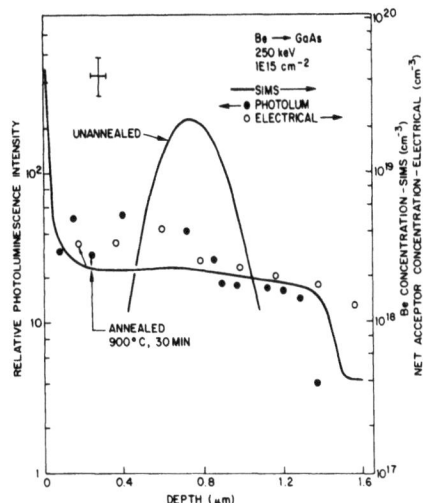

Fig. 5. Net acceptor concentration (HE) and Be atomic concentration (SIMS) versus depth for GaAs (VPE) samples implanted with Be to $6 \times 10^{14} cm^{-2}$. The anneals were performed at 900°C for 30 min.

Fig. 6. Relative photoluminescence intensity (PL), Be atomic concentration (SIMS) and net acceptor concentration versus depth for GaAs samples implanted with Be to $1 \times 10^{15} cm^{-2}$. The anneals were performed at 900°C for 30 min. Silicon-doped substrates were used for the SIMS and electrical measurements and Te-doped substrates were used for the photoluminescence study.

active Be profiles from samples implanted at 250 keV to a fluence of $6 \times 10^{14} cm^{-2}$ are shown in Fig. 5. The high atomic concentration (> $10^{19} cm^{-3}$) in the surface region, as compared to the electrical activity ($\sim 4 \times 10^{18} cm^{-3}$), suggests that a significant number of Be atoms in that shallow region are not active. Beyond $\sim 0.2 \mu m$ both profiles are relatively flat. The distinct edge observed in the atomic profile at $\sim 1.4 \mu m$, at which the Be concentration decreases sharply, is also observed in the electrical profile.

Electrically active and SIMS atomic profiles obtained from samples implanted to fluences of $1 \times 10^{15} cm^{-2}$ were similar to those shown in Fig. 5. Such profiles are illustrated in Fig. 6, along with the measured profile of optically active Be. The relative luminescence intensity, electrically active, and atomic profiles show the same general behavior. The distinct edge near $1.4 \mu m$ appears for all three profiling methods.

SUMMARY

In summary, the electrical and SIMS results indicate that for samples implanted to fluences $< 10^{14} \text{cm}^{-2}$ and annealed, there are no major Be losses or distribution changes. The electrical activity achieved after the anneal is $> 80\%$, and the hole mobility is typical of bulk p-type GaAs.

Significant redistribution effects are observed for samples implanted to fluences $\geq 6 \times 10^{14} \text{cm}^{-2}$ after a 900°C, 30 min anneal. The atomic, electrically active, and optically active profiles are relatively flat, extending to approximately twice the depth of the unannealed peak position. The redistribution during annealing results in accumulation of Be at the encapsulant/substrate interface and diffusion of Be into the encapsulant. In these experiments, the maximum Be concentration attainable in the flat region of the distribution after a 900°C, 30 min anneal appears to be in the range $1-3 \times 10^{18} \text{cm}^{-3}$. These results suggest that the Be redistribution may be associated with mechanisms analogous to solubility limit effects.

ACKNOWLEDGEMENTS

The authors wish to thank the Monsanto Co. and Ms. E. Tarrents of the Air Force Materials Laboratory for material used in this study.

REFERENCES

1. R. G. Hunsperger, R. G. Wilson, and D. M. Jamba, J. Appl. Phys. **43**, 1318 (1972).
2. P. K. Chatterjee, W. V. McLevige, K. V. Vaidyanathan, and B. G. Streetman, Appl. Phys. Letts. **28**, 509 (1976).
3. J. P. Donnelly, F. J. Leonberger, and C. O. Bozler, Appl. Phys. Letts. **28**, 706 (1976).
4. P. K. Chatterjee, K. V. Vaidyanathan, W. V. McLevige, and B. G. Streetman, Appl. Phys. Letts. **27**, 567 (1975).
5. "Ion Implantation of Wide Bandgap Semiconductors," L. Anderson Hughes Research Laboratory Report, 1975, Contract No. N00014-74-C-0158, Naval Electronic Systems Command.
6. J. Comas and L. Plew, J. Elect. Mat. **5**, 209 (1976).
7. J. M. Morabito and R. K. Lewis, Anal. Chem. **45**, 869 (1973).
8. W. S. Johnson and J. F. Gibbons, <u>Projected Range Statistics in Semiconductors</u>, Stanford Univ. Press, 1969.
9. P. K. Chatterjee, Ph.D. Thesis, University of Illinois (1976). Available from NTIS, Springfield, Va. (Report #ADA0-25-607).

DUAL SPECIES ION IMPLANTATION INTO GaAs

C. A. Stolte

Hewlett-Packard Laboratories
1501 Page Mill Road
Palo Alto, California 94304

ABSTRACT

Dual species implants of Se in conjunction with Ga or As implants and Si implants in conjunction with Ga or As implants into <100> GaAs have been investigated. Data on the doping efficiency and doping profiles are presented. Although other models are possible and should not be ruled out, the gross features of the Se results are explainable using a simplified model which postulates that the Ga implant produces donors via As vacancies and that the As implant produces acceptors via Ga vacancies. To explain the Si results, it is necessary to invoke the effect of stoichiometry on the amphoteric nature of the silicon dopant.

INTRODUCTION

The use of ion implantation into GaAs to produce n-type layers of device quality is well established.[1,2] The conditions required during implantation and annealing to produce high doping efficiencies and high carrier mobility for moderately doped layers, $10^{17} cm^{-3}$, with implants of Se, S and Te are known. For high doping levels, in the $10^{18} cm^{-3}$ and greater range, the doping efficiency decreases both for doping in the crystal growth process [3] as well as for ion implantation.[4] In the case of Si implants into GaAs, the doping efficiency and mobility are lower, presumably due to the amphoteric nature of the Si dopant.

Several reasons for the decrease in the measured doping efficiency with increasing concentration have been proposed. These include the formation of compensating centers such as the Ga vacancy-donor complex, [3,4] and degeneracy [5] which is said to lead to a misleading determination of the doping efficiency. However, the latter explanation is believed invalid due to the over simplifications made in the theory and also from evidence in the literature [3] which is in contradiction with the calculated results. The main effect reducing the doping efficiency is generally believed to be associated with a deviation from stoichiometry.

Heckingbottom and Ambridge [6] have proposed a theory which takes into account the deviation from stoichiometry and in this way predicts the decrease in doping efficiency with increasing dose. This theory has been partially verified by several experiments [7,8,9] but the experimental results indicate a much more complex situation than predicted by the simple theory.

In this paper we present experimental results for dual species implants, namely, Si and Ga, or As and Se, and Ga or As dual species implants into GaAs. The electrical properties of the implanted and annealed layers are evaluated. As detailed below, the results do indicate promise for improved properties via stoichiometry control but they also illustrate the complexity of the phenomena operative during the implant and anneal process.

EXPERIMENTAL PROCEDURE

The matrix of experiments performed in this investigation is indicated in Table I where the species, the implant energy and the implant dose are indicated. The sample numbers in the Table refer to the samples for which profile data is presented as described below.

Table I. Matrix of dual species implants

Secondary Implant \ Primary Implant	None	Se 500 keV 3.0×10^{12} (cm^{-2})	Se 250 keV 1.5×10^{12} (cm^{-2})	Si 250 keV 3.0×10^{13} (cm^{-2})	Si 100 keV 1.5×10^{13} (cm^{-2})
None	X	Se-28	Se-78	Si-9	X
As 250 keV 1.0×10^{13}cm^{-2}	X	Se-91	X	Si-7	X
As 250 keV 1.0×10^{14}cm^{-2}	X	X	X	Si-32	X
As 250 keV 1.0×10^{15}cm^{-2}	X	X	X	Si-10	X
Ga 250 keV 1.0×10^{12}cm^{-2}	X	Se-86	Se-79		
Ga 250 keV 1.0×10^{13}cm^{-2}	Ga-5	Se-87	Se-80	Si-8	X
Ga 250 keV 1.0×10^{14}cm^{-2}	Ga-3	Se-88	Se-81	Si-27	X
Ga 250 keV 1.0×10^{15}cm^{-2}	Ga-2	X	X	Si-12	X

The crosses indicate samples for which only total charge data are presented. The primary species, Si or Se was implanted with the substrate at 350°C. The secondary species, Ga or As, was implanted at room temperature prior to the primary species. In all cases the implant substrate was oriented for "random" direction implants.

The substrates used in the investigation were high purity Liquid Phase Epitaxial (LPE) layers grown on <100> semi-insulating GaAs. These layers, grown in this laboratory, have room temperature mobilities of 8000 $cm^2V^{-1}s^{-1}$ and liquid nitrogen temperature mobilities of greater than 100,000 $cm^2V^{-1}s^{-1}$. The inferred carrier concentrations are $N_D < 2 \times 10^{14} cm^{-3}$ and $N_A < 1 \times 10^{14} cm^{-3}$.

After ion implantation the samples were annealed at 850°C for 15 minutes in an open tube Ar flow furnace. The surface of the GaAs was protected with a chemical vapor deposited Si_3N_4 layer formed at 650°C in a silane-ammonia reactor. This protective cover yields excellent results with no significant change in electrical properties observed for unimplanted control samples and the cover maintains the good surface morphology of the LPE layers.

The implanted and annealed samples were characterized by Hall mobility and sheet resistance measurements using the Van der Pauw geometry. In addition to the determination of the effective Hall mobility, u_e, and the total charge, N_s, in the layers, the doping profile, N_x, and the mobility profile, μ_x, were obtained using the differential technique with a chemical etch layer removal procedure. Doping profiles were also obtained using standard Schottky barrier C-V techniques on etch step samples. With this technique the doping profiles are routinely obtained over large depths in the material.

EXPERIMENTAL RESULTS

The results for Se implanted samples are summarized in Fig. 1 where the total charge, N_s, in the implanted and annealed layers is plotted as a function of the secondary, As or Ga, implant dose. The total charge in the layers decreases with increasing As implant dose and the layers become p-type for an As dose of $10^{14} cm^{-2}$ for the deep, 500 keV, Se implant and for an As dose of $10^{13} cm^{-2}$ for the shallow, 250 keV, Se implant. The addition of the Ga implant increases the total charge and as seen produced total charges in the layers which are greater than the Se implant dose. The data for the Ga only implanted layers indicate that the Ga implant contributes a significant total charge, greater than the Se implanted charge, for high Ga implant doses. The relatively low charge in the shallow Se implanted samples is an artifact of the measuring procedure since for the doping levels used in these samples the depletion effect produces significant errors in the measurement.[10]

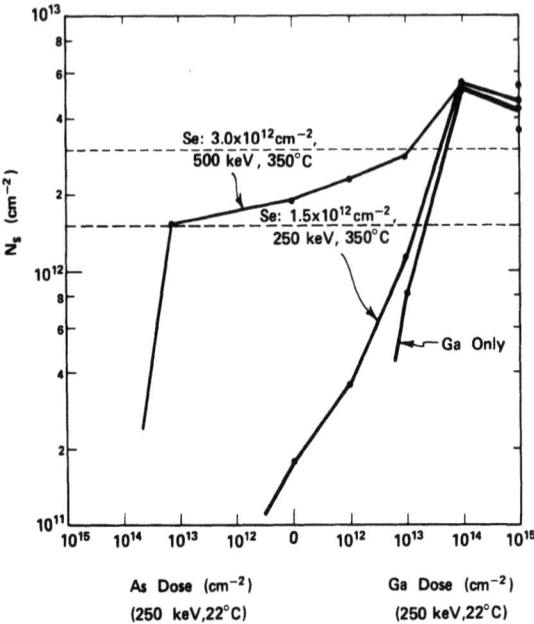

Fig. 1. Total charge, N_S, vs. the secondary species implant dose, As or Ga, for Se implants into GaAs. The Se implant conditions are indicated on the figure; the dashed lines represent the Se implant dose.

The doping profiles for the Se implanted layers are shown in Fig. 2. The profiles shown in Fig. 2a were obtained for the deep, 500 keV, Se implant and shallow 250 keV, Ga or As implants. The implant condition for each sample is indicated in Table I. The addition of a low As dose implant, $10^{13} cm^{-2}$, results in only a slight profile change, as illustrated by sample Se-91, compared to the sample, Se-28, which had the Se only implant. Higher As implant doses resulted in p-type layers. The addition of a Ga implant to the Se implant increases the profile magnitude and shifts the peak of the profile toward the predicted peak of the Ga implant profile. The profiles shown for the Ga and Se dual implants, Se-87 and 88, are approximately the superposition of the Se only profile, Se-28, and the Ga only profile, Ga-3 and Ga-5, respectively. The profiles shown in Fig. 2b for matched Se and Ga implants again show the superposition of the Ga only and Se only profiles to produce the dual species profiles.

The results for the Si implanted samples are shown in Fig. 3 when the total charge, N_S, is plotted as a function of the secondary implant species dose. In the Si implants the effect of As implants is more complex than in the Se implant case discussed above. For the deep Si implant, 250 keV, the addition of the As implant increases the total charge in the layer in contradiction to the effect in Se implanted layers and also in contradiction

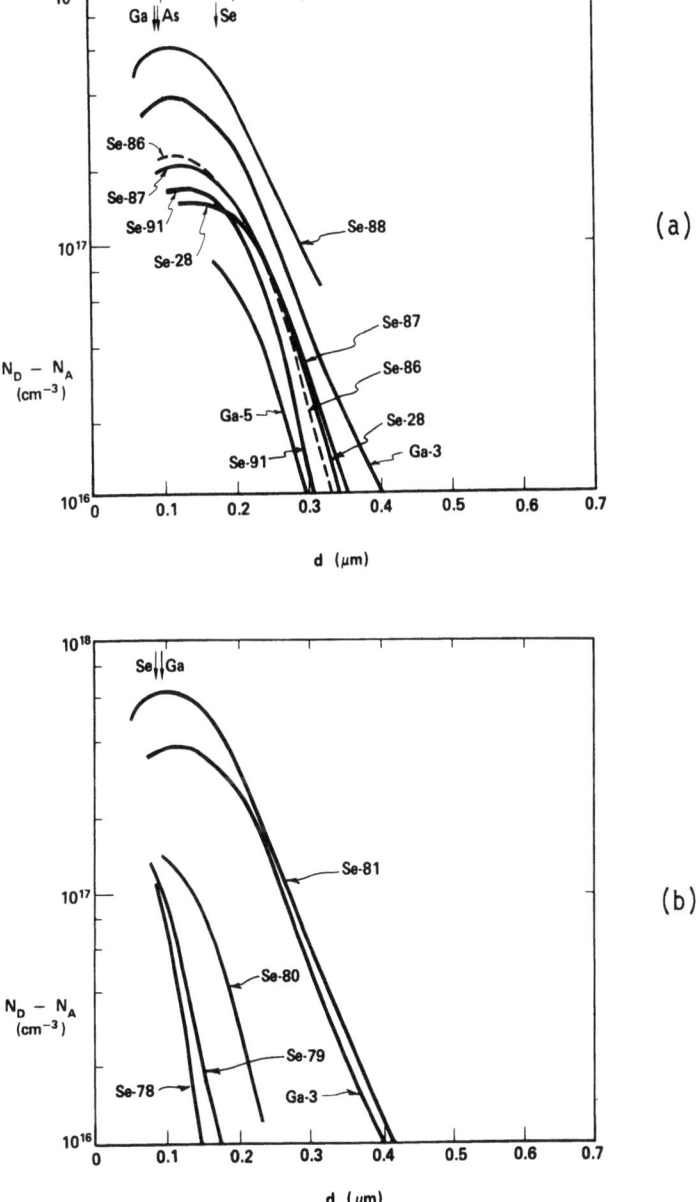

Fig. 2. Doping profiles for dual species implants into GaAs. (a) Deep Se plus shallow Ga or As implants; (b) shallow Se plus shallow Ga implants. The implant conditions for each sample are given in Table I.

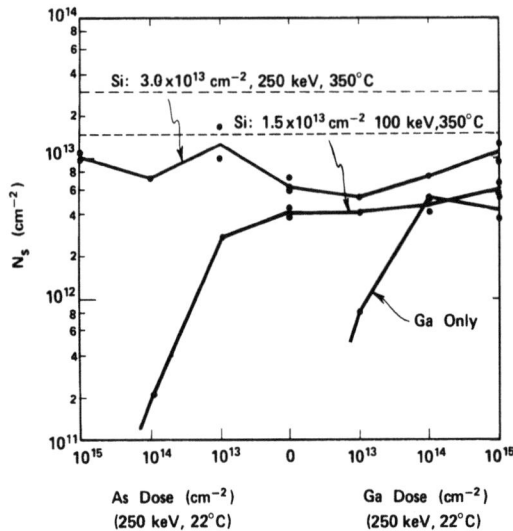

Fig. 3. Total charge, N_s, vs. the secondary species implant dose, As or Ga, for Si implants into GaAs. The Si implant conditions are indicated on the figure; the dashed lines represent the Si implant dose.

to the effect in the matched profile implants for Si (100 keV) shown in Fig. 3. For the dual implants of Ga with Si the total charge increases but the effect is not as dramatic as in the case of the Se implanted layers. This is due in part to the higher total charge in the Si implanted samples, the implant dose used is 10 times that used in the Se implant experiments.

The more complex behavior of dual species As and Si implants is illustrated in Fig. 4a. With increasing As dose the carrier concentration near the peak of the Si profile increases. However, the profile shape is modified near the surface, the region corresponding to the predicted peak of the As implant. This decrease in concentration at the high concentration region of the As profile is in agreement with that seen in the Se plus As implant.

The profile data for Si and Ga dual species implants illustrated in Fig. 4b again show the superposition effect of the Ga only and Si only profiles in agreement with the effect seen with the Se and Ga dual species implants. For the high dose Ga implant case, $10^{15} cm^{-2}$, there is a profile change at the surface, corresponding to the peak of the Ga implant. This effect is further illustrated in Fig. 5 where the profile data obtained from differential Van der Pauw measurements are shown along with profile data reproduced from Fig. 4. The agreement between the profiles obtained using the two techniques is good. The mobility for the Si only implant is relatively constant through the layer. The mobility for the Si plus

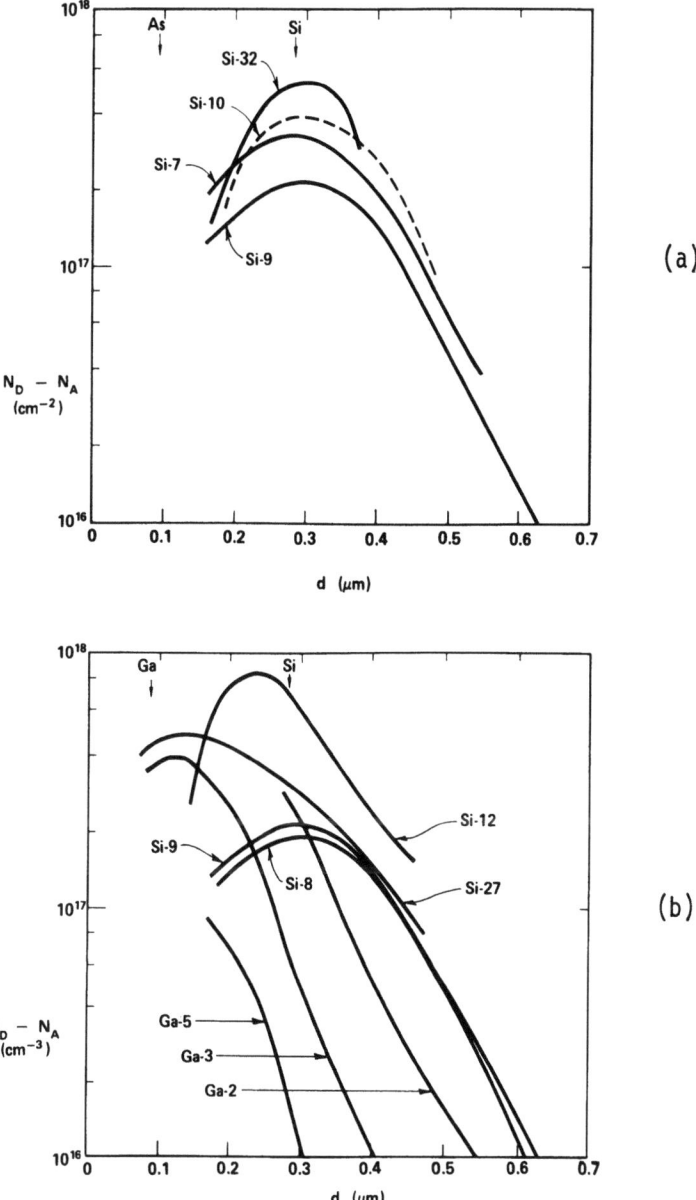

Fig. 4. Doping profiles for dual species implants into GaAs. (a) Deep Si plus shallow As implants; (b) deep Si plus shallow Ga implants. The implant conditions for each sample are given in Table I.

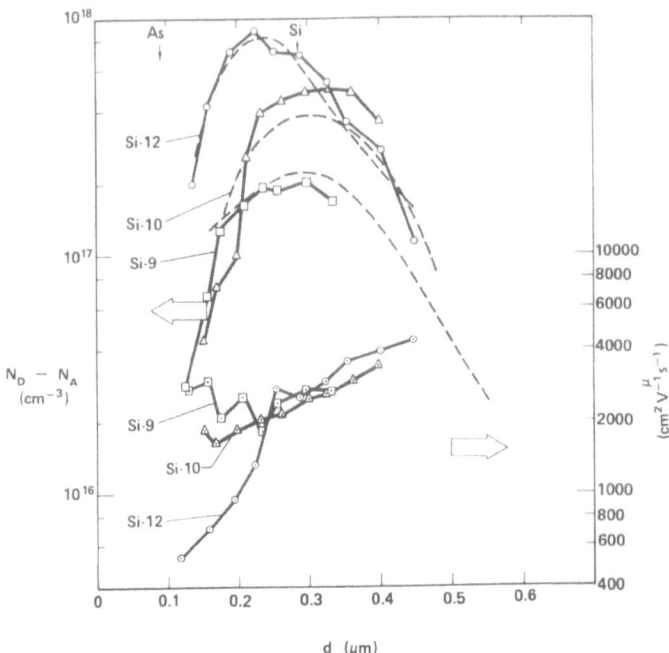

Fig. 5. Doping and Hall mobility profiles for dual species, deep Si plus shallow As or Ga, implants into GaAs. The dashed profiles are reproduced from Figs. 4a and 4b.

As dual implant is low near the surface and increases with depth. The same characteristic is seen for the Ga plus Si dual implant but in this case the mobility near the surface is much lower. This very low mobility is postulated to be due to residual damage remaining in the layer as a result of the high secondary species implant dose. This is discussed in detail in the next section.

DISCUSSION OF RESULTS

The data shown above indicates that several effects occur as a consequence of the secondary, Ga or As, implant into GaAs. The dominant effect for the conditions used in these experiments is the production of donors by the Ga implant and the production of acceptors by the As implant. These secondary species profiles are superimposed on the primary species donor profile. In the case of the Se implants, for the conditions used, this was the only effect observed.

The second effect of the secondary implant is most apparent in the Si implanted samples. In this case there is evidence that the stoichiometry of the sample is changed to yield an increase in the donor concentration of the Si dopant when implanted with As. This effect, however, is compensated by the As induced acceptor. The stoichiometry change induced by As and Ga implants should be further investigated for higher dose Se implants where stoichiometry effects would be more apparent.

In addition to these two effects there is evidence that for the higher dose As and Ga implants the results are influenced by the residual damage remaining after the anneal procedure. These damage effects do not, however, appear to be significant for the As or Ga doses below $10^{14} cm^{-2}$.

The results presented can be interpreted by a simplified model in which it is postulated that the Ga implant produces As vacancies, V_{As}, which act as donors and that the As implants produce Ga vacancies, V_{Ga}, which act as acceptors. These assumptions are an obvious over simplification of the actual complex behavior, but they do form a self-consistent model as described below. These vacancy sites, in addition to acting as donors or acceptors, serve as sites for the Si and Se species location. This control of site location is most apparent in the Si implant data where the addition of V_{Ga} sites by As implantation is postulated to increase the number of Si on V_{Ga} sites which act as donors. This effect is observed only in regions of low As concentration since at higher concentrations the acceptors produced via the V_{Ga} sites dominate. For the Se dose used in these experiments the site control is masked by the donors produced via the V_{As} site. Preliminary data from higher Se dose implants, to be reported at a future time, indicate an enhanced doping efficiency via an Se on a V_{As} site in addition to the superposition of effects seen in the experiments reported above.

CONCLUSIONS

The addition of a secondary Ga or As implant to the primary Si or Se implant into GaAs has a pronounced effect on the doping efficiency and carrier profile properties of the resulting layers. This effect is due in part to the additional donors formed by the Ga implant or to the acceptors formed by the As implant, in part due to the influence of the change in the primary species site location as a result of the secondary species implantation, and in part due to the residual damage produced by the high dose secondary implant. Hence the use of Ga or As implants in conjunction with donor species implants to control the stoichiometry of the GaAs substrate is more complex than initially believed but does offer promise for improved properties of ion implanted n-type layers in GaAs.

ACKNOWLEDGMENTS

The author wishes to thank C. A. Bittmann, T. W. Sigmon and P. E. Greene for their many discussions and suggestions, J. Hansen and D. Burriesci for their assistance in the implantation procedures, S. Malcolm for the GaAs LPE layer growth and V. Bitsch for the sample processing and electrical measurements of the implanted and annealed samples.

LATERAL SPREAD OF THE PROTON ISOLATION LAYER IN GaAs

Hideki Matsumura, Selcuk Gecim, and Kenneth G. Stephens

Department of Electronic and Electrical Engineering

University of Surrey, Guildford, Surrey, England

ABSTRACT

When ions are implanted into a tilted target, information on the lateral spread of ions and damage is contained in the depth distribution of the target. Using this principle, the lateral spread of the proton isolation layer in GaAs was measured by the Copeland method. It is found that the lateral spread of the isolation layer is largest near the depth at which the carrier removal rate is a maximum and that it increases as the incident energy increases. However, the ratio of the lateral to the longitudinal spreads decreases with increasing energy, and there is no significant change in this ratio after annealing.

INTRODUCTION

Knowledge of the lateral distribution of the isolation layer produced by protons in GaAs has become important for integrated optics [1] and for fabrication of high frequency diodes [2]. However, the simple direct measurement of the lateral spread under a mask may be ambiguous because of proton scattering from the mask edge.

Following earlier work by Furukawa and Matsumura (3), when ions are implanted into a tilted target, information on the lateral spread of ions and damage is contained in the depth distribution normal to the target surface. So similarly, without using masks it should be possible to estimate the lateral distribution of carrier removal of the proton isolation layer in GaAs from several depth distributions, formed by implantation with different angles of tilt of the samples.

Several GaAs samples were bombarded at room temperature with protons of 300 keV, 400 keV and 500 keV at incident angles of 5°, 45° and 60°. The depth distributions of carrier removal were measured by the Copeland method and the variation of the shape of these distributions after annealing was studied.

From our studies we were able to show that: (1) The lateral spread of the carrier removal distribution is largest near the depth at which the carrier removal rate is a maximum, i.e., at the estimated projected range of the protons [4,5]. (2) The magnitude of the lateral spread of carrier removal increases as the incident proton energy increases but the ratio of the lateral to the longitudinal spreads decreases. (3) This ratio does not change significantly for a low dose implant after annealing.

THEORETICAL BASIS OF THE EXPERIMENT

Let us denote the depth distribution of carrier removal $F(z)_\theta$ when protons are implanted in GaAs at an angle θ on the x-z plane as shown in Fig. 1, where the x-y plane is set on the surface of a GaAs target and the positive z axis is taken into the sample normal to the surface. If we can assume a linear relation between the carrier removal rate and the number of the implanted protons (assumption (a)) and if we imagine a carrier removal distribution $f(x,y,z)$ formed by only one single proton implant in GaAs, $F(z)_\theta$ should be expressed by the spatial superpositon of $f(x,y,z)$'s as shown by the following equation [6,7].

$$F(z)_\theta = N_\square \cdot \cos\theta \cdot R(\theta) \cdot \int_{-\infty}^{\infty}\int_{-\infty}^{\infty} f(X,y,Z) \cdot dxdy$$

where

$$X = x \cdot \cos\theta - z \cdot \sin\theta, \quad Z = x \cdot \sin\theta + z \cdot \cos\theta$$

 .... (1)

Here, $N_\square$ and $R(\theta)$ express the area density of a proton beam and the ratio of protons implanted into the GaAs to the total protons in the beam, respectively.

If we can estimate this imaginary $f(x,y,z)$, we can evaluate

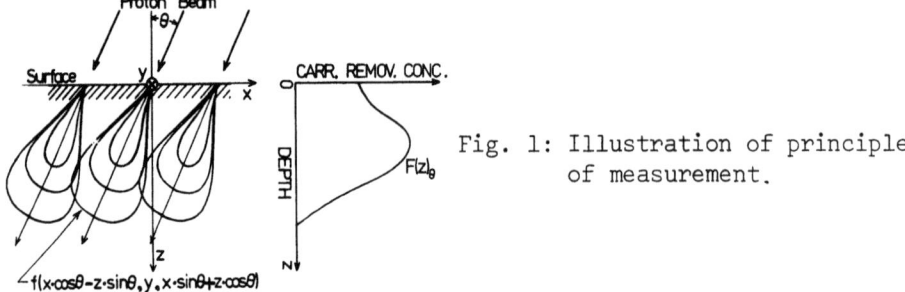

Fig. 1: Illustration of principle of measurement.

the lateral spread of carrier removal beneath any shape of a mask by superposing f(x,y,z)'s spatially. However, even if we measure several $F(z)_\theta$'s for different θ's, it is still mathematically difficult to estimate f(x,y,z) directly from Eq. (1). Thus for mathematical simplification we assume that f(x,y,z) is expressed as follows (assumption (b)),

$$f(x,y,z)=A(z)\cdot\exp\left\{-\frac{x^2+y^2}{2\cdot<\Delta L(z)>^2}\right\} \quad\quad\quad (2)$$

where $<\Delta L(z)>$ denotes the standard deviation of the lateral spread. It has been shown theoretically that the lateral distribution of damage formed by a single implanted ion can be approximated to a gaussian [7,8]. In addition the validity of the assumption (b) is shown experimentally later.

Using Eqs. (1) and (2) the following relation between $F(z)_{0°}$ and $F(z)_\theta$ is derived [6].

$$F(z)_\theta=\frac{\cos\theta\cdot R(\theta)}{R(0°)}\cdot\int_{-z\cdot\mathrm{ctn}\theta}^{\infty}\frac{F(Z)_{0°}}{\sqrt{2\pi}\cdot<\Delta L(Z)>}\cdot\exp\left\{-\frac{x^2}{2\cdot<\Delta L(Z)>^2}\right\}\cdot dx \quad (3)$$

The value of $\cos\theta\cdot R(\theta)/R(0°)$ can be calculated by comparing the total carrier removal rate for an implant at $\theta=0°$ with that at $\theta\neq 0°$. It is possible to solve $<\Delta L(z)>$ from Eq. (3) using a computer iteration method, thus, we can estimate f(x,y,z) from measured values of $F(z)_{0°}$ and $F(z)_{\theta(\neq 0°)}$.

VALIDITY OF THE ASSUMPTIONS OF THE METHOD

In this section we investigate the validity of the fundamental assumptions (a) and (b) which are essential to our method.

Assumption (a): Fig. 2(a) shows several $F(z)_{50}$'s measured by the Copeland method for a 400 keV proton implant. In our experiment $F(z)_{50}$ is assumed to be equal to $F(z)_{0°}$ to avoid channelling errors. The Copeland method using 5.7 MHz as the first harmonic is used to measure $F(z)_\theta$'s because it gives the carrier distribution excluding carriers emitted from deep traps, and this carrier distribution is directly useful for application of the isolation technique to high frequency devices [6]. The experimental conditions for fabricating the GaAs schottky barrier diodes for the Copeland measurements are listed in Table 1.

As shown in Fig. 2(a) the shapes of the carrier removal distributions are similar for all doses up to 3×10^{10} ions/cm^2. Fig. 2(b) shows the linearity of the carrier removal rate versus proton dose at the different depths A,B,C, and D of the distributions of Fig. 2(a). These two figures show that the assumption (a) holds for doses up to 3×10^{10} ions/cm^2 in our GaAs samples.

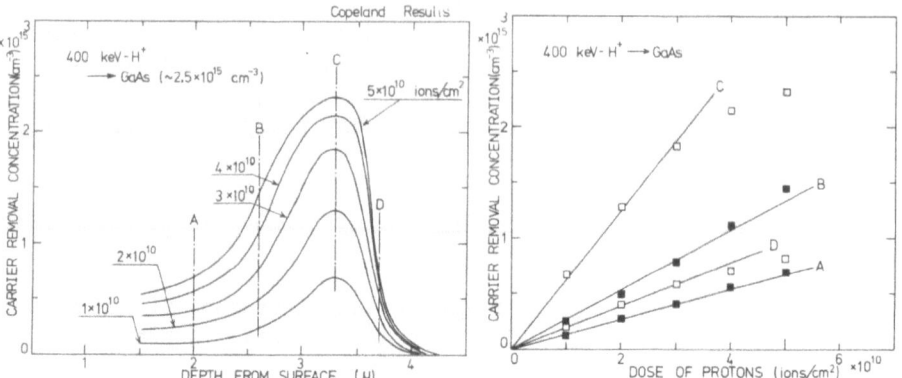

Fig. 2(a): Distributions of carrier removal of various proton doses.

Fig. 2(b): Relation between proton dose and carrier removal.

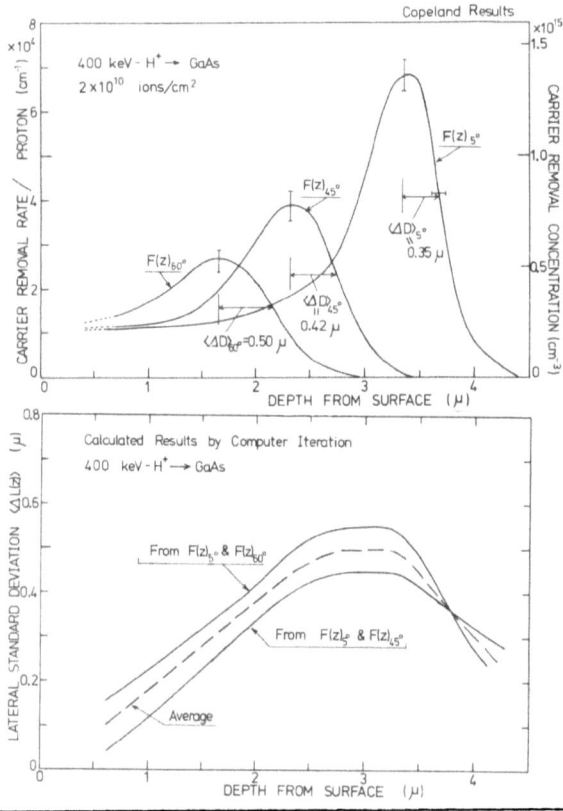

Fig. 3: Experimental results of carrier removal distributions.

Fig. 4: Results of $\langle \Delta L(z) \rangle$ calculated by an iteration method.

Table 1

Material	epitaxial n($\sim 7\mu$ thickness, $\sim 2.5 \times 10^{15} cm^{-3}$) on n^+GaAs
Ohmic Contact	Sn dot on back n^+ layer.
Schottky Electrode	Al of 1mm diameter

LATERAL SPREAD OF THE PROTON ISOLATION LAYER

Assumption (b): Our experimental results for 400 keV protons are shown in Fig. 3 for incident angles θ of 5°, 45° and 60°. The samples were bombarded simultaneously and for θ=5° the dose of protons was 2×10^{10} ions/cm^2. The error bar in the carrier removal concentration results from the error of the area of the Schottky electrode, and the error bar in the depth axis results from the ambiguity due to the Debye length [9] in addition to the error of the Schottky area. From these results it is possible to evaluate $<\Delta L(z)>$ from the combination of $F(z)_{5°}$ and $F(z)_{45°}$ or from the combination of $F(z)_{5°}$ and $F(z)_{60°}$ following Eq. (3). Additionally Fig. 4 shows calculated results of $<\Delta L(z)>$ versus depth from the experimental results of Fig. 3 using a computer iteration method.

The validity of this computer iteration method was checked by calculating a value of $<\Delta L(z)>$ from modelled values of $F(z)_{0°}$ and $F(z)_{60°}$ which are composed starting with an arbitarily assumed value of $<\Delta L(z)>$. As shown in Fig. 4, the shape of $<\Delta L(z)>$ derived from the combination of $F(z)_{5°}$ and $F(z)_{60°}$ is similar to the shape of $<\Delta L(z)>$ derived from $F(z)_{5°}$ and $F(z)_{45°}$. This fact shows the validity of our assumption about the shape of $f(x,y,z)$, that is the assumption (b).

LATERAL SPREAD FROM A MASK EDGE

The imaginary carrier removal distribution $f(x,y,z)$ can be readily calculated from $<\Delta L(z)>$ and $F(z)_{0°}$. However, for actual application it is more important to estimate the real lateral distribution of the carrier removal near the edge of a mask, which is composed by the superposition of $f(x,y,z)$'s.

Let us consider that the GaAs for x<0 is covered by a mask, that is, the edge of the mask is parallel to the y axis at x=0 and that all protons are implanted normal to the GaAs surface. In this case $N_D=0$ for x<0 and θ=0° in Eq. (1). Thus, the real carrier removal distribution $F(x,z)$ is derived from Eqs. (1) and (2) as follows,

$$F(x,z)=F(z)_{0°}\cdot\mathrm{erfc}\left\{\frac{x}{<\Delta L(z)>}\right\}, \quad \mathrm{erfc}=(2\pi)^{-\frac{1}{2}}\cdot\int_{-\infty}^{\infty}\exp(-S^2/2)\cdot dS \quad .. \quad (4)$$

This shows that the lateral carrier removal distribution is expressed by a complementary error function near the edge of the mask.

As an example, in Fig. 5 we show the lateral carrier removal distribution for a 400 keV proton implant. The contours in this figure are normalized to the maximum value of the carrier removal. This figure shows that the lateral spread is largest at the depth where $F(z)_{5°} \simeq F(z)_{0°}$ is the maximum reducing to a small value near the surface.

Fig. 6 shows the variation of the lateral spread of these contours due to the proton energy. The depth of the maximum carrier removal D_{max} is also shown together with the result by Pruniaux et al. [10] and compared with the projected range of protons which is derived from the theory of Lindhard et al [5] using a modified electronic stopping power [4]. Since in our experiments electronic stopping dominates over most of the range of the protons, the agreement between the projected range of protons and the depth of the maximum carrier removal is reasonable.

Using the above range theory the lateral standard deviation of 400 keV protons is estimated to be 0.51μ at a depth equal to the projected range. The average value of 0.50μ for the lateral standard deviation of the carrier removal spread in Fig. 4 agrees well with the theoretical estimate of proton spread and supports the validity of our experimental results.

Moreover, this figure shows that the lateral spread L from the mask edge increases as the incident energy increases but that the ratio of the lateral to the longitudinal spreads D decreases. For example, this ratio varies from ∼0.19 for a 300 keV proton implant down to ∼0.15 for a 500 keV proton implant. This difference is significant in spite of errors of about ±10%.

THE SHAPE OF THE ISOLATION LAYER AFTER ANNEALING

In order to apply the proton isolation technique we need to know the variation of the shape of the isolation layer (Fig. 5) after annealing.

Fig. 7 shows how the peak rate of carrier removal recovers during annealing for different doses of 400 keV protons. As we

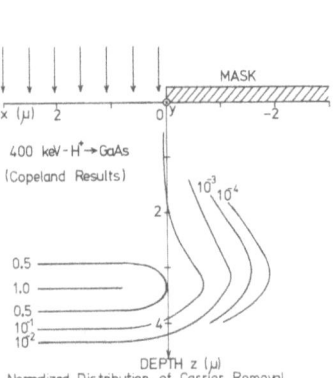

Fig. 5: Normalized lateral distribution of carrier removal.

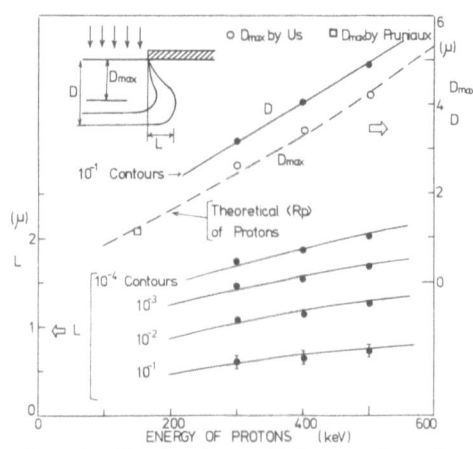

Fig. 6: Variation of lateral and longitudinal spreads with various proton energy.

LATERAL SPREAD OF THE PROTON ISOLATION LAYER

mentioned above, the carrier removal rate, as implanted, varies linearly with proton dose up to 3×10^{10} ions/cm^2 in our GaAs samples. Fig. 7 shows clearly that this linear relation between the dose and the carrier removal rate holds even after annealing, within the same dose limitations.

Thus for doses below 3×10^{10} ions/cm^2 we would expect the shape of any carrier removal distribution to be the same before and after annealing. To demonstrate this point we plot in Fig. 8 the variation of $F(z)_{50}$ curves with annealing temperatures for a 400 keV proton implant at a dose of 3×10^{10} ions/cm^2. Also plotted are the annealing recovery ratios at different positions A, B and C on the $F(z)_{50}$ curves.

In this figure, as we expected, it is found that the shapes of the profiles are similar during annealing for temperatures up to 385°C. Furthermore, from Figs. 7 and 8, we can conclude that the shapes of the normalized lateral distributions do not vary significantly after annealing nor is there a significant change in the ratio of the lateral to the longitudinal spreads.

One may notice that the dose of protons is very low in our experiment. However, one should understand that the carrier removal rate, as implanted, tended to saturate for proton doses over 3×10^{10} ions/cm^2 because we used GaAs samples having a carrier concentration of $2-3 \times 10^{15}$ cm^{-3}. Pruniaux et al [10] reported that the linearity between the proton dose and the carrier removal rate holds for doses up to 1.2×10^{11} ions/cm^2. Moreover, following the resistivity measurements by Kato et al [11], this linearity appears to be valid for doses of $10^{15} \sim 10^{16}$ ions/cm^2, which is more than enough for actual applications.

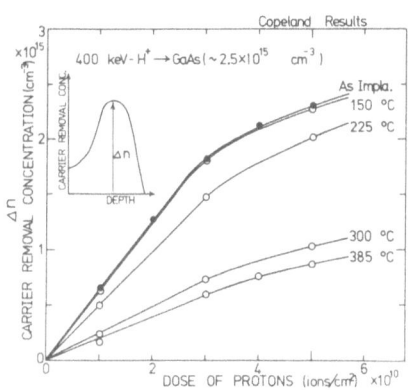

Fig. 7: Relation between proton dose and maximum carrier removal rate during annealing.

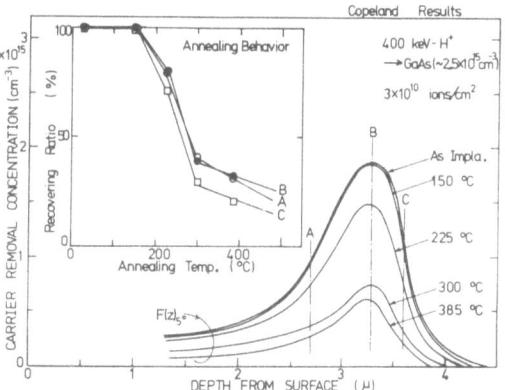

Fig. 8: Carrier removal distributions and recovery ratios at depths A, B and C during annealing.

CONCLUSIONS

We conclude as follows: (1) By comparing several depth distributions formed by implantation with different tilting angles to the target surface, we can get the lateral distribution of the carrier removal as well as the lateral distribution of the implanted ions. (2) The depth of the maximum carrier removal coincides with the theoretical projected range of protons, and the lateral spread of the carrier removal is largest near to that depth. For example, the removal rate goes down to 10% of the maximum at about 0.7μ from the mask edge for 400 keV protons, where the projected range of protons is about 3.3μ. (3) The lateral distribution of the carrier removal is expressed by a complementary error function near the mask edge. The lateral spread of the carrier removal increases as the energy of protons increases but the ratio of the lateral to the longitudinal spreads decreases. (4) The shape of the lateral distribution of carrier removal does not change significantly with annealing temperature. There is also little change in the ratio of the lateral to the longitudinal spreads of the carrier removal after annealing for doses up to 3×10^{10} ions/cm^2. We give reasons why we expect this ratio to be the same for doses as high as 10^{16} ions/cm^2.

ACKNOWLEDGEMENT

The authors are very grateful to Dr. B.J. Sealy for his discussions, to Dr. P.L.F. Hemmet for his advice on the experiment and the staff of the accelerator laboratory of the University of Surrey. This work is supported by the Science Research Council of the United Kingdom.

REFERENCES

[1] H. Stoll et al., Appl. Phys. Lett., 23 (1973) 664.
[2] J.D. Speight et al., Electr. Lett., 10 (1974) 98.
[3] S. Furukawa et al., Appl. Phys. Lett., 22 (1973) 97.
[4] M. Nagatomo et al., (private communication).
[5] J. Lindhard et al., K. Danske Widensk, Selsk. mat-fys. Medd. 33 (1963) 1.
[6] H. Matsumura, et al., (unpublished).
[7] H. Matsumura et al., J. Appl. Phys., 47 (1976) 1746.
[8] H. Matsumura et al., Japan. J. Appl. Phys., 14 (1975) 1783.
[9] W.C. Johnson et al., IEEE Electr. Devices, ED-18 (1971) 965.
[10] B.R. Pruniaux et al., Proc. of 2nd Int. Conf. on Ion Implantation (1972) 212.
[11] Y. Kato et al., J. Appl. Phys., 45 (1974) 1044.

METALLURGICAL APPLICATIONS OF ION IMPLANTATION*

S. M. Myers

Sandia Laboratories

Albuquerque, New Mexico 87115

ABSTRACT

Ion implantation can be used to produce atomic mixtures of controlled composition, independent of thermodynamic constraints, and to introduce lattice damage at rates exceeding 1 dpa/sec. These unique capabilities are being exploited for the study of intermetallic alloys in two different temperature regimes. At lower temperatures, where atomic mobilities are nil, the technique is being used to investigate new metastable phases. These have included substitutional solid solutions of immiscible elements, amorphous alloys, and improved superconductors of nonequilibrium compositions. At higher temperatures, where appreciable atomic diffusion occurs over experimental times, implantation is being employed to study equilibrium alloys. Previously unavailable information about phase diagrams and diffusion rates has been obtained at much lower temperatures and in complex alloy systems.

I. GENERAL DISCUSSION

Ion implantation is emerging as a powerful tool for the study of metals, and its usefulness stems primarily from two properties. First, the technique can be used to create an intimate atomic mixture of controlled composition, without many of the limitations of thermal alloying. For example, one is not constrained by solid solubilities or atomic diffusion rates. In addition, ion implantation permits lattice damage to be introduced at a high rate. Thus,

---

*This work was supported by the United States Energy Research and Development Administration, ERDA, under Contract E-(29-1)789.

atomic displacement rates exceeding 1 dpa/sec are readily achieved, which is orders of magnitude above the levels attainable with electron or neutron irradiation. These unique properties have been exploited in three areas of metals research: 1) radiation damage; 2) behavior of gases in metals, notably He; and 3) properties of intermetallic alloys. Implantation studies of intermetallic alloys is the newest and least developed of these areas, and this will be the subject of the present paper. Ion-induced radiation damage and gases in metals have been covered in detail elsewhere.[1,2]

The qualitative behavior of metal-implanted metals can usually be divided into two temperature regimes, which are distinguished on the basis of whether there is appreciable atomic diffusion over experimental times. Thus, Section II of this paper will deal with a low-temperature regime, defined approximately by

$$\sqrt{D_A(T) \cdot t} < a , \qquad (1)$$

while Section III will discuss a high-temperature regime where

$$\sqrt{D_A(T) \cdot t} > a . \qquad (2)$$

Here T is the temperature, t is the duration of the experiment, and a is the lattice constant. The parameter D_A is the diffusion coefficient for the implanted atom after its thermal spike has died out,[3] but it includes any subsequent diffusion enhancement due to the ion bombardment.[4,5] As a practical matter, it should be noted that room temperature falls within the low-temperature regime in almost all cases.

In the low-temperature regime, the absence of atomic diffusion means that a metastable condition produced by implantation can persist. Precipitation of second phases which would be expected from the equilibrium phase diagram does not occur. More generally, the laws of equilibrium thermodynamics are not applicable: for example, solid solubilities can be exceeded and the Gibbs phase rule violated. With regard to point defects, the low-temperature regime usually extends in temperature from absolute zero to above Stage III in the isochronal annealing profile, where the vacancies become mobile. This upper limit may be inferred by noting first, that most intermetallic diffusion proceeds via the vacancy mechanism, and second, that the vacancy concentration in units of atomic fraction is usually much less than one. As a consequence, the vacancy mobility must be relatively high for appreciable atomic diffusion to occur. As an implanted alloy moves through defect Stages I-III within the low-temperature regime, its properties may be expected to change significantly. While these changes are likely to be less dramatic than those between the low- and high-temperature regimes, their investigation is an important field for future research.

In the high-temperature regime, atomic diffusion permits the metal-implanted metal to move from the implanted condition toward thermodynamic equilibrium. As a result, such an alloy system is usually comprised of one or more equilibrium phases, and the laws of thermodynamics are applicable. Furthermore, precipitation of second phases tends to be quite rapid, since implantation produces an atomic mixture in which the components are separated by only a few atomic spacings. Thus, a relatively simple picture emerges, and one which is consistent with existing data on metal-implanted metals in the high-temperature regime. This contrasts with the observed behavior of inert gases in implanted metals[6] and various impurities in implanted semiconductors.[7] There, the binding of the implanted species to lattice defects can be strong enough to retard the evolution toward equilibrium, resulting in a more complex, non-equilibrium state. However, while such effects certainly could arise for metal-in-metal implants, the probability is less because of the lower metal-defect binding energies.

The remainder of this paper will be concerned with exploiting the above properties for the study of intermetallic alloys. Representative examples of research will be discussed which illustrate the capabilities, and which serve to indicate the directions of current work. Section II will deal specifically with the low-temperature regime, while the high-temperature regime will be covered in Section III.

II. LOW-TEMPERATURE REGIME

Ion implantation has been used in the low-temperature regime primarily to create and study new metastable phases. In such work, this technique provides a unique capability for the production of atomic mixtures and the introduction of lattice damage. Specifically, implantation extends and complements both rapid quenching and film co-deposition, by virtue of its effectively instantaneous quench, its complete independence of the properties of the corresponding melt, its high degree of control over composition and damage, and its wide applicability. In this section, current research will be exemplified by discussions of substitutional solid solutions of insoluble metals in Cu, of amorphous alloys of Cu and Ni, and of high-T_c compositions in the Pd-H-noble metal superconductors. Entirely new metastable phases have been produced in each of these three categories.

A series of binary alloys were created in single-crystal Cu by Borders and Poate[8] and by Sood and Dearnaley[9] using ion implantation. These implantations were done at room temperature or below, so that the system was within the low-temperature regime. The local concentration of the implanted species was typically several at.%. Ion back-scattering-channeling was then used to determine the fraction of the implanted atoms which occupied substitutional sites on the host lattice. Some representative channeling data for Au and W implants in Cu with peak concentrations of ~3 at.% from the Borders-Poate work are shown in Fig. 1. The [110] dips are nearly the same for the host and the implanted atoms, indicating high substitutionality: 100% for Au and 90% for W. Such a result is not unexpected for Au because of its high equilibrium solubility in Cu. Tungsten, however, is negligibly soluble, so that the latter alloy necessarily constitutes a metastable solution. Similarly substitutional implants were obtained for a number of other insoluble elements in Cu.[8,9]

The successful production of metastable substitutional solutions suggests the possibility of a wide array of new alloys with improved properties. It is therefore desirable to understand the physical processes which control the final state of such a system: most importantly, what determines the lattice location of the implanted atoms. This question has been discussed recently by several authors,[8-10] but at present there is neither a complete picture or a full consensus. Factors which have been considered include: the replacement collision probability, which deals with the atomic collisions by which the implanted atom comes to rest; the atomic size and electronegativity, as related to the ease of accommodation within the host lattice; the mobility of point defects, which might be trapped at the impurity site and so stabilize a

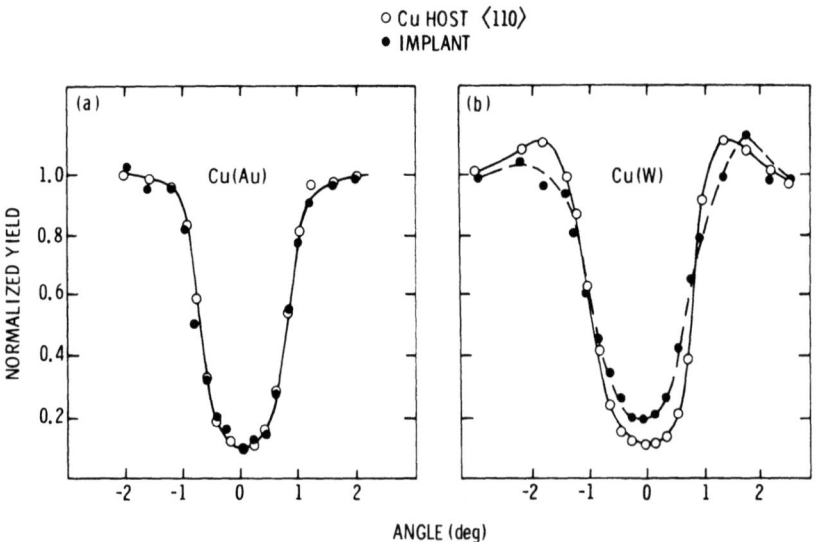

Fig. 1. [110] channeling for 1.8 MeV He backscattered from Cu implanted with Au or W (Ref. 8).

particular configuration; and atomic rearrangement during the lifetime of the thermal spike produced by the implantation cascade. Existing data indicate that the replacement collision probability is not the most important parameter, since theoretically calculated values for this quantity[11] do not correlate well with the experimentally observed substitutional fractions in Cu.[8] Among the remaining factors listed above, the atomic size appears most directly related to substitutionality. An example of this is shown in Fig. 2, where all of the elements implanted into Cu[8,9] are plotted as (atomic volume)$^{1/3}$ versus atomic number. These elements can be divided into groups of less than or greater than 50% substitutionality by a diagonal line.

Two papers have recently reported the creation of apparently amorphous intermetallic alloys by ion implantation. Andrew and co-workers[12] implanted Dy into Ni at room temperature to a local concentration of about 30 at.%, and then analyzed the resulting alloy by transmission electron microscopy. From a very diffuse ring in the electron diffraction pattern, as well as other data, it was concluded that there were essentially no ordered regions larger than about 10 Å. At about the same time, Cullis, Poate and Borders[13] reported a "disordered" layer in Cu which had been implanted with W to about 10 at.%. There was again a diffuse ring in the transmission electron diffraction pattern, and in addition, the channeled backscattering spectrum showed a damage peak. Both the Ni-Dy and the Cu-W alloys were shown by Andrew et. al.[12] to be consistent with the theoretical criteria for amorphous behavior given by Nagel and Tauc.[14] Thus, it appears probable that the

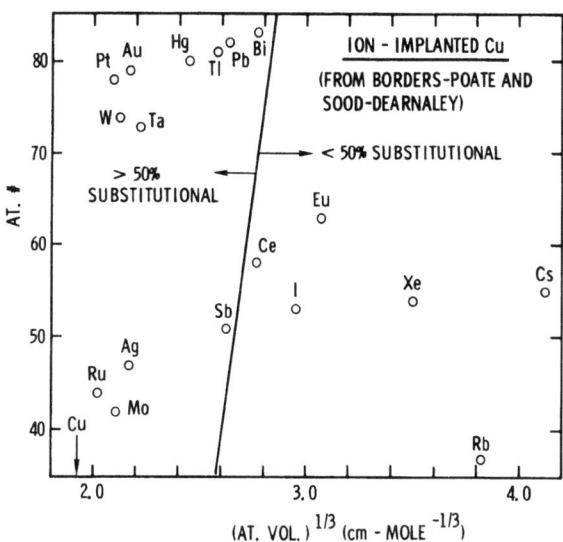

Fig. 2. Lattice Location in Cu (Refs. 8 and 9).

technologically important amorphous metals will be accessible to study by ion implantation. Subsequent papers in these proceedings will discuss this in more detail.

The properties of superconducting alloys are generally quite sensitive to composition, and in at least some instances, the optimum properties occur for a metastable state. In such cases, ion implantation can be used to produce compositions which are not attainable by other means. An example is the work of Stritzker, Heim, and Buckel[15] on the Pd-H-noble metal superconductors. Here the maximum transition temperature T_c requires roughly one H per metal atom, a concentration which is far above the H solubility. These authors circumvented this difficulty by implanting H into the Pd-noble metal alloys at 4 K, and were thus able to reach concentrations in excess of one H per metal atom. Some of their results are shown in Fig. 3, where T_C is plotted versus noble metal concentration. At each point, the H concentration was adjusted to the optimum value. Quite high transition temperatures were obtained, approaching 17 K in the case of the Pd-Cu alloy. Based on their data, Stritzker et. al. argued that the paramagnetism of the Pd ion is suppressed by both the H and the noble metal, and they suggested further that the peaks in T_c shown in Fig. 3 correspond to the appearance of soft phonon modes.

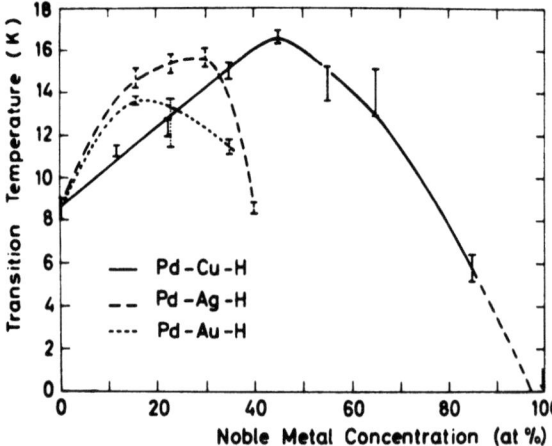

Fig. 3. Superconducting transition temperatures in the Pd-H-noble metal alloys (Ref. 15).

III. HIGH-TEMPERATURE REGIME

In the high-temperature regime, where by definition there is appreciable atomic diffusion over experimental times, a metal-implanted metal is usually made up of equilibrium phases. Where this simple picture appears to be valid, ion implantation is being used to obtain information about diffusion rates and phase diagrams which was previously unavailable. The general approach is to use implantation to create an alloy and then to follow its evolution during annealing by means of ion backscattering-channeling and transmission electron microscopy. Some of the advantages of such a method are: the microscopic scale of the experiment, which allows measurements at lower temperatures where diffusion rates are small; the possibility of using ion irradiation to enhance the atomic diffusion rates, thereby permitting data to be obtained at still lower temperatures; production of an intimate atomic mixture of tailored composition, so that the desired phases precipitate without the formation of intermediate phases; and direct introduction of the solute into the host matrix, which avoids difficulties with interfacial effects such as those due to oxide layers.

The above methods are illustrated by simple experiments which measured the diffusion rate[16] and the solubility[17] of Cu in Be. Diffusion couples were formed by implanting Cu into single-crystal bulk Be at room temperature, where the atomic diffusion rate is negligible. The samples were then annealed isothermally at temperatures of interest, and the composition-versus-depth profile

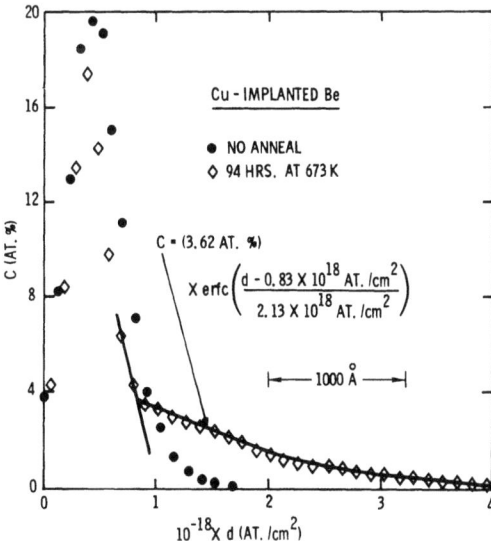

Fig. 4. Effect of Annealing on Cu-implanted Be (Ref. 17).

of Cu was monitored by 2 MeV He backscattering. Two qualitatively different types of behavior were observed, depending on whether the peak Cu concentration in the implanted profile was greater than the solid solubility at the annealing temperature. An example of the higher-concentration case is seen in Fig. 4, where the Cu profile is shown before and after annealing. The anneal has caused a diffused tail to emerge from the implanted peak, but this peak has not become broader. By contrast, when the initial Cu concentration did not exceed the solubility, the initial peak simply increased in width while retaining an approximately Gaussian shape. This difference in behavior results from the fact that there is precipitation locally wherever the Cu solubility is exceeded. Thus, in Fig. 4, the peak region contains both precipitates and Be phase with Cu in solution, while at greater depths only the latter is present. Analysis of the diffusion kinetics is straightforward in both the low- and high-concentration cases: for example, the diffused tail in Fig. 4 is seen to have the expected complementary error function shape.[17,18] The time dependence of such profiles then gives the diffusion coefficient, while the solid solubility is equal to the concentration at which the tail intersects the peak.[17,18]

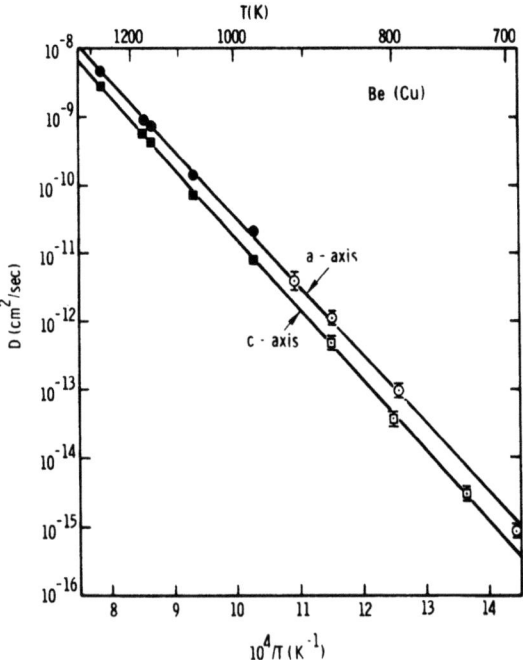

Fig. 5. Diffusion coefficients for Cu in Be (Ref. 16). The open symbols are ion beam data, while the solid points give previous conventional results.

The diffusion coefficient D and the solid solubility C_0 for Cu in Be are shown respectively in Figs. 5 and 6. There are two values of D at each temperature, corresponding to the two inequivalent directions of diffusion in the hcp Be lattice. Data from the ion beam experiments are given by the open symbols throughout, while the solid points are from conventional metallurgical measurements reported in the literature. There is good agreement among the various methods, but the ion beam results extend to much lower temperatures for the reasons discussed above. The solubility value at 593 K is especially noteworthy, in that it was obtained using enhanced diffusion. After the Cu implantation, the sample was irradiated at room temperature with 50 keV Ne to a fluence of 9×10^{16} at./cm^2, in order to damage the region through which the Cu would subsequently diffuse. The requisite penetration of the Cu tail, approximately 10^3 Å, was then reached after about 100 hours of annealing at 593 K. By comparison, approximately six months of annealing would have been necessary in the absence of the enhancement. Furthermore, if the conventional electron microprobe methods had been used, at least 10^4 Å of diffusion would have been required, and the experiment would have been the work of a long career. Possible mechanisms for the enhancement effect in this system will be discussed elsewhere. However, it should be noted here that a consistency check was made on the solubility value obtained using this effect: at 673 K, the solubility was measured both with and without enhancement, and the results were in excellent agreement.

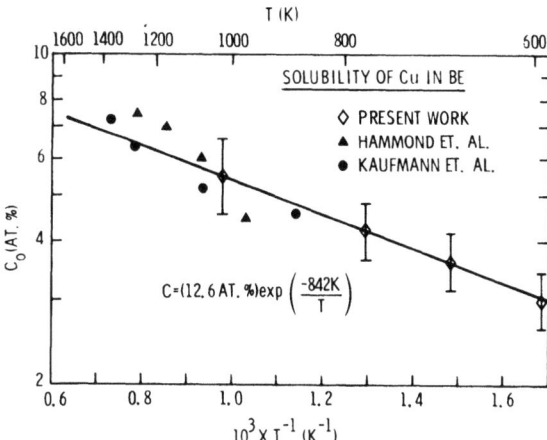

Fig. 6. Solubility of Cu in Be (Ref. 17). The open diamonds are the ion beam data, and the solid symbols are previous conventional results.

The approach used to study the phase diagram of the simple binary system Be-Cu is readily extended to handle ternary or higher order alloys. An example is provided by recent work on the Be-rich region of the Be-Al-Fe system,[18] for which a schematic isotherm is shown in Fig. 7. In this experiment both Al and Fe were implanted into Be, in amounts which located the implanted layer at selected points in the two- and three-phase fields about the Be phase. Then the measured solubilities of Al and Fe gave the coordinates of corresponding points on the boundary surrounding the Be phase. Data were obtained at the numbered points, with the results given in the table on the figure. Precipitation of the expected phases was confirmed by transmission electron microscopy at the three points labeled "E."[19] These results are new in the following respects: the Fe binary solubilities extend 200 K lower in temperature than previously; the upper bound on the Al binary solubility is about one-third the previous upper bound; and the results in the ternary regime are unique.

Similar ion beam experiments[20] have recently yielded the solubility of Zn in Al, where irradiation-enhanced diffusion was employed, and the solubility of Sb in Fe. In both cases there was good agreement with conventional results, which suggests that the ion beam methods outlined above have rather general applicability.

IV. CONCLUSION

The limited goal of this paper is to illustrate how the unique

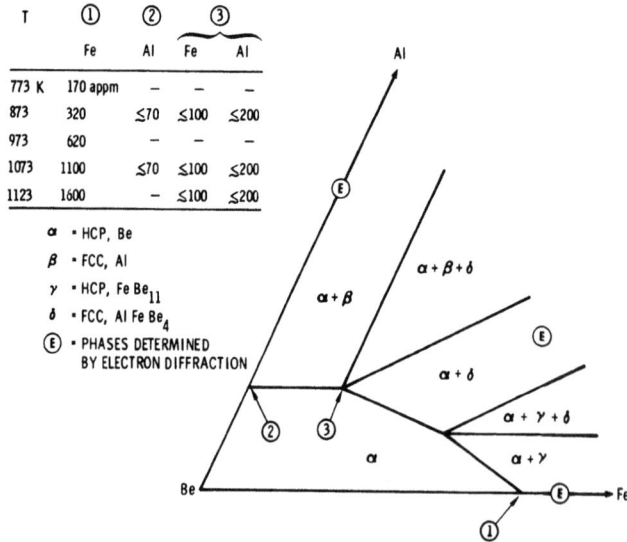

Fig. 7. Schematic isotherm of the Be corner of the Be-Al-Fe phase diagram (Ref. 18).

properties of ion implantation are currently being used in the study of intermetallic alloys. The examples given have included only a small fraction of the work in this area. Moreover, no mention has been made of a number of experiments designed specifically to understand the individual physical processes which occur during implantation in metals, such as enhanced diffusion, radiation-induced precipitation and dissolution, solute-defect interactions, and thermal spike effects. Finally, it may be said that, if the last few years are indicative, additional new and novel developments in metals implantation lie ahead.

The author is indebted to a number of colleagues for enlightening discussions of implantation into metals. Among these are J. A. Borders, W. A. Grant, R. A. Kant, E. N. Kaufmann, S. T. Picraux, J. M. Poate, and J. S. Williams.

REFERENCES

1. *Applications of Ion Beams to Materials, 1975*, ed. G. Carter, J. S. Colligon, and W. A. Grant (Inst. of Phys., London, 1976).
2. *Fundamental Aspects of Radiation Damage in Metals*, ed. M. T. Robinson and F. W. Young (Nat. Tech. Info. Ser., U. S. Dept. of Commerce, Springfield, Va., 1976), Rept. No. CONF-751006.
3. P. Sigmund, Appl. Phys. Lett. $\underline{25}$, 169 (1974).
4. Y. Adda, M. Beyeler, and G. Brebec, Thin Solid Films $\underline{25}$, 107 (1975), and references therein.
5. S. M. Myers, D. E. Amos, and D. K. Brice, J. Appl. Phys. $\underline{47}$, 1812 (1976), and references therein.
6. E. V. Kornelsen, Rad. Eff. $\underline{13}$, 227 (1972).
7. T. W. Sigmon, L. Csepregi, and J. W. Mayer, J. Electrochem. Soc. $\underline{123}$, 1116 (1976), and references therein.
8. J. A. Borders and J. M. Poate, Phys. Rev. B $\underline{13}$, 969 (1976).
9. D. K. Sood and G. Dearnaley, in Ref. 1, pp. 196-203.
10. E. N. Kaufmann, P. Raghaven, R. S. Raghaven, E. J. Ansaldo, and R. A. Naumann, Phys. Rev. Lett. $\underline{34}$, 1558 (1975).
11. D. K. Brice, in Ref. 1, pp. 334-339.
12. R. Andrew, W. A. Grant, P. J. Grundy, J. S. Williams, and L. T. Chadderton, Nature $\underline{262}$, 380 (1976).
13. A. G. Cullis, J. M. Poate, and J. A. Borders, Appl. Phys. Lett. $\underline{28}$, 314 (1976).
14. S. R. Nagel and J. Tauc, Phys. Rev. Lett. $\underline{35}$, 380 (1975).
15. B. Stritzker, Z. Physik $\underline{268}$, 261 (1974); G. Heim and B. Stritzker, Appl. Phys. (Germany) $\underline{7}$, 239 (1975); and references therein.
16. S. M. Myers, S. T. Picraux, and T. S. Prevender, Phys. Rev. B $\underline{9}$, 3953 (1974).
17. S. M. Myers and J. E. Smugeresky, Met. Trans., to be published.
18. S. M. Myers and J. E. Smugeresky, Met. Trans. $\underline{7A}$, 795 (1976).
19. G. J. Thomas, to be published.
20. S. M. Myers, to be published.

SOLUBILITY ENHANCEMENT IN ION IMPLANTED Cu ALLOYS

J. M. Poate

Bell Laboratories, Murray Hill, N. J. 07974

J. A. Borders*

Sandia Laboratories, Albuquerque, N. M. 87115

A. G. Cullis

Royal Radar Establishment, Great Malvern

WR 14 3 PS England

J. K. Hirvonen

Naval Research Laboratories, Washington, D. C.

We have been studying[1,2] the lattice site location and microstructure of metal atom implants in Cu by channeling and transmission electron microscopy measurements. At low atomic concentration, 1 at.%, the implanted metallic species reside primarily on substitutional Cu lattice sites. It was found from the previous study that high-dose implants of Au and W in Cu (10 at.%) were in general accordance with equilibrium phase diagram criteria in that they were 100% and 0% substitutional respectively.

---

*This work was supported by the United States Energy Research and Development Administration, ERDA, under Contract E-(29-1)789.

In this work, we report measurements on high-dose implants of Ag in Cu. Although the maximum reported solubility of Ag in Cu is 0.1 at.% at 200°C, we have observed complete substitutionality for 10 at.% Ag implants in Cu at room temperature. This is an intriguing result as complete solid solutions of Ag and Cu would indeed be expected from simple size and electronegativity arguments. These results will be correlated with previously reported measurements on splat-cooled samples where non-equilibrium alloys were obtained.[3]

REFERENCES

1. J. A. Borders and J. M. Poate, Phys. Rev. B **13**, 969 (1976).
2. A. G. Cullis, J. M. Poate and J. A. Borders, Appl. Phys. Lett. **28**, 315 (1976).
3. P. Duwez in Progress in Solid State Chemistry, edited by H. Reiss (Pergamon Press, Oxford, 1967), Vo. 3.

DOSE RATE EFFECTS IN A PRECIPITATION HARDENED NICKEL-ALUMINUM ALLOY

J.E. Westmoreland*, P.R. Malmberg*, J.A. Sprague[†],
F.A. Smidt, Jr.[†]
Naval Research Laboratory, Washington, D.C. 20375
L.G. Kirchner[+]
University of Wisconsin, Madison, Wisconsin 53706

ABSTRACT

A Ni-14 at. % Al alloy containing γ' (Ni$_3$Al) precipitates was irradiated at 725°C with 2.8-MeV ^{58}Ni$^+$ ions to determine precipitate stability under irradiation. The doses examined were 0.81, 2.5, and 8.1 displacements per atom (dpa) at both of two dose rates: 4.4×10^{-2} dpa/sec and 4.4×10^{-4} dpa/sec. Specimens were examined by transmission electron microscopy. The precipitates in the lower-dose-rate samples were seen to develop complex contrast features after irradiation but to retain their original size. The original precipitates in the higher-dose-rate samples developed a ragged appearance around the edges as if pieces were being dissolved, and a second smaller class of precipitates appeared in the matrix, and by 8.1 dpa the entire original precipitate size distribution had been converted to the new precipitates. These differences due to dose rate are not explained by current models of the behavior of precipitates under irradiation.

INTRODUCTION

The precipitate microstructure influences the mechanical properties of materials; hence, for materials operating in a radiation environment the stability of precipitates under irradiation is important. Heavy-ion irradiations for the simu-

---

* Materials Modification and Analysis Branch
[†] Thermostructural Materials Branch
[+] Nuclear Engineering Department

lation of neutron damage requires an awareness of any dose-rate effect because of the compressed time frame of these irradiations, and previous high temperature, high-dose-rate irradiations[1] suggest that dose rate is likely to be an important variable. Nickel ion irradiation of γ' precipitates in the same alloy employed in this experiment was previously studied[2] by transmission electron microscopy as a function of irradiation temperature and as a function of dose. At the high dose rate (see Experimental Procedures) of this experiment the largest effect on the precipitate microstructure was seen at 725°C, the highest irradiation temperature employed. where the original 400 Å size precipitates were replaced by 80 Å size precipitates. At lower temperatures the precipitate structures were less well defined and took on a ragged appearance with a wide spread in sizes and a smaller precipitate formed in the matrix between the original ones. These observations were in contrast to an earlier study[3] in which the γ' precipitates were observed to decrease in size in aged material and grow to the same size in solution-treated material. A model was proposed[3] which predicted that an equilibrium precipitate size which would be a function of dose rate would be achieved after a dose of a few dpa. This work was recently reviewed by Hudson.[4] The present study was undertaken to investigate the influence of dose rate on the γ' precipitate size distribution at 725°C and to examine more closely the early stages of the effects of heavy-ion irradiation on these precipitates.

EXPERIMENTAL PROCEDURES

Preparation of the Ni-Al alloy used in this investigation has been described previously.[2] The nickel-ion irradiations were performed with a 2.8-MeV ^{58}Ni$^+$ beam from the NRL 5-MV Van de Graaff accelerator. The specimens were irradiated at 725°C to doses equivalent to 0.81, 2.5, and 8.1 displacements per atom (dpa), and at peak displacement rates of 4.4×10^{-2} dpa/sec-high dose rate (HDR) and 4.4×10^{-4} dpa/sec-low dose rate (LDR). The deposition of initial damage energy by the nickel ions as a function of distance into the target foils was calculated with the E-DEP-1 computer code.[5] For 2.8-MeV ^{58}Ni$^+$ ions, this calculation yielded a peak energy deposition for elastic collisions of 1.16 MeV/μm at a depth of 5400 Å. A Kinchin-Pease secondary displacement model with an efficiency of 0.8 and a displacement energy of 40 eV was used to obtain dpa values. After ion bombardment, 4000 ± 500 Å of the front surface of each sample was removed using a laser interferometric polisher.[6] The front face was then masked off, and the sample was polished to perforation from the rear surface with one jet of a semiautomatic dual-jet electro-polisher. Specimens were examined with transmission electron microscopy in a JEM-200A electron microscope operated at 200 kV by using the {100} class of superlattice spots for imaging.

RESULTS

Definite dose-rate effects on the precipitate microstructure were observed in this experiment. The unirradiated precipitate microstructure will be described, followed by that of the HDR samples for each dose, and then the LDR samples for each dose.

The Ni_3Al precipitates of this investigation prior to irradiation are cuboidal with {100} faces and a mean cube edge of about 400 Å. The standard deviation of the size distribution is about 100 Å. Figure 1 shows these precipitates prior to irradiation.

Figure 2 shows micrographs of the HDR 2.5 dpa sample. The precipitate microstructure is typical of the appearance of that in both the 0.81 and 2.5 dpa samples. The precipitates in the HDR cases developed somewhat irregular shapes and contrast

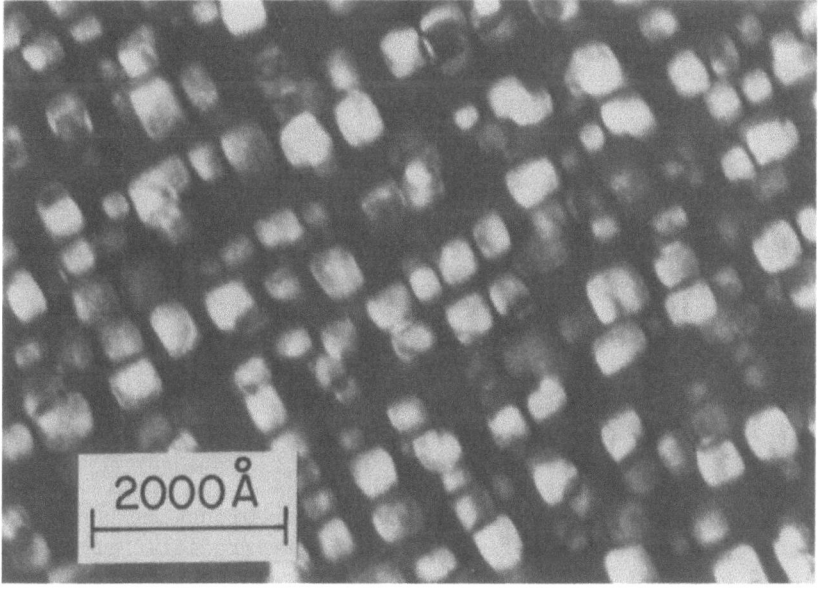

Figure 1. Ni_3Al precipitate microstructure in Ni-Al alloy before irradiation. g = {001} class.

This method of imaging was chosen because the small mismatch of γ' and the matrix does not yield sufficient contrast to produce well defined images. A particle size analyzer was used to characterize the precipitate size distribution.

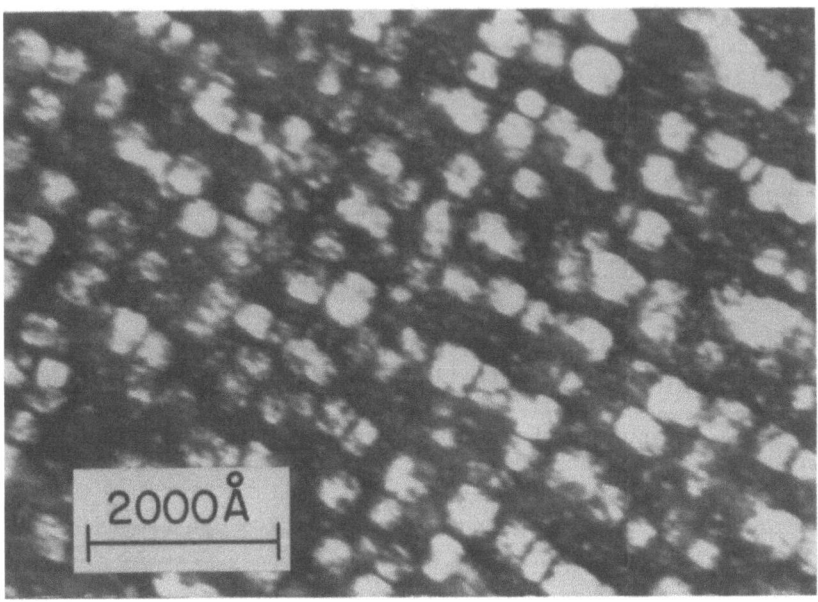

Figure 2. Ni$_3$Al precipitate microstructure of Figure 1 after irradiation with 2.8-MeV ^{58}Ni$^+$ ions to a dose of 2.5 dpa at a dose rate of 4.4 x 10^{-2} dpa/sec. g= {001} class.

features giving them a 'fractured' or 'ragged' appearance as if they were 'dissolved' around the edges. However, within the experimental error no size change was seen in the larger size precipitates in the 0.81 and 2.5 dpa HDR samples. In the HDR-irradiated samples a fine precipitate structure appeared between the larger precipitates even at the lowest fluence (0.81 dpa). With increasing HDR fluence these small precipitates became more evident until at 8.1 dpa the original 400 Å precipitates had been completely replaced by a smaller and higher density precipitate structure, as shown in Figure 3, where the precipitate mean size is about 50 Å. One may recall the results of Kirchner et al.[2] at 20 dpa where a result similar to Figure 3 was obtained but with a precipitate mean size of about 80 Å.

Figure 4 shows micrographs of the LDR 2.5 dpa sample. The precipitate microstructure is typical in appearance of those in both the 0.81 and 2.5 dpa samples. The LDR-irradiated samples developed contrast features during the irradiation, but retained their original size. No fine precipitation was observed in the

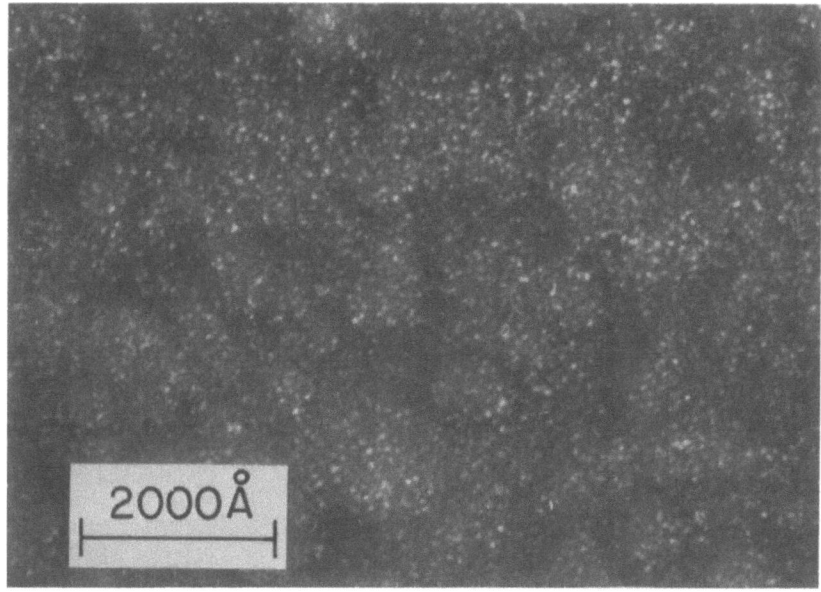

Figure 3. Ni$_3$Al precipitate microstructure of Figure 1 after irradiation with 2.8-MeV ^{58}Ni$^+$ ions to a dose of 8.1 dpa at a dose rate of 4.4×10^{-2} dpa/sec. g = {001} class.

matrix of the LDR-irradiated samples, even at 8.1 dpa; the appearance of the precipitate microstructure of the 8.1 dpa LDR specimen was similar to Figure 4.

Measurements of the Ni$_3$Al precipitate microstructural parameters before and after irradiation are summarized in Table 1. The scatter in the mean size is believed to be due both to the relatively small number of precipitates counted and to the difficulty in sizing the precipitates.

DISCUSSION

There is little direct information in the literature on the effect of dose rate on precipitate stability under irradiation. That information which might perhaps be useful will be summarized briefly and contrasted with the quite different behavior of the precipitate microstructure observed here under the factor of one hundred difference in heavy-ion dose rate employed. Some possible implications of these results will then be considered.

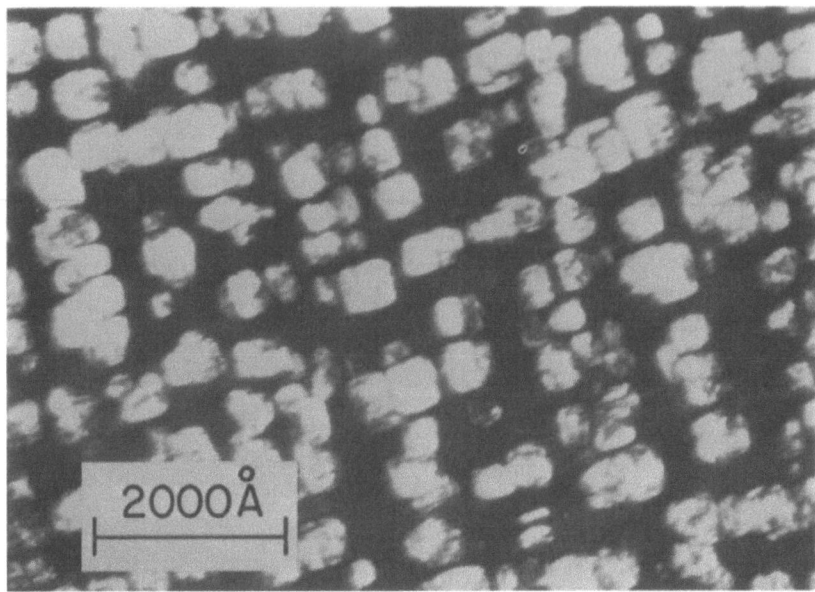

Figure 4. Ni₃Al precipitate microstructure of Figure 1 after irradiation with 2.8-MeV ^{58}Ni$^+$ to a dose of 2.5 dpa at a dose rate of 4.4×10^{-4} dpa/sec. g = {001} class.

The relationship of the HDR-type irradiations employed in this experiment to other work has been discussed in more detail elsewhere.[2] The first reported study[3] of the response of Ni₃Al precipitates to irradiation showed that in an alloy similar in composition to the one used in the present study the precipitates (γ'-Ni₃Al) were observed to break up during heavy-ion irradiation at all temperatures above 325°C to form a population of smaller precipitates. Solution-treated material was observed following the same heavy-ion irradiation to have formed precipitates of the same size. Based on these data, a model was proposed[3] in which an equilibrium size precipitate was formed as a consequence of a kinetic equilibrium between the dissolution of volumes of precipitate near the precipitate surface when disordered by a displacement cascade or replacement collision sequence and the growth of the precipitates as a consequence of radiation enhanced diffusion. The occurrence of precipitation during the irradiation and the effect of precipitate density on the final equilibrium size was acknowledged[3] but not incorporated into a predictive model.

TABLE 1

Characteristics of Precipitate Size Distributions in an Irradiated Ni-Al Alloy

Dose (dpa)	Dose Rate	Mean Size (Å)	Standard Dev. of Size Distribution (Å)	Number of Precipitates Counted
0[a]	Thermal Control	396	109	573
0[a]	Thermal Control	402	103	628
0.81	HDR	415	98	590
0.81	LDR	411	118	495
2.5	HDR	436	95	642
2.5	LDR	420	107	744
8.1	HDR	50	20	642
8.1	LDR	405	100	525

[a]Two micrographs of one control sample held at 725°C for approximately seven hours, the time to irradiate the 8.1 dpa LDR sample.

The results of the present experiment show that in this case the behavior of precipitates under irradiation is more complex than the proposed model[3] as the following effects due to dose rate alone are observed:

1. In the HDR samples examined new precipitates nucleate and grow in the matrix between the original precipitates.

2. The HDR irradiation causes areas at the "edge" of the image of the original precipitate to go completely out of contrast, that is, to appear to be dissolved. That this is the case is confirmed by the complete conversion of the precipitate size distribution by the 8.1 dpa HDR dose.

3. In the LDR samples examined complex contrast features associated with the precipitates are observed.

Whether one or more dissolution processes are operating is not yet clear. A model which will explain the present experimental results very likely must include one or more radiation enhanced nucleation mechanisms, homogeneous or inhomogeneous, as well as various dissolution processes.

CONCLUSIONS

A study has been performed of dose-rate effects observed in the irradiation at 725°C of a Ni-14 at. % Al alloy containing Ni_3Al (γ') precipitates. A factor of 100 difference in the rate of 2.8-MeV $^{58}Ni^+$ ions was employed to give dose rates of 4.4×10^{-2} dpa/sec (HDR) and 4.4×10^{-4} dpa/sec (LDR).

1. The precipitate size distribution in samples HDR-irradiated decreased from an average value of about 400 Å before irradiation to a value of about 50 Å after a dose of 8.1 dpa.

2. The precipitates in HDR-irradiated samples developed with increasing dose an increasingly ragged appearance. Fine scale precipitation occurred in the matrix between the larger precipitates. The precipitates in LDR-irradiated samples developed contrast features which were distinctly different in appearance from the HDR-irradiated precipitate microstructures. No fine scale precipitation was observed in the LDR-irradiated matrix.

REFERENCES

1. J.E. Westmoreland, J.A. Sprague, F.A. Smidt, Jr., and P.R. Malmberg, Rad. Effects 26, 1 (1975).

2. L.G. Kirchner, F.A. Smidt, Jr., L.G. Kulcinski, J.A. Sprague, and J.E. Westmoreland, Effects of Radiation on Structural Materials, proceedings of ASTM Symposium, May 1976, St. Louis, MO, to be published.

3. R.S. Nelson, J.A. Hudson, and D.J. Mazey, J. Nucl. Mat. 44, 318 (1972).

4. J.A. Hudson, J. Br. Nucl. Energy Soc. 14, 127 (1975).

5. I. Manning and G.P. Mueller, Computer Phys. Comm. 7, 85 (1974).

6. J.A. Sprague, Rev. Sci. Inst. 46, 1171 (1975).

AℓSb PRECIPITATE EVOLUTION DURING Sb IMPLANTATION IN Aℓ:

EXPERIMENT AND THEORY*

R. A. Kant
University of New Mexico, Albuquerque, N. M. 87106

S. M. Myers and S. T. Picraux
Sandia Laboratories, Albuquerque, N. M. 87115

ABSTRACT

Precipitate evolution during ion implantation has been studied in the model system Sb-implanted Aℓ. Transmission electron microscopy was used to determine the size distribution of AℓSb crystallites as a function of fluence, flux, and sample temperature. There was a dramatic increase in average size, accompanied by a decrease in number density, for increasing fluence, for decreasing flux, and for increasing temperature. A new theoretical model for the evolution of the precipitate size distribution has been developed which incorporates both thermal processes and implantation effects. Numerical solutions for the AℓSb system using physically realistic parameters agree qualitatively with the experimentally observed evolutions.

I. INTRODUCTION

Precipitation plays a central role in determining the properties of ion-implanted metals.[1] However, because of physical processes unique to the implantation environment, the factors which determine the final state of alloys produced in this way are complex. Various investigators have reported precipitate growth and dissolution under ion bombardment, and have suggested mechanisms to account for the observed behavior (see for example, Refs. 2-5).

---

*This work was supported in part by the United States Energy Research and Development Administration, ERDA, under Contract E-(29-1)789.

However, a detailed understanding of precipitate evolution under ion bombardment has not been available.

In this paper, we report the results of a systematic experimental and theoretical study of precipitate evolution in ion-implanted metals. Transmission electron microscopy was used to examine the precipitate size distribution in a model system, Sb-implanted Aℓ, as a function of flux, fluence, and temperature. The Aℓ-Sb system was selected for two reasons: first, the phase diagram is simple in that there is only one stable intermetallic phase, AℓSb; second, the solubility of Sb in Aℓ is so small that thermal (Ostwald) ripening[6] is insignificant. This allows one to separate irradiation effects unambiguously from the normal thermal aging. The experimental results are interpreted on the basis of a new mathematical model which incorporates the dominant physical process affecting the precipitate size distribution. This model is unique in that, for the first time, the solute concentration, the precipitate number, and the precipitate size are allowed to vary with time. Furthermore, the present treatment includes an ensemble of precipitates of differing size, instead of an array of a single size. This more general approach is required because two precipitates of different size will evolve differently, but interdependently.

II. EXPERIMENTAL RESULTS

Electropolished samples of 99.99% pure, single-crystal Aℓ were implanted with Sb at 50 keV to a mean depth of 235 Å. Implantations were carried out for ion fluences ranging from 5×10^{15} to 2×10^{17} cm^{-2}, producing average Sb concentrations of ~ 1 to 15 at.%, which are large compared to the solubility ($\leqslant 0.03$ at.%). Ion fluxes were varied between 6.25×10^{11} and 2.25×10^{13} cm^{-2} sec^{-1}, and the temperature ranged from room temperature to 300°C. Analysis throughout the implanted depth was carried out by transmission electron microscopy at 100 kV. In most cases, the samples were thinned for microscopy prior to implantation.

We begin our survey of the experimental results with a comparison of the precipitate size distribution obtained by implanting the Sb at 300°C, with that obtained from room temperature implantation followed by a 60-minute anneal at 300°C (Fig. 1). In both cases, the Sb fluence was 5×10^{15} cm^{-2}. However, it is apparent from the micrographs that the size distributions are quite different. Electron diffraction analysis of the room temperature implant prior to annealing indicated no evidence of AℓSb formation. However, upon annealing to 300°C, a diffraction pattern consistent with the AℓSb structure began to emerge. Within less than 10 min the system had stabilized, and no further AℓSb precipitation or growth was observed

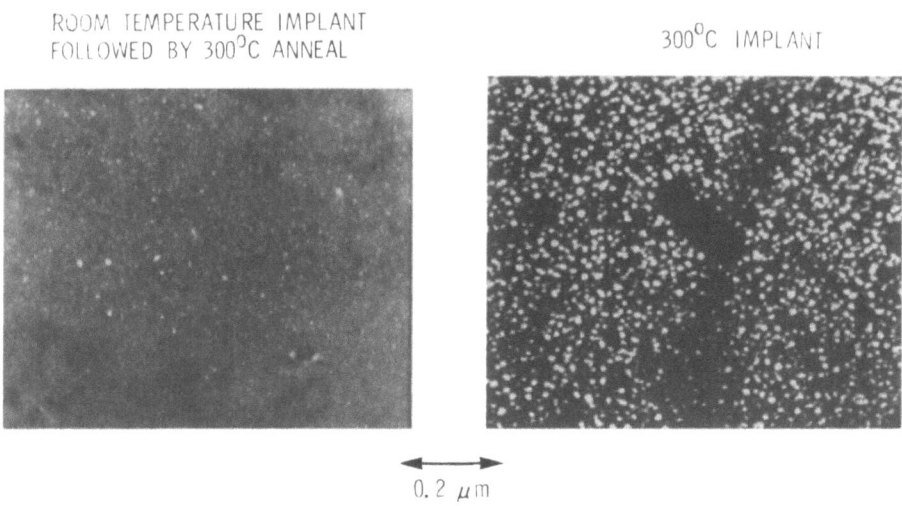

Fig. 1. Dark-field micrographs of AℓSb precipitates resulting from (a) post implantation annealing at 300°C, and (b) implantation at 300°C.

throughout the remainder of the 60-min. anneal. From Fig. 1 it is evident that the AℓSb precipitates grown in the absence of irradiation are considerably smaller (≤ 30 Å) and greater in number density than those which formed during implantation (~ 250 Å). We therefore conclude that while AℓSb forms whenever the Sb mobility is high, the final size distribution is strongly dependent on the implantation and annealing conditions.

Additional evidence for the important influence of irradiation effects on the precipitate size distributions can be obtained by examining intermediate stages of a high fluence implant. In Fig. 2 is shown the AℓSb precipitates for three different Sb fluences. With increasing fluence, the number of precipitates decreases while the mean size increases. This indicates that during implantation the larger precipitates are growing at the expense of the dissolution of the smaller ones. Such behavior was demonstrated to be the result of the irradiation by a subsequent experiment in which a precipitated sample was aged for more than an hour at 300°C, and no change in the size distribution was observed. The latter result is expected, since no thermal ripening should take place in a system with negligible solubility.[6]

At temperatures where precipitation can occur during implantation, the precipitate size distribution is also a strong function

IMPLANT FLUENCE DEPENDENCE

Al (50 KeV Sb, 2×10^{13} Sb/cm^2 - sec, 300°C)

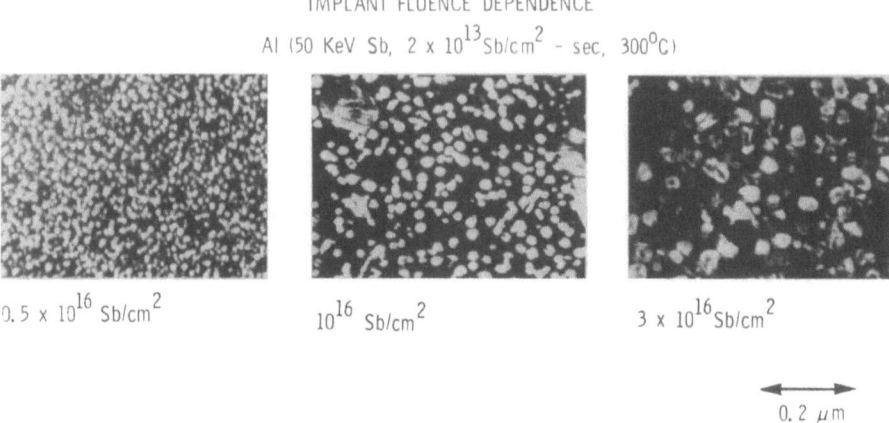

0.5×10^{16} Sb/cm^2 10^{16} Sb/cm^2 3×10^{16} Sb/cm^2

0.2 μm

Fig. 2. Evolution of precipitate size distribution with increasing Sb fluence.

FLUX DEPENDENCE

Al (50 KeV Sb, 5×10^{15} Sb/cm^2, 300°C)

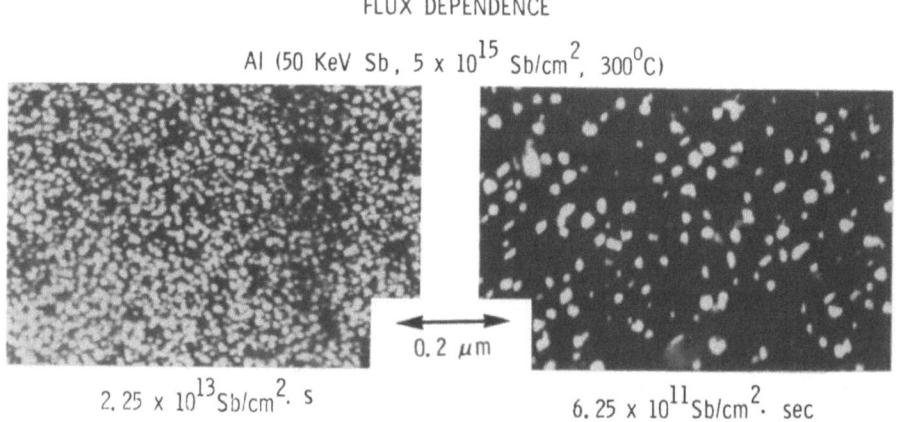

0.2 μm

2.25×10^{13} Sb/cm$^2 \cdot$ s 6.25×10^{11} Sb/cm$^2 \cdot$ sec

Fig. 3. Effect of a different Sb flux at constant fluence and temperature.

of flux and temperature. Figures 3 and 4 illustrate the effects of changing these parameters. In each case the Sb implant fluence was 5×10^{15} cm^{-2}. Decreasing the flux by a little more than a factor of ten, or increasing the temperature by 100°C, shifts the

IMPLANT TEMPERATURE DEPENDENCE

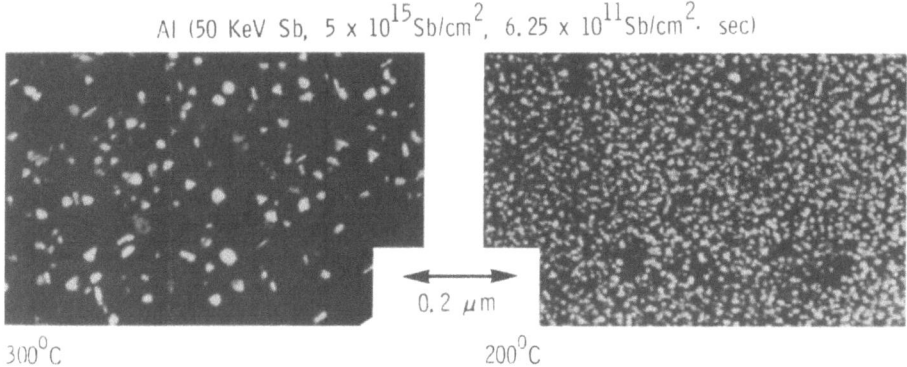

Fig. 4. Effect of a different sample temperature at constant fluence and flux.

distribution to a lower density of larger precipitates. Thus, these parameters have significant effects on the precipitates and must be accounted for in any theoretical model.

III. THEORETICAL MODEL AND DISCUSSION

In order to predict the behavior of the precipitate size distribution, the rate of change of the radius of a single precipitate must be determined. Then a time evolution of the size distribution is described by a continuity equation, which can be solved subject to the constraint that the total amount of impurity is conserved. In the present paper we will outline the formulation, and a detailed derivation will appear elsewhere.[7] The rate of change of the radius R of a single precipitate is given by approximately

$$\dot{R} = \frac{D}{nR}\left[C - C_\infty \exp\left(\frac{2\sigma v}{RkT}\right)\right] - \frac{FS}{4n}\left[1 - \frac{R\exp(-\epsilon/\xi)}{R+\epsilon}\right]$$

$$+ \frac{F}{4n}\left[1 - \exp(-R/a)\right], \qquad (1)$$

where D is the diffusivity, F the flux, T the sample temperature, C_∞ the solid solubility of a precipitate of infinite radius, σ

interfacial energy, k the Boltzmann constant, S the number of atoms ejected per ion incident on a precipitate, ϵ the effective range of an atom ejected from a precipitate, v the atomic volume of impurity in the precipitate, $\bar{\xi}$ the mean distance between precipitates, n is the number of solute atoms per unit of precipitate, and a the characteristic stopping length of an implanted atom within a precipitate.

The first term of Eq. (1) describes the purely thermal processes.[6,8] The rate of diffusion to, or away from a precipitate, depends on the difference between the solute concentration, C, and the solubility. The latter is given here as the solubility for a precipitate of infinite radius, C_∞, modified by the exponential factor which accounts for the increased relative interfacial energy as the precipitate size becomes small. This interfacial energy, σ, is responsible for the normal Ostwalt ripening in which a precipitated system lowers its free energy through the growth of large precipitates at the expense of the dissolution of smaller ones. Even for the purely thermal case, a correct description of precipitate growth requires that the entire distribution of precipitate sizes be considered, since in general, both growth and dissolution are present and the concentration, C, in Eq. (1) depends upon the behavior of the entire distribution.[6]

Our experiments have shown that the implantation parameters play an important role in determining the final state of the system. Effects arising from the implantation process are described approximately by the last two terms of Eq. (1). The second term accounts for dissolution due to recoil collisions, in which solute atoms are ejected from the precipitate. We obtain a functional dependence for the recoil dissolution term different from that originally given by Nelson[4] by noting that, since the range of these atoms is small, only an R-dependent fraction of the recoiling atoms diffuse away from the parent precipitate. While an incident ion may knock atoms out of a precipitate, it may also come to rest within the precipitate. This effect is taken into account by the final term in Eq. (1). A small precipitate will capture the incident atoms in proportion to its volume, whereas a large one will intercept the incident flux in proportion to its area. In this way, a precipitate may grow even in the absence of diffusion from the surrounding matrix.

The basic physics of the model is contained in Eq. (1), which describes a single precipitate. To determine the time evolution of an ensemble of precipitates, we then substitute this equation into the continuity equation.

$$\frac{\partial}{\partial t} N(R,t) = \frac{\partial}{\partial R}\left[\dot{R}N(R,t)\right], \qquad (2)$$

with the boundary condition $\partial^2[\dot{R}N(0,t)]/\partial R^2 = 0$.

The experimentally observed evolution to fewer and larger precipitates with increasing fluence (see Fig. 2) suggests that once the initial distribution is formed, further nucleation of precipitates is suppressed. Therefore, in the present calculations, an experimentally determined initial distribution is used and nucleation phenomena are not included. However, nucleation could be included in the above model by appropirate modification of the boundary conditions at R = 0, provided that information on the nucleation rate were available.

The value of $\dot{R}$ in Eq. (1) depends explicitly on the concentration C which at any given time t is given by

$$C(t) = \frac{\frac{Ft}{w} + \frac{4}{3}\pi n\left[\int_0^\infty N(R,0)R^3 dR - \int_0^\infty N(R,t)R^3 dR\right]}{\left[1 - \frac{4}{3}\pi \int_0^\infty N(R,t)R^3 dR\right]}, \qquad (3)$$

where w is the effective width of the implanted region. For the present case, the Sb concentration in solution in Aℓ is given by the amount of Sb implanted, corrected for any net gain or loss to precipitates, divided by the volume of the Aℓ phase.

Equations (1-3) are solved numerically using the method of lines and the stiff integrator due to Hindmarch.[9] While not all of the parameters are available in the literature, physically reasonable values were used in order to examine the qualitative behavior predicted by the model. We consider three cases for which numerical calculations have been performed which demonstrates the influence of fluence, flux and temperature on the precipitate size evolution. The same initial precipitate size distribution is used throughout, and is labeled as the "initial distribution" in Figs. 5-7. This corresponds to the distribution experimentally observed after a 5×10^{15} cm^{-2} implant at 300°C with an Sb of 2×10^{13} cm^{-2} sec^{-1}. In Fig. 5 is shown the ripening which results from the additional Sb implantation of 5×10^{15} cm^{-2}. The distribution is shifted to the right indicating growth, while at the same time the area under the curve is reduced, showing that the total number of precipitates has decreased. This is in agreement with the qualitative behavior observed experimentally, as shown in Fig. 2.

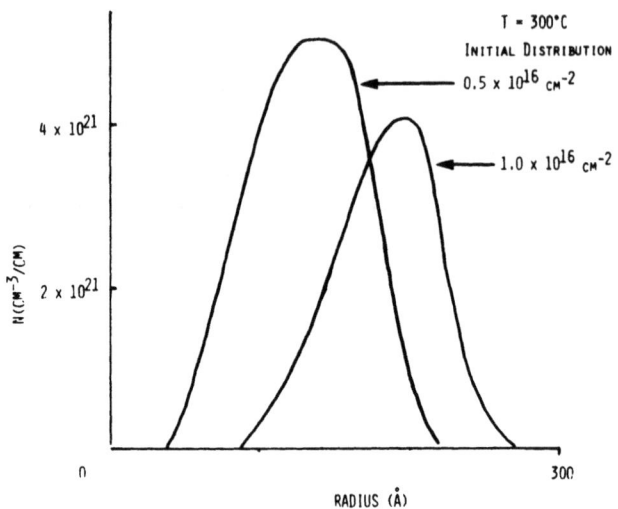

Fig. 5. Predicted evolution of the precipitate size distribution with increasing fluence for an Sb flux of 2.3×10^{13} cm^{-2} sec^{-1}.

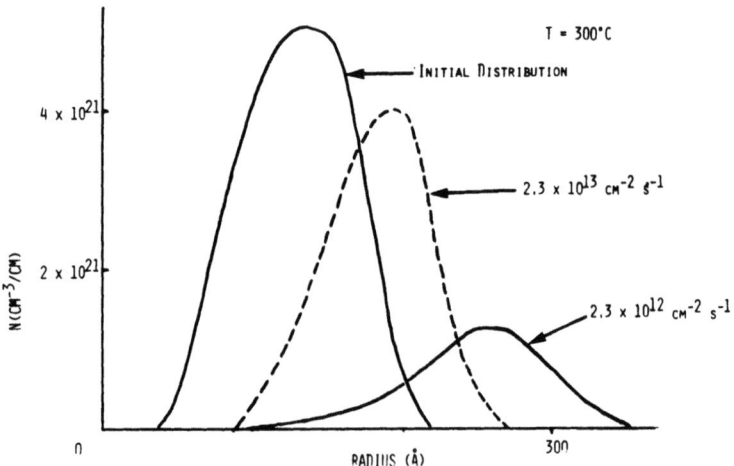

Fig. 6. Effect of changing the flux on the evolution of the theoretical size distribution for a total Sb fluence 1×10^{16} cm^{-2}. The dashed line gives the final distribution at the flux of Fig. 5.

Numerical calculations also demonstrate that the precipitate size distribution is strongly dependent on the ion flux. In Fig. 6, the final distribution from Fig. 5 (dashed line) is compared to that obtained at 1/10 of the Sb flux, with all other parameters held constant. In each case, the model predicts precipitate growth with a reduction in the total number of precipitates during Sb implantation, but at the lower flux the effect is more pronounced. This is consistent with the experimental observation (Fig. 3) of fewer and larger precipitates at the lower Sb flux.

Finally, we consider the effect of changing the temperature. A calculation has been done in which the flux and fluence are the same as in Fig. 5, but the temperature is lower by 100°C. The results are given in Fig. 7, and indicate that reducing the temperature drastically slows the growth of precipitates and suppresses precipitate dissolution. This is consistent with the experimentally observed difference between the precipitate size distributions for Sb implantation at 200 and 300°C (see Fig. 4).

A degree of insight into the above flux and temperature dependences can be obtained by considering the concentration C of Sb in solution in the $A\ell$ phase. In particular, we find that, at constant fluence decreasing C is accompanied consistently by larger and fewer precipitates. The reduction of C can, in turn, result either from lower flux or from higher temperature, the latter because of the greater diffusivity.

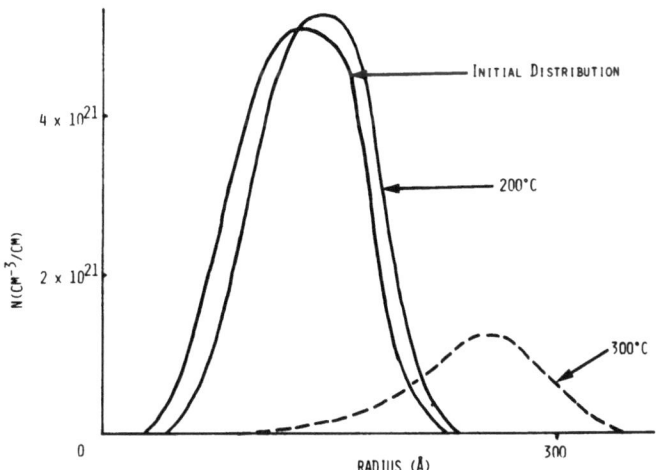

Fig. 7. Effect of implantation at different temperatures on the evolution of the theoretical size distribution for a total Sb fluence 1×10^{16} cm^{-2} and flux 2.3×10^{12} cm^{-2} sec^{-1}. The dashed line gives the final distribution at the temperature of Fig. 6.

IV. CONCLUSIONS

Experiments on AℓSb precipitation during Sb implantation into Aℓ have demonstrated the strong dependence of precipitate evolution on the implantation temperature, flux and fluence. Furthermore, the first theoretical model describing the evolution of the precipitate size distribution during implantation has been developed. Predictions by this model have been shown to correctly describe the qualitative behavior observed for the AℓSb.

The theoretical approach used here should be applicable to a wide range of precipitated systems under irradiation. For example, by including nucleation one should be able to describe the observed[4,5] changes in precipitate size distribution under self-ion bombardment. Other potential applications include the formation of gas bubbles and voids.

The authors gratefully acknowledge helpful discussions with T. J. Headley on electron microscopy; with D. E. Amos on the numerical calculations; and C. T. Fuller for his assistance with the implantations.

REFERENCES

1. See, for example, these proceedings.
2. P. A. Thackery and R. Nelson, Phil. Mag. 19, 169 (1969).
3. G. J. Thomas and S. T. Picraux in *Applications of Ion Beams to Metals*, Ed. by S. T. Picraux, E. P. EerNisse and F. L. Vook (Plenum Press, New York, 1974), p. 257.
4. R. S. Nelson, ibid, p. 221.
5. J. E. Westmoreland (these proceedings).
6. I. M. Lifshitz and V. U. Slyozov, J. Phys. Chem. Solids 19, 35 (1961).
7. R. A. Kant, S. M. Myers and S. T. Picraux (to be published).
8. F. S. Ham, J. Phys. Chem. Solids 6, 335 (1958).
9. A. C. Hindmarch, Lawrence Livermore Laboratory Report UCID-30059, "GEARB: Solution of Ordinary Differential Equations Having Banded Jacobian," May 1973.

FORMATION OF CORROSION-RESISTANT SURFACE ALLOYS BY IMPLANTATION OF LOW-ENERGY NICKEL AND CHROMIUM IONS INTO POLYCRYSTALLINE IRON

B. D. Sartwell, A. B. Campbell, and P. B. Needham, Jr.

College Park Metallurgy Research Center, Bureau of Mines
U.S. Department of the Interior, College Park, Md. 20740

ABSTRACT

The depth profiles for 25-keV nickel ions implanted into polycrystalline iron samples to a dose of 3×10^{16} ions/cm^2 have been determined using proton-excited X-ray analysis (PEX) combined with 1-keV argon-ion sputtering. Profiles obtained for the implanted samples were compared to profiles calculated on the basis of the theory of Lindhard, Scharff, and Schiott (LSS) and were found to substantially disagree with their theory. The method of using LSS theory was modified to more accurately take into account the sputtering of the iron by the incident 25-keV nickel ion beam. This resulted in good agreement between the experimental results and LSS theory.

The effects of heat treatment of the implanted samples were studied by obtaining the implant profiles after annealing at 500° C in ultra-high vacuum. The redistribution of the implanted nickel was found to disagree with Fick's Laws of Diffusion, with the centroid of the distribution shifting away from the sample surface. This behavior implied diffusion of the nickel in the surface region against the concentration gradient.

The corrosion characteristics of surface alloys were studied electrochemically under conditions of anodic polarization. Fe-6.6(s)Cr, Fe-13.3(s)Cr, and Fe-18(s)Cr alloys were studied in a buffered borate solution containing 0.4 wt-pct NaCl. Their corrosion characteristics have been found to be substantially the same as those obtained for bulk Fe-Cr alloys under the same conditions. In addition, similar polarization studies were carried out for Fe-6.6(s)Ni and Fe-6.6(s)Al, and the results are interpreted as indicating that the corrosion characteristics of the surface alloys are due to the formation of an alloy structure rather than to defect production.

INTRODUCTION

The Bureau of Mines, several years ago, initiated a program to develop alloy structures using the techniques of ion implantation. Justification for this program, in 1976 as it was in 1968, is based on the ever-increasing demand for materials constructed from stainless steels and other high-alloy steels. Virtually all of the alloying metals used in these steels, such as chromium, nickel, and cobalt, are imported by the United States, a reliance that contributes to a large balance-of-payments deficit and could, in the future, lead to critical domestic shortages of these metals. The ion implantation program at the Bureau of Mines was designed to provide one alternative method that would drastically reduce the consumption of these metals through the use of ion implantation techniques to form corrosion-resistant alloys in the surface regions of metal structures; hence the name "surface alloy."

The use of proton-excited X-ray (PEX) analysis in conjunction with ion implantation, and the use of ion implantation as a technique to alter the surface properties of metals through the formation of surface alloys, has been described previously[1,2].

This paper reports the results of studies to determine (1) the distribution of the implanted 25-keV nickel ions as a function of depth into the iron substrate, and (2) the effect of heat treatment at 500° C on this distribution. In addition, we report a brief summary of the aqueous corrosion characteristics of Fe-6.6(s)Cr, Fe-13.3(s)Cr, Fe-18(s)Cr, Fe-6.6(s)Ni, and Fe-6.6(s)Al surface alloys obtained under conditions of anodic polarization.

The concentration of the alloying element is defined here as the average concentration within two standard deviations ($\pm 2\Delta R_p$) of the mean projected range R_p of the ion. Thus, "Fe-12(s)Ni" denotes a surface alloy formed by implanting nickel ions into the surface region of iron to an average concentration of 12 atomic percent.

EXPERIMENTAL

Preparation of high-purity (99.995%) polycrystalline iron samples used in these studies consisted of mechanical polishing through 600 grit, stress relieving in an argon atmosphere at 500° C, and electropolishing in a 90% acetic acid-10% perchloric acid solution at a current density of 0.15 amp/cm^2. Following electropolishing, the samples were mounted in the ion implanter ultra-high vacuum chamber and implanted with 25-keV nickel, chromium, or aluminum ions to a predetermined dose.

Details of the experimental facility used for profiling have been described elsewhere[2,3] and only a brief description is given here. Figure 1 shows the UHV chamber in which the implant profiles were obtained and effects of heat treatment were studied. A 180-keV proton beam was collimated to 0.3 cm diameter by apertures S_1 and S_2 with aperture S_3 biased for secondary electron suppression. The emitted characteristic X-rays passed through two 2μ aluminized mylar windows, W_1 and W_2, into a gas-flow proportional counter or through one 2μ aluminized mylar window, W_3, and a 1.0-mil Be window, W_4, into a Si(Li) solid state detector. The X-ray filters are used to selectively filter out interfering X-ray lines by the introduction of gases, however, for the experiments reported here, they were maintained at 10 microns pressure. The samples were biased to +200 V for secondary electron suppression, and the proton beam current was measured by a current digitizer. For heat treatment, the surface alloys were annealed in situ by radiant heating from a tungsten filament placed behind the sample. The sample temperature was measured with a chromel/alumel thermocouple welded to the sample. The X-ray spectra were accumulated in a multichannel pulse height analyzer gated with a preset scaler connected to the output of the current digitizer.

The experimental X-ray data was obtained in terms of X-ray photons per microcoulomb of incident protons and was converted to the absolute X-ray yield, Y, expressed as X-rays/incident proton, by use of appropriate geometrical and X-ray absorption factors[2]. The X-ray yield can be converted to the number of atoms/cm^2, for very thin films, according to the equation[4]:

$$N = Y/\sigma_x \qquad (1)$$

where N is the number of atoms/cm^2, Y is the experimentally-determined X-ray yield in X-rays/proton, and σ_x is the thick-target X-ray production cross section in cm^2/atom.

For depth-profiling, the Fe-12(s)Ni surface alloy samples were incrementally sputtered away by bombardment with 1.0-keV argon ions at an angle of 30° with respect to the sample surface, with the Ni-L X-ray yield measured after each sputtering increment. The effective analyzing depth of 180-keV protons on iron is approximately 2,500 Å[5] whereas the mean projected range of 25-keV Ni$^+$ on iron is 75 Å[6]. Thus, the total amount of nickel remaining in the iron substrate following each sputtering interval can be determined using equation (1), and integral distributions of the implanted nickel can be obtained. These integral distributions can then be differentiated to obtain the depth profiles.

The corrosion characteristics under conditions of anodic polarization of both surface and bulk alloys has been studied using standard Greene cells modified to permit maximum agitation of the solution in front of the working electrode. The solution was sodium

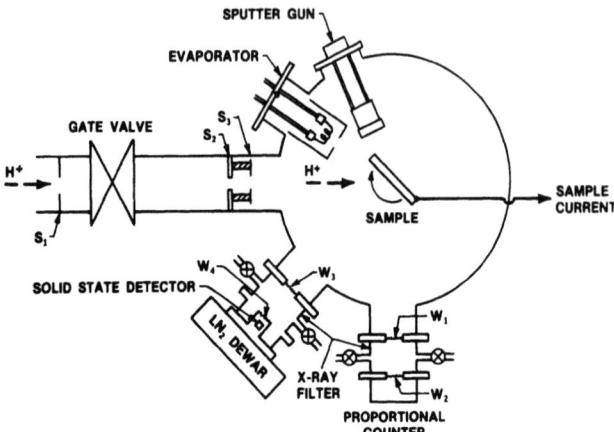

Figure 1. Schematic diagram of vacuum chamber used to profile and anneal the Fe-12(s)Ni surface alloys.

tetraborate ($Na_2B_4O_2$) buffered to a pH of 8.5 with 0.4 wt-pct NaCl added to enhance pitting corrosion in the transpassive region. The solution was deaerated with high-purity helium that flowed at a rate of 6 liters per minute. Potentials were measured with respect to a saturated calomel/luggin probe reference electrode but are reported here on the normal hydrogen electrode scale. The polarization was initiated at 100 millivolts negative with respect to the open circuit potential, and a scan rate of 100 mV/min was used.

RESULTS

For the Fe-12(s)Ni alloys, the total implant dose determined by current integration during fabrication was 2.5×10^{16} ions/cm² whereas subsequent PEX measurements (using $\sigma_x = 1.84 \times 10^{-22}$ cm² for 180-keV protons on nickel[7]) showed that the actual total implant dose was 3.0×10^{16} ions/cm². Uniformity of the implant was measured to be ±10% across the sample.

Prior to profiling the Fe-12(s)Ni alloys, the thickness of the residual air-formed oxide films and the extent of carbon surface contamination were determined by PEX. These analyses showed that the average oxide layer was 6.5 Å (±2.5 Å) on all samples. Monitoring of the C-K X-ray yield during profiling indicated that carbon was distributed throughout the implant region and was less than 6 atomic percent concentration.

Figure 2 shows the integral yield distributions obtained for Fe-12(s)Ni alloys before and after vacuum annealing at 500° C for 2,400 seconds in an ambient pressure of 1×10^{-8} torr. Both sets of

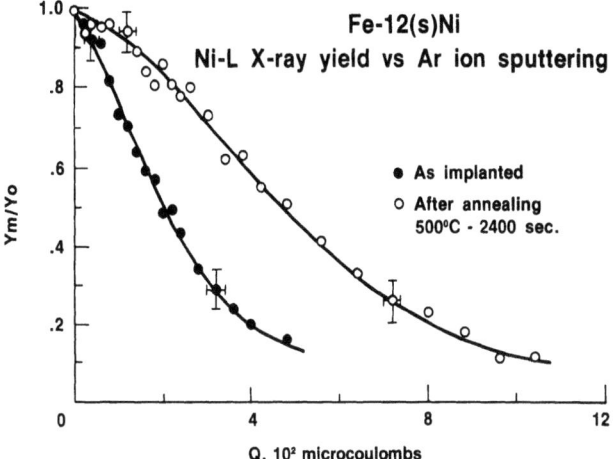

Figure 2. Experimental integral yield distributions for Fe-12(s)Ni surface alloys before and after annealing at 500° C for 2,400 seconds.

data, representing the average results for three samples, are plotted as the ratio of the Ni-L X-ray yield, Y_m, expressed in X-rays/incident proton after a total sputtering charge, Q, to the initial (zero sputtering) Ni-L X-ray yield, Y_o. The curves drawn through the data points represent fourth-order polynomial least-square-fits to the data points. The imprecision in the measurement of the X-ray yield and the sputter-beam charge is estimated to be ±7% and ±3%, respectively, with an estimated systematic error in the X-ray measurement of ±10%.

Figure 3 shows average anodic polarization curves obtained for six samples each of Fe, Fe-5Cr, Fe-12Cr, and Fe-18Cr bulk alloys, and Fe-6.6(s)Cr, Fe-13.3(s)Cr, and Fe-18(s)Cr alloys. The iron and the bulk alloys exhibit typical electrochemical behavior for this type of environment, with the addition of chromium reducing the active corrosion current (between -0.6 and -0.2 V vs. N.H.E.) and increasing the potential (pitting potential) where the transpassive region is initiated (e.g. -0.25 V for Fe-5Cr).

DISCUSSION

The argon-ion sputtering charge, Q, axis (figure 2) was converted to an absolute depth scale using a sputtering yield for iron recently obtained in this laboratory. This sputtering yield has been determined to be 4.1 atoms/ion[8] with an estimated uncertainty of ±12%. With this value, the charge axis of figure 2 was converted from microcoulombs of sputtering to depth in angstroms. Figure 4A shows the resulting integral distribution with the nickel concentration in atoms/cm^2 as a function of depth, X, into the iron. The

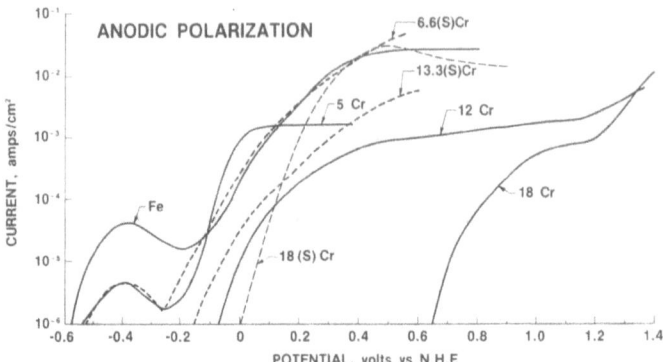

Figure 3. Results of anodic polarization studies of several bulk and surface alloys in buffered borate solutions.

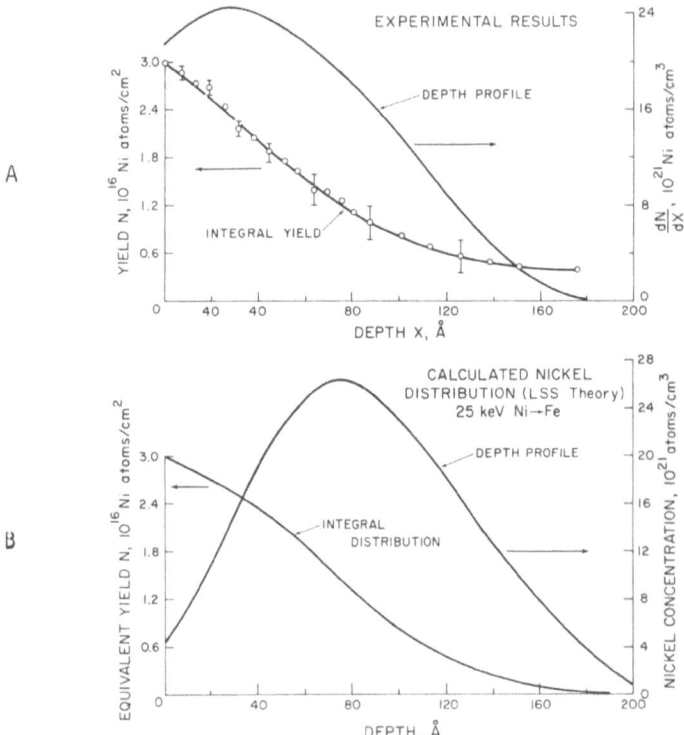

Figure 4. A. Integral yield distribution for Fe-12(s)Ni surface alloys and nickel depth profile obtained by taking the derivative. B. Theoretical[6] distribution of 25-keV Ni^+ implanted into polycrystalline iron.

absolute value of the differential, dN/dX, is the experimental depth profile of the nickel implant.

A comparison was made of the depth profile obtained for the Fe-12(s)Ni (figure 4A), with the depth profile calculated by use of LSS theory[6] for a total ion dose of 3.0 x 10^{16} ions/cm^2 (figure 4B). The comparison shows that the implanted nickel is substantially closer to the final iron surface than is predicted by theory. R_p is approximately 75 Å for the theoretical profile, whereas it is only 30 Å for the experimental profile. In order to explain this discrepancy, the effect of sputtering of the iron sample during the implantation, an effect not totally accounted for in the theoretical treatment, was considered in detail. Since we were unable to find published experimental data for the sputtering yield of iron by 25-keV Ni$^+$, we have estimated values using the self-sputtering yields for 45-keV Ni$^+$ incident on nickel (3.8 atoms/ion) and 45-keV Fe$^+$ incident on iron (3.0 atoms/ion)[9] and the fact that the sputtering yield is relatively constant for ion energies ranging from 10 to 50 keV[10]. Using these data, we have extrapolated a sputtering yield of 3.4 atoms/ion for 25-keV Ni$^+$ on iron and have calculated that approximately 100 Å of the iron sample surface could be expected to be removed during the implantation of 3 x 10^{16} Ni ions/cm^2.

To determine the effect of this sputtering on the final depth profile, we adopted a technique for using LSS theory in which the total implant dose was divided into 5 equivalent, sequential implants of 6.0 x 10^{15} atoms/cm^2, with 20 Å of the iron substrate having been sputtered away at the end of each implant. The resulting partial-implant distributions, shown in figure 5, were thus summed to yield the final distribution. There are several sources of imprecision in the final distribution that this model produces. The uncertainty in the sputtering yield for 25-keV Ni$^+$ on iron is estimated to be ±25%. In addition, with each sequential implant, the value of the instantaneous sputtering yield may change as the surface region stoichiometry changes. Finally, the effect of the total defect production at any point in the implantation sequence may appreciably affect the value of the instantaneous sputtering yield.

In figure 6, the LSS profile corrected for sputtering has been compared to the experimental profile. The agreement, clearly, is quite good, with the maximum in the theoretical profile occurring at a final sample depth of 30 Å. The difference in the amplitudes of the distribution are considered to be due to enhancement of the total uncertainty in the integral distribution through the derivative process.

For the Fe-12(s)Ni alloy annealed at 500° C for 2,400 seconds, the integral yield distribution from figure 2 was differentiated to obtain the experimental depth profile shown as the solid line in

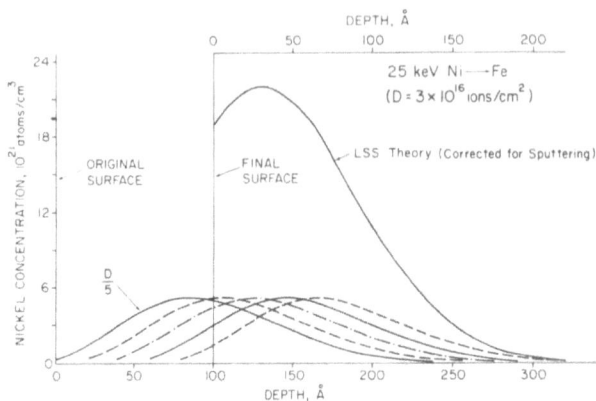

Figure 5. Method of constructing the modified theoretical distribution assuming 100 Å of iron surface sputtered away during the implanting of 3×10^{16} ions/cm² of 25-keV Ni⁺.

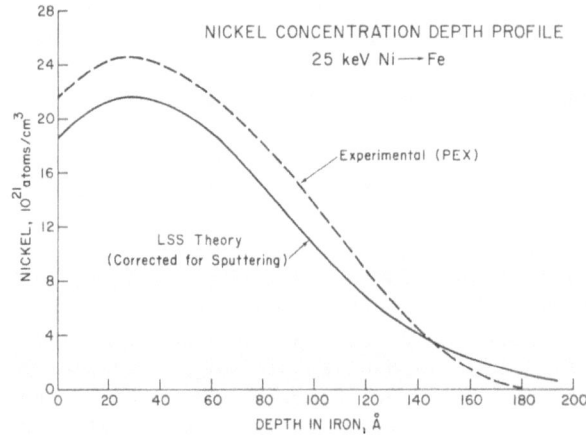

Figure 6. Comparison of modified theoretical profile with the experimental depth profile.

figure 7. Fick's Law[11] was used with the experimental unannealed profile to estimate the expected redistribution of the nickel implant. Assuming a constant diffusion coefficient, $D = 2 \times 10^{-16}$ cm²/sec[12], the second order differential equation of Fick's Law was solved using the method of Churchill[13] for a semi-infinite solid with an impenetrable surface at depth x=0. The solution is shown as the "annealed (Theor.)" curve in figure 7. We presently have no explanation for the difference between the experimental and theoretical annealed results; however, it is possible that application of a different set of boundary conditions to the solution or by developing a variable diffusion coefficient that has some dependence on radiation damage may result

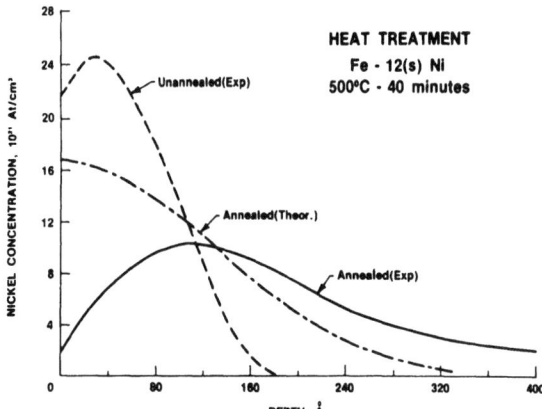

Figure 7. Comparison of experimental depth profile for heat-treated surface alloys with profile expected on the basis of Fick's Law.

in better agreement between Fick's Law and the experimental results. This is under further investigation.

The most important aspect of the annealing studies is that the experimental results clearly show a shift of the nickel profile-maximum away from the sample surface. This shift implies, at least in the outermost surface regions, a mechanism that allows diffusion against the concentration gradient.

The anodic polarization curves of figure 3 show that the Fe and Fe-5Cr exhibit active corrosion currents greater than 1 $\mu A/cm^2$, the minimum current density recorded in these experiments. The Fe-6.6(s)Cr alloy shows an active region that is essentially identical in current-density-magnitude and range-of-potential as the Fe-5Cr. The chromium surface and bulk alloys with higher chromium concentrations may have small active corrosion regions, but they are less than 1 $\mu A/cm^2$. The Fe-6.6(s)Cr and Fe-13.3(s)Cr alloys show shifts in values of the potentials in their polarization curves that are very close to those obtained for equivalent bulk alloys. In terms of the potential at which the transpassive region is initiated, the Fe-18(s)Cr is superior to the Fe-12Cr alloy; however, the subsequent transpassive dissolution currents for the surface alloys are substantially higher than either the Fe-12Cr or Fe-18Cr. The major difference between the surface and bulk alloys is their behavior in the high-potential transpassive regions where dissolution resulting from pit nucleation at defects and grain boundaries has largely destroyed the nearby regions of the implant, with resulting dissolution kinetics approximating those of iron. These and other electrochemical studies have been reported elsewhere in detail [14].

Finally, since other laboratories have demonstrated that corrosion behavior can be altered by radiation damage due to ion implan-

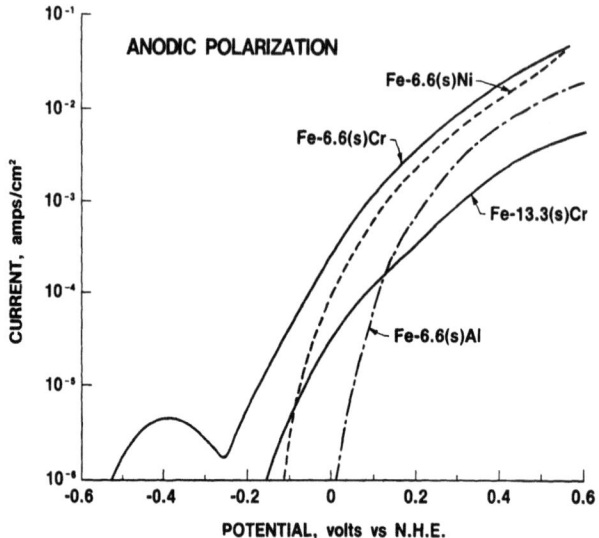

Figure 8. Anodic polarization curves for several different surface alloys.

tation[15], it was desirable to determine whether the results obtained in these experiments are due to the formation of a specific alloy structure and/or to the production of defects in the implant region. Figure 8 compares anodic polarization data obtained for two other types of surface alloys, Fe-6.6(s)Ni and Fe-6.6(s)Al, to the Fe-6.6(s)Cr and Fe-13.3(s)Cr data. If the shift in the potential at which the transpassive region begins and the reduction in dissolution current in the active region resulting from the implantation process were due to defect production, the Fe-6.6(s)Cr and Fe-6.6(s)Ni data would be expected to be essentially identical since the defect production rates for chromium and nickel ions of the same energy are approximately the same[6]. Clearly, the chromium and nickel surface alloys exhibit corrosion behavior so different as to be attributed primarily to differences in alloy characteristics. On the other hand, the defect production rate for aluminum ions could be expected to be less than that for chromium and nickel ions. Again, the anodic polarization behavior of the Fe-6.6(s)Al compared to that of the Fe-6.6(s)Cr and Fe-6.6(s)Ni is contrary to what would be expected on the basis of enhanced corrosion resistance due to defect production.

CONCLUSIONS

Depth profiles have been reported for 25-keV Ni^+ implanted into polycrystalline iron. These profiles were obtained using proton-excited X-ray analysis with 1-keV argon ion sputtering.

The experimentally-determined profile showed substantial disagreement with the profile predicted by LSS theory, principally that the maximum in the implant profile occurred at 30 Å whereas LSS predicted 75 Å. A simple correction for sputtering of the iron substrate by the nickel ions was made to sequential LSS calculations, with the result that good agreement was obtained between theory and experiment.

The redistribution of the implanted nickel ions following high-vacuum heat treatment at 500° C for 2,400 seconds was shown to disagree with Fick's Laws of Binary Diffusion. There was a pronounced shift of the profile maximum away from the sample surface with implications of diffusion against the concentration gradient. This phenomenon is under further investigation.

Anodic polarization studies have shown that the chromium surface alloys have corrosion resistance comparable to that of bulk alloys and that the corrosion resistance of the surface alloys is the result of alloying rather than defect production due to the implantation process.

REFERENCES

1. P. B. Needham, Jr., Proc. of Second Oak Ridge Conf. on the Use of Small Accelerators for Teaching and Research, CONF-700322, USAEC, 155 (1970).
2. P. B. Needham, Jr., and B. D. Sartwell, Advances in X-Ray Analysis, 14, 184 (Plenum Press, 1971).
3. P. B. Needham, Jr., H. W. Leavenworth, and T. J. Driscoll, J. Electrochem. Soc., 120, 778 (1973).
4. L. J. Christensen, J. M. Khan, and W. F. Brunner, Rev. Sci. Instr., 38, 20 (1967).
5. R. G. Musket and W. Bauer, Thin Solid Films, 19, 69 (1973).
6. K. Bruce Winterbon, Ion Implantation Range and Energy Deposition Distributions, Vol. 2 (Plenum Press, 1975).
7. P. B. Needham, Jr., and B. D. Sartwell (unpublished).
8. P. B. Needham, Jr., and T. J. Driscoll, App. Phys. Lett. (to be published).
9. G. Dearnaley, J. H. Freeman, R. S. Nelson, and J. Stephen, Ion Implantation, (American Elsevier, 1973), p. 209.
10. G. Carter and J. S. Colligan, Ion Bombardment of Solids, (American Elsevier, 1968).
11. J. Crank, The Mathematics of Diffusion, (University Press, 1957).
12. The Handbook of Chemistry and Physics, 53rd Edition (Chemical Rubber Company, 1973).
13. R. V. Churchill, Fourier Series and Boundary Value Problems, (McGraw-Hill, 1963).
14. P. B. Needham, Jr., B. D. Sartwell, and B. S. Covino, Jr., J. Electrochem. Soc. (to be published).
15. B. L. Crowder, ed., Ion Implantation in Semiconductors and Other Materials, (Plenum Press, 1973), pp. 405-442.

ION IMPLANTATION INDUCED DISORDER IN Ni STUDIED BY RUTHERFORD

BACKSCATTERING AND ELECTRON MICROSCOPY

J. S. Williams, R. Andrew, C. E. Christodoulides,
W. A. Grant, P. J. Grundy* and G. A. Stephens*

Department of Electrical Engineering, University of Salford
Salford M5 4WT, U.K.

ABSTRACT

The disordering effects of Dy^+, Pb^+ and Xe^+ ions implanted into Ni have been investigated by Rutherford backscattering and transmission electron microscopy. Following room temperature implantation of Ni with Dy, RBS analysis indicates a disorder structure characteristic of implanted semiconductors which are rendered amorphous by ion bombardment. TEM confirms the formation of a non-crystalline Ni phase following Dy ion bombardment. No such amorphous Ni phase was observed for Pb implantations. The annealing of the implantation-induced Ni disorder has been examined in some detail and the results emphasise the powerful approach that a combination of RBS and TEM offers for analysis of ion implanted metals.

INTRODUCTION

The wide application of implantation in device technology has prompted extensive investigation into ion-implantation-induced damage in semiconductors and the disordering processes which ultimately lead to the production of an amorphous layer after prolonged bombardment are relatively well understood. Ion implantation effects in metals have not been widely studied since only recently has interest been stimulated by possible applications.

---

*Department of Pure and Applied Physics, University of Salford, Salford M5 4WT, U.K.

As a consequence, the current understanding of this field is not comprehensive, but it is nevertheless clear that ion-induced disorder in metals is quite different to that in semiconductors: the mobility of point defects in metals and the non-directionality of bonding can promote annihilation of defects or segregation into dislocation loops and clusters and thus preserve the basic crystallinity of the lattice during ion bombardment.

A recent study[1] of Dy-implanted Ni by Rutherford backscattering and channelling (RBS) has shown the existence of a channelling disorder peak which is uncharacteristic of metals and usually associated with amorphization in semiconductors. The preparation of non-crystalline metal films, and Ni in particular, is possible by employing well established processes such as evaporation onto liquid-nitrogen-cooled substrates.[2] These non-crystalline metal phases are stable to above room temperature which leads to speculation as to the properties of the damaged layer observed by RBS for Dy-implanted Ni.

This paper presents the results of RBS and TEM investigations into the nature of ion implantation-induced disorder in Ni. The disorder arising from implantation with Dy^+, Pb^+ and Xe^+ ions at energies up to 200 keV has been monitored as a function of anneal temperature and it is shown that under certain circumstances non-crystallinity can be induced in Ni by ion bombardment.

EXPERIMENTAL

Single crystal Ni slices of <100>, <110> and <111> orientations were employed for RBS and channelling analysis. These were prepared for implantation by mechanical and chemical polishing to remove surface damage. Samples which were to undergo TEM examination were thinned from large-grain polycrystalline Ni foils (jet electropolished to perforation using Lenior's solution) prior to implantation. The most common orientations with respect to the foil plane were <110>, <112> and <013>. A few Ni evaporated films (deposited on NaCl at room temperature) were also prepared for implantation and subsequent TEM analysis.

All specimens were implanted at room temperature with either Dy^+, Pb^+ or Xe^+ ions at doses ranging from 5×10^{14} cm^{-2} to 5×10^{16} cm^{-2} and at energies up to 200 keV. The flux was kept below 1 µA cm^{-2} to minimize beam heating effects during implantation. The Dy and Pb implanted samples were analysed either by 2 MeV He^+ backscattering or with 100 keV electrons in a transmission microscope following ½ hour isochronal annealing cycles in a dry Ar atmosphere in the temperature range 25-850° C.

RESULTS AND DISCUSSION

The RBS measurements of Ni disorder for 5×10^{15} Dy cm^{-2} (20 keV) implantations have been previously reported.[1] These results, together with the more extensive data obtained in the present study, are summarised below.

(i) For Dy, Pb and Xe implantations the post-bombardment aligned spectra exhibited a pronounced near-surface disorder peak. The build-up of disorder with increasing dose was measured for Dy implantations, but this data is not particularly relevant to the present study and shall be reserved for a later, more detailed account of the backscattering results.[3] It is, however, important to note here that the magnitude of the Dy-implant disorder was measurably greater than that for Pb (i.e. comparing similar 20 keV, 5×10^{15} cm^{-2} implants).

(ii) A reproducible 20% increase in the measured disorder was observed for Dy-implanted Ni annealed to 300°C. No change in disorder was observed in Pb-implanted Ni in this temperature range.

(iii) For both Dy and Pb implants, the disorder peak was observed to decrease to the near-virgin level in the temperature range 300-600°C. This annealing was observed to initiate from the bulk and progress towards the surface.[1]

(iv) A marked reduction in the dechannelling level (rate) behind the disorder peak was noted for Dy-implanted Ni in the temperature range 650-850°C (insufficient data for Pb implants).

(v) For 20 keV Dy implants, the disorder and implant profiles had a similar depth distribution,[1] whereas high energy implants (80 keV Dy and 200 keV Xe) analysed in the present work, appear to result in a disorder peak located within 100 Å of the Ni surface and thus much shallower than the peak of the implant distribution.

Low angle RBS was employed to provide increased depth resolution[4] and showed that the initially Guassian-shaped Dy and Pb implant profiles were substantially modified with annealing temperature. Pb was observed to migrate (in apparent entirety for doses > 5×10^{15} cm^{-2}) to the target surface at a temperature of ~400°C and Dy was found to outdiffuse gradually in the temperature range 650-800°C. Variation in profile shape depending on channelled or random implants and corresponding differences in substitutionality were also detected, but such effects will be reported more fully elsewhere.[3]

Low angle channelling measurements[5] were also made to permit investigation of the disorder with enhanced depth resolution and provided the observations in (v) above. This type of analysis also helped to identify important differences between 20 keV Dy and Pb induced disorder in Ni. This is illustrated in Fig. 1, where low angle channelled and random profiles are shown for Dy and Pb implantations. A shift to higher backscattered energies can be noted for some aligned distributions (indicated by a ΔE separation in the tail of the aligned/random profiles) and this results from a lower stopping power for channelled He^+ ions. This effect is accentuated for low angle channelling, since the geometry is such that[4] the incident (channelled) path length is much longer than the scattered(random)path length. In Fig. 1 it is most interesting to note the absence of a stopping power effect in the room temperature Dy aligned profile, but a marked shift in the corresponding Pb aligned profile. This would appear to imply a basic difference between the Dy and Pb induced disorder, although both implanted species give rise to the same observation of a disorder peak (in (i) above). The Dy profile effect would be expected if the implanted Ni layer was totally disordered (c.f. amorphicity in semiconductors) or distorted such that no channelling resulted, whereas the Pb profile shift indicates a substantial channelling effect and implies that the implanted Ni layer is still basically (single) crystalline. Note that, by $600°C$, the aligned Dy profile exhibits the ΔE shift, consistent with the reduction in disorder peak area identified in (iii).

TEM analysis has revealed important differences between the disordering behaviour of Dy and Pb implanted into Ni. Treating the more interesting Dy implantations first, the typical as-implanted features are shown in Fig. 2 for a 1×10^{16} Dy^+ cm^{-2} (20 keV). The bright field micrograph (Fig. 2a) apparently shows a high density of damage clusters. The most interesting feature to note is the appearance of a single diffuse scattering maximum in the diffraction pattern (Fig. 2b) in addition to Ni matrix reflections (<123> orientation) due to undamaged crystal lying beyond the ion range. This diffuse ring is $\sim 3\%$ smaller in diameter than $\{111\}$ Ni and is indicative of a non-crystalline or microcrystalline state. The width of the ring suggests very small scattering centres, which is confirmed by the high resolution dark field (h.r.d.f.) micrograph (Fig. 2b) of the same region, taken from the diffuse ring. Other as-implanted samples resulted in up to 5% contraction of the inner diffuse ring (from $\{111\}$ Ni) and in such cases no resolvable diffracting centres could be obtained by h.r.d.f., indicating centres < 10 Å in size. Such samples were unstable under the conditions of observation and after a few minutes the more stable situation represented by Fig. 2 was reached. The above observations, together with more detailed results reported elsewhere,[6,7] are consistent with TEM observations of thin, splat-cooled Ni films which have been defined as non-crystalline.[2] Thus, we are led to interpret our data as

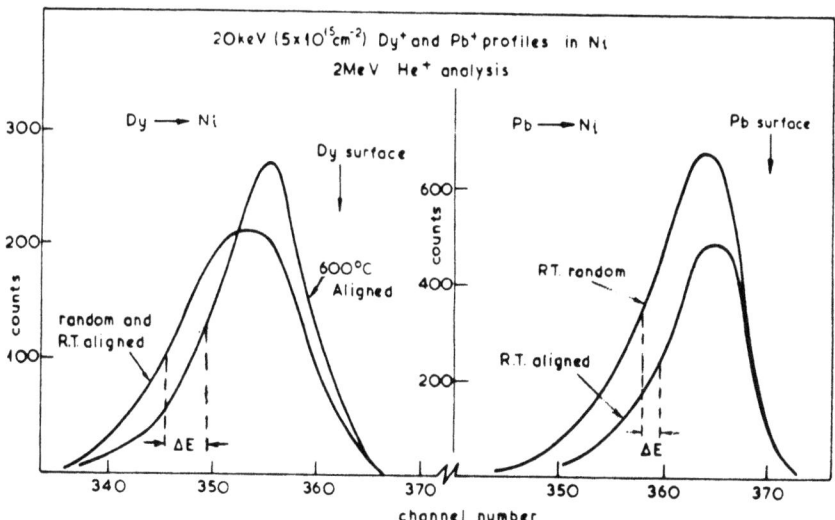

Fig. 1 Low angle (∅ = 80°) random and aligned profiles for Pb and Dy in Ni

evidence for the production of quasi-amorphous Ni surface layers by ion implantation.

Heating of Dy implanted samples resulted in the diffuse ring becoming wider and sharper, until by 300°C a good {111} Ni fit was obtained with the added appearance of {200} and {220} Ni rings. This behaviour is illustrated in Fig. 3 which gives TEM data of the same sample as in Fig. 2, following annealing at 300°C. The bright field micrograph (Fig. 3a) shows some coarsening in appearance, but the diffraction pattern (from a near <123> region) in Fig. 3b illustrates the sharpening of the diffuse Ni ring. This is indicative of a possible quasi-amorphous to microcrystalline transition which is confirmed in the h.r.d.f. micrograph where larger diffracting centres (∼50 Å) are clearly resolved.

Further annealing of Dy implanted Ni indicated a restoration of basically single crystal Ni by 600°C as revealed by the gradual disappearance of Ni rings in the diffraction pattern. Some evidence for Dy-Ni compound formation[7] was obtained at 600-700°C, but the main observations of heating to 900°C were that the large density of extended defects annealed out and clearly identifiable Dy_2O_3 rings began to appear.

TEM observations of Pb implanted Ni showed no evidence for amorphicity or even microcrystallinity resulting from Pb bombardment for doses up to 10^{16} Pb cm^{-2} (20 keV). As-implanted Ni revealed a basically single crystal Ni lattice with Ni spot

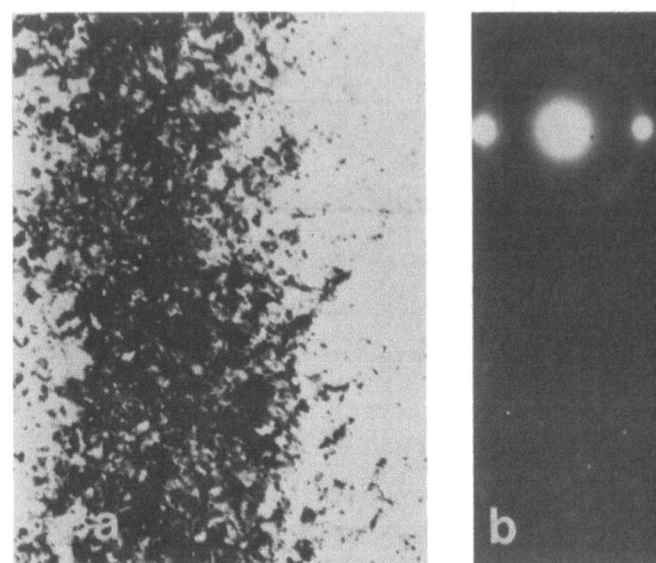

Fig. 2 Bright field micrograph (a) and corresponding diffraction pattern and dark field micrograph (b) for 20 keV Dy implant in Ni to a dose of 1×10^{16} cm^{-2}. Analysis immediately following implantation.

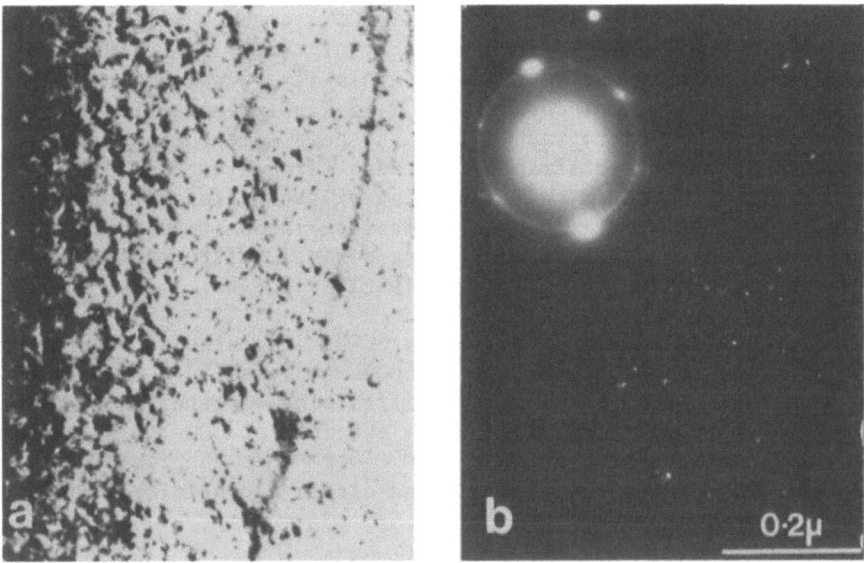

Fig. 3 Bright field micrograph (a) and corresponding diffraction pattern and dark field micrograph (b) for same sample as previous figure following annealing at 400°C.

streaking and splitting in the diffraction patterns, presumably
arising from 'strain' effects. Some evidence for Pb precipitation
was observed. Heating to 400°C resulted in the appearance of
dislocation loops and regular patches in the bright field micrograph (as illustrated in Fig. 4a) whereas only ill-defined 'strain'
centres had appeared previously. The corresponding diffraction
pattern showed the appearance of extra spots which could be
readily identified as Pb. This is illustrated in Fig. 4b, where
the diffraction pattern of a Ni <013> region reveals extra spots
arising from Pb <112>. Thus, Pb would appear to be growing
epitaxially with $[013]$ Ni $\parallel$ $[112]$ Pb. Identification of the
patches in Fig. 4a as Pb is confirmed in the h.r.d.f. (Fig. 4b)
which has been taken from the $[11\bar{1}]$ Pb reflection. Note that
only some of the Pb patches appear in this reflection - others are
in the twin orientation. Further annealing results in the loss of
Pb and in the annealing out of extended defects in the Ni by 850°C.

Preliminary TEM investigations of Dy and Pb implanted into
evaporated Ni substrates indicated a tendency for a non-crystalline

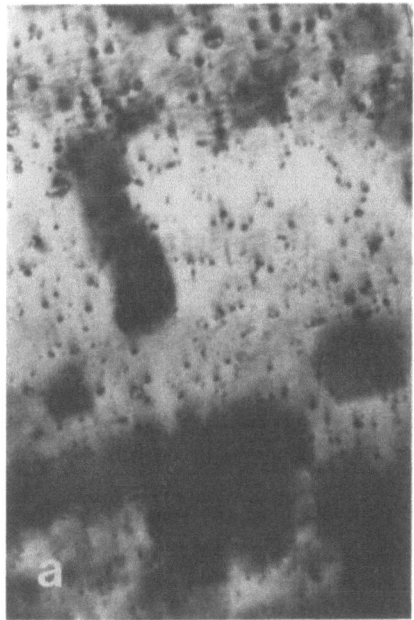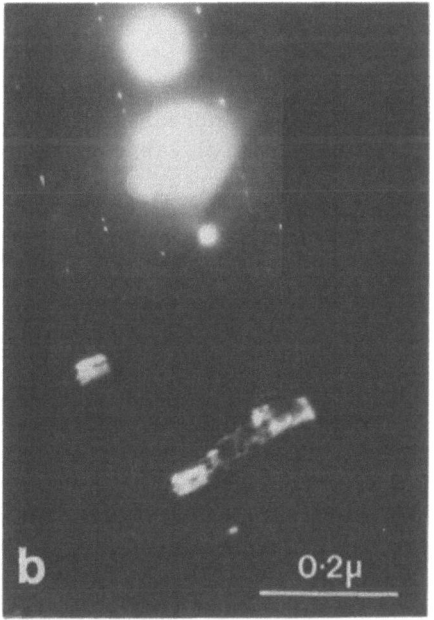

Fig. 4   Bright field micrograph (a) and corresponding diffraction
         pattern and dark field micrograph (b) for Pb-implanted Ni
         (1 x 10^{16} cm^{-2}, 20 keV) following annealing at 400°C.

phase to be formed for Dy-implantation, but no loss of crystallinity was observed for Pb implantations.

When viewed as a whole, the RBS and TEM data reveal the excellent correlation between the two methods of analysis. This is clearly shown from the summary table below (Table I) where the major RBS and TEM results for the annealing of Dy and Pb implanted Ni samples are presented. TEM analysis shows that the RBS disorder peaks obtained following implantation for both Dy and Pb implanted Ni arise from totally different forms of Ni lattice disruption (quasi-amorphous in the case of Dy and highly-strained single crystal in the case of Pb). This emphasises that considerable care should be exercised when using RBS analysis alone to monitor lattice disorder,[8] although, with the aid of low-angle channelling (Fig. 1), it was possible to identify

TABLE I A comparison of the most significant RBS and TEM observations as a function of annealing temperature for Dy and Pb implanted Ni

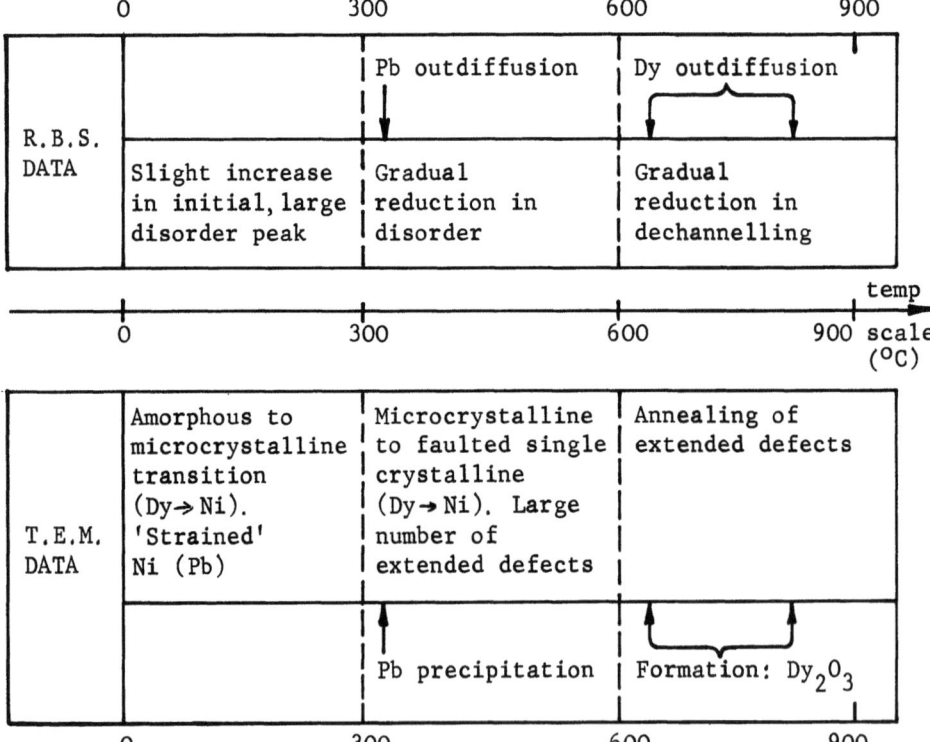

apparent differences in the Ni lattice disorder following Dy and Pb implantations. Good correlation of the reduction in RBS disorder peak area with a microcrystalline-to-single crystal Ni transition (400-600°C) and of a reduction in the dechannelling level with annealing of extended defects (600-850°C) is further evidence of the usefulness of combined RBS and TEM analysis.

An attempt has been made to understand the difference between the as-implanted Ni disorder for Dy and Pb bombardment by an examination of existing models[9,10] for non-crystallinity. It would appear that the models can be divided into two types: those based upon the random packing of hard spheres with no appreciable order beyond the first neighbour and those which suggest that a quasi-amorphous solid is a microcrystalline arrangement with localized energy barriers preventing adjoining small crystallites from recrystallisation. The former models usually involve the presence of an impurity to stabilize amorphicity and a recent attempt at providing a theoretical framework for such a model has been made by Nagel and Tauc.[9] When applied to our Dy-Ni and Pb-Ni systems Nagel and Tauc's model predicts, as we have previously discussed,[6] the observed implantation effects (quasi-amorphicity for Dy implanted Ni, but no non-crystalline phase for Pb implanted Ni). A recent proposal of a microcrystallinity model by Wang and Merz[10] suggests that small crystallites may take up any one of a number of polymorphic forms and have correlated bonding anisotropy with polymorphicity. Following this approach Ni and Fe are considered to be the metals most likely to be stable in a non-crystalline phase - no requirement of the presence of a stabilizing impurity is assumed here. The interpretation of our results in terms of models for amorphicity is discussed in more detail elsewhere.[6]

CONCLUSIONS

(i) The production of a quasi-amorphous Ni phase following Dy ion implantation has been identified and the subsequent annealing characteristics examined in some detail.

(ii) Room temperature stability of this amorphous Ni phase is in some way dependent upon the implanted species employed to create the disorder.

(iii) The combination of RBS and TEM techniques has been shown to be a powerful approach for investigating ion-implantation-induced disorder in metals.

(iv) Considerable care is needed when interpreting RBS (channelling) disorder peaks and the employment of another comparative technique (e.g. TEM) is most desirable.

The authors are indebted to Dr E Johnson and Professor L T Chadderton at the University of Copenhagen for the use of facilities and invaluable discussion and to Professor G Carter for stimulating this study by suggesting the possibility of metal amorphization by ion implantation. Both NATO and the SRC are gratefully acknowledged for financial support.

REFERENCES

1. G. A. Stephens, E. Robinson and J. S. Williams, Proc. Int. Conf. Ion Implantation in Semiconductors, Osaka (1974) Publ. Plenum Press, N.Y. (1975) p. 375
2. L. B. Davies and P. J. Grundy, Phys. Stat. Sol. (a) $\underline{8}$,189 (1971)
3. J. S. Williams and G. A. Stephens, to be published.
4. J. S. Williams, Nucl. Instr. Meth. $\underline{126}$, 205 (1975)
5. J. S. Williams, Proc. Int. Conf. Ion Beam Surface Layer Analysis Karlsruhe, Germany (1975) Publ. Plenum, N.Y. (1976) p. 223
6. R. Andrew, W. A. Grant, P. J. Grundy, J. S. Williams and L. T. Chadderton, Nature, to be published (1976)
7. R. Andrew, Phys. Stat. Sol. (a), to be published (1976)
8. Y. Quere, Rad. Effects, $\underline{28}$, 253 (1976)
9. S. R. Nagel and J. Tauc, Phys. Rev. Letters, $\underline{35}$, 380 (1975)
10. R. Wang and M. D. Merz, Nature $\underline{260}$, 35 (1976)

NITROGEN IMPLANTATION INTO MOLYBDENUM:

SUPERCONDUCTING PROPERTIES AND COMPOUND FORMATION

G. Linker and O. Meyer

Institut für Angewandte Kernphysik

Kernforschungszentrum Karlsruhe

ABSTRACT

The implantation of nitrogen ions into evaporated molybdenum layers caused an increase of the superconducting transition temperature, T_c from $T_c < 1.2K$ up to T_c (max) = 9.2 K with increasing N concentration up to about 23 at %. Besides the Mo bcc phase, an amorphous-like phase as well as a fcc and a fct phase have been observed in the implanted layers. The results indicate that the T_c enhancement is due to the disordered Mo phase. This disordered phase is stabilized by the nitrogen atoms probably forming radiation induced impurity-defect complexes.

INTRODUCTION

Bulk molybdenum has a bcc structure and is a superconductor with T_c = 0.92 K. Thin films of the element often exhibit different structural and superconducting properties to the bulk material depending on the preparation processes. Vapour deposition on cryogenic substrates (1) leads to the formation of an (quasi-) amorphous structure and to an increase of T_c. Mo films produced by sputtering with an intense beam of inert gas ions (2) also revealed an enhancement of T_c, accompanied by a sizable lattice expansion but no additional phases were detected. A fcc phase however has been observed in evaporated and sputtered Mo films (3) the enhanced T_c-values ($T_c \simeq 6.2$ K) in these films were attributed to a polymorphous modification of molybdenum. Structural transformations in Mo from bcc to fcc have been observed after argon and nitrogen ion bombardment (4,5) and an increase of T_c in Mo layers implanted with different ions has been reported (6,7).

The aim of the present paper is to find out how the observed increase of T_c is correlated with radiation induced disorder and with structural transformations.

EXPERIMENTAL

The Mo-layers were deposited onto heated ($\sim 900°C$) quartz substrates by electron beam evaporation at a rate of about 100 Å/sec and a vacuum of 10^{-7} Torr was maintained during evaporation; these deposition conditions were necessary to produce high purity layers. Layer thickness, which was monitored during evaporation with a quartz oscillator, was typically 1500-3000 Å.

Implantations were performed at room temperature and into liquid helium cooled layers to maximum fluences corresponding to 33 at. % nitrogen. In order to obtain a constant concentration of nitrogen over the range of interest several different ion energies were used; the fluences for each energy were determined by the calculation of added depth distributions using values of projected range and straggling from Johnson and Gibbons (8). The actual distribution, layer thickness and purity have been analyzed by Rutherford backscattering (9) of 2 MeV $^4He^+$ ions while structural information has been obtained by diffraction measurements with a thin film X-ray camera. The superconduction transition temperatures were determined resistivity using a standard four point probe arrangement. The lowest temperature attainable in our cryostat was 1.2 K.

RESULTS

All the as-evaporated Mo-layers had the bcc structure. A residual resistance ratio of 4 or better was found for all layers and has been considered as a measure of the layer purity which is the essential condition for obtaining reproducible results as has been shown earlier (7).

The nitrogen implantations lead to an enhancement of the transition temperature of the layers. This enhancement is shown in Fig. 1 where T_c values are plotted as a function of nitrogen concentration for both room temperature and low temperature implantations. For both sets of implantation conditions an increase in T_c with increasing nitrogen concentration is observed until saturation occurs at about 23 at. % nitrogen. The curves show a similar shape up to a dose of 12 at. % nitrogen, at which point the saturation region starts for the room temperature implants. An additional T_c-increase is observed for the low temperature implantations between 12-20 at. % nitrogen. The maximum detected T_c-values were 7.0 K and 9.2 K for the room and low temperature

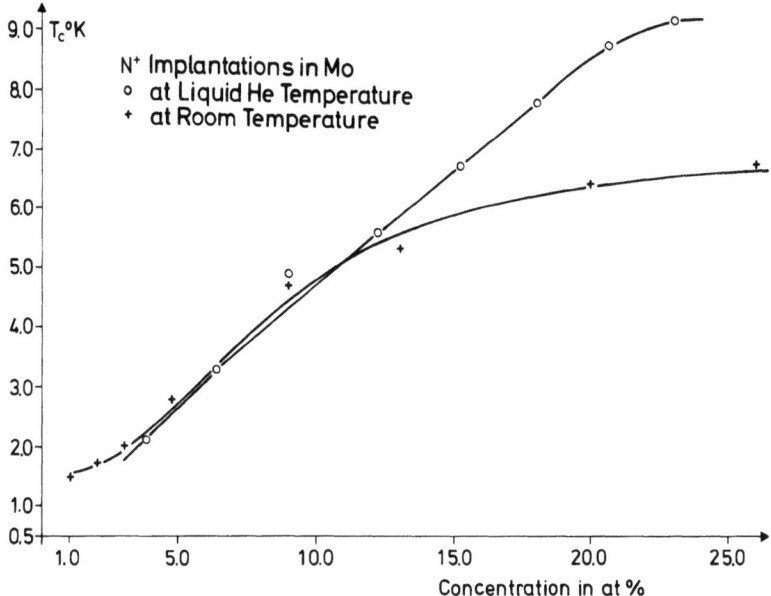

Fig. 1 Increase of the superconducting transition temperature T_c as a function of implanted nitrogen concentration at liquid He- and room temperature implantations.

implantations respectively. Synchronously with the observed T_c-increase the residual resistivity ratio of the implanted Mo-layers decreases. Structural modifications were found to depend on the nitrogen concentration, on the implantation temperature and on the temperature in a subsequent annealing procedure.

Room temperature implantations with small nitrogen concentrations (up to 5 at. %) lead only to a line broadening of the reflections from the bcc-Mo phase. Above a dose of 10 at. % nitrogen, however, additional lines attributable to a fcc Mo phase were detected in the diffraction patterns. With increasing nitrogen concentration not only does the number of lines from this fcc-phase increase but also the lattice parameter expands from 4.15 Å (10 at. %) to 4.21 Å (33 at. % nitrogen). Structural investigations for layers implanted at liquid He temperature were restricted to concentrations where T_c-saturation had been achieved and were performed for samples already annealed to room temperature. Here only the strongest Mo-bcc line (110) has been observed. This line exhibits considerable broadening when compared with the equivalent line from an unbombarded layer; a result which indicates that highly disordered, amorphous-like structures stable up to room temperature are produced at the low temperature bombardments.

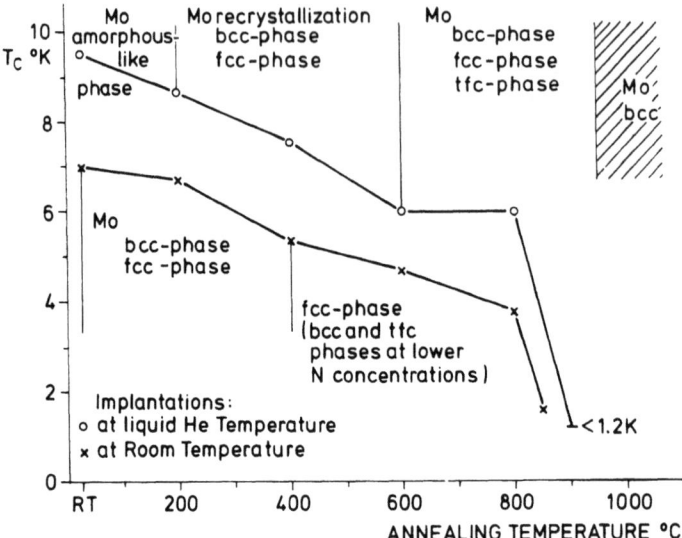

Fig. 2 Decrease of the superconducting transition temperature T_c and structural transformations as a function of temperature in an isochronal (1 h) annealing process. Samples were implanted till T_c-saturation at liquid He temperature and with 33 at. % nitrogen at room temperature respectively.

Annealing of the implanted samples effects the transition temperature and the structure of the layers. Some representative results are illustrated in Fig. 2 where the T_c-values as a function of annealing temperature are shown in an isochronal annealing process (the structural alterations are included in a parametric presentation). The results stem from two samples, one implanted at liquid He temperature until T_c-saturation was reached (~ 23 at.%) and the other at room temperature with 33 at. % nitrogen. For both samples T_c decreases continuously as a function of annealing temperature up to about 900°C where a rapid reduction to below 1.2 K is observed. The amorphous-like structure generated at low temperature is conserved up to room temperature. It recrystallizes between 200 and 600°C whereby lines from bcc Mo and fcc-phases emerge in the diffraction patterns. Above 600°C the bcc-phase is recrystallized and a weakening of the fcc-lines is observed, however some additional lines attributed to a molybdenum face centered tetragonal (tfc) phase are detected. Above 1000°C only bcc-Mo is observed. The disappearance of the face centered phases above 1000°C is related to nitrogen outdiffusion from the samples; an effect that has been detected from backscattering measurements.

For room temperature implantations bcc- and fcc-phases are present simultaneously in the as-implanted layers. In the sample shown in Fig. 2 above 400°C only reflections of a pure fcc-phase are observed; for samples with lower nitrogen content than that presented, above 400°C similar to the results described for the liquid He implanted layers also bcc and depending on concentration, fct-lines are observed. Above about 1000°C again the face centered phases disappear together with nitrogen loss from the samples. It should be noted that T_c decreases below 1.2 K in the temperature region where the fcc as well as the fct phase are still present.

Channelling experiments performed on nitrogen and neon implanted Mo single crystals support the idea of a characteristic lattice-distortion closely related to the chemically active dopant as has been suggested previously (7). The results from these measurements are illustrated in Fig. 3. Here backscattering spectra of a pure Mo-crystal aligned in [100]-direction and randomly oriented are shown together with spectra of aligned crystals implanted with nitrogen and neon ions. The spectrum from the Ne implanted sample shows an almost linear increase of dechannelling in the range of the implanted ions indicating the presence of long range order distortions like e.g. dislocations. The spectra from the nitrogen implanted samples however reveal well resolved disorder peaks originating from direct backscattering of He ions from displaced Mo-atoms probably bound to the displaced positions by the nitrogen through complex formation. The behaviour of the disorder peak for the 8 at. % nitrogen implant in an isochronal (1h) annealing process is shown in Fig. 4. Here [100]-aligned backscattering spectra are shown for three different temperatures. After annealing at 400°C nearly no change in the defect distribution is visible, at 800°C however the defect peak decreases whereas a strong peak emerges at the crystal surface. At 900°C the defect peak disappeared whereas a small surface peak is still present.

DISCUSSION

The results described above demonstrate that the T_c-enhancement in the implanted Mo-layers correlates with the formation of disorder- and fcc-phases, and the lattice constant of the fcc-phases is found to be in the range of Mo_2N.

Several arguments however can be given to support the idea that nitride formation is not the primary reason for the T_c increase. The highest T_c values were found for low temperature implants, where only the disordered fcc phase was found to exist. Further, T_c was found to decrease continuously as a function of annealing temperature while the fcc phase emerges more distinctly in the X-ray patterns. Finally T_c decreases below 1.2 K while the fcc and fct phases are still present. The latter results are supported by the channelling measurements where we believe that

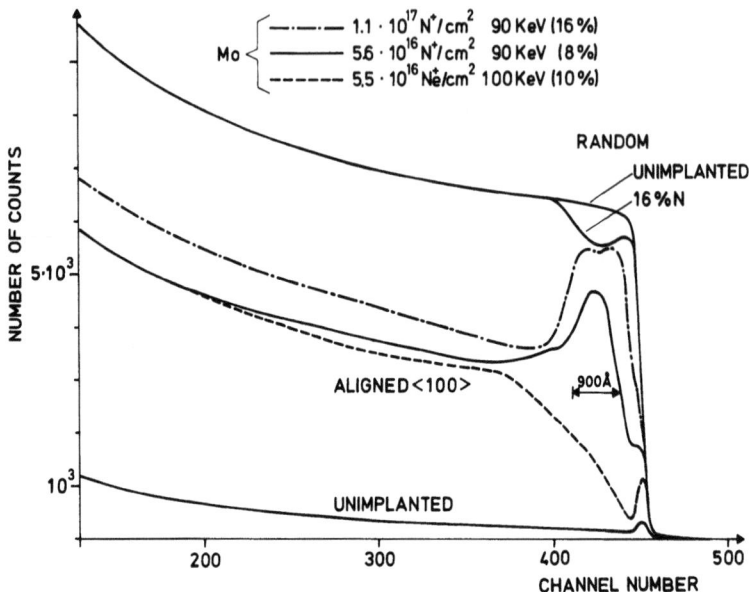

Fig. 3 Backscattering spectra with 2 MeV ^{4}He$^+$ ions from aligned and randomly oriented Mo single crystals; the aligned spectra from nitrogen and neon implanted samples clearly reveal the different dechannelling behaviour in the region of the implanted ions.

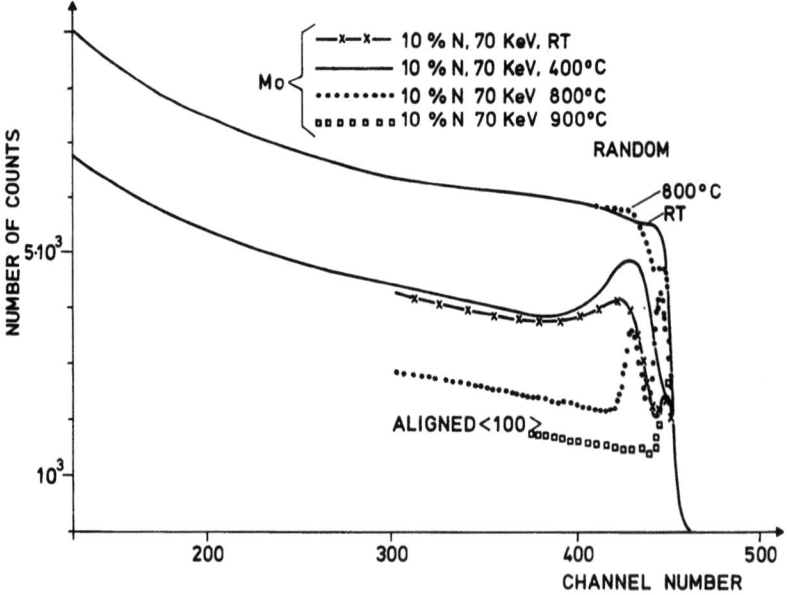

Fig. 4 Backscattering spectra with 2 MeV ^{4}He$^+$ ions from a nitrogen implanted Mo single crystal after three annealing stages in an isochronal annealing process.

a nitride phase has been formed at 800°C at the surface and where the disorder peak is found to anneal out between 800° and 900°C coinciding with the T_c decrease below 1.2 K in the implanted layers.

From these results we conclude that the T_c-enhancement in the nitrogen implanted Mo-layers is due to the disordered Mo phase, this phase being stabilized by the chemically active nitrogen possibly by the formation of Mo-N complexes. An indication that such a radiation induced formation of complexes occurs has been already found in oxygen contaminated Mo-layers where a T_c increase has been observed by noble gas ion bombardment at low temperatures (4). In layers having a small amount of contamination (<0.5 %) however the generation of intrinsic defects by either implanting noble gas atoms or by bombarding through the layers did not increase the T_c above 1.2 K. Also no structural transformations were observed in these experiments with Ne or Ar irradiation a result which is not in agreement with the observations of Pavlov (4). We believe that the formation of these radiation induced defect complexes might explain the T_c enhancements reported for Mo layers formed or treated with energetic ion beams.

ACKNOWLEDGEMENTS

The authors would like to thank M. Kraatz and R. Smithey for their help in performing the experiments.

REFERENCES

1. Collver, M.M. and Hammond, R.H., Phys. Rev. Lett. __30__, 92(1973).

2. Schmidt, P.H., Castellano, R.N., Barz, H., Cooper, A.S., and Spencer, E.G., J. Appl. Phys. __44__, 1833 (1973).

3. Chopra, K.L., Randlett, M.R. and Duff, R.H., Phil.Mag. __16__, 261 (1967).

4. Pavlov, P.V., Tetelbaum, D.I., Pavlov, A.V. and Zorin, E.I., Dokl. Adad. Nauk. SSR __217__, 330 (1974).

5. Bykov, V.N., Troyan, V.A., Zdorovtseva, G.G. and Khaimovich, V.S., phys. stat. sol. (a), __32__, 53 (1975).

6. Meyer, O., Mann, H., and Phrilingos, E., Application of Ion Beams to Metals, p. 15, Plenum Press, New York and London (1973).

7. Meyer, O., Application of Ion Beams to Materials, p. 168. The Institute of Physics, Conf. Ser. Numb. 28, London and Bristol (1976).

8. Johnson, W.S. and Gibbons, J.F., Projected Range Statistics in Semiconductors, Stanford, California (1970).

9. Mayer, J.W., Eriksson, L. and Davies, J.A., Ion Implantation in Semiconductors, p. 136, Academic Press, New York and London (1970).

ION TRAPPING, SPUTTERING AND STRUCTURAL CHANGES IN O_2^+ AND N_2^+ BOMBARDMENT OF POLYCRYSTALLINE ALUMINUM FILMS

O. Auciello, R. A. Baragiola, E. R. Salvatelli and

J. L. Spino. Centro Atómico Bariloche[**] and Instituto

de Física "Dr. J. A. Balseiro",[†] 8400 Bariloche, Argentina

ABSTRACT

We have studied ion trapping and sputtering during bombardment of polycrystalline aluminum films with 30 keV O_2^+ and 26 keV N_2^+ ions by using a quartz-crystal resonator technique; and the resulting compounds, Al_2O_3 and AlN, by transmission electron microscopy.

I - INTRODUCTION

Ion implantation of oxygen and nitrogen ions into metals and semiconductors leads usually to the formation of compounds. These compounds are important in secondary-ion mass spectrometry since they give a high yield of sputtered ions[1]. Interest in the formation of layers of oxides and nitrides comes also from their possible application in microelectronics; this has led to studies of the electrical properties of ion implanted compounds of Cu[2], Al[2-10], Si[2,11,12], Ta[2,13-16] and Ti[6,8,17].

The formation of the layers results from a competition between ion trapping, sputtering and diffusion[18]. In order to determine the importance of these parameters in the process of fabrication of ion implanted resistors[10], we have studied the dose dependence of the mass change of polycrystalline aluminum films during bombardment with 30 keV O_2^+ ions and 26 keV N_2^+ ions. The resulting structural transformations have been examined by transmission electron microscopy.

II - MEASUREMENTS OF ION TRAPPING AND SPUTTERING

IIa - Experimental Methods

The changes in the areal mass ($\Delta M/A$) of the sample due to bombardment were measured by means of the quartz-crystal resonator technique(19-21), in which changes in the resonance frequency of the crystal caused by variations of the mass of thin films deposited onto it, are measured. The frequency shifts Δf are proportional to $\Delta M/A$ if this last quantity is small as it is in our case. These shifts Δf can also result from stress effects(22). In our experiments, the beam energy was low (~ 30 keV) and the ion doses used sufficiently high so that the small induced stresses would have saturated at the beginning of the implantations. Frequency shifts also arise from temperature variations; therefore the frequency was allowed to stabilize after each bombarding step before taking a reading. This introduces the penalty that it is not possible to observe mass changes due to the release of implanted ions after the bombardment is stopped.

We have used Toyo 5MHz AT-cut quartz-crystals mounted on a water cooled holder. This holder forms part of a suitably shielded Faraday cage which allowed accurate measurements of the ion beam currents. The targets were made by evaporating 99.999% Al films onto the crystals. The mass-analyzed beams of 30 keV O_2^+ and 26 keV N_2^+ ions were uniformly scanned over the target surface. The dose was measured with a 1% accuracy and the changes in areal mass were reproducible to within 15% among many different experiments. The results were insensitive to variations in the initial film thickness between 650Å and 2500 Å, in the beam current density between 4.5 and 16 $\mu A/cm^2$ and in the pressure in the target chamber between 3 and 20 x 10^{-6} torr, the independence on the latter two factors indicating stable gas-covered surface conditions of the targets.

IIb - Results

Figure 1 shows the variations in the areal mass, $\Delta M/A$, of the Al films with dose of 30 keV O_2^+ and 26 keV N_2^+ ions. At low doses, ion trapping is more important than sputtering and out-diffusion, and the areal mass increases with dose. At higher doses, the areal mass decreases with dose, indicating the predominance of sputtering and outdiffusion, and that the quantity of gas trapped saturates.

The sputtering yields can be determined at low doses (trapping probability $P_t \sim 1$) and at high doses ($P_t \sim 0$). The results are given in amu/ion and in atoms/ion in table 1. To obtain values of atoms/ion it is assumed that the mean mass of sputtered atoms corresponds, at low doses, to that of Al_2O_3 (20.8 amu) and at high doses to that of Al_2O_3 for oxygen bombardment and of AlN (20.5 amu) for nitrogen bombardment.

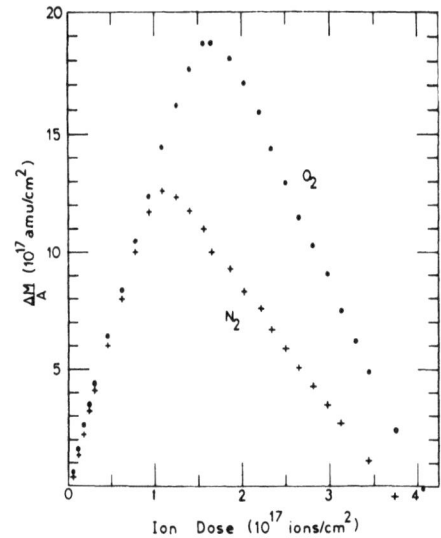

Fig. 1) Variation of the areal mass ($\Delta M/A$) of Al films with dose of implanted 30 keV O_2^+ and 26 keV N_2^+. The results shown are representative of data taken over many different samples.

TABLE I

Ion	Energy (keV)	S (low dose)		S (high dose)	
		amu/ion	atoms/ion	amu/ion	atoms/ion
O_2^+	30	16.3	0.78	8.4	0.40
N_2^+	26	12.9	0.62	4.8	0.23

III - ELECTRON MICROSCOPIC EXAMINATIONS

IIIa - Specimen Preparation

Thin films ($\sim$ 1000 Å thick) were deposited onto glass discs previously coated with a detergent. The discs were masked during implantation in order to obtain virgin and implanted regions in the same specimen, for direct comparison under the microscope. After implantation, the films were floated free by dissolving the detergent layer, and fragments were then placed onto a copper grid.

Transmission electron microscopy and electron diffraction were then carried out on a Philips EM 300 electron microscope using an accelerating potential of 80 KV. For the analysis of diffraction patterns, the camera constant was determined using the unimplanted aluminum zone as a standard.

IIIb - Results

Aluminum films bombarded with 30 keV O_2^+ ions have been thoroughly studied with the electron microscope (5,8) and it has been shown that a structure of oxide precipitates is formed, embedded in the Al matrix. These precipitates increase in size with increase in bombarding dose forming a completely oxidized surface layer at sufficiently high O_2^+ doses. We have obtained similar results and they will not, therefore, be presented here.

A typical result for N_2^+ bombardment is shown in the micrographs of Fig. 2, for virgin Al and N_2^+ implanted Al (2×10^{17} N_2^+/cm^2) before and after in situ heating to ~ 500 °C. The implanted zone gives diffraction rings of Al and other rings with d=2.687 Å, 1.822 Å, 1.511 Å and 1.413 Å corresponding closely with the values of Jeffrey et al. (23) for AlN ($10\bar{1}0$), ($10\bar{1}2$), ($11\bar{2}0$) and ($10\bar{1}3$) respectively. We note that the formation of AlN by implantation has been identified previously by Pavlov et al.(24) using 40 keV N_2^+ ions but, surprisingly, could' not be detected by Bykov et al.(25) using 30-50 keV N^+ ions even at doses as high as 10^{18} ions/cm^2.

In Fig. 2 it is seen that at these doses, the AlN precipitates have a size similar to the grain size of the virgin Al. Upon heating, the precipitates grow to a size ~ 1.5 larger while the Al grains grow by a factor of ~ 1.5 in the implanted region and ~ 6 in the virgin region. This difference can be explained by the high thermal stability of AlN at 500 °C(26), the precipitates fixing the grain boundaries and inhibiting the growth of crystallites. A similar inhibition was observed by Dasarathy and Hudd(27) in cold rolled aluminum containing embedded AlN precipitates.

The additional observation was made that after heating the specimens to 500 °C at the 10^{-5} torr pressure of the electron microscope column, Al_2O_3 was formed in the virgin area as determined by electron diffraction while no oxidation was apparent in the implanted region. The inhibition of oxidation induced by nitrogen bombardment was previously observed by Naguib et al.(28) in copper and by Wada and Ashikawa(29) in silicon.

IV - DISCUSSION

IVa - Sputtering Yields

We have shown in table 1 that the sputtering yields depend on dose. The higher value of $S(O_2, Al_2O_3)$ at low doses could be due to the neglect, in the calculation, of two processes which would make the initial trapping efficiency to be less than one: the reflection of incident ions and gas release during bombardment. The first pro-

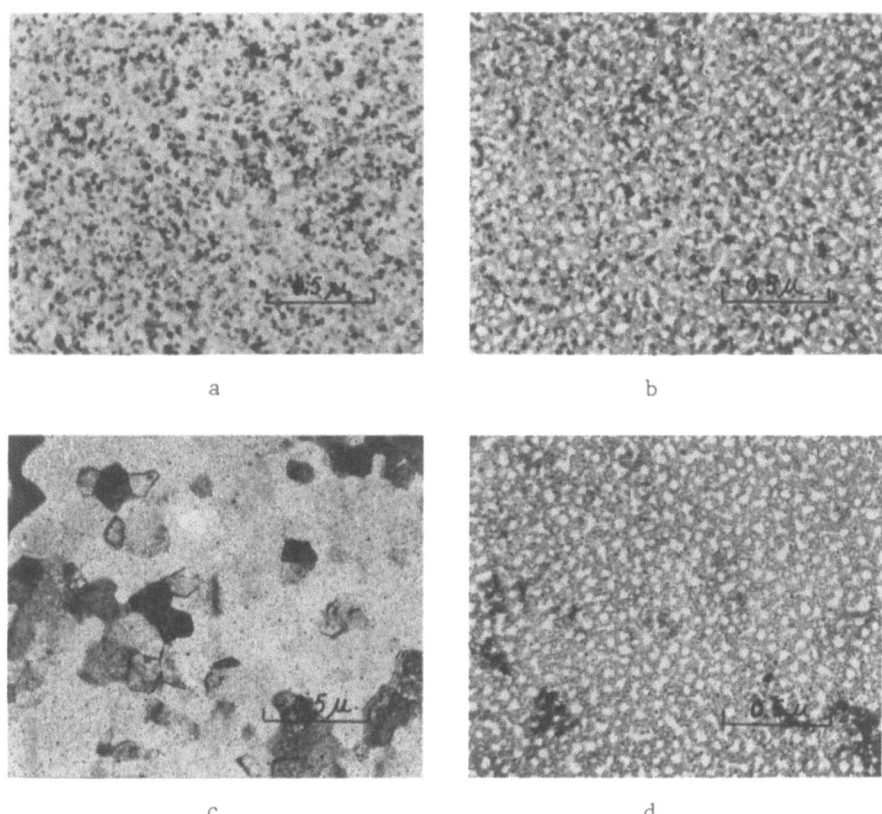

Fig. 2) Electron micrographs of Al films. a) Virgin zone, no heat treatment. b) Implanted zone, no heat treatment. c) and d) idem to a) and b) respectively but after heating to $\sim$ 500 °C in the "hot stage" of the microscope.

cess occurs with low probability (a few percent) as can be inferred from the data of Bøttiger et al.(30), while oxygen release is improbable since due to its great chemical affinity to aluminum, it would remain "frozen" in the lattice where it comes to rest(31). The out-diffusion or "gas sputtering" of gases buried in the film during evaporation, as observed by Andersen and Bay(21) is more probable. The yield at high doses is lower than could be expected from extrapolation of the values of Kelly(32) to higher energies. However, Kelly calculated S from the dose necessary to perforate a very thin anodic oxide film. In this case, forward sputtering(33) would cause S to appear to be larger. N_2^+ ions also give a lower S at high doses than at low doses. In this case, we have the additional fact that two different compounds are bombarded, Al_2O_3 at low doses and AlN at high doses; AlN given a lower S due to its higher surface binding energy, 10.8 eV(34) compared to 6.4 eV for Al_2O_3(35).

IVb - Ion Trapping

As stated above, the quantity of gas trapped results from a competition between the increase due to the penetration of the ions and the decrease caused by sputtering and out-diffusion. The majority of the work reported in the literature has been analyzed with the sputtering model of saturation(36-40) in which diffusion is neglected. The model predicts saturation to be reached at doses D_s at which the target is sputtered to a depth $\sim (R_p + \Delta R_p)$, where R_p and ΔR_p are the mean projected range and its standard deviation, respectively. Using our sputtering yield and values of R_p (285 Å) and ΔR_p (169 Å) derived from theory(41) for O on Al, obtains $D_s \simeq 7 \times 10^{17}$ ions/cm^2, clearly too high (it would still be too high if one uses the higher value of S for O_2 on Al). The disagreement is even greater for N_2^+ bombardment.

This model has been extended by Biersack(42) and Collins and Carter(43) by including a constant diffusion coefficient D. However, as pointed out by Andersen(44), a constant D is rather unrealistic since if diffusion does take place, it will be determined mainly by the depth-dependent radiation damage. A further shortcome of the theories is the neglect of mixing caused by knock-on effects, of precipitation of the implanted atoms and of the dose dependence of the stopping power and sputtering yield. Kräutle(45) has recently introduced the last two effects in a theory, but his treatment is rather simplified and incorrect as can be clearly seen from the excellent review of sputtering by Andersen(44).

In the case of insulating layers, a further mechanism of gas release can occur which is postulated here. Implanted gas atoms lying in the insulator-metal interface may capture an electron and move to the surface under the influence of the electric field in the insulator produced by the irradiation, and then be desorbed. That bombardment induced fields can cause ion migration has been shown to occur at low energies by McCaughan et al.(46). An additional complication is the strong binding between implanted ions and target atoms; it is possible that ions are trapped only if there are trapping sites available, this would explain the higher value of the quantity of trapped oxygen at saturation since there are $\sim$ 50% more trapping sites possible than in the case of nitrogen for nearly equal ranges as it happens in our work. However, the occurrence of bubble formation may render this simple picture invalid. We must conclude that due to the complex nature of the overall process, and the fact that several parameters are involved which cannot be evaluated a priori, it is not possible at present to give a quantitative analysis of the complete trapping-sputtering curves.

AKNOWLEDGMENTS

We thank E. V. Alonso and R. Rapacioli for their help during the experiments.

REFERENCES

* Work supported in part by CITEFA and by the Multinational Program in Physics of the Organization of American States.
** Comisión Nacional de Energía Atómica, † Universidad Nacional de Cuyo.

1. J.S. Colligon, Vacuum 24, 373 (1974).
2. M. Balarin, G. Otto, I. Storbeck, M. Schenk and H. Wagner, Thin Solid Films 4, 255 (1969).
3. J.G. Perkins and L.E. Collins, Thin Solid Films 5, R59 (1970).
4. L.E. Collins, P.A. O'Connell, J.G. Perkins, F.R. Pontet and P.T. Stroud, Nucl. Instrum. Methods 92, 455 (1971).
5. J.G. Perkins, Ph.D. Thesis, University of Surrey, 1971.
6. J.G. Perkins, Thin Solid Films 9, 257 (1972).
7. P.T. Stroud, Thin Solid Films 11, 1 (1972).
8. P.T. Stroud, H.M. Lindsay and J.G. Perkins, Vacuum 23, 125 (1972).
9. P.A. O'Connell, Ph.D. Thesis, University of Surrey, 1975.
10. J. Arancibia, R.A. Baragiola, E.R. Salvatelli, O. Auciello and P. Mendoza, Comunicaciones AFA (in press).
11. M. Watanabe and A. Tooi, Japan J. Appl. Phys. 5, 737 (1966).
12. J.H. Freeman, G.A. Gard, D.J. Mazey, J.H. Stephen and F.B.Whiting, European Conf. on Ion Implantation(P. Peregrinus:London,1970)p74.
13. M. Deery, K.H. Goh, K.G. Stephens and I.H. Wilson, Thin Solid Films 17, 59 (1973).
14. I.H. Wilson, K.H. Goh and K.G. Stephens, Application of Ion Beams to Metals (Plenum: New York, 1974) p 269.
15. K.H. Goh, K.G. Stephens and I.H. Wilson, Ion Implantation in Semiconductors (Plenum: New York, 1975) p 325.
16. I.H. Wilson, K.H. Goh and K.G. Stephens, Thin Solid Films 33, 205 (1976).
17. P.T Stroud, Thin Solid Films 10, 205 (1972).
18. G. Carter and J.S. Colligon, Ion Bombardment of Solids (Heinemann: London, 1968).
19. D. McKeown, Rev. Sci. Inst. 32, 133 (1961).
20. E.P. EerNisse, Rep. Sandia SC-RR-70-377 (1970).
21. H.H. Andersen and H. Bay, Rad. Eff. 13, 67 (1972).
22. E.P. EerNisse, J. Appl. Phys. 43, 1330 (1972); Ion Implantation in Semiconductors and other Materials(Plenum:New York,1973)p531.
23. G.A. Jeffrey, G.S. Parry and R.L. Mozzai, J. Chem. Phys. 25,

1024 (1956).
24. P.V. Pavlov, E.I. Zorin, D.I. Tetelbaum, V.P. Lesnikov, G.M. Ryzhkov and A.V. Pavlov, Phys. Stat. Sol. (a)19, 373 (1973).
25. V.N. Bykov, V.A. Troyan, G.G. Zdorovtseva and V.S. Khaimovich, Phys. Stat. Sol. (a)32, 53 (1975).
26. R.P. Elliot, Constitution of Binary Alloys, First Supplement (McGraw-Hill: New York, 1965).
27. C. Dasarathy and R.C. Hudd, Acta Met. 15, 1665 (1967).
28. H.M. Naguib, R.J. Kriegler, J.A. Davies and J.B. Mitchell, J. Vac. Sci. Technol. 13, 396 (1976).
29. Y. Wada and M. Ashikawa, Japan J. Appl. Phys. 15, 389 (1976).
30. J. Bøttiger, J.A. Davies, P. Sigmund and K.B. Winterbon, Rad. Eff. 11, 69 (1971).
31. E. Ruedl and R. Kelly, Modern Developments in Powder Metallurgy, Vol. 2: Applications (Plenum: New York, 1966) p 145.
32. R. Kelly, Can. J. Phys. 46, 473 (1968).
33. P. Mertens, Nucl. Instr. Meth. 132, 307 (1976).
34. R.K. Willardson and A.C. Beer, Eds., Semiconductors and Semimetals (Academic Press: New York, 1968) p 151.
35. O. Kubaschewski, E.L. Evans and C.B. Alcock, Metallurgical Thermochemistry (Pergamon Press: London, 1967) p 303.
36. O. Almén and G. Bruce, Nucl. Instr. Meth. 11, 257 (1961), Nucl. Instr. Meth. 11, 279 (1961).
37. G. Carter, J.S. Colligon and J.H. Leck, Proc. Phys. Soc. 79, 299 (1962).
38. M.T. Robinson, Appl. Phys. Letters 1, 49 (1962).
39. J.C.C. Tsai and J.M. Morabito, Surf. Sci. 44, 247 (1974).
40. F. Schulz and K. Wittmaack, Rad. Eff. 29, 31 (1976).
41. J. Lindhard, M. Scharff and H.E. Schiøtt, Mat. Fys. Medd. Dansk. Vid. Selsk 33, 1 (1963); K.B. Winterbon, Rep. AECL-3194 (1968).
42. J.P. Biersack, Rad. Eff. 19, 249 (1973).
43. R. Collins and G. Carter, Rad. Eff. 26, 181 (1975); G. Carter and R. Collins, Rad. Eff. 28, 123 (1976).
44. H.H. Andersen, Physics of Ionized Gases, 1974, ed. V. Vujnović (Inst. Phys. Univ. Zagreb, 1975) p 361.
45. H. Kräutle, Nucl. Instr. Meth. 134, 167 (1976).
46. D.V. McCaughan, R.A. Kushner and V.T. Murphy, Phys. Rev. Letters 30, 614 (1973).

IMPLANTATION OF Co INTO ALUMINUM: DAMAGE AND LATTICE LOCATION STUDIES

R. Kalish[*], Rutgers University, New Brunswick, N.J. 08903

and Bell Laboratories, Murray Hill, N.J. 07974

L. C. Feldman, Bell Laboratories, Murray Hill, N.J. 07974

I. INTRODUCTION

Radiation damage in metals induced by ion bombardment has been studied extensively over the last few years. Swanson and Maury (1) have shown that irradiation in aluminum at 40-70°K with 0.3-1.0 MeV He$^+$ ions to doses of 10^{15}-10^{16}/cm^2 causes the impurity atoms (Mn, Zn, and Ag) to be displaced from lattice sites by the trapping of self-interstitial Al atoms into $\langle 100 \rangle$ dumbells. These dumbells are observed to anneal out at temperatures from 180° to 220°K, at which vacancy-interstitial annihilation takes place.

The behavior of cobalt impurities in aluminum was studied by the Munich Mössbauer group (2,3) using the Mössbauer effect. The conclusions drawn from the Mossbauer studies on neutron and electron irradiated CoAl samples are consistent with the formation of cobalt-aluminum interstitial dumbells, the annealing behavior of which is similar to that observed by Swanson and Maury (1).

The present channeling work on Co implanted Al crystals was motivated by the above studies and was meant to be a bridge between the Mössbauer-effect experiments and the lattice location studies. The present results seem to confirm the $\langle 100 \rangle$ dumbell configuration observed previously; furthermore, interesting information regarding the radiation damage induced by 100 keV Co implantation into Al crystals at 15°K and at 300°K was obtained. The cold and room temperature implantation exhibit rather different

---

[*]Permanent Address: Physics Department, Technion, Technion City, Haifa, Israel.

damage patterns: In the former case one observes a "knee" in the backscattering yield corresponding to dechanneling deep inside the crystal, which anneals out at annealing stage III ($\sim 200°K$); the room temperature implantation shows a clear damage peak at about the Co range, which anneals at $\sim 200°C$.

II. EXPERIMENT AND RESULTS

1. Room Temperature Implantation

An aluminum single crystal was implanted along a random direction, with $9 \times 10^{14}/cm^2$, 100 keV cobalt ions at room temperature. The implantation dose rate was about 30 nA. 1.8 MeV He$^+$ ions, collimated to $\sim .04°$, were used to examine the crystal using the backscattering channeling technique. The backscattered He spectra taken along the $\langle 110 \rangle$ channel in an annular surface barrier detector (resolution ~ 15 keV) show an oxide peak, a pronounced damage peak, a typical surface peak, some impurities (introduced during the cleaning procedure of the crystal), and the Co peak. Analysis of the spectra show that the oxygen peak is on the surface of the crystal, the damage peak is about 770 Å inside the crystal, while the Co peak corresponds to He backscattered from Co at a depth of about 570 Å inside the crystal (the calculated range of 100 keV Co in Al is ~ 575 Å).

Angular scans, through the $\langle 110 \rangle$ axis display a minimum yield for the Al (taken immediately beyond the surface peak) of $\chi_{min}^{(Al)} = 0.080$ with an angular width of $2\psi_{1/2}^{(Al)} = 1.00°$, while the corresponding numbers for the Co impurity were $\chi_{min}(Co) = 0.61$ and $2\psi_{1/2}(Co) = .90°$ (see fig. 1). As can be seen in fig. 1, a slight peaking in the center of the Co channel may be suggested by the data. Planar scans in the $\{100\}$ direction yielded $\chi_{min}(Al) = 0.52$, $2\psi_{1/2}(Al) = 0.35°$, and $\chi_{min}(Co) = 0.65$, $2\psi_{1/2}(Co) = 0.38$.

The behavior of the damage peak was studied as a function of anneal temperatures up to 600°C. While the site of the Co remains constant to about 300°, the Al damage peak anneals out very rapidly over the first 200°C. Figure 2 shows the annealing of the damage peak and the associated growth in the surface peaks. The damage decreases from 3×10^{16} scattering centers per cm^2 at 15°C to 0.55×10^{16} centers per cm^2 at 200°C. The change in the surface peak accompanying this annealing is from 4.1×10^{16} scattering centers per cm^2 at 15°C to 4.8×10^{16} centers per cm^2 at 200°C. No significant change in the oxide peak related to the annealing could be observed.

IMPLANTATION OF Co INTO ALUMINUM

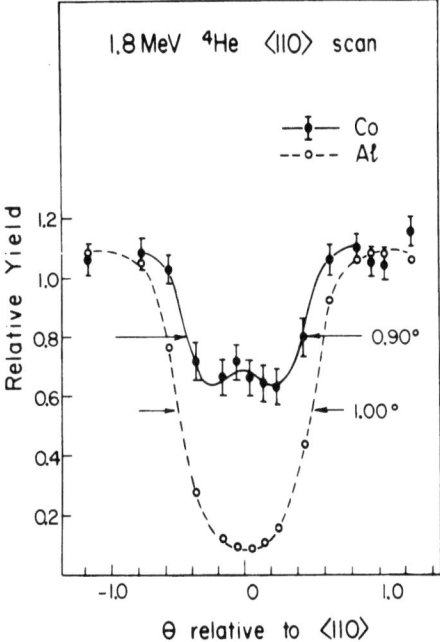

Fig. 1 ⟨110⟩ angular scan for the Al host and implanted Co impurity (See text for experimental details).

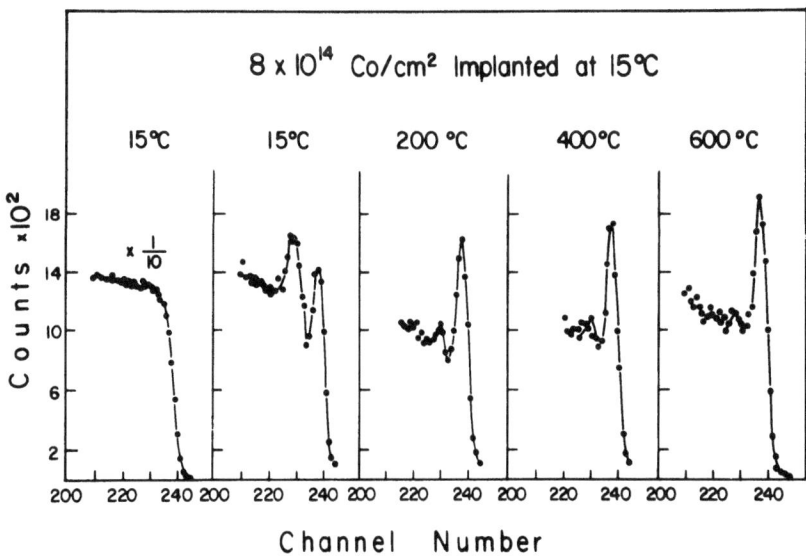

Fig. 2 Leading edge of the backscattered energy spectra for 1.8 MeV He$^+$ incident in a a) random and b) -e) ⟨110⟩ channeling direction after Co implantation and subsequent annealing.

2. Cold Implantations

An aluminum single crystal was mounted on a goniometer with provision to cool the sample to temperatures close to liquid He temperature. Angular scans for undamaged Al $\langle 110 \rangle$ at 300°K and 12°K are shown in Fig. 3.

The implantation of $4 \times 10^{14}/cm^2$ 100 keV Co ions was carried out, without warming up the Al crystal, along a random direction. The average Co beam current was 20 nA and it was swept across the sample at ~100 c/sec × 1000 c/sec. The results of the implantation are shown in Fig. 4. While the minimum yield along the $\langle 110 \rangle$ direction did not change by more than a factor of 2 immediately after the surface peak (χ_{min} = 0.096), it went up by about of factor of 8 deep inside the crystal (χ_{min} = 0.32). The effect of the implantation becomes evident by the increase in dechanneling the deeper the probing beam penetrates into the crystal, giving rise to the observed "knee" in the backscattered spectra.

The annealing behavior of this "knee" has been studed by gradually raising the temperature of the sample at ~50° steps, for 30 minutes/step. The backscattering spectra taken at the various temperatures are shown in Fig. 5. While no observable changes occur in the spectra up to 100°K, a clear change in the slope of the backscattering yield is evident from the data at 200°K.

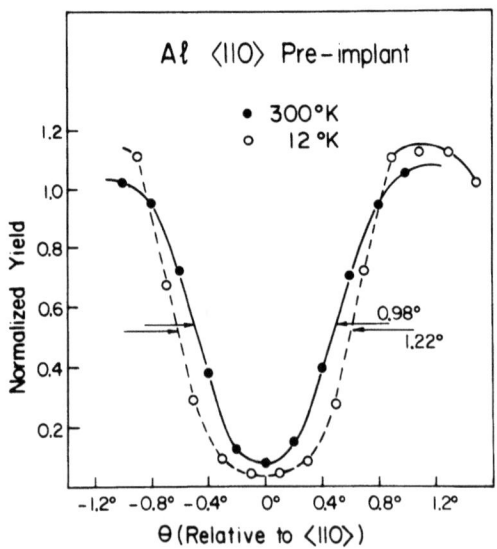

Fig. 3   Angular scans for undamaged Al along the $\langle 110 \rangle$ direction at 12°K and 300°K.

IMPLANTATION OF Co INTO ALUMINUM

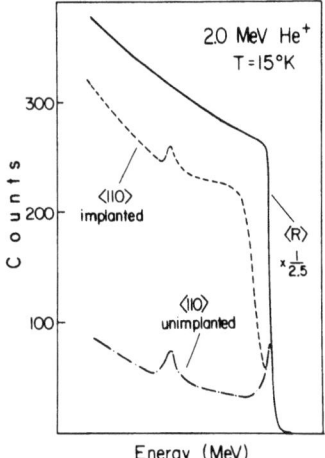

Fig. 4 Energy spectra in the random and aligned directions before and after implantation of 4×10^{14} Co/cm^2 at ~15°K.

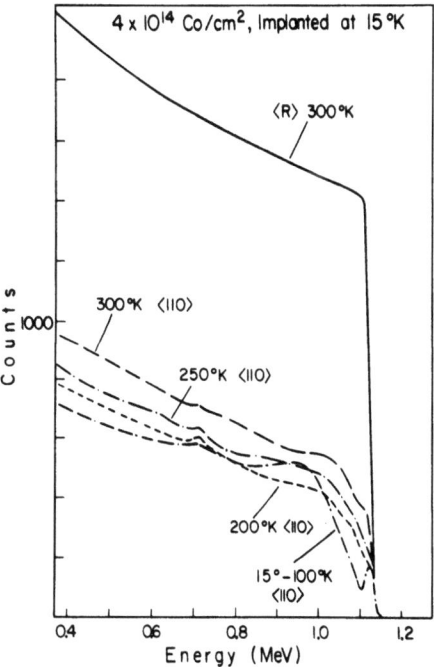

Fig. 5 Energy spectra as a function of anneal temperature following implantation at 15°K.

III. DISCUSSION

1. Lattice Location of Co

As noted previously the "S" factors ($S=1-X_{min}(Co)/1-X_{min}(Al)$) for the shadowing of the Co were measured along the $\langle 110 \rangle$ axial and $\{100\}$ planar symmetry directions. The measured values are $S\langle 110 \rangle = .45$ and $S\{100\} = .73$. Impurity atoms in $\langle 100 \rangle$ dumbell configurations should indeed show a larger S factor in the $\{100\}$ plane. Following Swanson the simplest interpretation of the data is a 45% substitutional component and the remainder in $\{100\}$ dumbell sites. Since 2/3 of these sites are shadowed along the $\{100\}$ plane this model "predicts"

$$S\{100\} = .45 + 2/3 \cdot .55 = .81$$

in reasonable agreement with the measured value. The angular width data also support this conclusion.

2. Damage

The understanding of heavy-ion induced radiation damage in metals, particularly as observed by channeling has recently been reviewed by Behrisch (4). It is clear that as compared to the semiconductor case which show well defined damage peaks, with quantitatively understandable depth profiles, the problem in metals is poorly understood. The difficulties are most probably connected to the nature of the damage clusters in metals. For the specific case of Al, Rimini et al. (5) recently studied the energy dependence of the dechanneling at room temperature and concluded that in the case of high dose Zn implants, the damage profile is governed by the local curvature of the lattice near dislocations. Such a model explains the absence of prominent damage peaks.

For the low dose Co implant described here, a small but unmistakeable damage peak is observed. This is one of the few cases where such a distinct peak is observed. It should be noted that the damage peak anneals almost completely by 200°C and is accompanied by some growth in the surface peak. It is as though some of the defects, once released, get trapped at the surface and contribute to the scattering yield of the surface peak. (As stated before, the oxygen scattering peak does not increase during the anneal assuring that the surface peak increase does not simply correspond to surface oxidation.)

In the case of the low temperature implantation the damage profiles and annealing behavior are even more striking. The initial spectrum at 15°K following an implant of 4×10^{14} Co/cm^2 shows a well-defined surface peak of the same intensity as the virgin spectrum and a very abrupt dechanneling, rising to approximately 1/3 of the random spectrum. It is interesting to compare this to the work of Rimini et al. who studied high dose (1.2×10^{16}/cm^2) Zn implants into Al at room temperature. Although there is a factor of 30 in heavy ion dose, the spectra are comparable; but the features at low temperature are considerably sharper.

It is difficult to understand such strong dechanneling in the context of a point defect model. Furthermore, angular scans (at 2.0 MeV) in the damage region do not indicate any narrowing. In fact, the angular width in the damaged case is slightly larger than for the unimplanted crystal; in agreement with Rimini et al. Queré (6) and Rimini et al. (5) have put forth a model in which the dechanneling is due primarily to strain from extended defects in the host lattice, e.g., dislocations and vacancy loops. The dechanneling cross section is related to the local curvature of the lattice near the dislocations. Thus the data presented here show that for implantation at low-temperature the strain-type dechanneling is observed for very low dose implants. Experiments in Al which require radioactive implantation or high neutron dose irradiation (as in Refs. 1-3) may also find the active impurities in strong strain fields. The large strain observed here, even for a low dose implant, further indicates that the strain does not result from the impurity concentration, but is intrinsic to the damaged crystal.

As discussed previously the dechanneling curve undergoes significant changes as the sample is warmed to 300°K. The most abrupt changes occur at approximately 200°K, generally called "Stage III Annealing", and believed to be associated with the release of trapped interstitials. It is clear that the nature of the damage is different above and below 200°K and that the mechanism of release of interstitials may be connected to the relaxation of crystal strain.

IV. ACKNOWLEDGMENTS

We gratefully acknowledge the technical expertise of W. M. Augustyniak and W. F. Flood.

REFERENCES

1. M. L. Swanson and F. Maury, Can. J. Phys. $\underline{53}$, 1117 (1975).

2. W. Mansel, G. Vogl and W. Koch, P.R.L. $\underline{31}$, 359 (1973).

3. G. Vogl, W. Mansel and P. H. Dederichs, P.R.L. $\underline{36}$, 1497 (1976).

4. R. Behrisch and J. Roth, Ion Beam Surface Layer Analysis, ed. by O. Meyer, G. Linker and F. Kappeler, p. 539, Plenum Press, New York (1976).

5. E. Rimini, S. U. Campisano, G. Foti, P. Baeri and S. T. Picraux, Ion Beam Surface Layer Analysis, ed. by O. Meyer, G. Linker and F. Kappeler, p. 597, Plenum Press, New York (1976).

6. Y. Queré, Ann. Phys. (N.Y.) $\underline{5}$, 105 (1970).

DECHANNELING BY DISLOCATIONS IN Zn-IMPLANTED Al

G. Foti, S. T. Picraux*[†] S. U. Campisano,
E. Rimini and R. A. Kant #[†]

Instituto di Struttura della Materia
Corso Italia, 57 - Catania - Italy

*Sandia Laboratories, Albuquerque, N. M. 87115
#U. of New Mexico, Albuquerque, N. M. 87106

ABSTRACT

The influence of dislocations on the dechanneling of energetic channeled ions has been investigated. Channeling effect and transmission electron microscopy (TEM) techniques were used to measure the disorder in 1.2 to $2.6 \times 10^{16}/cm^2$ Zn-implanted Al crystals. Channeling analysis along the <110> axis was made with He beam energies ranging between 1.0 and 3.0 MeV. From TEM measurements, a dense network of dislocations is found to be the predominant type of defect in the implanted layer. From our analysis of the energy dependence of the dechanneling, together with the TEM determination of the total projected dislocation length, we obtain the first experimental measurement by backscattering of the dechanneling cross section, λ, for a dislocation. The experimental value, $\lambda(Å) = 20[E(MeV)]^{1/2}$, agrees closely in magnitude and in energy dependence with theoretical prediction. From these studies, we estimate that the minimum dislocation density for detection by channeling is 10^9-10^{10} cm length of line/cm^3.

[†]This work was supported by the United States Energy Research and Development Administration, ERDA, under Contract E-(29-1)789.

I. INTRODUCTION

Previous measurements[1] of implantation damage in Aℓ have shown a different channeling behavior for N and Zn implants. For the N implant the dechanneling decreased with increasing analysis energy. This is consistent with the presence of a random distribution of displaced atoms, as observed for implantation disorder in semiconductors,[2] or local rearrangements of host atoms, as may be the case for these high N concentrations ($\sim$ 10 at.%). However, for the heavier Zn ions the energy dependence of the dechanneled fraction was proportional to $\sqrt{E}$. It was proposed that this was due to a type of implantation-induced disorder which resulted primarily in local distortions for the lattice rows, rather than to randomly displaced atoms. Strains in the Aℓ crystal, such as would result from dislocation lines, dislocation loops, or small defect clusters, could give rise to such an effect.

In this paper the Zn-implanted Aℓ system has been studied in more detail. New dechanneling measurements have been correlated with transmission electron microscopy (TEM) observations. An analysis of the experimental results has been developed, utilizing the energy-dependence of the dechanneling and the defect density measured by TEM. This study provides the first direct check by ion backscattering of the mechanism proposed by Quéré for dechanneling by dislocations.[3]

II. EXPERIMENTAL PROCEDURE

Aluminum single crystals of (110) orientation were prepared by electropolishing, and for some cases were subsequently vacuum annealed at 400°C. Implantations of Zn ions were performed at 150 or 200 keV to fluences of 1.2×10^{16}/cm^2, respectively. All implants were done at room temperature along a nonchanneling direction 12 to 15° off the <110> direction with the Zn$^+$ beam intensity $\sim$ 1 μA/cm^2/sec.

Channeling effect analysis was carried out using He ion backscattering at 165° in the energy range 1-3 MeV. In each case, the same sample was analyzed along the <110> direction in both implant-damaged and unimplanted regions. TEM measurements at 100 kV were done in pre-thinned samples which were implanted under the same conditions as those for the channeling analysis, except that the sample surface was tilted 45° from the beam normal to decrease the depth of the implanted region.

III. EXPERIMENTAL RESULTS

In Fig. 1 are shown the backscattering energy spectra for unimplanted and 150 keV Zn-implanted Aℓ to a fluence of $1.2 \times 10^{16}/cm^2$. The <110> and random spectra were

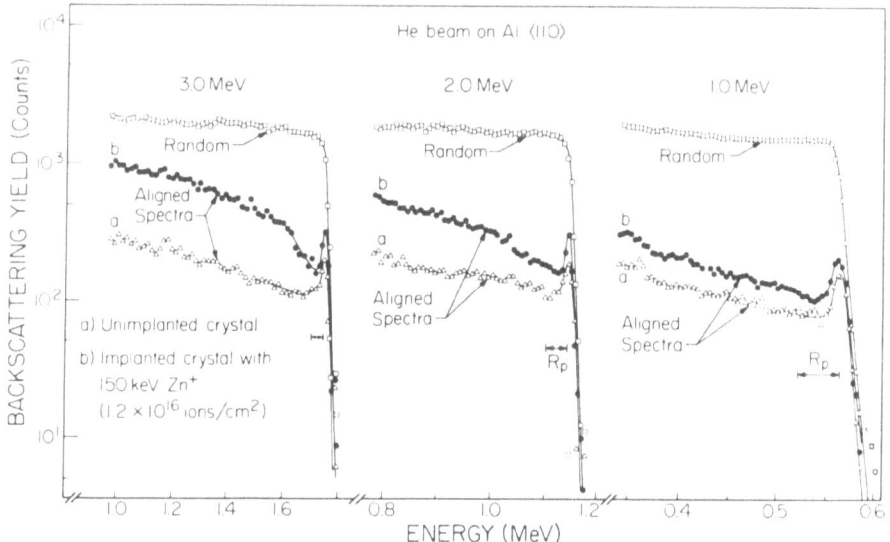

Fig. 1. Backscattering energy spectra of 3.0, 2.0 and 1.0 MeV He beam incident along the <110> direction of Aℓ. Aligned spectra are taken both for the unimplanted and 150 keV, $1.2 \times 10^{16}/cm^2$ Zn implanted crystal.

obtained for 1.0, 2.0 and 3.0 MeV He analysis beam energies. In all cases the minimum yield of the unimplanted crystal was < 5%. For the implanted sample the minimum yield is ∿ 5 to 8% near the surface and increases appreciably at greater depths. The arrows show the distance from the surface to the theoretical projected range of the implanted Zn (R_p ∿ 790 Å).[4] The aligned yield increases rapidly up to a depth of about twice the projected range, indicating the presence of high disorder in the implanted region. Beyond this depth the rate of dechanneling decreases, as is seen from the "knee" in the 3.0 MeV <110> spectrum. Similar results have been obtained from 200 keV Zn implantation to a fluence of 2.6 × $10^{16}/cm^2$.

The most important aspect of Fig. 1 is the increase in the normalized aligned yield with increasing energy of the He probing beam. This is shown more quantitatively in Fig. 2, where the dechanneled fraction is given by the ratio of the aligned-to-random yield. The depth scale is obtained from the known He ion stopping power and geometry of scattering. The same stopping power has been assumed for random and aligned directions. The upper part of Fig. 2 shows the results for the implanted crystal and the lower part for the unimplanted crystal. The decrease in

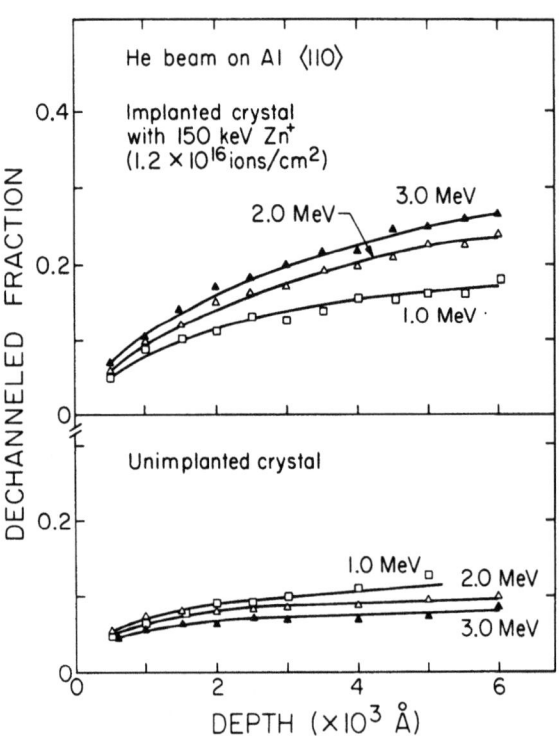

Fig. 2. Dechanneled fractions vs. depth obtained from the data of Fig. 1 are shown for the implanted crystal in the upper part and for the unimplanted crystal in the lower part.

dechanneling with increasing energy for the unimplanted crystal is consistent with theoretical and experimental results for a perfect crystal.[5] This dechanneling behavior arises from the multiple scattering of the channeled ions by electrons and thermally vibrating lattice atoms. The opposite energy dependence for the implanted crystals indicates that the disorder giving rise to dechanneling cannot be described in terms of randomly-distributed displaced atoms. This is consistent with the absence of a disorder peak in the aligned yield. Certain types of disorder can give rise to small displacements or distortions of the lattice rows, and thereby result in appreciable dechanneling without sufficient direct scattering to cause a disorder peak.

To better characterize the nature of the disorder TEM measurements were done on samples implanted together with those used for the channeling effect studies. A representative dark field micrograph is shown in Fig. 3 after 200 keV Zn implantation at 45° to the surface normal to a fluence of $2.6 \times 10^{16}/cm^2$. A dense network of dislocations is observed and this appears to be the predominant type of defect present after Zn implantation in Aℓ.

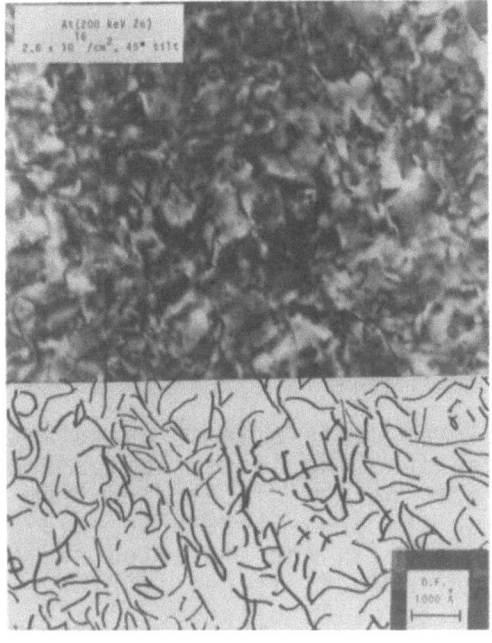

Fig. 3.

Dark field micrograph using g = 111 for 200 keV, $2.6 \times 10^{16}/cm^2$ Zn-implanted Aℓ. Example of the interpreted dislocation network is shown in the lower part of the figure.

The entire depth of the disordered region is included in the micrograph shown in Fig. 3. Selected area diffraction shows that the implanted layer remains single crystal Aℓ and that there is no indication of Zn precipitation.

For quantitative comparison with the channeling measurements we have carried out estimates of total projected length of dislocation lines in the plane approximately normal to the <110> direction. A portion of the map of the dislocation lines used in this evaluation is shown in the lower part of the figure. We obtain a projected length of $\approx 3.5 \times 10^5$ cm/cm^2, which after correction for sample titling during implantation and the fraction of dislocations out of contrast gives $\approx 9.6 \times 10^5$ cm/cm^2. From geometric considerations, the actual length is greater than the projected length by a factor ≈ 1.27. The absolute accuracy of this dislocation length measurement is estimated to be within a factor ≈ 2, and is primarily limited by the overlapping strain fields of the dislocations.

IV. ANALYSIS AND DISCUSSION

The standard treatment[6] for an approximate analysis of channeling disorder measurements in implanted semiconductors describes the normalized aligned yield, χ_D, at depth z by

$$\chi_D = \chi_R + (1 - \chi_R) \frac{n_D}{n}, \qquad (1)$$

where χ_R is the dechanneled component of the aligned beam at depth z, n_D is the density of displaced atoms and n is the crystal atom density. The second term on the right-hand side of Eq. (1) represents direct scattering of channeled ions by defects and the first term is given by

$$\chi_R = \chi + (1 - \chi)[1 - \exp(- \sigma N_D)], \qquad (2)$$

where χ is the aligned yield at depth z in an unimplanted crystal, σ is the defect dechanneling cross-section and N_D the total number of defects per cm^2 integrated from the surface to the depth z. For the present case we have seen from the lack of a disorder peak in the implanted region that the direct scattering by defects is negligible. This means σ is not related to n_D by coulomb interaction as would be the case for randomly distributed displaced atoms.

Thus $\chi_D \sim \chi_R$, and from Eq. (2) we obtain

$$\frac{1 - \chi_D}{1 - \chi} = \exp(-\sigma N_D) \quad . \tag{3}$$

In the present case $N_D \sigma = \ell \lambda$, where ℓ is the total projected length of dislocation lines[3] per unit area and the channeling width λ is the cross-section per unit length.

Based on theoretical predictions discussed below we would expect the functional dependence $\lambda \propto E^{1/2}$, where E is the analysis beam energy. The dechanneling data for the 2.6 x 10^{16} Zn/cm^2 implant is therefore plotted in Fig. 4 as the left-hand side of Eq. (3) versus $E^{1/2}$.

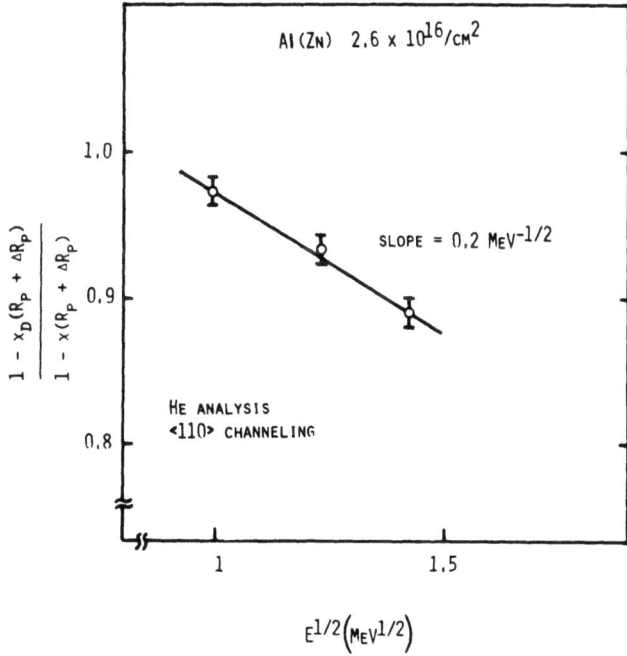

Fig. 4. The ratio between the channeled fraction in the implanted and in the unimplanted crystal at a depth $R_p + \Delta R_p$ = 1450 Å is plotted vs. the square root of the analysis beam energy for 200 keV, 2.6 x 10^{16}/cm^2 Zn-implanted Aℓ.

Here, we have used $\exp(-\ell\lambda) \simeq 1 - \ell\lambda$. The dechanneled fractions are taken at a depth $z = R_p + \Delta R_p$, which for 200 keV Zn in Aℓ is $\simeq$ 1450 Å.[4] Good agreement with the predicted functional dependence is observed, both in Fig. 4 and for the lower fluence Zn implant of $1.2 \times 10^{16}/cm^2$ where measurements were extended over a greater energy range of E = 1 to 3 MeV. From the slope of the line in Fig. 4, together with the measured value of ℓ, we obtain $\lambda = 20 \, E^{1/2}$, where λ is given in Å and E in MeV.

This is the first experimental determination of the dechanneling cross-section for dislocations by backscattering. This result may be compared with a distortional steering model by Quéré for dechanneling by dislocations.[3] For axial channeling this model predicts

$$\lambda_{theor.} = \left[\frac{bdaE}{\alpha Z_1 Z_2 e^2} \right]^{1/2}, \qquad (4)$$

where b is the Burgers vector, d the interatomic spacing along the channeling axis, a the Thomas-Fermi screening radius, $Z_1 e$ and $Z_2 e$ the atomic charge of the projectile and target atoms and α a number equal to 4.5 for edge dislocations and 12.5 for screw dislocations. Using $\alpha = 8.5$, i.e., assuming equal numbers of edge and screw dislocations, we obtain for the <110> direction in Aℓ $\lambda = 21 \, E^{1/2}$, where λ is given in Å and E in MeV.

In conclusion the energy dependence of dechanneling is shown to give information on the nature of disorder. For dislocations we have demonstrated that the dechanneling cross section, λ, is consistent with an $E^{1/2}$ dependence, in agreement with the theoretical prediction. In addition for 1 MeV He along the Aℓ <110> direction we obtain $\lambda = 20$ Å, which compares extremely well with the calculated value of 21 Å, considering the accuracy of the measured dislocation length by TEM. A high density of dislocations is required for detection by dechanneling measurements. In our case the dislocation density in the implanted region is $\simeq 7 \times 10^{10}$ cm length of line/cm^3 (= lines/cm^2). This density is comparable to that found in cold-worked metals. We estimate a minimum dislocation density of 10^9 to 10^{10} lines/cm^2 is required for detection by channeling effect measurements.

Assistance with implantations by C. Fuller and G. Harper is acknowledged.

REFERENCES

[1] E. Rimini, S. V. Campisano, G. Foti, P. Baeri, S. T. Picraux, Ion Beam Surface Layer Analysis, Vol. 2, Edited by O. Meyer, G. Linker, and F. Käppeler (Plenum Press, New York, 1976), p. 597.
[2] F. H. Eisen in Channeling: Theory, Observations and Applications, Edited by D. V. Morgan (John Wiley, New York, 1973), p. 415.
[3] Y. Ouéré, Phys. Stat. Sol. 30, 713 (1968).
[4] D. K. Brice in Ion Implantation Range and Energy Deposition Distributions, Vol. 1 (Plenum Press, New York, 1975), and private communication.
[5] F. Grasso in Channeling: Theory, Observations and Applications, Edited by D. V. Morgan (John Wiley, New York, 1973), p. 181.
[6] E. Bøgh, Can. J. Phys. 46, 653 (1968).

ION IMPLANTATION IN PIEZOELECTRIC SUBSTRATES*

P. Hartemann and M. Morizot

THOMSON-CSF - Laboratoire Central de Recherches

Domaine de Corbeville - 91401 ORSAY (France)

ABSTRACT

The elastic and piezoelectric properties of substrates used in acoustic surface wave devices are changed by ion implantation as a consequence of the crystalline structure perturbation produced by ions. Quartz, lithium niobate, lithium tantalate and bismuth germanium oxide substrates were exposed to a beam of light ions. The influence of this bombardment on density, acoustic surface wave velocity, electromechanical coupling coefficient, static capacitance, d.c. conductivity and temperature coefficient of delay has been studied.

As examples of application, the density and velocity change induced by ion implantation is used for reflecting surface waves. Guides and resonators are realized by these technics.

INTRODUCTION

Piezoelectric substrates used in acoustic surface wave devices have been exposed to an ion beam. The properties of these substrates are changed by ion implantation (1, 2). In a first section of this paper, main changes induced by implanted ions are reviewed, in the second section, some devices realized using ion implantation are described.

---

* Work supported by the Direction des Recherches et Moyens d'Essais - Paris (France).

The ions perturb the crystalline structure and a more or less amorphous layer is obtained at the surface of substrates as shown by X-ray topographies and electron diffraction patterns. Quartz, lithium niobate, lithium tantalate and bismuth germanium oxide substrates were bombarded by different kinds of ions. However, quartz and lithium niobate were more particularly studied. Light ions as helium, lithium, beryllium, boron have been implanted because they are the most efficient.

Channelling effects were avoided by using an ion-beam 7° off the perpendicular to the surface of substrates.

PERTURBATIONS INDUCED BY ION IMPLANTATION

Density change

For quartz, the volume of the implanted region increases as a consequence of a density decrease, and the height (e) of the resulting step at the boundary of this region has been measured using a "Talystep". The height (e) versus ion dose is shown in figure 1, for unannealed substrates of quartz. It is relatively large and reaches 950 Å (Y cut) for a dose equal to 1.5×10^{16} He^+/cm^2 at 100 KeV. At 52 KeV the height corresponding to the saturation is about 630 Å for $^9Be^+$ and 480 Å for $^{11}B^+$ (ST cut quartz).

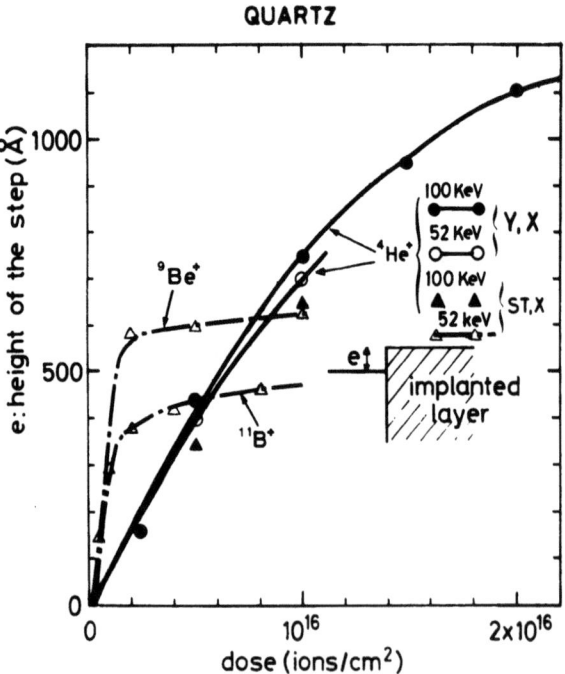

Fig.1 - Step height at the boundary of an implanted quartz surface versus ion dose.

Inversely, the large variation of the step height for doses between 10^{15} and 10^{16} ions/cm^2 can be used to calibrate an unknown implanted dose.

The thickness of the perturbed layer was determined by etching this layer using an hydrofluoric acid solution. For helium ions at 100 KeV it is about 7750 Å (Y cut). At 52 KeV, it is about 5100 Å (Y cut), 4220 Å (ST cut) and 3060 Å (ST cut) for helium, beryllium and boron ions respectively. Therefore, the density of the implanted layer is close to that of amorphous silica which is 17 % smaller than that of quartz. A small increase (80 Å) of the step height was found after annealing of the substrate at 560°C during 56 hours or at 800°C during 1 hour, the dose being 1.5 x 10^{16} He/cm^2 at 100 KeV.

No step with a significant height was observed with lithium niobate and tantalate and bismuth germanium oxide.

Acoustic surface wave velocity change

The acoustic surface wave velocity is increased by implantation in quartz (see figure 2) mainly as a consequence of the density decrease. This velocity increase is a function of the frequency because the implanted layer thickness is smaller than the acoustic wavelength in the used frequency range and the wave propagates in a stratified material. The curves are straight lines and a 3 % relative increase in velocity has been measured at 500 MHz.

The behaviour of lithium niobate and tantalate is strikingly different. For these materials, ion implantation induces a decrease of acoustic surface wave velocity. The curves for niobate (see figure 3) show a saturation effect : the relative velocity decrease is practically constant for frequencies higher than 450 MHz. This saturation effect seems to show that the thickness of the perturbed layer is larger than the penetration depth of the ions in the niobate substrate.

The normalized profile of beryllium atoms implanted in niobate was determined **using a Cameca ion analyzer, the dose being 1.5 x 10^{16} Be$^+$/cm^2 at 100 KeV (see figure 4).** The projected range and the standard deviation are about 2340 Å and 800 Å respectively.

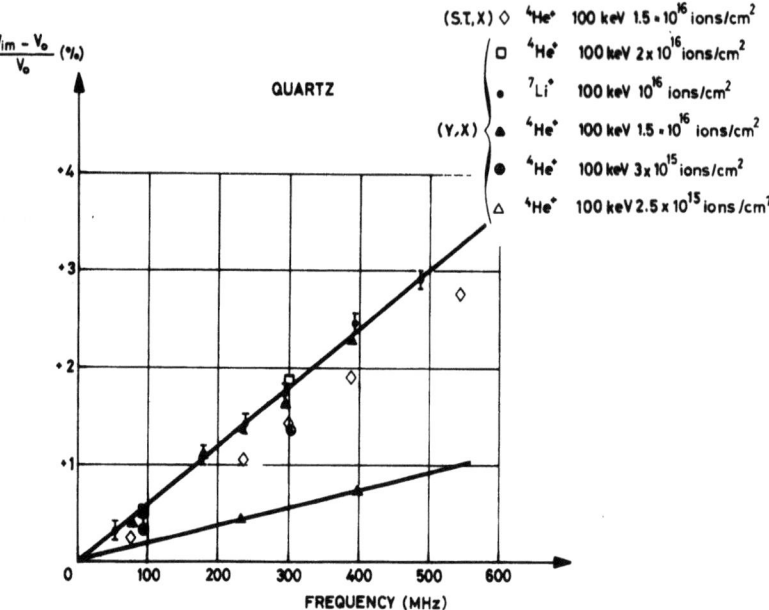

Fig. 2 - Relative acoustic surface wave velocity change induced by ion implantation in quartz versus frequency.

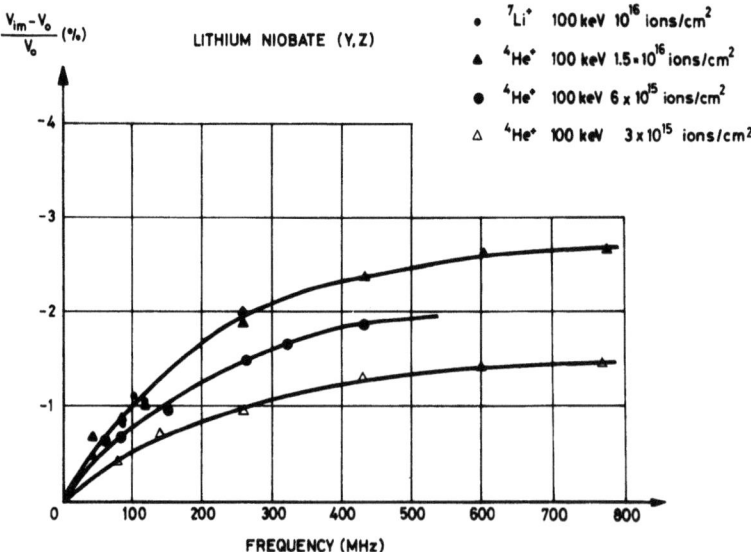

Fig. 3 - Relative acoustic surface wave velocity change induced by ion implantation in lithium niobate versus frequency

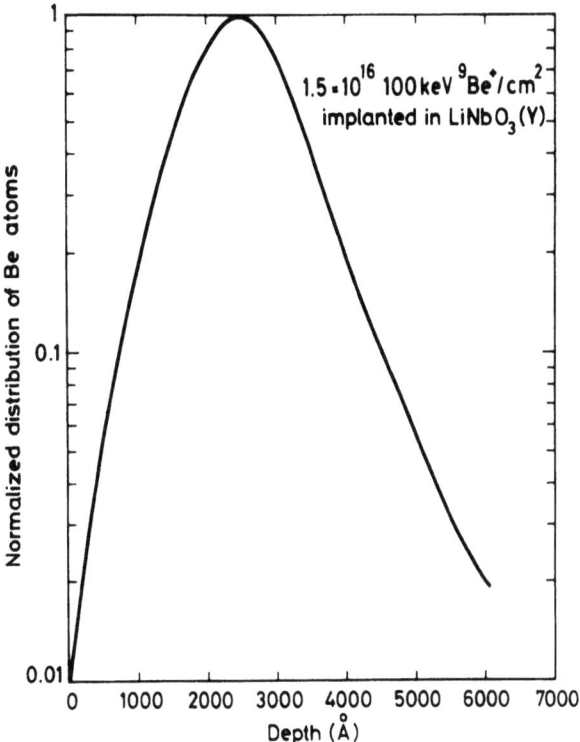

Fig. 4 - Normalized distribution of beryllium implanted in lithium niobate.

Electromechanical Coupling Coefficient Change

The electromechanical coupling coefficient of the implanted layer is decreased by implantation as a consequence of the crystalline-to-amorphous transition. This decrease has been measured for lithium niobate (Y cut, Z propagating). In fact, an effective coupling coefficient K_i for the depth probed by the acoustic wave is deduced from the conventional formula :

$$K^2 = 2(V - V')/V'$$

where V is the velocity of Rayleigh waves propagating on a free surface, and V' is the velocity on the surface with a massless conductor coating.

K_i^2 versus frequency is shown in figure 5 for 2 ion doses. It may be deduced from these curves that the main cause of the velocity decrease produced by implantation is the decrease of the electromechanical coupling coefficient.

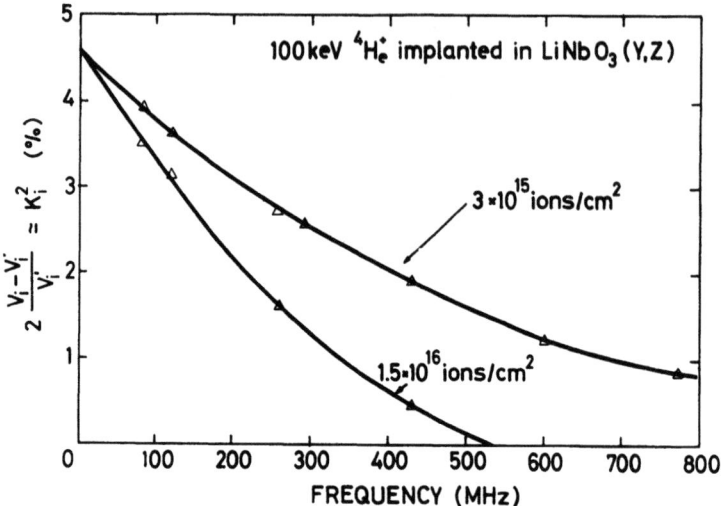

Fig. 5 - Squared effective electromechanical coupling coefficient of ion implanted lithium niobate versus frequency.

Static capacitance and d.c. conductivity change

Interdigital transducers centered at 138 MHz were deposited on niobate implanted with different doses of ^{4}He$^+$ ions at 100 KeV. The number of fingers and the interdigital spacing are equal to 20 and 6.3 µm respectively. The static capacitances of the transducers were measured at 1 MHz and the capacitance F_0 of a finger pair has been determined in picofarad per meter. As shown in the table I, a 18 % increase of capacitance was found in a 1.5 x 10^{16} ions/cm^2 range.

The d.c. resistance of these transducers was also measured by using an Ohm-meter. The results are reported in the table I. The d.c. resistance decreases versus ion dose.

Li Nb O$_3$ (Y, Z)

Dose ^{4}He$^+$/cm^2 100 KeV	0	3×10^{15}	6×10^{15}	1.5×10^{16}
F_0 (pF/m)	235	243	258	277
R (kΩ)	∞	∞	1300	500

Table I - Finger-pair capacitance F_0 and d.c. resistance R of 138 MHz interdigital transducers deposited on an ion implanted lithium niobate substrate.

Temperature coefficient of delay change

The delay introduced by a propagation length is equal to T and the temperature coefficient of delay (T.C.D.) is $\Delta T/T$. It is equal to the difference of the thermal expansion coefficient and relative velocity variation. For the implanted layer, this coefficient may be different from that of the unimplanted substrate. The T.C.D. of an implanted substrate depends upon the frequency because the implanted layer is generally thinner than the penetration depth of the wave.

The curves of the figure 6 show the relative frequency variation versus temperature for delay line oscillators using the same implanted substrate. This frequency variation is equal to the opposite of the effective temperature coefficient of delay including the effect of inductances matching the transducers. For unimplanted ST cut quartz the coefficient of delay is quadratic with a turn-around temperature close to 20°C. This temperature is 47°C at 78 MHz, 76°C at 133 MHz and 113°C at 139 MHz for a propagation length implanted by 1.5×10^{16} He/cm^2 at 100 KeV. This turn-around temperature shift is due to the T.C.D. of the implanted layer which is linear and negative. Moreover the quadratic coefficient of delay variation is reduced by ion implantation from about $34.8 \times 10^{-9}/(°C)^2$ to $24 \times 10^{-9}/(°C)^2$ at 139 MHz.

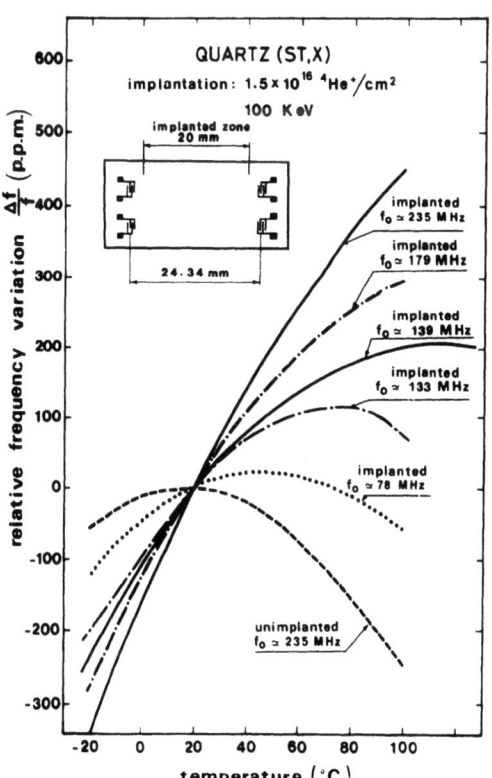

Fig. 6

Relative frequency variation for delay line oscillators. This variation is equal to the opposite of the effective temperature coefficient of delay

No significant change of T.C.D. was observed for implanted lithium niobate and tantalate substrates.

APPLICATION TO DEVICES

The acoustic impedance variation at the boundary of an ion implanted region is not only due to the velocity and density changes, but also to the surface step as a consequence of the density decrease. Surface wave guides and resonators have been realized using this acoustic impedance mismatch induced by ion implantation in quartz.

Guides

For quartz, the guide has a slot structure and the main part of the acoustic power propagates in an unimplanted channel. 100 MHz and 400 MHz slot guides were realized by implanting 1.5×10^{16} He$^+$/cm^2 at 100 KeV, the unimplanted channel being 148 µm and 18 µm wide respectively (3).

Resonators

Reflective array may be obtained using the acoustic impedance mismatch produced by ion implantation. For quartz, the reflection coefficient is close to 100 % at 125 MHz with a periodic grating made of 630 lines (6.3 µm wide) implanted with 1.5×10^{16} He$^+$/cm^2 at 100 KeV. A surface wave resonator consists of a conventional interdigital transducer deposited between two ion-implanted gratings (4). This device operates as a Fabry-Perot cavity made of distributed mirrors. At the resonance frequency the impedance of the transducer varies and a Q value relatively large (19000) has been obtained at 125 MHz.

CONCLUSION

Several effects of ion implantation in piezoelectric substrates have been shown up. They are consequences of the crystalline-to-amorphous transition induced by ions. But further investigations must be made in order to specify the origins of these effects. Nevertheless, surface acoustic wave devices can be advantageously realized using ion implantation.

REFERENCES

(1) T.R. LARSON, W.H. WEISENBERGER and W.H. LUCKE, Appl. Phys. Letters 22, 617 (1973)
(2) P. HARTEMANN and M. MORIZOT, Electr. Lett. 9, 497 (1973)
(3) P. HARTEMANN, Electr. Lett. 10, 110 (1974)
(4) P. HARTEMANN, Ultrasonics Symposium Proceedings IEEE cat 75 CHO 994, 4SU p. 303 (1975)

ASSOCIATION OF THE 6-eV OPTICAL BAND IN SAPPHIRE WITH OXYGEN VACANCIES

B.D. Evans
Naval Research Laboratory, Material Sciences Division
Washington, D.C. 20375

H.D. Hendricks
NASA Langley Research Center
Hampton, Virginia 23365

F.D. Bazzarre
Technics, Inc.
Alexandria, Virginia 22310

J.M. Bunch
Los Alamos Scientific Laboratory
Los Alamos, New Mexico 87545

INTRODUCTION

Arnold and Compton[1] have demonstrated that the prominent 6-eV optical absorption band observed in reactor irradiated crystalline sapphire by Levy and Dienes[2,3] and Mitchell et al.[4] is due to atomic displacement events and not to ionizing radiation alone, as x-ray and γ-irradiation will not produce this band. The orientationally averaged displacement threshold energies associated with the production of the 6-eV band were determined to be 90±5 and 50± eV when the absorption is associated with respectively anion or cation vacancies by a series of low temperature 0.6-to-1.8 MeV electron irradiations. However, they were unable to determine which damaged sublattice, anion or cation, is responsible for the uv optical absorption. Arnold[5,6] et al. and Evans[7] have subsequently shown that energetic light ions such as 50-keV and 200-keV

H^+, 100-keV D^+ (and 100-keV $^3He^+$ to a lesser degree) at fluences of 10^{16} cm^{-2} also produce the intense 6-eV band, whereas bombardment to the same fluence with heavier ions such as 220-keV O^+, 500-keV A^+ and 200-keV Xe^+ results in virtually none of these characteristic bands. Arnold[5] has offered an explanation of these observations based on defect center charge decoration. It is proposed that the electron capture cross-section for heavier ions near the end of their trajectory is greater than for light ions so that Frenkel defects produced near the Bragg peak in nuclear stopping power are denied compensating charge and quickly self-annihilate driven by strong, long range Coulombic forces.

An alternative explanation is presented in this work for the origin of the characteristic uv absorption bands at 6.0, 5.4 and 4.8 eV in terms of local stoichiometric imbalance in favor of excess aluminum or, alternatively, oxygen vacancies. Al_2O_3 is a material particularly well suited for studying stoichiometric upset in favor of excess aluminum in that only two out of three sites on the cation sublattice are normally occupied whereas all of the anion sites are filled. The concept of insulator defects based on local stoichiometric imbalance is not new; it has formed the basis for interpreting additive coloration and ionizing radiation effects in alkali halide crystals for many years[8]. More recently, stoichiometric upset has been shown to be a fruitful approach to interpreting radiation induced charge build-up near the oxide semiconductor interface in MOS structures[9,10]. By comparing optical absorption and emission spectra from aluminum and oxygen implanted sapphire with those from proton irradiated, and pile and fusion neutron irradiated material it is concluded that stoichiometric imbalance in favor of oxygen vacancies has a stronger influence on the intensity of the characteristic uv bands than the previously offered explanations, especially for self-implants.

EXPERIMENTAL

Samples of crystalline Al_2O_3 (Adolf Meller Co., substrate grade, c-axis parallel to sample plane) were implanted at ambient temperature with 200-keV ions of H^+, O^+, Ne^+, Al^+ and Xe^+ at fluxes respectively of 0.8, 0.8, 1.0, 0.8-to-2.0, and 0.4 µA/cm^2. Other samples (Union Carbide, uv grade, c-axis perpendicular to sample plane) were irradiated at room temperature with 14-MeV neutrons at the Lawrence Livermore Laboratory ICT rotating target Intense Neutron Source to a fluence of 10^{17} n/cm^2 at a flux of approximately 10^{12} n/cm^2 sec[11]. Absorption spectra were taken with a Cary 14 spectrophotometer and a McPherson model 225 spectrometer. Luminescence and excitation spectra were obtained with Bausch and Lomb 1/4-meter monochromators fitted with an S-20 photomultiplier tube. This combination was calibrated with an Electro Optics Associates model L-101 spectral irradiance standard quartz-iodide lamp. Damage profiles were obtained by sputtering with a Technics model MIM-TLA 15 ion milling machine. Sputtered layer thicknesses were measured with a Sloan Dektak.

RESULTS AND DISCUSSION

Figure 1 shows the elastic stopping power profiles $S_D(x)$ for 200-keV Xe^+, Al^+, Ne^+, and O^+ into sapphire obtained from an E-DEP-1 code[12] for the theory of Lindhard, Scharff and Schiott[13]. The stopping power integral is listed as an inset. Assuming the uv coloration is proportional to the number of lattice defects and to the energy into elastic processes, which a priori is not inconsistent with the early results of Arnold and Compton[1] and the 3-MeV Ne^+-induced optically active lattice damage measurements in crystalline magnesia[14], the intensity of the 6-eV band in sapphire resulting from Al^+ implant should be no more than 1.5 times that due to O^+ implant, about 1.2 times that due to Ne^+ bombardment, and about 10^3 times that due to H^+ irradiation when all implants are performed with 200-keV ions to the same fluence. On the other hand, later work by Arnold et al.[6] suggests the uv coloration is proportional to the energy deposited through inelastic processes. The average energy density into elastic (ρ_d) and inelastic (ρ_e) processes and the average dpa level, listed in Table I, are not very different between 200-keV O^+ and 200-keV Al^+ implants, and yet the optical response varies greatly, as shown below. For comparison, the dpa level at 10^{17} 14-MeV neutrons/cm^2 is approximately 1.3×10^{-4} based on the damage energy calculations for aluminum of Robinson[15].

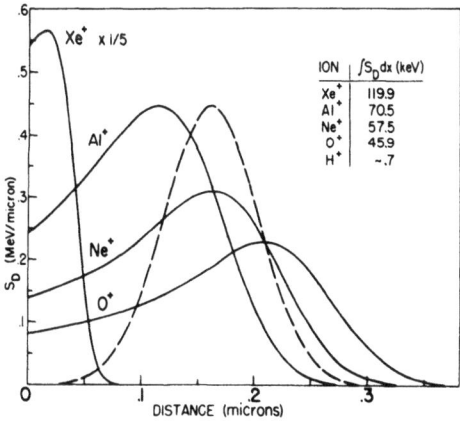

Fig. 1. Spatial profile of elastic stopping power, $S_D(x)$, for 200-keV ions into sapphire. The insert lists the integral energy deposited through elastic events.

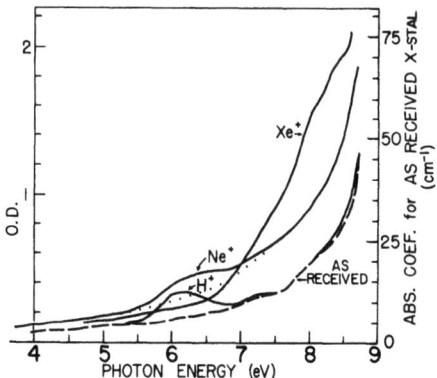

Fig. 2. Optical absorption spectra of crystalline sapphire unimplanted and implanted with 200-keV H^+, Ne^+ and Xe^+ to 10^{16} cm^{-2}.

Figure 2 shows the room temperature optical absorption before and after ion implantation at ambient temperature to 10^{16} cm^{-2} of 200-keV H$^+$, Ne$^+$ and Xe$^+$. Both the H$^+$ and Ne$^+$ implants result in a weak absorption band near 6 eV while the Xe$^+$ implant only produces a broad band near 8 eV. The dotted line in the Ne$^+$ spectrum represents a reasonable estimate of the underlying implant-induced background absorption which decrease monotonically to lower photon energies. Similar background absorption has been observed in Ne$^+$ implanted crystalline MgO and may be related to perturbed excitons[14]. Figure 3 compares induced absorption after implanting to 10^{16} cm^{-2} of 200-keV Al$^+$, H$_+^+$ and O$^+$. The Al$^+$ implant results in an intense uv band while the O$^+$ implant produces only the nondescript background absorption; furthermore, no 6-eV band is present in the O$^+$-implanted sample after 10^6R exposure to Co^{60} irradiation. From Figures 3 and 4 and Table I, it is not clear that the uv coloration is strictly proportional to energy density into either elastic or inelastic processes; a complex interaction of both these including saturation effects is more likely at this dpa $\gtrsim 1$ level. These results do not imply that O$^+$ and other medium-to-heavy-mass ion implantations do not create defects associated with these uv bands but rather they do not to the same concentration as Al$^+$ implants. O$^+$ implants may very well result in vacancy concentrations near 10^{18-19} cm^{-3} in a thin submicron layer (as 3-MeV Ne$^+$ implantation of MgO dose[14]), but would not be detected optically, whereas the centimeter thick, uniform damage resulting from fast-neutron irradiaton allows detection as low as 10^{16} cm^{-3} as shown below.

Al$^+$ implantation produces a weak 4.8-eV band observed in pile[3,4] and fusion-[11] neutron-irradiated sapphire. A band at 5.4 eV is also present, however it is obscured by background absorption and by either a broad band(s) centered above 6 eV underlying the characteristic 6-eV band observed after neutron and proton irradiation, or by the low-energy tail of a broader version of the characteristic 6-eV band. Such broadening could arise from perturbations caused by a high degree of lattice debris accompanying these dpa

TABLE I. Energy Density and dpa for 200-keV Ion Implantation to 10^{16} cm^{-2}.

Ion	ρ_d(keV/cm^3)	ρ_e(keV/cm^3)	dpa	OD(6eV)	$\int$OD(E)dE (arb. units)
Xe+	2.8x10^{23}	1.9x10^{23}	17	0.03	---
Al$^+$	4.5x10^{22}	8.2x10^{22}	2.7	1.11	766
Ne$^+$	3.1x10^{22}	7.7x10^{22}	1.9	0.12	75
O$^+$	2.3x10^{22}	7.7x10^{22}	1.4	0.03	---
H$^+$	5.1x10^{20}	1.3x10^{23}	0.03	0.17	70

ASSOCIATION OF THE 6-eV OPTICAL BAND

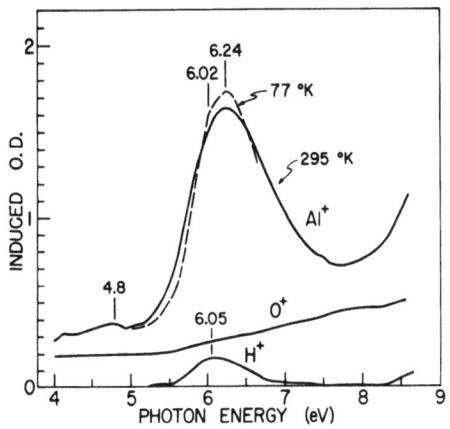

Fig. 3. Room temperature (solid line) and 77°K induced spectra (dashed line) after implantation with 200-keV H^+, O^+ and Al^+ to 10^{16} cm^{-2}.

levels. However, there is some evidence for the former case in the 77°K spectra shown as the dashed line in Fig. 3 where a low-energy shoulder coincides with position of the 6.02-eV band while the peak absorption occurs at 6.24 eV. In any event, Al^+-and Ne^+-implant-produced uv absorption appears broader and shifted to higher energy than when induced by protons and neutrons. Table II summarizes these results.

Using photoexcitation techniques we have observed a band at 5.34 eV which we believe is the same band as the 5.4 eV band observed in neutron[3,4,11] and proton[16,17] irradiation sapphire. Figure 4 includes the emission spectrum of Al^+-implanted sapphire when exposed to uv light. A single band is observed at 3.78±.05 eV of FWHM 0.41±.04 eV. The excitation spectrum of this emission band

TABLE II. The prominent uv absorption band at 295°K.

Energy/particle	Peak Position (eV)	FWHM (eV)
E > .1-MeV neu.[a]	6.02	.60
E > .1-MeV neu.[b]	6.1	N.A.
14-MeV neu.(1.7x10^{17}cm^{-2})	6.02±.02	.79±.05
14-MeV neu.(.3x10^{17}cm^{-2})[c]	6.03±.02	.68±.05
200-keV H^+ [d]	6.05±.03	.80±.03
200-keV Al^+	6.24±.05	1.33±.05
100-keV Al^+	6.24±.05	1.33±.05
200-keV Ne^+	6.3 ±.1	.95±.08
200-keV O^+	---	---
200-keV Xe^+	---	---

[a] Ref. 2.   [b] Ref. 4.   [c] Fig. 5.   [d] Ref. 16.

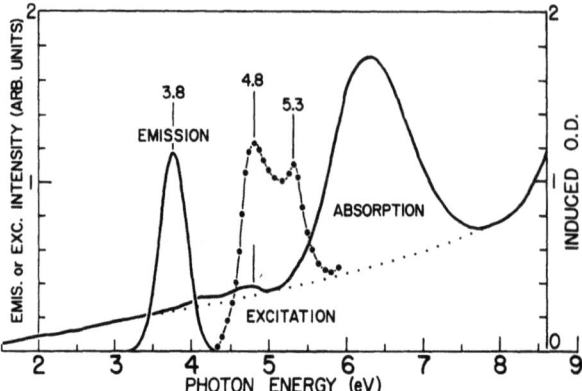

Fig. 4. A comparison of room temperature absorption, uv-stimulated photoemission (solid) and the corresponding excitation spectrum (dot-dash) after 10^{16} cm^{-2} 200-keV Al$^+$ implantation. The dotted line is an estimate of the background absorption.

is shown as the dot-dashed line. It clearly shows two peaks at 4.82 and 5.34 eV. This excitation spectrum has been corrected for both variations with energy of the intensity of the incident pump light and for changes in self-absorption of the excitation light throughout the absorption spectrum[18]. The emission spectrum has been corrected for detector response. The 3.8-eV emission is not observed in the unimplanted portion of these Al$^+$ implanted crystals, nor is it detected in either implanted or unimplanted halves of O$^+$ implanted samples. However, similar emission and excitation spectra are observed in proton[16] and neutron irradiated samples. Figure 5 shows the 77°K spectra for 3×10^{16} cm^{-2} 14-MeV neutron irradiated material. Here the 4.8-and 5.4-eV bands in the excitation spectrum are more pronounced and clearly mimic similar bands found in absorption. As in Fig. 4, the excitation and emission spectra have been suitable corrected.

These spectra then suggest that there is a characteristic 3.8-eV emission band in particle irradiated crystalline sapphire stimulated by pumping the three uv absorption bands near 4.8, 5.4, and 6.0 eV. Furthermore, that the defects associated with at least the two lower energy bands (and possibly all three bands) may be related is shown by the fact that they emit at the same energy, i.e., 3.8 eV. Either the assumption must be made that energy is transferred from the defect responsible for the 5.4-eV band to the defect associated with the 4.8-eV absorption or that the two defect wavefunctions interact strongly and overlap. The latter is not unreasonable in the ion implanted case where a high density ($\sim 10^{21}$ cm^{-3}, see below) of damage is created in a thin 0.1μ layer. However

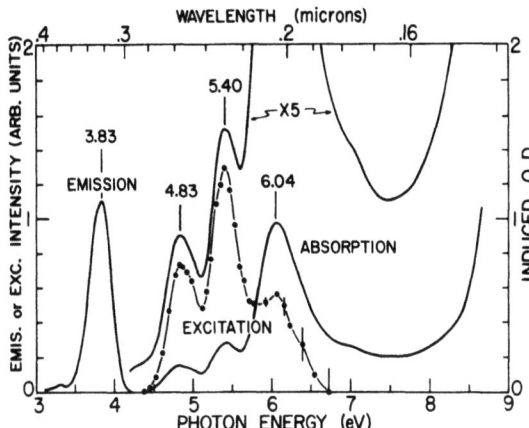

Fig. 5. A comparison of 77°K absorption, uv stimulated photoemission (solid) and the corresponding excitation spectrum (dot dash) after 3×10^{16} cm^{-2} 14-MeV neutron bombardment. A x5 absorption spectrum was shifted vertically 0.1 O.D. for convenient comparison with the excitation spectrum.

this is not the case in low-fluence (10^{17} n/cm^2), fast-neutron irradiated crystals where the optically active defect concentration is typically 10^{17} cm^{-3} (Fig. 5). The fact that both high and low-damage concentration samples show similar emission and excitation spectra rules out the case for long range spatial energy transfer unless the unlikely assumption is made that energy transfer operates at low damage concentration but not at high. The two centers could be very close, within a lattice spacing, or an extreme case, where they are in fact spatially coincident and represent different excited states of the same localized defect.

The number N of stable optically active defects per implanted Al$^+$ may be estimated from Fig. 4 with the use of Samkula's equation[19]: $Nf = 0.87 \times 10^{17} (n/n^2+2)^2 \alpha \Delta E$, where n is the refractive index, α the maximum absorption coefficient, ΔE the band FWHM and f the oscillator strength. α at 6 eV is calculated from the induced O.D. above the background (dotted line Fig. 4), 1.10 O.D. ΔE is taken from the values determined from neutron and proton irradiations (Table II), 0.80 eV, rather than the larger value observed in Fig. 4 because, as mentioned above, there is some evidence of another band above 6 eV which makes this band appear broader. f is assumed unity. Then, at a fluence of 10^{16} cm^{-2}, approximately one optically active damage center is found per implanted Al$^+$. In spite of the lower energy into elastic processes ($\int S_D dx = 26.7$ keV) this is much lower than the 120 oxygen displacements per incident 40-keV Kr$^+$ determined by a Rutherford back-scattering-channeling technique reported by Naguib[20] in Al$_2$O$_3$.

The simplest of displacement defect structures to assume are charges trapped at cation or anion vacancies. As the characteristic optical spectra are produced with cation and not anion self-implants, anion vacancies appear more likely. La et al. use a point-ion potential model to calculate the position of optical transitions associated with a single electron trapped at an O^{-2} vacancy. These are 2.26, 3.39 and 5.15 eV.[21] Intense uv

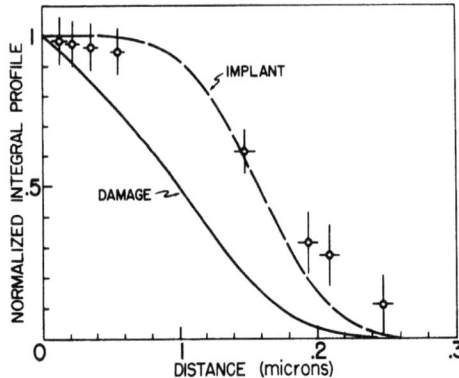

Fig. 6. A comparison of the integral of the spatial damage (solid) and implant (dashed) profiles shown in Fig. 1 for 200-keV Al$^+$ with the etch-back profile (circles) of absorption remaining at 6 eV after a 200-keV Al$^+$ implant to 10^{16} cm^{-2} at 2×10^{-6} A/cm^2.

absorption bands in other oxides, albeit different crystal structures, have been assigned to electrons trapped at anion vacancies: the 5.75-eV band in SiO$_2$[22], the 4.95-eV band in MgO[23], the 3.7-eV band in CaO[24], etc. The Stokes shift, $\Delta E'=1.05$ eV, observed in particle irradiated Al$_2$O$_3$ associated with the 4.78-eV absorption and the 3.83-eV emission lies intermediate between the case for the single electron trapped at an anion vacancy (F$^+$ center) in MgO, where $\Delta E'=1.9$ eV[25], and in CaO where $\Delta E'=0.4$ eV at room temperature[26].

As further evidence that the uv coloration is to be associated with a stoichiometric upset, Fig. 6 shows the result of etch-back experiments. The solid and dashed lines represent respectively the integral of the damage and aluminum implant distributions shown in Fig. 1. The data points are the (normalized) absorption remaining at 6 eV with the baseline (dotted line in Fig. 4) subtracted after material to the indicated depths has been removed by ion milling. The uv coloration spatial profile is closer to the implant than the damage profile, again supporting the thesis that excess aluminum plays a strong role in stabilizing displacement damage in sapphire.

Now that the thickness of the damaged layer has been measured, $0.10\pm.02\mu$, Smakula's equation with Fig. 4 may be used to esimtate the average concentration of optical damage; $N \sim 1\times10^{21}$ cm^{-3}, which is nearly 1% of the atomic density. The average calculated dpa in the displacement damage profile is about 2.7 (Table I). This is a much higher damage level than estimated from optical results. The optically active damage is stabilized at a lower concentration similar to the excess implanted aluminum density. However, this dpa level is about an order of magnitude higher than that corresponding to the maximum induced absorption at 6 eV observed by Luera et al. after heavy ion irradiation[27]. In fact we have observed no saturation or decrease in 6-eV absorption with increasing fluence of 200-keV Al$^+$ implants from 3×10^{14} to 10^{16} cm^2; this fluence range corresponds to an average dpa range of 0.092-to-2.7 and a range of average energy density into elastic processes of 1.5×10^{21}-to-4.5×10^{22} keV/cm^3. Again, excess aluminum plays an important defect stabilization role in Al_2O_3.

CONCLUSION

We have shown that there is a characteristic 3.8-eV photoemission band associated with the particle-irradiation-induced coloration bands at 4.8, 5.4 and 6.0 eV in crystalline sapphire. Furthermore, excess aluminum has a strong influence in stabilizing the optically active lattice damage produced by energetic particles. The fact that excess anions produced by oxygen implantation do not result in uv absorption or emission bands suggests that these bands are to be associated with anion vacancies.

ACKNOWLEDGEMENTS

The authors gratefully acknowledge John W. Burgess for performing the implants, and I. Manning and G.P. Mueller for assistance with the energy deposition code.

REFERENCES

1. G.W. Arnold and W.D. Compton, Phys. Rev. Lett. 4, 66 (1960).
2. P.W. Levy and G.J. Dienes, Report of Bristol Conf. on Defects in Crystalline Solids, July, 1954 (The Physical Society, London, 1955), p. 256.
3. P.W. Levy, Phys. Rev. 123, 1226 (1961).
4. E.W.J. Mitchell, J.D. Rigden and P.D. Townsend, Phil. Mag. 5, 1013 (1960).
5. G.W. Arnold, Proc. First Topical Meeting on the Technology of Controlled Nuclear Fusion, CONF-74042, (NTIS, Washington, D.C.), Vol. II, p. 500.
6. G.W. Arnold, G.B. Krefft, and C.B. Norris, Appl. Phys. Letters 25, 540 (1974).
7. B.D. Evans, "Ion Beam Simulation of Fast-Neutron Damage in Crystalline Magnesia and Sapphire", presented at Int. Conf. on the Application of Ion Beams to Materials, University of Warwick, U.K., 8-12 September 1975.
8. J.H. Schulman and W.D. Compton, Color Centers in Solids, (The MacMillan Co., New York, 1962), Ch. 1.
9. C.L. Marquardt and G.H. Sigel, Jr., IEEE Trans. on Nucl. Sci. NS-22, 2234 (1975).
10. M.H. Woods and R. Williams, J. Appl. Phys. 47, 1082 (1976.

11. These are companion samples to those reported in J.M. Bunch and F.W. Clinard, Jr., J. Amer. Ceram. Soc. 57, 279 (1974).
12. I. Manning and G.P. Mueller, Comput. Phys. Commun. 7, 85 (1974).
13. J. Lindhard, M. Scharff and H.E. Schiott, Denske Videnskab. Selskab 33, No. 14 (1964).
14. B.D. Evans, Phys. Rev. B9, 5222 (1974).
15. M.T. Robinson, in "Nuclear Fusion Reactors", Proc. of Brit. Nucl. Energy Soc. Conf. on Nucl. Fusion Reactors, UKAEA Culham Laboratory, 17-19 September 1969, p. 364.
16. B.D. Evans, H.D. Hendricks, and J.M. Bunch, Amer. Ceram. Soc. Bull. 55, 458 (1976).
17. D.W. Muir and J.M. Bunch, Proc. Int. Conf. on Rad. Effects and Tritium Tech. for Fusion Reactors, Gatlinburg, TN., 1-3 Oct. 1975, USERDA CONF-750989, (NTIS, Washington, D.C.), Vol. II, p. 517.
18. D. Curie, *Luminescence in Crystals*, (John Wiley and Sons, Inc., New York, 1963).
19. D.L. Dexter, Solid State Phys. 6, 353 (1958).
20. H.M. Naguib, J.F. Singleton, W.A. Grant, and G. Carter, J. Matl's. Sci. 8, 1633 (1973).
21. S.Y. La, R.H. Bartram, and R.T. Cox, J. Phys. Chem. Solids 34, 1079 (1973).
22. R.A. Weeks, J. Appl. Phys. 27, (1956).
23. Y. Chen, W.A. Sibley, F.D. Srygley, R.A. Weeks, E.B. Hensley, and R.L. Kroes, J. Phys. Chem. Solids 29, 863 (1968).
24. J.C. Kemp, W.M. Ziniker, J.A. Glaze, and J.C. Cheng, Phys. Rev. 171, 1024 (1968).
25. Y. Chen, J.L. Kolopus, and W.A. Sibley, Phys. Rev. 186, 865 (1969).
26. B.D. Evans, J.C. Cheng, and J.C. Kemp, Phys. Letters 27A, 506 (1968).
27. T.F. Luera, J.A. Borders, and G.W. Arnold, presented at V Int. Conf. on Ion Implantation in Semiconductors and Other Materials, Univ. of Colorado, Boulder, CO., 9-13 August 1976, IV-4; also appearing in this volume.

THERMOLUMINESCENCE OF ION-IMPLANTED SiO_2*

G. W. Arnold

Sandia Laboratories

Albuquerque, New Mexico 87115

ABSTRACT

Thermoluminescence (TL) has been measured from room temperature to 500°C for ion-implanted fused silica glasses, crystalline synthetic quartz and rf-sputtered SiO_2 films. Measurements of the TL spectra for widely varying values of electronic and atomic energy depositions, along with the known impurity concentrations of the various systems, has allowed some of the TL features to be identified. In particular, 1) a TL peak at 150°C in fused silica has been identified with defects formed by structural modification, 2) a 330°C peak in crystalline quartz and relatively impure fused silica is tentatively assigned to a center involving Al, 3) a 100°C peak, common to all silicas may be related to oxygen vacancies, and 4) an ~200°C peak may be the analog of the 245 nm impurity absorption band seen in some fused silica glasses.

INTRODUCTION

Ion implantation has been used in the present experiments to study the induced thermoluminescence (TL) in well-characterized commercial fused silica glasses as well as in rf-sputtered SiO_2 films and crystalline synthetic quartz. Theoretical calculations by Brice[1] were used to determine the partitioning of ion energy into electronic (ionization and excitation) and atomic (displacement and lattice heating) processes. Experimental variations in ion mass, fluence, and energy allow the partitioned energies to be varied over a wide range of values. Corresponding systematic variations in the growth, quenching, and intensity of the TL peaks, combined with the knowledge of the impurity content and growth conditions of the SiO_2

system under investigation, allow inferences to be drawn concerning the nature of the defects involved in the TL process. This approach permits the great sensitivity of TL techniques to be used to probe the defect structure of silica, yet avoids having to rely on detailed analysis of the individual TL curves by any of the various theoretical treatments which have been critically reviewed by Kelly and Bräunlich[2] and found to be inadequate.

EXPERIMENTAL

The commercial fused silica glasses used in this investigation (Suprasil W1, Suprasil 1, Corning 7940, Optosil 1, and Infrasil 1) can be categorized using the nomenclature introduced by Hetherington et al.[3] This systemization is on the basis of method of manufacture and leads to a characterization in terms of impurity content (see Fig. 1). Measurements were also made on rf-sputtered SiO_2 films[4] ($\sim$2 μm thick) deposited on Corning 7940 substrates at a rate of 40 Å/min in argon at a pressure of $\sim$3 μm and on crystalline synthetic (Sawyer) quartz samples. The glasses and SiO_2 films were annealed in flowing N_2 gas for 1 hour at 600°C prior to implantation. The quartz samples were annealed at 550°C (in N_2) for 1 hour.

The implantations were made with an Accelerators, Inc. machine in the ion energy range of 50 - 250 keV. All implantations were made at room temperature using continuous beam currents ranging from $\sim$100 nA/cm^2 to 2 $\mu A/cm^2$. To minimize charging effects, a metal ion-beam mask was in close contact with the samples and the beam was over-scanned. After implantation the samples were stored at temperatures ($\sim$0-5°C) below room temperature until TL measurements were made. These measurements were made over the range from room temperature to 500°C. The measurements were made with reference to the non-thermoluminescent rerun of the same sample using a modified Harshaw Model 2000 Thermoluminescence Analyzer at a rate of 0.5°C/sec. Optical absorption measurements were made with a Cary 14 spectrophotometer.

RESULTS AND DISCUSSION

Figure 1 shows the TL spectra induced by 150 keV H^+ ions, at various fluence levels, in the fused silica glasses categorized as types I through IV according to Hetherington et al.[3] Similar results are obtained with other ions at other energies and fluences. The common dominant TL characteristic is the peak centered near 150°C. Other peaks are regularly observed, in the glasses with relatively large metallic impurity concentrations, at about 100, 220 and 330°C. The 330°C peak is especially prominent, in a particular low fluence range, in the Infrasil 1 glass. A weak 100°C peak is common to all the SiO_2 systems examined. Note that the TL intensity for all the

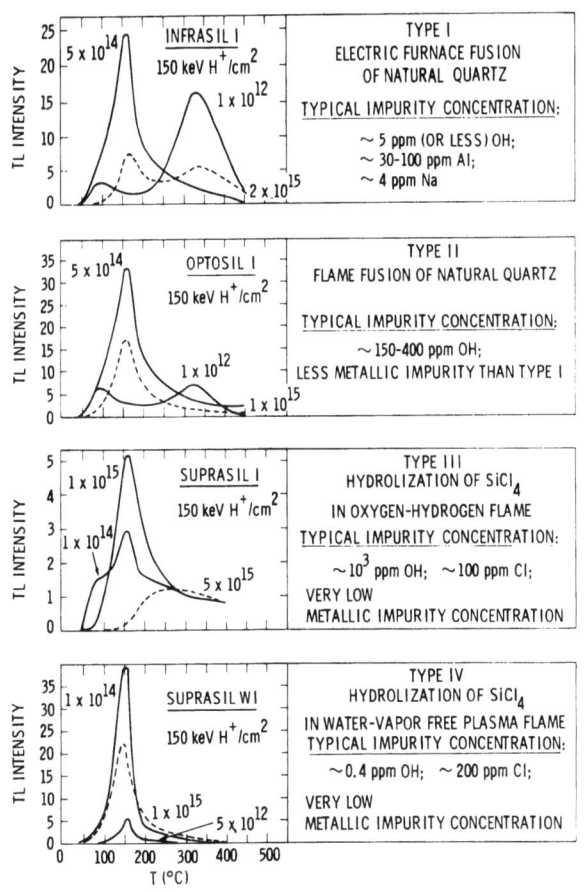

Figure 1. TL intensity vs. temperature for fused silica Types I-IV as represented by Infrasil 1, Optosil 1, Suprasil 1 and Suprasil W1 for various 150 keV H^+ ion fluences. The typical impurity concentrations are due to Bruckner (J. Non-Crystall. Solids 5, 123 (1970)). The units of TL intensity are arbitrary but the relative intensities between the various glasses are scaled properly.

glasses is quenched at high fluence levels.

The growth of the 150°C integrated TL peak intensity in Suprasil W1 for 150 keV H^+, 157 keV He^+ and 50 keV Xe^+ ion implantation is shown in Fig. 2 as a function of ion fluence, energy into electronic processes (E_e^i keV/cm^2) and energy into atomic processes (E_n^i keV/cm^2). As can be seen in Fig. 2, the individual growth curves can be brought into rather good agreement, in the linear growth region, only when the integrated TL intensity for the 150°C peak is plotted as a function of energy (E_n^i keV/cm^2) into atomic processes. It is appropriate to attempt the correlation of the TL data with incident energies per cm^2 rather than as a function of energy per cm^3 because the integral measurement of TL output is proportional to the incident energy. Figure 3 shows that the correlation of the linear growth of the 150°C TL peak integrated intensity with incident ion energy into atomic processes holds for a wide range of ion masses. A comparison is made in Table 1 of the integrated 150°C peak TL intensity in Suprasil W1 for a common value of 10^{12} keV/cm^2 into electronic and atomic processes for H^+, He^+,

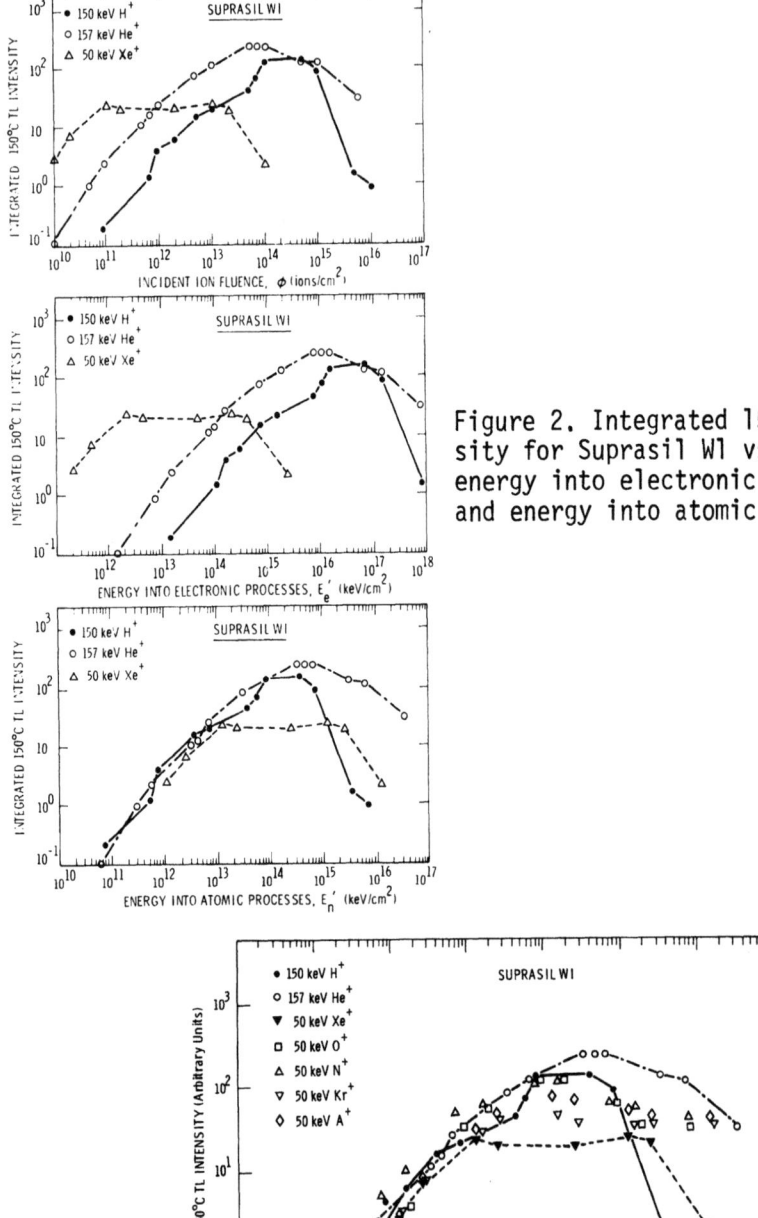

Figure 2. Integrated 150° TL intensity for Suprasil W1 vs. ion fluence, energy into electronic processes, and energy into atomic process.

Figure 3. Integrated 150°C intensity for Suprasil W1 for various ions and energies as indicated vs. energy into atomic processes.

Table I. 150°C TL Peak Integrated Intensity for 10^{12} keV/cm^2 into Atomic (E_n') and Electronic (E_e') Processes in Suprasil W1.

ION	E(keV)	EQUAL E_n'		EQUAL E_e'	
		ϕ(ions/cm^2)	TL PEAK AREA (cm^2)	ϕ(ions/cm^2)	TL PEAK AREA (cm^2)
H$^+$	150	1.27×10^{12}	4.28	6.76×10^9	0.20
He$^+$	157	1.54×10^{11}	5.47	6.70×10^9	0.32
Ne$^+$	50	5.33×10^{10}	5.39	3.47×10^{10}	2.54
Xe$^+$	50	3.81×10^{10}	4.60	4.77×10^{10}	11.46

Ne$^+$ and Xe$^+$ ions.

The departure from linearity (Fig. 3) occurs at lower energy deposition levels for increasing ion mass. The saturation value of integrated 150°C TL intensity occurs at about the same volume energy deposition for the various implanted ions. This value is of the order of 10^{19} keV/cm^3. The decrease in integrated TL intensity per unit energy deposition past the saturation value is most marked for H$^+$ ions. The quenching of TL intensity past the saturation value is also greatest for Suprasil W1(low-OH) than for Suprasil 1 or Corning 7940 (high-OH) as seen in Fig. 4. The radiation-quenching of TL intensity could occur because of competitive trapping of charge-carriers by accumulated concentrations of other traps with larger capture cross sections. The common volume energy deposition value for saturation, however, indicates that the quenching may be due to a structural transformation, such as the saturation of compaction observed by EerNisse,[4] which decreases the radiative transition rate. Figure 4 also shows that the more open-structured silicas (Suprasil 1 and 7940) require a higher energy deposition value before quenching of the TL commences. This is reasonable, if TL quenching is due to re-orientation under stress, because the loose-structured glasses can accommodate greater stresses before short-range order is significantly altered. The data for Suprasil 1 of Fig. 4 shows that the defects generated by the initial implant are annealed out during the first TL measurement since the results for the same samples after the second implantation are very nearly the same.

Figure 5 shows a comparison of TL spectra from Suprasil W1, Infrasil 1, and synthetic crystalline quartz. The band near 330°C is found in Infrasil 1, Optosil 1, and crystalline quartz, but not in Suprasil W1. The other bands for crystalline quartz at 150°C and at ~220°C are also found in the two fused silica glasses. Analyses of Infrasil 1, Optosil 1 and the quartz samples show Al to be present

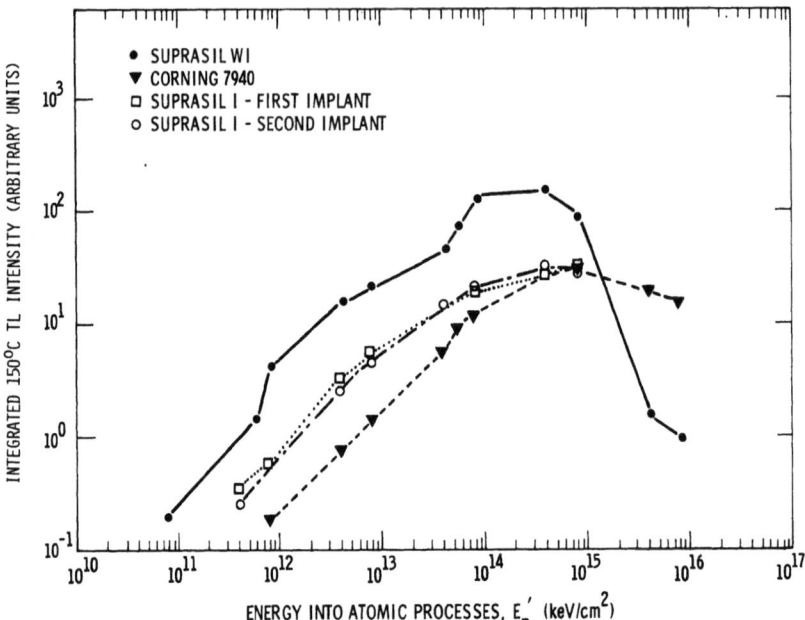

Figure 4. Integrated 150°C TL intensity for 150 keV H^+ ions implanted into Suprasil W1, Corning 7940, and Suprasil 1 vs. energy into atomic processes.

in concentrations of 10, 10, and 3 ppm, respectively. While these observations do not constitute proof that the 300°C TL peak is to be associated with Al impurities, Al is common to both the silica glasses and to quartz but absent in Suprasil W1.

Figure 6 shows a comparison of thermoluminescence from rf-sputtered 7940 silica, in the form of a 2 μm film,[5] and bulk 7940 silica glass for 5×10^{12} 150 keV H^+ ions/cm². The peak of the TL distribution for the sputtered SiO_2 film occurs at a temperature (~100°C) for which a peak is observed in bulk 7940 and in the other silica glasses as well. The growth and quenching of the TL spectra occurs in this thin film system as well as in the bulk glasses studied. Hickmott[6] has studied the intrinsic TL intensity in X-irradiated rf-sputtered films and found a similar TL spectra. His results, in conjunction with ESR measurements,[7] led him to conclude that some of the defects in as-sputtered films were to be associated with Si-H bonding. His work also indicated that the as-grown films are oxygen-deficient. Oxygen vacancies produced in the ion-bombarded films of the present experiment might account for the increased 100°C thermoluminescence at low fluence levels.

Figure 7 shows a comparison of TL intensity induced in various unimplanted silica glasses by ultraviolet radiation from a 150 watt high-pressure Hg-lamp. For this purely ionizing subbandgap source of

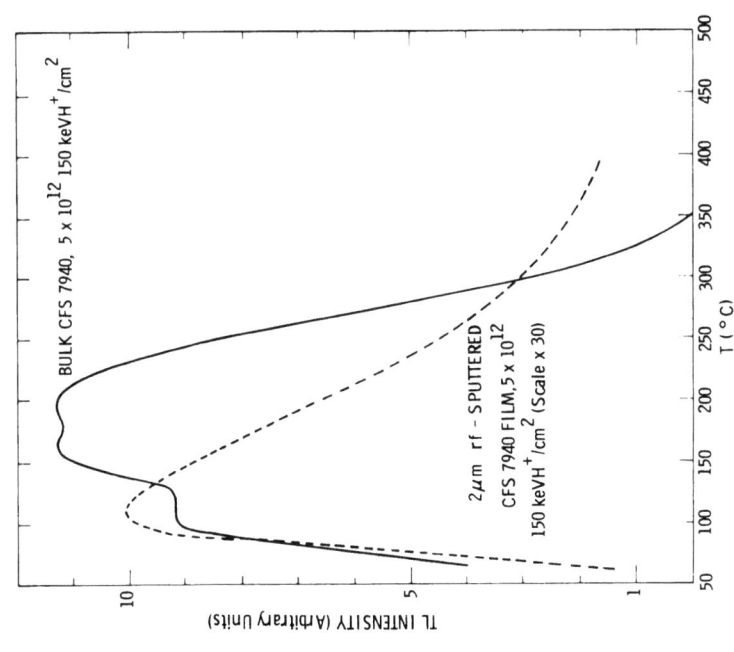

Figure 6. TL intensity vs. temperature for a 2 μm rf-sputtered 7940 film on 7940 substrates and bulk 7940 for 5×10^{12} 150 keV H$^+$ ions/cm^2.

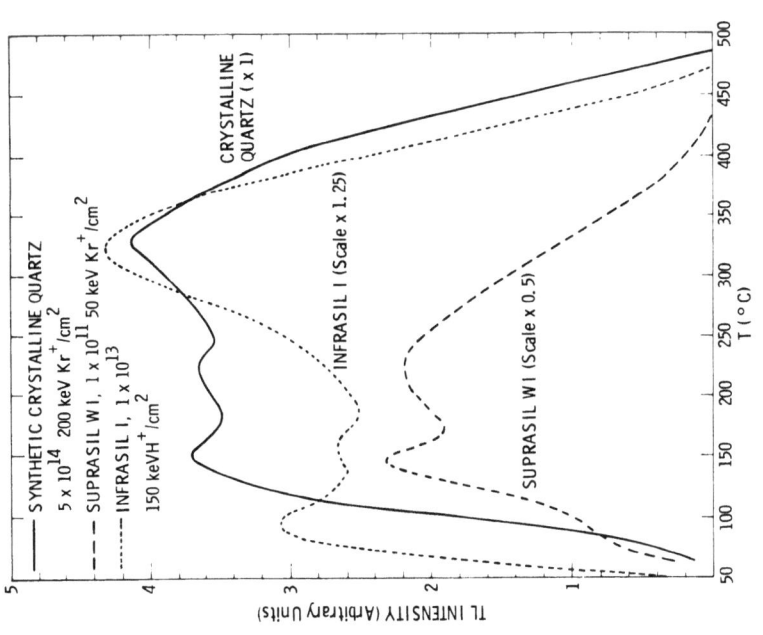

Figure 5. TL intensity vs. temperature for synthetic crystalline quartz, Suprasil W1 and Infrasil 1, for the ions, energies, and fluences indicated.

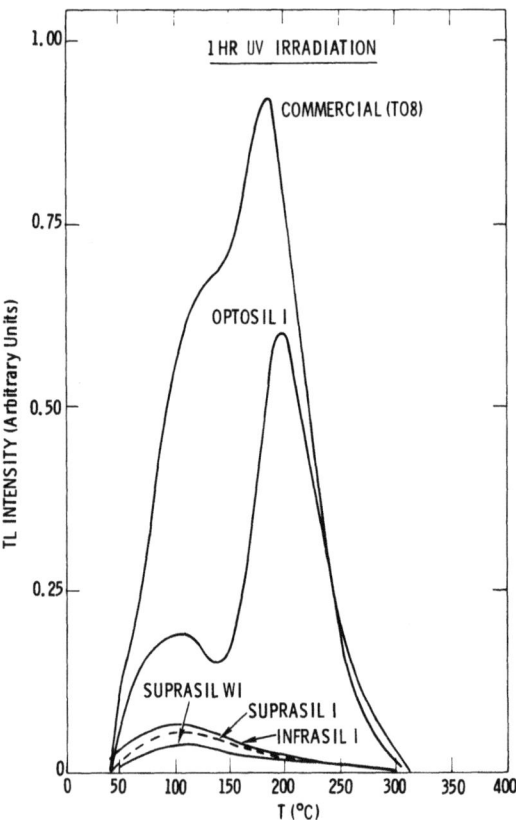

Figure 7. TL intensity vs. temperature for various fused silica glasses after 1 hour UV irradiation.

irradiation, the TL peaks at ~100°C and ~200°C are especially prominent only for the Optosil 1 and T08 silica glasses. These glasses are both Type II silicas, i.e., they are prepared from natural quartz by flame fusion. They also both have, in the as-fabricated form, an absorption band at ~245 nm which has been assigned to a trivalent impurity (Al, Ge) substituting for Si^{4+} adjacent to an oxygen vacancy.[8] This type of center may be responsible for the ~200°C TL peak for the two glasses in Fig. 7.

CONCLUSIONS

The sensitivity of TL techniques and the availability of accurate calculations[1] of ion energy partition, together with well-characterized experimental materials has allowed the following conclusions and inferences to be drawn:

1) The 150°C TL peak in fused silica is formed by that part of the ion energy which goes into atomic processes. The structural changes brought about by this type of energy deposition are known to be primarily responsible for compaction of the implanted volume.[4,9]

2) The 330°C TL peak observed in fused silica glasses of relatively high metallic impurity concentration, and in crystalline synthetic quartz, is tentatively assigned to electronic transitions involving Al impurity.

3) A 100°C TL peak, observed in most silica systems, may be related to oxygen vacancies on the basis of correlations of the TL spectra of rf-sputtered Corning 7940 films with bulk 7940 material.

4) A near 200°C TL maximum may arise from the center seen in absorption at 245 nm which has previously been assigned[8] to a trivalent impurity substituting for Si adjacent to an oxygen vacancy.

ACKNOWLEDGEMENTS

The author thanks Roger Shrouf for his assistance in collecting the experimental data.

*This work was supported by the United States Energy Research and Development Administration (ERDA) under Contract E(29-1)789.

REFERENCES

1. D. K. Brice, Rad. Effects 6, 77 (1970), and unpublished calculations on damage in SiO_2.
2. P. Kelly and P. Bräunlich, Phys. Rev. 1B, 1587 (1970).
3. G. Hetherington, K. H. Jack, and M. W. Ramsay, Phys. Chem. Glasses 3, 129 (1962).
4. E. P. EerNisse, J. Appl. Phys. 45, 167 (1974).
5. Prepared by G. Kominiak, Sandia Laboratories.
6. T. W. Hickmott, J. Appl. Phys. 45, 1060 (1974).
7. T. W. Hickmott, J. Appl. Phys. 45, 1050 (1974).
8. G. Hetherington, K. H. Jack, and M. W. Ramsey, Phys. Chem. Glasses 6, 6 (1965).
9. H. M. Presby and W. L. Brown, Appl. Phys. Lett. 24, 511 (1974).

STUDIES OF RADIATION DAMAGE PRODUCED BY ION IMPLANTATION IN SAPPHIRE[*]

T. F. Luera
White Sands Missile Range, NM 88002

J. A. Borders and G. W. Arnold
Sandia Laboratories, Albuquerque, NM 87115

ABSTRACT

Radiation damage produced by room temperature ion implantation into single crystal Al_2O_3 has been studied by a combination of optical absorption spectroscopy, Rutherford backscattering-channeling (RBC) and channeled proton-induced x-ray (CPIX) measurements. For implantations with light ions, optical absorption at 204 nm is a linear function of fluence below $10^{16} cm^{-2}$. For heavy ions the intensity of the 204 nm absorption band increases linearly for low fluences, reaches a maximum for displacing energy densities of about 10^{20}-10^{21} keV cm^{-3}, and decreases for higher energy depositions. The displacement damage curves obtained with RBC and CPIX techniques are linear at low fluences and saturate at higher fluences. The quenching of the 204 nm absorption occurs in the linear growth portion of the displacement damage curves. Samples implanted with Xe or Ar ions showed no measurable decrease in displacement damage by radiation annealing when irradiated with 50 keV protons.

---

[*] Work supported by the U.S. Energy Research and Development Administration (ERDA) under Contract E(29-1)789 and the Department of the Army (TRMS #7-CO-000-PB6-WS1-005).

INTRODUCTION

Sapphire is a technologically important insulator because of its high melting point, good thermal conductivity, and excellent dielectric properties. Apart from its application in the fabrication of solid state devices it is a candidate for controlled thermonuclear reactor insulator applications. Because of its importance, several investigators have studied radiation effects on this material through optical absorption,[1-6] lattice dilatation,[7-10] electron spin resonance,[11-14] and channeling techniques.[10,15] Recent papers have attmepted to correlate implantation-induced lateral stress with optical effects,[9] and to identify specific defects.[16] The optical absorption band at 204 nm has been shown to result from displacements of atoms from lattice sites.[1,2] A charge state dependence of this center has also been established.[3] The defect responsible for this abosrption has not been identified, but some evidence[16] points to an electron trap such as interstital Al^{3+}, an oxygen vacancy, or a simple defect cluster. In the present work, RBC and CPIX are used to provide direct information on lattice damage for comparison with 204 nm optical absorption measurements and the results of earlier lateral stress measurements.[8,9]

EXPERIMENTAL

Implantations were made 7° off-axis at room temperature with an Accelerators Inc. 300 keV accelerator. Dose rates were kept below 1 µA cm^{-2} to minimize the effects of beam heating. Adolf Meller UV quality, 1 mm thick, sapphier samples were used for most of the optical measurements. Some optical measurements were made with Union Carbide UV grade material, and the results were consistent with those obtained with the Adolf Meller samples. A different sample was used for each fluence level. Measurements were made relative to the ion implanted sample with a Cary 14M spectrophotometer. The calculations of Brice[17,18] were used to determine ion ranges and energy partition. Energy densities for electronic and atomic processes were calculated with the ion range and twice the rms damage spread, respectively.

Samples for the channeling experiments were Union Carbide UV grade material furnished by F. Clinard of LASL. Samples were annealed at 1200°C in vacuo for 2 hours. The effects of charge buildup in CPIX measurements were eliminated by depositing 50 Å of carbon on the sapphire surface. RBC analyses were made with 2 MeV He ions on samples oriented with either the <0001> or <11$\bar{2}$0> axis normal to the surface (c- and a-axis, respectively). Analysis beam currents were 6 nA (random) and up to 65 nA (aligned) and the total charge per analysis was 40 µC in a beam spot of 1.0 mm^2. The values of χ_{min} (ratio of aligned to random yields) obtained before implantation were approximately 2 percent. CPIX analysis employed 250 keV protons with a beam current of 15 nA in a beam spot of 1.3 mm^2. Charge increments of as low as 0.05 μC were used in the analysis in order to investigate the effects of ionization stimulated annealing. The χ_{min} obtained in CPIX, through the carbon layer and before implantation, were 15 percent for channeling along either the a- or c-axis.

RESULTS

The 204 nm optical absorption resulting from light ion implantation is shown in Fig. 1. These data were obtained on samples implanted with H and He ions to provide equal energy into atomic processes at the same depth. For these conditions, it is seen in Fig. 1 that the optical density is proportional to the energy into electronic processes. Earlier optical experiments[3] suggested a charge state dependence of the defect responsible for absorption. at this wavelength. The defect charge state dependence was also seen in measurements of lattice stress by a cantilever beam technique.[9] Lateral stress in the sapphire lattice induced by heavy ions

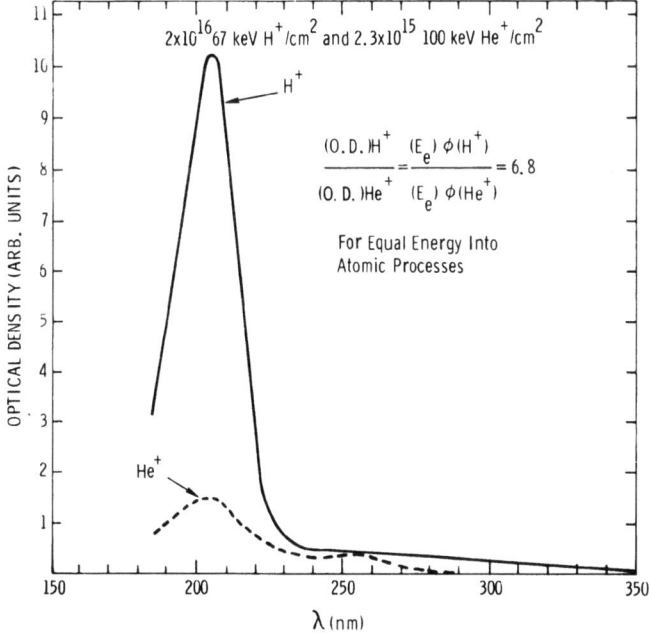

Fig. 1   Measured optical density vs. wavelength (λ) in namometers for H+ and He+ implantations of 5.5 x 10^{20} keV cm^{-3} into atomic processes at a depth of 3000 Å.

was decreased by subsequent implantation of protons or electrons of subthreshold energy. It was found that very little 204 nm absorption is produced by heavy ion fluences near $10^{16} cm^{-2}$, while implantation-induced lateral stresses are quite high at these fluences.[9] The results of low fluence implantation of 500 keV Xe and 180 keV Ar are shown in Fig. 2. The absorption is less than for light ions, and it is not observed at high fluences because of a radiation quenching phenomenon. These heavy-ion low-fluence data do not fit the simple model proposed earlier. That is, for a fixed value of E_n, the ratio of energies into electonic processes does not equal the ratio of optical densities.

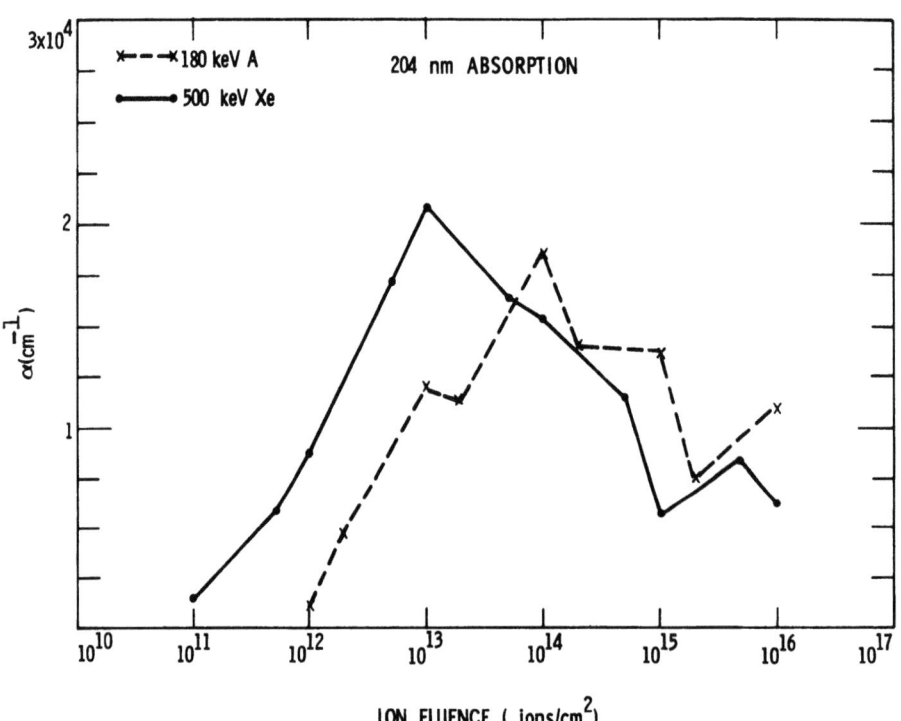

Fig. 2 204 nm opitcal absorption coefficient vs. fluence of Xe^{++} and Ar^+ ions implanted to a depth of 1000 Å.

Rutherford backscattering channeling (RBC) along the <0001> and <11$\bar{2}$0> directions was used to measure the lattice damage introduced by 500 keV Xe implantation. RBC spectra, for an a-axis sample implanted with Xe up to a fluence of $4.5 \times 10^{15} cm^{-2}$, are shown in Fig. 3. In contrast to RBC measurements of 40 keV Kr ion damage in c-axis sapphire,[15] well defined Aℓ and O surface and damage peaks were obtained even at the lowest fluences. Scattering from a

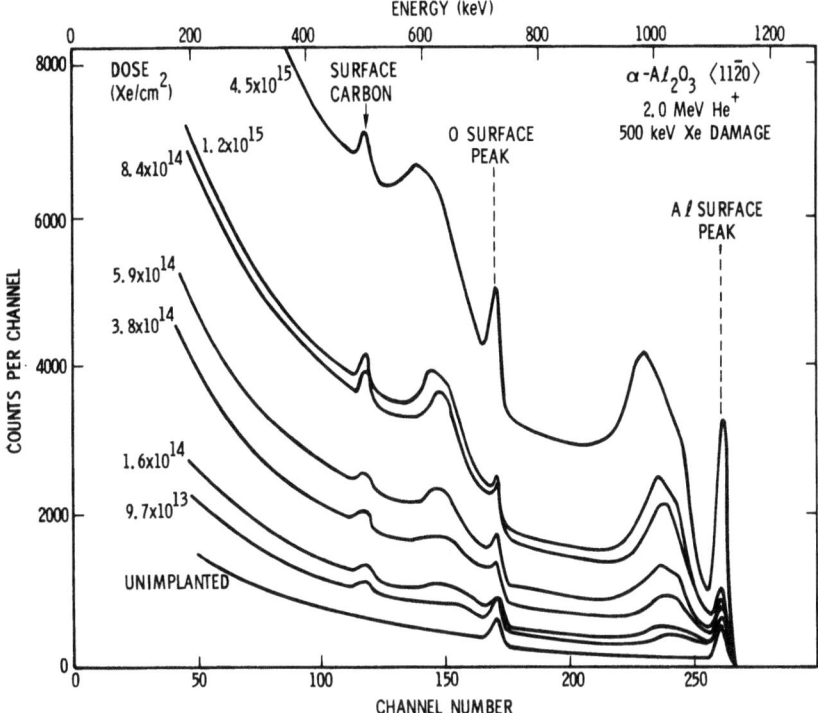

Fig. 3 a-axis aligned RBC spectra for 500 keV Xe^{++} fluences between 9.7×10^{13} and 4.5×10^{15} cm^{-2}.

surface layer begins to increase rapidly between 1 and 5×10^{15} ions cm^{-2}. At a fluence of 2×10^{16} cm^{-2} (not shown in Fig. 3) the Aℓ and O peaks reach the random level, and indicate a completely disordered surface layer. The damage peak did not reach the random level for the highest fluence analyzed with RBC. At the higher fluences shown in Fig. 3 the damage distributions are skewed away from the surface. The number of displaced Aℓ and O atoms, calculated from the areas under the damage peaks, are linear functions of Xe fluence up to about 10^{15} cm^{-2}. Above this fluence the relation becomes sublinear. In the linear region, the a-axis data yeild 180 and 125 displaced O and Aℓ atoms per incident ion, or about the proper ratio for the stoichiometric Aℓ$_2$O$_3$ lattice. For channeling along the c-axis, the measured number of Aℓ displacements is less by a factor of two. This result indicates that displaced Aℓ atoms preferentially reside in positions along the c-axis rows.

Channeling along the <11$\bar{2}$0> direction is sensitive to Aℓ atoms which might be displaced into the normally vacant Aℓ lattice sites in the stoichiometric lattice. Along the <0001> direction these sites are shadowed by rows of Aℓ atoms. An angular scan about the <0001> direction, after a 500 keV Xe implantation of 6 x 10^{14}cm^{-2}, did not show the flux peaking expected for displacements into well defined sites. On this basis it is estimated that $\leq$ 10% of the displaced Aℓ atoms occupy the normally vacant sites. A repeat of an RBC aligned spectrum analysis revealed the measured damage was stable with respect to additional ionization.

Channeled proton induced x-ray yield along the <0001> and <11$\bar{2}$0> directions was used as a measure of lattice damage introduced by 500 keV xenon implantation. Lattice disorder, as monitored by proton-induced Aℓ K$_\alpha$ x-ray intensity, is shown in Fig. 4. Disorder as viewed along the a-axis is observed to increase more rapidly and remains higher at all fluence levels. These results are in agreement with our RBC measurements, and similar to those obtained in measurements of lattice dilatation upon neutron irradiation[7] and lateral stress induced by ion implantation.[8] In those measurements lattice expansion and stress parallel to the c-axis were shown to exceed those perpendicular to this axis. At a fluence of 7 x 10^{16}cm^{-2} (not shown in Fig. 4), CPIX revealed no long range crystallininty along the a-axis.

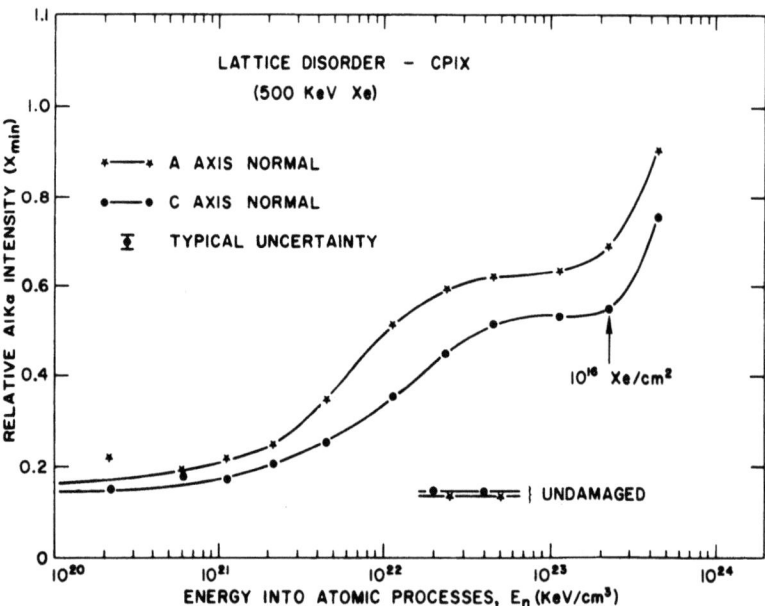

Fig. 4 Relative lattice disorder measured by aligned CPIX Aℓ K$_\alpha$ x-ray yield vs energy into atomic processes for 500 keV Xe^{++} implantations.

Ionization-induced annealing, which has been found to relieve lateral stress in sapphire,[9,10] did not have a measurable effect on CPIX yield. After damage was introduced, the crystal was returned to the aligned orientation and the analysis beam applied in 4 μC cm^{-2} increments for a minimum of 12 points. No trend toward improved channeling was seen in these data. Additional implantations of 500 keV Ar and 250 keV Xe, after annealing with a 50 keV proton beam up to 10^{16}cm^{-2}, revealed no improvement in channeling. These experimental conditions duplicated those under which lateral stress has been shown to be relieved by ionizing radiation.[10]

DISCUSSION AND CONCLUSIONS

The contrasting behavior of 204 nm optical absorption and lattice displacement damage is shown in Fig. 5. The optical absorption undergoes radiation quenching in the linear growth portion of the displacement damage curve. Therefore, 204 nm absorption does not reflect the total defect concentration for damage produced by heavy ions, except during the linear growth of the optical absorption curve.

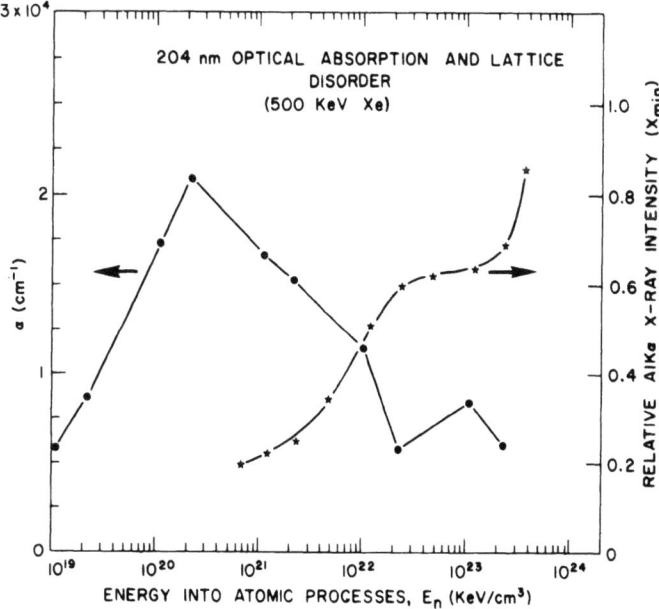

Fig. 5 204 nm optical absorption coefficient and lattice disorder determined by CPIX vs energy into atomic processes for 500 keV Xe^{++} implantations.

A comparison (Fig. 6) of RBC and CPIX data for a-axis samples, shows substantial agreement. The CPIX measurements provide a measure of the number of displacements in the linear region, and are advantageous in evaluating ionization-induced annealing because they require much less energy deposition by the analysis beam. The RBC measurements, however, reveal more information on the depth distribution of the damage.

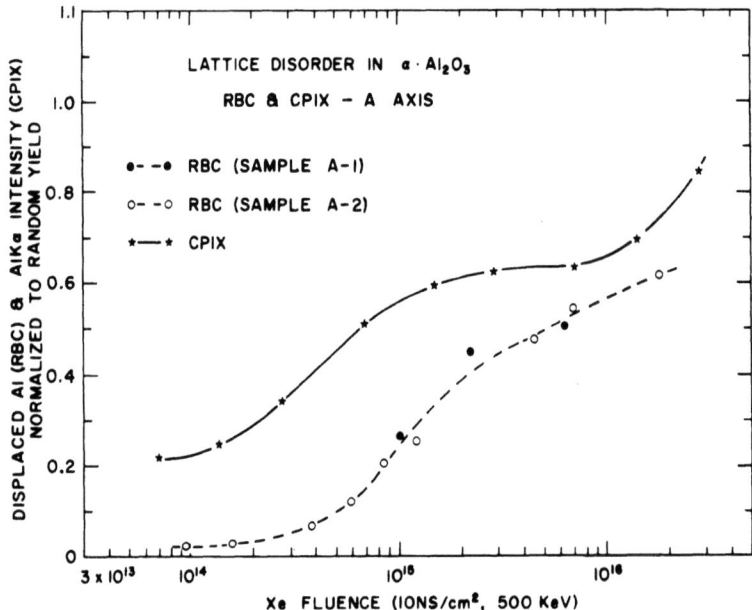

Fig. 6 Lattice disorder in a-axis sapphire measured by CPIX and RBC vs 500 keV Xe^{++} fluence.

Comparison of optical absorption for Xe with that produced by Ar implantations (Fig. 2) suggests a series of maxima at increasing fluence as the ion mass (and, therefore, the ratio of energies into atomic and electronic processes) decreases. Baranova, et al,[19] have seen behavior similar to the radiation damage quenching of the 204 nm absorption in the 1.8 μm divacancy absorption band of silicon. These investigators proposed a model in which the damage quenching is dependent on the relative volume of locally amorphous regions produced by the ion implantation. Such a model might be applicable in interpreting the present results.

Previously published ion-implantation-induced lateral stress measurements show that, for heavy ions, a maximum stress is reached at fluences corresponding to about 6×10^{21} keV cm^{-3} into atomic processes.[9] Our CPIX (Fig. 5) and RBC results show that, at this level, the disorder begins to deviate from the linear relationship found at lower fluences.

From our results it can be concluded that: (1) Optical absorption at 204 nm is produced by low fluence heavy ion implantation; (2) Optical absorption scales directly as the energy into electronic processes for light ion implantation, but not for heavy ion bombardment; (3) Lattice atom displacements are a linear function of 500 keV Xe fluence up to $10^{15} cm^{-2}$ above which saturation begins to occur; and (4) Optical absorption at 204 nm is not a general measure of lattice damage because radiation quenching of this absorption occurs where RBC and CPIX measurements show damage concentrations still to be increasing.

To a large extent the relationships among the major ion implantation damage effects in sapphire remain unresolved. Some of the important unknowns are: (1) the identity of the defect responsible for 204 nm absorption; (2) the mechanism of radiation quenching of this 204 nm absorption; and (3) the mechanism of lateral stress relaxation by ionization.

ACKNOWLEDGEMENTS

The helpful suggestions of E. P. EerNisse, P. S. Peercy and G. B. Krefft, and the assistance of J. Snelling, N. Wing and J. Smalley in preparing samples and making the measurements are gratefully acknowledged.

REFERENCES

1. G. W. Arnold and W. D. Compton, Phys. Rev. 116, 802(1950).
2. W. D. Compton and G. W. Arnold, Discuss. Faraday Soc. 31, 130 (1961).
3. G. W. Arnold, Proceedings of the First Topical Meeting on Technology of Controlled Nuclear Fusion, San Diego, CA, 1974, CONF-74042.
4. P. W. Levy, Phys. Rev. 123, 1226(1961).
5. E. W. J. Mitchell, J. P. Rigden, and P. D. Townsend, Phil. Mag. 5, 1013(1960).
6. B. D. Evans and H. D. Hendrix, Proceedings of the Fifth International Conference on Ion Implantation in Semiconductors and Other Materials (1976) (to be published).
7. D. G. Martin, Phys. and Chem. of Solids 10, 64(1959).
8. G. B. Krefft and E. P. EerNisse, Am. Ceram. Soc. Bull. 53, 621 (1974).
9. G. W. Arnold, G. B. Krefft, and C. B. Norris, Appl. Phys. Letters 25, 540 (1974).

10. G. B. Krefft, W. Beezhold, and E. P. EerNisse, IEEE Trans. Nucl. Sci. NS-22, #6 (1975).
11. F. T. Gamble, R. H. Bartram, C. G. Young, O. R. Gilliam, and P. W. Levy, Phys. Rev. 138, A577(1967).
12. R. T. Cox and A. Herve, C. R. Acad. Sci. (Paris) 261, 5080 (1965).
13. R. T. Cox, Phys. Letters 21, 503(1966).
14. S. Y. La, R. H. Bartram and R. T. Cox, J. Phys. Chem. Solids 34, 1079(1973).
15. H. M. Naguib, J. F. Singleton, W. A. Grant, G. Carter. J. Matl. Sci. 8, 1633(1973).
16. T. J. Turner and J. H. Crawford, Jr., Phys. Rev. B 13, #4(1974).
17. D. K. Brice, Radiat. Eff. 11, 227(1971).
18. D. K. Brice, J. Appl. Phys. 46, 3385(1975).
19. E. C. Baranova, V. M. Gusev. Yu.v. Martynenko, C. V. Starnin and I. B. Hailbullin, Ion Implantation in Semiconductors and Other Materials, B. L. Crowder ed., (Plenum Press, N.Y. 1973).

THE STRUCTURE DAMAGE, PHASE FORMATION AND Si DEPTH DISTRIBUTION IN THE IMPLANTED NATURAL DIAMOND

V.V.Krasnopevtsev, Ju.V.Milyutin, V.S.Vavilov

P.N.Lebedev Physics Institute, Moscow, USSR

A.E.Gorodetsky, A.N.Khodan, A.P.Zakharov

Institute of Physical Chemistry, Moscow, USSR

The implantation of tens keV heavy ions brings about appearance of the displacement spikes (disordered regions) in the crystal subsurface layer. The displacement spikes are able to stimulate reactions in the solid state and the new phase formation at high doses of implanted atoms.

The purpose of the present work is to study structure changes and phase transformations in diamond by Si ion implantation. It is known that increase of the ion dose to the certain critical value results in the amorphization of diamond /1,2/. It was also found the graphite is formed during the implantation at the 40 keV Ar ion doses several times more than the critical one (more 5×10^{14} Ar^+/cm^2) /3/. On the other hand, the Si ion implantation into diamond at doses more 10^{17} Si^+/cm^2 leads to the SiC formation /4/. Of particular interest are both the phase formation kinetics and the phase depth distribution in the Si implanted diamond layer.

Natural diamond crystals were implanted with 30 keV Si ions at room temperature. Doses, D, between $1,2 \times 10^{13}$ to 2×10^{17} ions/cm^2 were used. The samples used were plane spinel twins characterized with both edge length about 2 mm along $\langle 110 \rangle$-direction and the pronounced (111) face.

The implanted samples were studied by RHEED technique with 65 keV electrons. Diffraction patterns of the initial samples etched in perchloric acid $HClO_4$ (200 C,

10 min) show Kikuchi-lines, -bands and -envelopes indicating a perfect structure of the surface layers (Fig.Ia).
The implantation of $1,2\times 10^{13}$ Si ions/cm² results in the appearance of both weak haloes and a background in the diffraction pattern taken at the glancing angle

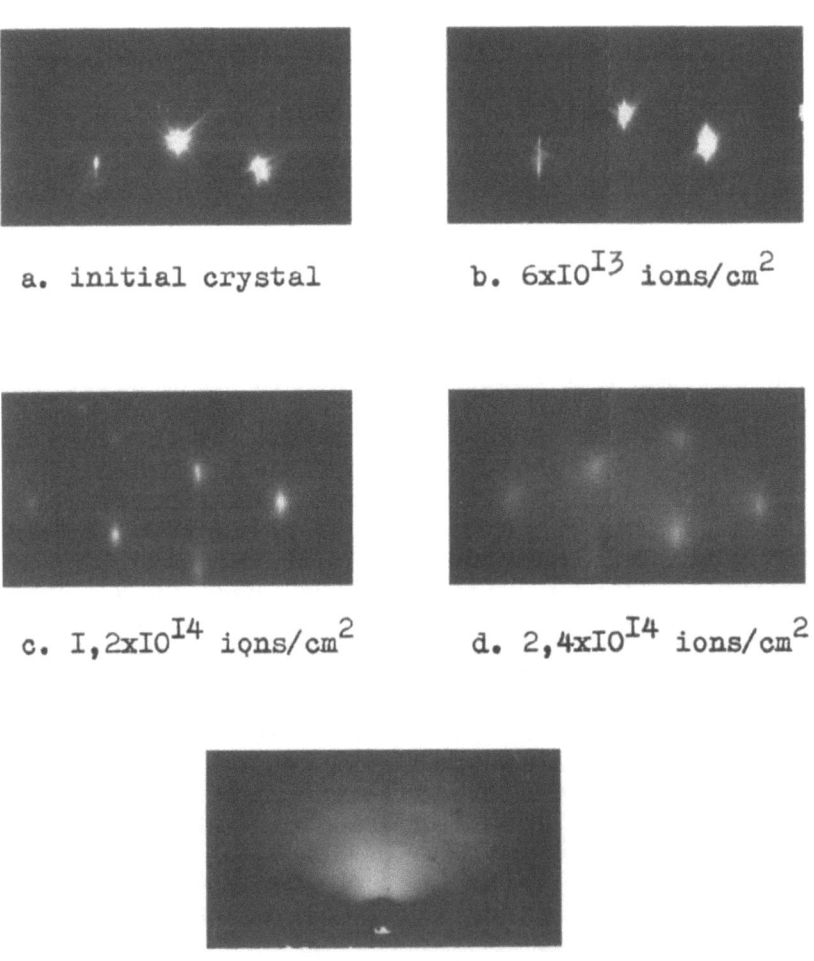

a. initial crystal

b. 6×10^{13} ions/cm²

c. $1,2\times 10^{14}$ ions/cm²

d. $2,4\times 10^{14}$ ions/cm²

e. 2×10^{17} ions/cm²

Fig.I. The electron diffraction patterns for the diamond subsurface layer implanted by 30 keV Si ions. The electron beam direction along ⟨110⟩, the angle, α, $1,3°$ for a.–d. and $0,2°$ for e.

$\alpha = 0° \pm 0,2°$ (an angle between the incident electron beam and the sample face). The background present is due to the partial amorphization of the surface layer I-5 Å thick. The complete amorphization of the layer is observed at $D=6 \times 10^{13}$ ions/cm^2. The amorphization is followed by the formation of another phase nuclei (see Fig.I,b and 2,a-c). In this case the deeper layers remain crystalline and sufficiently perfect. As in the Ar ion implanted diamond /3/ the dominant damage at D between $10^{13}-10^{14}$ ions/cm^2 are the displaced C atom clusters having the shape of discs parallel to the (III) plane 2-I5 Å thick with diameter 40-80 Å. The discs are located at 50-I00 Å beneath the crystal surface (see Fig.2,a-c). The discs sizes were determined from smearing of reflections observed in the pattern taken along ⟨IIO⟩ and ⟨II2⟩ directions at the angle α between 0°-5°. As a result of ion implantation, the length of all reflections increases along the direction normal to the (III)

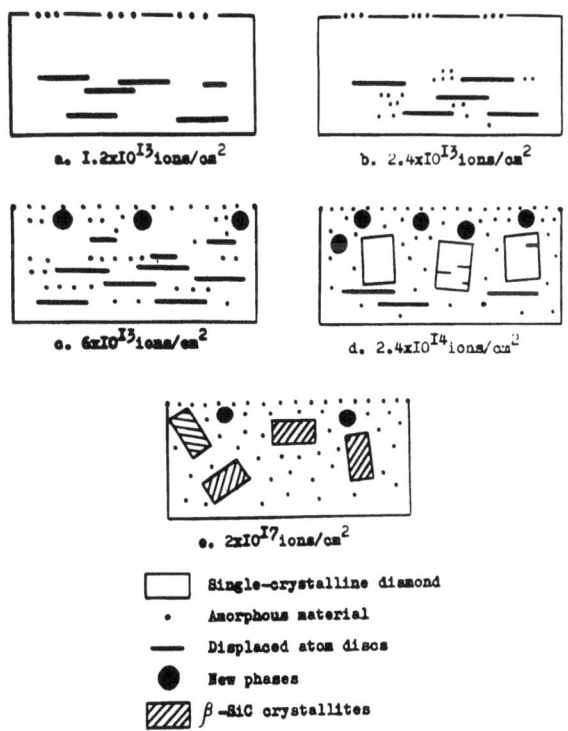

Fig.2. Schematic picture of structural changes in diamond implanted by various doses.

plane. Such reflection smearing is generally due to diffused (non-Bragg) coherent electron scattering on special shape inclusions present in the crystal. The number of clusters-discs rises with increasing dose. This effect results in a simultaneous increase of both intensity and length of the smeared-out reflections adjacent to diamond reflections. The density of discs reaches a maximum at D $(1,2-2,4) \times 10^{14}$ ions/cm^2. Further Si implantation as in case of Ar ions leads to the break-up of the crystal into single subgrains with size of 1000-3000 Å (Fig.1,c,d and 2,d). At the same time the lattice disordering and the accumulating of amorphous material occur. At D about $9,6 \times 10^{14}$ ions/cm^2 the subsurface layer becomes amorphous up to 300 Å. The amorphization process occurs in two directions: downwards from the sample surface and upwards from the depth of 100 Å /5/. These two amorphous layers overlap at D $(4-10) \times 10^{14}$ ions/cm^2. It should be emphasized that the diamond amorphization is followed, in case of the Ar ion implantation, by the formation of turbostratic carbon having graphite network with typical interplanar spacing of 3,5; 2,1 and 1,2 Å. No graphite structures form in the Si implanted diamond at all doses used. It is supposed that the tetrahedral Si-C bond existance prevents formation of the hexagonal carbon networks. However an amorphous material includes the highly dispersed precipitates of new phases. So, ranging from D=$4,8 \times 10^{14}$ ions/cm^2, there are three clear haloes in the pattern taken at $\alpha \approx 0$; the halo intensity maxima correspond to the interplanar spacing of 4,4; 2,3 and 1,3 Å. It is believed that about 10 Å nuclei of the corresponding phase are as if embedded into amorphous layer. This new phase revealed up to D=2×10^{17} ions/cm^2 is most likely non-stoichiometric SiC. At the dose range $6 \times 10^{13}-3,6 \times 10^{16}$ ions/cm^2 one can see as many as five very weak diffractions lines caused by another phase formation (see Table I); the corresponding interplanar spacing are best attributed to Si_3N_4 or SiO_2. Finally, as the patterns show the implantation of 2×10^{17} Si ions/cm^2 into a sample results in the formation of highly dispersed β- SiC in a shape of 30-50 Å crystallites oriented in random way (Fig.1,e and 2,e). The strong background observed is due to the electron scattering in an amorphous phase. Electron beam scanning along the sample surface shows that regions containing SiC alternate with regions including the new phases mentioned.

In addition, the diamond structure and new phase distribution through the all implanted layer depth was of interest. The samples implanted both with the high

Table I. The interplanar spacing, d, and diffraction line intensity, I, for the phases formed in the Si implanted diamond.

β-SiC					Si_3N_4			
Present Experiment		Reference data			Pres. Exper.	Reference data		
d, Å	I	d, Å	I	hkl	d, Å	d, Å	I	hkl
2,60	mean	2,51	100	111	3,81	3,85	40	002
2,21	mean	2,17	20	200	3,30	3,34	40	400
1,54	strong	1,54	63	220	2,87	2,87	95	030
1,30	mean	1,31	50	311	2,18	2,15	40	040
1,24	weak	1,26	5	222	2,05	2,07	60	041
						1,93	20	004

(2×10^{17} Si+/cm^2) and relatively low (4×10^{15} Si+/cm^2) doses were studied using successive removing of thin layers.

Successive layer etching of the implanted diamond was performed in boiling $HClO_4$ (200 C). Using the interference microscope the thickness of the layer removed by etching (i.e. a height of the step between the implanted part of the crystal and nonimplanted one) was measured (the accuracy about 50 Å). The RHEED technique was used after each stage of etching.

Using the microscope the implanted surface part was shown markedly expand above the nonimplanted one (1000-1500 Å) at D about 10^{17} ions/cm^2. Up from the dose about 10^{17} Si+/cm^2 the initial sample surface moves markedly towards the ion beam during the implantation.

The structure of the surface layer of about 300 Å depth was described above (see Fig.1,e and 2,e). The region just below the upper layer extends down to 1200-1500 Å and includes crystal grains in size of 500 Å and more. Both the SiC phase and completely disordered material are located on the boundaries of these grains.

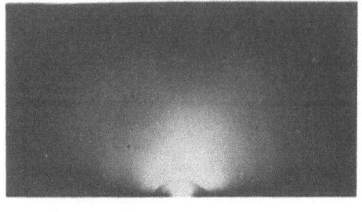 

a.   b.

Fig.3. Diffraction patterns obtained after the removal of the implanted layer 500-600 Å thick (dose of 2×10^{17} Si+/cm^2):
a. SiC phase ($\alpha = 0,2°$, the beam direction along $\langle 110 \rangle$);
b. diamond crystal grains ($\alpha = 2,3°$, the same beam direction).

The reflections of the SiC phase formed during the implantation appear in the patterns taken at the angle α of $0°-0,2°$ (Fig.3,a). The diamond reflections, Kikuchi-lines and diffused background due to the occurence of disordered regions are seen in the diffraction patterns taken at $\alpha = 2,3°$ (Fig.3,b). The SiC crystallites have the shape of needles (in diameter of 20-40 Å) stretched normal to the sample face. The smeared-out reflections (see Fig.3,a) adjacent to the newly formed phase reflections are due to the lattice strain in diamond crystal grains which is parallel to the substrate plane.

The third 100-200 Å thick layer located below the second one consists of both the SiC phase and disordered material (see Fig.4). In addition to the crystallites-needles, mentioned above, the regions of the SiC phase in the layer have a shape of the planar discs 5-10 Å thick and in diameter of about 100 Å. No diamond structure appears in the layer.

Finally, there is the fourth region located at the depth more 1500-2000 Å from the surface of the sample implanted part; the SiC phase is absent in this region. The corresponding diffraction patterns show only the diamond structure (Fig.5). However, the total number of implanted Si atoms is found by X-ray microanalysis to be 10^{16} atoms/cm^2 in the region. One can easily see the smeared-out reflections adjacent to the diamond reflections again (Fig.5); these reflections are the

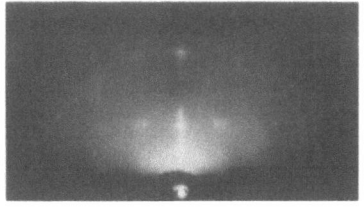 

Fig.4.   Fig.5.

Fig.4. SiC phases in the third layer (see the text). $\alpha = 0,2°$; the beam direction along $\langle 110 \rangle$.

Fig.5. Diamond substrate structure after the removal of the implanted layer 2000 Å thick. Total number of Si remained is 10^{16} atoms/cm^2. $\alpha = 2,3°$; the beam direction along $\langle 110 \rangle$.

evidence that aggregates of both implanted Si atoms and displaced C atoms in shape of discs cause the strain (expansion) of the substrate lattice. The diffused background observed also in the patterns arises from an elastic electron scattering by defect clusters in size of 5-10 Å. The Si atom migration up to 1500-2000 Å depth does not result from the projectile slowing-down but from another mechanism (radiation-controlled diffusion, migration along crystal grain boundaries etc.). Fig.6 shows a schematic picture of the structure and phase depth distribution in Si implanted diamond.

Successive etching of the sample implanted at 4×10^{15} Si$^+$/cm^2 allows to distinguish three typical layers: I. the amorphous layer about 200-300 Å thick, II. the crystalline one without essential defects and III. the diamond substrate region containing radiation damage in shape of displaced atom discs which is located at a depth more than 500 Å from the surface. All latter facts mentioned permit to suppose that just this layer up to 500 Å thick changes its thickness approximately three times during further implantation of Si into diamond up to 2×10^{17} ions/cm^2. The main volume of the SiC phase is formed in the upper subsurface layer about 300 Å thick while the lower layers consist generally of the carbon phase (amorphous or crystalline) which includes SiC precipitates. The total thickness (about 500 Å) of all three layers is in agreement with He ion back-scattering data /6/ as well as with LSS theory values of averaged projected range, Rp, and straggling, $\triangle$Rp,

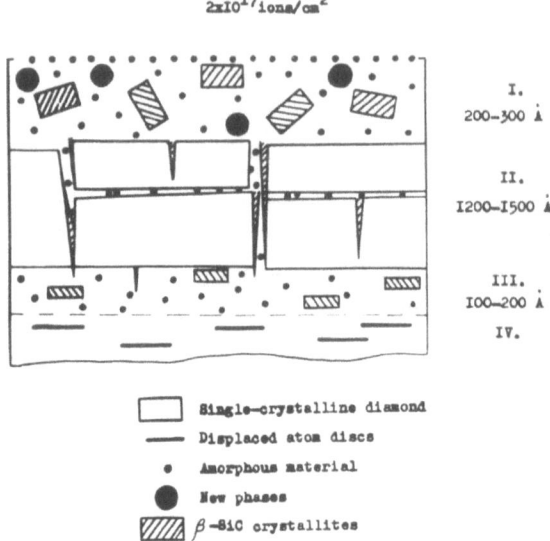

Fig.6. Schematic picture of the structure profile in implanted diamond; 30 keV Si ions.

for 30 keV Si ions: $R_p=207$ Å, $\Delta R_p=45$ Å in case of diamond and $R_p=362$ Å, $\Delta R_p=79$ Å for graphite if density is 3,5 and 2,0 g/cm³, respectively /7/.

Samples implanted by more than 10^{17} Si⁺/cm² were annealed in vacuum of 10^{-6} Torr. No structural changes of the subsurface diamond layer occur up to 600-700 C. However the annealing at higher temperature is accompanied by the crystalline SiC phase growth in the implanted diamond. The SiC phase crystallization yield

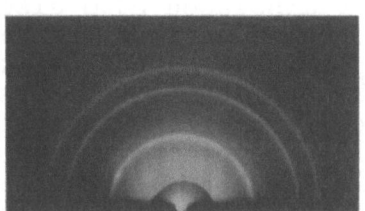

Fig.7. The diffraction pattern of polycrystalline β-SiC layer which is synthesized in diamond by Si implantation followed by annealing.

and structural characteristics of the layer synthesized depend on dose and also conditions of both bombardment and following anneal treatment. Fig.7 shows the pattern for the crystal implanted under rather favourable conditions (distributed ion energy of 24 and 40 keV, total dose of $4,2 \times 10^{17}$ ions/cm^2, target temperature below 300 C, annealing at 1200 C). In this case one sees the lines due to weakly textural β-SiC (Table 2) characterized by the typical averaged crystallite size about 200 Å. In addition, weak smeared halo observed corresponds to an amorphous phase with interplanar spacing of 3,3 - 3,5 Å.

Table 2. Interplanar spacing, d, and diffraction line intensity, I, for β-SiC synthesized in diamond by Si implantation and subsequent annealing at elevated temperature.

Present experiment		Reference data		
d, Å	I	d, Å	I	hkl
2,54	100	2,51	100	111
2,23	10	2,17	20	200
1,55	40	1,54	63	220
1,32	30	1,31	50	311
1,26	10	1,26	5	222
1,11	4	1,09	6	400
1,01	10	1,00	18	331
0,97	5	0,97	6	400

The amorphous phase occupies less than 10 per cent of the crystalline SiC amount.

Thus, new phase synthesis as well as radiation damage and substrate structure disordering occur by the Si ion implantation in diamond. Lattice strain appeared at the boundary between the implanted layer and diamond substrate changes the substrate structure to the considerable depth (to 10000 Å) enhancing thereby migration of Si atoms into the crystal. The yield of SiC crystalline phase synthesis by Si ion implantation in diamond with subsequent anneal treatment depends strongly on experiment conditions.

The authors wish to thank Dr. I.P.Akimchenko for helpfull discussions of the experimental results, Kh.R.Kasdajev and I.K.Ten for taking part in some measurements.

References

1. A.I.Gerasimov, E.I.Zorin, P.V.Pavlov, D.I.Tetelbaum. Proceedings of the Conference, Gorky, 1971.
2. V.V.Galkin, F.I.Zorin, V.V.Krasnopevtsev, Yu.V.Milyutin. Proceedings of the Symposium, Krivoj Rog, 1971.
3. V.S.Vavilov, V.V.Krasnopevtsev, Yu.V.Milyutin, A.E. Gorodetsky, A.P.Zakharov. Radiation Effects, $\underline{22}$, 141 (1974).
4. I.P.Akimchenko, V.S.Vavilov, V.V.Galkin, V.S.Ivanov, V.V.Krasnopevtsev, Yu.V.Milyutin. Soviet Physics-Semiconducters, $\underline{6}$, 1182 (1972).
5. A.I.Gerasimov, E.I.Zorin, P.V.Pavlov, D.I.Tetelbaum. Phys. status solidi(a), $\underline{12}$, 679 (1972).
6. I.P.Akimchenko, V.V.Krasnopevtsev, Yu.V.Milyutin, V.S.Vavilov, J.Gyulai, G.Mezey, T.Nady. Intern. Conf. on Radiation Effects in Semiconductors. Dubrovnik, Yugoslavia, 1976.
7. W.S.Johnson, J.F.Gibbons. Projected Range Statistics in Semiconductors. U.S.A., 1970.

EXPANSION OF THERMALLY GROWN SiO$_2$ THIN FILMS

UPON IRRADIATION WITH ENERGETIC IONS

D. W. Ormond, E. A. Irene, J. E.E. Baglin, B. L. Crowder

IBM Thomas J. Watson Research Center

Yorktown Heights, New York, 10598

Thin films (~1000 to 7000Å) of SiO$_2$ were prepared by several methods of the thermal oxidation of single crystal silicon substrates. The films were irradiated with argon ions with energies ranging from 40 to 500 KeV and ion fluences from 5×10^{13} to 1×10^{16} ions/cm^2. Ellipsometry was used to monitor changes in the films' thickness and refractive index that took place as a direct result of being irradiated. Confirmation of some of the ellipsometric measurements was established with the use of an interference microscope. In addition, infrared spectroscopy was utilized to observe possible changes in the Si-O stretching vibration (~9μm) that may result due to the irradiation. Expansion of the thin films was found to depend on how the films were prepared. "Ultradry" oxides (<1ppm H$_2$O) and HCl grown oxides were found to expand upon implantation with a decrease in refractive index; however wet grown oxides (~10,000ppm H$_2$O) and Na contaminated steam grown oxides were found to remain essentially unchanged to within the experimental accuracy of about 2 percent. Also, infrared analysis showed that the Si-O stretching vibration (~9μm) shifted ~50cm^{-1} to lower energy in the case of the former films, while in the latter films, using identical implant conditions, the shift was reduced by approximately 50 percent. Post-implantation annealing at the SiO$_2$ films' preparation temperature restores the films to their original condition. Density measurements, performed on the as prepared, thin SiO$_2$ films, revealed that the density of the original films varied according to their method of preparation. High purity oxides, of MOSFET gate insulator quality, grown in a clean room facility at 1000° C in "ultradry" O$_2$ ambient (less than 1ppm H$_2$O), were found to have a density of 2.30±.02 g/cm^3; "wet" oxides (10,000 ppm H$_2$O) had a density of 2.16±.02 g/cm^3; and oxides grown in a Na contaminated furnace at 1050°C with a steam-O$_2$ mixture (about 1:1) had a density of 2.20±.02 g/cm^3.

A simple picture has emerged that shows that the high purity, high density films can only become less dense upon irradiation; however, the low purity, low density films remained essentially unchanged within the experimental accuracy of approximately 2 percent.

INTRODUCTION

The interaction of energetic ions with silicon dioxide has been the subject of many studies [see for example Ref. (1-7)]. Primak, Fuchs, and Day[8] reported that bulk vitreous silica compacted (increased in density) when exposed in a nuclear reactor, an effect which they attributed to energetic neutrons. Investigating neutron radiation damage further in crystalline quartz and amorphous silica, Primak[9] found that the quartz decreased in density, while the vitreous silica increased in density. Weismann and Nakajima[10] demonstrated that the neutron irradiation of both the quartz and fused silica resulted in an quasi amorphous film that contained roughly 20 percent by volume α-quartz. These preceding studies indicated that an equilibrium state may exist between the ordering and disordering processes.

A review of the more recent literature in this field uncovered some apparent discrepancies. Fritzsche and Rothemund[11], while implanting ionic species of argon and helium into thin, thermally grown films, found by ellipsometry and interference microscopy that the films expanded (in the perpendicular direction to the plane of the film) up to 6 percent. They also noted a decrease in the refractive index upon irradiation. Furthermore, they found by transmission infrared spectroscopy a shift in the Si-O stretching band to lower energy, in addition to a broadening of this band. On the other hand, EerNisse and Norris [12,13], using a cantilever plate technique[14,15] (a method which is based on the stress exerted by the thin film on the substrate) and stylus measurements, observed a compaction of bulk vitreous silica, as well as a compaction of thin, thermally grown SiO_2 films when these samples were irradiated with a large variety of ionic species from hydrogen and helium up to argon. The amount of compaction reported was 2.7 percent. The common thread between these preceding papers was the use of argon implantation.

In view of the previously mentioned discrepancies, it appeared that there were either differences in the SiO_2 films used, or errors in the measurements, or both. The present study was aimed at resolving these differences. With this purpose in mind, we chose to examine thin silicon dioxide films, thermally grown under various conditions, to determine whether or not processing could alter the properties of the films. Implantation conditions were selected to be comparable with the previous studies, argon being used as the implantation species. Ellipsometry was used to monitor any changes in the films' thickness and refractive index that might have resulted due to the ion irradiation. In addition, confirmation of the ellipsometric measurements was established with the use of interference microscopy. Infrared spectroscopy was utilized to observe any shift in vibrational frequencies which may occur during irradiation, as further proof of expansion or compaction of the SiO_2 tetrahedral network. Furthermore the density of the SiO_2 films was measured as a function of processing to correlate, as Primak had done with bulk material, the initial density of the material with the final density after irradiation. Finally, the existence of an equilibrium state produced by the ordering and disordering processes of irradiation was explored.

EXPERIMENTAL

Wafer Preparation

The silicon substrates used in this study were Czochralski grown, <100>, high purity semiconductor grade, single crystal silicon, 1 1/4 inch dia., 8-10 mils thick, 2Ω-cm resistivity, doped p-type with boron. All the wafers were chemmechanically polished on both sides. Prior to oxidation, the wafers were chemically cleaned using ultrasonic cavitation in peroxide solutions of ammonia and hydrochloric acid, followed by a concentrated hydrofluoric acid dip of 15 seconds duration. Deionized water of 18 Megaohm-cm resistivity was used between cleaning steps for rinsing the wafers. Clean silicon substrates are hydrophobic following the HF dip and do not need to be "blown dry" with purified, filtered gas after the water rinse. Most of the oxidations were performed in a standard ultra-clean semiconductor pilot-line. The oxides used in this study were determined to be of MOSFET quality by C-V and dielectric breakdown[16,17] measurements.

Oxidation Conditions

The wafers were placed vertically in a quartz boat and inserted into a quartz furnace at approximately 1000°C. The oxidation ambients were varied: some were "HCl grown" using an ambient of 4.5 percent hydrogen chloride gas, 95.5 percent oxygen, while others were grown as "wet" oxides by varying the water content. In one experiment the oxides were grown using a sodium contaminated quartz boat in a high water content ambient produced by mixing steam and O_2 together. In another experiment, "ultradry" oxides[18,19] were grown in an "ultradry" O_2 ambient which was produced by removing methane, CH_4, from the oxygen supply. (Methane is known to be in most liquid oxygen tanks in ~20 ppm concentration. Our method of purification was as follows: oxygen was passed through a high temperature furnace which cracks the methane into CO_2 and H_2O. The gas was then passed through a cold trap to condense the H_2O. This method yields an oxidation ambient with <1 ppm H_2O.) Most of the high quality oxides produced were annealed at the growth temperature for five minutes in dry nitrogen (to remove oxide fixed charge and lower surface states) and then subsequently quenched to room temperature by quickly removing them from the hot zone of the furnace tube. Oxide growth conditions were chosen so as to produce two sets of oxide film thicknesses: one set to be comparable with the previously mentioned work of Fritzsche and Rothemund[11], namely thin oxides ~1000Å, and a thicker set ~6 to 7,000Å, to be comparable with the previously mentioned work of Eer-Nisse and Norris[12].

Ion Implantation Conditions

Argon was chosen as the implantation species for this present work, in order that a one to one comparison could be made with the earlier studies.

The first set of implantation conditions, designated Set I in Table I, placed the peak of the argon ion concentration near the center of the film, far away from the

Table I

	Ion Implantation Conditions		R_p	σ_p
Set I	for ~1000 Å Oxides	at 40KeV Ar$^+$	404 Å	170 Å
	for 6000 Å and 7000 Å Oxides	at 340KeV Ar^{++}	3478 Å	943 Å
Set II	for ~1000 Å Oxides	at 80KeV Ar$^+$	791 Å	304 Å
	for 6000 Å and 7000 Å Oxides	at 500KeV Ar^{++}	5077 Å	1203 Å

SiO_2-Si interface. For example in the thinner films, ~1000Å thick, $^+$Ar was implanted at 40 KeV with a fluence of $5\times10^{15}/cm^2$ and similarly for the thicker films, ~6000Å to 7000Å thick, $^{++}$Ar was implanted at 340 KeV with a fluence of $5\times10^{13}/cm^2$. The second set of implantation conditions, designated Set II in Table I, placed the peak of argon ion concentration closer to Si-SiO_2 interface than in Set I. To be specific, in the thinner films, ~1000Å thick, $^+$Ar was implanted at 80 KeV with a fluence of $1\times10^{16}/cm^2$ and likewise in the thicker films, ~6000Å thick, $^{++}$Ar was implanted at 500 KeV with a fluence of $5\times10^{13}/cm^2$. The projected range, R_p, of argon in SiO_2 for the two sets of implantation conditions is listed in Table I along with the projected standard deviation, σ_p. The values of R_p and σ_p found in Table I were taken from LSS calculation by Gibbons, et.al.,[20] of argon in silicon which closely simulate the implantation of argon into amorphous silicon dioxide. Notice that for Set I the amount of SiO_2-Si interfacial damage was minimal since R_p was approximately three times σ_p from the interface duplicating the study by Fritzsche and Rothemund[11]. However in Set II, R_p was approxi-

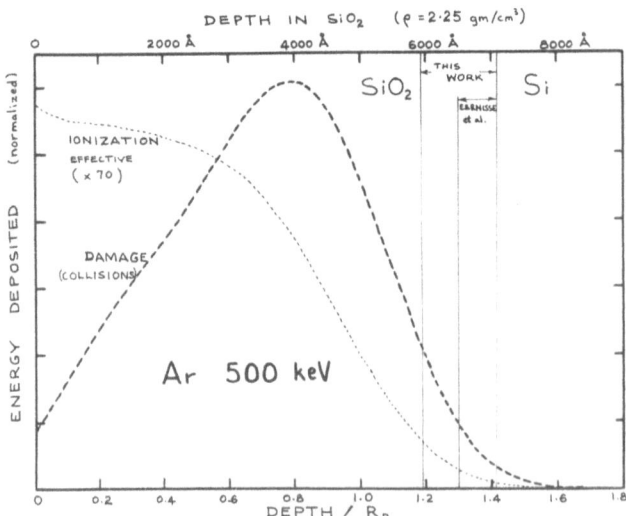

Figure 1. Plot of implantation damage vs. depth in silicon dioxide. Comparison of collision damage and effective ionization damage for Argon, 500 KeV.

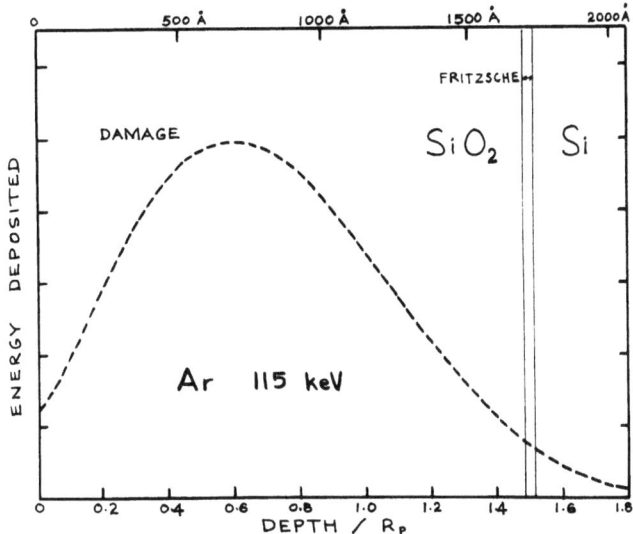

Figure 2. Plot of implantation damage vs. depth in silicon dioxide for Argon, 115 KeV.

mately one σ_p distance away from the SiO_2-Si interface duplicating the study by EerNisse and Norris[12].

Most of the literature on compaction emphasizes the importance of nuclear (atomic) stopping processes, as opposed to the other type, namely electronic (ionic) energy interactions. For compaction of SiO_2 to occur, nuclear processes or collisions have been shown to be 200 times more effective than are electronic or ionization interactions[7]. Figures 1 and 2 show plots of the damage profiles of argon implanted into SiO_2 using Brice's calculations[21]. Energy deposited was plotted on a normalized scale. Figure 1 illustrates one of the implantation conditions specified in Set II, namely those used by this study and also by EerNisse and Norris[12]; note the fact that the "effective" electronic damage is negligible. Figure 2 illustrates only the collision damage caused by the implantation conditions used by Fritzsche and Rothemund[11]; note the minimal damage to the interface and recall that the Set I implantation conditions were modelled after that research work.

Film Thickness Measurements

Film thicknesses and refractive indices were measured by ellipsometry before and after ion implantation. An automatic ellipsometer[22] operated with light having a wavelength of 6328Å was used for the measurements. For the ellipsometric measurements it was assumed that the SiO_2 films had zero absorption of the 6328Å light and that the Si surface was undamaged by the ion beam. However, there was damage in the SiO_2 which may have resulted in optical absorption. Therefore, in order to ascertain whether the results from ellipsometry were correct

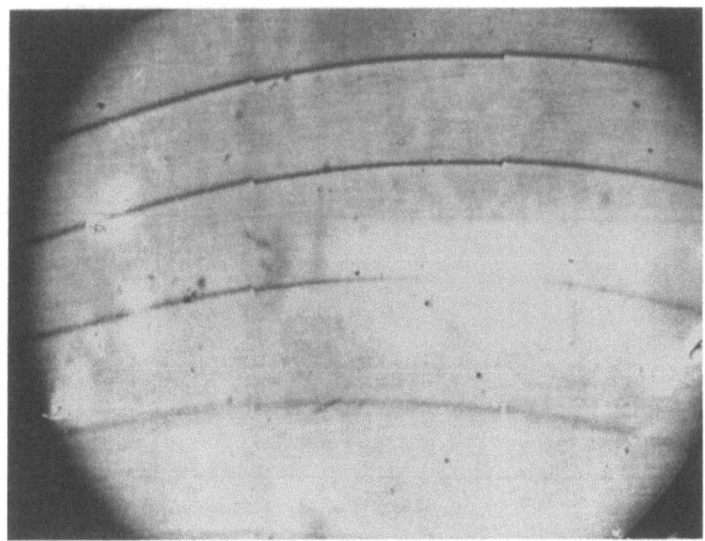

Figure 3. Tolansky interference micrograph showing ~200 Å step between implanted and non-implanted region. Distance between lines is 3164Å.

in direction (expansion and/or compaction) and magnitude, several representative samples were checked by interference microscopy. A Tolansky interference measurement was obtained in Figure 3 by measuring the implanted versus non-implanted region. The step size of 200Å calculated from this method agreed with the ellipsometric value. Moreover, there was a definite step in the upwards direction when proceeding from the non-implanted to the implanted region. Therefore, we concluded that the amount of absorption due to damage in the SiO_2 must have been small and that the thickness results from ellipsometry were valid.

Infrared Analysis

The infrared spectrum of the thin films was monitored using a Beckmann 12, IR spectrometer which was operated in the transmission mode. The Si-O stretching vibration with an absorption near 9μm was the most pronounced feature of the spectrum and it was a shift in this peak after irradiation that was reported by Fritzsche and Rothemund[11]. Thick silicon substrates (>30 mils) were used to eliminate interference fringe effects in the IR spectrum. Most of the spectra were taken with a bare silicon substrate placed in the reference beam to eliminate substrate absorption effects such as the Si-Si stretching vibration (16-17μm). IR analysis was generally performed before and after ion implantation without any post-implantation anneal except where specifically indicated.

Density Measurements

Direct density measurements were performed on some samples of commercially available, high purity, fused quartz (vitreous silica). These measurements yielded a range of densities from 2.0 gm/cm^3 to 2.2 gm/cm^3. Direct density measurements were also made on the thin SiO$_2$ films by utilizing photo-lithographic methods to define an oxide pad of known area. Thickness of the pad was measured by ellipsometry. Thus, using photo-lithography to define area and ellipsometry to measure thickness, the volume of the oxide pad was calculated to better than 1 percent accuracy. Knowing the volume of the oxide pad to within 1 percent, the wafer was weighed by a microbalance before and after etching in a buffered hydrofluoric acid solution which preferentially etches away the SiO$_2$ (~1200 Å/min at room temperature) while leaving the substrate essentially unetched. Density measurements performed in this manner yielded an accuracy of about 1 percent. Oxidation ambients were varied in order to obtain films of varying densities as will be discussed in more detail later on.

RESULTS

Table II summarizes the density measurements and it also illustrates the oxide pad geometry used to obtain the density data. As one can discern from the data presented, the initial density of the SiO$_2$ film was found to be strongly dependent upon the conditions of thermal oxidation.

All the oxides which would have been used in this laboratory as gate oxides in MOSFET's, namely the high density, ultra-dry oxides and HCl oxides, showed expansion after implantation with argon, as indicated in Table III. Recall that the

Table II

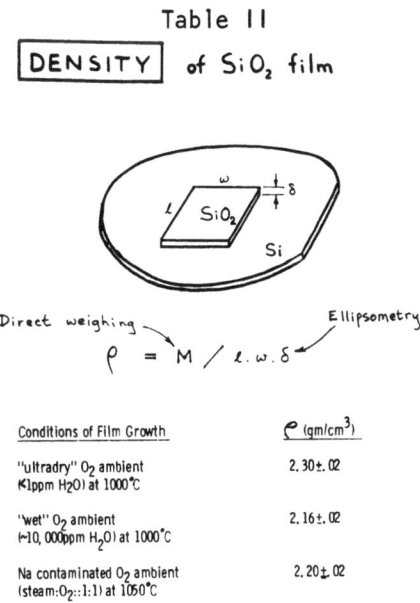

Conditions of Film Growth	ρ (gm/cm^3)
"ultradry" O$_2$ ambient (<1ppm H$_2$O) at 1000°C	2.30 ± .02
"wet" O$_2$ ambient (~10,000ppm H$_2$O) at 1000°C	2.16 ± .02
Na contaminated O$_2$ ambient (steam:O$_2$::1:1) at 1050°C	2.20 ± .02

Table III

Type of Oxide	Implantation Condition	Average Initial d_{ox}	Average Initial n_{ox}	Δd_{ox} %	Δn_{ox} %
Set I { HCl	40KeV Ar$_2^+$ 5x10^{15}/cm^2	984 Å	1.46	+6	-1
Set I { Dry	340KeV Ar$_2^{++}$ 5x10^{13}/cm^2	5921 Å	1.47	+6	-3
Set II { HCl	80KeV Ar$_2^+$ 1x10^{16}/cm^2	996 Å	1.46	+17	-3
Set II { Dry	500KeV Ar$_2^{++}$ 5x10^{13}/cm^2	6000 Å	1.47	+9	-4

Table IV

Type of Oxide	Implantation Condition	Average Initial d_{ox}	Average Initial n_{ox}	Δd_{ox} % Δn_{ox} %
Set I { Na Contam. Steam	340KeV Ar$_2^{++}$ 5x10^{13}/cm^2	7081 Å	1.46	NO Change
Set I { Wet	340KeV Ar$_2^{++}$ 5x10^{13}/cm^2	6018 Å	1.47	NO Change
Set II { Na Contam. Steam	500KeV Ar$_2^{++}$ 5x10^{13}/cm^2	7081 Å	1.46	NO Change
Set II { Wet	500KeV Ar$_2^{++}$ 5x10^{13}/cm^2	6018 Å	1.47	NO Change

Set I type implantation placed the implanted species near the center of the film, in close approximation to the experiment of Fritzsche and Rothemund[11], while the Set II type followed closely the implantation conditions set forth by EerNisse and Norris[12]. Note that the ellipsometer in all these cases showed expansion of the thin films usually accompanied by a decrease in refractive index in agreement with Fritzsche and Rothemund.

All the oxides which were grown as wet oxides and Na contaminated oxides, specifically the low density oxides, showed neither expansion nor compaction after implantation (Set I and Set II) with argon, as shown in Table IV; thus these low density films were found to remain unchanged to within the experimental accuracy of about 2 percent.

Infrared analysis of the high density, ultra-dry oxides and HCl oxides after implantation (Set I and Set II) showed a pronounced broadening and shift of the Si-O stretching vibration to lower energy, as shown in Figure 4. Isochronal annealing of 1 hour duration in helium gas was performed on the sample after implantation. Figure 5 illustrates the shifting back of the Si-O peak to its pre-implant position and the decrease in the half-width of the peak with annealing. The insert in Figure 5 shows graphically the agreement of the ellipsometric data with the IR data before and after annealing. Almost complete restoration of the film was accomplished after annealing at the film's growth temperature of 1000°C for 1 hour. More quantitative data are presented in Table V for the isochronal annealing behavior of the Si-O stretching vibration, which is located at ~1078 cm^{-1} before ion implantation and then shifts to ~1035 cm^{-1} after ion implantation.

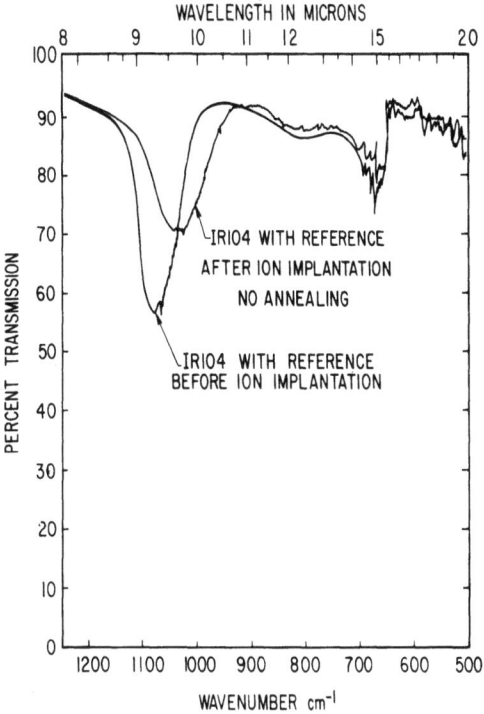

Figure 4. Transmission infrared spectrum showing broadening and shift of the Si-O stretching vibration after Argon implantation for the case of HCl, MOSFET quality oxide.

Infrared analysis of the low density, wet and Na contaminated oxides after implantation (Set I and Set II) showed less broadening and shift of the Si-O stretching vibration to lower energy, as shown in Figure 6. The shift was reduced by approximately one-half the value of that reported earlier for the dense film irradiation. No annealing experiments were conducted with these samples, although their annealing properties would be expected to be similar to those of the more dense films.

DISCUSSION AND CONCLUSIONS

To find a possible explanation of our results, let us go back to some earlier radiation studies on quartz and fused silica. Figure 7, a compilation taken from Primak's data[9], plots the change in the density of quartz and fused silica as one irradiates the samples with a neutron flux. Note the behavior of the crystalline material which goes to a less dense phase, conversely the low density amorphous phase becomes more dense. Observe that the same qualitative changes that were taking place in Primak's bulk material were taking place in our thin films. The most dense thin films, $2.30 \pm .02$ g/cm^3, became less dense; while the less dense

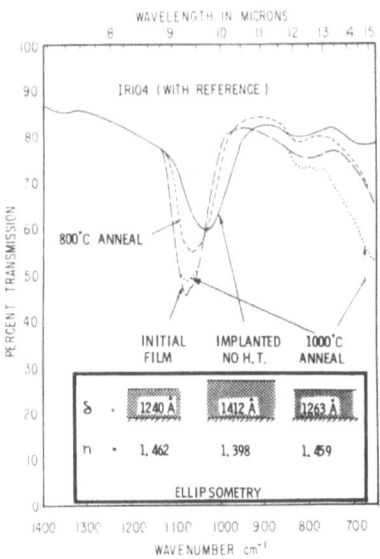

Figure 5. Annealing behavior of HCl, MOSFET quality oxides in terms of transmission infrared spectroscopy and ellipsometry.

Table V

Infrared Analysis of Si-O Stretch

Sample #IR104

Condition	Absorption Peak (cm^{-1})		Δ (cm^{-1})
Before II	1078	→ sharp	
After II	1035	→ broad	~43
Isochronal Anneals of 1 hr. duration at specified temp. (°C)			
100	1035		0
200	1035		0
300	1035		0
400	1040		5
500	1045		5
600	1050		5
700	1055		5
800	1060		5
900	1065		5
1000	1070	→ sharp	5
	Total		Δ ~35 cm^{-1}

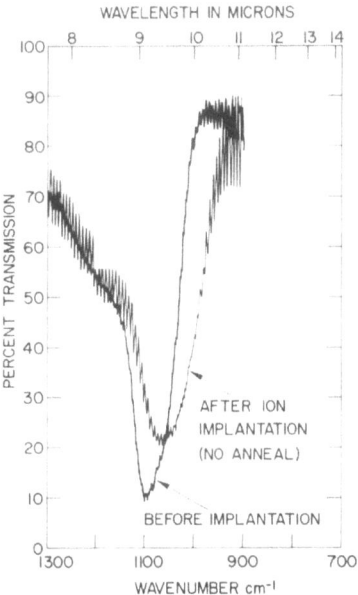

Figure 6. Transmission infrared spectrum showing only a slight broadening and shift of the Si-O stretching vibration after Argon implantation which is typical of wet, steam and Na contaminated oxides.

films, $2.16 \pm .02$ g/cm^3, remained the same within the experimental accuracy of ~2 percent. (Note that the ellipsometric measurements are indicative of a change in the film's thickness, perpendicular or normal to its surface, and that any lateral changes are assumed to be negligible since the film is "pinned" to the substrate's surface. Therefore, a change in the film's thickness can be assumed to be a direct change in the film's density.) Utilizing our observation that dense films expand and less dense films remain near a density of ~ 2.16 g/cm^3, it would appear that an equilibrium state may exist between the ordering and disordering processes. The infrared data reinforces the density observations in that one interprets the broadening of the Si-O band and shifting to lower energy as a disordering of the SiO$_4$ tetrahedra and also as a "loosening up" of the Si-O bonds[23].

The fact that we found various densities for the thermal silicon dioxide films, depending upon film preparation, is in good agreement with a paper by Deal[25]. Deal reported that the dry oxides prepared in his laboratory had a density of 2.27 g/cm^3, while the wet oxides had a density of 2.18 g/cm^3. Deal also found that the steam grown oxides varied in density within a range of 2.00 g/cm^3 to 2.20 g/cm^3.

Though our study is in disagreement with the compaction study of EerNisse and Norris[12], it is our belief that the starting silicon dioxide films may have been

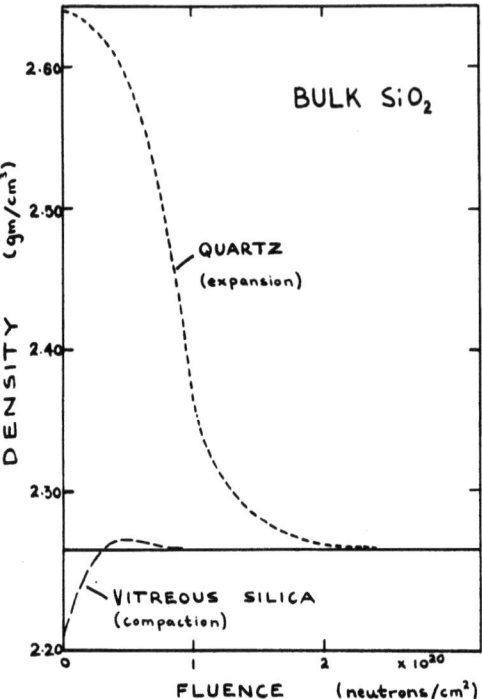

Figure 7. Composite of Primak's data plotting density vs. fluence. Shows crystalline quartz becoming less dense, while vitreous silica becomes more dense. Solid line at 2.26 g/cm^3 seems to be an equilibrium position for irradiated bulk silicon dioxide.

different, and thus, exhibited different properties. EerNisse and Norris stated in their paper that, "All wet samples came from one wafer and all dry samples came from another wafer." Their sample size was admittedly small. They also reported an "anomalous expansion effect" that occurs for dry oxides only. Precisely what we observed. In conclusion, we feel that a simple picture has emerged in that very dense silicon dioxide films can only become less dense upon irradiation; while less dense silicon dioxide films remain unchanged within 2 percent. Furthermore, annealing of the irradiated films in He at their growth temperature, restored the films to their pre-implantation condition.

REFERENCES

1. W. Primak, *The Compacted States of Vitreous Silica*, Gordon and Breach Science Publishers, New York, 1975.

2. S. Weissman and K. Nakajima, J. Appl. Phys., *34*, 611(1963).

3. T.Nishimura, H. Aritome, K. Masuda and S. Namba, ibid, *13,* 1317 (1974).

4. E. Kooi, Philips Res. Repts., *20,* 595 (1965).

5. W. Primak and R. Kampwirth, J. Appl. Phys., *39,* 6010 (1968).

6. S. Namba, H. Aritome, T. Nishimura, and K. Masuda, J. Vac. Sci. Technol., *10,* 936 (1973).

7. C. B. Norris and E. P. EerNisse, J. Appl. Phys., *45,* 3876 (1974).

8. W. Primak, L. H. Fuchs, and P. Day, Phys. Rev., *92,* 1064 (1953).

9. W. Primak, ibid., *110,* 1240 (1958).

10. S. Weissmann and K. Nakajima, J. Appl. Phys., *34,* 3152 (1963).

11. C. R. Fritzsche and W. Rothemund, J. Electrochem. Soc., *119,* 1243 (1972).

12. E. P. EerNisse and C. B. Norris, J. Appl. Phys., *45,* 5196 (1974).

13. E. P. EerNisse, ibid., *45,* 167 (1974).

14. E. P. EerNisse, Appl. Phys. Let., *18,* 581 (1971).

15. A. E. Hill and G. R. Hoffman, Brit. J. Appl. Phys., *18,* 13 (1967).

16. C. M. Osburn and D. W. Ormond, J. Electrochem. Soc., *119,* 591 (1972).

17. C. M. Osburn and D. W. Ormond, ibid., 597 (1972).

18. E. A. Irene, ibid., *121,* 1613 (1974).

19. E. A. Irene and Y. J. van der Meulen, ibid., *124,* 1380 (1976).

20. J. F. Gibbons, W. S. Johnson and S. W. Mylroie, *Projected Range Statistics: Semiconductors and Related Materials,* Dowden, Hutchinson and Ross, Stroudsburg, Pennsylvania, Second Edition, 1975.

21. D. K. Brice, *Ion Implantation Range and Energy Deposition Destributions, I,* Plenum Press, N.Y., 1975.

22. Y. J. van der Meulen and N. C. Hien, J. Opt. Soc. Am., *64,* 804 (1974).

23. W. A. Pliskin and H. S. Lehman, J. Electrochem. Soc., *112,* 1013 (1965).

24. B. .E. Deal, ibid., *110,* 527 (1963).

RECOIL IMPLANTATION

R. A. Moline

Bell Laboratories

Reading, Pennsylvania 19604

ABSTRACT

This paper reviews previous work on recoil implantation of atoms from a thin target. Some examples of effects produced by recoil implantation are described.

Thin target recoil implantation yields have been calculated based on collision cascades and by direct calculation of the primary recoil yield. These models are discussed and compared. A general feature of calculations based on collision cascades is that the yield is inversely proportional to a minimum escape energy. The primary recoil yield does not diverge at low escape energies.

Recoil implantation profiles are shallow and sharply peaked at the target-substrate interface. This creates experimental difficulty since the target is generally removed prior to recoil yield measurements. Experimental techniques used to measure the recoil implantation yield are discussed. Typical recoil implantation yields found in the literature range from 2 to 10 recoils per projectile when the target thickness is approximately the same as the projectile range.

INTRODUCTION

Recoiling target atoms play an important role in the stopping of low energy, heavy projectiles. Multiple scattering and range straggle are heavily influenced by the energy distribution of atomic recoils. The final resting place of recoils is also of considerable interest since it determines sputtering rates, changes in stoichiometry and radiation damage distributions.

This paper will review a specific effect caused by recoiling target atoms, namely recoil implantation. While it is to be expected that the local stoichiometry of compound targets will be altered by bombardment, due to the mass dependence of target atom scattering cross sections, these effects are outside the scope of this paper. Only effects which are caused by target atoms recoiling from one layer to another will be considered.

Recoiling atoms can be either advantageous or detrimental to the application of ion implantation. Recoil implantation can be used to implant species of which beams are not readily obtainable. It also will cause recoiling of surface layers, such as oxides, into the substrate. This effect is unavoidable if patterned implantations are used since the masking material (e.g. SiO_2) will have tapered edges. Even if the windows are absolutely clean, the window edges will have a zone where recoil implantation can occur. The relative importance of recoil implantation to a specific situation must be judged on an individual basis.

After reviewing examples where recoil implantation has resulted in beneficial and detrimental effects, this paper describes some models and calculations relevant to recoil implantation. Experimental techniques and results are then reviewed.

EXAMPLES OF EFFECTS

Figure 1 schematically illustrates the most common configuration for recoil implantation experiments. Typically, a thin, uniform target is formed on the substrate and the structure is then bombarded with projectiles which may, but need not, have a projected range larger than the target thickness.

Although only forward scattered recoil atoms are shown in this figure, clearly many atoms will be sputtered in a backward direction. Stroud, et al.[1] have shown that if a thin Al film is formed on a SiO_2 substrate and bombarded with Ar ions energetic enough to penetrate into the SiO_2, the Al is doped with oxygen.

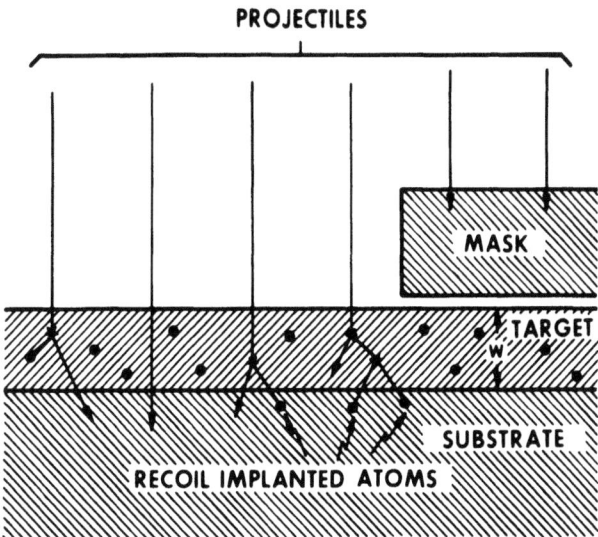

Figure 1. Typical configuration for recoil implantation experiments.

Collins, et al.[2] have demonstrated greater than a factor of 60 increase in the adhesion of Al to glass following recoil implantation of Al past the Al-glass interface following a bombardment of 10^{15}, 120 keV Ar/cm². Similar, although less dramatic, effects have been observed for gold [3]. Cermets have been formed [4] by bombarding an Al-SiO$_2$-Al sandwich with $\sim 10^{16}$, 80 keV Ar/cm². Both forward recoil implantation from the first Al layer and backward sputtered atoms from the bottom Al layer contribute to doping the SiO$_2$. The Au-Si Schottky barrier height has been lowered by the formation of a thin, Sb doped surface layer using recoil implantation [5]. The shallow profile produced by recoil implantation is important for this application.

Some effects produced by recoil implantation are not beneficial. Cass and Reddi [6] observed residual defects in Si produced by implantation of 10^{16} As/cm² through a thin SiO$_2$ layer. Transmission electron micrographs exhibited heavily dislocated regions in the Si. This damage has been shown to be the direct result of recoil implanted oxygen [7,8]. Other authors,[9,10] studying implanted rare-earth atoms in iron, have suggested recoil implanted oxygen might be influencing their results.

The ability to accurately estimate recoil implantation yields is important. Cass and Reddi anticipated recoil implanted oxygen might produce the residual damage they observed. They implanted 0.1 oxygen/As and did not see the residual damage effect they had observed after As was implanted through SiO_2. Recently it has been shown [11] that the oxygen dose threshold for gross residual damage in Si is $1\text{-}3 \times 10^{15}/cm^2$ for 10-50 keV ions. Thus, since the recoil yield is 1-2 per projectile for their experimental conditions, [12] if a more accurate estimate had been used the authors would have confirmed their initial concern about recoil implantation.

Additional effects caused by an "insert" ion beam may cause complications when using recoil implantation. Not only must radiation damage be dealt with but also effects specificit to the ion. Seidel, et al. [13] have shown that Ar atoms remain in Si even following an 1150°C anneal. Considerable residual lattice damage is caused by high levels of Ar.

THEORY

Nelson [14] has developed a theory based on collision cascades for recoil implantation yields. As an approximation he has assumed heavy projectiles, isotropic cross sections, and a target sufficiently thin so the projectile energy loss can be ignored. The final recoil spectrum, shown in Eq. (5) of Ref. 14 can be rewritten as:

$$dS = 0.2078 \; \pi a^2 \; DN \frac{T_M}{\varepsilon} \left[\frac{1}{T^2} - \frac{1}{T_M^2} \right] dT, \qquad (1)$$

where
- S is the forward sputtering yield per projectile,
- D is the mean interatomic spacing,
- T the energy transfer to the recoil,
- T_M the maximum recoil energy,
- N the atomic density,
- ε the projectile energy in dimensionless units, [15] and
- $a = a_0 * 0.8853 \; (Z_1^{2/3} + Z_2^{2/3})^{-1/2}$.

This results in a recoil energy spectrum diverging as $1/T^2$ at low energy. The total recoil yield is then obtained by integrating from some minimum energy to T_M. If this minimum energy, u, is much less than T_M,

$$S \approx 0.2078 \; \pi a^2 \; \frac{DN}{\varepsilon} \frac{T_M}{u}. \qquad (2)$$

Sigmund [16] has developed a more general theory of sputtering, containing (forward) transmission sputtering as a special case. This model is also based on collision cascades. When the recoiling atoms have energies within the elastic collision region, he estimates the thin foil transmission sputtering yield, S, as [16]

$$S = \alpha' \Lambda N S_n, \qquad (3)$$

where
- α' is a constant ($\sim 1/2$ for Rutherford scattering),
- Λ is a material constant which can either be measured or eliminated as shown below, and
- S_n is the elastic stopping cross section of the target.

Sigmund identifies

$$\Lambda = \frac{\Delta x}{\pi^2 u}, \qquad (4)$$

where
- Δx is the escape depth of sputtered atoms and
- u is the escape energy.

Thus Eq. (3) can be rewritten as

$$S = \frac{\alpha'}{\pi^2} \frac{\Delta x \, N \, S_n}{u} \qquad (5)$$

or

$$S \approx 0.05 \frac{\Delta E}{u} \qquad (6)$$

where ΔE is the elastic energy loss in the escape depth. Sigmund calculates estimates of Δx for various experimental conditions but cautions against arbitrary use since, for very massive projectiles, significant deviations can occur.

When the projectile is massive there is a statistically significant chance of obtaining energetic recoils. These recoils can penetrate considerable distances, resulting in a large effective escape depth.

Another complication comes from the fact that for a heavy projectile, the projected range of the most energetic recoils exceeds the projected range of the projectile. Figure 2 demonstrates this effect for a thin Al target on Si. The relative range of the highest energy recoil is shown as a function of relative projectile mass for 150 keV projectiles. Range data was taken from Ref. 17

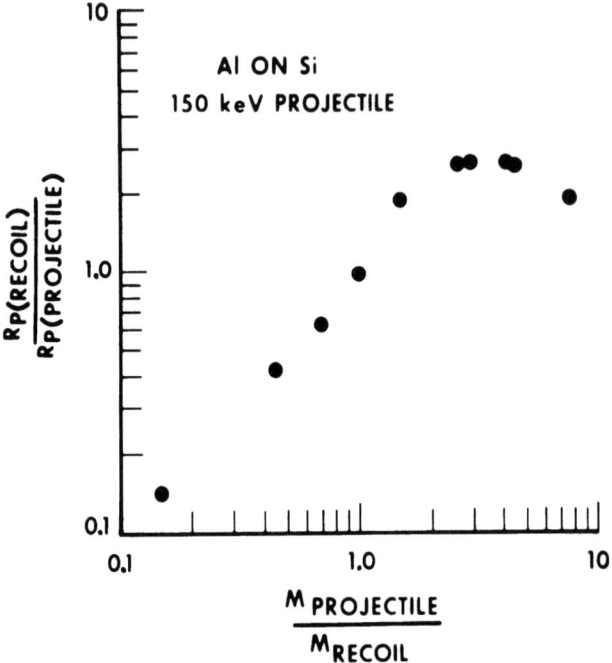

Figure 2. The range of the maximum energy recoil, relative to the projectile range, is shown for various projectile masses. LSS ranges are assumed.

except LSS [15] values for the electronic stopping were assumed for B and C in Si. (This assumption results in a smoother curve.) Clearly, for $M_{Projectile} > M_{Target}$ some recoils will penetrate completely through the target providing the projectile is energetic enough to penetrate the target.

Neglecting this problem for the moment, if the elastic energy loss is assumed constant [15] (in general a poor approximation) Eq. (6) can be rewritten as

$$S \sim 0.0165 \, \pi a^2 \, \frac{\Delta x N}{\varepsilon} \, \frac{T_M}{u}. \tag{7}$$

This yield estimate differs from Eq. (2) only by a constant factor which can be absorbed in the minimum escape energy.

As mentioned above, for massive projectiles, the primary recoil yield might contribute significantly to the transmission sputtering yield. The number of primary recoils, S_p, which escape the target has been calculated for a $1/r^2$ potential [12]. For a thin target of thickness

W, and a constant stopping power for the recoil,

$$S_p \sim 0.33 \ \pi a^2 \ \frac{WN}{\varepsilon} \ [\ \frac{3}{2} \ (\frac{R_M}{W})^{1/3} - 1], \qquad (8)$$

where R_M is the projected range of the maximum energy recoil. This yield is independent of escape energy since at low energy the primary recoil yield does not diverge as rapidly as the cascade yield does. The yield does depend almost linearly on W. Numerical integration of the recoil yield for thicker targets has also been made for the $1/r^2$ and $1/r^3$ potential [12].

If $R_M \gg W$, Eq. (8) yields

$$S_p \sim 0.50 \ \pi a^2 \ \frac{WN}{\varepsilon} \ (\frac{R_M}{W})^{1/3}. \qquad (9)$$

Littmark and Sigmund [18] have calculated the momentum deposited in a semi-infinite substrate as a function of depth for a $1/r^3$ potential. Recoil atoms were included in their calculations. They found a net negative momentum deposition near the surface and significant positive momentum transfer even at depths greater than twice the projectile projected range. They argue that since recoils go predominately in the direction of momentum transfer, more surface atoms will be backward sputtered than will be recoil implanted deeper into the substrate. At greater depths in the substrate, more recoils will be going forward than backwards.

Brice [19] has included recoil atoms in the damage energy density distribution and found for heavy projectiles significant errors were introduced if the recoil atoms were neglected. Brice [20] has also calculated the energy-weighted average projected range for all recoils within a thick target, for various ions incident on Si. For 100 keV As ions, the weighted average recoil range is 220Å! Even for B ions, the weighted average range is 100Å. This large average recoil range underscores the complexity associated with the application of equations which depend on an escape depth to recoil implantation.

With the success of numerical integration techniques for ion range and damage distributions, it is reasonable to assume similar techniques can be developed to calculate recoil implantation yields. This task will be complicated by the fact that a layered structure is used and thus calculations will not be general. Compound targets and substrates will add additional complexity since recoil implantation induced by recoiling atoms of other species will be important. Also, cross sections will be tested at very low energies since recoils which have ranges of only a few Å can have significant impact on the recoil yield, a factor not important for damage distributions.

EXPERIMENTAL RESULTS

One feature all theories predict is an extremely shallow profile of the recoil implanted atoms in the substrate. This complicates experiments since typically the target shown in Fig. 1 is chemically removed prior to measurement of the recoil implantation yield. It is important that this etch stop exactly at the interface. An error of a few atomic layers might significantly alter the result.

Many factors contribute to uncertainty in the amount of substrate removed. If the bombardment dose exceeds the amorphous dose, the interface is essentially gone. Atoms from the substrate will be backward sputtered into the target, tending to alter the location of the interface. Chemical effects between the target and the substrate can also be important.

Nishi, et al. [21,22] have observed that when M_O is evaporated on Si and then chemically removed, approximately one mono-layer of M_O remained on the Si even without bombardment. This effect was not observed when SiO_2 was used as a substrate. They then bombarded a 300Å thick M_O target with 150 keV Ar projectiles. ($M_{Projectile}/M_{Target} = 0.4$). A non-linear dependence of the residual M_O yield per projectile was observed for a Si substrate. For sub-amorphous doses the number of residual M_O per projectile was approximately three times greater than for amorphous projectile doses. This non-linear yield behavior was not observed when SiO_2 was used as the substrate. Similar non-linear effects were observed for Si substrates when a Cr target was bombarded with As ions. They attribute this non-linear behavior to bombardment - enhanced chemical effects between the target and Si.

The variation in recoil implantation yields with target thickness was also measured [21,22]. Yields were found to increase with increasing target thickness until a significant fraction of the projectiles did not penetrate to the interface. Maximum recoil yields were found to be 4 for Ar→M_O and 10 for As→Cr. Using estimates of Δx [16] and S_n [20], good agreement is obtained with

Eq. (3) if u is assumed to be 10 eV for M_O on Si and 5 eV for Cr on Si.

Perkins and Stroud [23] have studied Ar projectiles incident on Al films. ($M_{Projectile}/M_{Target}$ = 1.5). They studied two experimental configurations, namely free standing Al targets and Al targets formed on fused silica or polyethylene [1]. They observe a factor of four higher yield for the free standing target (transmission sputtering yield) when compared to the target-substrate combination (recoil implantation yield). They have fit their results to Sigmund's theory with values of u = 4 eV for a free standing target and u = 16 for the target substrate combination. They argue that for the free standing target, u is related to surface binding energies while for the substrate case, bulk binding energies are relevant.

Moline, et al. [12] studied recoil implantation of ^{18}O from SiO_2 targets on Si substrates. For Kr projectiles ($M_{Projectile}/M_{Target}$ = 4.7), they found excellent agreement with the primary recoil yield calculated using Eq. (8) for small oxide thicknesses. For larger thicknesses, a multiplication factor of $1 + (W/W_o)^{2/3}$, where $W_o = 50$ Å, was needed to fit their data. This agreement is shown in Figure 3.

Brice [20] has calculated damage energy density curves for various projectiles in Si. Estimates for As damage profiles in Si are shown as points in Figure 4. Also shown as solid lines is the primary oxygen recoil yield (Eq. 8) for Kr in SiO_2, replotted from Ref. 12. Note that for 24 keV projectiles the shape is nearly identical for the two parameters. To the extent that Kr and As are the same mass and Si and SiO_2 stopping power are similar, these curves can be compared. For this particular case Eq. (3) and Eq. (8) yield nearly identical results if $\alpha'\Lambda = 0.01$. Assuming $\alpha' = 0.3$, [16] u = 17 eV. This agreement is not generally true as can be seen from the data plotted for 48 keV projectiles.

Since Eq. (3) already includes cascade type effects, the multiplication factor needed to make Eq. (8) fit the data would not apply to Eq. (3). Also, the measured decrease in yield with increased projectile energy [12] is underestimated using Eq. (3). This disagreement is not surprising since many assumptions which were made in deriving Eq. (3) were stated to be invalid for the case of heavy projectiles [16].

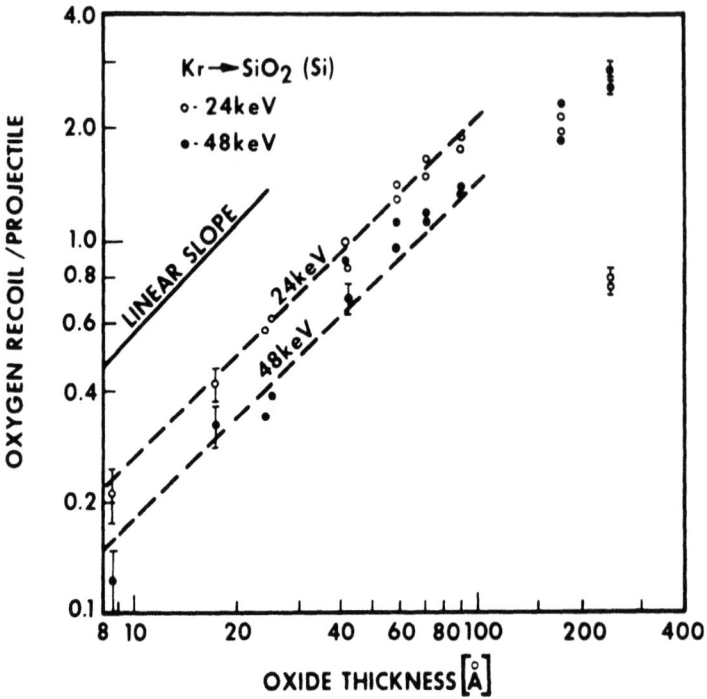

Figure 3. Yield of ^{18}O recoiled into Si from a SiO_2 target. The dashed line is the primary recoil yield calculated from Eq. (8) multiplied by $1 + (W/W_0)^{2/3}$.

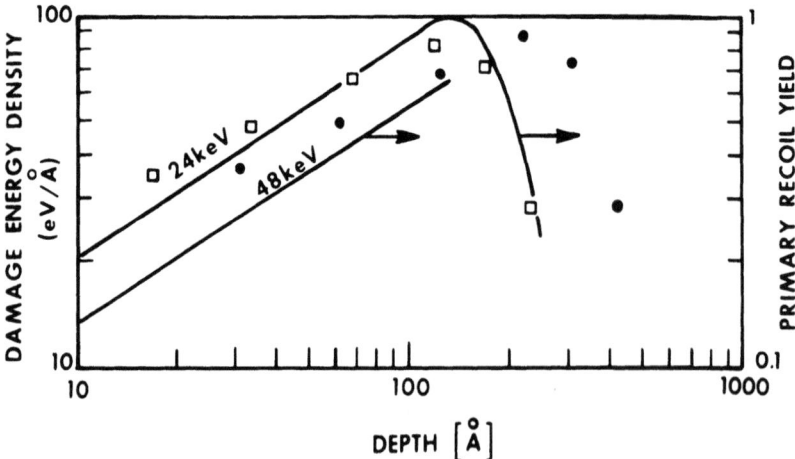

Figure 4. Solid lines: primary oxygen recoil yield from Eq. (8) for Kr in SiO_2. Points: damage energy density estimated from Ref. 20 for As in Si.

DISCUSSION AND SUMMARY

Although recoil implantation can result in yields of ~ 10, only careful experimental methods will establish its importance for new situations. For example, Lee, et al. [24] studied bombardment enhanced interactions between Al and Si. Their control experiments determined that radiation damage, rather than recoiling Al atoms, was the dominant factor responsible for the enhanced interaction.

Recoil implantation yields can be estimated using Eqs. (3) or (8) depending on the projectile mass. However, for accurate predictions, covering arbitrary cases, complete numerical integration methods will have to be developed. Also, the importance of doping in the first few atomic layers of the target must be established.

Increased yield has been described for free standing targets. Equation (5) can be used to anticipate this effect if u is assumed to vary and S_n is assumed constant. However, S_n will depend on the target-substrate structure. In cases where recoil atoms contribute significantly to S_n, S_n will decrease if no substrate back-scattered recoils enter the target. Thus, if the increased yield is due only to decreased u, the decrease must be more than the observed factor in yield increase. Additional experimental evidence is required to establish the importance of substrate composition, target and substrate thickness and projectile mass.

Additional experimental measurements of recoil implantation profiles will provide needed opportunities to test models and calculations. Also, the impact of experimental techniques and structures on recoil yields must be established.

The preceding discussion indicates the difficulty in making precise estimates of recoil implantation yields. Estimates can be based on the fact that measured maximum yields range from 2 to 10 for typical ion implantation conditions. Extrapolation to other situations can be made using Eq. (3) when cascade effects dominate or Eq. (8) when primary recoil effects dominate the yield.

REFERENCES

[1] P. T. Stroud, L. E. Collins, J. G. Perkins and K. G. Stephens, European Conference on Ion Implantation, (Peter Peregrinus Ltd., England, 1970), 116.
[2] L. E. Collins, J. G. Perkins and P. T. Stroud, Thin Solid Films 4, 41 (1969).
[3] P. T. Stroud, Thin Solid Films 11, 1 (1972)
[4] L. E. Collins, P. A. O'Connell, J. G. Perkins, F. R. Pontet and P. T. Stroud, Nucl. Instr. Methods 92, 455 (1971).
[5] Hiroshi Ishiwara and Seijiro Furukawa, this conference (V International Conference on Ion Implantation).
[6] T. R. Cass and V. G. K. Reddi, Appl. Phys. Lett. 23, 268 (1973).
[7] W. K. Chu, H. Müller, J. W. Mayer and T. W. Sigmon, Appl. Phys. Lett. 25, 297 (1974).
[8] R. A. Moline and A. G. Cullis, Appl. Phys. Lett. 26, 551 (1975).
[9] R. L. Cohen, G. Beyer and B. Deutch, Phys. Rev. Lett. 33, 518 (1974).
[10] L. Thomé, H. Bernas, J. Chaumont, F. Abel, M. Bruneaux and C. Cohen, Phys. Lett. 54A, 37 (1975).
[11] A. G. Cullis and R. A. Moline, "Implantation of Oxygen into Silicon" talk to be presented at the 6th European Congress on Electron Microscopy in Jerusalem, Israel, September 14-20, 1976.
[12] R. A. Moline, G. W. Reutlinger and J. C. North, proceedings of V Int. Conf. on Atomic Collisions in Solids (1973), Atomic Collisions in Solids, Vol. 1, Edited by S. Datz, B. R. Appleton and C. D. Moak, (Plenum, New York, 1975) 159.
[13] T. E. Seidel, R. L. Meek and A. G. Cullis, J. Appl. Phys. 46, 600 (1975).
[14] R. S. Nelson, Radiation Effects 2, 47 (1969).
[15] J. Lindhard, M. Scharff and H. E. Schiøtt, Kgl. Danske Videnskab, Mat.-Fys. Medd. 33, No. 14 (1963).
[16] Peter Sigmund, Phys. Rev. 184, 383 (1969).
[17] James F. Gibbons, William S. Johnson and Steven W. Mylroie, Projected Range Statistics, (Halsted Press, 1975).
[18] Uffe Littmark and Peter Sigmund, J. Phys. D:Appl. Phys. 8, 241 (1975).
[19] David K. Brice, Ion Implantation in Semiconductors, Ed. Susumu Namba (Plenum, New York, 1975) 399.
[20] David K. Brice, J. Appl. Phys. 46, 3385 (1975).
[21] H. Nishi, T. Sakurai and T. Furuya, Ion Implantation in Semiconductors, ed. Susumu Namba (Plenum, New York 1975) 347.

[22] H. Nishi, T. Sakurai, T. Akamatsu and T. Furuya, Appl. Phys. Lett. 25, 337 (1974).
[23] J. G. Perkins and P. T. Stroud, Nucl. Instr. Methods 102, 109 (1972).
[24] D. H. Lee, O. J. Marsh and R. R. Hart, Ion Implantation in Semiconductors, Ed. I. Ruge and J. Graul (Springer-Verlag Berlin-Heidelberg - New York 1971) 262.

APPLICATION OF THE BOLTZMANN TRANSPORT EQUATION TO THE CALCULATION OF RANGE PROFILES AND RECOIL IMPLANTATION IN MULTILAYERED MEDIA

D. H. Smith and J. F. Gibbons

Stanford University

Stanford, California 94305

ABSTRACT

A straightforward approach to calculations of range profiles in multilayered media is developed based on numerical integration of the Linearized Boltzmann Transport Equation,

$$\frac{\partial f(E,x)}{\partial x} = N \int [\partial\sigma(E' \to E) f(E',x) - \partial\sigma(E \to E') f(E,x)]$$

where x is the path length traveled by the implanted species. For heavy ions in light substrates, such as As in a Si_3N_4-SiO_2-Si layered target, the path length is nearly equal to the depth below the surface. Therefore these calculations provide direct estimates of the range profiles for this situation.

The calculation thus performed necessarily involves calculation of probability for energy transfer, T, to a substrate atom at each depth within the target. As a result, values for the energy, recoil angle and rate of generation of recoil atoms at a given depth within the target are also available. By following the path of the recoiling species we can construct, in addition to the range profile for the implanted ion, a range profile for recoiling atoms deposited in the multilayer substrate.

I. INTRODUCTION

The application of ion implantation to device fabrication often involves processing in which surface coatings are present on the target substrates [1-4]. Two questions of interest that arise in cases such as these are as follows: (1) what is the range distribution of the primary projectile in the coated (or multilayer) target, and (2) what is the range distribution of the atoms that are recoil implanted from the surface coatings into the target.

The answers to both of these questions can be obtained when the energy distribution of the primary projectile is known as a function of depth in the multilayer target. In particular, if an As ion is implanted into a Si_3N_4-SiO_2-Si target as illustrated in Figure 1, then a knowledge of the As energy distribution is sufficient to calculate both the recoil energy and recoil angle for each N and/or O recoil that is produced in a slab of width Δx_p at the position x_p. A knowledge of the As energy distribution at the Si_3N_4-SiO_2 interface is also sufficient to predict the relative concentration of As ions on the two sides of the interface.

It is of course possible to calculate energy distributions for semi-infinite targets using either the moments method of Winterbon et al [5] or the direct construction technique of Brice [6]. How-

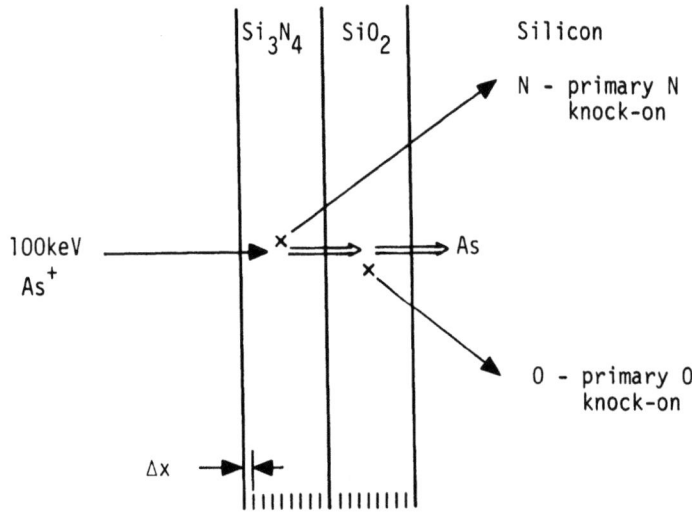

Fig. 1   Arsenic implantation though a Si_3N_4-SiO_2 layer. Important Secondary Processes are the recoil implantation of Oxygen and Nitrogen.

APPLICATION OF THE BOLTZMANN TRANSPORT EQUATION 335

ever these methods are not readily applicable in a multilayer target and it is therefore natural to resort to the Boltzmann transport equation, where the evolution of the energy distribution with distance is a quantity that may be calculated directly.

Use of the Boltzmann equation to calculate the energy distribution for the primary ion is in principle straightforward when each layer of the target is assumed to be a random stopping medium. However even for this case the analysis requires that several simplifying assumptions be made to construct numerical solutions in particular cases. Fortunately, results show good agreement with experiments where such comparison is possible, and the analysis provides a useful basis for additional calculations.

The technique is similar in concept to the Quasi-Monte-Carlo calculations of Furukawa and Ishiwara [7].

II. THE BOLTZMANN EQUATION

Consider a collection of a large number of particles with different velocities, $(\vec{v})$ located at different points in spaces $(\vec{x})$. We describe the collection by an average number,

$$dN_p = F(\vec{v},\vec{x}) \, d^3\vec{v} \, d^3\vec{x} \quad (1)$$

for each differential element of phase space $d^3\vec{v} \, d^3\vec{x}$. We further assume that particles can undergo transitions from one region in phase space to another by means of collisions. The probability for a particle with velocity $\vec{v}$ to be scattered into the energy interval $d^3\vec{v}'$ about $\vec{v}'$ during a time dt is given by

$$K(\vec{v} \rightarrow \vec{v}') \, d^3\vec{v}' \, dt = N_s |\vec{v}| d\sigma \, (\vec{v} \rightarrow \vec{v}') \, dt \quad (2)$$

where $K(\vec{v} \rightarrow \vec{v}')$ is the transition rate, N_s is the density of scattering centers, and $d\sigma(\vec{v} \rightarrow \vec{v}')$ is the differential cross section for scattering from $\vec{v}$ into $d^3\vec{v}'$. Note that the spatial dependence has been omitted for simplicity. A consideration of the average number of particles scattered into and out of a differential element of phase space leads to an integro-differential equation governing the density $F(\vec{v}, \vec{x})$

$$\frac{\partial F(\vec{v})}{\partial t} + \vec{v} \cdot \frac{\partial F(\vec{v})}{\partial \vec{x}} = N_s \int \left\{ d\sigma(\vec{v}' \rightarrow \vec{v}) |\vec{v}'| |F(\vec{v}')| - d\sigma(\vec{v} \rightarrow \vec{v}') |\vec{v}| |F(\vec{v})| \right\} + Q(\vec{v}) \quad (3)$$

The quantity $Q(\vec{v})$ is a source term which describes the generation of new members of the ensemble. For example if the particles under consideration are generated in collisions (i.e. recoils), then $Q(\vec{v})$ would be calculated by consideration of these collisions. The spatial dependence of the quantities in equation (3) has again been omitted for simplicity. Of course if more than one type of particle is being considered (e.g. primary ions and recoils) approximate subscripts should be added to the various quantities. Equation (3) is the standard form of the Boltzmann equation.

Equation (3) in conjunction with appropriate boundary conditions provides a basis for a complete description of the complex processes which result from energetic ion moving in an inhomogeneous media. A numerical solution in six dimensional phase space, while possible in principle, is unfortunately considerably beyond the reach of even the most energetic programmer equiped with the most sophisticated, modern computer. However, simplifying assumptions can be made which allow the extraction of a considerable amount of information in certain cases.

III. ONE DIMENSIONAL PENETRATION

For heavy ions incident on light targets, as for example As on SiO_2, angular scattering of the incident ion is relatively small and the ion travels in nearly a straight line. The density function is then a function only of the depth of penetration x and the energy E of the ion. With this simplifying assumption equation (3) can be rewritten,

$$\frac{\partial f(E,x)}{\partial x} = N_s \int \left((d\sigma(E' \to E) f(E',x) - d\sigma(E \to E') f(E,x) \right) \quad (4)$$

where $f(E,x)$ is the total flux of ions at depth x,

$$f(E,x) = \int dt \, |\vec{v}| \, F(E,x,t) \quad (5)$$

$f(E,x)$ dAdE is then total number of particles with energy E to E + dE which cross a element of area dA perpendicular to the direction of incidence. Equation (4) has a simple physical interpretation, as in figure 2. The first term on the RHS represents the rate for scattering from E' into dE, while the second term represents scattering out of the energy interval dE. Integration of equation (4) is carried out starting from x = 0 (the sample surface) and integrating inward, the starting condition being

$$f(E,0) = D \, \delta(E-E_0) \quad (6)$$

where D is the total dose in ions/cm^2, E_0 is the energy of the incident ion, and $\delta(E-E_0)$ is the Dirac delta function. It is tacitly understood that the terms on the right hand side must be summed over all collision types (i.e. elastic scattering via nuclear collisions as well as inelastic or electronic scattering).

IV. SCATTERING CROSS SECTIONS AND SEPERATION OF LOW ENERGY COLLISIONS.

For numerical analysis we have used an analytical form for the nuclear cross section suggested by Sigmund (8).

APPLICATION OF THE BOLTZMANN TRANSPORT EQUATION

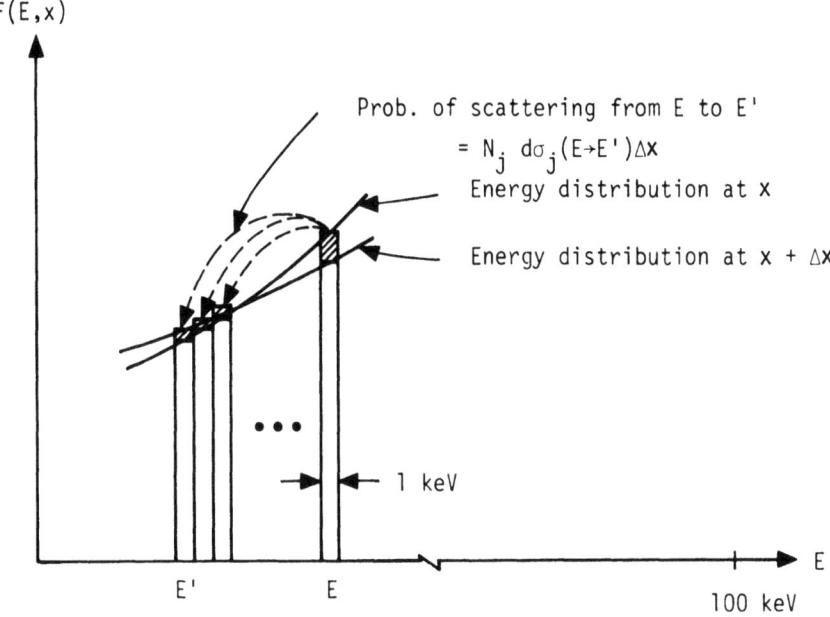

Figure 2 Schematic showing the process by which the quantity F(E,x) evolves.

$$d\sigma(E' \to E) = \left(\frac{\pi a^2}{4}\right) \left\{\frac{2\lambda t^{-1-m} dt}{\left[1 - (2\lambda t^{1-m})^6\right]^{1/8}}\right\} \quad (7)$$

where λ, q, and m are constants and

$$a = .8853 a_0 \, (Z_1^{2/3} + Z_2^{2/3})^{-1/2}$$

$$t = TE \left\{\frac{M_2}{4M_1} \left(\frac{a}{Z_1 Z_2 e^2}\right)^2\right\}$$

$a_0 = \hbar/me^2 = .529$ Å

$T = E' - E =$ transferred energy

$M_1 =$ ion mass

$M_2 =$ target atom mass

$Z_1 =$ ion atomic number

$Z_2 =$ target atom atomic number

$e =$ electronic charge

$\hbar =$ plank's constant

Equation (7) is understood to apply only for $T \leq T_{max} = \dfrac{4M_1 M_2}{(M_1+M_2)^2} E$

For $T \geq T_{max}$, $d\sigma = 0$.

By a proper choice of λ, q, and m this expression provides a good analytical approximation to scattering cross sections derived by various workers. A number of examples, given by Sigmund, are tabulated in Table I. This expression provides a power law scattering for small t, and Rutherford scattering at higher values of t.

As will be shown below the inelastic (electronic) scattering does not require a knowledge of the differential cross section for this process, but can be properly accounted for if the stopping power,

$$S_e(E) = \int T d\sigma_e \tag{8}$$

is known. For this purpose we use the expression,

$$S_e(E) = CE^p \tag{9}$$

where C and p are constants depending on the ion-target atom combination.

V. SEPARATION OF LOW ENERGY AND ELECTRONIC COLLISIONS, NUMERICAL INTEGRATION.

It is convenient to rewrite the collision integral of equation (4) as follows:

$$I = \int \left[d\sigma(E' \to E) F(E') - d(E \to E') F(E) \right] \tag{10}$$

$$= -\frac{\partial}{\partial E} J(E')$$

where

$$J(E) = N_s \int_{E'=E}^{E_{max}} K(E',E) F(E') dE' \tag{11}$$

and

$$K(E',E) = \int_{E''=0}^{E''=E} d\sigma(E'' \to E') \tag{12}$$

$$= \pi p^2$$

where p is the impact parameter for the transition $E' \to E$. For distributions $F(E')$, where $F(E')$ does not vary appreciably over the energy interval where $K(E',E)$ is non-zero (small energy or soft collisions) we can take $F(E') \approx F(E)$ in equation (11) and write

$$J(E) \approx N_s S(E) F(E) \tag{13}$$

APPLICATION OF THE BOLTZMANN TRANSPORT EQUATION 339

Table I Parameters for use in the cross section expression eq. (7) (From Sigmund (8)).

Screening function	m	q	λ
Thomas-Fermi	.333	.667	1.309
Thomas-Fermi-Sommerfeld	.331	.588	1.70
Lenz-Jensen	.191	.512	2.92
Moliere	.216	.530	3.07
Bohr	.103	.570	2.37

where we have used the fact that

$$\int d E' \, K(E',E) = \int T \, d\sigma$$
$$= S(E) \qquad (14)$$

(10) and (14) are proven by partial integration.

The technique for numerical integration of equation (4) will now be described. Let E_0 be the energy of the incident ion beam. Then the range $o \geq E \geq E_0$ is divided into energies $E_1, E_2, \text{---}, E_2, \text{---} E_n = E_0$. The flux f_j is defined by

$$f_j = \int_{E=E_{j-1}}^{E=E_j} f(E) \, dE \qquad (15)$$

For electronic scattering and for nuclear events such that

$$T < \frac{E_j - E_{j-1}}{2}$$

the collision integral is represented by terms of the type shown in equation (10) with $J(E)$ given by equation (13). The following describe the development of the energy spectrum on passing from x to $x + \Delta x$.

$$x \to x + \Delta x \qquad (16a)$$

$$f_j \to f_j + N_s \, \Delta x \, \Sigma_{j'} \left[f_{j'} d\sigma_{j' \to j} - f_j d\sigma_{j \to j'} \right] \qquad (16b)$$

$$E_j \to E_j - N_s \Delta x \, S_\Delta(E_j) \qquad (16c)$$

Here,

$$d\sigma_{j' \to j} = \int_{E=E_{j-1}}^{E=E_j} d\sigma \, (E_{j'} \to E) \qquad (17)$$

and S_Δ accounts for electronic and small energy nuclear scattering events. This scheme can be shown to be mathematically rigorous, and as Δx and ΔE approach zero, one obtains an increasingly accurate approximation to the exact solution for $f(E)$. The details of proof for this assertion will not be given here.

VI. AN EXAMPLE

We have used the algorithm outlined above as the basis for a computer program, which calculates the function $f(E)$. Results for 100 kev As incident on a SiO_2 target are shown in figure 3. This calculation uses Lenz-Jensen cross sections and 20% of the electronic stopping given by the Linhard formula.

Special considerations are required at very low energies and this region is shown shaded in the figure. One can show however that the total number of particles that pass into the shaded region between x and $x + \Delta x$ is equal to $J(E_1) \Delta x$, where E_1 is the lowest energy considered. Since these ions have very little energy they can be considered to have stopped in this interval.

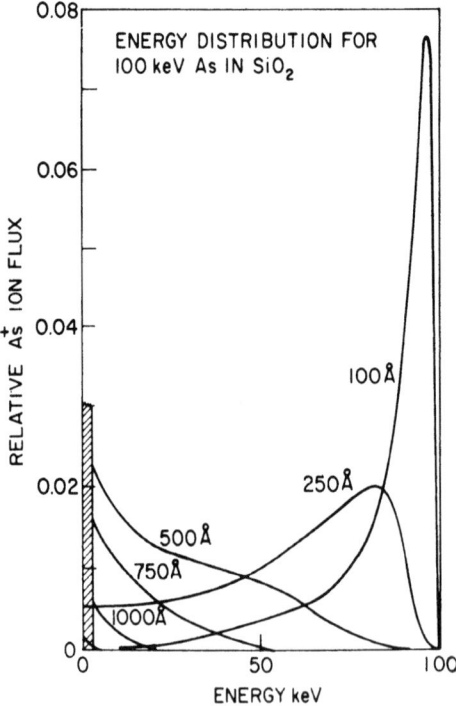

Fig.3. Evolution of As ion flux with depth of penetration for 100 kev As → SiO_2.

Figure 4 shows the range profile calculated using this technique. This profile is essentially identical to the one which would be calculated from LSS Theory using a joined half-Gaussian distribution and the stopping powers selected for the Boltzmann analysis.

VII. KNOCK-ONS

A natural consequence of the calculation outlined above is the source density $Q(\vec{v},x)$ for oxygen knock-ons produced in energetic collisions (figure 5). Thus an As-oxygen collision involving energy transfer T, produces a recoiling oxygen with energy T traveling in the direction

$$\cos \theta = \sqrt{\frac{T}{\gamma E}} \qquad (18)$$

By following the penetration of recoiling oxygen atoms one can construct the distribution of recoil implanted oxygen atoms.

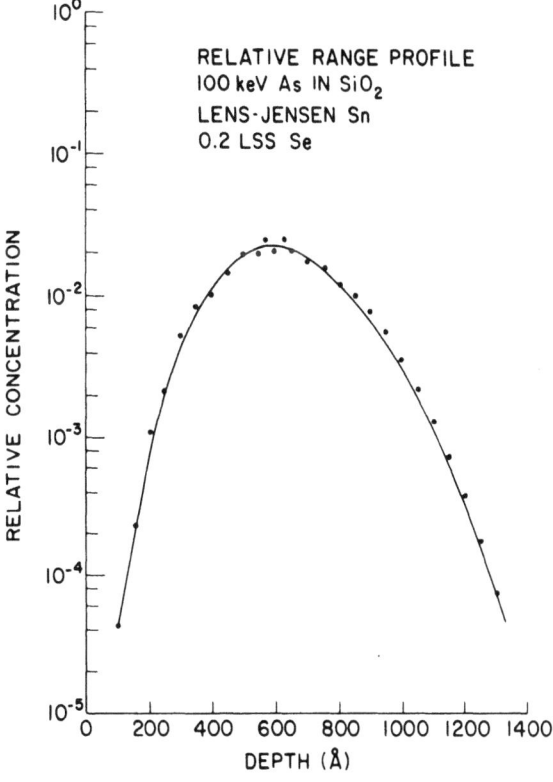

Fig. 4 Range profile for 100 kev As → SiO_2.

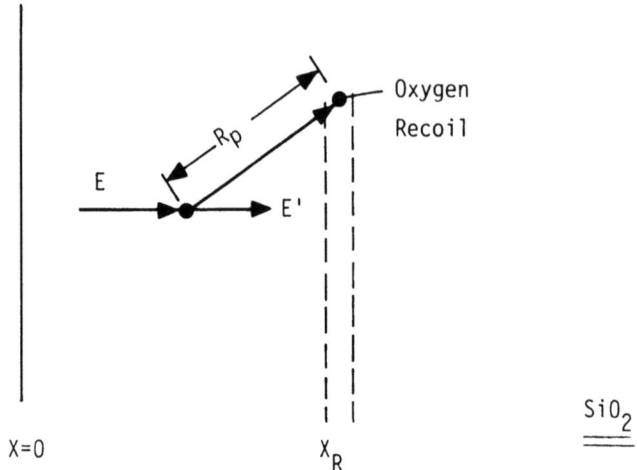

Calculation of Oxygen Recoil Distribution in SiO_2: Method 1.

Fig. 5 Production of an energetic oxygen recoil.

A simple way to do this is illustrated in figure 5. The recoiling atom is assumed to stop at a distance r determined by the stopping powers of the materials involved. For this purpose we use an effective stopping power defined by

$$S_{eff} = \frac{dE}{dRp} \qquad (19)$$

where Rp is the projected range calculated using the Linhard approach to range calculations. Values of this parameter for a number of species can be obtained from the range tables published by Gibbons, Johnson and Mylroie [9]. This approach therefore provides a first order correction for effects due to straggling.

We have used this technique to calculate the total doses due to recoil implantation of oxygen, when 100 kev As is incident on a silicon target covered with a SiO_2 layer. The thickness of the SiO_2 layer is taken as a variable and the total oxygen dose implanted into silicon is calculated as a function of this thickness. For this purpose we consider only recoiling oxygen atoms which penetrate more than 10 Å below the interface. The results of this calculation are denoted by "method 1" in figure 6. Also shown are experimental data of Goetzberger et al [10]. As can be seen the agreement is quite good except in the tail of the distribution where straggling becomes important. The curve labelled "method 2" was obtained using a fairly rough estimation of the effects of straggling of the recoil ions on the final recoil distribution.

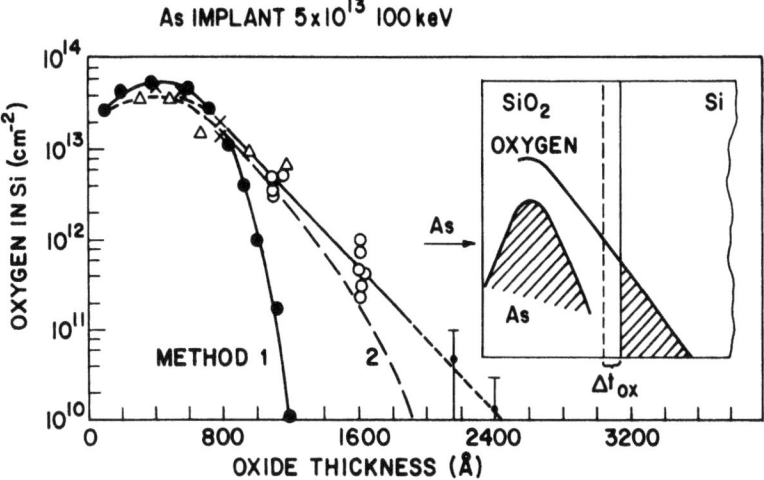

Fig. 6   Oxygen recoil implantation. The vertical axis shows the total dose of recoil implanted oxygen atoms as a function of oxide thickness. Experimental data is from Goetzberger et al [10]. The curves labeled method 1 and method 2 are calculated by techniques explained in the text.

VIII. DISCONTINUITY AT AN INTERFACE

Because of the differences in stopping power associated with various substances, a concentration discontinuity is expected at an interface between two layers [7]. This result appears naturally in range calculations using the technique outlined above. It is also possible to obtain an analytical expression which should be fairly accurate even for light ions, or ions crossing the interface at an angle.

The number of ions scattering into energies below any given energy E is proportional to $J(E)$ defined in equations (10) - (12). For $E \to 0$ this then gives the number stopping at depth x. If we use the approximation of equation (13) for J we can write

$$\frac{n_1}{n_2} \approx \left(\frac{S_1}{S_2}\right)_{E \to 0} \qquad (20)$$

Here C_j is the concentration and S_j is the stopping power, in substrates 1 and 2 respectively. Using equation (7) one can take the limit indicated in (20) and obtain

$$\frac{n_{(1)}}{n_{(2)}} = \frac{\left(\Sigma_j \, C_\rho^{\,j} \, (C_\varepsilon^{\,j})^{-2m}\right)_1}{\left(\Sigma_j \, C_\rho^{\,j} \, (C_\varepsilon^{\,j})^{-2m}\right)_2} \tag{21}$$

where C_ρ and C_ε are given by

$$C_\rho^{\,j} = N_j \pi a_j^2 \gamma_j$$

$$C_\varepsilon^{\,j} = \left[\frac{a_j M_2^{\,j}}{Z_1 Z_2 j c^2 (M_1 + M_2^{\,j})}\right]$$

and the superscript j refers to different target atom type. The quantities C_ρ and C_ε have been tabulated for a number of target-ion combinations in the book by Gibbons, Johnson, and Mylroie.

Values of the calculated interface discontinuity rates for three interface types are tabulated in table 2. Note that the values are fairly insensitive to the choice of the parameter m. Experimental measurements of Combosson et al are included in column 3 of the table.

IX. CONCLUSIONS

The approach using the Boltzmann equation is seen to be a useful one for calculation of ion implantation effects in multilayered targets. In particular, the energy distributions for a heavy primary ion can be calculated as a function of depth in a multilayer target, and from this information range distributions for both the primary ion and the recoils can be constructed.

Tabel II Comparison Between Theory and Experiment for Boron Range Profile Discontinuity. Tabulated Quantities are C_1/C_2 as given by eq. 21. The last columns are experimental results of Combasson, et al [11].

Interface	T - F (m = .333)	L - J (m = .191)	Combasson, et al (experiment)
SiO_2 on Si	1.68	1.86	0.7 - 2.2
Si_3N_4 on Si	2.18	2.42	2.1
Si_3N_4 on SiO_2	1.3	1.3	1.1

Computer programs which can provide useful information about range and knock-on effects in multilayer targets of specific interest in semiconductor processing are being developed and will be described in a subsequent publication.

REFERENCES

[1] T. R. Cass and V. G. K. Reddi, Appl. Phys. Lett. 23, 268 (1973).

[2] E. H. Bogardus and M. R. Poponiak, Appl. Phys. Lett. 23, 553 (1973).

[3] W. K. Chu, H. Muller, and J. W. Mayer, Appl. Phys. Lett. 25, 297 (1974).

[4] R. A. Moline, G. W. Reutlinger, and J. C. North, Proc. of the Fifth International Conf. on Atomic Collisions in Solids, Gatlinburg, Tenn., Plenum Press, New York (to be published).

[5] K. B. Winterbon, P. Sigmund and J. B. Sanders, Mat. Fys. Medd. Dan. Vid. Selsk. 37, No. 14 (1970).

[6] D. K. Brice, Proc. First International Conf. on Ion Implantation (Thousand Oaks, CA), pp 101-111, F. Eisen and L. Chadderton, Eds., Gordon and Breach, London, 1971.

[7] S. Furukawa and H. Ishiwara, J. Appl. Phys. 43, 1268 (1972).

[8] Peter Sigmund, Rev. Roumanian Physics, Vol. 17, pp 283 ff, 1972.

[9] J. F. Gibbons, W. S. Johnson, S. Mylroie, Projected Range Statistics, (Dowden, Hutchinson and Ross, Inc., Stroudsburg, Penn., 1975).

[10] A. Goetzberger, D. J. Bartelink, J. P. McVittie, J. F. Gibbons, Appl. Phys. Lett., Nov. 1976 (in press).

[11] J. L. Combasson, J. Bernard, G. Guernet, N. Hilleret, M. Bruel, Proc. International Conf. on Ion Implantation in Semiconductors and other Materials, edited by Billy L. Crowder (Plenum Press, New York, 1973) p 285.

PREFERENTIAL SPUTTERING AND RECOIL IMPLANTATION DURING DEPTH PROFILING*

D.K. Murti and Roger Kelly

Institute for Materials Research, McMaster University

Hamilton, Ontario, Canada, L8S 4M1

ABSTRACT

Most of the methods which are currently used to obtain depth-composition profiles in solids rely on the continuous removal of surface layers by sputtering with heavy ions. When a solid containing more than one element is profiled in this way, initially the various constituents will be sputtered at different rates, producing a layer of altered composition at and beneath the surface. At the same time recoil implantation will cause further alterations in the composition. In the present work we consider three effects which interfere with profiling, including brief reviews of experimental results which demonstrate the effects and of theories which rationalize the effects. We conclude by discussing one example where these problems were avoided and two examples where the problems were not recognized and where the interpretations must therefore be reconsidered.

1. INTRODUCTION

When a solid containing more than one element is bombarded with ions, initially the various constituents will be removed at different rates, producing a layer of altered composition at and beneath the surface. Such preferential removal effects are particularly marked with oxides, where they can be considered to arise in three ways:

(a) A number of oxides show preferential sputtering of oxygen from the extreme outer surface. For example, bombardment of PdO

―――――――――
*Supported by a grant from the National Research Council of Canada, Ottawa.

with 1 keV Ar^+ to a dose of 2×10^{15} ions/cm^2 results in a surface layer of Pd with a thickness of ∼2 nm [1].

(b) An important variation concerns instances where preferential sputtering combines with diffusion (or an equivalent process) and leads to an atomically thick layer of altered composition. For example, bombardment of Nb_2O_5 with 35 keV Kr^+ or O_2^+ to a dose of 2×10^{17} ions/cm^2 results in the formation of an altered layer which has the stoichiometry NbO and a thickness of ∼31 nm [2,3]. Examples with substances other than oxides include PtSi, which develops an altered layer with a composition approaching Pt_2Si and having a thickness of ∼20 nm [4].

(c) Concurrent with preferential sputtering is recoil implantation. Depending on the experiment, this can manifest itself as an unusually long time-constant, as for the desorption of oxygen from Al [5], or else as skewness in a depth profile. Examples of the latter, which should be regarded as a further case of altered composition, include skewness at the oxide-metal interface when oxide films are being profiled [6] or at the interface between two metal films when evaporated layers are being profiled [7]. In principle, recoil implantation combined with diffusion can be considered as a variant of (c), just as (b) is a variant of (a), but will not be discussed here.

The following sections will serve to demonstrate and explain these effects more explicitly, including one example where they were avoided and two examples where they probably occurred but were not recognized.

2. REVIEW OF EXPERIMENTAL RESULTS

Work in the past few years using surface probes such as photoelectron or Auger-electron spectroscopy has revealed a remarkable tendency for oxides to become oxygen deficient in the extreme outer surface when subjected to high-dose ion impact. The work of Kim et al. [1], in which the conversion of the outer ∼2 nm of PdO to Pd was revealed by X-ray photoelectron spectroscopy, has already been referred to. A result of a different sort was obtained with MoO_3, where exposure to 0.4 keV Ar^+ ions at various doses led to the intermediate oxide MoO_2 rather than to the metal as the final product. A partial list giving the chemical state of target surfaces following high-dose ion impact is presented in columns 1-3 of Table I. The total number of such examples which revealed surface alteration is now at least 24 [1,11,15].

Studies with oxides subjected to high-energy ion bombardment have revealed the formation in particular cases of a much thicker layer of altered composition which is detectable by high-energy electron diffraction or even visually. As an example, the results obtained in the bombardment of Nb_2O_5 and Ta_2O_5 will be compared

TABLE I

Examples of the chemical state of target surfaces following high-dose ion impact

Target	Ion and energy	State of extreme outer surface	Ion and energy	State beneath outer surface	Thickness (nm)	References
Al_2O_3	0.4 keV Ar^+	Al_2O_3	10 keV Ar^+	amorphous Al_2O_3	~10	1,8
Fe_2O_3	0.4 keV Ar^+	Fe	30 keV Kr^+	Fe_3O_4	-	1,9
MoO_3	0.4 keV Ar^+	MoO_2	40 keV Kr^+	MoO_2	~115	1,10
MoS_2	1 keV Ar^+	~$MoS_{0.8}$	-	-	-	11
Nb_2O_5	-	-	35 keV O_2^+	NbO	~31	2,3
PdO	0.2-1 keV Ar^+	Pd	-	-	-	1
PtSi	-	-	20 keV Ar^+	~Pt_2Si	~20	4
Ta_2O_5	0.4 keV Ar^+	~Ta_2O_5	35 keV Kr^+	mostly amorphous Ta_2O_5**	~24	1,12
TiO_2	0.4 keV Ar^+	(Ti_2O_3)*	30 keV·Kr^+	Ti_2O_3	11±2	1,13
V_2O_5	-	-	40 keV Kr^+	V_2O_3	~115	10

*TiO_2 was not studied explicitly but Ti_2O_3 was shown to be unaltered in both low and high-energy bombardments [1,13]. This implies strongly that TiO_2 would have evolved to Ti_2O_3.

**Bombardment of Ta_2O_5 leaves it mainly in the amorphous state though with small amounts of crystalline δ-Ta_2O_5 [14] also present.

[2,3,12]. Specimens of Nb and Ta were anodized at 20 V in 0.25% KF, resulting in anodic films of Nb_2O_5 and Ta_2O_5 with thicknesses of about 60 nm. They were then coated with parlodion, which after drying was peeled off to remove the anodic films. The films were supported on 200 mesh grids and the parlodion was removed by treating with amyl acetate. The as-prepared films had an ill-defined structure as shown by bright-field images and the corresponding diffraction patterns consisted of the usual halos characteristic of amorphous materials. Bombarding Nb_2O_5 with 35 keV O_2^+ to a dose of 5×10^{15} ions/cm^2 revealed a few highly diffracting regions which were dark in bright field and bright in dark field. They were evidently nuclei consisting of microcrystallites with sizes of 8-15 nm. A similar dose left Ta_2O_5 largely unchanged. The effect on Nb_2O_5 of a dose of 5×10^{16} ions/cm^2 is shown in Fig. 1. The highly diffracting regions already noted are now in an intermediate state of impingement and the corresponding diffraction pattern reveals well-defined rings spaced as for NbO (except for having intensities appropriate to f.c.c.). Ta_2O_5 showed much less evidence for alteration, including the persistence of amorphous halos (Fig. 2). These experiments demonstrate that bombardment of Nb_2O_5 with 35 keV O_2^+ ions leads to the formation of NbO whereas Ta_2O_5 is largely, though not completely, unchanged. The altered regions of Ta_2O_5 which appeared at still higher doses have been tentatively identified as δ-Ta_2O_5, i.e. Ta_2O_{5-x}, by comparison with the diffraction pattern of Terao [14].

Bombardment-induced stoichiometry changes can in some cases also be followed by monitoring the electrical conductivity with a four-point probe since a slight deviation from stoichiometry of transition-metal oxides often leads to marked changes in conductivity [16-18]. In fact, bombarded Nb_2O_5 pellets showed a catastrophic increase in electrical conductivity (Fig. 3). In the region of low dose the conductivity remained similar to that for unbombarded specimens, whereas beyond a threshold dose of about 2×10^{15} ions/cm^2 (for Kr^+) or 4×10^{15} ions/cm^2 (for O_2^+) it increased rapidly towards a saturation level. This conductivity change is easily understood since at room temperature NbO exhibits metallic conductivity [19]. Ta_2O_5, by contrast, remained largely insulating when bombarded (Fig. 3). Such a lack of change correlates (a) with such phases as TaO being unstable in the solid state, (b) with the crystallites of δ-Ta_2O_5 failing to impinge, and (c) with the results of Kim et al. [1] as included in Table I.

The total number of examples investigated by high-energy electron diffraction or conductivity and which revealed the presence of a thick altered layer is now at least 12 [20], of which a number are summarized in columns 4-6 of Table I.

An example of a composition change due to recoil implantation (perhaps, but not necessarily, combined with diffusion) follows from the work of Braun et al. [7], who studied the intensity of

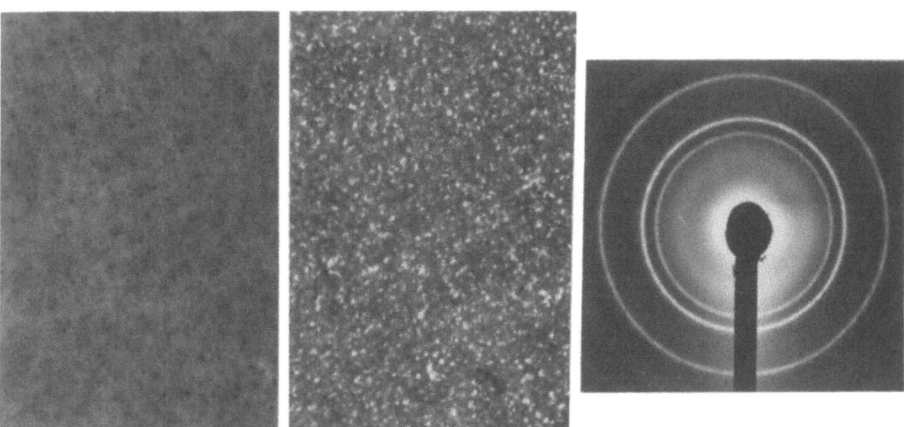

Fig. 1. Transmission electron microscopy at 80 kV of anodic Nb_2O_5 films which have been bombarded with 35 keV O_2^+ to a dose of 5×10^{16} ions/cm^2. (a) — bright-field image. (b) — dark-field image. (c) — diffraction pattern. The diffraction pattern corresponds to NbO, except that the intensities imply an f.c.c. symmetry instead of the expected simple-cubic symmetry. Due to Murti and Kelly [3].

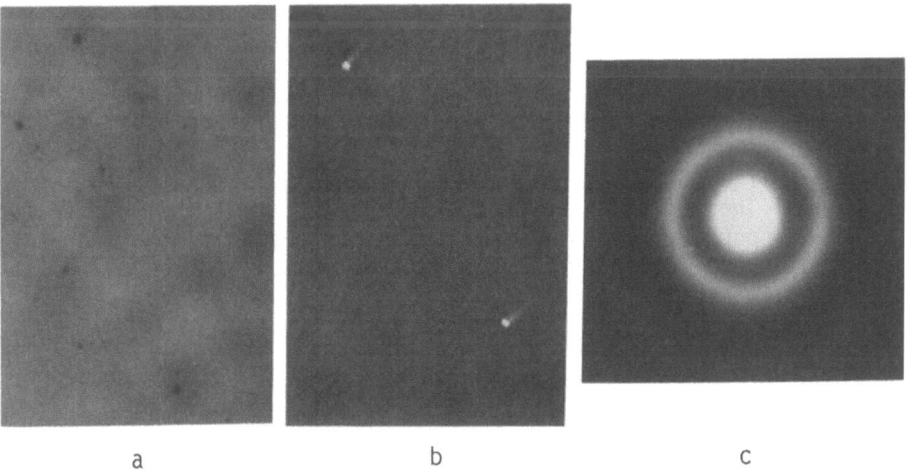

a   b   c

Fig. 2. Transmission electron microscopy at 80 kV of anodic Ta_2O_5 films which have been bombarded with 35 keV O_2^+ to a dose of 5×10^{16} ions/cm^2. (a) — bright-field image. (b) — dark-field image. (c) — diffraction pattern. Due to Murti [12].

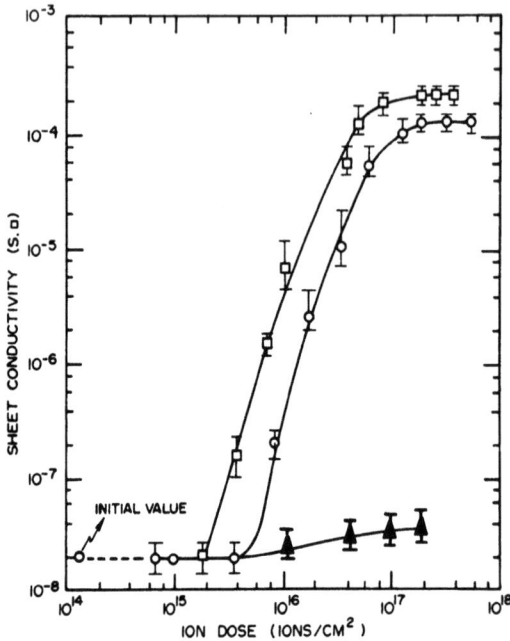

Fig. 3. Effect of ion dose on the sheet conductivity of Nb_2O_5 pellets bombarded with 35 keV Kr^+ (□) or O_2^+ (O) ions and of Ta_2O_5 pellets bombarded with 35 keV Kr^+ (▲) ions. Note that the units of conductivity are here Siemens x square. Due to Murti and Kelly [2,3,12].

secondary photons as a function of ion dose for a specimen initially consisting of an Ag film of thickness 100 nm on an Mo substrate. The specimen was bombarded with 17 keV Ar^+ ions and the intensities of an AgI line (328.0 nm) and an MoI line (379.8 nm) were followed (Fig. 4). The tails of the photon intensity curves indicate that the Ag-Mo boundary region is smeared out extensively, the long tail of Ag being explicitly ascribed by the authors to the recoil implantation of Ag atoms.

Fig. 5 shows similar results for an anodic Al_2O_3 film as obtained by Good [6]. Anodic films are believed to have particularly abrupt boundaries and the tails must therefore be attributed to recoil implantation combined to some extent with an imperfect beam profile. Still a further example is found in work by Naguib [10] and Arora [21] on V_2O_5. They showed that, whereas bulk V_2O_5 is converted to V_2O_3 by ion impact, an anodic V_2O_5 film is converted first to V_2O_3 and then VO. Evidently, the metal substrate acts as a sink for recoil-implanted O atoms. There are relatively few

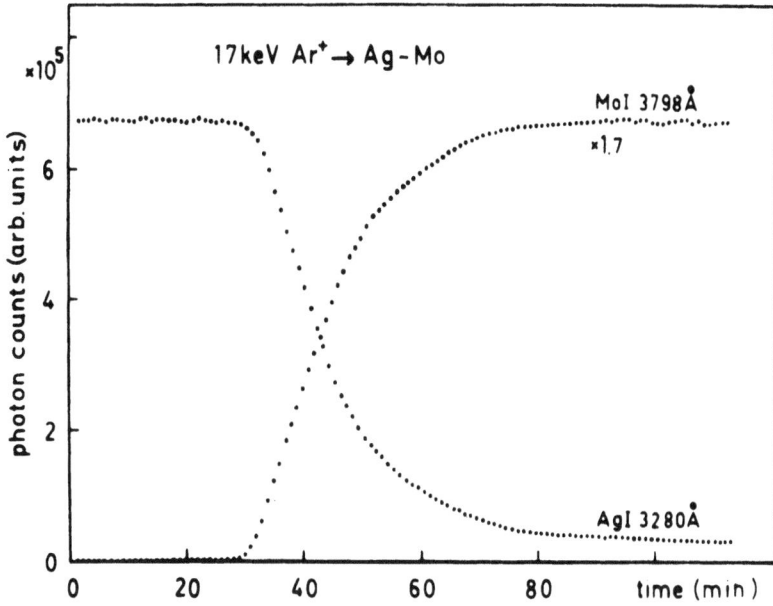

Fig. 4. The intensity of secondary photons as a function of ion dose (17 keV Ar^+) for a specimen consisting of a 100 nm thick Ag film on an Mo substrate. Due to Braun et al. [7].

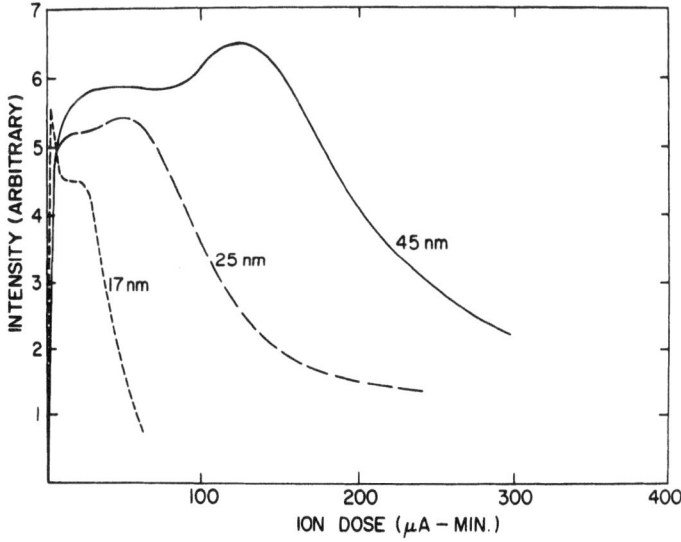

Fig. 5. The intensity of secondary photons (AlI 308.2 nm) as a function of ion dose (12 keV Kr^+) for specimens consisting of anodic Al_2O_3 films on Al substrates. The initial rise of intensity is an artifact due to surface dirt, especially of organic origin. The thicknesses of the anodic films are indicated in nanometers. Due to Good [6].

other examples in which effects as in Figs. 4 and 5 or as found with V_2O_5 were recognized explicitly.

3. REVIEW OF THEORY

The results considered here indicate that ion bombardment of compounds in general, and oxides in particular, leads to layers of altered composition ranging from the extreme outer surface to depths of as much as 115 nm below the surface, depending on the bombardment conditions. At the same time, recoil implantation causes altered compositions at interfaces beneath the surface. Specific models to explain these results have been treated elsewhere and can be summarized as follows.

Effects at the extreme outer surface, as in columns 1-3 of Table I, can be envisaged in the case of oxides to take place due to the preferential sputtering of oxygen with the resultant formation of nuclei of lower oxides. According both to Sigmund's theory of collisional sputtering [22] as well as to the work of Winters and Sigmund [23] on the sputtering of chemisorbed gas, the most material-dependent quantity controling S, the sputtering coefficient, is E_b, the surface binding energy,

$$S \alpha 1/E_b \quad \text{or} \quad S \alpha 1/(E_b)^m,$$

where E_b can be approximated by the heat of atomization, ΔH_a, expressed in units of energy per gas atom, and m is 1/3 or 1/2. Preferential oxygen sputtering will in principle occur whenever E_b for partial oxygen removal is less than E_b for the removal of an average atom [20]. For the specific cases of the solids included in Table I the appropriate values of ΔH_a are given in Table II (mostly from Ref. [20]). The information in Table II suggests that all entries except Al_2O_3 would lose O or S to the extent shown in column 4 when bombarded and this is borne out with one exception (Ta_2O_5 evolves to a certain extent to δ-Ta_2O_5 but not to Ta). An even greater degree of self consistency is obtained if thermal sputtering is considered [24,25], as thermal sputtering is able to explain the differing response to impact of certain closely related pairs such as ZnO and CdO, or MoO_2 and WO_2, or CoO and NiO, for which an argument based on ΔH_a fails. A preliminary application of thermal sputtering to oxygen loss is given in Ref. [26].

The results obtained in the bombardment of Nb_2O_5 and Ta_2O_5 with Kr^+ or O_2^+ ions, for example the formation in the former case of a layer of NbO having a thickness of ~31 nm, can be explained on the basis of a model [3] which combines preferential oxygen sputtering at the surface, diffusion of the relevant point defects, and random nucleation of a phase with lower stoichiometry. The governing equation is an extended form of the diffusion equation,

TABLE II

Heats of atomization (ΔH_a) relevant to the sputtering of certain compounds [20]

Target[*]	ΔH_a for congruent sputtering (eV/gas atom)	ΔH_a for incongruent sputtering (eV/gas atom)	Residual solid phase assumed in column 3[*]
$Al_2O_3(\ell)$	6.2	8.0	$Al(s)$
$Fe_2O_3(\ell)$	<5.0	<<5.0	$Fe_3O_4(s)$
$MoO_3(\ell)$	5.5	3.8	$MoO_2(s)$
$MoS_2(\ell)$	<5.1	<4.2	$Mo(s)$
$Nb_2O_5(\ell)$	6.7	5.9	$NbO(s)$
$PdO(\ell)$	3.6	3.3	$Pd(s)$
$Ta_2O_5(\ell)$	7.1	6.7	$Ta(s)$
$TiO_2(\ell)$	6.4	5.1	$Ti_2O_3(s)$
$V_2O_5(\ell)$	5.6	4.0	$V_2O_3(s)$

[*]The symbols "ℓ" and "s" mean respectively "liquid", i.e. amorphized, and "solid", i.e. non-amorphized.

$$\partial C/\partial t = D\partial^2 C/\partial x^2 + v\partial C/\partial x - DC/L^2, \qquad (1)$$

where C is the differential free concentration, v is the velocity of surface recession due to sputtering, and L is the diffusion length for trapping. If eq. (1) is solved with the boundary condition $C = C_0$ at x=0, which is formally equivalent to recognizing the surface alterations discussed in the preceding paragraph, the result for steady-state conditions is

$$\text{differential trapped concentration} = -(D/vL^2) \int C \, dx$$
$$\simeq (DC_0/vL) \exp(-x/L).$$

The model thus accounts for the altered layer having a final mean thickness (L) which is atomically large, while a more detailed analysis also accounts for the dose dependence, i.e. time constant, of the alteration. (The dose dependence has also been considered by Winters and Coburn [27], though with a somewhat different starting point than eq. (1).)

The role of recoil implantation in altering the composition

of a bombarded solid has been considered by Kelly and Sanders [28, 29]. The authors argued that, provided the recoil source is of monolayer thickness, the number of atoms implanted beyond a depth x is given by

$$H(x) = I\theta\lambda N \int d\sigma \, F(x,\psi) \text{ cm}^{-2}\text{s}^{-1},$$

where I is the incident ion current, θ is the fractional surface coverage characterizing the recoil source, λ is the mean atomic spacing of the substrate, $N = \lambda^{-3}$ is the atomic density of the substrate, and $d\sigma$ is the differential scattering cross-section for ions incident on the recoil source. $F(x,\psi)$ is the integral distribution function describing the slowing down of the recoil-implanted atoms and ψ is the angle of implantation with respect to the direction of the incident ions.

For a recoil source having a thickness y, where y is sufficiently small that the incident ions do not slow down or change direction to a "significant" extent, the number of atoms implanted at low doses is given approximately by [28]

$$H(x,y) \approx (1/\lambda) \int_0^y H(x+y')dy'$$

with $\theta = 1$. In the limiting case when y is large enough we have

$$H(x,\infty) \approx (1/\lambda) \int_0^\infty H(x+y')dy'$$
$$= IC_{123}\text{Int}^{(2)}(\alpha),$$

where C_{123} and α are calculable and $\text{Int}^{(2)}(\alpha)$ is tabulated [29].

Numerical values of $H(x,\infty)/I$, which has the units as well as the characteristics of a sputtering coefficient, are given in Table III for x=0.5 nm. We find, for example, that each 10 keV Xe$^+$ ion should cause 1.46 0 atoms to be driven into Al from a surface layer of Al_2O_3. We would note first of all that 1.46 is within a factor of two of the value 0.8±0.3 inferred from Fig. 5 by attributing the tails to recoil implantation [30] as well as of the value 2.5 obtained by Moline et al. [31] for 24 keV Kr$^+$ incident on SiO_2. Secondly, we would point out that since the coefficient for backward sputtering of oxides lies typically in the interval 1.5-4 atoms/ion [32], backward sputtering and recoil implantation occur to a similar extent. Evidently recoil implantation will constitute a major perturbation in virtually all attempts to profile.

TABLE III

Numerical values of $H(x,\infty)/I$ for various ions incident on oxygen on the surface of Be, Al, Mo, or W when $x = 0.5$ nm*

Ion	Energy (keV)	Result for Be	Result for Al	Result for Mo	Result for W
He	2	.0365	.0260	.0232	.0574
Ne	2	.733	.495	.439	.981
	10	.349	.233	.228	.476
	20	.305	.203	.202	.410
Xe	2	1.31	.945	.786	2.04
	10	2.22	1.46	1.33	2.79
	20	2.51	1.62	1.51	3.03
	100	1.57	1.04	1.06	2.10

*Taken from Table 3 of Ref. [29]. Note that $H(x,\infty)/I$ as used in the present work is identical with $H(x,\infty)S_{13}/\theta\lambda N$ of Ref. [29], namely $C_{123}Int^{(2)}(\alpha)$ in both cases.

4. PREVENTION OF ALTERED COMPOSITIONS

Most of the methods which are currently used to obtain depth-composition profiles in solids rely on the continuous removal of surface layers by heavy-ion sputtering. It is of interest then to minimize or suppress changes due to preferential sputtering or recoil implantation. Specifically, experiments have been done [12] with Nb_2O_5 pellets using O_2^+ and Kr^+ beams at various energies from 1 to 35 keV and at a constant dose of 1×10^{17} ions/cm². Following bombardment, the electrical conductivity was measured with a four-point probe. As seen in Fig. 6 incident Kr^+ caused a significant conductivity increase, thence an inferred reduction to NbO, even at the lowest energy. By contrast, O_2^+ ions showed a well-defined threshold for conductivity increase at 5 keV. Evidently reduction to NbO does not occur below this energy. These results were explained as follows.

Bombardment with O_2^+ ions may be regarded, in the case of Nb_2O_5 or in general MO_a, as creating a surface layer of NbO or in general MO_b having a thickness R_a. This is basically the same effect, namely sputtering combined with diffusion, as summarized in columns 4-6 of Table I. In addition, O_2^+ ions play a chemical role such that a differential concentration profile is set up which

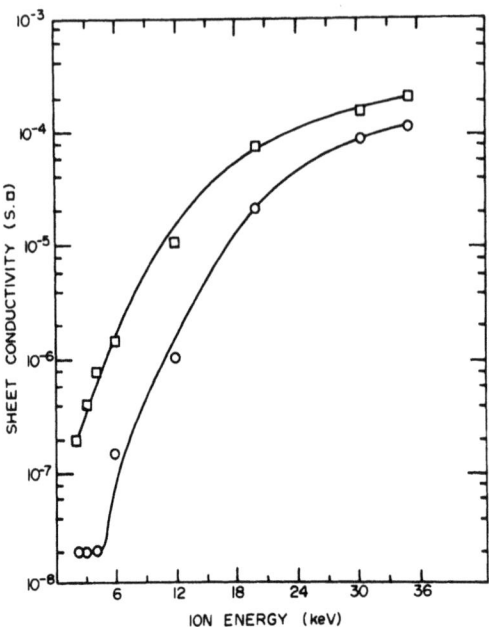

Fig. 6. Effect of ion energy on the sheet conductivity of Nb_2O_5 pellets bombarded with Kr^+ (□) or O_2^+ (○) ions. Due to Murti [12].

for large doses of a diatomic ion takes the approximate form (eq. (7) of [13]):

$$c^{diff}(x)dx \approx \{Ndx/S\}\ \text{erfc}\{\frac{x-<x>}{(2\mu_2)^{1/2}}\}\ \text{atoms/cm}^2. \qquad (2)$$

Here S is the sputtering coefficient, $<x>$ is the mean projected ion range, and μ_2 is the mean square projected ion straggling. This oxygen distribution can be regarded as being superimposed on the stoichiometry which is developed by sputtering effects, namely MO_b. The net result due to O_2^+ ion bombardment is that MO_a is altered to $MO_{b+\Delta b}$, where Δb is given by eq. (2) multiplied by λ_c^2 and with λ_c substituted for dx, λ_c being the mean <u>cation</u> spacing. Since $N^c = \lambda^{-3} = (1+b)\lambda_c^{-3}$, we have

$$\Delta b \approx \{(1+b)/S\}\ \text{erfc}\{\frac{x-<x>}{(2\mu_2)^{1/2}}\}.$$

If $\Delta b > a-b$ for $x = R_a$ then reduction is avoided. We finally note that S is an increasing function of energy and that there will, therefore, be a critical energy below which the condition $\Delta b > a-b$ is met. Fig. 6 suggests that, for O_2^+ incident on Nb_2O_5, the critical energy is 5 keV, while a more detailed analysis shows that the argument is numerically self-consistent [12].

In the light of the discussion presented here, we would suggest that some aspects of recent studies on depth-composition profiles in oxides [33-35], using techniques based on ion bombardment, need reconsideration. For example, oxygen profiles in anodized Nb, obtained with ion-scattering spectrometry, were interpreted as follows [33]:

adsorbed gases/Nb_2O_5/NbO,

where the NbO was regarded as being a pre-existing constituent of the anodic film. Based on the work of Refs. [2,3] we would suggest as an alternative

Nb_2O_5/NbO/O in Nb.

The NbO is adequately explained by preferential sputtering combined with diffusion, while the O in Nb can be accounted for in terms of recoil implantation alone or recoil implantation combined with diffusion. Thus neither is necessarily a pre-existing constituent of the anodic film.

In another study [34], ion-scattering spectrometry was used to obtain depth-composition profiles in oxide films formed on Fe in air at room temperature. The results were interpreted as follows:

excess O_2/Fe_2O_3/oxides of continually lower stoichiometry.

Based on recent work by Thomson [9] we would suggest as an alternative

Fe_2O_3/Fe_3O_4/O in Fe,

where the Fe_3O_4 and O in Fe are to be regarded mainly as artifacts rather than pre-existing constituents of the film. Similar problems occur with the work of Ref. [35].

5. CONCLUSIONS

The overall conclusion that we wish to make is that some compounds, whether in bulk or thin-film form, cannot be meaningfully profiled using an ion beam due to the preferential sputtering of a component, in some cases combined with diffusion. Even with systems which are completely resistant to the loss of a component, such as an Al_2O_3 layer on Al, concentration profiles will be skewed and compositions, therefore, effectively altered owing to the substrate acting as a sink for recoil-implanted or implanted and diffused atoms. In so far as preferential loss of oxygen is concerned, it has been shown both experimentally and analytically that the effect can be minimized by the use of an O_2^+ beam at energies below a threshold. We are not aware, at present, of an explicit technique to suppress effects due to recoil implantation and would re-iterate the previous comment that recoil implantation will often occur to a comparable extent to backward sputtering and will, therefore, constitute a major perturbation to profiling.

ACKNOWLEDGEMENTS

The authors thank Dr. M. Braun (Research Institute for Physics, Stockholm) and Ms. C.J. Good (Xerox Research Center of Canada Ltd., Mississauga) for permitting reproduction of Figs. 4 and 5.

REFERENCES

1. K.S. Kim, W.E. Baitinger, J.W. Amy, and N. Winograd, J. Electron Spect. and Related Phenom., 5, 351 (1974).
2. D.K. Murti and R. Kelly, Surface Sci., 47, 282 (1975).
3. D.K. Murti and R. Kelly, Thin Sol. Films, 33, 149 (1976).
4. J.M. Poate, W.L. Brown, R. Homer, W.M. Augustyniak, J.W. Mayer, K.N. Tu, and W.F. van der Weg, Nucl. Instr. Methods, 132, 345 (1976).
5. R. Kelly and C.B. Kerkdijk, Surface Sci., 46, 537 (1974).
6. C.J. Good, M.Sc. Thesis (McMaster University, 1976).
7. M. Braun, B. Emmoth, and R. Buchta, Rad. Effects, 28, 77 (1976).
8. C. Jech and R. Kelly, J. Phys. Chem. Sol., 31, 41 (1970).
9. B.A. Thomson and R. Kelly, to be published.
10. H.M. Naguib and R. Kelly, J. Phys. Chem. Sol., 33, 1751 (1972).
11. A. Preisinger, T. Tortschanoff, P. Viehböck, and W. Weissmann, Japan. J. Appl. Phys., Suppl. 2, Pt. 2, 791 (1974).
12. D.K. Murti, Ph.D. Thesis (McMaster University, 1975).
13. T.E. Parker and R. Kelly, J. Phys. Chem. Sol., 36, 377 (1975).
14. N. Terao, Japan. J. Appl. Phys., 6, 21 (1967).
15. L.I. Yin, S. Ghose, and I. Adler, J. Geophys. Res., 77, 1360 (1972).
16. R.F. Janninck and D.H. Whitmore, J. Chem. Phys., 37, 2750 (1962).
17. J.M. Berak and M.J. Sienko, J. Sol. State Chem., 2, 109 (1970).
18. N. Kimizuka, M. Saeki, and M. Nakahira, Mat. Res. Bull., 5, 403 (1970).
19. C.N.R. Rao and G.V. Subba Rao, Phys. Stat. Sol. (a), 1, 597 (1970).
20. H.M. Naguib and R. Kelly, Rad. Effects, 25, 1 (1975).
21. M.R. Arora, Ph.D. Thesis (McMaster University, 1974), p. 234.
22. P. Sigmund, Phys. Rev., 184, 383 (1969).
23. H.F. Winters and P. Sigmund, J. Appl. Phys., 45, 4760 (1974).
24. R. Kelly, Rad. Effects (submitted, 1976).
25. R. Kelly, to be published.
26. R. Kelly, Proc. Int. Conf. on Phys. Met. of Reactor Fuel Ele-

ments (The Metals Soc., London, 1974), p. 275.
27. H.F. Winters and J.W. Coburn, Appl. Phys. Lett., $\underline{28}$, 176 (1976).
28. R. Kelly and J.B. Sanders, Nucl. Instr. Methods, $\underline{132}$, 335 (1976).
29. R. Kelly and J.B. Sanders, Surface Sci., $\underline{57}$, 143 (1976).
30. C.J. Good and R. Kelly, to be published.
31. R.A. Moline, G.W. Reutlinger, and J.C. North, Proc. 5th Int. Conf. on Atomic Collisions in Solids (Plenum, New York, 1975), p. 159.
32. R. Kelly and N.Q. Lam, Rad. Effects, $\underline{19}$, 39 (1973).
33. K.E. Gray, Appl. Phys. Lett., $\underline{27}$, 462 (1975).
34. R.P. Frankenthal and D.L. Malm, J. Electrochem. Soc., $\underline{123}$, 186 (1976).
35. M. Seo, J.B. Lumsden, and R.W. Staehle, Surface Sci., $\underline{50}$, 541 (1975).

THE INFLUENCE OF RECOIL IMPLANTATION OF ABSORBED OXYGEN

ON THE ENTRAPMENT OF XENON IN ALUMINUM AND SILICON

K. Wittmaack and P. Blank

Gesellschaft für Strahlen- und Umweltforschung mbH
Physikalisch-Technische Abteilung
D-8042 Neuherberg, Germany

ABSTRACT

Xenon collection in polycrystalline aluminium and amorphized silicon has been studied by means of Rutherford backscattering. Implantations were carried out at energies between 10 and 160 keV, current densities between 0.07 and 5 $\mu A/cm^2$, and oxygen partial pressures between $< 10^{-8}$ and 10^{-5} Torr. It was found that the saturation areal densities of collected material strongly increased with increasing oxygen partial pressure during implantation. Simultaneously, pronounced accumulation of oxygen in the target was observed. This effect is attributed to recoil implantation of adsorbed oxygen. Enhanced collection of xenon is explained by assuming that oxygen acts as a trap for implanted xenon. Based upon a simple model a consistent interpretation of recoil implantation from adsorbed oxygen layers could be obtained.

1. INTRODUCTION

Effects due to recoil implantation of atoms from thin films into the bulk of the underlying substrate have attracted increasing interest during the last few years. A review of recent experimental results and the current state of theory has been given at this conference /1/.

Most frequently recoil implantation has been studied in "through-oxide" implantation of various ions into oxide covered silicon /2-5/. From the experiments one finds that the cross sections for recoil implantation amount to several $10^{-15} cm^2$ /1,4,5/.

Accordingly one would expect recoil implantation of adsorbed oxygen to be observable. Pronounced effects should occur in case that the oxygen pressure is high enough to cause steady replacement of sputtered or recoil implanted atoms in the adsorbed layer.

Indications for recoil implantation of adsorbed oxygen were found in studies of the oxygen enhanced secondary ion yield /6,7/ and photon yield /8/. More recent depth profiling measurements by means of secondary ion mass spectrometry (SIMS) showed that the range distribution of oxygen recoil implanted from adsorbed layers into silicon is peaked at the surface and extends to distances in excess of 1oo Å for typical SIMS bombardment conditions (1o keV argon ions) /9/. Since secondary ion yield enhancement has been observed already at oxygen partial pressures as low as 10^{-8} Torr /6/ one must suspect that recoil implantation from adsorbed layers has also affected those irradiation studies which have been carried out at moderate vacuum conditions. Pronounced effects might have occurred in high fluence experiments because the surface concentration of implanted atoms can become very large /1o/.

We have recently investigated xenon collection in silicon under UHV conditions /11/. It was found that the saturation distributions of xenon exhibit pronounced gradients towards the surface. Moreover, up to energies of about 5oo keV the maximum xenon concentrations were considerably smaller than the theoretical limit. Without specifying details of the mechanism leading to loss of implanted material these effects may be ascribed in part to insufficient trapping of xenon in silicon. Since oxygen can be expected to act as an effective trap we have extended our collection experiments to investigations at enhanced oxygen pressure. In addition to silicon we used aluminium as a target material because UHV implanted xenon distributions in this metal exhibited even more pronounced loss towards the surface than in silicon (see below).

2. EXPERIMENTAL

The investigations were carried out in a UHV target chamber in which a total background pressure of 10^{-8} Torr or better was achieved routinely. For the present study a variable oxygen leak was added which allowed controlled increase of the oxygen partial pressure in the target chamber. The pressure was monitored by an ionization gauge. A calibration for oxygen has not been carried out.

Implantation and analysis by Rutherford backscattering were done in situ without breaking the vacuum. Details of the procedure have been described elsewhere /11/. Polished silicon crystals and polycrystalline aluminium films were used as backing materials. In the range of implantation fluences discussed in this study silicon is known to become amorphous due to heavy ion bombardment. The crystal orientation is of no importance, therefore. The aluminium films (thickness 3 µm) were prepared by vapour deposition onto polished silicon crystals. The total pressures prior to and during deposition were 10^{-6} and 4×10^{-6} Torr, respectively (evaporation rate ~ 200 Å/S).

3. RESULTS

First indications for an effect possibly due to recoil implantation of oxygen were found in collection experiments at relatively low energies. In the 300 kV accelerator used for the present studies the current densities achieved in the target chamber at a distance of about 5 m downstream from the acceleration stage are only about 0.1 µA/cm^2 for 10 keV xenon ions. At a pressure of 10^{-8} Torr the flux density of residual gas atoms and molecules hitting the target surface is about one order of magnitude larger than the primary beam flux density at 0.1 µA/cm^2. Therefore one has to worry about the effect of adsorbed material unless the sticking coefficient of residual gas particles is very small.

Fig. 1 shows the collection curve for 10 keV xenon bombardment of aluminium. At very small fluences one finds the expected linear increase in the amount of material collected. Deviations from proportionality, however, occur already above about 2×10^{15} ions/cm^2.

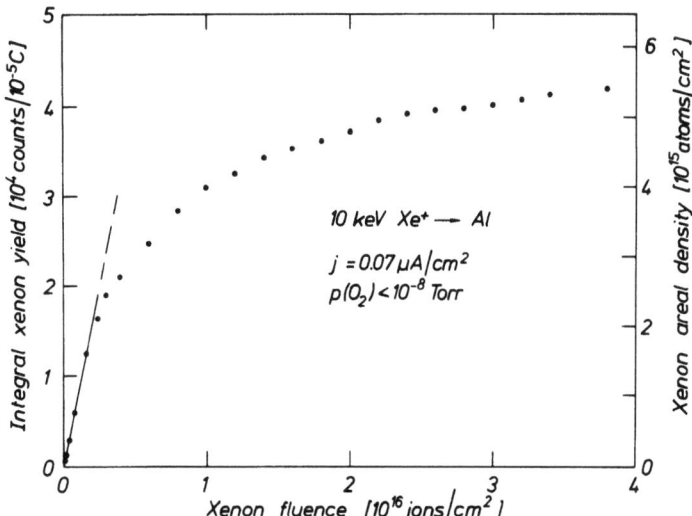

Fig. 1. Collection curve for 10 keV xenon implanted in aluminium.

Different from the behaviour at higher energies and larger beam current densities saturation is not achieved by additional small fluence increments. In fact the areal density of collected xenon increases even after having implanted a fluence which exceeds the proportionality limit by as much as a factor of 2o. Consequently the evaluation of a saturation quantity becomes a questionable task under the above conditions.

Much better defined experimental conditions were achieved at higher energies. Figs. 2 and 3 show backscattering spectra after high fluence 4o keV xenon implantation in aluminium and silicon. The profiles obtained for implantation at base pressure are saturation data, i.e. they reflect true steady state, whereas the profiles affected by oxygen recoil implantation were taken at adequately large fluences. Comparison of the "clean" spectra indicates pronounced differences in profile shape. The xenon distributions in aluminium are much broader and exhibit considerable tails on either side. More detailed investigations show that this effect is due to irradiation enhanced diffusion of xenon.

According to Figs. 2 and 3 implantations at elevated oxygen pressure result in an enhanced xenon entrapment both in aluminium and silicon. Simultaneously the backscattering spectra reveal recoil implantation of large amounts of oxygen. Note that the cross section for backscattering is proportional to the square of the atomic number. Accordingly the sensitivity for oxygen is 46 times smaller than for xenon. In view of this fact even a rough estimate indicates that in Figs. 2 and 3 the quantities of recoil implanted oxygen and implanted xenon are nearly equal.

Enhanced trapping of xenon in silicon is observed at all depths except for a small surface region. Similar results are observed in aluminium, the loss region being larger. The most peculiar effect with aluminium is the drastic change in the xenon profile shape due to oxygen recoil implantation. The main characteristics of the oxygen induced profile form - long range peak and short range shoulder - are observed also at other implantation conditions and energies (Fig. 4). Comparison of Figs. 2 and 4 indicates that higher ratios of oxygen pressure versus beam current density mainly result in an increase in height of the long range peak whereas the short range shoulder is only slightly affected.

Another interesting result of Fig. 4 is that xenon loss in near surface regions increases with increasing implantation energy. From this one may conclude that there exists an energy dependent mechanism producing release of xenon from traps.

The energy dependence of xenon collection in silicon has been discussed in detail elsewhere /11/. A compilation of "saturation"

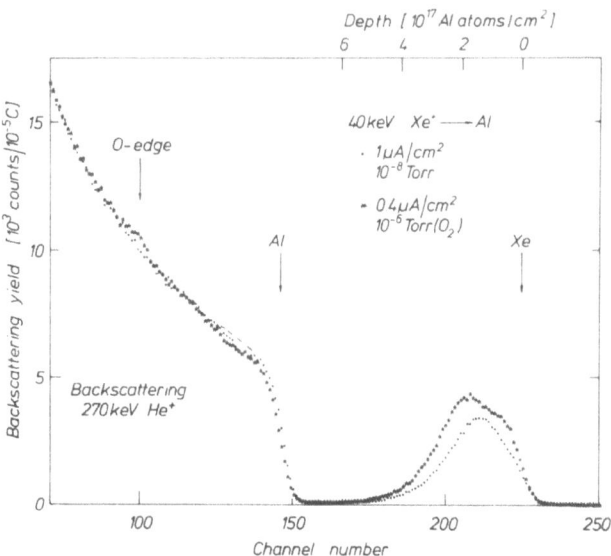

Fig. 2. Backscattering spectra of 4o keV xenon saturation distributions in aluminium. Parameter is the oxygen pressure during implantation.

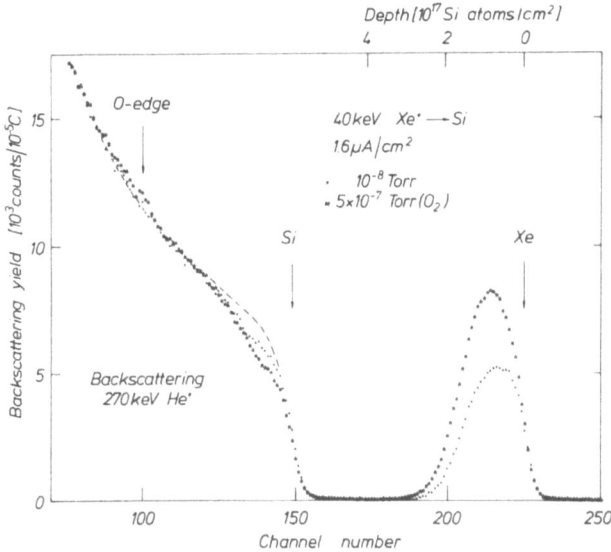

Fig. 3. Backscattering spectra of 4o keV xenon saturation distributions in silicon. Parameter is the oxygen pressure during implantation.

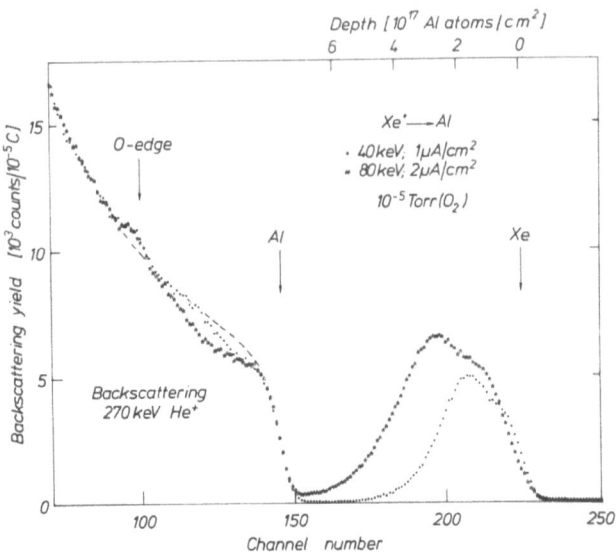

Fig. 4. Backscattering spectra of oxygen enhanced saturation distributions of xenon in aluminium. Parameter is the xenon energy.

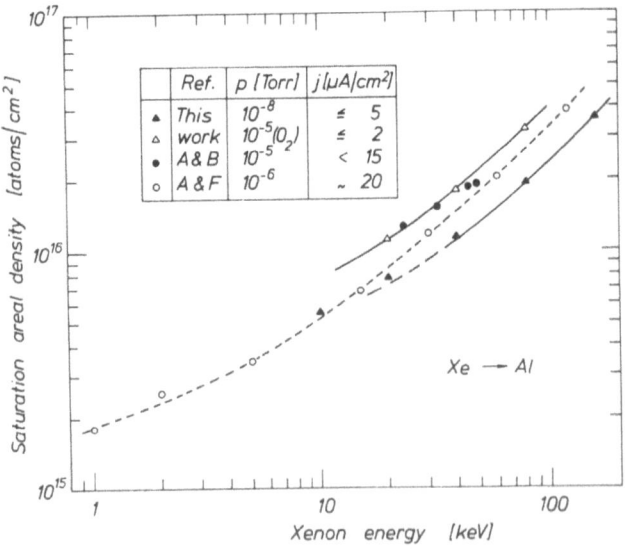

Fig. 5. Energy dependence of xenon saturation areal densities in polycrystalline aluminium. A&B /12/, A&F /13/.

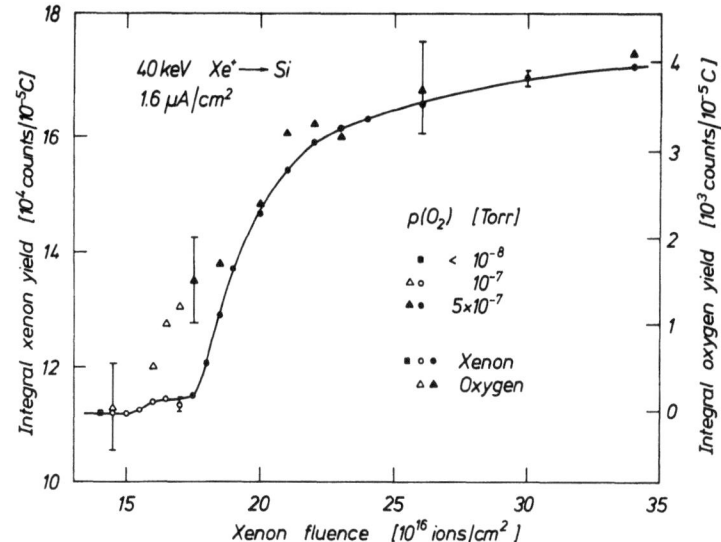

Fig. 6. Fluence dependence of oxygen recoil implantation and oxygen enhanced xenon collection in silicon.

areal densities of xenon in aluminium is presented in Fig. 5. Also shown are results of other authors /12,13/ obtained at moderate or poor vacuum conditions. Comparison with the results of this study clearly indicates that the data of Almén and Bruce /12/ and Arminen and Fontell /13/ were markedly affected by recoil implantation of adsorbed residual gas. Note that Arminen et al. /14/ have observed a pronounced oxygen induced enhancement of copper collection in aluminium.

To achieve a better understanding of the effect of oxygen on the entrapment of xenon we measured the fluence dependence of the collection enhancement at different oxygen pressures. The results are shown in Fig.6. We start with a silicon sample saturated at base pressure with 4o keV xenon. The oxygen content is below the limits of detectability. Further xenon implantation at an oxygen pressure of 10^{-7} Torr results in considerable oxygen recoil implantation. The effect on xenon entrapment is small. Strongly enhanced xenon collection is observed, however, at 5×10^{-7} Torr. Moreover the oxygen content increases. Within the limits of accuracy the saturation oxygen content is proportional to the oxygen pressure. Pronounced enhancement in xenon collection seems to require that the oxygen content exceeds a certain threshold. Since enhanced entrapment has been found only at a certain distance from the surface one may conclude that the oxygen concentration on the long range side of the xenon peak should exceed a certain limit to cause effective trapping.

For completeness we mention that sample storage at elevated oxygen pressure without xenon bombardment does not change the oxygen content of the sample (within the limits of accuracy).

A conversion of the oxygen backscattering yield, Y_o, to the oxygen areal density, N_o, can be achieved by use of the Rutherford cross section and either the calibrated xenon conversion factor or the silicon yield per channel and the energy loss function /15/. Somewhat accidentally we obtain exactly the same result in either case, $N_o/Y_o = 6 \times 10^{12}$ atoms cm^{-2}/count, for an integrated helium current of 10^{-5} C. The "saturation" areal density of oxygen in Fig. 6 is thus 2.4×10^{16} atoms/cm^2, which corresponds to the amount of oxygen contained in a 55 Å layer of silicon dioxide. This comparison clearly indicates that recoil implantation from adsorbed layers may introduce very large amounts of impurities.

4. DISCUSSION

A quantitative interpretation of the results of this study requires the knowledge of the cross section for recoil implantation. Recently reported models /1,5,16,17/ suffer from the fact that the calculated cross sections are inversely proportional to an unknown minimum escape energy. In the following discussion we will therefore use cross sections estimated from available experimental data.

The present investigations differ markedly from all the other recoil implantation studies reported previously. In this study high fluence bombardment of continuously renewed surface layers was applied to produce the effect rather than low fluence bombardment of thin films deposited on bulk substrates /1-5/. We will try to explain the observed phenomena by using the results of a model calculation of ion collection in the presence of sputtering /10/. In the so-called "zero order approximation" it was assumed that (1) the range distribution of injected ions is not affected by atoms collected previously, (2) the target sputtering is characterized by a constant yield, (3) the rest position of collected atoms in the target is not changed by further implantation and (4) diffusion does not occur. These assumptions seem to be fulfilled for oxygen implantation in silicon. The results of the model calculation can thus be used directly if the quantities defined for direct implantation are redefined for recoil implantation. For example, the recoil implantation yield, S_r, can be handled as a synonym for the accommodation coefficient, α,

$$\alpha = \int_0^\infty p(x)dx, \tag{1}$$

INFLUENCE OF RECOIL IMPLANTATION OF ABSORBED OXYGEN

$$\alpha \doteq S_r = dN_r/d\phi \big|_{\phi \to 0} = \sigma N_s, \qquad (2)$$

where $p(x)$ is the range distribution of recoil implanted atoms, x the distance from the surface, N_r the areal density of recoil implanted atoms, ϕ the bombardment fluence of primary (xenon) ions, σ the cross section for recoil implantation and N_s the areal density of the adsorbed surface layer, i.e. of the recoil implantation source.

The most important quantities to be deduced from the model calculation /10/ are the saturation surface concentration of recoil implanted atoms, c_s^∞, and the respective saturation areal density N_r^∞,

$$c_s^\infty = n_s^\infty/n_o = S_r/S_{eff}, \qquad (3)$$

$$N_r^\infty = S_r n_o <x>/S_{eff} = <x>n_s^\infty, \qquad (4)$$

where S_{eff} is the effective target sputtering yield at elevated oxygen pressure, n_o and n_s the number density of target atoms and recoil implanted atoms, respectively, and $<x>$ the mean range of recoil implanted atoms.

Backsputtering of adsorbed atoms does not enter explicitly into the calculation. It is known that the corresponding sputtering yield may be very large /18/. The effect is implicitly contained in N_s, a quantity which is determined by a dynamical equilibrium between sputtering and adsorption. The equilibrium coverage has been estimated recently by assuming the adsorption probability to be proportional to the fractional surface area uncovered /8/. In case of bombarded, highly damaged surfaces this picture is questionable.

Alternatively we try to determine the areal density of adsorbed material by use of the above model. From Eq. (2) we have

$$N_s = S_r/\sigma. \qquad (2a)$$

S_r may be determined from the initial slope of the oxygen collection curve in Fig. 6. At an oxygen pressure of 5×10^{-7} Torr one finds $S_r \simeq 0.3$ atoms/ion. Due to the large scatter in the data points the accuracy of this result is poor. The cross section for recoil implantation of oxygen is estimated from the work of Moline et al /5/. For 40 keV xenon bombardment one finds $\sigma \simeq 7 \times 10^{-15} cm^2$. Introducing these data into Eq. (2a) one ends up with an areal density of adsorbed oxygen of 4×10^{15} atoms/cm^2 which corresponds to about two monolayers of adsorbed atoms. This seems to be a realistic result.

Further calculations require the knowledge of the target

sputtering yield at the respective oxygen pressure. It is well known that bombardment at elevated oxygen pressure reduces the sputtering yield considerably (6,19,2o/. For argon bombardment of silicon a reduction factor of four has been observed for similar ratios of oxygen to primary ion flux /6/. With the experimentally determined UHV sputtering yield for 4o keV xenon bombardment of silicon, S = 3.3 atoms/ion /21/, one has $S_{eff} \simeq$ o.8 (silicon) atoms/ion. Application of Eq. (3) thus yields a steady state surface concentration of oxygen of o.4. This is too low a concentration to allow silicon dioxide formation.

In view of the strong reduction in target sputtering yield at elevated oxygen pressure one could be misled to consider this as the reason for the increase in xenon areal density. We have shown, however, that up to energies of about 5oo keV the saturation quantities are determined by stress and bombardment induced release and not by the target sputtering yield /11/.

Continuing the quantitative analysis we determine the mean range of recoil implanted atoms. According to Eq. (4) one has

$$<x> = N_r^\infty \, S_{eff} / n_o S_r . \qquad (4a)$$

Introducing the above data we end up with $<x> \simeq$ 13o Å. For comparison we note that the most probable range of oxygen atoms which have received the maximum energy transfer in collisions with 4o keV xenon atoms is x_p (16 keV) $\simeq$ 34o Å /22/. The estimated mean range is not unrealistic, therefore. In fact recently reported calculations indicated that the mean range of recoils can be quite large /23/.

It is clear that the above results can be considered only as very rough estimates. A detailed check of the model of recoil implantation from adsorbed layers would require a more accurate determination of the areal density and the depth distribution of implanted oxygen. Moreover a control of the other quantities involved would be desirable. Experimentally, a combination of SIMS depth profiling with an absolute measurement of implanted quantities by nuclear reactions would be a very promising approach. Comparison of the results with adequate theories might allow a determination of the minimum escape energy.

Finally we make some remarks concerning oxygen enhanced trapping of xenon in aluminium and silicon. In pure targets dislocations and grain boundaries as well as bombardment induced defects /24-26/ may act as traps. At high concentrations of noble gas atoms bubble formation has to be taken into account. Room temperature implantation of heavy noble gas ions in metals causes, if any /24/, only the build up of very small bubbles (diameter 2o Å /26/). The configuration of trapped gases in silicon seems to be unknown.

Release of xenon from the solid can occur as a result of thermal or bombardment induced activation /27/. The latter mechanism is likely to be responsible for the steady state gradient in the xenon concentration close to the surface. Enhanced entrapment due to the presence of oxygen might indicate that the binding between xenon and oxygen in both aluminium and silicon is quite strong. It is interesting to note that recoil implantation of 2.4×10^{16} oxygen atoms/cm^2 causes the xenon areal density to increase by 0.8×10^{16} atoms/cm^2 (Fig. 6). Since the oxygen distribution is peaked in a region where pronounced bombardment induced xenon loss occurs one may conclude that at larger depths xenon trapping by oxygen is very effective. Most likely the presence of oxygen becomes the more important the smaller the bombardment induced defect density is. In agreement with this interpretation the variation in concentration has been found to be most pronounced at the trailing edge of the xenon distribution.

5. CONCLUSIONS

The present study has shown that recoil implantation from adsorbed layers can result in high impurity concentrations in near surface regions. Although the possible consequences have been demonstrated only with respect to ion collection it is clear that this effect should always be taken into consideration. Of course, the best way to avoid detrimental effects is to improve the vacuum conditions. Base pressures of the order of 10^{-8} Torr seem to be required in most of the implantation experiments currently discussed.

Acknowledgements. The authors like to thank G. Herzog for carefully operating the accelerator and E. Schneider for assistance in data analysis. Computer programs for data processing were kindly provided by F. Schulz.

REFERENCES

/1/ R.A. Moline, this conference.
/2/ T.R. Cass and V.G.K. Reddi, Appl. Phys. Lett. 23 (1973) 268.
/3/ W.K. Chu, H. Müller, J.W. Mayer and T.W. Sigmon, Appl. Phys. Lett. 25 (1974) 297.
/4/ R.A. Moline and A.G. Cullis, Appl. Phys. Lett. 26 (1975) 551.
/5/ R.A. Moline, G.W. Reutlinger and J.C. North, Proc. Fifth Int. Conf. Atomic Collisions in Solids, Vol. 1, ed. by S. Datz, B.R. Appleton and C.D. Moak (Plenum Press, New York 1975) 159.
/6/ J. Maul and K. Wittmaack, Surface Sci. 47 (1975) 358.
/7/ K. Wittmaack, Int. J. Mass Spectrom. Ion Phys. 17 (1975) 39.
/8/ C.B. Kerkdijk and R. Kelly, Surface Sci. 47 (1975) 294.
/9/ K. Wittmaack, to be published.
/10/ F. Schulz and K. Wittmaack, Radiation Eff. 29 (1976) 31.

/11/ P. Blank, K. Wittmaack and F. Schulz, Nucl. Instr. Meth. 132 (1976) 387.
/12/ O. Almén and G.Bruce, Nucl. Instr. Meth. 11 (1961) 257.
/13/ E. Arminen and A. Fontell, Ann. Acad. Sci. Fennicae Ser.A VI, No. 357 (1971).
/14/ E. Arminen, A. Fontell and V.K. Lindroos, Phys. Stat. Sol.(a) 4 (1971) 663.
/15/ W.K. Chu, J.W. Mayer, M.-A. Nicolet, T.M. Buck, G. Amsel and F. Eisen, Thin Solid Films 17 (1973) 1.
/16/ R.S. Nelson, Radiation Eff. 2 (1969) 47.
/17/ R. Kelly and J.B. Sanders, Surface Sci. 57 (1976) 143.
/18/ H.F. Winters and P. Sigmund, J. Appl. Phys. 45 (1974) 7460.
/19/ J.-F- Hennequin, Comptes Rendus Acad. Sci. Paris 264, Serie B (1967) 1127.
/20/ W.O. Hofer and H. Liebl, Proc. Sec. Int. Conf. Ion Beam Surface Layer Analysis, Vol. 2, ed. by O. Meyer, G. Linker and F. Käppeler (Plenum Press, New York, 1976) 659.
/21/ P. Blank and K. Wittmaack, to be published.
/22/ J.F. Gibbons, W.S. Johnson and S.W. Mylroie, Projected Range Statistics, 2nd Edition (Halsted Press, 1975).
/23/ D.K. Brice, J. Appl. Phys. 46 (1975) 3385.
/24/ R.S. Nelson, Phil. Mag. 9 (1964) 343.
/25/ L. Hendriksen, A. Johansen, J. Koch, H.H. Andersen and R.M.J. Cotterill, Appl. Phys. Lett. 11 (1967) 136.
/26/ B. Hertel, J. Diehl, R. Gothardt and H. Sultze, Proc. Int. Conf. Applications of Ion Beams to Metals, ed. by S.T.Picraux, E.P. Eer Nisse and F.L. Vook (Plenum Press, New York) 507.
/27/ G. Carter and J.S. Colligon, Ion Bombardment of Solids (Heinemann, London, 1968).

FORMATION OF HIGHLY-DOPED THIN LAYERS BY USING KNOCK-ON EFFECT

H. Ishiwara and S. Furukawa

Tokyo Institute of Technology

Nagatsuda, Midoriku, Yokohama 227, Japan

ABSTRACT

Optimum conditions in the recoil implantation to form highly-doped thin layers on the semiconductor surfaces have been discussed by using He and N ions backscattering techniques and electrical measurements. A thin n-type layer with sheet resistivity of 300 $\Omega/\square$ was obtained on a p-type Si wafer by implantation of 100 keV Si ions through an As film at a dose of 3×10^{14} cm^{-2}. In case of an Sb film the sheet resistivity was about 10 times higher than that in the As film.

INTRODUCTION

Recently, it has become important to form highly-doped thin layers on the surface of solids in metallurgical applications, semiconductor fabrications, and so on. Especially, applications to Ohmic contacts, Schottky barrier contacts, solar cells, tunnel junctions, photo- and secondary electron-emitting surfaces, etc. are interesting in the semiconductor field.

One of the methods of forming such thin layers is considered to use the knocked-on atoms in implantation into double-layer substrates (the recoil implantation) [1-3]. This method will have the following merits over other doping methods; (1) Formation of the thin (< 100 A) layers, (2) Maximum concentration at surface, (3) Low dose (In suitable conditions, an amount of the knocked-on atoms is known several times greater than that of implanted ions.), and (4) Simple process (Ion beams to form other doping layers can also be used for this process.).

Usefulness of this method has recently been reported by J.M. Shannon [4] where the barrier height of Schottky diodes are controlled by the recoil implantation at relatively lower doses. However, it will be also interesting in the other applications [8] stated above to know electrical properties of the layers at the higher doses. In this paper, crystallographical and electrical properties of the layers formed by the recoil implantation of As and Sb atoms into Si are discussed by using He and N ions backscattering and channeling techniques and by using electrical measurements.

PREPARATION OF SAMPLES

Samples were prepared and measured as shown in Fig.1. In order to form n^+ layers on the surface of Si, $n(\sim 5\,\Omega\,cm)$-n^+ and $p(\sim 5\,\Omega\,cm)$ type <111> wafers were chemically cleaned. As films ~ 600 Å thick and Sb films $100 \sim 1500$ Å thick were evaporated at pressures less than 1×10^{-5} Torr. and less than 5×10^{-6} Torr., respectively. For the former samples, Sb films ~ 300 Å thick were successively coated in the same vacuum in order to prevent the As films from oxidation. Si ions with an energy of 100 keV were implanted through the films at doses between 3×10^{12} and 1×10^{15} cm^{-2}. The theoretical values of the projected average range of 100 keV Si in As and Sb are 820 Å and 830 Å, respectively.

The surface films of the implanted samples were twice dip-etched in 1:1 $H_2O:HNO_3$ for 10 min. For a sample implanted at a dose of 1×10^{15} cm^{-2}, remarkable increase of adhesion of the film [5]

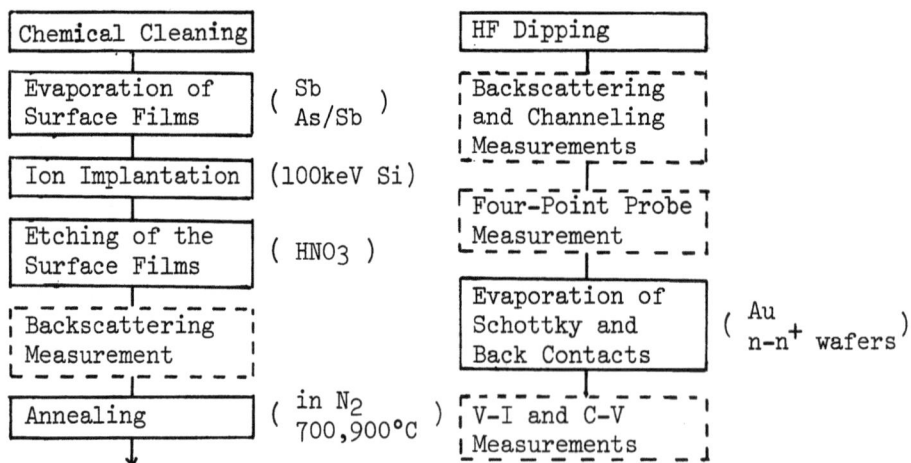

Fig.1 Preparation of samples

FORMATION OF HIGHLY-DOPED THIN LAYERS

was observed and it was dipped for about one hour. The etched samples were annealed at temperatures ranging from 700 to 900°C for 10 min. in N_2 atmosphere. After the annealing process, the samples were dip-etched in HF and Au dots with a diameter of 300 μm were evaporated on the n-n$^+$ wafers to form Schottky barrier contacts.

BACKSCATTERING AND CHANNELING ANALYSIS

Backscattering and channeling effect measurements were performed at the 4 MeV Van de Graaff accelerator at Tokyo Institute of Technology. 1.5 MeV ^{4}He$^+$ ions and ^{14}N$^+$ ions were used to measure thickness of the evaporated films and the recoil yield, respectively. Typical beam currents were 3 nA for both ions. N ions were used to avoid the pile-up effect and to obtain better separation between Si and the impurity atoms as shown in Fig.2.

Variation of the recoil yield with thickness of the Sb films is shown in Fig.3. This figure shows that about 1.5 Sb atoms are knocked-on into Si by a 100 keV Si ion at the optimum thickness of the Sb film, which corresponds to about 0.8 Rp of 100 keV Si in Sb. Dose dependences of the recoil yield are shown in Fig.4 for cases of the Sb and As films. From this figure and Fig.2, the following conclusions can be derived; (1) Dose dependences are linear up to 3×10^{14} cm^{-2} for both cases. (2) Recoil yield from the As films is about 3 times greater than that from the Sb films. This difference will be enhanced by choosing the optimum thickness of the As film, since in case of the Sb film the nearly optimum thickness is used as shown in Fig.3. And (3) an amount of Sb atoms recoiled into Si through the As film 580 Å thick is about 0.1 atoms per a Si ion.

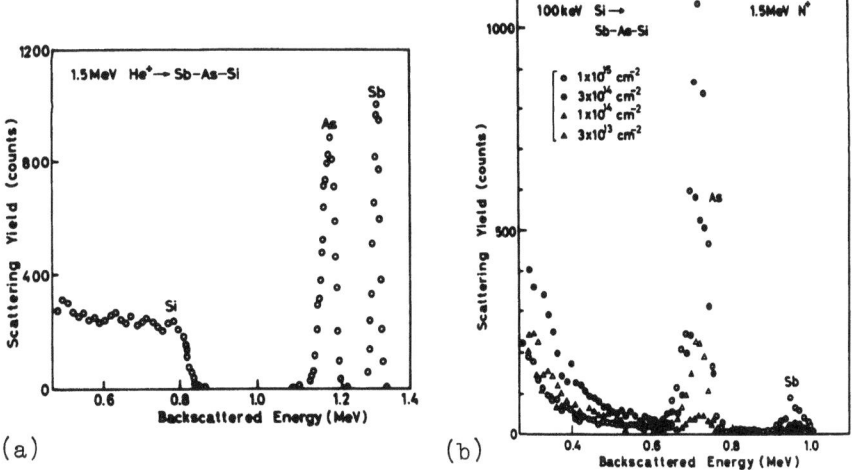

Fig. 2 (a) Composition of the evaporated film measured by 1.5 MeV He ions. (b) Recoil yields measured by 1.5MeV N ions.

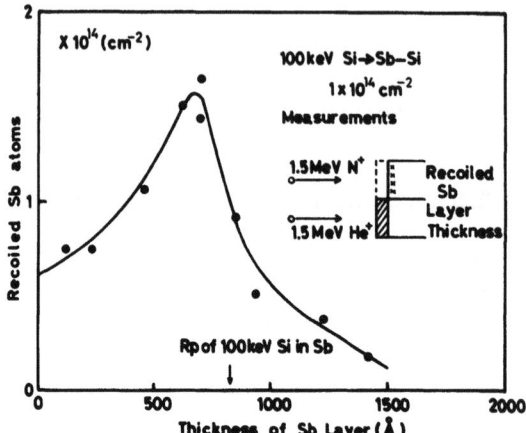

Fig.3 Variation of the recoil yield with thickness of Sb films.

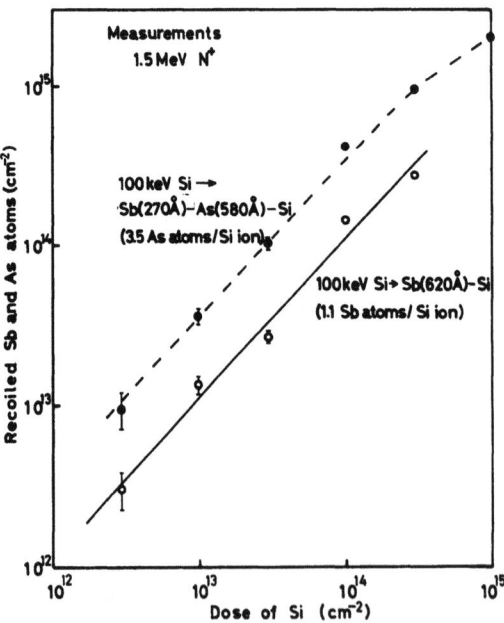

Fig. 4 Dose dependences of the recoil yield

FORMATION OF HIGHLY-DOPED THIN LAYERS

Next, in order to investigate the lattice location of the knocked-on atoms, the channeling effect measurement by N ions was performed. Samples were aligned to the beam by using He ions, since the backscattered energies of N ions from Si atoms were too low to obtain effective information on the channeling effect. Typical results are shown in Fig.5, in which the aligned spectrum was measured by translating the sample in every 0.5 µC in order to minimize the beam effects by N ions. However, since an amount of N ions for which the beam effect can be neglected has not yet been checked, we can only say that these results give the lower limit of As atoms existing in the substitutional and/or tetrahedral interstitial sites. From Fig.5, it can be said that (1) by annealing at 900°C and HF dipping processes, about 80 % of the recoiled As atoms are lost, (2) about a half of the residual As atoms occupy the substitutional and/or interstitial sites in the Si crystal and this amount corresponds to 4×10^{13} cm^{-2}, and (3) the fraction occupying these sites seems to increase as the dose increases, since in the sample at a dose of 3×10^{13} cm^{-2} the amount was less than 8×10^{12} cm^{-2}.

ELECTRICAL MEASUREMENTS

Conductivity type of the surface layers formed on the p-type wafers was checked by using Seebeck effect. Type conversion to n-type was ascertained for both cases of the Sb and As films at higher doses than 3×10^{13} cm^{-2}. Sheet resistivity of the layers was measured by using the four-point probe method. Typical results are shown in Fig. 6, which derives the following conclusions; (1) For both cases the sheet resistivity of the layers annealed at 900°C is 15∿30 % of that at 700°C. (2) The saturated resistivities at 900°C are 200 Ω/◻ and 2 kΩ/◻ for the As and Sb layers, respectively.

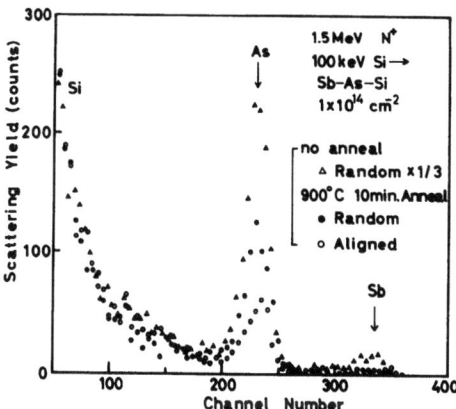

Fig. 5 <111> axis channeling effect measurement of the knocked-on As atoms. The spectrum of the non annealed sample is reduced.

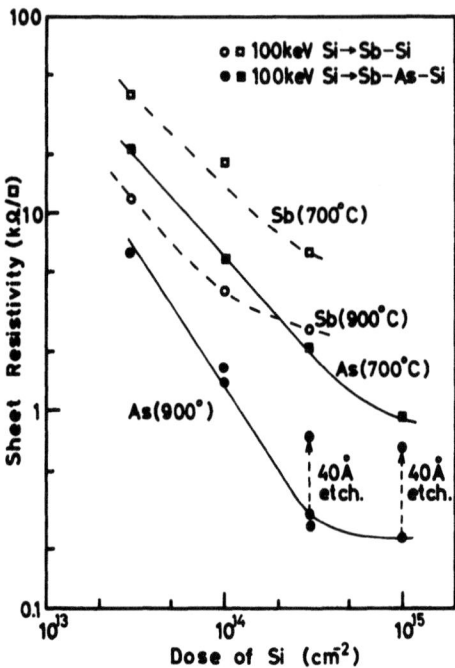

Fig. 6 Sheet resistivities of the As and Sb layers formed by recoil implantation of 100 keV Si.

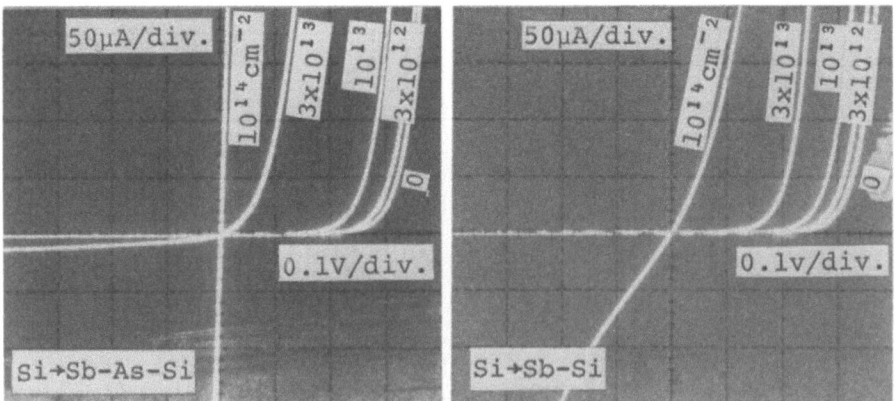

Fig. 7 Decrease of the Schottky barrier height of Au-Si diodes by recoil implantation of Si. Thickness of the As film is thinner ($\sim$10 Å) than other samples.

This result will be due to the difference of the solid solubility of As and Sb atoms in Si. (3) Dose dependence of the sheet resistivity is different for both samples, especially at 900°C. That is, in case of the Sb layers, the resistivity tends to saturate above a dose of 1×10^{14} cm^{-2}. While in case of the As layers, it deceases super-linearly up to a dose of 3×10^{14} cm^{-2}. The latter result suggests that the fraction of electrically active As atoms increases with the dose under the assumption of a constant mobility, and it corresponds to the results by the channeling measurement. These phenomena may be associated with the defects produced by the implanted ions and recoil atoms. Number of electrically active As atoms is presumed to be about 4×10^{13} cm^{-2} for the sample implanted at a dose of 1×10^{14} cm^{-2} and annealed at 900°C, assuming that $\mu =$ 100 cm^2/Vs. Comparison with the result by the channeling measurement suggests that decrease of the backscattering yield on the <111> axis is mainly due to occupancy of the substitutional site by As atoms.

Next, in order to estimate the depth distribution of the recoil atoms, the surface As layers were etched by about 40 Å using anodic oxidation. The sheet resistivity of the remaining layers increased by about 2.5 times as shown in Fig.6. This result shows that more than 60 % of the electrically active As atoms exist in this range and its average impurity concentration is presumed more than 3×10^{20} cm^{-3}.

Electrical properties of the layers formed by implantation at lower doses were investigated by measuring the decrease of the Schottky barrier height of Au-Si diodes [6]. Typical results are shown in Fig.7, in which better doping efficiency in the As layers was shown. Details will be published elsewhere [7].

CONCLUSIONS

Optimum conditions in the recoil implantation to form highly-doped thin layers on the semiconductor surface were experimentally discussed by using He$^+$ and N$^+$ ions backscattering and channeling effect techniques and by using electrical measurements. For samples implanted by 100 keV Si ions through the As and Sb films on Si wafers, the following conclusions are obtained.
(1) There exists an optimum film thickness for which the recoil yield becomes maximum. For the Sb films implanted by Si ions it is about 0.8 Rp.
(2) In case of the As films, the recoil yield is at least 3 times greater than that of the Sb films.
(3) The recoil yield increases linearly with the implanted dose up to 3×10^{14} cm^{-2} for both the As and Sb films.
(4) The sheet resistivity of the layer formed by the recoil implantation of As atoms does not saturate up to a dose of 3×10^{14} cm^{-2}

and the saturated value is about 200 $\Omega/\square$. On the other hand, in case of Sb atoms it tends to saturate at higher doses than 1×10^{14} cm^{-2} and the saturated value is about 2 k$\Omega/\square$.
(5) More than 60 % of the electrically active As atoms exist in the region within 40 Å from the surface.
(6) The maximum transfer coefficient from a Si ion to electrically active As atoms is about 0.7 at a dose of 3×10^{14} cm^{-2}, assuming that $\mu = 100$ cm^2/Vs. It will be able to improve by using the optimum film thickness and by minimizing the loss of As atoms in the annealing and HF dipping processes.

ACKNOWLEDGEMENT

The authors gratefully acknowledge assistance of the members of the steering committee and the staffs of the Van de Graaff accelerator at Tokyo Institute of Technology. They are also grateful to Prof. K.Takahashi and F.Kitagawa for their advice and technical supports in evaporation of arsenic, and to H.Wakabayashi for his assistance in the measurements.

REFERENCES

(1) R.A.Moline, G.W.Reutlinger and J.C.North; Proc. 5th Intern. Conf. on Atomic Collisions in Solids, Galtinburg, edited by S.Datz, B.R.Appleton and C.D.Moak (plenum, N.Y. 1975) Vol.1, p.159
(2) H.Nishi, T.Sakurai, T.Akamatsu and T.Furuya; Appl. Phys. Letters, 25, 337 (1974)
(3) R.A.Moline and A.G.Cullis; Appl. Phys. Letters, 26, 551 (1975)
(4) J.M.Shannon; Proc. Intern. Conf. on Applications of Ion Beams to Materials, edited by G.Carter, J.S.Colligon and W.A.Grant (Institute of Physics, London, 1976) p.37
(5) L.E.Collins, P.A.O'Connell, J.G.Perkins, F.R.Pontet and P.T. Stround; Nucl. Instrum. Meth., 92, 455 (1971)
(6) J.M.Shannon; Appl. Phys. Letters, 24, 369 (1974)
(7) H.Ishiwara and S.Furukawa; to be published in Proc. 2nd Intern. Conf. on Solid State Devices, Tokyo, 1976
(8) O.Christensen and H.L.Bay; Appl. Phys. Letters, 28, 491 (1976).

DAMAGE PRODUCTION AND ANNEALING IN ION IMPLANTED Si-SiO$_2$ STRUCTURE AS STUDIED BY EPR

T. Izumi, T. Taku, and T. Matsumori

Department of Electronics, Faculty of Engineering

Tokai University, Shibuya, Tokyo, Japan

ABSTRACT

This report shows annealing behavior on the depth distribution of two kinds of paramagnetic centers[P_A; g=2.0013, $\Delta Hmsl$=4 Oe (electron trapped in an O-vacancy), P_C; g=2.008, $\Delta Hmsl$=16 Oe (hole associated with non-bonding oxygen atoms)] in SiO$_2$ films and effects of knocked-on oxygen-atoms on the Si-SiO$_2$ interface, using EPR technique. The ratio of the spin density of P_A and P_C was 1:0.3 for the specimens as implanted. A very weak line (P_I; g= 2.005) which was produced in the silicon substrate close to the interface by a knocked-on oxygen was observed. The P_I center lies near the interface and its properties depend strongly on the location of the profile of implanted atoms near the interface rather than on the dose implanted into silicon through the interface.

INTRODUCTION

A considerable amount of works on properties of the Si-SiO$_2$ interface states created by ion implantation have been reported. It is well known that ion implantation into Si-SiO$_2$ structure induces displacement damage in the oxide films, and creates the fast surface states and positive fixed charge near the Si-SiO$_2$ interface. The structure of these radiation damage is usually composed of some kind of paramagnetic centers with unpaired electrons. Although paramagnetic defects in Si-SiO$_2$ systems have been observed by Nishi (as P_A, P_B and P_C) [1] and Caplan et al (as P_A, P_B and P_C) [2], EPR data on ion-implanted Si-SiO$_2$ structure are so little as to be neglected till now. We have already found three kinds of paramagnetic damage centers in the ion-implanted Si-SiO$_2$ structures by the EPR method and have named them P_A, P_B and P_C centers in the order of increasing the g-values [3]. P_A and P_C centers with the g-values of 2.0013 and 2.008, respectively, are located in the SiO$_2$ films and P_B with the g-value of 2.006 is located in the silicon. We have suggested in the previous work that P_A arose from an electron trapped in an oxygen-vacancy, P_C arises from a hole associated with a non-bonding oxygen and the properties of P_B are similar to those of the amorphous center [4] in the implanted silicon. P_A and P_C and effects of knocked-on oxygen-atoms on the interface, were studied using EPR techniques.

EXPERIMENTAL PROCEDURES

Specimens used in this study were made of p-type silicon with the specific resistivity of 3--20 Ω-cm. The main surfaces were oriented to (100), and were chemically polished to a mirror finish. Ne$^+$- and P$^+$- ions were implanted at energies of 40--170keV and with doses of 1×10^{13}--2.1×10^{15} ions/cm^2 into SiO$_2$ films thermally grown in dry oxygen at 1050°C on silicon substrates. Isochronal anneals were carried out for 10 min. in an Ar ambient at each temperature with 50°C steps up to 600°C. The EPR of these specimens were measured at liq. N$_2$ using an x-band microwave spectrometer.

RESULTS AND DISCUSSION

1. Annealing Behavior on Depth Distribution of P_A and P_C

Figure 1 shows the derivative curve of the EPR spectrum observed in the 50 keV Ne$^+$-ion implanted Si-SiO$_2$ structure with doses of 1.0×10^{13}/cm^2. P_A and P_C are observed, but P_B is not observed in this figure. The projected range R_p was 770 Å and the thickness of the SiO$_2$ films was 1900 Å; thus the location of the profile was in the SiO$_2$ layer.

Figure 2 (a) and 2 (b) show the depth distributions of the P_A and P_C centers, respectively, and their variations with annealing. The peak position of the distribution curve of the P_A and P_C center were located at the depth of 700 and 900 Å, respectively, from the SiO$_2$ surface. The peak position of the P_A distribution moved from the far-surface to the near-surface of the SiO$_2$ and from the near-surface to the far-surface alternately with each annealing step of 100°C. The P_A distribution decreased for anneals above 200°C and was not observed after annealing at 450°C. The P_C distribution separated into two peaks after 100°C anneal, one at 400 Å and the second at 800 Å. The total spin quantity remained unchanged for anneals below 200°C and decreased slightly for anneals above 200°C. The activation energies for the annealing process are 0.2eV for P_A and 0.05eV for P_C 3. The ratio of the spin density of P_A and P_C

DAMAGE PRODUCTION AND ANNEALING

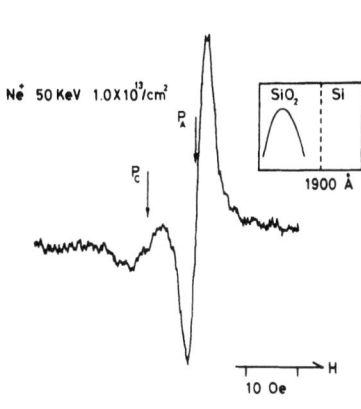

Fig. 1 Typical example of the EPR spectrum of damage centers in ion-implanted Si-SiO$_2$. Ne$^+$ ions were implanted into Si-SiO$_2$ structure with doses of 1.0×10^{13}/cm^2 at an energy of 50 keV, in this case. The SiO$_2$ film was 1900 Å and a schematic diagram of the profile of the implanted atoms is shown in the insert at the right-hand in the figure. The DC magnetic field for the EPR measurements was perpendicular to the SiO$_2$ surface.

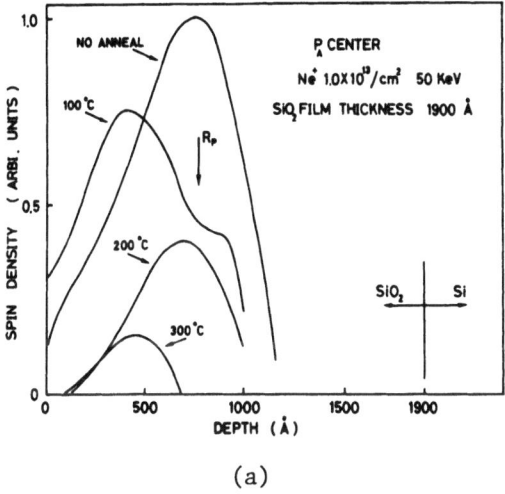

(a)

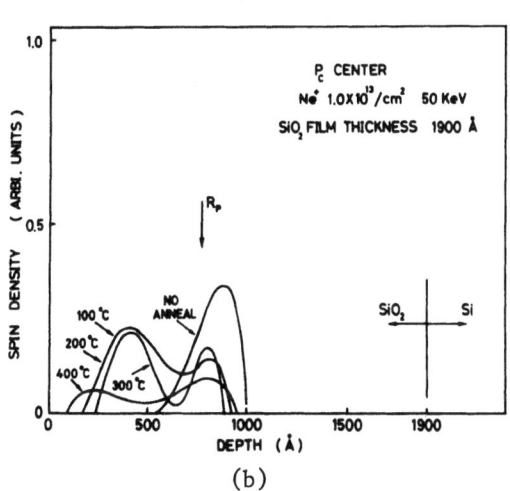

(b)

Fig. 2 (a) Annealing behavior of the depth distribution of P$_A$ centers in the SiO$_2$ film.

(b) Annealing behavior of the depth distribution of P$_C$ centers in the SiO$_2$ film.

Ne$^+$ ions were implanted into Si-SiO$_2$ structure with doses of 1.0×10^{13}/cm^2 at an energy of 50 keV in either case.

was 1:0.3 for the specimens as implanted. We have already suggested that P_C is a signal from a non-bonding oxygen and comparing this consideration with the results in Figure 2(a) and (b) may indicate that the P_C is a signal from a non-bonding oxygen which was knocked-on by the implanted ion. The reason that the spin density of P_C is much lower than P_A is that the knocked-on oxygen atoms are easy to move through the SiO_2 structure and bond with other atoms or vacancies.

2. Effect of Knocked-on Oxygen-atoms on the Si-SiO$_2$ Interface

When 1800Å of the SiO_2 films were removed after implants of 50 keV Ne$^+$ into 1900 Å SiO_2 on Si, P_A and P_C spectra disappeared and we could detect a very weak line with a g-value of 2.005, named P_I by us, which was not observed in unimplanted Si-SiO$_2$ samples. Then we found that the distribution of P_I was located within 100 Å from the Si-SiO$_2$ interface in the silicon substrate. Whereas the implanted atoms didn't reach the interface, P_I was observed. Therefore, it may be considered that P_I is produced by a knocked-on oxygen which penetrates the interface. In order to investigate the origin of the P_I center, we prepared the samples which were implanted with phosphorus ions under the following conditions (see Table I).

The samples were classified into three, that is, No.1, No.2 and No.3. In the sample No.1, the peak position of the profile of the implanted phosphorus atoms is in the oxide layers, while in the No.2, the peak position is close to the interface, and in the No.3, the peak is in the silicon substrates. Moreover, we prepared three kinds of samples in each No., that is, A, B and C, in order to realize the same profile in spite of different oxide thicknesses and implantation energies. In all the samples the doses of phosphorus atoms implanted into the silicon substrates through the oxides were fixed at $3\times10^{14}/cm^2$. When we made the EPR measurements, the oxide layers were removed completely.

Figure 3 shows spin quantities of defects centers created in the silicon substrates. Spin quantity in A-3 is much larger than those in A-2 or A-1, as can be seen in Figure 3. On the other hand, spin quantities in the same No. samples were the same, that is, the same in A-1, B-1 and C-1, and also in A-2 and B-2, and in A-3 and B-3. These results indicate that damage production in the silicon substrates depends strongly on the location of the profile of implanted atoms near the interface rather than on the dose implanted into silicon through the interface. When these samples were annealed at 300°C for 10 min. in dry Ar, the spin quantities in A-3 and B-3 decreased rapidly compared with those in A-1, B-1 and C-1.

Figure 4 shows annealing characteristics of the paramagnetic resonance spectra in the sample B. In the sample B-1, we could observe the spectrum due to the amorphous center with the g-value of 2.0061 for the sample as implanted. With increasing the anneali-

DAMAGE PRODUCTION AND ANNEALING

Table 1 Preparations of samples were made in order to investigate the effects of knocked-on oxygen atoms on the Si-SiO$_2$ interface. Samples were implanted with P$^+$ ions under the conditions as table 1 and classified into three, that is No.1, No.2 and No.3. In the sample No.1, the peak position of the profile of the P atoms is in the oxide layer, while in the No.2, the peak position is close to the interface, and in the No.3, the peak is in the silicon substrate. In all the samples the doses of P atoms implanted into the silicon substrates through the oxide were fixed at $3\times10^{14}/cm^2$.

SAMPLE	SiO$_2$ FILM THICKNESS (Å)	ACCELERATION ENERGY (keV)	TOTAL DOSE (/cm^2)
A-1		40	1.7×10^{15}
A-2	524	54	6.0×10^{14}
A-3		80	3.6×10^{14}
B-1		75	1.7×10^{15}
B-2	1206	100	6.0×10^{14}
B-3		147	3.5×10^{14}
C-1	1719	130	2.1×10^{15}
C-2		170	6.0×10^{14}

$Q_{Si} = 3.0\times10^{14}/cm^2$ fixed

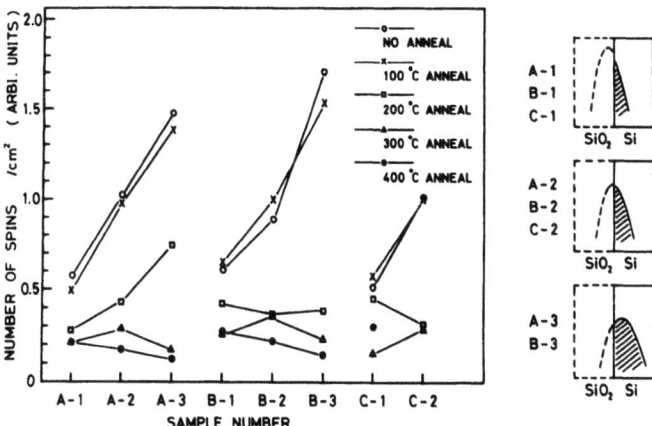

Fig. 3 Spin quantities in the silicon substrate in each sample.

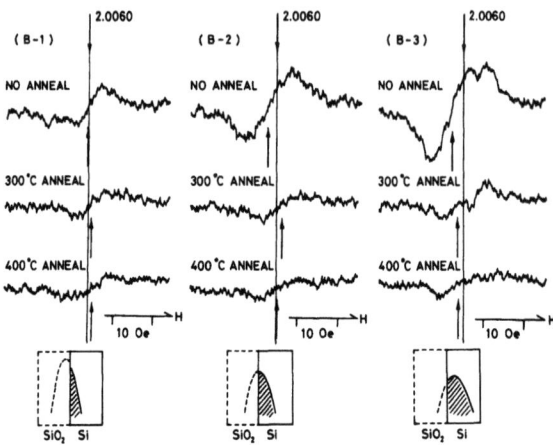

Fig. 4 Annealing characteristics of the EPR spectra in the sample B as an example.

ng temperature, its g-value decreases to 2.005 which is the g-value of the P_I center that we have mentioned above.

On the other hand, we could observe two kinds of spectra due to the amorphous center and the P-3 center. With increasing the annealing temperature, the spectrum due to the P-3 center tends to diminish and the g-value approached 2.0061, which is the g-value of the amorphous center.

As for the results, it may be suggested that in the sample B-1 the damaged layers introduced in the near-interface in the silicon substrate are stable and uniform amorphous layers[5]. In sample B-2 and B-3, however, there are highly disordered but non-amorphous layers surrounding a number of isolated amorphous regions, since the width of the profile in the silicon substrate of the implanted ion in B-3 is larger than in B-1. In this damaged layers, the Si-P3 center inclusive of the amorphous center was observed and this Si-P3 center disappeared after 250—300°C annealing. Brower et al reported that the Si-P3 center was observed in 400 keV O^+ ion-implanted silicon at room temperature [6]. The annealing properties of the P3 centers observed in the sample B-2 and B-3 agreed with Bower's results [6]. The isochronal annealing properties of the sample (A-1, A-2, A-3) and (C-1, C-2) were the same as that described for sample (B-1, B-2, B-3).

From the above mentioned results, it is satisfactory to consider that the defect - creation mechanisms are as follows:

In the samples (A-1, B-1 and C-1) which have the peak position of the profile of the implanted phosphorus atoms in the oxide layers, a large number of knocked-on atoms penetrate into the silicon substrates through the interface, so that a continuous amorphous layer will be formed in this region and at the same time the P_I center with the g-value of 2.005 will be formed. The P_I center dominates upon annealing to 300—400°C and disappears upon the 600°C annealing.

On the other hand, in the samples (A-3, B-3 and C-3) which have the peak position in the silicon substrates, isolated amorphous layers accompanied with complexes of defects such as the P-3 center are formed in this region. A large majority of the P-3 center are decomposed easily by low temperature annealing of 200—300°C.

In conclusion, we reported the properties of five kinds of defect centers, P_A, P_B, P_C, P_I and P-3, in ion implanted Si-SiO$_2$ structures and obtained the distribution profiles of P_A and P_C, and their annealing properties, and moreover, the effects of knocked-on atoms on the interface. We wish to make further experiments on the P_I center, and to make clear its physical structure.

ACKNOWLEDGMENTS

The authors would like to thank Dr. T. Tokuyama and M. Miyao of the Central Reseach Laboratory Hitachi Ltd, for suppling the samples and for helpful discussions. We also thank Prof. T. Sakata of Tokai University for his encouragement and advice.

REFERENCES

[1] Y. Nishi, Japan J. Appl. Phys. 10, 52 (1971)

[2] P. J. Caplan, J. N. Helbert, B. E. Wagner and E. H. Poindexten, Surface Science 54, 33 (1976)

[3] T. Izumi and T. Matsumori, Japan J. Appl. Phys. 14, 1067 (1975)

[4] B. L. Crowder, et al, Appl. Phys. Letters 16, 205 (1970)

[5] L. C. Kimerling and J. M. Poate, Inst. Phys. Conf. Ser. No. 23, 126 (1975)

[6] K. L. Brown, F. L. Vook and J. A. Borders, Appl. Phys. Letters 13, 208 (1969)

ANOMALOUS RESIDUAL DEFECTS IN SILICON AFTER ANNEALING OF THROUGH-OXIDE PHOSPHORUS IMPLANTATIONS

M. Tamura, N. Natsuaki, M. Miyao and T. Tokuyama

Central Research Laboratory, Hitachi Ltd.

Kokubunji, Tokyo, 185, Japan

ABSTRACT

Crystal defects formed in (111) silicon substrates implanted with phosphorus ions through SiO_2 layers are investigated as a function of implantation dose, energy, temperature, and oxide thickness using a transmission electron microscope. The density of these defects becomes maximum for samples with SiO_2 film thicknesses slightly smaller than the projected range of phosphorus ions in SiO_2, and they are not detected in samples having a SiO_2 film thicker than 2 times the projected range. These results are independent of implantation energy in the range between 50 and 350 keV. Comparing these results and observations of depth distribution of defects with Monte Carlo calculations of the yield of recoil oxygen atoms into silicon, the variation of defect density is closely related to the quantity of recoil oxygen atoms. Defect elimination methods are also discussed.

INTRODUCTION

Several authors [1-5] have recently reported that anomalous residual defects (ARD's) are formed in silicon after 1000°C annealing of "through-oxide" arsenic implanted samples. Moreover, it has been shown [4,5] that the recoil oxygen atoms from SiO_2 during through-oxide implantation strongly influence the formation of ARD's. However, the effect of implantation energy, dose and temperature during implantation is not clearly understood.

In the present paper, it is shown that ARD's are also produced by through-oxide phosphorus implantation which has a mass much lighter than that of arsenic. The structural changes in ARD's were observed as a function of implantation conditions and oxide thickness using a transmission electron microscope (TEM). The results are discussed by considering the concentration of knocked-on oxygen atoms calculated by Monte Carlo simulations.

EXPERIMENTAL PROCEDURES

Silicon wafers of (111) orientation and resistivity of 10 ohm·cm (boron doped) were used throughout the experiments. Oxide layers of 50 to 7000 Å were grown by both thermal oxidation and chemical vapor deposition techniques. Elipsometry and step height measurements were used to determine the thickness of the oxide films. Phosphorus implants were made at 50, 180 and 350 keV, with doses ranging between 1×10^{15} and 2×10^{16} ions/cm^2. The temperature during implantation was varied between room temperature and 1000°C. Oxygen implants were carried out to compare ARD's with oxygen-implantation-induced defects. Samples were annealed after implantation in dry N_2 or wet O_2 atmospheres from 5 to 240 min in the temperature range between 800 and 1200°C. However, the standard sample treatment employed was 1000°C, 30 min annealing in dry N_2.

An examination of defects by TEM was made for chemically thinned samples. The TEM observations and selected area diffractions were performed with HITACHI HU-11E electron microscope, operated at 100 keV, using a specimen tilting and rotation mechanism. All TEM micrographs shown herein were taken under low-beam bright field conditions with a deviation parameter of $s \geq 0$.

ARD FORMATION CONDITION

In order to determine the SiO_2 film thickness where the density of ARD's reaches a maximum, ARD's were observed for various SiO_2 film thickness. Phosphorus implantation through various film thickness was performed as a function of implantation energy at a fixed dose of 2×10^{16} ions/cm^2. Selected TEM micrographs reveal structural changes of ARD after annealing at 1000°C for 30 min in dry N_2 as shown in Fig. 1. Micrographs are shown for reduced oxide thickness, T, for each implantation energy; here T is the ratio of SiO_2 film thickness to the projected range of phosphorus ions in SiO_2 [6]. For comparison, dislocation network structures generated by implantation into bare silicon and annealing are also shown.

It is clearly seen that the defects generated even for samples with T of 0.1 are different from those formed by direct implantation. In particular, the TEM micrographs indicate that precipitate-like defect agglomerates are produced together with irregular and dense dislocation lines for the 180 keV implantation energy. The density of these agglomerates increases with increasing T and became maximum for samples with T of 0.9, irrespective of implantation energy. For T greater than 0.9, the density of defects under the SiO_2 layers begins to decrease, and no defects are observed in any TEM micrograph with T of 2 as seen in Fig. 1.

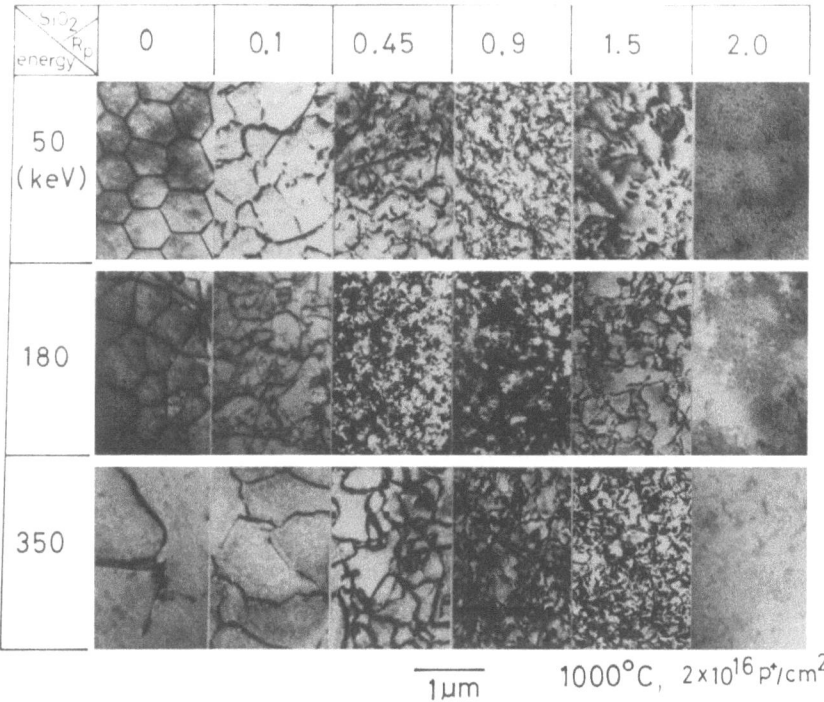

Fig. 1 The change of defects in silicon after 1000°C, 30 min anneal of 2 x 10^{16} P$^+$/cm^2 through-oxide implantation. T is the ratio of SiO$_2$ thickness to the projected range of P$^+$ in SiO$_2$.

An example of ARD variations observed as a function of ion dose is shown for 1500 Å through-oxide 180 keV implants in Fig. 2. Also shown are the results of implantation into bare silicon after 1000°C, 30 min dry N$_2$ annealing. The transformation of dislocation loops to dislocation lines is noted with increase of ion dose for bare silicon implants. On the other hand, precipitate agglomerates increase with dose for through-oxide implanted silicon. Considerable defect coagulations were observed for doses greater than 7 x 10^{15} ions/cm^2 in the 180 keV case. However, even for a 1 x 10^{15} ions/cm^2 implanted sample in which no coagulation occurs, the size of the dislocation loops is clearly different from that in bare silicon implantation. Namely, while the average loop size in bare silicon is about 3000 Å in diameter, the size in oxide covered silicon is less than 1500 Å in diameter. This result suggests that knocked-on oxygen atoms also have a strong effect on the growth of dislocation loops. In the experiments, these ARD's were observed even for samples annealed at 1200°C in dry N$_2$, although the density decreased.

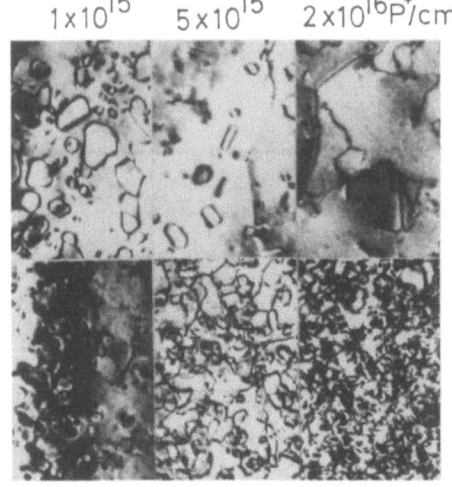

Fig. 2 The change of defects in silicon implanted through 1500 Å-oxide and into bare silicon as a function of ion dose. Ion energy is 180 keV. Annealing conditions are 1000°C, 30 min in dry N_2.

DEPTH DISTRIBUTION OF ARD's

The ARD distribution was examined as a function of depth from the specimen surface. First, carrier concentration profiles were determined by anodic oxide stripping and sheet resistivity measurements using Irvin's data [7] for 2×10^{16} ions/cm^2 implantation through 1500 Å oxide followed by annealing for 30 min at 1000°C in dry N_2. The profile curve is shown together with the one obtained for implantation into bare silicon in Fig. 3. Next, TEM observations were carried out at different depths indicated by arrows on the profile curves and the results are shown in Fig. 4. In bare silicon, dislocation networks were observed at depths between 2500 and 3000 Å from the surface. At slightly deeper depths, the networks were broken, and at deeper depths, short dislocations and faulted loops were seen as shown in Fig. 4. However, in through-oxide implants, the depth of the observed defects approached the surface side, and defect density was higher than for bare silicon. The high defect density region of through-oxide implants is confined to the range from the surface to 1000 Å depth. At still deeper depths, irregular and dense dislocation networks are observed. The cell width in the networks is about 2200 Å about 1/2 the width of the nets formed in bare silicon implants. Also, small agglomerates are seen on the dislocation lines. These dislocations are broken into smaller lines when 1650 Å of the surface layer is etched off. At deeper depths, smaller dislocations and faulted loops exist the same as in bare case.

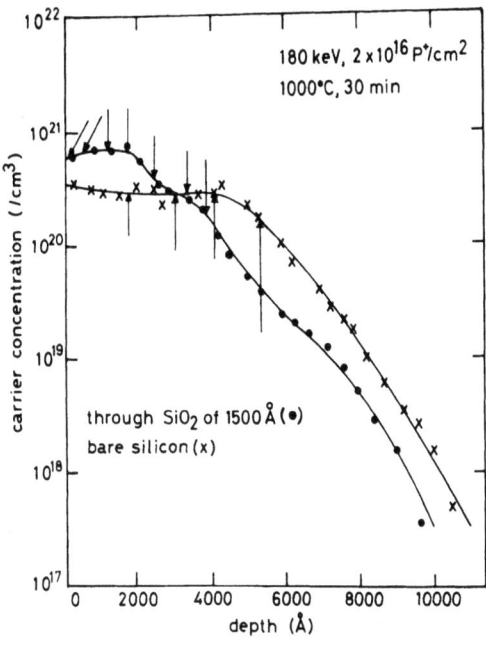

Fig. 3 Carrier concentration profiles in silicon substrates resulting from implantation of 2×10^{16} P^+/cm^2 at 180 keV through a SiO_2 film of 1500 Å and into bare silicon after 1000°C, 30 min, dry N_2 anneal. The levels observed by TEM are indicated by arrows on the profile curves.

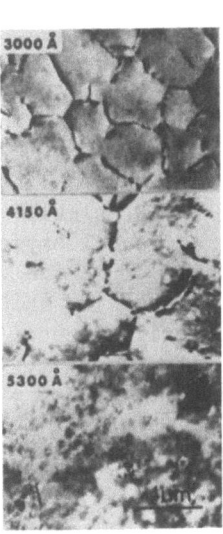

Fig. 4 TEM micrographs of depth distribution for defects generated by bare silicon and through-oxide implants after 1000°C, 30 min dry N_2 anneal. The depth observed correspond to the indications in Fig. 3.

From these phosphorus profiles and defect distributions, the following results are deduced: the region where ARD's form range from the surface to a 1000 Å depth, in this region the carrier concentration is high compared with bare silicon implantation. The results suggest that if ARD formation is a side effect of knocked-on oxygen atoms, the penetration depth of knocked-on oxygen should be confined to the thin layer near the surface. Hirata et al [8] observed that the substitutional phosphorus vacancy pair (E-center) anneals by migrating to an oxygen center, producing an oxygen-vacancy pair (A-center) leaving substitutional phosphorus atoms. Therefore, if this mechanism is applied to the present results, substitutional phosphorus concentration will be higher in the region where ARD formation occurs. In addition, high density A-centers or A-center complexes may be produced.

CALCULATIONS OF RECOIL OXYGEN YIELD

The results of Monte Carlo calculations are shown in Fig. 5 for recoil oxygen yield as a function of SiO_2 film thickness for various energies of phosphorus ions corresponding to the conditions in Fig. 1. Primary recoil yield including the results of an analytical method [9] and total yield are plotted. The total yield includes a multiplication process for the recoil collision cascade. The error bars represent statistical error. The thickness of SiO_2 is shown as normalized by the projected range of phosphorus ions in SiO_2 indicated in Fig. 1. Thus, the recoil yield becomes maximum when the SiO_2 thickness is slightly less than the phosphorus range, irrespective of ion energy. This agreed well with the observed results of the change in ARD density shown in Fig. 1, and it may be concluded that the formation of ARD's directly related to the existence of knocked-on oxygen atoms.

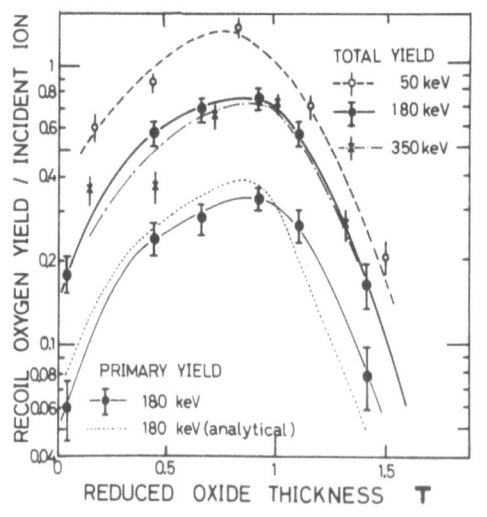

Fig. 5 Monte Carlo calculations of recoil oxygen yield as a function of reduced oxide thickness. Primary recoil yield (two lower curves including a dotted line curve obtained by an analytical method [9]) and total yield (three upper curves) are plotted.

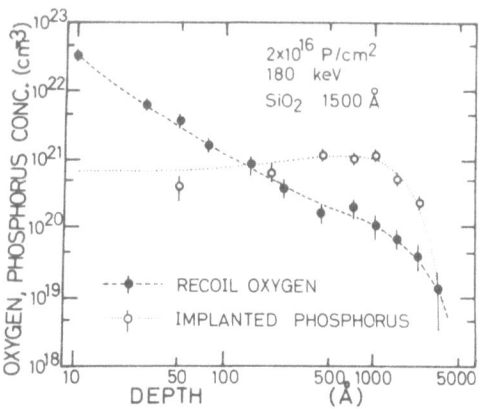

Fig. 6 Calculated concentration profiles of recoil oxygen (closed circles) and implanted phosphorus atoms (open circles) in the silicon substrates below the interface of SiO_2/Si for $2 \times 10^{16} P^+$ ions/cm^2 implantation at 180 keV through 1500Å thick SiO_2.

In Fig. 6, the calculated concentration profiles of recoil oxygen together with implanted phosphorus atoms are shown for 2×10^{16} P^+ ions/cm^2 implants at 180 keV through 1500 Å SiO_2. Closed circles (recoil oxygen) and open circles (implanted phosphorus) show the results of Monte Carlo calculations. Near the SiO_2/Si interface, the oxygen concentration is more than 10^{22} atoms/cm^3. At depths below 1000 Å, the oxygen concentration decreased rapidly to values below 1×10^{20} atoms/cm^3. By comparing the calculation with the observed depth distribution of ARD's shown in Fig. 4, the minimum oxygen concentration necessary for ARD formation may be considered as 1×10^{20} atoms/cm^3. However, since the solubility of oxygen in silicon is 2×10^{17} atoms/cm^3 at 1000°C, oxygen concentration at depths below 1000 Å largely exceeds the solubility limit. This suggests that oxygen atoms in excess of the solubility level interact strongly with implantation-induced defects or phosphorus atoms, and leave a stable defect formation. Also, it was shown that stable defects were actually formed in silicon layers implanted at 25 keV with oxygen ion doses above 3×10^{15} ions/cm. The maximum oxygen concentration for this condition exceeds 1×10^{20} atoms/cm^3.

ELIMINATION OF ARD's

When phosphorus or arsenic ions are implanted in the region limited by the oxide mask, ARD's are inevitably generated at the periphery of the oxide cuts. The existence of such ARD's results in the degradation of junction characteristics [10], and it is important

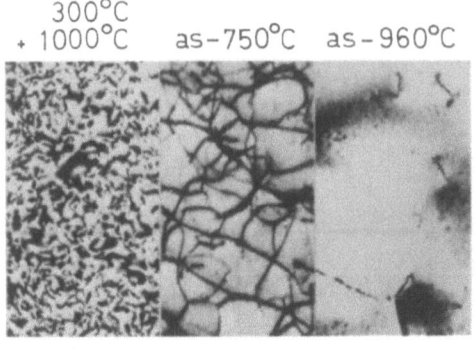

Fig. 7 The effect of temperature during implantation on defect structures. 1500 Å-thick through-oxide implantations of 180 keV P^+ ions at a dose of $2 \times 10^{16}/cm^2$.

to discuss methods of eliminating or preventing ARD formation. Two methods of eliminating ARD are considered here. One is to convert silicon layers with ARD's into SiO_2 films by annealing in oxidizing atmospheres. The other is to perform hot implantation to avoid the formation of stable defect structures during implantation. The ARD's generated at the oxide mask edge were eliminated by annealing at 900°C for 30 min in wet O_2 [10]. However, more care is needed for generation of outside dislocations propagating into unimplanted regions [11], which usually occur during wet O_2 annealing of high-dose phosphorus and arsenic implanted silicon.

A substrate temperature of about 200°C during implantation is known to prevent continuous amorphous layer formation [12]. Since vacancies generated during implantation easily migrate at this temperature, it was expected that knocked-on oxygen atoms might also migrate into crystals with vacancies as A-center and ARD's would not form at this substrate temperature. However, as shown in Fig. 7, although the ARD density decreased, they still formed at implantation temperatures of 300°C.

In order to completly avoid ARD formation, it was ascertained by further experiments that high substrate temperatures, above 750°C, were required for phosphorus implants. In Fig. 7, micrographs of defects generated under SiO_2 layers are shown for 1500 Å-thick-oxide covered samples implanted at various substrate temperatures. It can be seen from the figure that, in samples implanted at temperatures above 750°C, dislocation generation occurs but ARD's are not formed. Particularly, in 960°C implants, the density of long dislocation lines decreases, and inclined and curved dislocations were observed instead. This indicates that the interaction of the dislocations with vacancies generated during implantation occured actively.

CONCLUSION

The structural changes of anomalous residual defects (ARD's) in (111) silicon formed by through-oxide phosphorus implantations after annealing was investigated as a function of implantation energy, dose and temperature. These ARD's were clearly observed in 180 keV samples implanted with doses above 7×10^{15} ions/cm^2 after dry N_2 annealing at temperatures between 800 and 1200°C. The density of ARD's was maximum for samples with SiO_2 film thicknesses slightly smaller than the projected range of phosphorus ions in SiO_2, independent of ion energy. No defects were observed in samples with SiO_2 films thicker than 2 times the phosphorus ion range in SiO_2. It was shown by Monte Carlo calculations that these experimental results have a strong correlation with the amount of oxygen atoms recoiled from SiO_2 into silicon. The depth distribution of ARD's was confined to the region from the surface to 1000 Å, and this agreed well with the calculations of high concentration regions for knocked-on oxygen atoms. These ARD's can be eliminated by both annealing in oxidizing atmospheres and hot implantation at substrate temperatures above 750°C.

REFERENCES

1. T. R. Cass and V. D. K. Reddi: Appl. Phys. Lett., 23, 268 (1973).
2. E. H. Bogardus and M. R. Poponiak: Appl. Phys. Lett., 23, 553 (1973).
3. W. J. Chu, H. Muller, J. W. Mayer and T. W. Sigmon: Appl. Phys. Lett., 25, 297 (1974).
4. T. W. Sigmon, W. K. Chu, H. Muller and J. W. Mayer: Proc. 4th Int'l Conf. Ion Impl. in Semicon. and Other Materials, Osaka, 1974 (Plenum Press, New York, 1975) p. 633.
5. R. A. Moline and A. G. Cullis: Appl. Phys. Lett., 26, 551 (1975).
6. W. S. Johnson and J. F. Gibbons: Projected Range Statistics in Semiconductors (Stanford Univ. Bookstore, 1969).
7. J. C. Irvin: Bell Syst. Tech. J., 41, 387 (1962).
8. M. Hirata, M. Hirata and H. Saito: J. Phys. Soc. Japan, 27, 405 (1969).
9. R. A. Moline, G. W. Reutlinger and J. C. North: Proc. 5th Int'l Conf. Atomic Collision in Solids, Gatlinburg, 1973 (Plenum Press, New York, 1975) p. 159.
10. N. Natsuaki, M. Tamura, M. Miyao and T. Tokuyama: Proc. 1976 Int'l Conf. Solid State Devices, Tokyo (to be published as a supplement to Japan J. Appl. Phys.).
11. T. Ikeda, M. Tamura, N. Yoshihiro and T. Tokuyama: Proc. 6th Conf. Solid State Devices, Tokyo, 1974, p. 311.
12. F. F. Morehead and B. L. Crowder: Rad. Effects, 6, 27 (1970).

DIVACANCY FORMATION BY POLYATOMIC ION IMPLANTATION*

H. J. Stein

Sandia Laboratories

Albuquerque, New Mexico 87115

ABSTRACT

The production of the neutral divacancy absorption band (1.8 μm at 300 K) by the polyatomic series C^+ (70 keV), CO^+ (163 keV), CO_2^+ (256 keV), and by O_1^+ (85 keV), O_2^+ (170 keV), O_3^+ (255 keV) has been investigated. The oxygen series is emphasized because equal total energy and impurity deposition can be achieved simultaneously. For fluences significantly less than those required for amorphous layer formation, divacancy formation by 255 keV O_3^+ implantation at 300 K is ~1.5 times that for an equal atomic dose introduced by 85 keV O_1^+ implantation. Divacancy formation at 80 K followed by heating to 300 K, is ~1.3 times that for an equivalent implantation at 300 K. An enhanced probability for divacancy formation with increasing initial defect density is suggested to explain the polyatomic and temperature effects. For polyatomic implantation the initial defect density is increased by simultaneous collisions within a cascade, while suppressed annealing allows accumulation of initial defects for low temperature implantation. Defect annealing for energy deposition near the crystalline-to-amorphous transition is especially important in determining the nature of the disorder. The results are compared with previous channeling-backscattering measurements of disorder produced by polyatomic implantation.

INTRODUCTION

Davies et al[1] have used polyatomic implantations of low-mass ions to study the dependence of displacement damage on the energy density within a collision cascade. Using ion channeling-backscattering (CBS) measurements, they found that the damage produced and retained in Si per implanted atom at 300 K is nearly a factor of 10 larger for CO_2^+ than for C^+ implantation when the ion energies and currents are selected to give equal energy C atoms and equal atomic dose from each molecular species. The disorder peaks observed in CBS were attributed to grossly disordered crystalline or amorphous regions.

Divacancy (V-V) defect production by ion implantation can be monitored in crystalline Si by a 1.8 μm infrared absorption band; however this band is not observed in amorphous Si or in regions of high crystalline disorder.[2,3] Therefore, if polyatomic implants produce more grossly disordered regions than single atom implants, the observed V-V formation may decrease as the number of atoms per molecule increases. We report here on an investigation of V-V production by C^+ (70 keV), CO^+ (163 keV), and CO_2^+ (256 keV) implantations, and by O_1^+ (85 keV), O_2^+ (170 keV), and O_3^+ (255 keV) implantations into Si at 300 K. Instead of a decrease, we observe a small increase in V-V formation with an increase in the number of atoms per molecule for low fluences (fluences much less than those required to form an amorphous layer). The intensity of the V-V band produced by O_1^+ and O_3^+ implantations at 80 K followed by heating to 300 K is also larger than that for comparable low fluence implantations at 300 K. The results are compared and contrasted with those obtained previously by CBS measurements.[1]

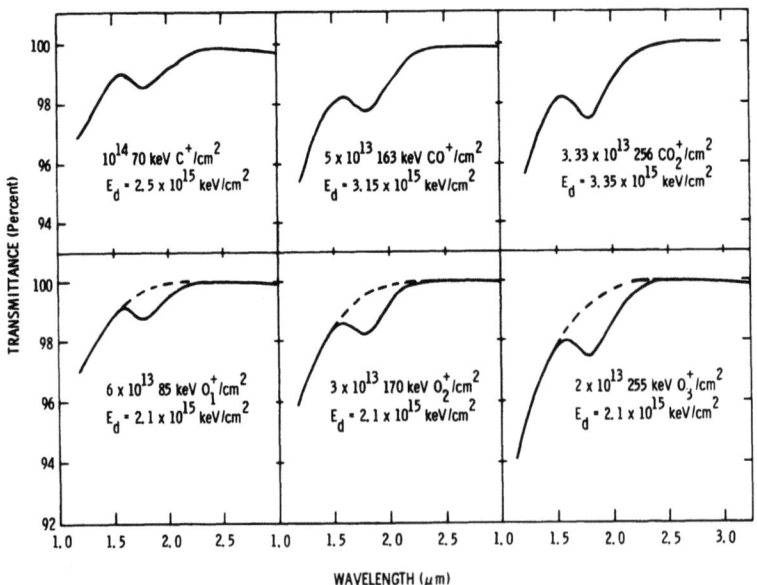

Figure 1. Production of the 1.8 μm divacancy by polyatomic implantations. Atom dose was maintained constant within one series and the energy depositions are listed in the figure (two side implants).

EXPERIMENTAL DETAILS

Implantations were performed on a 300 keV Accelerators Inc. machine. Ions were extracted from an RF ionized O_2 or CO_2 source, accelerated, magnetically selected, and electrostatically scanned across a beam defining aperture and onto the sample. A shallow Faraday cup combined with secondary electron suppression was used to determine ion fluence. Both optical faces of the samples were implanted. Optical transmittance data were obtained on a Perkin Elmer model 221 (large source, CaF prism) spectrophotometer. Optical measurements at 300 K were made differentially with an unbombarded Si blank in the reference beam. Measurements at 80 K were made with only purging nitrogen gas in the reference beam. The data presented in this report were obtained on (100) 1 ohm-cm, n-type Si with commercially polished surfaces, but additional data were taken on (111) p-type Si with etched surfaces to determine material independence of observed effects.

EXPERIMENTAL RESULTS AND DISCUSSION

Implantations at 300 K

The top row of transmittance versus wavelength data shown in Fig. 1 represents production of the 1.8 μm V-V band by C^+, CO^+, and CO_2^+ implantation at 300 K. The atomic dose and implantation time (~5 min) are the same for each ion species. The ion energies are scaled up from those used by Davies et al,[1] to achieve greater sample thickness for the optical measurements. The results show a small increase in the magnitude of the V-V band with the number of atoms in a molecule. Part of this increase can be attributed to an increase in energy deposition into displacement processes for the larger molecules as given in Fig. 1. In addition, the presence of carbon[4] and oxygen[5] can affect the V-V formation rate.

To eliminate variations associated with different energy deposition and different atomic species, O_1^+, O_2^+, and O_3^+ implantations were performed, and the results are shown in the bottom row of Fig. 1 where a molecular effect is still apparent. The dashed lines over the 1.8 μm band in Fig. 1 are estimated baselines for determining peak absorption-coefficient-layer-thickness products (αd) for the band. To obtain a V-V density/cm^2, αd is multiplied by a previously established calibration of 7.7×10^{16} V-V/αd.[2] V-V density results obtained in this way are shown in Fig. 2 as a function of atomic oxygen dose implanted as either O_1^+, O_2^+, or O_3^+ ions. The V-V density is only weakly dependent upon the fluence. This fluence dependence has been previously attributed to quenching in high disorder density crystalline regions.[2,3] The energy deposition per unit volume

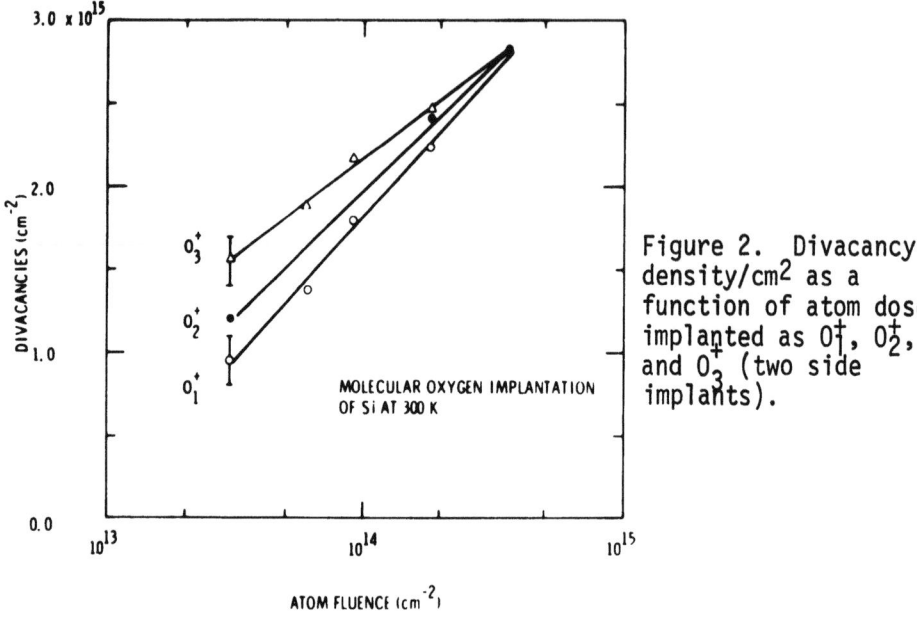

Figure 2. Divacancy density/cm² as a function of atom dose implanted as O_1^+, O_2^+, and O_3^+ (two side implants).

represented by the highest fluence implant is approximately the same as that for the CBS experiments of Davies et al[1] where a large polyatomic effect in the displacement damage was observed.

Comparisons of Results from Implantations at 80 and 300 K

Results from 3×10^{13} 85 keV O_1^+/cm^2 and 10^{13} 255 keV O_3^+/cm^2 implantations performed in a cryostat system at 300 K and at 80 K are shown in Fig. 3. The shallow Faraday cup, and secondary electron suppression bias are illustrated in the lower right of Fig. 3. The transmittance was adjusted to 100 percent at 5 μm, and the V-V band appears at 1.7 μm rather than 1.8 μm because the measurements were made at 80 K. The left column again shows the polyatomic effect for implantation at 300 K. Results in the right column of Fig. 3 are for O_1^+ and O_3^+ implantations at 80 K followed by heating to 300 K, and then recooling to 80 K. These results indicate that a larger divacancy band is formed by implantation at 80 K followed by heating to 300 K, than for implantation at 300 K. Therefore, both higher instantaneous damage density provided by polyatomic implantation, and accumulation of damage by implantation at low temperature enhance V-V formation at low fluences. An increased V-V formation rate with increasing ion mass for equal energy deposition can be deduced from the work of Baranova et al,[3] and stress measurements by EerNisse[6] showed larger damage production for equal energy deposition by heavy ions than for light ions.

It should be realized that neither the present low fluence

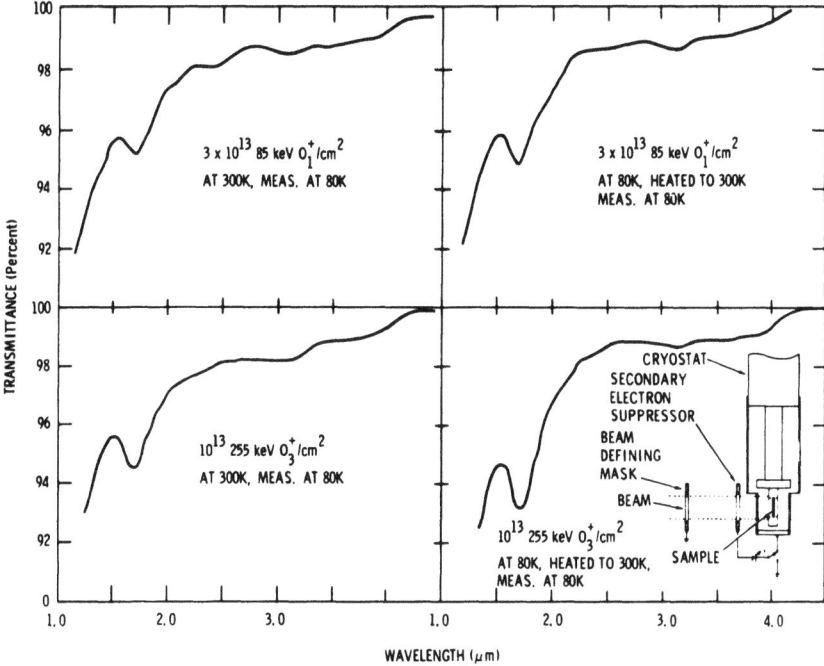

Figure 3. Implantation temperature dependence for divacancy formation by O_1^+ and O_3^+ (two side implants).

results nor the CBS results of Davies et al[1] represent isolated track effects because the damage volumes are such that each volume is written over hundreds of times by the fluences used in the experiments. Vook[7] reviewed results from many experiments which demonstrate that the energy deposition required for amorphous layer formation at 80 K is ~6×10^{23} eV/cm^3. Presumably, this energy density in a collision cascade would also cause amorphous zone formation. Plotted in Fig. 4 are calculated energy depositions as a function of ion mass for 40 keV ions.[8] The values are probably high because the damage is assumed to be contained in a volume of $(\Delta R_p)^3$ where ΔR_p is the standard deviation in the projected ion range. Average energy densities within the collision cascade decrease with increasing ion energy. Since all energy densities within a collision cascade fall well below the energy required to form an amorphous layer in the mass range under consideration, amorphous zones are most likely formed by disorder accumulation.[2,3] It is interesting to note that the average energy density deposited in the experiments of Davies et al[1] was ~5×10^{23} keV/cm^3, or very close to the energy density needed to form an amorphous layer if all the damage were retained. Therefore, annealing could be an important aspect of high fluence polyatomic or ion mass effects at room temperature.

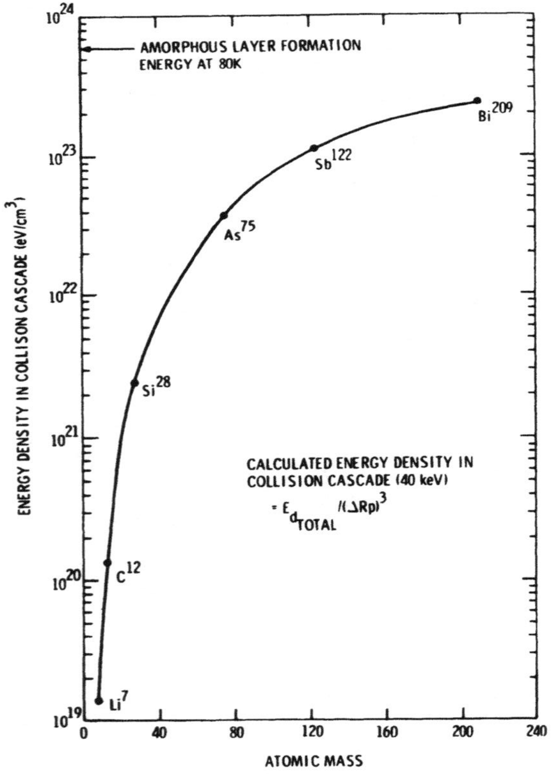

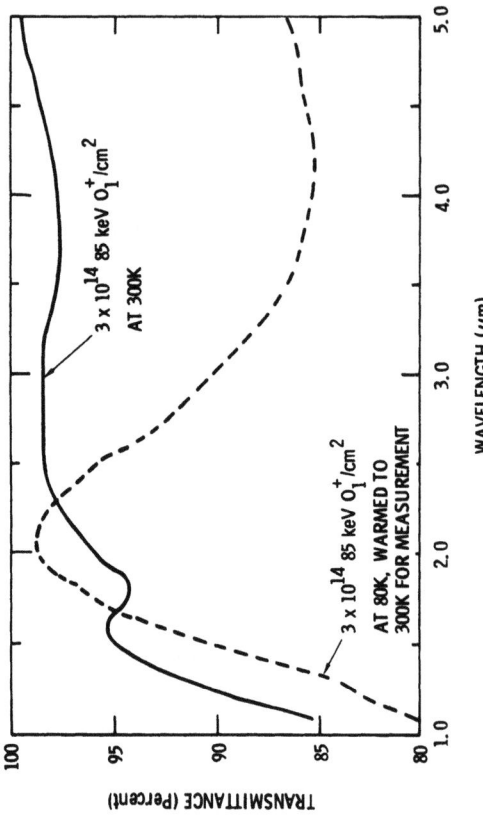

Figure 5. Effect of implantation temperature on the nature of the disorder for high fluence implantation (two side implants).

Figure 4. Calculated energy deposition per unit volume for 40 keV ions as a function of ion mass. Energy deposition for amorphous layer formation at 80 K is indicated.

To demonstrate the effects of accumulated high density disorder on V-V production, we have performed implants with 3 x 10^{14} 85 keV O_2^+/cm^2 at 300 K and at 80 K followed by annealing to 300 K. The accumulated damage in the implanted layer for these implants is also ≈5 x 10^{23} eV/cm^3. Results in Fig. 5 show large interference effects appear after implantation at 80 K, indicative of a refractive index change due to amorphous material formation,[3] and the 1.8 μm band is hardly resolvable. In contrast, a large V-V band is observed for 300 K implantation with only weak interference effects. Consequently, the degree of reordering is temperature-dependent and plays a very important role in determining the nature of the damage at these high fluences.

SUMMARY AND CONCLUSIONS

Divacancy formation is apparently enhanced by an increase in the initial disorder density for densities significantly below that required to produce an amorphous film. An increase in the initial disorder can be achieved by either an increased energy density in the collision cascade or by implanting at low temperature to suppress annealing. Neither formation nor quenching of V-V centers at 300 K correlate with the disorder measured by CBS.[1] This observation is consistent with self-quenching of V-V centers.[2,3] produced by implantation at 300 K since vacancy centers are not considered to be significant in the production of disorder peaks observed in CBS.[1]

ACKNOWLEDGEMENTS

The author is indebted to F. L. Vook and K. L. Brower for helpful discussions, and to R. H. Baxter and N. D. Wing for assistance with the experiments.

REFERENCES

1. J. A. Davies, G. Foti, L. M. Howe, J. B. Mitchell, and K. B. Winterbon, Phys. Rev. Letters 34, 1441 (1975).

2. H. J. Stein, F. L. Vook, D. K. Brice, J. A. Borders, and S. T. Picraux, Rad. Effects 6, 19 (1970).

3. E. C. Baranova, V. M. Gusev, Yu. V. Martynenko, C. V. Starinin, and I. B. Hailbullin, in *Ion Implantation in Semiconductors and Other Materials*, edited by Billy L. Crowder (Plenum Press 1973) p. 59.

INVESTIGATION OF ION IMPLANTATION DAMAGE WITH X-RAY DOUBLE REFLECTION

D.P. Lecrosnier, G.P. Pelous, J. Burgeat

Centre National d'Etudes des Télécommunications
22301 Lannion, France

ABSTRACT

The lattice parameters of boron implanted silicon layers have been measured by an X-ray diffraction technique. The rocking curves show an expansion volume which anneals in several stages : 100°C, 280°C and 400-600°C. Around the 500°C anneal, a stress-free crystal is found. This result suggests the formation of defect complexes having a compensating effect on the crystal lattice.

INTRODUCTION

When rocking curve experiments are performed on surface doped silicon, the X-ray diffracted intensity shows two peaks : a main one produced by the substrate and a secondary one originating from the doped region where the lattice parameter is different from the bulk. The position of the secondary peak relative to the main one depends on the nature and amount of the dopant species. For example, in the case of boron whose covalent radius is less than that of silicon, the secondary peak is expected to be on the high angle side. Nevertheless, on boron implanted silicon, we have found a secondary peak located on the low angle side. Similar results have been obtained by Burgeat and Colella (1) on silicon irradiated with α-particles. These authors have attributed this secondary peak to the damage region where defects produce a variation of lattice parameter. In the present study, rocking curve experiments were used to determine the lattice parameter change of a boron implanted layer as a function of post-implantation heat treatment.

EXPERIMENTAL

Crucible grown, low dislocation grade, n type, < 100 > orientation, 3 mm thick silicon samples were implanted with 500 keV or 1 MeV B^{11} ions in a random direction. Such thick samples were chosen in order to make sure that all the stresses induced in the damaged layer do not release and so, do not produce any elastic deformation as already observed on thin samples (2, 3, 4). Moreover, high energy implantation gives well-suited specimens for X-ray techniques, the thickness of the implanted layer being close to the penetration of Cu $K\alpha_1$ radiation here used.

The convoluted profiles have been measured with a double crystal spectrometer in the (n, -n) arrangement with a perfect (unirradiated) silicon sample on the first axis. All the rocking curves were obtained with Si 400 Cu $K\alpha_1$ reflection. Such experimental conditions allow us to measure less than 1 part in 10^{+5} variation in lattice parameter.

RESULTS

Figure 1 shows the diffracted intensities of samples as implanted in various conditions when rocking around the Bragg angle. For lower implanted doses ($1 - 2 \times 10^{15}$ ions cm^{-2}) a well defined secondary peak appears on the low angle side while the main peak shows only slight depression and broadening (The subpeaks may probably be attributed to "pendellösung" effects). According to Burgeat et al. (1) we may consider that the damage region has still a quite good crystalline structure with a lattice parameter slightly larger than the bulk. On the other hand, for higher doses, in such conditions where an amorphous layer is thought to be created (5), we noted a large depression of the main peak, indicating the presence of an absorbing layer, while the secondary peak becomes very broad and small, suggesting the existence of a region of poor crystallinity.

The relative position of the peaks show a volume expansion of the damaged crystal ; using the formula,

$$\frac{\Delta d}{d} = - \frac{1}{tg\ \theta_{Bragg}} \times \Delta\theta$$

we find an expansion of $1,2 \times 10^{-4}$ for 10^{15} cm^{-2} implanted ions. This result indicates that implanted boron introduces a very much larger volume variation than light particles as found by Burgeat (1) ($\Delta d/d = 5 \times 10^{-6}$ for 10^{16} α cm^{-2}) and Moyer (6) ($\Delta d/d \simeq 10^{-7}$ for electrons). The volume change, previously seen by several authors (2, 4) may be attributed to long range effects of lattice point defects and/or, of interstitial boron atoms since it is

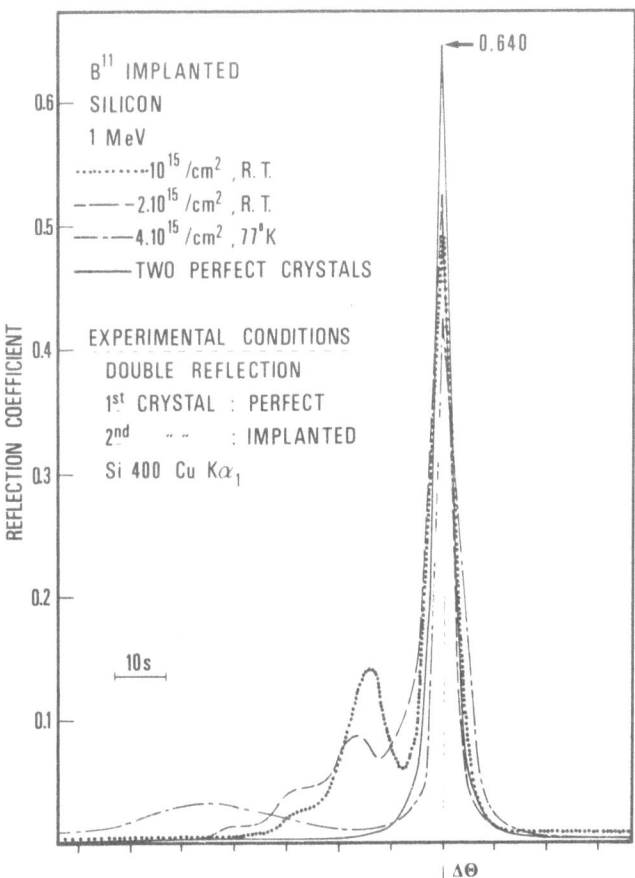

Fig. 1. Diffracted intensities of Si 400 Cu Kα₁ radiation for various boron implanted samples. The Bragg angle is $\theta_B = 34°\ 33'\ 55''$.

known that substitutional boron causes a volume reduction (7, 8).

After isochronal (30 minutes) heat treatment in an inert atmosphere, we observe an evolution of the diffracted profiles as it is shown on Figure 2. The position of the secondary peak and the height and width of the main peak have been measured for 10^{15} cm^{-2} borons implanted at 500 keV. The results are plotted in Figures 3 and 4. In Figure 3, several annealing stages are clearly seen. The first one appears at about 100°C and corresponds to a large change in lattice parameter. This can be well correlated

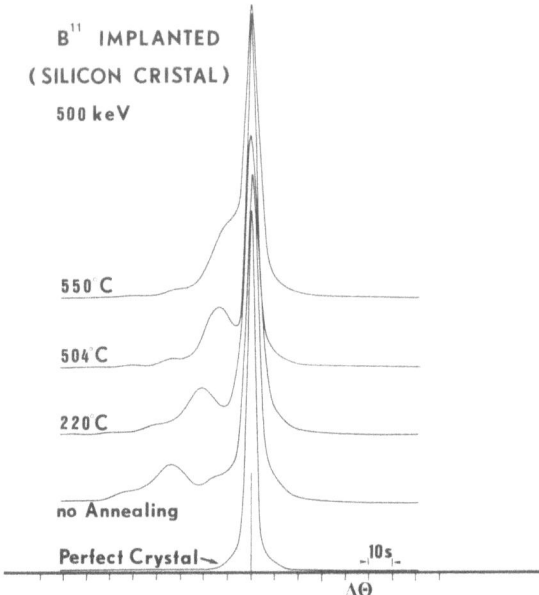

Fig. 2. Diffracted profiles at various anneal temperatures.

with the annealing of < 110 > split silicon self interstitials as proposed by Tan, Berry and Frank (9) from internal friction measurements. In the same way, the small lattice variation observed at 280°C may also result of the reordering from several interstitial defects as seen previously by photoconductivity measurements (10) and in (9) experiments.

The main anneal stage gradually occurs from 400°C to 600°C. At this temperature the secondary peak has completely vanished, and so the damaged region has recovered a lattice parameter very close to that of the bulk.

For higher temperatures, the secondary peak reappears, but on the high angle side, indicating the substitutional position of boron atoms.

However, the most striking feature of these experiments, as shown in Figure 4, is that around 500°C, the sample looks like a

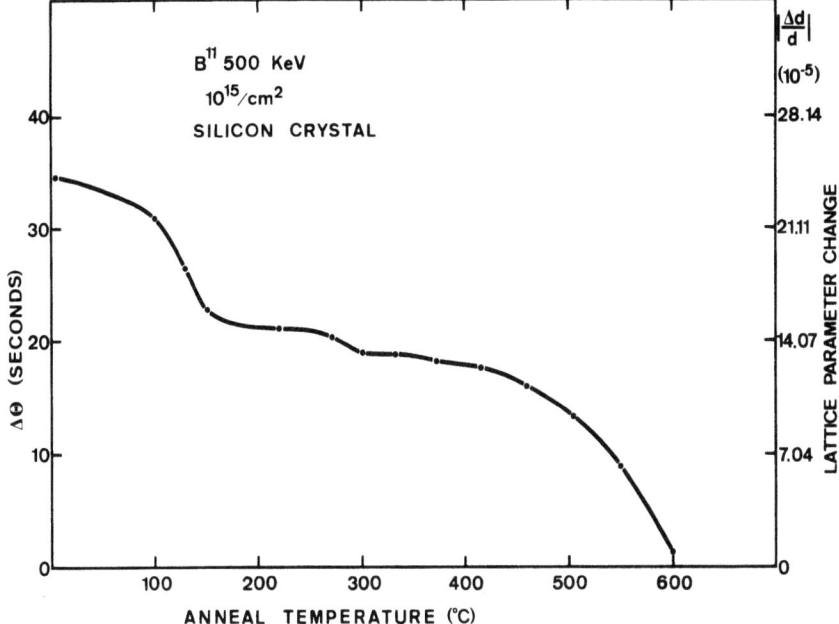

Fig. 3. Angular displacement of the secondary peak as a function of annealing temperature.

stress-free crystal. At this temperature, the values of the height and width of the main peak are very close to those of perfect crystals. It must also be noted that primary and secondary defects, which are present at respectively lower and higher temperature, introduce about the same order of broadening and depression of the main peak. So it seems to indicate, that at a temperature around 500°C which corresponds to the well know "reverse annealing" effect in boron implanted silicon, the defects would be arranged in such complexes so that their influence on the lattice parameter becomes negligible.

DISCUSSION

As a tentative explanation of a stress-free implanted layer, we are searching for defect complexes having a compensating effect on lattice strain. Since during the 500°C anneal, about 90 % of the boron atoms are electrically inactive, we may assume that this

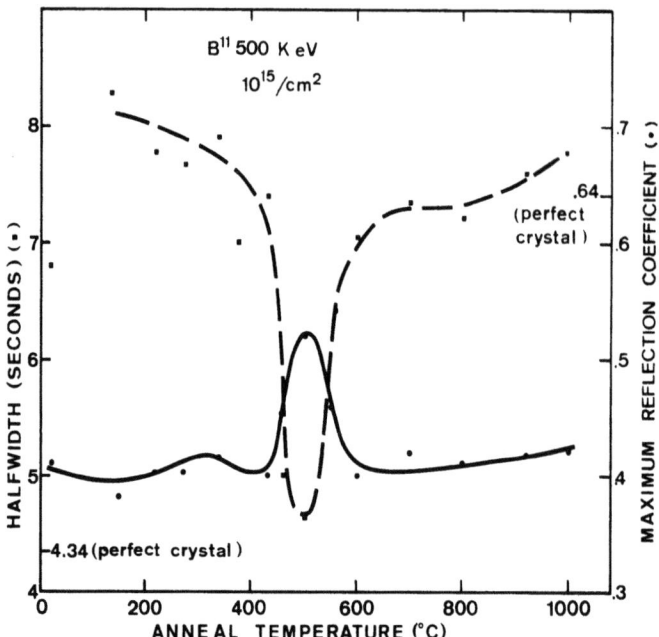

Fig. 4. Halfwidth (■) and maximum reflection coefficient (●) of the main peak as a function of annealing temperature. For an unirradiated sample the halfwidth and the maximum reflection coefficient are respectively 4,34 sec. and 0,64.

fraction occupies interstitial sites and introduces a volume expansion. Consequently, it appears necessary to associate boron with defects that contracts the lattice.

A possible configuration would be a boron and vacancy-type defect as suggested by Frank (11). If we assume, as did Frank, that this complex is under an "extended" form, the damaged volume of crystal should be not negligable, so that the main diffraction peak would probably be reduced in height. A more "contracted" complex seems to be more convenient. For example, if we reconsider the Masters' semivacancy pair (12), we may suppose that the interstitial silicon, located between the two unoccupied sites of this defect configuration , is replaced by a boron atom. Such an arrangement may produce a local compensating effect on the lattice and would suggest that divacancies, which disappear in the 250 - 300°C range (13), are trapped by boron atoms. This boron-semivacancy pair would be consistent with the resulting secondary defects observed after high temperature anneal (14, 15). In this model, when a boron atom jumps into a vacancy site, a vacancy is released. These vacancies may agglomerate into clusters and produce vacancy-types dislocation loops (14, 15). In this assumption, the density of loops would be strongly correlated to the boron concentration, as experimentally pointed out in (15) showing the identity of boron and loop depth distribution.

An alternative approach can be suggested in considering substitutional-interstitial complexes of boron atoms only. In fact, channeling experiments (16, 17) indicate that during a 500°C anneal, a large fraction of boron is displaced from a regular substitutional site. In order to compensate the low radius of boron, we are tempted to associate with such a particular site, one (or several) interstitial boron atoms. This model would exclude a long range interraction between primary defects and implanted ions, and would also mean that the dislocation loops result only from the effect of lattice strain induced by the low atomic size of boron. This last remark agrees with the previously mentionned correspondence between boron and loop depth distribution.

In any cases, the existence of ordered complexes may explain their high thermal stability since all boron atoms are found on substitutional sites around 1000°C (16, 18).

CONCLUSION

Rocking curve experiments appear as a useful technique for investigation of damage in implanted layers. Information has been obtained over a large range of anneal temperature. Although the nature of defects occuring in boron implanted silicon after 500°C anneal has not been entirely elucidated, the evidence of an almost stress free crystal suggest that ordered complexes having a compensating effect on lattice strain are present.

ACKNOWLEDGMENTS

Discussions with Pr P. Baruch, Dr W. Frank and Dr E.V.K. Rao are gratefully acknowledged.

REFERENCES

(1) J. Burgeat, R. Colella, J.A.P. Vol. 40, N° 9 (1969)

(2) S. Kishino, A. Noda, Proceedings of the 4th Conference on Solid State Devices, Tokyo (1972)

(3) E.P. Eer Nisse, A.P.L., Vol. 18, N° 12 (1971)

(4) R.L. Meek, W.M. Gibson, J.P.F. Sellschop, A.P.L., Vol. 18, N° 12 (1971)

(5) F.F. Morehead, J.R. and B.L. Crowder, "Ion implantation" edited by F.H. Eisen and L.T. Chadderton (Gordon and Breach, 1971)

(6) N.E. Moyer, R.C. Buschert, "Radiation Effects in Semiconductor", Santa Fe, Oct. 3-5, 1967

(7) J. Burgeat, C.R. Acad. Sc. Paris, t. 260, pp. 1915-1917 (1965)

(8) M. Mihara, T. Hara, M. Arai, M. Nakajima, S. Nakamura, A.P.L. Vol. 29, N° 1 (1976)

(9) S. Tan, B.S. Berry, W. Frank, Proceedings of the Intern. Conf. on Ion Implantation in Semiconductors and others Materials. Yorktown Heights (1972)

(10) B. Nétange, M. Cherki, P. Baruch, A.P.L. 20, 349 (1972)

(11) W.F.J. Frank, B.S. Berry, Radiation Effects, Vol. 21, pp. 105-111 (1974)

(12) B.J. Masters, Solid State Communications, Vol. 9, pp. 283-286 (1971)

(13) G.D. Watkins, J.W. Corbett, Phys. Rev. 138 A 543 (1965)

(14) R.W. Bicknell, Pro. Royal Soc., A, Vol. 311 (1969)

(15) D.P. Lecrosnier, G.P. Pelous, P. Henoc, Conf. on Lattice Defects in Semiconductors, Freiburg, July 1974

(16) J.C. North, W.M. Gibson, A.P.L. 16, 126 (1970)

(17) G. Fladda, K. Björkqvist, L. Eriksson, D. Sigurd, A.P.L. 16, 313 (1970)

(18) T.E. Seidel, A.U. Mac Rae, Rad. Effects, 7, 1 (1971)

EPR STUDY OF OXYGEN-IMPLANTED SILICON

P. R. Brosious

IBM Thomas J. Watson Research Center

Yorktown Heights, New York 10598 U.S.A.

ABSTRACT

A new spin-1/2 anisotropic EPR spectrum (Si-I5) described by a spin-Hamiltonian having a Zeeman term with nearly <011> axial symmetry and a hyperfine term with <111> axial bonding symmetry has been observed in float-zoned silicon after implantation with oxygen. The experimental parameters are compatible with the identification of the I5 spectrum with a high concentration of vacancy-oxygen defects in crystalline regions of high damage density. The dependence of lattice damage production and annealing on oxygen fluence has been studied by monitoring the intensity of the amorphous resonance (Si-a) at g=2.0055.

INTRODUCTION

Knowledge of the atomic structure of a simple point defects is necessary to the understanding of the lattice damage produced in ion-implanted silicon. While many experimental methods have been used to study lattice damage [1,2], relatively few have been successful in identifying radiation-induced defects in silicon. EPR is an extremely effective probe for studying point defects in solids produced by electron and neutron irradiation [3] as well as ion implantation [4-7]. In this study, EPR is used to investigate the fluence dependence and annealing kinetics of the lattice damage produced in silicon following implantation with oxygen ions.

EXPERIMENTAL PROCEDURE

The electromagnetically analyzed beam from the ion implantation facility at IBM/Yorktown was utilized to implant $^{16}O^+$ at room

temperature into silicon at seven degrees off the <100> with 3/5 of
the total dose at 200 keV followed by 2/5 at 100 keV. The fluence
rate was held constant at $0.5\mu a/cm^2$ over the fluence range 3×10^{14}-
6×10^{16} O^+ ions/cm^2. The silicon samples were oriented by x-ray
diffraction to within an accuracy of ±1° and then cut and polished
to a size of 2.03x10.2x.127mm with orientation <110>x<110>x<001>
respectively. The float-zone grown silicon was phosphorus doped
to an n-type resistivity of $10^3 \Omega\cdot cm$. The annealing of samples was
conducted in a quartz tube with a He gas environment and the temperature was monitored by a thermocouple located inside the tube.
The EPR measurements were made at x-band frequencies over the
temperature range 25-300°K.

RESULTS AND DISCUSSION

1. EPR Spectra

Figure 1 shows the actual EPR signal which was observed at
25°K in the O^+-implanted silicon after a fluence of 3×10^{14} ions/cm^2.
The spectra is a superposition of the isotropic Si-a spectrum corresponding to the amorphous state [8] at g=2.0055 and a new anisotropic spectrum labelled Si-I5.

The rotation pattern of the I5 spectrum, taken in the (110)
plane in 5° increments, is shown in Fig. 2. The relevant geometry
and coordinate system of the principal g-tensor components are
given in Fig. 3. In Fig. 2, the solid lines show the results of
the computed fit of the Zeeman Hamiltonian $\mathcal{H}=\beta\bar{H}\cdot g\cdot\bar{S}$ for $\bar{S}=1/2$ to
the I5 rotation pattern in the (110) plane using the principal g-
values given. The I5 g-tensor symmetry is characteristic of the
bent-pair [9] bond in silicon. A similar g symmetry is seen in the
B1 [9], SL1 [10] and I3 [11] spectra of the negative, neutral and
positive charge states of the vacancy-oxygen defect.

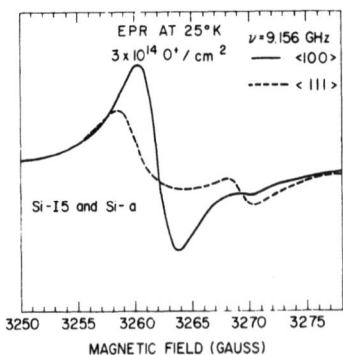

Fig. 1. Si-I5 and Si-a EPR spectra with the magnetic field parallel
to the <100> and <111> crystal directions.

A hyperfine interaction with nearby Si^{29} nuclei was also observed for the I5 spectrum and the results are summarized in Table 1, where the similar parameters for the B1 spectrum are shown for comparison. The I5 hyperfine interactions are axially symmetric and half of the axes for each set are closely along the [111] axis, the other half along the [$\bar{1}\bar{1}1$] axis for the defect depicted in Fig. 3. The experimental parameters shown in Table 1 are compatible with the identification of the I5 spectrum with a high concentration of vacancy-oxygen defects in crystalline zones of high damage density which results in the spreading out of the electronic wave function to overlap other regions.

The complete spin Hamiltonian for the I5 defect can now be described by the Zeeman term plus a hyperfine perturbation of the form $\Sigma \bar{I}_j \cdot A_j \cdot \bar{S}$ which describes the magnetic interaction of the spin with nearby Si^{29} nuclei (4.7% abundant, I=1/2). A_j is the corresponding hyperfine tensor for the jth lattice site. A simple LCAO (Linear Combination of Atomic Orbitals) analysis on I5 can be written in terms of the wave function for the unpaired electron given by $\Psi = \Sigma_j \eta_j \psi_j$, where ψ_j is approximated at each atom site j as a hybrid 3s3p orbital of the form $\psi_j = \alpha_j (\psi_{3s})_j + \beta_j (\psi_{3p})_j$. It is found that the paramagnetic electron on the I5 defect is localized in orbitals with enhanced p-like character (α^2=13% 3s and β^2=87% 3p) over normal sp^3 hybrids. Approximately 50% ($2\eta^2$) of the total wave function is localized on the two silicon atoms neighboring the bent-pair bond as shown in Fig. 3. This represents a nearly 22% decrease of the total wave-function localized at these sites as compared to the completely isolated B1 defect spectrum where the corresponding α^2, β^2 and $2\eta^2$ probabilities are 37%, 63% and 72% respectively.

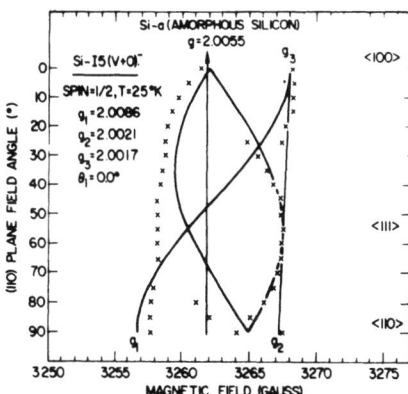

Fig. 2. Angular variation of the Si-I5 spectrum in the (110) plane at 25°K. The solid lines are from theoretical calculation.

Table 1. Coupling parameters of the negative vacancy-oxygen defect.

Spectrum	g_1	g_2	g_3	θ_1	$A_{\parallel}$ $\times 10^{-4}$ cm^{-1}	$A_{\perp}$ $\times 10^{-4}$ cm^{-1}
Si-B1	2.0093	2.0025	2.0031	0.0	153.0	128.8
Si-I5*	2.0086	2.0021	2.0017	0.0	49.1	40.6

*Vacancy-oxygen defects at high concentration (>10^{18} defects/cm^3) in regions of high damage density.

The atomic model for the vacancy-oxygen defect, and a simple one-electron LCAO description of the electronic configuration of the defect in its various charge states is shown in Fig. 3. The I5 spectrum essentially derives from the same electronic structure that is responsible for the negative charge state B1 spectrum with the small spectral perturbations in the Zeeman and hyperfine interactions arising from high defect concentration and high damage density effects.

A third EPR spectrum, identified as Si-P2 [12] was observed at 300°K after the 3×10^{14} O$^+$ ions/cm^2 implant. The spin-1 P2 spectrum has been identified [13] as a divacancy having two oxygen atoms trapped in its bent-pair bonds. Previously the P2 spectrum had been observed [12] after room temperature neutron irradiation of Czochralski silicon followed by annealing at 400°C.

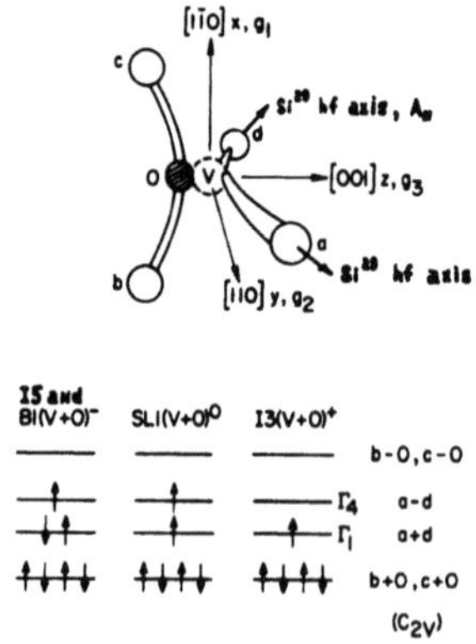

Fig. 3. Atomic configuration of the vacancy-oxygen defect, with the principal coordinates of the EPR coupling parameters shown above and the electronic arrangements thought to be responsible for the EPR spectra of the various charge states shown below.

2. Lattice Damage

To achieve a more uniform distribution of implanted ions and energy deposition into atomic processes we chose a double energy oxygen implant in the proportions given in Fig. 4 which also shows the theoretical damage density profile according to the Brice theory [14]. The energy deposition into atomic processes gives rise to three resolvable EPR spectra. In Fig. 5 we show the number of paramagnetic defects/cm^2 for the "a", I5 and P2 spectra as a function of oxygen fluence.

At the lowest oxygen fluence the I5 spin concentration of $\sim 7.5 \times 10^{13}$ spins/cm^2 agrees well with the concentration of oxygen-vacancy defects, determined by internal friction measurements [15] which implies that all the oxygen-vacancy defects are paramagnetic. Compared to the 3×10^{14} O^+ ions/cm^2 implanted at this fluence only one in every four oxygen ions produce oxygen-vacancy defects. The I5 spin concentration decreases very rapidly with increasing oxygen fluence and is not observable beyond 3×10^{15} O^+ ions/cm^2. Since the oxygen-vacancy defect concentration is known [15] to increase with oxygen fluences to 3×10^{15} O^+ ions/cm^2, we attribute the decrease in I5 spin concentration to a decreasing Fermi level combined with probable averaging effects at high defect and damage concentrations which may destroy the anisotropy of the I5 spectrum and contribute to the spin concentration of the isotropic Si-a amorphous signal.

The spin concentration of the P2 spectrum, which is only seen at the lowest fluence of 3×10^{14} O^+ ions/cm^2, is about three orders of magnitude higher than one would expect from the random statistical probability that two oxygen atoms will occur together as close neighbors.

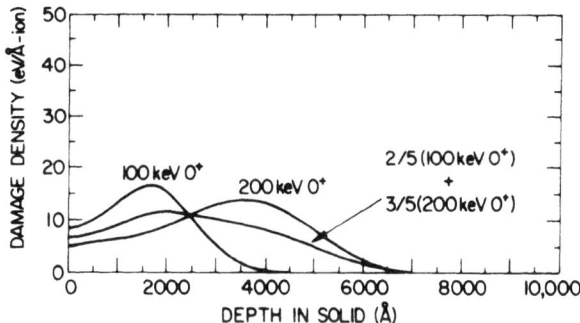

Fig. 4. Oxygen damage density profile into silicon [14] for 3/5 fluence at 200 keV followed by 2/5 at 100 keV.

It is suggested that radiation-enhanced diffusion of the implanted oxygen species may account for the high spin concentrations of the P2 spectrum.

The Si-a amorphous silicon resonance has a slightly sublinear production rate over the fluence range $3 \times 10^{14} - 10^{15}$ O^+ ions/cm^2 where TEM measurements [16] show that no continuous amorphous layer exists. Thus, in this range we believe the amorphous spin density arises from regions surrounding the incoming oxygen ions where the localized energy deposition into atomic processes is high enough to convert the regions from crystalline to amorphous. There may also be a contribution from the previously mentioned averaging of the I5 spectrum. In the range $10^{15} - 3 \times 10^{15}$ O^+ ions/cm^2 the localized regions begin to overlap and the amorphous resonance production rate changes from slightly sublinear to vary nearly as the square of oxygen fluence. Over the range $3 \times 10^{15} - 10^{16}$ O^+ ions/cm^2 the amorphous resonance production rate continues to vary approximately as the square of oxygen fluence. It is in this range that complete overlap of the high damage density regions occurs. At 7×10^{15} O^+ ions/cm^2 TEM measurements [16] reveal a 1000Å thick buried amorphous layer centered at a depth of 2000Å. The buried amorphous layer can be explained by the damage energy density profiles shown in Fig. 6, which are obtained after multiplying the composite profile in Fig. 4 by the various ion fluences. To agree with the TEM measurements [16] a damage energy density into atomic processes of $\sim 6 \times 10^{24}$ eV/cm^3 is required to convert the silicon from crystalline to a continuous amorphous layer with a room temperature oxygen implant. This agrees well with Beezhold's [17] oxygen implantation measurements where he observed a sharp increase in lattice expansion at 2.8×10^{24} eV/cm^3. These values are about an order of magnitude higher than that which is required to convert crystalline silicon

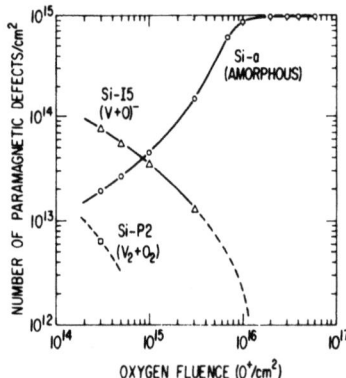

Fig. 5. Paramagnetic defect production versus oxygen fluence with EPR measurements at 300°K.

to an amorphous condition at temperatures below 77°K [18]. Thus a considerable amount of thermally activated recovery is inferred to take place during the room temperature implantation. Finally, over the range $10^{16}-6\times10^{16}$ O^+ ions/cm^2, the amorphous resonance remains saturated at approximately 10^{15} spins/cm^2. It is in this fluence range the TEM measurements [16] reveal a continuous amorphous layer from the surface (except at 10^{16} O^+ ions/cm^2 where a 250–500Å thin crystalline surface layer still exists) to a depth characteristic of the implanted ion range. The damage energy density profiles shown in Fig. 6 again explain very well the TEM measurements [16] for the depth of the continuous amorphous layer, which reaches a maximum depth of $\sim$ 6000Å. Thus, expressed per unit volume, the amorphous resonance spin concentration in the continuous saturated layer is found to correspond to $\sim 1.7\times10^{19}$ spins/cm^3.

3. Anneal of Lattice Damage

This section will be concluded with several remarks on the annealing of the amorphous lattice damage which is dealt with in more detail elsewhere [16]. An isochronal anneal in 100°C steps at 30 min per step was carried out for three oxygen fluences of 5×10^{14}, 10^{16} and 3×10^{16} O^+ ions/cm^2 respectively and the intensity of the amorphous resonance was monitored. The annealing kinetics of the amorphous lattice damage were observed to slow down with increasing oxygen concentration into the implanted layer.

At the low oxygen fluence of 5×10^{14} O^+ ions/cm^2, where no continuous amorphous layer exists, the amorphous resonance anneals at a temperature near 250°C. The annealing kinetics may proceed faster in this case because the localized amorphous regions are completely embedded within a crystalline matrix.

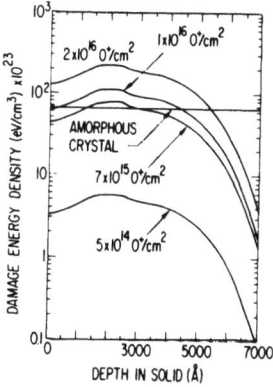

Fig.6. Damage energy density profiles for oxygen in silicon at the various oxygen fluences.

At an oxygen fluence of $10^{16}O^+$ ions/cm^2, where a $\sim$ 4500Å continuous amorphous layer exists beneath a thin $\sim$ 250Å crystalline surface layer, the amorphous resonance has a sharp annealing stage at $\sim$ 600°C. In this case the TEM measurements [16] show the loss of the amorphous resonance spins at 600°C correlates to epitaxial recrystallization of the implanted layer. Finally, at the high oxygen fluence of $3 \times 10^{16}O^+$ ions/cm^2, where a continuous amorphous layer exists from the surface down to $\sim$ 6000Å, the amorphous resonance has a sharp annealing stage near 700°C. However, after an 800°C anneal, a residual amorphous spin density of 0.6% of the original unannealed amorphous spin density still remains. In this case the TEM measurements [16] show the loss of amorphous resonance spins at $\sim$ 700°C correlates to polycrystallization of the implanted layer. The residual amorphous spin density remaining after 800°C anneal may be explained by the dangling bonds on the surfaces of the randomly oriented microcrystallites which make up the polycrystalline layer.

In conclusion, increasing the oxygen concentration in the implanted layer stabilizes the amorphous structure to higher temperatures. This is probably related to the ability of oxygen to form a hierarchy of oxygen-vacancy defects. Such defects are known from the observation of a series of Spin-1 EPR spectra. The spectra Si-I1 [19,20], Si-A14 [13] and Si-P2 [12] have been identified with divacancies having none, one and two oxygen atoms respectively in their bent-pair bonds. Similarly, the spectra Si-P4 [12], Si-P5 [12] and Si-A15 [13] have been identified with trivacancies having one, two and three oxygen atoms respectively in their bent-pair bonds. There exists a critical oxygen concentration between a fluence of 10^{16} and $3 \times 10^{16}O^+$ ions/cm^2 beyond which the recrystallization mode changes from epitaxial to polycrystalline. This effect may be related to the oxygen concentration and formation of an interface boundary region near the end of the implanted ion range which eliminates the connection to the crystalline lattice.

SUMMARY AND CONCLUSIONS

Three EPR spectra have been observed from float-zoned silicon implanted at room temperature with 100 and 200 keV oxygen ions. One spectra is Si-a, the isotropic resonance at g=2.0055 arising from amorphous silicon. Another is Si-P2, the spin-1 spectrum arising from a divacancy having two oxygen atoms trapped in its bent-pair bonds. The third is Si-I5, a new spin-1/2 anisotropic spectrum which has been identified with a high concentration of vacancy-oxygen defects in crystalline regions of high damage density. The amorphous resonance has been used to study the fluence dependence of lattice damage production and annealing. Increasing the concentration of implanted oxygen in the amorphous layer has been found to retard the annealing kinetics of the amorphous to crystalline transformation.

REFERENCES

1. J. W. Mayer, L. Eriksson, and J. A. Davies, Ion Implantation in Semiconductors (Academic, New York, 1970).
2. F. H. Eisen and L. T. Chadderton, Ion Implantation (Gordon and Breach, New York, 1971).
3. G. D. Watkins, in Radiation Damage in Semiconductors, edited by P. Baruch (Dunod, Paris, 1965), p. 97.
4. J. A. Borders and K. L. Brower, Rad. Eff. 6, 135 (1970).
5. F. F. Morehead and B. L. Crowder, Rad. Eff. 6, 27 (1970).
6. K. L. Brower and W. Beezhold, J. Appl. Phys. 43, 3499 (1972).
7. Y. H. Lee, P. R. Brosious, L. J. Cheng and J. W. Corbett, in Ion Implantation in Semiconductors, S. Namba, Ed., Plenum Press, New York (1975), p. 519.
8. B. L. Crowder, R. S. Title, M. H. Brodsky and G. D. Pettit, Appl. Phys. Lett. 16, 205 (1970).
9. G. D. Watkins and J. W. Corbett, Phys. Rev. 121, 1001 (1961).
10. K. L. Brower, Phys. Rev. B4, 1968 (1971).
11. P. R. Brosious, Appl. Phys. Lett. 29, 265 (1976).
12. W. Jung and G. S. Newell, Phys. Rev. 132, 648 (1963).
13. Y. H. Lee and J. W. Corbett, Phys. Rev. B 13, 2653 (1976).
14. D. K. Brice, Sandia Laboratories Research Report, SC-RR-71-0599.
15. B. S. Berry and W. C. Pritchet, Internal Friction Study of Vacancy-Oxygen Centers in Ion-Implanted Silicon, this conference, p.
16. B. S. Berry, P. R. Brosious and L. D. Glowinski, (to be published).
17. W. Beezhold, Bull. Amer. Phys. Soc. 16, 835 (1971).
18. F. L. Vook, in Radiation Damage in Semiconductors, edited by J. E. Whitehouse (Ins. of Phys., London, 1973).
19. P. R. Brosious, (to be published).
20. P. R. Brosious and B. S. Berry, Bull. Am. Phys. Soc. 20, 318 (1975).

EPR OF THE LATTICE DAMAGE FROM ENERGETIC Si IN SILICON AT 4 K*

K. L. Brower

Sandia Laboratories

Albuquerque, New Mexico 87115

ABSTRACT

An electron paramagnetic resonance (EPR) study of the lattice damage produced by 14.2 MeV neutrons in p-type silicon at 4 K is presented. The EPR measurements were made at 5 K without any intermediate warmup of the sample. Our EPR spectra indicate that each damage region, which is produced by a Si recoil of energy <1.89 MeV, is characterized by a high density of localized defects. A significant fraction of the lattice damage consists of distorted {110} 4-vacancies (Si-P3) embedded in a quasi-crystalline environment. Although a search for isolated vacancies was made, none was found. Even though the defects are complex and overlap, there is no evidence that a <1.89 MeV Si recoil produces amorphous regions at 4K. Upon annealing the lattice damage to 50 K, a trace of the Si-G6 spectrum due to $(V + V)^+$ was observed. After annealing to 500 K, the Si-B3 center, which has recently been identified as a <001> Si split interstitial, emerged.

INTRODUCTION

One of the fundamental questions of ion implantation is concerned with the nature of the lattice damage produced in a solid by individual, energetic ions. Of course the answer to this question is unique to the material and the conditions of the irradiation. In silicon alone, various models and descriptions of the lattice damage have emerged from numerous ion implantation studies. In one extreme, the primary lattice damage has been characterized in terms of vacancies and interstitials which evolve during subsequent kinetic processes into more complex centers.[1,2] In the other extreme, the primary lattice damage has been characterized in terms of amorphous tracks.[2,3]

In studies dealing with the nature of the primordial lattice damage, certain experimental conditions and difficulties need to be recognized. First, the damage regions created by each energetic ion should ideally not overlap. Overlap effects in the lattice damage produced by ion implantation are negligible only at low fluences ($\lesssim 10^{13}$ ions/cm^2).[4] Obviously, this constraint presents serious signal-to-noise difficulties for most types of measurements. Second, the lattice damage should be examined before it anneals. In silicon any rearrangement in the lattice damage can be minimized by keeping the sample at liquid helium temperatures -- even then self-interstitials may migrate.[5] The difficulties of constructing a cryogenic system in which ion implants can be done at 4 K followed by electron paramagnetic resonance (EPR) measurements at 5 K are nontrival. Third, the character of the lattice damage needs to be qualified in terms of the incident ion and its energy since the energy into atomic processes varies considerably with respect to these parameters.

This paper presents the results of our EPR study on the lattice damage produced by individual, energetic Si ions in silicon at 4 K.

EXPERIMENT

Each damage region was produced by an energetic Si ion which had been knocked out of a normal lattice site by a 14.2 MeV neutron. This method has several advantages. First, each damage region is isolated. Second, the whole sample rather than just the surface region can be damaged; therefore, the signal-to-noise can be enhanced considerably.

In particular, a p-type, 10 Ω-cm, B-doped sample of silicon was irradiated with $\approx 2 \times 10^{13}$ n/cm^2. Since the total scattering cross section for 14.2 MeV neutrons in silicon is 1.8 barns,[6] ~4 $\times 10^{11}$ primary silicon recoils/cm^2 were produced in our sample, which is 2.34 mm thick. The mean free path for 14.2 MeV neutrons in silicon is 11 cm; therefore, multiple scattering effects are negligible. Approximately half of the n-Si collisions are elastic.[7] The energy distribution of these primary, elastically scattered Si recoils ranges uniformly from 0 to 1.89 MeV in the center-of-mass frame. The total energy distribution is altered somewhat due to contributions from inelastic scattering. The total volume of lattice damage, $V_{\ell d}$, is estimated to be

$$V_{\ell d} = \frac{\xi V_{FP}}{2E_d} \int_0^{1.89 \text{ MeV}} N(E) \, \nu(E) \, dE \qquad (1)$$

where $\xi \approx 0.8$, V_{FP} is the volume of a Frenkel pair, E_d is a displacement threshold energy (between 13 and 22 eV[8,9]), $N(E)$ is the number of primary Si recoils of energy E, and $\nu(E)$ is the total energy into atomic processes. Assuming an average energy of $<E> = 0.9$ MeV and $V_{\ell d} \approx \xi V_{FP}\nu(<E>)/2E_d$, the ratio of the lattice damage volume to sample volume is $\approx 5 \times 10^{-5}$ so that overlap in the lattice damage from different primary Si recoils is negligible.

In order to minimize any changes in the lattice damage due to annealing, the sample was submerged in liquid helium during irradiation. Producing the lattice damage at 4 K with neutrons rather than with accelerated ions was a major simplification in experimental procedure since our existing experimental apparatus is compatible with low-temperature neutron irradiation. As in electron irradiations, the sample was positioned in the irradiation chamber of our microwave probe directly behind the aluminum irradiation window of our dewar.[10]

RESULTS AND INTERPRETATION

The first <u>in situ</u> EPR measurements of the lattice damage produced by neutrons in silicon at 4 K are shown in Fig. 1. <u>The lack of any sharp resonances in this spectrum indicates that the paramagnetic lattice damage does not consist of isolated defects.</u> Isolated, paramagnetic defects in silicon give rise to narrow resonances (linewidth ≈ 1.5 Oe) whose angular dependence with magnetic field direction sharply reflect the symmetries of the defect and the host lattice. In order for a defect to be isolated as seen by EPR, the defect wave functions which encompass ≈ 200 lattice atoms must be unperturbed. Our concept of a defect includes the lattice imperfection (vacancy, interstitial, impurity, or some combination of these constituents) plus the surrounding lattice, called the outer region, which supports the defect wave functions. It is important to realize that relaxation in the positions of the neighboring lattice atoms is a natural part of the total defect.

At intermediate damage densities defects overlap. The degree of overlap can vary. In some cases, only the outer regions may overlap; in other cases, other lattice imperfections may be located within the outer region of a defect. The complexity of this situation may be compounded by variations in the kinds of defects which overlap. Variations in the degree of overlap account for variations in the interactions between the unpaired electrons and their environment and give rise to broadening in EPR spectra. As the amount of overlap increases, it becomes increasingly more ambiguous to think of the lattice damage in terms of a collection of specific kinds of defects; instead, one has a collection of weakly interacting defects. In this case the paramagnetic electrons, although perturbed, remain localized.

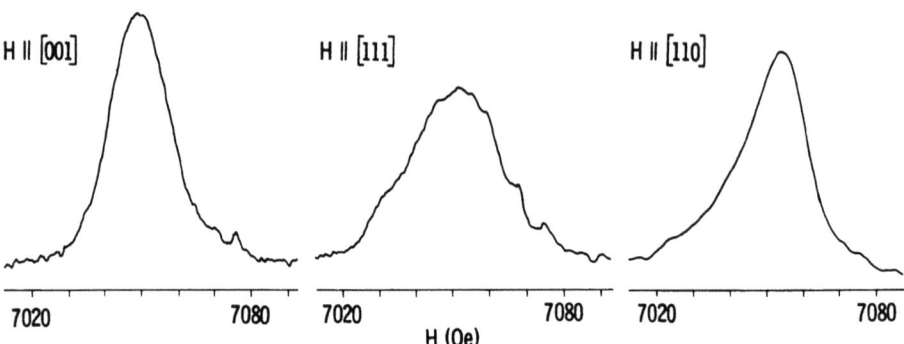

Figure 1. EPR spectrum observed from 10 Ω-cm, p-type, VFZ silicon irradiated at 4 K with 14.2 MeV neutrons to a fluence of 2×10^{13} n/cm^2. This spectrum was measured at 5 K without any intermediate warmup.

The spectrum illustrated in Fig. 1 is severely broadened due to overlap effects. Since the lineshape of the resonance varies with the direction of H, the defects still appear to be embedded within a quasi-crystalline lattice. Even though the defects overlap, the unpaired electrons are still localized. This conclusion is based upon the observations that the spectrum is anisotropic with respect to magnetic field direction and that the spectrum is observed in fast passage which indicates that the spin-lattice relaxation times are very long.

There is no evidence from our EPR measurements for the existence of amorphous zones within the damage regions. The presence of amorphous silicon is manifested in EPR spectra by an isotropic resonance at g = 2.0055 with a linewidth of 6.7 Oe.[11,12] This resonance arises from unpaired, delocalized electrons with very short spin-lattice relaxation times. Consequently, this resonance is observed in slow passage. Amorphous silicon produced by energetic ions is stable up to ≈850 K.[12] The amorphous silicon resonance, which is easily observed if present, is not present in Fig. 1 nor did it emerge at any point upon annealing to 800 K.

In order to better understand the nature of the paramagnetic damage, the EPR spectra of silicon irradiated under various conditions were compared. In Fig. 2, the top spectrum was observed from silicon irradiated with fission neutrons at 330 K. The resolved portion of this spectrum arises from isolated Si-P3 centers. The

EPR OF THE LATTICE DAMAGE FROM ENERGETIC Si

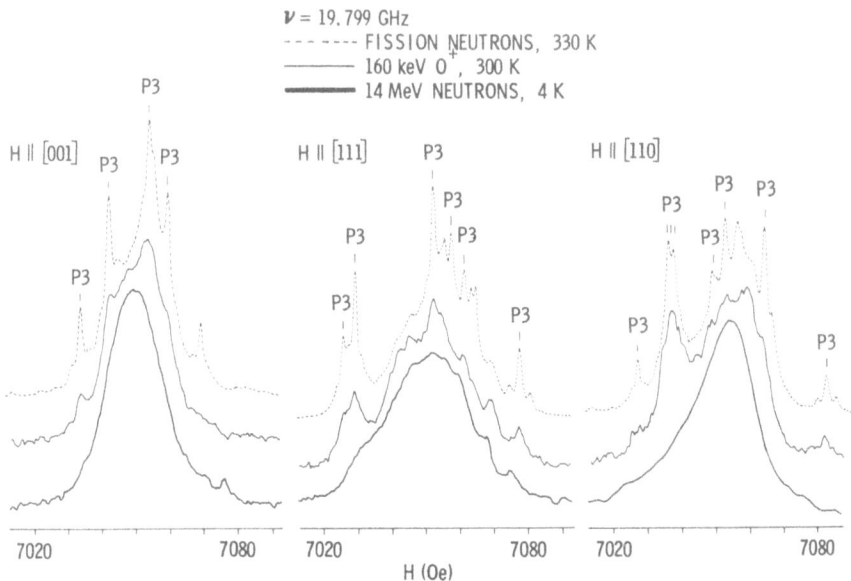

Figure 2. The top spectrum arises from lattice damage produced at 330 K by fission neutrons with a fluence of 4.16 x 10^{16} n/cm^2 and was measured at 15 K. The middle spectrum arises from lattice damage produced by 160 keV O$^+$ ions implanted at 300 K to a fluence of 2 x 10^{13} O$^+$/cm^2 and was measured at 5 K. In both cases, p-type, 1000 Ω-cm, VFZ silicon was used. The bottom trace comes from Fig. 1.

Si-P3 center has previously been identified as a {110} planar 4-vacancy.[13] The unresolved portion of this spectrum is attributed to Si-P3 centers which are distorted due to overlap effects. The middle spectrum in Fig. 2 was observed from silicon implanted with 160 keV O$^+$ ions at 300 K.[14] In this case the partially resolved Si-P3 spectrum is broadened due to overlap effects. It is also observed that as the sample is annealed overlap effects diminish and the resolution in the Si-P3 center recovers.[15] These results tend to suggest that as the density of the lattice damage increases, the Si-P3 spectrum converges to a spectrum like that shown in the bottom trace. Certainly the overall linewidth, lineshape, and line position agree very well with the unresolved portions of the other two spectra. The slight hump at 7028 Oe for H||[111] in the bottom trace appears to correlate with the Si-P3 spectrum. This extrapolation leads us to believe that <u>the paramagnetic lattice damage produced at 4 K in 10 Ω-cm, B-doped silicon by 14.2 MeV neutrons is dominated by Si-P3 centers which are severely distorted due to</u>

overlap effects.

The intensity of the spectrum in Fig. 1 corresponds within a factor of ±3 to ≈1.5 × 10^{14} distorted Si-P3 centers as determined by comparison with a previously calibrated[14] Si-P3 spectrum. This means that on the average ≈10^3 Si-P3 centers are created per 14.2 MeV neutron collision in silicon at 4 K. According to Sigmund,[16] the average number of Frenkel pairs, N, produced by an ion of energy E is given by the relationship

$$N = \frac{\xi \nu(E)}{2E_d} . \qquad (2)$$

Assuming that the energy of all primary Si recoils can be represented by a mean energy of 900 keV, Brice's calculations[17] indicate that $\nu(E) = 1.53 \times 10^5$ eV. Since estimates of E_d range from 13 to 22 eV,[8,9] between 2800 and 4700 Frenkel pairs are created on the average per primary Si recoil. This means that between 30 and 100% of all the vacancies produced at 4 K are contained within distorted 4-vacancies. It is reasonable to believe that other kinds of complex multiple-vacancy centers also exist within the damage region.

Since vacancies undergo thermally activated migration for temperatures ≳40 K and interstitials migrate below 4 K,[5] we irradiated our sample at 4 K so as to minimize any changes in the primary lattice damage due to annealing. Since interstitials are not observed at low temperatures, EPR sees only the vacancy component of the lattice damage. An attempt was made to observe the Si-G1 (V$^+$) and Si-G2 (V$^-$) spectra under light illumination, but no vacancy spectrum was observed. It is also significant that no divacancy spectra were observed initially. Although vacancies might have been mobile during the collision cascade due to ionization effects and a thermal spike, there is no indication that vacancies escaped from the damage region and formed positive divacancies outside of the damage region in p-type silicon. The lack of any resolved spectra indicates that all of the paramagnetic defects remained trapped within the dense defect cluster. The immediate formation of complex defects such as the 4-vacancy is not surprising. First of all, the energy density into atomic processes is more than sufficient to form complex multiple vacancy centers spontaneously. Second, the lattice damage is very dense -- defects overlap. Even if single vacancies were formed, only ∼1000 jumps would be required before the vacancy could combine with another defect. The presence of ionization, which is known to enhance vacancy diffusion,[18,19] and a thermal spike[20] could make the vacancy mobile. It is also worth noting that the momentum transfer to the lattice would shift the spatial position of the interstitial component of the lattice damage with respect to the vacancy component.

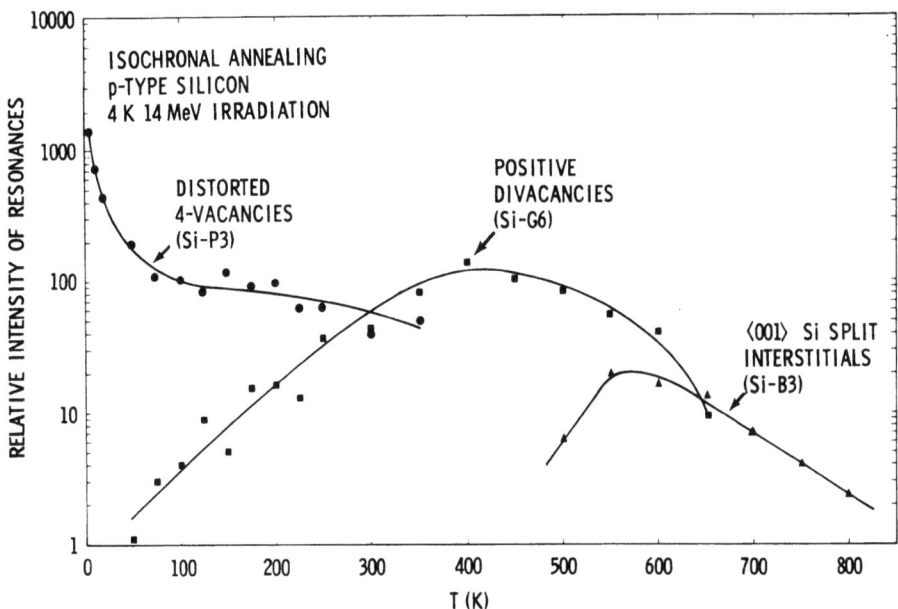

Figure 3. Isochronal annealing of the paramagnetic lattice damage in p-type, VFZ silicon irradiated at 4 K with 2×10^{13} n/cm^2. The annealing times were 20 min.

Upon annealing at 10, 20, and 50 K, there is a significant reduction in the intensity of the distorted Si-P3 spectrum (Fig. 3). This change indicates that the catacomb of overlapping defects is unstable and readjustments in molecular and electronic structure occur as the temperature is raised. At 50 K a trace of the Si-G6 spectrum corresponding to positive divacancies is observed. This spectrum is partially resolved and corresponds to divacancies in p-type silicon. It appears that as the damage density decreases within the cluster existing divacancies near the periphery of the defect cluster are left stranded and switch from neutral, diamagnetic to a positive, paramagnetic state as the Fermi level drops along the edges of the defect cluster. At higher temperatures vacancies could be liberated from the defect cluster and recombine in regions outside of the defect cluster.[21] This process would tend to disperse the lattice damage. After the anneal at 125 K, a stress of 1000 kg/cm^2 was applied along the [$\bar{1}$10] direction at 30 K. There was no apparent change in the Si-G6 spectrum. This result indicates that the Jahn-Teller distortion in these divacancies is stabilized by strain from neighboring defects. The gradual increase in the number of positive divacancies between 50 and 500 K as opposed to well-defined annealing stages associated with isolated defects is consistent with a defect cluster made up of overlapping defects. In this model one would expect the annealing stages of the lattice damage to be rather broad. At 350 K the EPR spectrum is dominated by a partially-resolved Si-G6 spectrum. At 500 K the Si-B3 center appears; this defect has recently been identified as a <001> Si split interstitial.[22] The annealing of this spectrum is like that observed previously.[22,23] Above 400 K, the EPR spectra are rather sharp and correspond to well-known, point-like defects.[24]

*This work was supported by the United States Energy Research and Development Administration (ERDA) under Contract E(29-1)789.

REFERENCES

1. F. L. Vook and H. J. Stein, Radiation Effects 6, 11 (1970).
2. F. F. Morehead, Jr. and B. L. Crowder, Radiat. Effects 6, 27 (1970).
3. E. C. Baranova, V. M. Gusev, Yu. V. Martynenko, C. V. Starinin, and I. B. Hailbullin, in *Ion Implantation in Semiconductors and Other Materials*, edited by B. L. Crowder (Plenum, New York, 1973), p. 59.
4. E. P. EerNisse, Appl. Phys. Letters 18, 581 (1971).
5. G. D. Watkins, in *Radiation Damage in Semiconductors*, edited by P. Baruch (Dunod, Paris, 1965), p. 97.
6. R. B. Schwartz, R. A. Schrack, and H. T. Heaton, II, *MeV Total Neutron Cross Sections* (U. S. Government Printing, Washington, 1974) NBS Monogr. 138, p. 61.
7. J. Höhn, H. Pose, and D. Seeliger, Nucl. Phys. A134, 289 (1969).
8. R. Bauerlein, in *Radiation Damage in Solids*, edited by D. S. Billington (Academic, New York, 1962), p. 358.
9. P. C. Banbury, in *Radiation Effects in Semiconductors*, edited by F. L. Vook (Plenum, New York, 1968), p. 280.
10. K. L. Brower, accepted for publication in Rev. Sci. Instru.
11. M. H. Brodsky and R. S. Title, Phys. Rev. Letters 23, 581 (1969).
12. B. L. Crowder, R. S. Title, M. H. Brodsky, and G. D. Pettit, Appl. Phys. Letters 16, 205 (1970).
13. K. L. Brower, Radiat. Effects 8, 213 (1971).
14. K. L. Brower and W. Beezhold, J. Appl. Phys. 43, 3499 (1972).
15. C. B. Norris, K. L. Brower, and F. L. Vook, Radiat. Effects 18, 1 (1973).
16. P. Sigmund, Appl. Phys. Letters 14, 114 (1969).
17. D. K. Brice, *Ion Implantation Range and Energy Deposition Distributions* (IFI/Plenum, New York, 1975).
18. H. H. Sander and B. L. Gregory, IEEE Trans. on Nucl. Sci., NS-13, 53 (1966).
19. B. L. Gregory and H. H. Sander, IEEE Trans. on Nucl. Sci., NS-14, 116 (1967).
20. K. Brack and G. H. Schwuttke, Phys. Stat. Sol. 5, 711 (1971).
21. R. E. Whan, J. Appl. Phys. 37, 3378 (1966).
22. K. L. Brower, Phys. Rev. B 14, 872 (1976).
23. D. F. Daly, J. Appl. Phys. 42, 864 (1971).
24. D. F. Daly and H. E. Noffke, Radiat. Effects 8, 203 (1971); 10, 191 (1971).

INTERNAL FRICTION STUDY OF VACANCY-OXYGEN CENTERS IN ION-IMPLANTED SILICON

B. S. Berry and W. C. Pritchet

IBM Thomas J. Watson Research Center

Yorktown Heights, New York 10598

ABSTRACT

Thin reeds of oxygen-implanted silicon vibrating near 400Hz exhibit an internal friction peak close to 200°K, which has been shown to originate from the stress-induced reorientation of vacancy-oxygen (V-O) centers. The peak has been used to study the production of these centers in 100 and 200 keV oxygen-implanted float-zoned silicon over the fluence range 10^{14} to 5×10^{16} O^+/cm^2. Peak broadening, present even at the lowest fluence, indicates a damage-induced perturbation of the crystalline environment containing the defects; however the average activation energy for defect reorientation (0.38 eV) is the same as that for less heavily damaged electron-irradiated samples. The strength of the peak increases sublinearly with fluence over the range 10^{14} to 10^{15} O^+/cm^2 and subsequently passes through a maximum near the fluence at which a continuous amorphous layer is formed. At a fluence of 10^{14} O^+/cm^2 about one-third of the implanted oxygen ions are detected as V-O centers. This fraction decreases rapidly in the fluence range where the amorphous layer forms and thickens, and is only two per thousand for a fluence of 5×10^{16} O^+/cm^2.

1. INTRODUCTION

One of the important problems associated with ion implantation is understanding the nature of lattice damage in terms of specific atomic models. This remains a major goal even in the case of silicon, where a considerable amount of work has already been reported. It is clear at least that a comprehensive characterization of lattice damage demands the use of a wide variety of experimental techniques, which together enable the details of the microstructure to be probed in different and complementary ways. This paper is concerned with an internal friction study of just one crystallographic point defect in ion-implanted silicon, namely the vacancy-oxygen (V-O) center shown in Fig. 1. This well-known radiation defect was originally identified in electron irradiated Czochralski silicon by EPR and infrared absorption measurements [1,2]. The ability to study the defect by internal friction measurements stems basically from its lower-than-cubic (orthorhombic) symmetry (Fig. 1), which endows the defect with a set of equivalent but distinguishable orientations and make it susceptible to stress-induced ordering, a process which occurs by thermally-activated reorientation jumps. This internal relaxation process gives rise to an internal friction peak whose characteristics are related to several important features of the defect [3,4]. The particular internal friction peak associated with the V-O defect in silicon is now known to occur close to 200°K for a measurement frequency near 400 Hz. This peak was first seen in boron-implanted Czochralski silicon [3], and was labelled Peak II. At that time, it was suggested that the peak may represent a negatively charged boron interstitial. Additional evidence gained since then has shown this suggestion to be incorrect. For example, it has been found that peak II is not specific to boron; it is also produced when Czochralski silicon is either hydrogen implanted or electron irradiated. On the other hand, a dependence on oxygen is indicated by the virtual absence of the peak in parallel experiments on float-zoned silicon. As will be reported in detail elsewhere, the kinetics of defect reorientation deduced from Peak II are in excellent agreement with the data reported by Watkins and Corbett [1] for the V-O center. This evidence, coupled

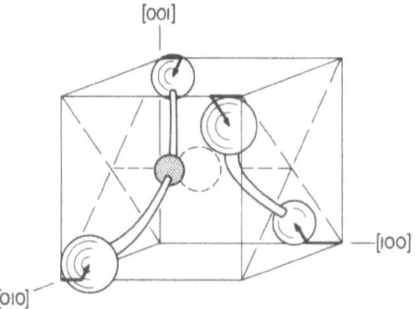

Fig. 1. The vacancy-oxygen defect in silicon. The vacancy is at the center of the cube; the oxygen atom (darkened circle) is displaced from it along the [100] direction. The displacements of the neighboring silicon atoms from their normal lattice sites are indicated schematically by the heavy lines.

with data concerning the strength, fluence-dependence, orientation-dependence and annealing behavior of the peak now provide a convincing case that peak II originates from the stress-induced ordering of V-O centers. The present investigation has been carried out to capitalize on this identification. Oxygen has been implanted to various fluences into float-zoned silicon, and the peak has been used to determine the fraction of incoming ions that appear as V-O centers. In this way some insight is obtained on the extent to which the lattice damage can be accounted for in terms of simple crystal defects.

2. PROCEDURE

Samples were prepared from float-zoned silicon of 1000 ohm-cm resistivity and n-type conductivity. Single crystal wafers of {100} orientation were thinned to a nominal thickness of 0.006 cm by a chemical lapping treatment applied to both sides of the wafer. Strips of <100> and <110> orientation, cut from these wafers, were bonded to supporting pedestals to provide vibrating-reed samples for the internal friction measurements. The internal friction, measured in terms of the logarithmic decrement of free decay, was observed as a function of temperature using either the fundamental mode or several higher overtones which together provided frequencies in the range 60-2000 Hz. Further details of the technique and apparatus have been given elsewhere [3,5].

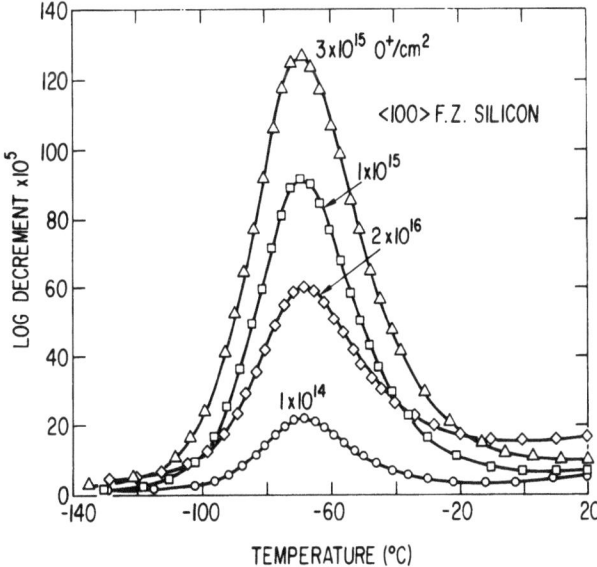

Fig. 2. Internal friction of a float-zoned silicon reed (0.006cm nominal thickness), after oxygen implantation on each face to the fluences indicated. Measurement frequency: 408 Hz.

The implantations were performed at room temperature with a swept beam of $^{16}O^+$ ions at a current density of $0.3\mu A/cm^2$. The angle of incidence was 7° from <100>. To improve the sensitivity of the measurements, the reeds were implanted on both faces, using for each side a two-step procedure in which 60% of the desired fluence was implanted at 200 keV, followed by the remaining 40% at 100 keV. This procedure produced a well-resolved peak over the whole fluence range employed, even though the implanted regions represent only about 2% of the total specimen volume. While such a small volume fraction is disadvantageous from the viewpoint of the peak resolution above background, the use of a sample whose thickness is many times larger than the implant range is highly advantageous from the viewpoint of quantitative interpretation of the data. With such a geometry, the peak strength is independent of the geometrical distribution of defects in the implanted layer, and is simply proportional to their total number. Furthermore, the peak strength obtained at a defect concentration of N_{o-v}/cm^2 can be shown to be just 3 times larger than that which would be exhibited if the same number of defects were uniformly dispersed throughout the full volume of the sample (i.e. to a volume concentration of $(N_{o-v}/d)/cm^3$, where $\underline{d}$ is the sample thickness). This relation provides an important link between the peak strengths observed in implantation and electron-irradiation experiments, since in the latter case the defect distribution is uniform and the calibration factor relating peak strength to defect concentration has already been determined with the help of EPR measurements.

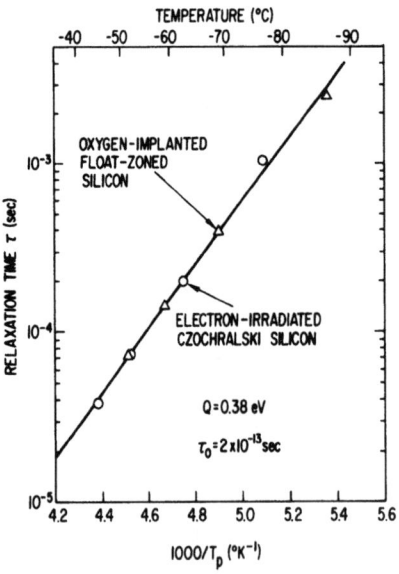

Fig. 3. Temperature dependence of the relaxation time for Peak II, as determined for oxygen-implanted float-zoned silicon and electron-irradiated Czochralski silicon.

3. RESULTS

Figure 2 shows the appearance of Peak II following oxygen implantation of a <100> oriented reed of float-zoned silicon. The peak, which is absent prior to implantation, increases to a maximum height for a fluence of 3×10^{15} O^+/cm^2 and thereafter decreases towards a final value of about one-half the maximum. Quantitatively, the important parameter describing the strength of the peak is not the peak height but the relaxation strength Δ, which is related to the area under the peak and which therefore also involves the peak width. In the present case, a satisfactory expression for the relaxation strength is

$$\Delta = 2r\delta_p/\pi \tag{1}$$

where δ_p is the peak decrement and $\underline{r}$ is the ratio which expresses the peak width relative to that of a standard Debye peak governed by a single relaxation time τ. Values of $\underline{r}$ greater than unity reflect the existence of a distribution of relaxation times. By calculating a fictitious or apparent activation energy Q_w from the peak width, the value of $\underline{r}$ may be found from the relation

$$r = Q/Q_w \tag{2}$$

where Q is the actual mean activation energy for defect reorienta-

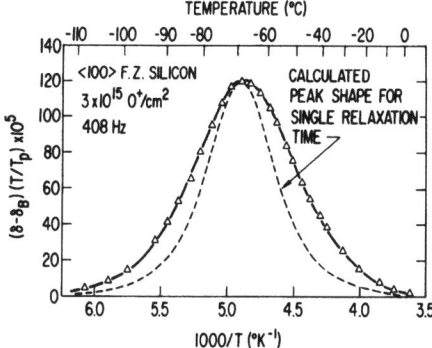

Fig. 4. The shape of Peak II in an oxygen-implanted sample of float-zoned silicon, compared with that calculated with the assumption of a unique relaxation time. The experimental data have been corrected for the background damping δ_B, and an expected 1/T dependence of the relaxation strength.

tion. Data yielding the value of Q are shown in Fig. 3. These results were obtained by determining the peak temperature T_p for several different vibration frequencies, and finding the corresponding relaxation time from the condition $\omega\tau=1$, where ω is the circular frequency of vibration. The data of Fig. 3 conform to the expression

$$\tau = \tau_0 \exp Q/kT \qquad (3)$$

with Q=0.38 eV and $\tau_0 = 2 \times 10^{-13}$ sec. Figure 3 also shows the excellent agreement obtained between the implanted sample and a sample of Czochralski silicon irradiated with 1.9 MeV electrons. In contrast to the electron-irradiated case, where no significant peak broadening was observed, the peaks of Fig. 2 exhibit a moderate degree of broadening. This is illustrated by Fig. 4. The width of a Gaussian distribution of activation energies producing such broadening is 0.027 eV, or 7% of the mean value. Values of the relative width factor $\underline{r}$ are shown as a function of oxygen fluence in the upper section of Fig. 5; the lower section shows the relaxation strength calculated from the height and width using Eq. (1). An obvious similarity exists between the two curves. Conversion of the relaxation strength to the number of V-O centers produced per unit area, N_{O-V}, is shown in Fig. 6. This reduction was performed with the aid of the calibration factor obtained from electron-irradiation experiments, modified as noted in Section 2 to allow for the different defect distribution in the implanted sample. For convenience, Fig. 6 also shows the results in the form N_{O-V}/N_O, where N_O is the implanted oxygen fluence. It is seen that the fraction of implanted oxygen ions detected as V-O centers varies very widely with fluence.

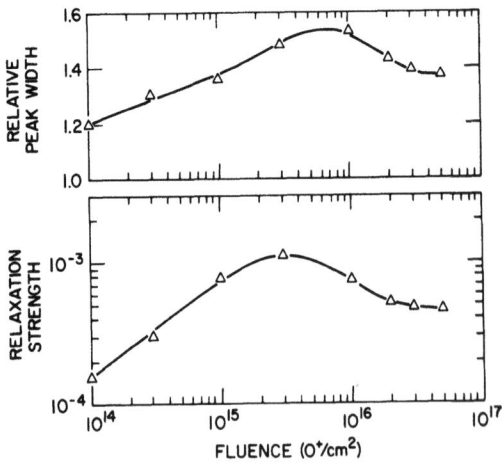

Fig. 5. Dependence of the width and strength of Peak II on the fluence of implanted oxygen ions.

At the lowest fluence one-third of the implanted ions are accounted for as V-O centers. This ratio continually decreases with increasing fluence and at the highest fluence is only two per thousand.

4. DISCUSSION

The principal feature of the present work is that it provides quantitative results on the production of V-O centers in oxygen-implanted silicon. This type of information has not been obtained from EPR measurements, even though spectra related to various electronic configurations of the V-O center have been observed [6,7]. On the other hand, the internal friction measurements give a very limited view of the overall state of the lattice damage. For this reason, the present measurements have been coordinated with EPR and TEM studies conducted by two of the authors' colleagues (P. R. Brosious and L. D. Glowinski), some of whose results are made use of below.

The finding that as many as one-third of the implanted oxygen ions are detectable as V-O centers, after implantation to a fluence of $10^{14} O^+/cm^2$, serves to emphasize the considerable importance of a point-defect description of lattice damage at low to moderate fluences. The perturbation of the crystallographic environment

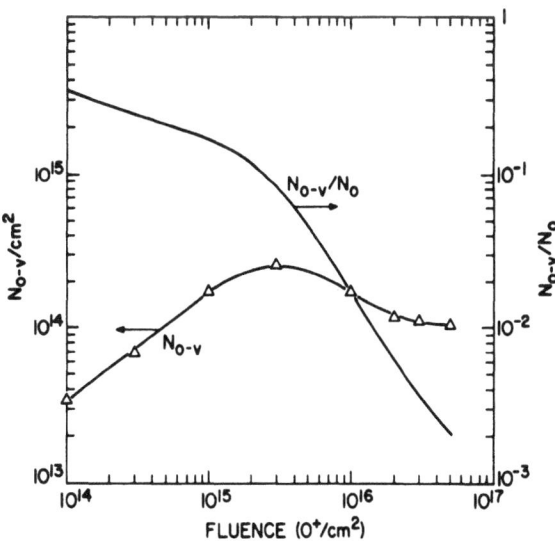

Fig. 6. The deduced absolute and relative numbers of V-O defects produced by oxygen implantation of float-zoned silicon.

containing these defects is evident from the broadening exhibited
by the peak. This broadening, which is ascribed to internal strain
and interactions associated with a relatively high total defect concentration in the implanted layer, is analagous to the EPR lineshape broadening described by Brower and Beezhold [6]. Evidence
that the defects sensed by the present measurements occur in essentially crystalline regions of the implantation zone is provided
by the fact that the peak temperature (and inferentially, the activation energy Q) remains the same at all fluences and is identical
to that observed in electron-irradiated material containing a much
lower density of damage. Inasmuch as Fig. 6 shows the number of
implanted ions which appear as V-O centers, it also draws attention to the large number of oxygen atoms whose fate is not revealed
by the internal friction measurements, and which are partitioned
between other point-defect species and amorphous zones. The existence of other strongly competitive defect configurations for the
implanted oxygen, or of overlap phenomena between damaged regions,
is indicated even at the lowest fluence by the sublinear ($N_o^{0.7}$)
change in the V-O concentration with the implanted fluence, N_o.
Possible other point-defect configurations include interstitial
oxygen, and the heirarchy of higher-order vacancy-oxygen complexes
known from work on neutron and electron-irradiated Czochralski
silicon [8]. Of these, the only one Brosious [7] has detected by
EPR measurements is the P2 center. Observation of this defect is
particularly surprising since it is thought to contain two oxygen
atoms (and two vacancies). In the present experiments there is a
very low statistical probability of two oxygen atoms appearing in
close proximity. Based on the rapid increase in the intensity of
the Si-a spectrum [7], it seems in any event that at fluences above
$10^{14} O^+/cm^2$ most of the oxygen is likely to be contained in heavily
disordered regions or amorphous zones. The amount of oxygen in such
regions appears to increase monotonically with fluence, as evidenced
by the continual drop in the ratio N_{o-v}/N_o, even though N_{o-v} itself
exhibits the maximum shown in Fig. 6. This maximum occurs near the
fluence where Brosious has observed the Si-a spectrum to be increasing most rapidly, and where Glowinski first detected a continuous amorphous layer by TEM. As the amorphous layer thickens at
higher fluences, it is noteworthy that the width of peak II shows
some narrowing (Fig. 5). This is interpreted to mean that the growth
of the amorphous layer removes the most heavily damaged crystalline
regions, improving the relative perfection of those which remain.
Since the highest fluences employed in this work cause the amorphous
layer to extend to the free surface, the eventual approach of the
peak strength to a final limiting value indicates a corresponding
saturation of the point defect damage in the transitional damaged
region below the amorphous layer.

Finally, it should be noted that the maximum present in the N_{o-v} curve of Fig. 6 clearly shows that conversion of a V-O center from a crystalline to an amorphous environment inactivates the defect as far as a contribution to peak II is concerned. It is interesting to speculate on why this is so. One possibility is that the transition from a crystalline to an amorphous environment destroys the V-O center as a recognizable local entity. On the other hand, since the terms amorphous and crystalline have to do more with long-range than short-range order, it is also possible that the V-O center is essentially preserved, but becomes locked into a particular orientation by strong local distortions. This latter viewpoint raises the interesting question of the extent to which identifiable point defects may persist in amorphous silicon, or indeed may be present as structural entities in amorphous materials in general.

ACKNOWLEDGMENTS

We thank P. R. Brosious and L. D. Glowinski for many useful discussions and permission to quote their results. We also thank R. Fiorio for performing the implantations, and E. Mendel for furnishing the thin polished wafers.

REFERENCES

1. G. D. Watkins and J. W. Corbett, Phys. Rev. 121, 1001 (1961).
2. J. W. Corbett, G. D. Watkins, R. M. Chrenko and R. S. McDonald, Phys. Rev. 121, 1015 (1961).
3. S. I. Tan, B. S. Berry and W. F. J. Frank, in Ion-Implantation in Semiconductors and Other Materials (edited by Billy L. Crowder), Plenum (New York) 1973, p. 19.
4. A. S. Nowick and B. S. Berry, Anelastic Relaxation in Crystalline Solids, Academic (New York) 1972.
5. B. S. Berry and W. C. Pritchet, IBM J. Res. Dev. 19, 334 (1975).
6. K. L. Brower and W. Beezhold, J. Appl. Phys. 43, 3499 (1972).
7. P. R. Brosious, this volume, p.
8. Y. H. Lee and J. W. Corbett, Phys. Rev. 13B, 2653 (1975).

ANNEALING OF DEFECTS IN ION-IMPLANTED LAYERS BY PULSED LASER RADIATION

G. A. Kachurin, V. A. Bogatyriov,
S. I. Romanov, and L. S. Smirnov

Institute of Semiconductor Physics
Novosibirsk, 90, USSR

The annealing of ion-implanted layers by laser pulses of high power is extremely interesting for several reasons. First of all, sufficiently short light pulses with photon energies exceeding the forbidden band of the semiconductor will heat only thin surface layers. As a result, it is possible to anneal radiation defects in the implanted layers without undesirable heating of the bulk. Secondly, use of focused light beams makes it possible to carry out local annealing at different temperatures all over the sample. Finally, pulsed laser irradiation is characterized by a time constant of $10^{-3} - 10^{-11}$ sec, and is accompanied not only by heating but also by effective ionization, shock waves, and quenching. All these factors affect the annihilation of defects and their interaction with impurities.

In this paper the action of short ($10^{-3} - 10^{-8}$ sec) laser pulses on implanted Ge, Si, GaAs, and InSb layers is investigated. The incident power of free generated pulses was $10^4 - 10^5$ W/cm^2, and for the Q-switched mode it was $10^6 - 10^7$ W/cm^2. We used lasers with wavelengths of 0.69 and 1.06 µm, i.e., ruby and Nd-glass lasers respectively. Special efforts were made to achieve uniform irradiation of a 0.25 - 0.5 cm^2 area. In order to prevent surface decomposition under the laser beam, we deposited protective films by reactive sputtering of silicon. Irradiation of InSb samples was performed both with such protective films on the surface and without them. Ge and Si were irradiated without films. As a rule, laser irradiation was carried out at room temperature. We used P- and As-implanted silicon samples, P-implanted germanium, GaAs implanted with Zn and Te ions, and InSb implanted with Zn.

It was found that a single laser pulse with a power density exceeding the threshold is sufficient for crystallization of amorphous layers and for electrical activation of an impurity. The threshold power for implanted layers of GaAs was $2 \cdot 10^6$ W/cm^2 in the case of the Q-switched ruby laser. Annealing of Te-implanted GaAs ($D = 3 \cdot 10^{15}$ cm^{-2}, $E = 35$ keV) with such pulses gives an electron concentration of the order of 10^{18} cm^{-3} and a mobility of about 1000 cm^2/V·sec. Ten additional laser pulses of the same power did not change the properties of the samples noticeably.[1] In the case of free generated ruby laser pulses (t = 5 msec) the threshold was higher than that for GaAs, and for InSb it was lower. For instance,

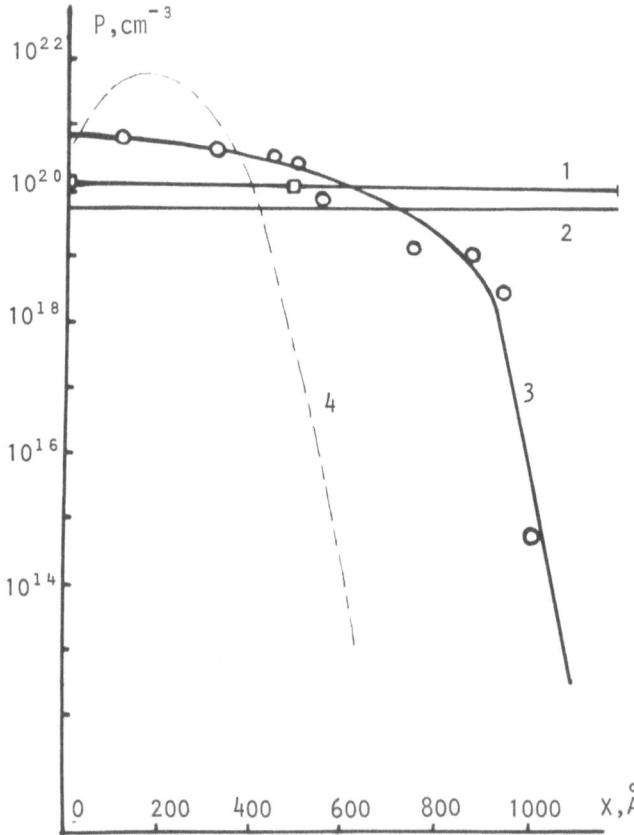

Fig. 1. Profile of electrically active Zn in GaAs. (1) Ion implantation and thermal annealing during 20 min at 750°C. The depth of the p-n junctions is $x_j = 0.7$ μm. (2) Theoretical curve for Zn diffusion in GaAs from an infinite thin layer during 20 min at 750°C; $x_j = 4.5$ μm. (3) Ion implantation and laser annealing. (4) Distribution of implanted Zn in GaAs according to LSS theory.

As implanted into silicon (E = 100 keV, D = 5·10^{15} cm^{-2}) becomes electrically active after irradiation with ruby pulses (5 msec) of 2·10^4 W/cm^2 power. It was found from Hall effect measurements that the electron concentration was 2.2·10^{15} cm^{-2}. Electronographic investigations show that it is possible to crystallize Ge layers that were amorphous after P$^+$ implantation (E = 100 keV, D = 2.5·10^{14} cm^{-2}) by irradiation with 2·10^3 W/cm^2 1-msec ruby laser pulses. However, in this experiment the laser irradiation was carried out at a sample temperature of 210°C.

We investigated the properties of p-n junctions obtained by ion implantation of Zn into n-GaAs and following free generated ruby laser annealing.[2] Zn ions were implanted at an energy of 35 keV with a dose of 10^{16} cm^{-2}. At first, by the use of Hall measurements and the stripping technique, we investigated the hole distribution in the layers after thermal and laser annealing. Figure 1 shows that the hole concentrations on the surface are the same after laser and thermal annealing. At the same time, as a result of the short duration of the laser pulse, no broadening of the profiles occurs. Some broadening of the doped layer may be a consequence of the channeling or the enhanced diffusion. Figure 2 represents the I-V characteristics of the p-n junctions obtained by thermal and laser

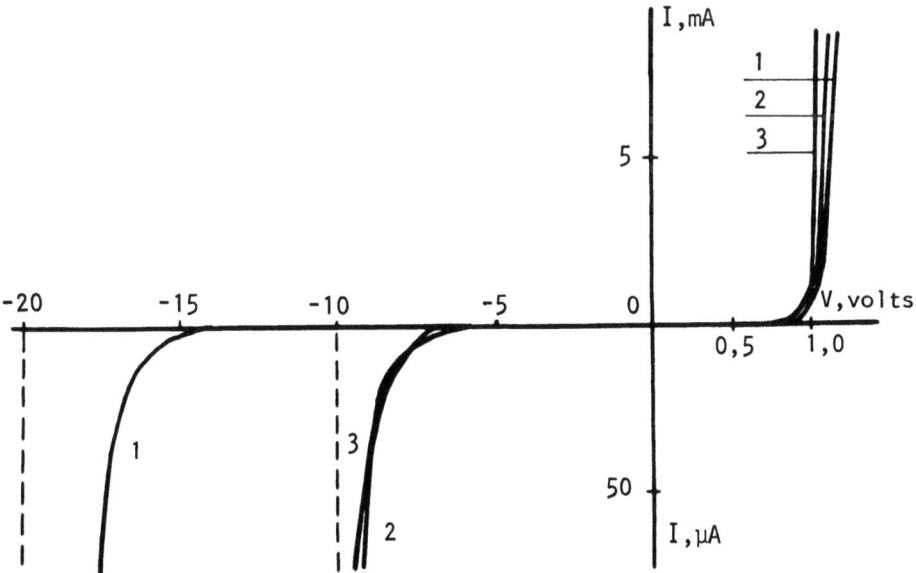

Fig. 2. I-V characteristics of mesa diodes fabricated by Zn$^+$ implantation into n-GaAs with electron concentration n = 6·10^{16} cm^{-3} (1) and N = 2·10^{17} cm^{-3} (2) with subsequent laser annealing. (3) Thermally annealed p-n junctions (n = 2·10^{17} cm^{-3}).

annealing. The resistance of p-n junctions obtained by laser annealing is greater than that after thermal treatment. The breakdown voltages in both cases coincide with the theoretical values of the breakdown voltages for sharp p-n junctions. From C-V measurements at 1 MHz the estimated width of the space charge layer after laser annealing was ~0.6 µm. The profiles of donor concentrations were calculated from the data of C-V measurements, and are shown in Fig. 3. It appears that there is a compensation of n-base material near the p-layer. It seems that the compensated layer is a result of incomplete annealing of defects in the depth. However the rectification of the laser-annealed p-n junctions was rather good. At 1 V the rectification coefficient was $10^5 - 10^6$.

An unexpected result was obtained in the laser annealing of Zn-doped InSb (E = 40 keV, D = $5 \cdot 10^{13}$ cm^{-2}). Instead of formation of a p-type layer, we found an increase of electron concentration at the surface up to 10^{14} cm^{-2}. The initial layer concentration of electrons in the sample was $2 \cdot 10^{13}$ cm^{-2}. All measurements were made at 77°C. The thickness of the layer with increased electron concentration was more than 1 µm.

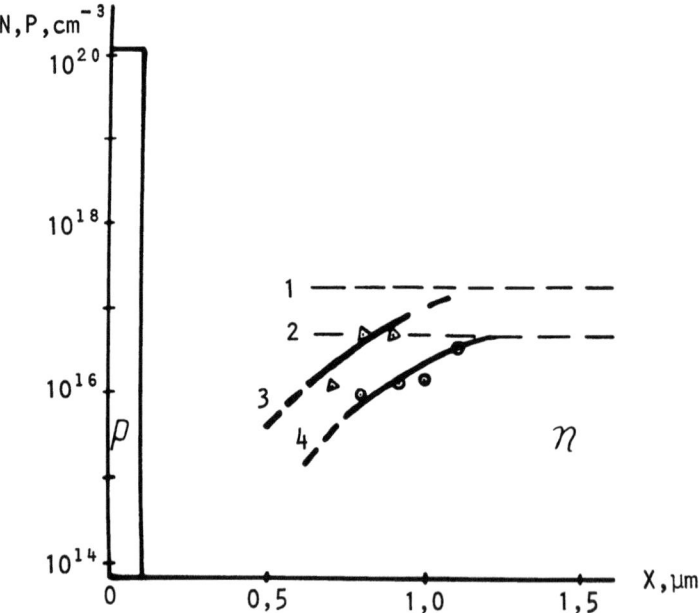

Fig. 3. Distribution of carriers in GaAs p-n junctions fabricated by ion implantation and laser annealing. (1,2) Initial concentration in n-bases; (3,4) results of calculations from C-V measurements.

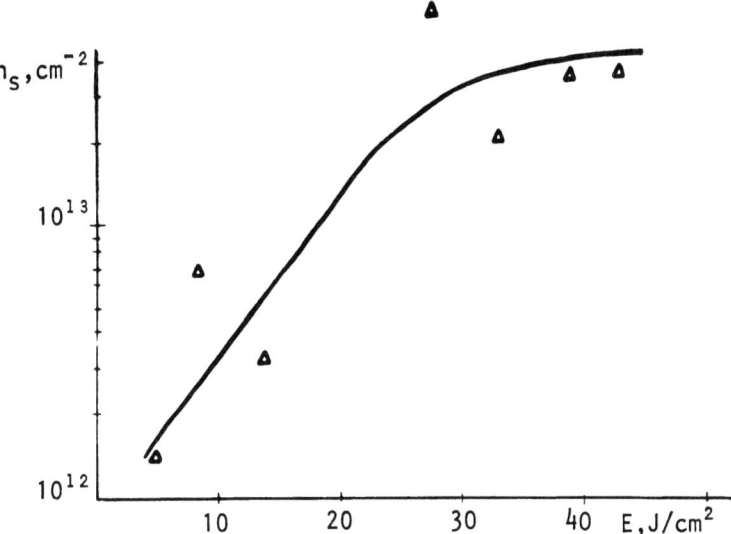

Fig. 4. Surface concentration of electrons in InSb as a function of ruby pulse energy.

The effect of formation of n-layers was investigated on p-InSb using the p-n junctions for isolation of a p-base in Hall effect measurements. In these experiments no ion implantation was carried out. For 5-msec ruby pulses the threshold energy of n-layer formation was 5 J/cm^2. The surface concentration of electrons increased with the laser beam energy and reached saturation for about 30 J/cm^2 (Fig. 4). This increase in electron concentration is mainly due to the broadening of the n-layer but not due to the growth of the volume concentration. Irradiation by the Q-switched Nd-glass laser (τ_{pulse} = 10 nsec) gives more efficient results. The electrical properties of the formed n-layers are stable up to 375 - 400°C.

The change in the conductivity type of the surface layer is not the result of thermal heating by the laser beam, since heat treatment of InSb usually introduces acceptors.[3,4] It is impossible to explain the formation of an n-layer by the decomposition of the surface and evaporation of Sb because the use of the protective films did not decrease the effect. Since we observed the formation of an n-layer well below the melting energies of the laser pulses, we concluded that the appearance of donors was not caused by melting. Also, n-layers were produced on InSb samples with different initial hole concentrations. This means that in this case we are not dealing with electrical activation of uncontrolled residual impurities. Moreover InSb permits very deep purification, and it is therefore difficult to assume the presence of an uncontrolled impurity concentration of up to 10^{18} cm^{-3}.

We assume that the formation of donor centers is a result of the quenching due to rapid cooling of the surface layer after laser

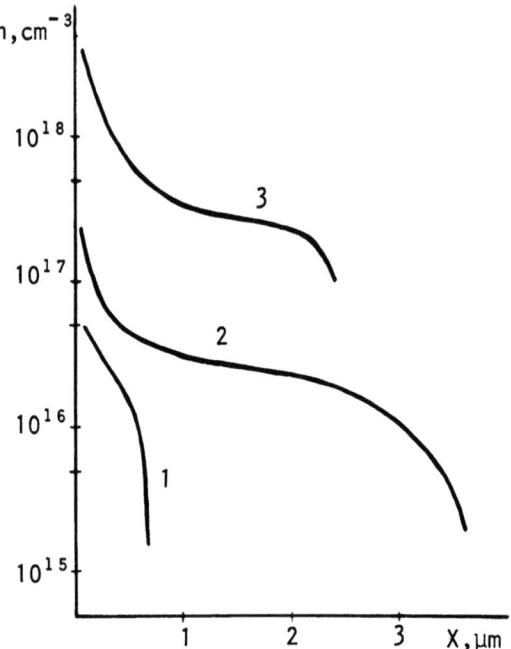

Fig. 5. Depth distribution of electron concentrations in the n-layers on p-InSb. The energies of the ruby laser were (1) 5 and (2) 40 J/cm^2. Curve 3 shows the results of Nd-glass laser irradiation.

irradiation. The formation of donor centers in InSb was observed earlier after bombardment by heavy particles.[5-7] The similarity of the carrier concentration (up to 10^{18} cm^{-3}) and mobility (10^4 cm^2/V·sec) after heavy-particle bombardment and after laser irradiation suggests the same origin of donor centers in both cases. We consider that structure defects resulting from rapid cooling of thermal spikes or laser-irradiated regions are responsible for the observed n-layer formation.

One can conclude that laser annealing of implanted semiconductors may be successfully used in semiconductor technology. The effect of short laser pulses on semiconductors and the mechanism of laser annealing of ion-implanted layers require further study. It is necessary to investigate the role of ionization, shock waves, the latent heat of crystallization, the migration of impurities, and the material composition of the matrix. The coherent interaction of light wave fronts with surface atoms seems to be a significant factor. Recent data prove that the major effect of millisecond laser pulses is ordinary thermal heating. However the role of specific laser effects in the annealing process will increase with decrease in pulse duration and with increase in the power.

References

1. G. A. Kachurin, N. B. Pridachin, and L. S. Smirnov, Fiz. Tekh. Poluprovod., 9:1428, 1975.

2. V. A. Bogatyriov, A. A. Gavrilov, G. A. Kachurin, and L. S. Smirnov, Fiz. Tekh. Poluprovod., 10:1392, 1976.

3. Lian Chi-Chao and D.N. Nazledov, Fiz. Tverd. Tela, 3:1458, 1961.

4. V. V. Gavrushko and O. V. Kosogov, Fiz. Tekh. Poluprovod., 4:2378, 1970.

5. J. W. Cleland and J. H. Crawford, Jr., Phys. Rev. 95:1177, 1954.

6. A. G. Foyt, W. T. Lindley, and J. P. Donnely, Appl. Phys. Lett., 16:395, 1970.

7. K. H. Wiedeburg, H. Betz, and H. Kranz, Phys. Stat. Sol. (a), 31:K69, 1975.

ANNEALING BEHAVIOUR OF PROTON BOMBARDMENT DAMAGE IN P-TYPE SILICON

Amitabh Jain*, B. J. Smith and J. Stephen

Electronics and Applied Physics Division,

Harwell, Oxfordshire, OX11 ORA, England

ABSTRACT

Capacitance-voltage measurements have been made on silicon n^+p diodes which have been implanted with protons at 400 keV. The position of the damage is such that the majority carrier profile in this region is obtained. Measurements made after $50°C$ annealing steps from $50°C$ to $700°C$ show four distinct types of defect. These appear in different temperature ranges. One of these defects which occurs in the annealing temperature range $100°C-300°C$ has been identified as the silicon divacancy. This has been deduced from estimates of the energy of the defect levels and their behaviour at room temperature and at $132°K$. The remaining three defects have been compared with those reported elsewhere but their physical structure is not clearly identified.

INTRODUCTION

In recent years there has been considerable interest in the various damage centres produced in ion irradiated silicon. This has been stimulated by the need to understand the role of damage in silicon devices doped by ion implantation. In this paper we present a study of defects produced in p-type silicon by proton irradiation and their behaviour during annealing. This work should be regarded as complementary to the published work on n-type silicon.

---

*On attachment from Physics Dept., Indian Institute of Technology, New Delhi, 110029, India.

WAFER PREPARATION

Boron doped Lopex silicon wafers oriented in the $\langle 111 \rangle$ direction were prepared by implanting the polished face with 5×10^{14} phosphorus ions/cm^2 at 40 keV. They were then coated with a 1000Å thick layer of r.f. sputtered silicon dioxide and annealed at 800°C for 30 minutes to form a n$^+$p junction approximately .2μ below the surface. The oxide was removed and an array of 500μ diameter mesas was etched over the surface. The wafers were then implanted with 1×10^{12} or 5×10^{11} protons/cm^2 at 400 keV at room temperature and annealed repeatedly at increasingly higher temperatures from 50°C up to 700°C. The normal to each wafer was tilted 7° to the proton beam. The temperature increments used were 50°C and the annealing time was 30 minutes at each temperature. Measurements were made after implantation and also after each anneal.

CAPACITANCE - VOLTAGE MEASUREMENTS

After each annealing stage the distribution of majority carriers in the wafer was measured by the capacitance-voltage (C-V) method. A 1 MHz Boonton type 75D capacitance-conductance bridge was used to measure capacitance. It was modified[1] to permit rapid automatic adjustment of the diode reverse bias voltage to equate the diode capacitance with the bridge setting. By this technique a complete profile could be measured in about two minutes. The voltages were recorded on punched tape, read by a digital computer and the majority carrier profile calculated using standard methods. Kennedy et al[2] have pointed out that the majority carrier distribution does not exactly reproduce the distribution of fixed charge centres in a medium; particularly under conditions of rapid changes of carrier concentration. This is due to departures from charge neutrality that occur under these conditions. Careful checks were made to ensure that charge neutrality is maintained. No serious violations occurred although the minimum of the donor dip discussed later in figure 2 should be lower than shown.

Defects producing a deep acceptor level not more than a few kT above mid gap change their charge state when the level is made to cross the quasi-Fermi level for holes by altering the bias. This produces a peak in the measured carrier profile. On the left hand side of figure 1 this is shown schematically. In this example the deep acceptor level changes its charge state at the distance y from the junction whilst the measured capacitance shows the depletion width to be x. Consequently the apparent position of the deep acceptor at y is displaced by an amount x-y deeper into the material. The amount of displacement is dependent on the energy level of the deep centre.

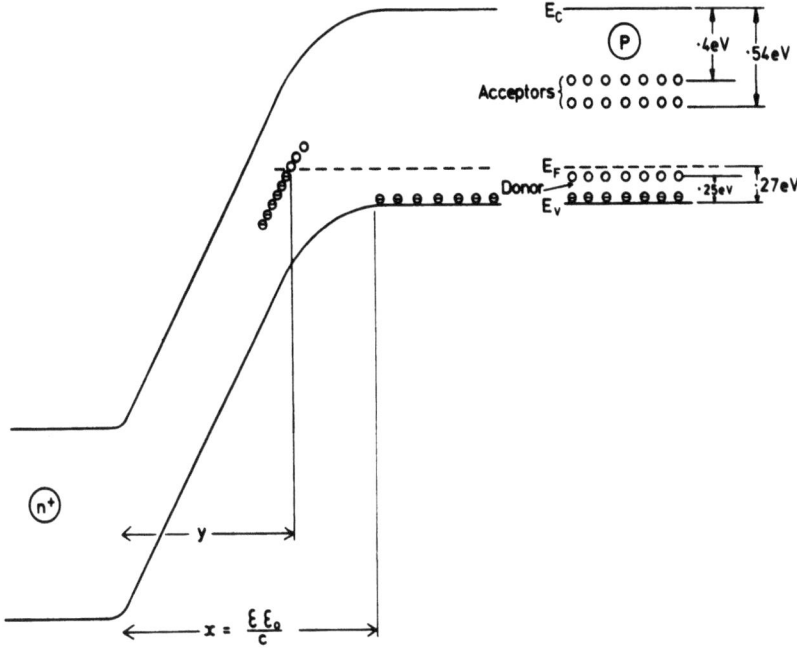

Figure 1. Left hand side: Band diagram of the n⁺p junction showing deep acceptors. Right hand side: Energy levels of the silicon divacancy.

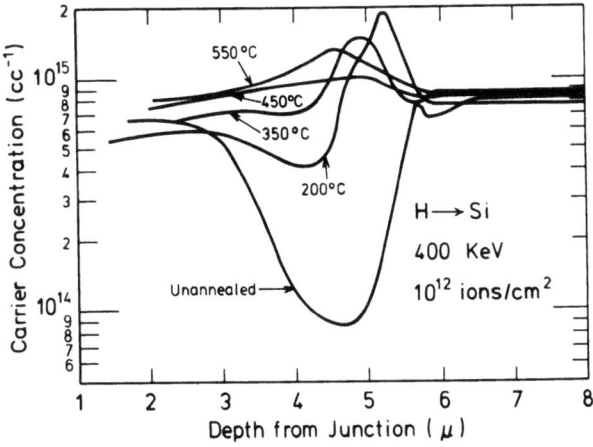

Figure 2. Carrier concentration profiles measured after various annealing steps.

In the case of a localised deep donor level in the bottom half of the band gap the capacitance-voltage profile shows a dip in the net carrier concentration as the positively charged defect enters the depletion region. Then as further depletion occurs the positive charge is neutralised when the level falls below the quasi-Fermi level giving rise to an anomalous peak in the profile. The origin of this peak is discussed in greater detail in the literature[3,4]. If the donor level is clearly in the top half of the band gap the emission probability of electrons to the conduction band is much greater than for holes to the valence band. Consequently the level is unoccupied by an electron even when the reverse bias is large enough for it to fall below the quasi-Fermi level for holes. Therefore the measured profile is simply the total background acceptor concentration minus the localised donor concentration and no anomalous peak occurs. Furthermore as no change of occupancy is required to show the level it does not appear displaced along the depth axis.

RESULTS AND DISCUSSION

The C-V profiles measured on wafers implanted with 1×10^{12} ions/cm^2 after annealing at various temperatures are shown in figure 2. A number of profiles at intermediate temperatures[5] have been omitted for clarity. The essential features of the annealing process can be summarised as follows:

i. In the unannealed state there is a dip at 4.6μ from the junction.

ii. On annealing at 100°C this dip reduces and a deeper peak begins to appear. By 200°C this peak is the predominant feature, and lies at 5.2μ.

iii. Above 200°C the 5.2μ peak diminishes and by 350°C is replaced by another peak at 4.9μ. Between 200 and 350°C both peaks are present together.

iv. The 4.9μ peak begins to anneal out at 400°C and by 450°C leaves an almost flat profile.

v. A new peak, slightly broader than the earlier ones, appears by 550°C at 4.5μ.

vi. At 600°C the size of the peak is reduced and it virtually disappears by 650°C.

The position of the dip in (i) agrees within experimental error[5] with the depth of the damage peak calculated by the method of Manning and Mueller[6] using the electronic stopping cross sections of Northcliffe and Schilling[7].

The interpretation of the profiles is in terms of four separate phenomena which appear in different annealing temperature ranges. The donor activity in unannealed specimens is similar to that observed by Kimerling[8]. On annealing at 100°C the centre is partially removed which coincides with the reported annealing temperature of the positively charged <110> split silicon self interstitial[9]. This defect has been postulated by Frank and Seeger[10] and assigned a donor level at E_v + 0.4 eV. However the centre responsible for the unannealed curve in figure 2 clearly has no anomalous peak associated with it and is therefore in the upper half of the band gap. Therefore it is unlikely to be the Frank-Seeger type of defect.

An alternative explanation is the presence of boron interstitials introduced by the Watkins interstitial conversion mechanism[11]. Boron interstitials appear to produce a deep donor level. However Watkins has also shown that such a defect is likely to be unstable at room temperature in the positively charged state. He also offers explanations for the apparent existence of this centre at room temperature in published studies. A third possibility is that the migration of silicon interstitials would cause some of them to eventually form clusters. Such clusters would have dangling bonds at the edges. These are thought to produce donor activity in p-type silicon[12]. States arising from strain fields at the edges could also remove majority carriers. At present there is insufficient data to conclusively determine which, if any, of these defects accounts for the observed profile.

The profile at about 200°C is attributed to a different doping centre. From the annealing behaviour of this centre the most likely explanation is in the form of a silicon divacancy. This centre appears only after annealing and is probably due to the break up of vacancy clusters. Such clusters would form due to the free migration of vacancies produced by the proton beam and are likely to be electrically inactive[13]. The centre disappears again at around 300°C. This temperature range is consistent with the annealing behaviour of the divacancy reported previously[14,15]. From the displacement in depth of the peak from the position of the damage, given by the dip in the unannealed profile, the position of the energy level in the band gap has been approximately calculated using a technique similar to that used by Schulz[3]. By this method the acceptor level is calculated to be 0.6 $\pm$ .1 eV below the conduction band compared with 0.54 eV reported previously for the divacancy (see for example ref. (12)).

Energy levels reported for the divacancy and the position of the Fermi level, for the doping concentration used, at room temperature are shown in the right hand side of figure 1. The (E_c - 0.54 eV) acceptor accounts for the 200°C profile of figure 2.

The higher acceptor being well above mid gap remains unoccupied by electrons and the lower donor level clearly remains occupied. In order to confirm the presence of the divacancy the wafer was cooled to 132°K in a specially designed cryostat. At the reduced temperature it would be expected that the Fermi level would fall to (E_v + 0.11 eV), therefore crossing the level of the deep donor. Under these circumstances the donor level would contribute to the profile.

Figure 3 shows profiles in wafers implanted with 5×10^{11} ions/cm² and annealed at 200°C. Curves A and B were from measurements at 132°K and curve C was measured at room temperature on the same diode. The lower dose used in this measurement does not produce a qualitative change in the room temperature profile. Curve A was obtained by biasing the p-n junction in the reverse direction, cooling and then measuring in the normal manner. In this case the measurements were made as the junction bias was reduced. It is noted that an anomalous peak is produced. This is clearly due to the donor level becoming occupied at a point closer to the junction

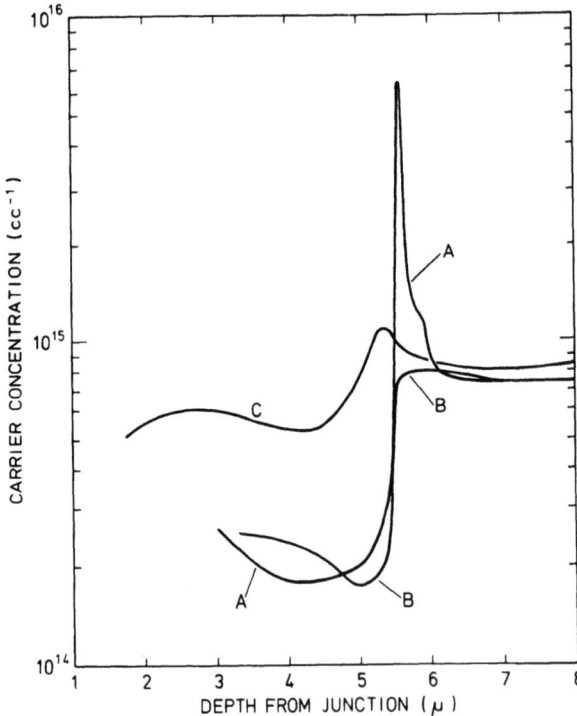

Figure 3. Carrier concentration profile after annealing at 200°C. A - Cooled to 132°K under maximum reverse bias and measured as bias is reduced; B - Measured at 132°K starting at forward bias; C - Measured at 300°K.

than the depletion edge, and is a classic example of case 2b in ref. (4). Curve B shows the effect of forward biasing the junction immediately before each measurement. The level quickly becomes unoccupied in forward bias but requires several minutes to become occupied again on applying reverse bias. Consequently the deep donor level remains unoccupied, and as there is no change in the occupancy no anomalous peak occurs. It would need further measurements at higher temperatures to determine the energy of the donor level. However the existence of such a level and its general position relative to the Fermi level at room temperature and at 132°K is consistent with the silicon divacancy. The acceptor peak is absent from curve B since the level would remain unoccupied during the measurement. The possible disappearance or reduction in size of this peak in curve A suggests that the energy level is in the top half of the band gap, but close to mid gap. Normally an acceptor level in the top half of the gap would be unoccupied, but within a few kT of mid gap it could have a reasonable probability of being occupied with sufficient reverse bias. At reduced temperature kT is smaller and therefore the probability of occupancy is reduced. The level would only contribute to the profile if it is occupied in reverse bias.

The peak at 350°C is less well understood. Clearly there is an acceptor state that begins to form at about 250°C. One possibility is that of a divacancy-oxygen complex. Such a defect has been reported by Kozlov et al[16]. They reported that divacancies can diffuse at temperatures around 300°C to produce a complex with oxygen that gives rise to a level at 0.35 eV above the valence band which is consistent with the estimate from the peak displacement in figure 2. The centre is reported as stable up to about 420°C when it anneals out. This is in good agreement with the annealing characteristics reported here. Some doubt must be expressed about this interpretation of the present results as the silicon used was of low oxygen concentration. However it is still likely that sufficient oxygen is present to produce this defect as the oxygen concentration was probably similar to that used by Kozlov.

The centre responsible for the peak in the 550°C curve is not understood. It may be related to the acceptor which has been observed to form at similar high temperatures (Lappo et al[17]), and then decay at slightly higher temperatures.

CONCLUSIONS

Proton irradiation produces defects in silicon which change their nature on annealing. Four distinct types of defect have been observed, each with its own electronic properties. One defect has been identified as the silicon divacancy. Other defects have been compared with those reported by other authors but not unambiguously identified. The damage finally anneals out at about 650°C.

ACKNOWLEDGEMENTS

We are grateful for helpful discussions with Drs G Dearnaley, J A Grimshaw, A H Harker, D W Palmer (and his research group), A M Stoneham and D H J Totterdell. We wish to express our thanks to Dr M D Matthews for help with calculations of damage profiles and to Mr M A Wilkins for the implants. One of us (AJ) is grateful to Prof A B Bhattacharyya for his interest and encouragement.

REFERENCES

1. B J Smith, J Stephen and P J Hammersley, Radiat. Effects 26, 17 (1975)
2. D P Kennedy, P C Murley and W Kleinfelder, IBM J. Res. Develop. 12, 399 (1968)
3. M Schulz, Appl. Phys. Lett. 23, 31 (1973)
4. L C Kimerling, J. Appl. Phys. 45, 1839 (1974)
5. A Jain, Ph. D. Thesis, Indian Institute of Technology, Delhi; in preparation
6. I Manning and G P Mueller, Comput. Phys. Commun. 7, 85 (1974)
7. L Northcliffe and R Schilling, Nucl. Data Tables 7, Nos. 3-4 (1970)
8. L C Kimerling and J M Poate, Lattice Defects in Semiconductors, 1974, ed. F A Huntley (The Institute of Physics), 126 (1975)
9. S I Tan, B S Berry and W F J Frank, Ion Implantation in Semiconductors and Other Materials, ed B L Crowder (Plenum), 19 (1973)
10. A Seeger and W Frank, Radiation Damage and Defects in Semiconductors, ed J E Whitehouse (The Institute of Physics, London), 262 (1973)
11. G D Watkins, Phys. Rev. B 12, 5824 (1975)
12. B Henderson, Defects in Crystalline Solids, (Arnold, 1972)
13. A J R de Kock, Appl. Phys. Lett. 16, 100 (1970)
14. L J Cheng, J C Corelli, J W Corbett and G D Watkins, Phys. Rev. 152, 761 (1966)
15. H J Stein, Radiation Effects in Semiconductors, eds J W Corbett and G D Watkins (Gordon and Breach), 125 (1971)
16. I P Kozlov, A G Litvinko, P F Lugakov, S V Mishuk and V D Tkachev, Sov. Phys. - Semicond. 6 1743 (1973)
17. M T Lappo and V D Tkachev, Sov. Phys. - Semicond. 5, 1411 (1972)

RESIDUAL DAMAGE IN SILICON IMPLANTED AND POST-ANNEALED SILICON

L. D. Glowinski, P. S. Ho and K. N. Tu

IBM Thomas J. Watson Research Center

Yorktown Heights, New York 10598

ABSTRACT

TEM was used to study lattice defects in silicon wafers which were first implanted with 10^{16} Si ions/cm^2 to produce an amorphous surface layer and then annealed to produce epitaxial regrowth. The effects of doping (n-type or p-type), ion energy and irradiation temperature have been studied. A highly damaged transition layer located between the amorphous layer and the undamaged silicon was observed in all samples. The density, distribution and characteristics of the lattice defects have been determined in the transition layer and the regrown layer. Some factors in controlling the recovery annealing of the implanted damage are discussed.

INTRODUCTION

In using ion implantation to fabricate silicon devices, heavy implantation at a level sufficient to form an amorphous layer are often required. The amorphous layer can be recrystallized epitaxially by thermal annealing but some residual lattice damage has been observed to remain in the implanted region [1-3]. The investigation of the annealing behavior and the structure of the residual damage has drawn considerable interest lately. In a recent publication [4], we reported a TEM study on the defect structure produced by implanting 270 keV Si ions at room temperature into a p-type Si wafer. The self ions were used to avoid impurity and precipitation effects caused by the implanted ions. Using stereomicroscopy, we observed the defect structure to consist of two regions, one an amorphous layer extending from the surface to a depth of about 5000Å and the other, located below the amorphous layer and about 500-600Å thick, was highly damaged but did not turn amorphous. The latter which we called as the transition layer has been found earlier by internal friction measurements [5]. Its existence was also confirmed recently by Csepregi et al. [6] using channeling measurements with He ion backscattering techniques. However, these authors reported that the transition layer was not detected in samples implanted at liquid nitrogen (LN_2) temperatures.

Upon annealing the epitaxial regrowth of the amorphous layer was observed to initiate from the transition layer. And the defects in the transition layer developed into a high density of interstitial loops which accounted for most of the residual damage. The transition layer is expected to be important for controlling the recovery annealing of the implanted damage. In this paper we report the results in a sequel study using TEM on the defect structures and their annealing behavior in self-ion implanted n-type silicon wafers. Here we study the effect of ion energy and irradiation temperature. By comparing the results with the previous study on p-type substrates we also observe effects due to the doping impurity in the substrate.

EXPERIMENTAL PROCEDURE

The samples were n-type 10 to 20 Ω-cm silicon wafers of (001) orientation grown by the Czochralski method. They were implanted with 270 keV or 80 keV Si ions to a dose of 1×10^{16} ions/cm^2. After implantation the wafers were cut into squares of 2mm×2mm. Some samples were kept for TEM study of the as-implanted condition or for in-situ annealing experiments; others were annealed in a purified helium atmosphere at either 600°C for 2 hr. or 800°C for 1 hr. These annealing conditions were chosen for observing the initial regrowth of the amorphous layer and the development of the defect structure in the transition layer respectively. TEM specimens were prepared by jet etching from the unimplanted side with a mixture of 1 part HF and 10 parts HNO_3.

The defect clusters were analyzed using image contrast techniques [7,8]. The depth distribution of the defects were determined by stereomicroscopy. The angle between the two pictures of the stereo pair has been calculated using a Kikuchi map. The accuracy of the measurement is estimated to be 10%.

EXPERIMENTAL RESULTS

1. Effect of the Irradiation Temperature

a. As-irradiated and 600°C annealing. To study the effect of the irradiation temperature, the substrate during implantation was kept at either room temperature or at LN_2 temperature and an ion energy of 80 keV was used in both runs. We found that the implantation produced a surface amorphous layer extending to 1600 Å at room temperature and 1900Å at LN_2 temperature. The transition layer was observed at both temperatures but the thickness was different. For RT implantation, it extended roughly 250Å below the amorphous layer while for LN_2 implantation its thickness was at most 50 Å. For the latter even though it was quite thin, its presence was clearly observed. (Figs. 1a and 1b)

After an anneal at 600°C for two hours, both amorphous layers recrystallized epitaxially over the range of implantation although the overall residual damage was very different as one can see by comparing the micrographs in Figs. 1c and 1d. For the RT implant, the transition region contained a high density of defect clusters which were too dense to be resolved. The recrystallized layer contained a density of 2 to 3×10^8 per cm^3 of large half loops with their apex at the transition layer and extending all the way to the sample surface. The nature of the half loops has been identified to be of the interstitial type and their characteristics were identical to that observed in p-type Si [4]. For the LN_2 implant, several TEM samples were examined but no defect clusters in the recrystallized layers were observed at this stage of annealing (Fig. 1d). There appeared to have some defect clusters remaining in the transition layer but their size was not sufficient for character identification.

b. **800°C annealing**. After annealing at 800°C for 1 hr., the defect structure in the transition layer of the RT implanted sample developed into a two-dimensional dislocation network parallel to the sample surface and its position remained at 1600Å below the irradiated surface (Fig. 1e). All six $\frac{1}{2}<110>$ Burgers vectors were found in this network and the loops can be identified to be interstitial in nature. In the recrystallized layer, the general configuration and the density of the half loops have remained virtually unchanged, so they appeared not to be affected by this stage of annealing. Some planes of precipitates can now be found in the recrystallized layer. The lattice parameter of the precipitates was determined using the Moire fringes obtained with the 220 and 400 diffraction vectors. It was found to fit the lattice parameter of the β-CuSi. The same type of precipitate has been found in non-implanted Si subjected to similar annealing and cooling sequences. and was attributed to the incorporation of Cu during wafer preparation [9]. They have been studied extensively and we will not discuss them further.

The samples implanted at LN_2 contained a low density (2×10^8 per cm^3) of loops (Fig. 1f). Their Burgers vectors are either $\frac{1}{2}<110>$ or $\frac{1}{3}<111>$. All loops identifiable were of the interstitial type. They were not confined to the transition layer but extended

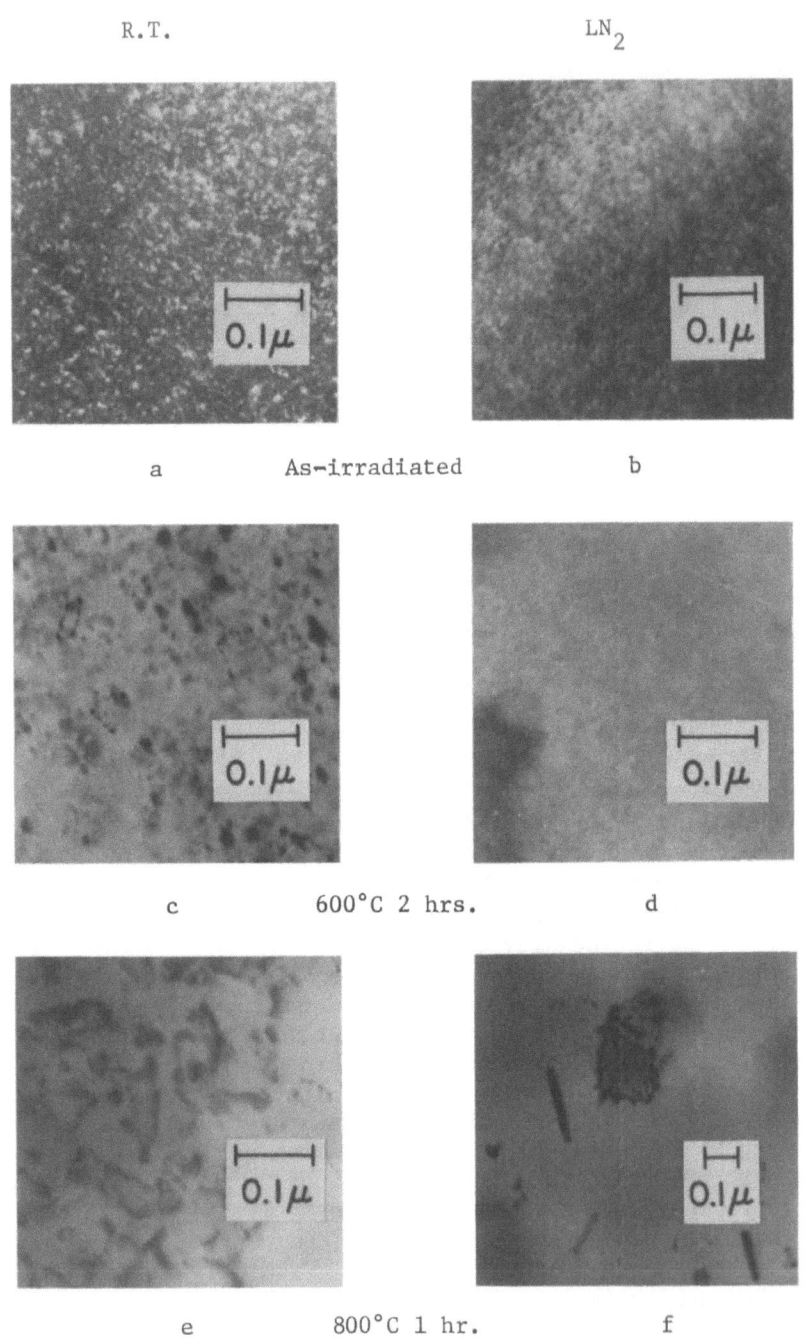

Fig. 1. Defect structure in n Si implated with 10^{16} cm^2 of 80 keV Si ion at R.T. and LN$_2$.

1500Å above and below a plane situated 2500Å below the irradiated surface. Most of the faulted loops were elongated along a <110> direction. Planes of β-CuSi precipitates were also present in those samples.

Since no defect clusters were seen in samples implanted at LN_2 and annealed at 600°C for 2 hrs., some samples were subjected to a 600°C anneal for 2 hrs. followed by an 800°C anneal for 1 hr. The results are identical to those obtained after a straight 800°C anneal.

2. Effect of the Ion Energy

This experiment was done by comparing the results from 80 keV implantation with those obtained by implanting with 270 keV ions; all other conditions were kept constant.

After an implantation of 270 keV at R.T., the surface amorphous layer was found to be about 5000Å thick and the transition region about 500Å thick. A 600°C anneal for 2 hrs. recrystallized the amorphous layer and induced the formation of a very high density of loops in transition regions (Fig. 2a). An 800°C anneal for 1 hr. did not change the defect structure in the recrystallized layer but developed the defect structure in the transition region into two planar networks of dislocations, both parallel to the irradiated surface. They were found to be 500Å apart and inter-connected by dislocation lines. (Fig. 2b)

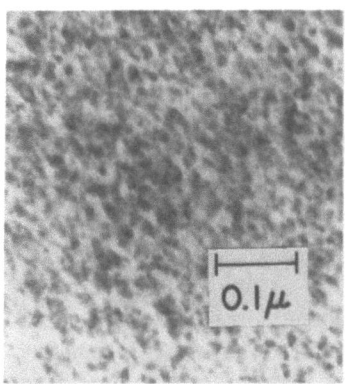
a. 600°C 2 hrs.

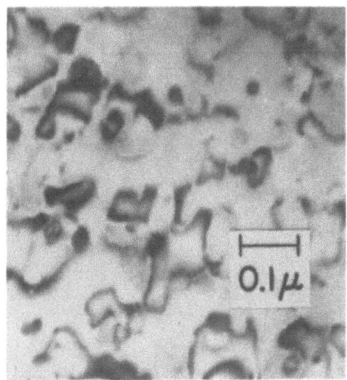
b. 800°C 1 hr.

Fig. 2. Defect structure in n Si implanted with $10^{16}/cm^2$ of 270 keV Si ions at room temperature.

DISCUSSION

We observed that changing the irradiation temperatue from RT to LN_2 has the effect of increasing the thickness of the amorphous layer while decreasing the thickness of the transition layer, a trend consistent with channeling results [6]. Increasing the ion energy from 80 keV to 270 keV has the effect of increasing the amorphous layer thickness from 1600Å to 5000Å and the transition layer thickness from 250Å to 500Å. The observed depth distribution of the implanted damage can be correlated to the energy released into atomic processes as calculated by Brice [10]. In Fig. 3, we plot the damage energy density computed according to Brice's calculation and the implanted ion intensity. The amorphous ranges observed are in good agreement with the predicted ion ranges and for the two implantations at RT, they correspond to an energy density of about 3.5×10^{24} eV/cm^3. This value exceeds 10^{24} eV/cm^3 which is generally believed to be sufficient for the conversion into an amorphous state [3]. In contrast, the amorphous range at LN_2 corresponds to a lower damage density of about 10^{24} eV/cm^3. The temperature effect observed here can be attributed to a higher degree of annealing of the lattice damage during implantation at RT. Judging from our results, the critical dose is about three times higher for RT, indicating that the kinetics of defect annihilation must be correspondingly faster at that temperature.

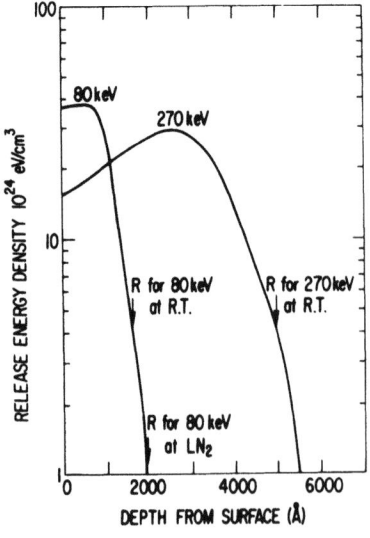

Fig. 3. Correlation of the theoretical energy density released into atomic processes [10] and the observed range R of the amorphous layer.

Since the observed range of the amorphous layer measures the depth within which the damage energy is sufficient for amorphous state conversion, the energy released beyond this range would be responsible for forming the transition layer. In this way, we conceive that the presence of the transition layer where the lattice is highly damaged but still crystalline will necessarily accompany the formation of the amorphous layer. In addition, the resultant thickness of the amorphous and transition layers would depend very much on the mobility of the defects. With higher defect mobility at RT, one expects a decrease in the thickness of the amorphous layer with a corresponding increase in the thickness of the transition layer. These predictions are consistent with our observations.

It should be noted that the channeling measurement [6] did not find the transition layer in a 100 keV Si implantation at LN_2 at doses up to a few times of 10^{16} ions/cm^2. We have no explanation for the discrepancy although there may be some question on the ability of the channeling technique in detecting a highly disordered layer of only about 50Å underneath an amorphous layer. More experiments would be needed to clarify this situation.

We observed that the irradiation temperature has significant effect on the residual defect structure after 600°C and 800°C anneals. While at RT, the defects after 600°C annealing segregate into numerous dislocation loops in the transition layer, which then develop into a two-dimensional dislocation network after subsequent annealing at 800°C. In contrast, annealing of the LN_2 implanted sample at 600°C reveals neither large half loops in the amorphous layer nor dislocation loops in the transition layer and a subsequent 800°C anneal produces faulted loops of quite different geometry. Mader and Michel [11] in a TEM study of lattice damage in 80 keV RT As^+ implanted Si found also the existence of the transition layer filled with small interstitial loops and large half loops similar to our observations. The half loops were attributed to be misfit dislocations formed for releasing the built-in compressive stress due to the implantation. These authors have considered the mechanism for formation of the 2-D dislocation network (meandering loops) which they observed in Si after annealing of "through oxide" As^+ implants. They reasoned that the oxygen atoms recoiled into the implanted layer can cause compressive stress to counteract the climb force of dislocations and also generate interstitial Si atoms during annealing by replacing them at the lattice sites. Under these conditions, the growth of the prismatic interstitial loops (with $\frac{1}{2}$<110> type Burgers vectors) is facilitated and as a result of lateral climb interaction, the loops will eventually develop into a network parallel to the surface and with only Burgers vectors parallel to the loop plane. This mechanism can be used to explain our results. Although we did not implant through an oxide layer, the implantation of Si ions, instead of As ions, generates numerous self interstititals and the Czokralski wafer contains oxygen concentration as high as 10^{18} cm^{-3}. It seems that the combination of these factors was sufficient for developing the meandering dislocation networks. The fact that we still observed loops with Burgers vectors not parallel to the surface is probably due to the lower annealing temperature (800°C instead of 1000°C) inhibiting complete dislocation reactions. It follows that the network formation is relatively independent of the ion energy and ion type as long as a sufficient amount of implanted damage and oxygen are present.

By comparing the results on 270 keV implantation in n-type substrate to our previous results on p substrate, we found that changing the substrate type has no observable effect on the formation and the recrystallization of the surface amorphous layer but does modify the rate of development of the defect structure within the transition region. For 270 keV implants, after the 600°C anneal, small resolvable loops are observed in n-type samples and only small defect clusters in p-type samples. After annealing at 800°C for 1 hr., dislocation tangles are observed in n-type samples while individual loops are observed in p-type. So the annealing kinetics of the defects seems to be faster in n Si than in p Si. This could be due to several causes: a) the nucleation of dislocation loops is easier in n than in p. b) The concentration and/or the diffusivity of the interstitials are higher in n than in p. c) The mobility of the vacancy is different due to the different charge states, thus affecting the overall balance in the annealing kinetics of interstitials and vacancies. d) The oxygen atom in Czokralski Si wafers can form a different defect complex with n and p dopants to influence the annealing kinetics. At present it is difficult to ascertain which of these factors or which combination is operating. It seems nevertheless that loop nucleation is easier in n than in p material. The very high density of loops observed in n type material during the first few minutes of annealing at 800°C made impossible any reliable measurement of the loop density but it is probably an order of magnitude higher than in p-type. This when combined with slower growth rate in p-type may explain the fact that loops do not conglomerate into a dislocation network in p Si but do so in n Si.

REFERENCES

1. B. L. Crowder, R. S. Title, M. H. Brodsky and G. D. Pettit, Appl. Phys. Lett. 16, 205 (1970).
2. J. W. Mayer, "Radiation Damage and Defects in Semiconductors," ed. by J. E. Whitehouse, The Institute of Physics, London, (1972), p. 72.
3. F. L. Vook, ibid, p. 60.
4. L. D. Glowinski, K. N. Tu and P. S. Ho, Appl. Phys. Lett. 28, 312 (1976).
5. S. I. Tan, B. S. Berry and B. L. Crowder, Appl. Phys. Lett. 20, 88 (1972).

6. L. Csepregi, E. F. Kennedy, S. S. Lau and J. W. Mayer, to be published in Appl. Phys. Lett. (1976).
7. P. B. Hirsch, A. Howie, R. B. Nicholson, D. W. Pashley and M. J. Whelen, "Electron Microscopy of Thin Crystals," Butterworths, Washington (1965).
8. D. H. Maher and B. L. Eyre, Phil. Mag. $\underline{23}$, 404 (1971).
9. E. Nes and J. Washburn, J. Appl. Phys. $\underline{42}$, 3559 (1972); ibid, 3562 (1972).
10. D. K. Brice, Rad. Effects $\underline{6}$, 77 (1970).
11. S. Mader and A. E. Michel, to be published in J. Vac. Sci. Technol. (1976).
12. S. Mader and A. E. Michel, to be published.

RECOVERY OF RADIATION DAMAGE BY PHOSPHORUS IMPLANTATION IN SILICON: T.E.M. AND PROTON BACK-SCATTERING ANALYSIS.

F.Cembali, R.Galloni, M.Servidori, F.Zignani[+]

Laboratorio di Chimica e Tecnologia dei Materiali

e dei Componenti per l'Elettronica, C.N.R.

Bologna, Italy

ABSTRACT

The radiation damage produced by room temperature implantations of 200 keV phosphorus ions into silicon targets and its recovery processes in the temperature range 100°C-900°C have been studied by comparison of proton back-scattering and transmission electron microscopy results. Experimental data obtained on samples implanted under channeling conditions along the [110] axis have been compared with those obtained on same energy random implantations.

Two stages are observed in the damage formation: at low implanted doses only a high density of point defect clusters are observed by the dark-field weak-beam technique. When the fraction of atoms displaced reaches values between 40-50%, in both aligned and random experiments, the appearance of amorphous zones is identified by the electron diffraction pattern. The crystalline-amorphous transition is obtained for implanted doses of about 7.5×10^{14} at/cm^2 in case of channeling implantations and about 2×10^{14} at/cm^2 in case of random implantations.

From the experimental results collected we can conclude that each implanted phosphorus ion produces point defects

---

[+] Istituto Chimico, Facoltà di Ingegneria, Università di Bologna

along its path in the crystal that cluster by annealing at room temperature; by increasing the dose of implantation amorphous zones begin to nucleate where the concentration of point defects reaches the critical value. The suggested model for the crystalline-amorphous transition is based on that by Breitling and Richter extended by Grigorovici relative to the amorphous silicon configuration. In case of implanted silicon, interstitial atoms and vacancies, due to radiation damage, make possible the formation of new bonds rotated with respect to the normal tetrahedral positions when their concentration is sufficiently high. Partial and total dislocation loops are formed by annealing of point defect clusters and amorphous zones.

INTRODUCTION

Our studies of the damage produced during ion implantation by the implanted species are mainly aimed to the end of obtaining the highest electronic performances of the implanted layers which are needed in device technology. Several studies have already been made[1,2,3,4,5] to correlate the annealing of the damaged crystals to the electrical activity, mobility, lifetime etc.; we think now that a more detailed knowledge of the kind of defects actually generated is essential if an almost defect free sample has to be obtained after annealing. Two major aspects can be therefore singled out in the study of radiation damage produced by ion implantation: a) dose dependence of the kind and amount of damage produced; b) evolution of damage during the annealing and identification of residual lattice defects. It is well known that the kind of damage is a function of dose, dose rate, implant temperature etc., but mainly of ion mass, heavy ions being responsible for the production of amorphous regions whereas light ions only produce crystalline damaged regions with a high concentration of point defects.

In this work we have analysed the evolution during annealing of the damage produced by the room temperature implantations of phosphorus ions into silicon, both in channeling and random conditions, because phosphorus is of fundamental importance in electronic applications and its mass, nearly equal to that of the Si target, cannot be

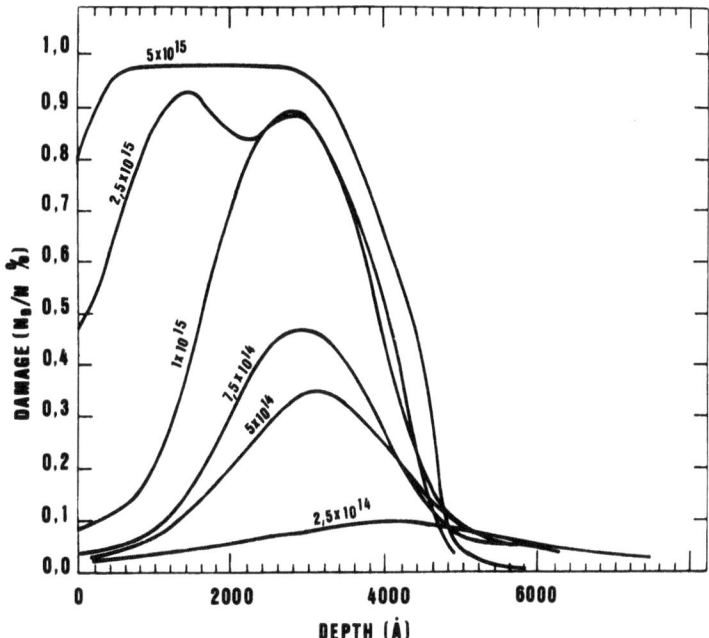

Fig. 1 - Damage distribution vs. depth for different doses of implantation of phosphorus ions/cm^2 along the [110] channel of silicon samples.

considered either heavy or light. To this end we have compared proton back-scattering analysis and transmission electron microscopy (T.E.M.) observations as these techniques can give complementary information: back-scattering can evaluate the total damage produced by the displacement of the atoms from their lattice sites, but does not give information on the nature of the defect; T.E.M. cannot instead detect isolated point defects but can only give quantitative information on extended defects as dislocations, stacking faults, dipoles or also point defect clusters, etc., that are normally formed as a consequence of point defect diffusion and interaction during annealing.

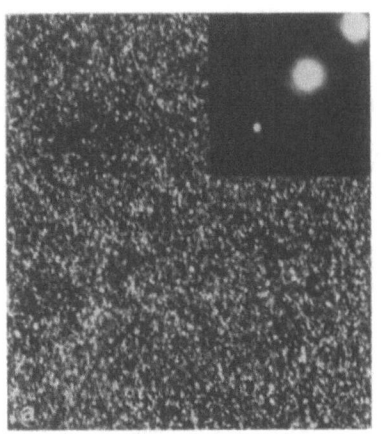
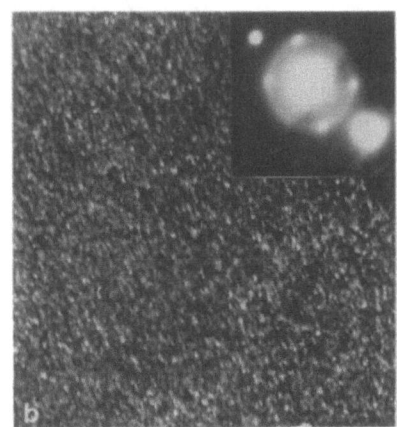

0,1 µm

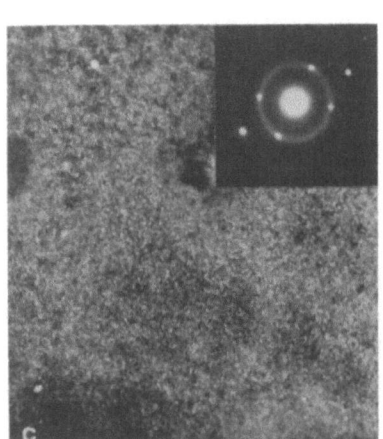
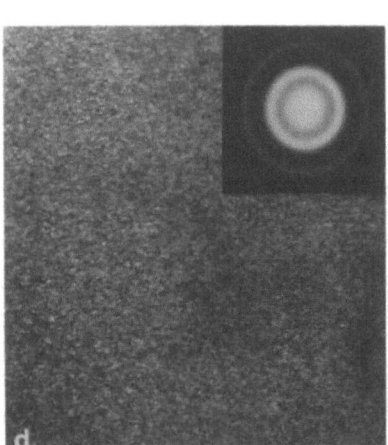

Fig. 2 - T.E.M. micrographs and electron diffraction patterns of Si samples implanted at room temperature with different doses of phosphorus ions along the [110] axis: a) 5×10^{14} P^+/cm^2; b) 7.5×10^{14} P^+/cm^2; c) 1×10^{15} P^+/cm^2; d) 5×10^{15} P^+/cm^2.

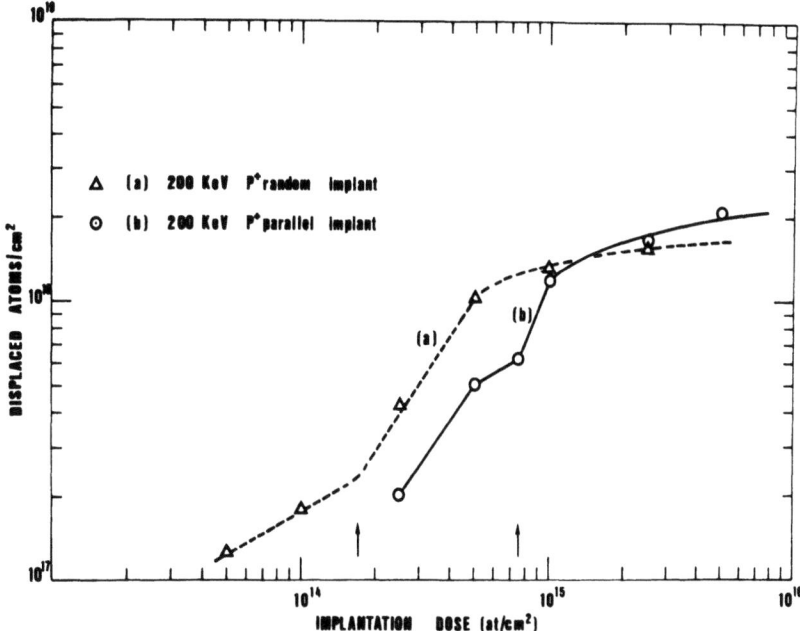

Fig. 3 - Total damage produced by room temperature implantations of P^+ ions in Si samples vs. dose: b) channeled implantations along [110] axis; a) random implantations.

EXPERIMENTAL TECHNIQUES

Silicon wafers, [110] orientation, resistivity ~ 500 ohm cm, floating zone, p-type, dislocation free by SMIEL-MONTEDISON (Merano Italy) have been used in our experiments. The samples have been chemically cleaned before implantation. The radiation damage has been produced by implantation of 200 keV phosphorus ions along the [110] axis, or random at several doses between 5×10^{13} and $5 \times 10^{15} P^+/cm^2$.

Implanted samples have been annealed in vacuum (10^{-7} torr) for 1 h. at increasing temperatures in the range 100°C-900°C. After each annealing treatment samples have been analysed by back-scattering and by T.E.M.

The back-scattering analysis has been made by 300 keV protons aligned along the [110] axis and the damage depth distribution has been obtained by elaboration of the

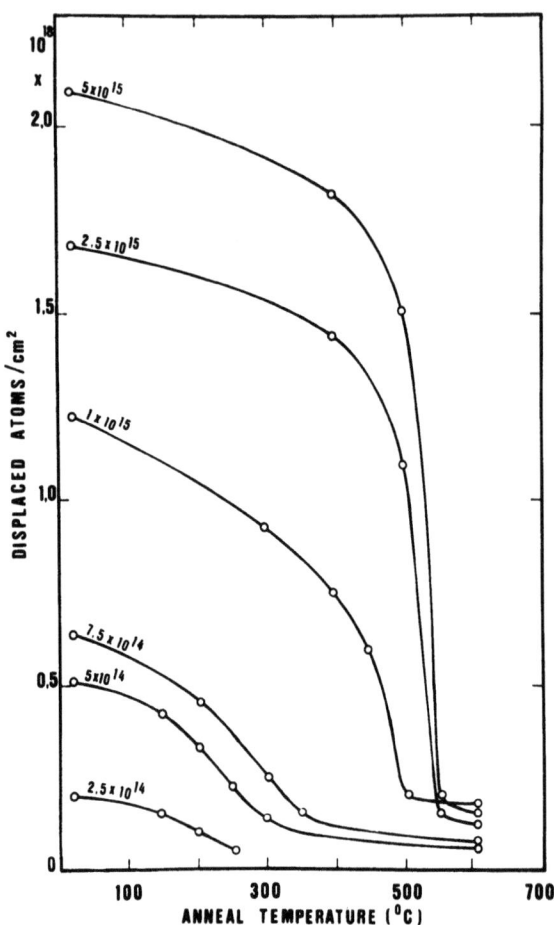

Fig. 4 - Channeling implantations of phosphorus along the [110] axis. Isochronal annealing of the total damage by integration of the damage depth distributions given by back-scattering analysis.

spectra by the usual Ziegler method[6].

The samples to be used in T.E.M. observations have been thinned on the back surface by standard jet thinning technique at low implant doses and by ion beam etching at high doses. In the samples with parallel implantations layers 2000 Å thick have also been stripped from the front surface to get the transparent layer approximately centered on the maximum of the damage distribution. T.E.M. observations have been made mainly with the dark-field or the dark-field weak-beam technique when amorphous regions or crystalline damaged regions had to be observed, respectively. Such techniques have proved to be the only ones able to give satisfactory results whenever defect structures so small had to be evidenced.

RESULTS AND DISCUSSION

Observations of a_s implanted samples

The damage distribution detected by back-scattering analysis as a function of the dose of phosphorus implanted along [110] axis is shown in Fig. 1. As the fraction of ions that stop near the surface grows by increasing the dose, due to the damage that builds up and acts as a de-channeling factor the detected damage distribution grows accordingly and extends to the surface. A similar trend for increasing doses has been observed in the case of same energy random implantations, but the damage peaks are shifted by about 1000 Å toward the surface and the doses needed to reach equal damage are lower; the differences between random and channeling almost disappear at higher doses as the de-channeling factor due to the crystal damage becomes dominant.

In case of channeling implantations electron diffraction patterns (Fig. 2) after the room temperature implantations show a transition in the crystalline structure of the sample between 5×10^{14} and 7.5×10^{14} P^+/cm^2: by increasing the dose diffraction spots produced by the perfect crystal lattice are replaced by continuous Debye rings. T.E.M. observations with the technique of dark-field weak-beam where crystal spots are present, and dark-field by the 1st Debye ring at higher doses, show the well known point-like microstructures. A similar trend for increasing doses has been observed in case of random implantations, but here the transition is at about 2×10^{14} P^+/cm^2 dose. As will be further evidenced, the microstructure shown is attributed to small clusters of point defects where diffraction spots are present, and to small amorphous zones (~ 30 Å diameter) where the diffraction pattern shows continuous Debye rings.

The total damage, measured by integration of the

damage distribution by back-scattering experiments, vs.the
dose of implantation is shown in Fig. 3. Two stages are
visible in the curves which we interpret as due to the increasing of point defect cluster density and to increasing
of amorphous zones going from the lower to the higher
implanted doses.The knee of both the curves, for random
and for channeling, corresponds in fact to the critical
doses for amorphisation (7.5×10^{14} P^+/cm^2 channeling,
2×10^{14} P^+/cm^2 random) in agreement with T.E.M. observations.
The difference in the saturation value of the total damage
of Fig. 3 is due to the fact that in the channeling case
a thicker crystal layer is damaged. The curve from the
random experiments in fig. 3 is also in very good agreement with the trend shown by the curves obtained by
Baranova et al.[7] by optical methods.

Damage Annealing

The isochronal annealing of the total damage, obtained
by integration of the damage depth distributions given by
back-scattering analysis, is shown in Fig. 4 for the channeling implantations. It can be observed that at low doses
$\lesssim 7.5 \times 10^{14}$ P^+/cm^2, at which amorphous zones have not been
formed yet, the recovery of the crystal is gradual and at
low temperature (between 200°C and 350°C). T.E.M. observations (Fig. 5) show that the point structure observed,
attributed to point defect clusters, is still present after
550°C annealing, but with point dimension sensibly grown.
After 700°C annealing, Frank dislocation loops are observed (Fig. 5) but they are still very small. Back-scattering
and T.E.M. observations are not in disagreement if the
following interpretation is given: the recovery of the
back-scattering yield is only partly produced by vacancy-
-interstitial recombination but mainly by the new configuration of the interstitial atoms in Frank dislocation
loops; it can be supposed in fact that the contribution to
back-scattering yield of the strain around a dislocation
line in a loop is much lower than that relative to the
clusters of interstitial atoms randomly distributed in the
crystal before loop formation.

At doses higher than 7.5×10^{14} P^+/cm^2 the total damage
isochronal annealing curves of Fig. 4 show the presence of

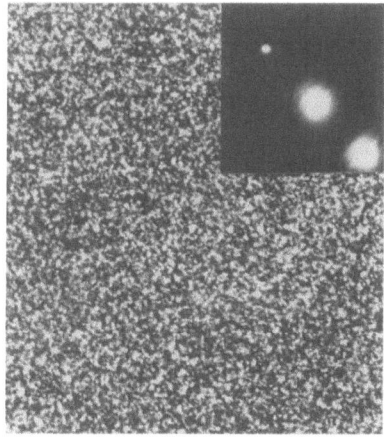
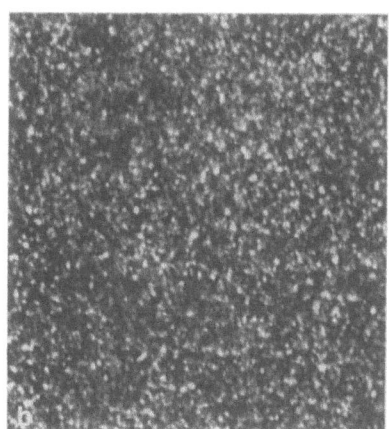

0.1 μm

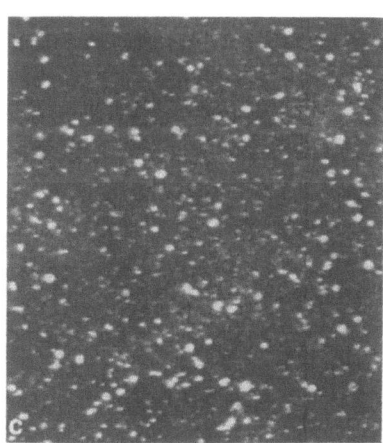
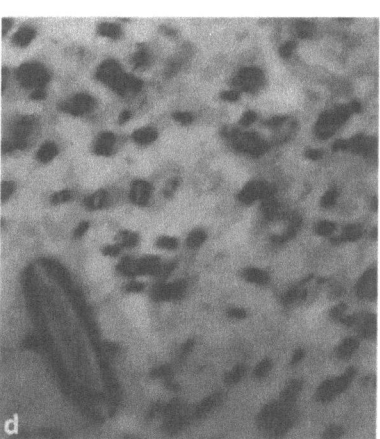

Fig. 5 - T.E.M. micrographs and electron diffraction patterns of Si samples implanted with $5 \times 10^{14} P^+/cm^2$ along the [110] axis for different annealing temperatures; a) 450°C; b) 550°C; c) 700°C; d) 900°C. Micrographs a, b, c, are obtained with the dark-field weak-beam technique, micrographed with the 220 bright-field.

a very steep decrease in damage between 500°C and 550°C
that can be correlated with the recovery of amorphous
regions. Due to the rapidity of the recovery process it
is not possible to confirm from our back-scattering data
whether or not the recovery follows as epitaxial regrowth
of the amorphous on the underlying perfect crystal.T.E.M.
observations shown in Fig. 6 confirm the rapid recovery
of the crystal between 500°C and 550°C. The amorphous
recrystallization leaves behind mainly Frank dislocation
loops, V shaped dislocations similar to those already
observed by Glowinski et al.[8], dipoles, etc. Between
700°C and 900°C annealing the diameters of the dislocation
loops increase, and then as usually observed, partial
loops become total.

The experimental results shown confirm that phosphorus ions implanted into silicon single crystals produce a damage that may be considered intermediate between that produced by heavy and light ions: in this case in fact we can see that the number of displacements in the single cascade is not high enough to destroy long range order of the lattice; only when the concentration of point defects reaches a critical value amorphous regions begin to be revealed. If we assume an average value for the nuclear stopping power at energies below 200 keV we find in fact, from a rough estimation[1,10], that only very few atoms are displaced by each primary recoil.

As the implantation is performed at room temperature, at which both self-interstitials and vacancies are highly mobile[11], it is reasonable to think that clusters are formed by the coalescence of point defects. By increasing the dose, point defect clusters transform into amorphous regions. Assuming the Breitling and Richter[12,13,14] model for amorphous silicon (in which the relative rotation of adjacent tetrahedra destroys long range order but preserves the distance involved in the first and second coordination) it can be suggested that in ion implanted silicon the eclipsed configuration is obtained by the formation of new bonds with interstitial atoms when a sufficient number of vacancies and interstitials have been produced by the ion bombardment. This hypothesis is supported by the fact that by 60° rotation of one tetrahedra, the

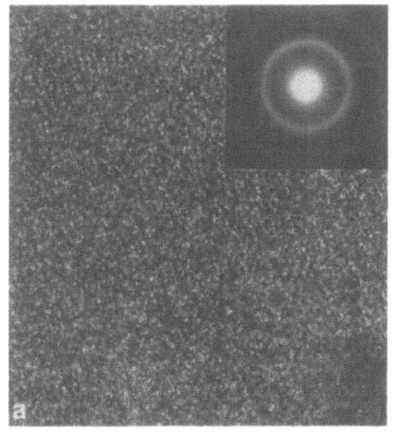
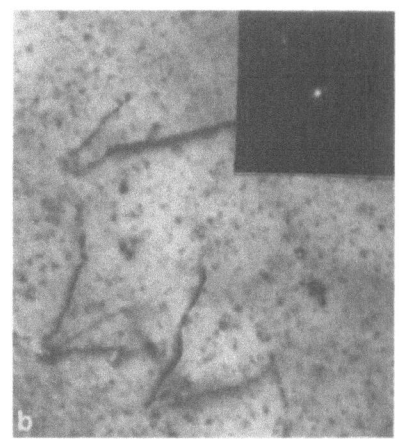
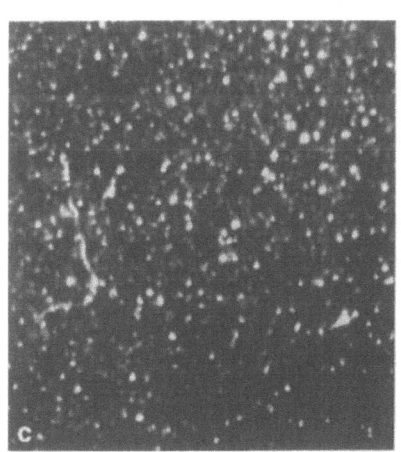
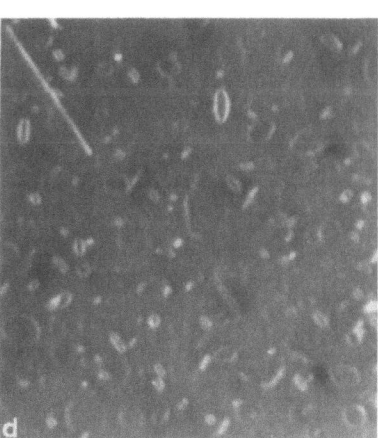

Fig. 6 - Idem as Fig. 5 for 5×10^{15} P^+/cm^2 dose. Annealing temperatures are: a) 500°C; b) 550°C; c) 600°C; d) 800°C. Micrographs a) is a first Debye ring dark-field; b) is 220 bright field; c), d) are weak-beam dark-field.

atoms whould occupy positions very near the tetrahedral interstitial sites[14].

AKNOWLEDGEMENTS

We should like to express our gratitude to Mr. R. Lotti and Mr. A. Piombini for their help in the samples implantations.

BIBLIOGRAPHY

1) J.E. Gibbons, Proc. of the IEEE, 9, 1062 (1972).
2) J.W. Mayer, L. Eriksson, J.A. Davies, "Ion Implantation in Semiconductors" Academic Press. N.Y. (1970).
3) G. Dearnaley, J.H. Freeman, R.S. Nelson, J. Stephen, "Defects in Cristalline Solids" N. Holland. Pu. Co. N.Y. (1973).
4) J.W. Mayer, Rad. Effects, 8, 269 (1971).
5) F. Cembali, R. Galloni, F. Zignani, Rad. Effects, 26, 161 (1975).
6) J.F. Ziegler, J. Appl. Phys., 7, 2973 (1972).
7) E.C. Baranova, V.M. Gusev, Yu. V. Martynenko, Proc. of the 3rd Int. Conf. on Ion Impl. in Semicond. and Other Mat. p. 59 (1972).
8) L.D. Glowinsky, K.N. Tu, P.S. Ho, Appl. Phys.Letters, 28, 312 (1976).
9) J.F. Gibbons, W.S. Johnson, S.W. Mylroie, "Projected Range Statistics" J. Wiley Ed. (1975).
10) A. Desalvo, R. Rosa, F. Zignani, Rad. Effects, 27, 89 (1975).
11) W. Frank, "Lattice Defects in Semiconductors" The Inst. of Phys. Conf. Series n°23, p.23 (1974).
12) P.A. Walley, Proc. Europ. Conf. on Ion Implantat., Reading. England, p. 219 (1970).
13) R. Grigorovici, R. Manaila, Thin Sol. Films, 1, 343 (1967/68).
14) Richter, H. and Breitling, G., Z. Naturforsch. 13a, 988 (1958).

RADIATION DAMAGE OF 50-250 keV HYDROGEN IONS IN SILICON

W. K. Chu, R. H. Kastl, R. F. Lever, S. Mader and B. J. Masters

IBM System Products Division, East Fishkill

Hopewell Junction, New York 12533

ABSTRACT

Damage distributions in [001] Si crystals bombarded with a fluence of 10^{16} to $10^{17}/cm^2$ at 50 keV to 250 keV H^+ ions at temperatures below 600°C were measured using high energy He^+ channeling. The nature of the damage was investigated by transmission electron microscopy (TEM). Different defect types are formed at different target temperatures, e.g. microblisters located on {111} planes exist between 100°C and 450°C. With higher fluences or post-implantation anneal the Si surface becomes visibly blistered.

INTRODUCTION

Hydrogen implantation into silicon is of interest in semiconductor technology for obtaining shallow donors[1] and for inducing radiation enhanced diffusion.[2-5] Prerequisites for these applications are data on the ranges of implanted H^+ and an assessment of the bombardment damage.

Recently Ligeon and Guivarc'h[6] determined the ranges in Si of H^+ ions with energies between 1.5 and 60 keV, while Baruch et al[5] reported values at 250 keV and 400 keV. In the present paper we concentrate on implantation energies between 50 keV and 250 keV. We measure the range and the distribution of damage in Si by channeling and backscattering of 2.0 MeV to 2.8 MeV $^4He^+$ ions and we will take the damage ranges as indicators for the H^+ ranges.[6]

Radiation damage, of course, strongly depends on the target temperature. In this paper we restrict ourselves to target temperatures below 600°C. For enhancement of diffusion one employs higher irradiation temperatures, but the ensuing residual lattice damage (to be described elsewhere[7]) is quite different from the "primary" radiation damage to be described here. Also, above 600°C implanted hydrogen seems to escape from Si (compare Stein[8]) whereas below 500°C it is retained and--under certain conditions--forms blisters and craters in the Si surface. Similar phenomena are observed in metals and have received broad attention[9,10] in connection with thermonuclear reaction environments. Here we can study an example of blistering in covalently bonded and brittle crystals.

HYDROGEN IMPLANTATION AND DAMAGE RANGES

[001] oriented wafers made from 20Ω-cm p-type Czochralski grown Si (doped with 10^{15} B/cm^3) were proton bombarded (7° off [001]) with scanned beams of typically 20μA/cm^2 obtained from a van de Graaff accelerator with an R.F. source. The target holder could be cooled or heated, and the wafers were mounted with heat conducting adhesives to ensure good thermal contact.

Typical examples of channeled backscattering measurements are shown in Fig. 1 for a fluence of 4×10^{16} H$^+$/cm^2 at 85, 125, 175 and 250 keV. A 2.4 MeV He$^+$ beam was directed along a [001] direction and the backscattering was measured at 170° in a solid angle of 4.11 msterad.

The damage region is well localized at a certain depth, and damage in the shallower depth portion of the samples is below the detectability of channeling. Every spectrum contains contributions of scattering and of dechanneling from the bombardment damage and the two contributions have to be separated in order to extract the damage profiles. For our analysis we modified the method of Ziegler[11] and combined both single scattering and multiple scattering in the dechanneling calculations. The depth scale is obtained using the energy loss table by Ziegler and Chu.[12] For the initial portion of the channeled beam, just inside the crystal, the energy loss along <100> channels is assumed to be 0.8 times the value of the random energy loss. Details of the analysis, an assessment of its accuracy, and complete damage distribution curves will be given elsewhere.[13]

The results of the backscattering analysis are damage ranges, R_D, and standard deviations of the damage distributions, ΔR_D. They are listed in Table 1 for room temperature bombardment at various energies. They interpolate well between the values of proton ranges in Refs. 5 and 6.

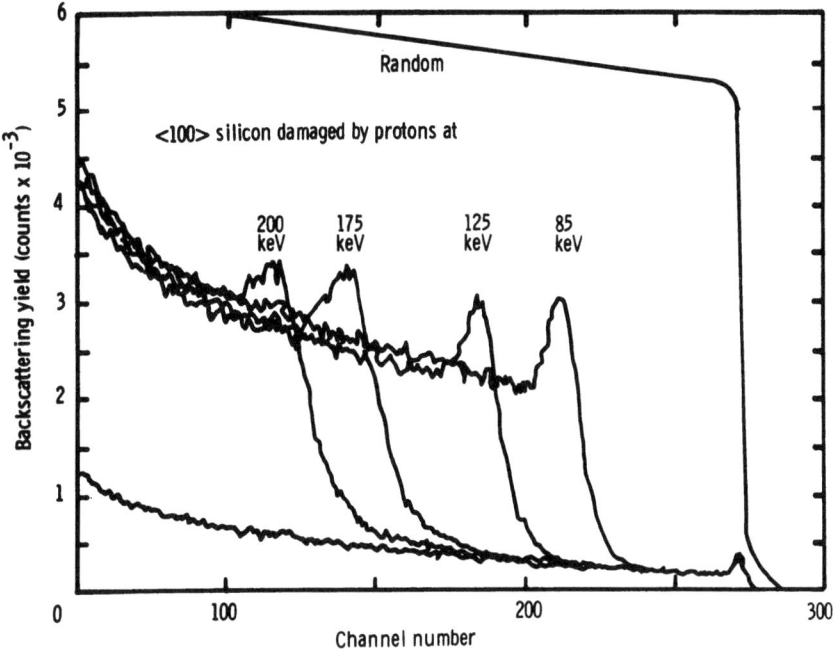

Fig. 1. Channeled backscattering spectra of 2.4 MeV He^+ from Si bombarded with 4×10^{16} H^+/cm^2 at room temperature.

Table 1. Defect Distribution in Silicon Due to Proton Bombardment

Proton Energy (keV)	Defect Depth R_D (Å)	Distribution ΔR_D (Å)
50	5000	400
75	6300	480
85	7200	520
100	8550	580
125	10180	610
150	12000	720
175	14700	750
200	17300	840
225	20000	900
250	23700	1000

Another parameter is the area under a damage distribution curve which can be converted into "equivalent displacements per atom." This is a relative measurement of the defect production per incident proton. Another parameter obtained from backscattering analysis is the "relative dechanneling efficiency" which is the ratio of dechanneling to scattering probability of a certain defect configuration divided by the same ratio for a buried layer of amorphous Si. Both of these parameters are given in Fig. 2 for implants of 4×10^{16} H^+/cm^2 at 50 keV into targets having different temperatures.

INFLUENCE OF TARGET TEMPERATURE

Figure 2 shows a strong temperature dependence of the amount of damage and of relative dechanneling. Three different temperature ranges can be seen: below 50°C, where damage is moderate; between 100°C and 450°C, where damage is high; and above 500°C, where the backscattering peak disappears but relative dechanneling increases. In the temperature range of 100°C to 450°C the damage profile becomes slightly shallower and broadens with increasing temperature to R_D = 4200 Å and ΔR_D = 680 Å at 450°C for 50 keV implants (compare Table 1).

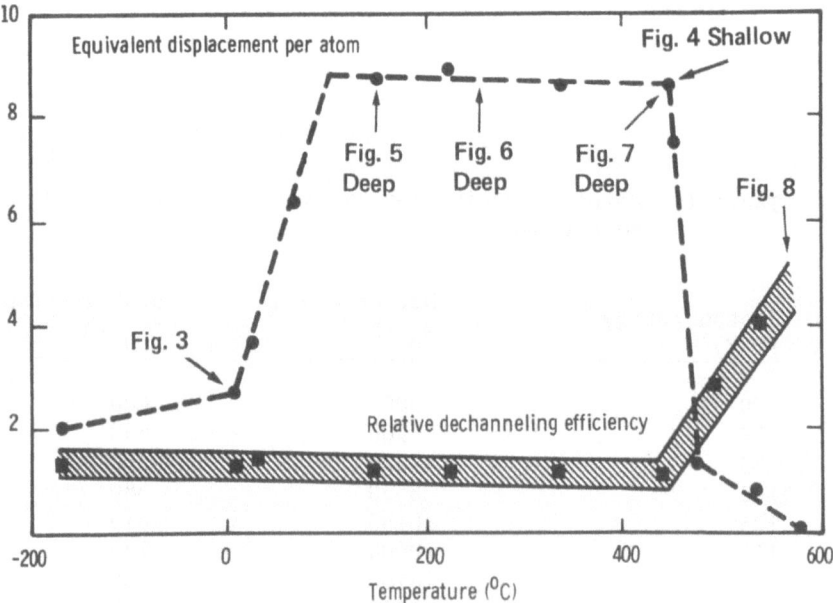

Fig. 2. Temperature dependence of damage and relative dechanneling for Si bombarded with 4×10^{16} H^+/cm^2 at 50 keV.

Different types of microscopic defects are observed in the three temperature ranges. The inset numbers in Fig. 2 refer to micrographs of the different defect configurations which will be described in the following section.

TEM STUDY OF THE NATURE OF THE DAMAGE

Electron transparent foils were prepared in the usual way by etching dimples and small holes from the (unbombarded) backside of the specimens. To obtain good TEM resolution of features located deeper than a few 1000 Å from the front surface, material was also removed from the front surface by repeatedly immersing the specimens in a Cu displacement solution (55g $CuSO_4$ + 10 mℓ HF + 1ℓ H_2O) and dissolving the Cu layer in HNO_3. Each step removes about 1100 Å of Si. A depth distribution of the damage can be obtained in this manner. The thin foil specimens become somewhat rough but they are adequate for dark field TEM analysis.

At low temperatures the damage consists of small clusters of mainly isotropic compression centers (as deduced from black-white contrast in suitable areas[14]). This is shown in Fig. 3. Out-of-focus contrast, characteristic of voids, was not observed. The cluster density is highest near the location of the backscattering damage peak, but a small density exists even very near the surface. The damaged layer contains high compressive strains which cause extensive bending at the edges of the foil. We interpret this damage as an agglomeration of bombardment-induced point defects where clusters of interstitials create larger strains in the Si matrix than clusters of vacancies.

Two types of defects exist between 100°C and 450°C. The first is located between the surface and the depth of maximum back-scattering. As shown in Fig. 4, these defects consist of clusters with anisotropic shapes and black-white contrasts. They are presumably small stacking faults and dislocation loops. Their density is highest near the depth of the backscattering damage peak. There they overlap with the second type of defects which are present between the peak location and the deep portion of the backscattering damage distribution, $(R_D + \Delta R_D)$.

The deeper lying defects are shown in Figs. 5-7, where the shallow damage has been removed. They are circular microblisters located on {111} planes. Their diameter increases from 250 Å at 150°C and to 2000 Å at 445°C. Figure 5 is a view along [001] which shows all four {111} planes equally inclined. Figures 6 and 7 are views along [112] showing "head on" images of features on (111) planes. Some of the circular features are truncated by the foil surfaces and they show weaker contrast than the untruncated

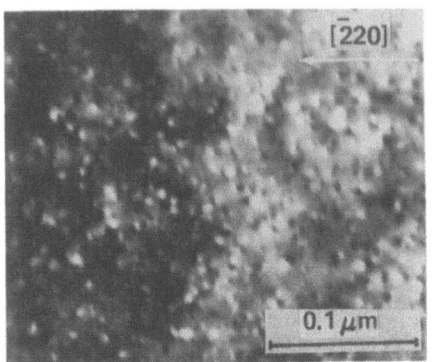

Fig. 3. Defect clusters target at 7°C. Darkfield.

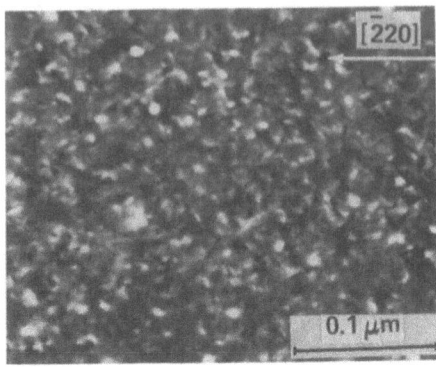

Fig. 4. Defect clusters target at 445°C. Darkfield.

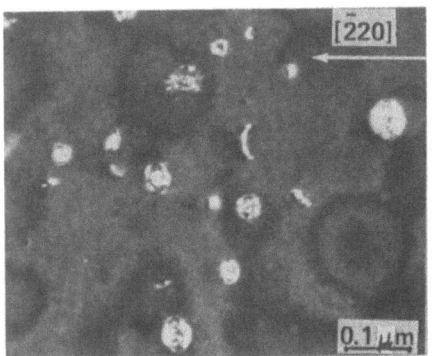

Fig. 5. Microblisters target at 150°C. Weak beam darkfield.

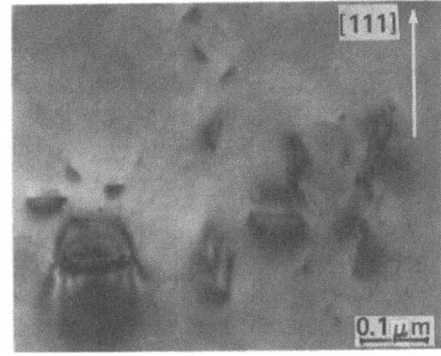

Fig. 6. Microblisters target at 250°C. Brightfield.

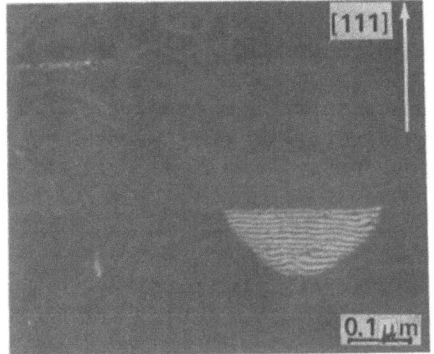

Fig. 7. Microblisters target at 445°C. Weak beam darkfield.

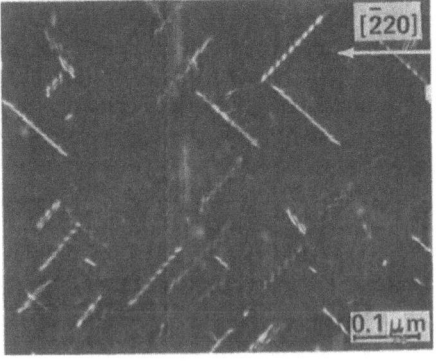

Fig. 8. Rod defects target at 575°C. Weak beam darkfield.

complete features (Fig. 5). The diffraction contrast of the complete features is compatible with displacements normal to the loop plane, but the displacements are not associated with a complete or partial Burgers vector of the Si lattice. The feature in the lower left corner of Fig. 6 is also a complete microblister viewed head-on. It is associated with multiple double-arc contrast, indicating that it transmits a large compressive strain into the Si matrix. All observations are consistent with the notion of microblisters with the inside hydrogen gas pressure which escapes when the blister is intersected by the foil surface during thinning of the specimen. The large blisters in Fig. 7 are all truncated. Their finite widths (of about 10 Å) in "head-on" views and their fringe contrast show that they have not collapsed completely and they presumably contain chemisorbed hydrogen. (The fringe contrast is always visible, unlike the contrast of ordinary stacking faults which can be made to disappear under certain diffracting conditions.) It is interesting to note that channeled backscattering measurements did not indicate the existence of these two defect structures in this temperature range.

No microblisters are formed with bombardment above $500°C$ where implanted hydrogen presumably diffuses out of the specimen. Instead, the bombardment creates another type of defect, namely rods in <110> directions, shown in Fig. 8. They occur in a layer between 2500 Å and 5000 Å below the surface. They are invisible when the diffraction vector is parallel to the rod direction. Rod-like defects have been observed before and they were analyzed by Madden and Davidson[15] as being narrow extrinsic ribbons. From inside-outside contrast observations we also find that the defects in Fig. 8 are ribbons which compress the surrounding Si matrix. But unlike Madden and Davidson[15] who inferred a {100} type habit plane for the ribbons, we observed that they are parallel to {113} planes and they are about 40 Å wide. These defects cause little backscattering but appreciable dechanneling of the He^+ beam (high temperature range of Fig. 2).

The above TEM results were obtained by 4×10^{16} H^+/cm^2 at 50 keV. An extensive study was also made using various fluences at 150 keV and targets between room temperature and $100°C$. Now the microblisters were located between 1.2 µm and 1.4 µm below the surface, but otherwise no new features were found.

SURFACE BLISTERS

At higher fluences, on the order of 10^{17} H^+/cm^2, the Si surface becomes visibly blistered. Upon further bombardment, or during post-implant annealing, these blisters erupt to form flat-bottomed craters. Figure 9 shows the beveled cross section of a <100>

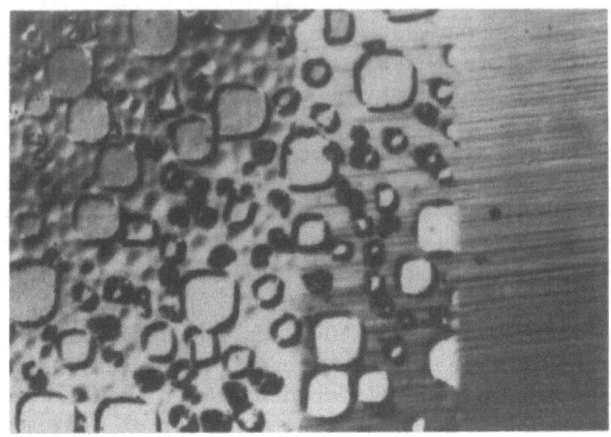

Fig. 9. Surface blisters and craters. Bevelled cross section. Magnification: 160 times. (Specimen prepared by E. Gorey)

oriented wafer which was implanted near room temperature with 2×10^{17} cm^{-2} of 150 keV protons and subsequently annealed for 30 minutes at 900°C. Both unbroken blisters ($\sim$ 10 μm diameter) and the resulting larger craters are evident. The crater depth obtained in this measurement, 1.30 μm, is in reasonable agreement with the values of Table 1.

A TEM examination of a similar implant revealed numerous truncated microblisters (similar to Figs. 5-7) around the periphery, and at the bottom, of the craters with the exception of the deepest areas at the center of the craters. This indicates that surface blisters are preceded by microblisters of the type described above. Their density increases with increasing fluence until some of them impinge and form a larger crack. The compressive stress at the depth region of maximum displacement damage, as well as the gas pressure, then aid in propagating the crack and causing the eruption of a crater.

DISCUSSION

We identified by channeled backscattering three temperature regions in which H$^+$ irradiation creates different damage in Si. These damage modifications were correlated with different microscopic defects in the Si lattice. Each of these is the result of the interplay during the irradiation of primary displacement damage, introduction of protons and self-annealing. The most prominent defects are microblisters which form between 100°C and 450°C. The transitions between the various regimes occur in narrow temperature ranges (Fig. 2). This indicates abrupt changes in the dominant self-annealing mechanisms. We refrain here from assigning specific Si point defect mechanisms to the three temperature ranges and offer only a conclusion on the behavior of hydrogen under the present irradiation conditions. The existence of microblisters indicates that hydrogen can diffuse short distances (a few 100 Å) above 100°C and longer distances (to escape from the specimen) above 450°C.

The range of implanted H^+ ions (as opposed to the damage range) is still somewhat uncertain. Ligeon and Guivarc'h[6] measured hydrogen distributions directly using the $^1H(^{11}B,\alpha)\alpha\alpha$ nuclear reaction; they compared them with He^+ backscattering curves and found hydrogen distributions and damage profiles to be very similar. On the other hand, Thompson and Robinson[16] concluded from a Monte Carlo calculation for H^+ at 10 keV to 40 keV that the H^+ range distribution peaks at about 20% greater depth than the damage distribution. This parallels our observation that microblisters are mainly located beyond the peak of the damage profiles.

The observation of surface blisters is similar to observations with metals although the morphology of microblisters in Si is different from voids observed in metals.[9,10] The microblisters consist of lenticular voids oriented parallel to {111} planes. These are cleavage planes of Si. The gas pressure presumably induces a separation along these planes.

REFERENCES

1. Y. Ohmura, Y. Zohta and M. Kanazawa, Solid State Com., 11, 263, 1972.

2. P. Baruch, C. Constantin, J. C. Pfister and R. Saintesprit, Discuss. Faraday Soc. 31, 76, 1961.

3. D. G. Nelson, J. F. Gibbons and W. S. Johnson, Appl. Phys. Lett., 15, 246, 1969.

4. Y. Ohmura, S. Mimura, M. Kanazawa, T. Abe and M. Konaka, Rad. Effects. 15, 167, 1972.

5. P. Baruch, J. Monnier, B. Blanchard and C. Castaing, Appl. Phys. Lett., 26, 77, 1975.

6. E. Ligeon and A. Guivarc'h, Rad. Eff., 27, 129, 1976.

7. S. Mader and B. J. Masters, to be published.

8. H. J. Stein, J. Electronic Materials, 4, 159, 1975.

9. "Application of Ion Beams to Metals," Edited by S. T. Picraux, E. P. EerNisse and F. L. Vook, Plenum Press, Proc. Intl. Conf. at Alburquereque, NM, 1973.

10. Proc. Intl. Conf. at Warwick, England, 1975.

11. J. F. Ziegler, J. Appl. Phys., 43, 2973, 1972.

12. J. F. Ziegler and W. K. Chu, Atom. Nucl. Data Tables, 13, 463, 1974.

13. W. K. Chu, et al., to be published.

14. M. R. Ruhle, In "Radiation Induced Voids in Metals," Ed. J. W. Corbett and L. C. Iannietto, Oak Ridge, 255, 1972.

15. P. K. Madden and S. M. Davidson, Rad. Effects, 14, 271, 1972.

16. D. A. Thompson and J. E. Robinson, Nucl. Inst. Methods, 132, 261, 1976.

ANOMALOUS ANNEALING BEHAVIOR OF SECONDARY DEFECTS IN Si
IMPLANTED WITH As IONS THROUGH DIELECTRIC LAYER

G. NAKAMURA and Y. YUKIMOTO
Kita-Itami Works, Mitsubishi Electric Corp.
4-1, Mizuhara, Itami 664, JAPAN

Y. AKASAKA and K. HORIE
LSI Development Lab., Mitsubishi Electric Corp.
Amagasaki 661, JAPAN

ABSTRACT

Defect observations and sheet resistance measurements were performed in Si crystals implanted by As ions with an energy of 200 keV-300 keV, a dose of 10^{15}-10^{16}cm^{-2}, and annealed at high temperature. Implantations were made on bare Si surfaces, through Si_3N_4 layers, and through SiO_2 layers. Annealings made at higher temperatures than 1000°C give good crystalline quality in the case of implantation through SiO_2 layer.

1. INTRODUCTION

Heavy As ion implantations are an attractive method for producing low resistivity emitters in Si bipolar devices, and source and drain regions in MOS devices. However, the residual damage has been observed in As ion implanted layers even after annealing at 1000°C for 30 min. by electron microscope and He backscattering analysis.[1-6] Several workers reported that the residual damage was enhanced when As ion implantations were made through SiO_2 layers due to the oxygen recoils knocked into the Si substrate by the As ions. Sigmon et al.[3] reported that the residual damage was reduced by using Si_3N_4 layers in place of SiO_2 layers in As$^+$ ion implantation.

This paper reports the transmission electron microscope observations of the defects in Si implanted with As ions on bare Si surfaces, through Si3N4 layers and through SiO2 layers. Special attention was paid to annealings at the temperature above 1000°C, because no experiment was reported hithertofore in this temperature range.

2. EXPERIMENTAL PROCEDURE

Silicon wafers of $\langle 111 \rangle$ orientation, p-type substrates, with a resistivity of 1-2 ohm-cm were used. Oxide layers of 100-1000 Å for through oxide implants, and Si_3N_4 layers of 300Å for through nitride implants were prepared by standard thermal oxidation and chemical vapor deposition techniques, respectively. The thickness of these layers were measured by ellipsometry. As^+ ions with energy of 300 keV and doses of $1 \times 10^{15} - 1 \times 10^{16} cm^{-2}$ were implanted through these layers.

All samples were implanted at room temperature at a dose rate below 2 $\mu C \cdot cm^{-2}$. Care was taken to avoid channeling of the implanted ion beam by tilting 8° off of the $\langle 111 \rangle$ axis. All samples were annealed in a dry N_2 atmosphere at 1000-1100°C for 20 min. Defects in the implanted layers were studied by transmission electron microscope (TEM) operating at 100 keV. Sheet resistance of implanted samples were measured by four point probe method.

3. RESULTS AND DISCUSSIONS

3.1 Isochronal annealing behavior of sheet resistance

Fig.1 shows sheet resistance of As ion implanted Si annealed at 1000-1100°C for 20 min. Sheet resistance of Si implanted on a bare Si surface decreases with annealing temperature until it saturates (by to 600-700°C) at which regrowth of the amorphous layer is nearly completed. Fair reported the good correlation between the dose of As^+ ions and the sheet resistance of Si annealed at 1000°C for 30 min.[8] The sheet resistances of Si implanted with doses of 5×10^{15} and 1×10^{16} cm^{-2} on bare Si surfaces for annealing at 1000°C agree well with the data reported by Fair.[8] Sheet resistance of Si implanted with a dose of 1×10^{15} cm^2 is about 1.6 times larger than that of Fair and increases by annealing slightly at 1050°C and rapidly at 1100°C, while the samples implanted with doses of 5×10^{15} and 1×10^{16} cm^{-2} show slight decreases at 1050°C and slight increases above 1100°C.

Samples implanted with As^+ ions through Si_3N_4 layers show large increases in sheet resistance by annealing above 1050°C. On the contrary, samples implanted with As^+ ions through SiO_2 layers show different annealing behavior from the samples implanted on bare Si surfaces and through Si_3N_4

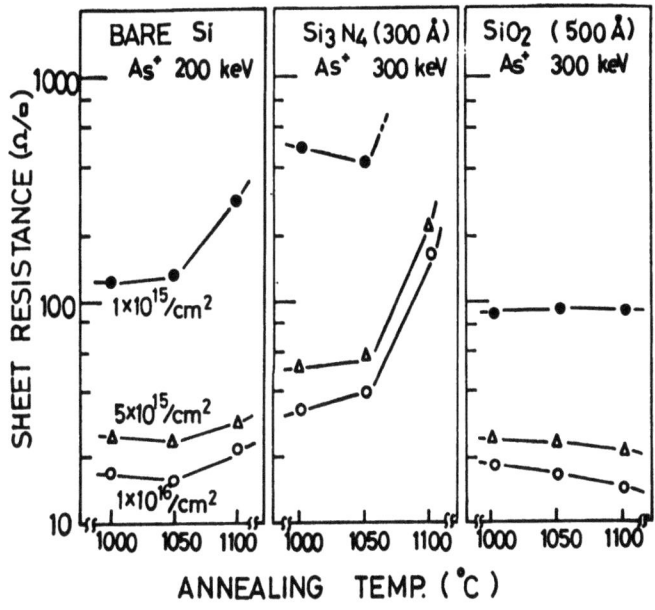

Fig.1 Isochronal anneal (20min) of sheet resistance of As ion implanted Si.

layers. They decrease slightly even after the annealing at 1100°C. The sheet resistance variation depends on the dose of As ions, and the minimum value of these samples coincides with the data reported by Fair.[8] The values of sheet resistance of Si implanted on bare Si surfaces show that the nearly all As ions are incorporated into Si (Rp for As with an energy of 300 keV in SiO2 is 2060 Å 10).

3.2 TEM observation

Fig.2 shows typical TEM images of the samples implanted with As ions on bare Si surfaces with an energy of 200 keV and doses of 5×10^{15} and $1 \times 10^{16} cm^{-2}$. Annealings were carried out at 1000 and 1100°C for 20 min. in an N_2 atmosphere. Fig.3 shows other TEM images of Si implanted with an energy of 60 keV and same doses and annealed at 1000°C for 20 min.

The structures of the defects may be classified into two types. The first type consists of small prismatic dislocation loops, as indicated by A in Fig.2c and Fig.3b, and three dimensional dislocation networks, as indicated by B. The nature of these dislocations coincides with that of the defects observed by Mader et al.[5,6] in (001) Si wafers. The small prismatic dislocation loops have an interstitial type character with an inclined Burgers vector to the surface, and the dislocation lines forming networks consist of pure edge dislocations with a surface parallel Burgers vector having an extra half plane at the Si surface.

The other type is a precipitated secondary phase,

(a) 5×10^{15} cm^{-2}, 1000°C (b) 5×10^{15} cm^{-2}, 1100°C

(c) 1×10^{16} cm^{-2}, 1000°C (d) 1×10^{16} cm^{-2}, 1100°C
Fig.2 TEM images of Si samples implanted by As$^+$ ions with an energy of 200 keV on bare Si surface.

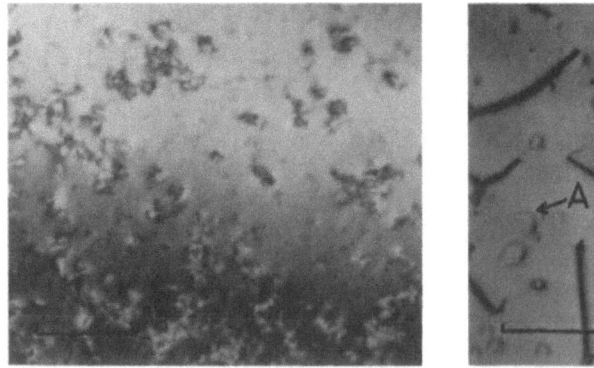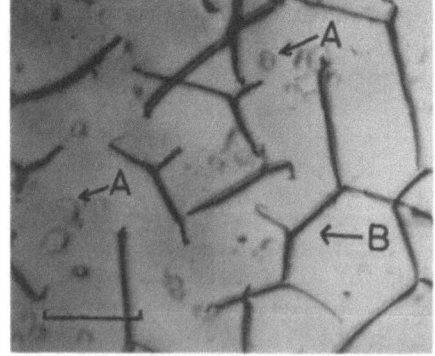

(a) 5×10^{15} cm^{-2}, 1000°C (b) 1×10^{16} cm^{-2}, 1000°C
Fig.3 TEM images of Si samples implanted by As$^+$ ions with an energy of 60 keV on bare Si surface and annealed in N_2 atmosphere at 1000°C for 20 min.

accompanied with misfit dislocations observed in low dose samples. Electron diffraction patterns have streaked spots around the 220 diffraction spots. Annealings at 1000-1100°C for 20 min. of the samples implanted with an energy of 200 keV and dose of $5 \times 10^{15} cm^{-2}$ cause the increases in the density of dislocation lines and in the size of the secondary phase, as well as the appearance of vapor etched holes, as shown in Fig.2b by C. As shown in Fig.2d, uniformly distributed liquid phase regions and vapor etched holes are observed in the samples implanted with an energy of 200 keV and dose of $1 \times 10^{16} cm^{-2}$ and annealed at 1100°C.

The difference in the nature between the two defect types may not be caused by the difference in the dose or energy of the implanted As^+ ions, but the difference in As^+ ion concentration at the depth of Rp.

The increase in sheet resistance occuring in the high temperature annealings as shown in Fig.1 is considered to be resulted from the decrease in As atom concentration in Si by the out diffusion effect, as revealed by the vapor etching phenomenon.

To study the effect of recoiled oxygen atoms on the recovery stage of As^+ ion implanted regions, the thickness of oxide layers was varied from 100 to 1000 Å. Fig.4 shows the isochronal annealing behavior of sheet resistance for the samples annealed at 1000°C -1100°C for 20 min., and Fig.5 shows defects observed by TEM after annealings at 1000°C-1100°C for 20 min. The As ions were implanted with an energy of 300 keV and a dose of 5×10^{15} cm^{-2}. Sheet resistance increases with the thickness of the oxide layer and decrease with the annealing temperature. The resistance changes are larger in the samples covered with thicker oxide layers. Annealings made at the temperatures above 1100°C increase the sheet resistance steeply, as seen in the case of bare Si implants. As^+ ions were out diffused, because the covered oxide layer was stripped off before annealing.

The characteristic features of the defects observed in As^+ ion implanted layers through SiO_2 with high ion doses are the appearance of high density prismatic dislocation loops with interstitial nature and their change into meandering dislocation lines having an extra half plane at the Si surface, as observed by Mader et al. 5,6) in (001) crystals.

After an annealing at 1000°C, the following defects are observed in As^+ ion implanted samples; isolated prismatic dislocation loops and meandering dislocation lines with high density in the case of implants through 100Å thick oxide, meandering dislocation lines and

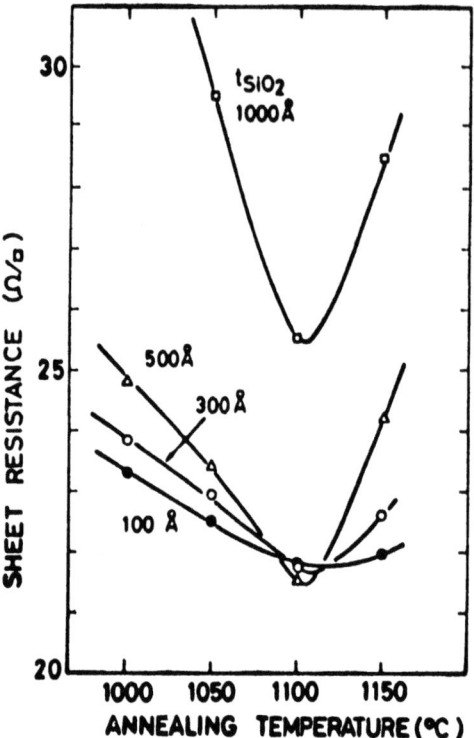

Fig.4 Isochronal anneal of sheet resistance of Si implanted by As$^+$ with various oxide thicknesses.

dislocation networks through 300-500 Å thick oxide, and microtwins suggested from the extra spots in electron diffraction patterns observed in the samples implanted through 1000 Å thick oxide. The microtwin structure remains after annealing at 1100°C.

The dislocation density decreases rapidly by an annealing at temperatures above 1000°C in the samples which have meandering dislocation lines and becomes nearly zero in the samples implanted through 500 Å thick oxide. The dislocation density does not change in the samples implanted through 100 Å thick oxide. Therefore the development of meandering dislocation lines may be considered to play an active role in the annihilation process of dislocation lines.

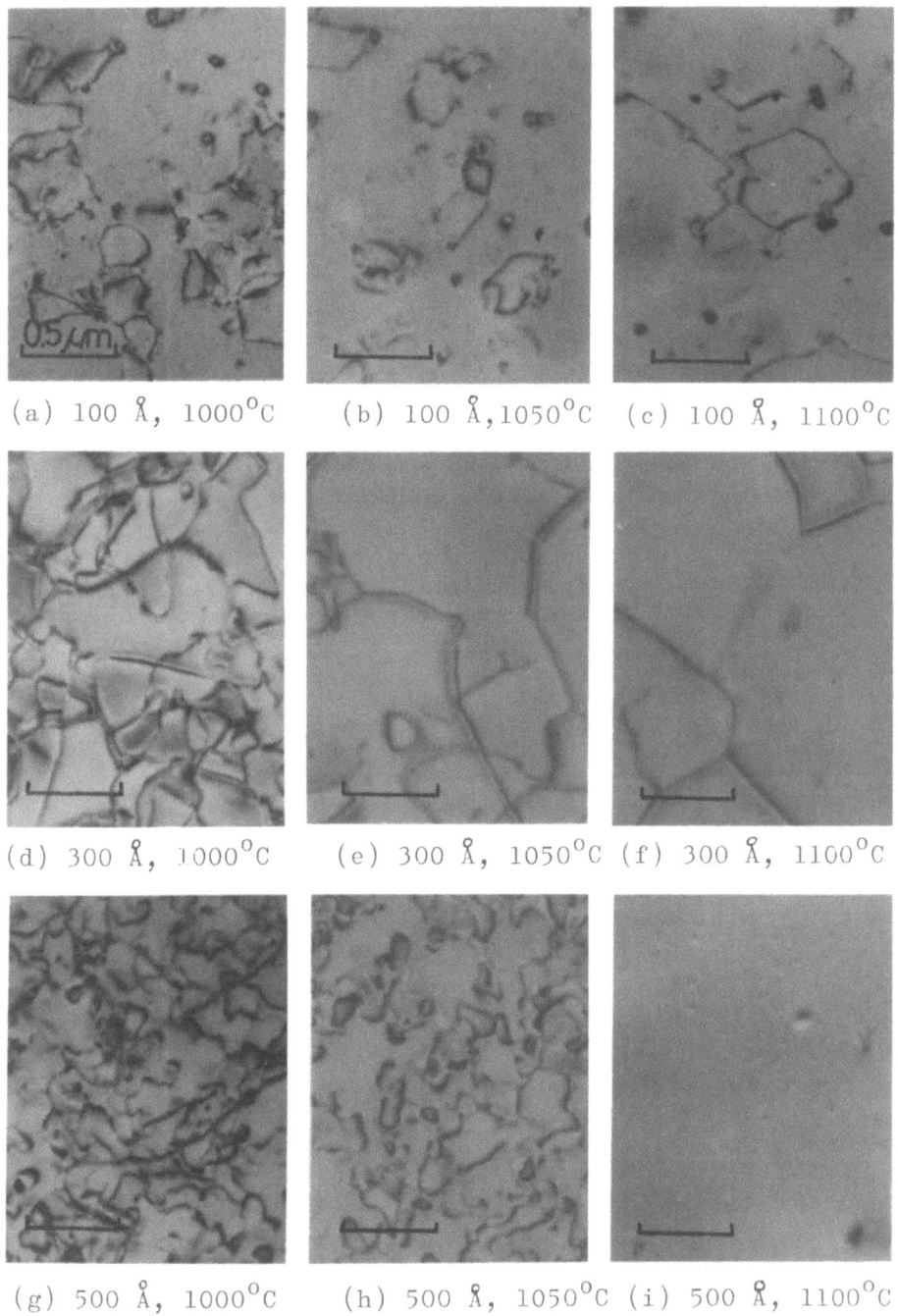

Fig.5 TEM images of Si samples implanted by As ions with an energy of 300 keV and a dose of $5 \times 10^{15} cm^{-2}$ through SiO2 layers of various thickness.

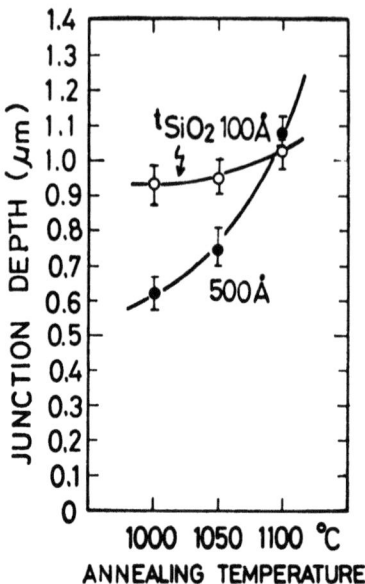

Fig.6 Junction depth variations in Si implanted with As^+ ions through oxide.

Fig.6 shows the junction depth variations in the samples implanted through 100 Å and 500 Å thick oxide layers. The latter sample shows a larger increase in junction depth with annealing temperature and this suggests the larger enhanced diffusion than the former sample.

The concentration of recoiled oxygen atoms is reported to increase with the oxide thickness in the heavy ion implantation. High density of dislocation lines caused partly by these recoiled oxygen atoms and partly by As^+

ions will interact with each other by the Lomer reaction process. This interaction may generate a high concentration of excess vacancies.[14] The excess vacancies will increase the electrically active As^+ ion concentration and also will enhance the climbing motion of dislacation lines which have an extra half plane at the Si surface. On the other hand, the samples with isolated dislocation loops and lines can not climb out after the high temperature annealing. In the samples in this stable steady state, the phenomena such as $VSiAs_2$[11] complex formation or vacancy undersaturation will be dominating and dislocation loops and lines will remain in the region near the depth Rp.

The active As^+ ion concentration in the samples implanted on bare Si surfaces shows a similar increase when the defects change from secondary phase into three dimensional dislocation networks.

4. CONCLUSIONS

Secondary defects observed in Si implanted with As ions on bare Si surfaces and through Si_3N_4 can not be annealed out even at temperatures above 1000°C. Vapor etched holes at the secondary phase can be seen and these may cause the sheet resistance increase.

In the case of As ion implantations through SiO_2 layers, many dislocation loops or meandering dislocation lines are observed first by annealing at 1000°C for 20 min. The density of these dislocations increases with dose density and oxide thickness. However, annealings at temperatures above 1050°C make these dislocations disappear from the implanted layers and give us good crystalline implanted layers.

5. REFERENCES

1. T.R.Cass and V.G.K.Reddi, Appl.Phys.Lett.23(1973)268.
2. W.K.Chu et al., Appl. Phys. Lett. 25(1974)297.
3. T. W. Sigmon et al., Ion Implantation in Semicon. and Other Materials, ed. S. Namba, (1974)633.
4. R. A. Moline et al., Appl. Phys. Lett. 26(1975)551.
5. S. Mader et al., J. Vac Sci. Technol.,13(1976)391.
6. S. Mader et al., Phys. Stat. Sol. (a)33 (1976)793.
7. H. Muller et al., Appl. Phys.,4 (1974)115.
8. R. B. Fair, Solid State Electrn.17 (1974)17.
9. W. K. Chu et al., Ion Implantation, (1974)177
10. W. K. Chu et al., Appl. Phys. Lett., 22 (1973)490.
11. R. B. Fair et al., J. Appl. Phys. 44 (1973)273.
12. R. B. Fair et al., J. Electrochem Soc.,122(1975)1689.
13. D. P. Kennedy et al Proc.IEEE, 59(1971)335.
14. T. J. Parker, J. Appl. Phys.,38 (1967)3471.

ANALYSIS OF DEFECT STRUCTURES IN RECRYSTALLIZED AMORPHOUS LAYERS
OF SELF-ION IRRADIATED SILICON BY CHANNELING AND TRANSMISSION
ELECTRON MICROSCOPY MEASUREMENTS*

P. P. Pronko and M. D. Rechtin
Materials Science Division, Argonne National Laboratory
Argonne, Illinois 60439
and
G. Foti, L. Csepregi, E. F. Kennedy and J. W. Mayer
California Institute of Technology
Pasadena, California 91125

I. INTRODUCTION

The purpose of this work is to study the structure and characteristics of residual defects in a <111> epitaxially regrown amorphous layer of ^{28}Si bombarded silicon by channeling and transmission electron microscopy (TEM). A secondary objective is to compare the channeling data with the TEM results in order to make critical judgments about the adequacy of standard channeling analysis[1] as it is applied to microdefects in diamond structure semiconductors. The standard approach assumes that randomly distributed interstitials produce direct back-scattering as well as forward scatter dechanneling. The total amount of dechanneling is assumed to be in strict proportion to the number of interstitials with no consideration given to the possibility that other defect types may be present which produce dechanneling with little or no direct scattering.

It has become increasingly evident that this standard channeling analysis cannot be used as a general technique to obtain quantitative information in implanted semiconductors or metals. It has been shown, for example, that the displaced atoms are not randomly distributed within the channel in light ion irradiated silicon.[2] This result is similar to that found for energetic ion induced defect clusters in gold.[3] It has also been found that the amount of experimental dechanneling in recrystallized layers

---

*Work supported in part by the U.S. Energy Research and Development Administration.

of silicon is higher than that obtained from calculations.[4] In addition, work[3,5] on heavy ion induced defect cluster strain fields has shown that little or no direct scattering is associated with the dechanneling from these clustered defects. It will be shown in what follows that dechanneling effects, over and above that contributed by interstitials, must be considered when analyzing a reordered layer of <111> silicon.

The amorphous layer on the specimens was prepared by implanting ^{28}Si ions into randomly oriented <111> Si substrates held near liquid nitrogen temperature. The silicon wafers had a resistivity of 2-10 Ω cm (low doped) to eliminate impurity effects in the regrowth process. The energies of the implanted ions ranged from 50 to 250 keV (in 50 keV steps) with fluences that started at 5×10^{13} cm^{-2} and ended at 10^{15} cm^{-2}. The total integrated fluence is 8×10^{15} ions/cm^2 which results in a well defined amorphous layer of about 4500 Å thickness. A vacuum furnace stabilized to within ±1°C was used to anneal the samples at 600°C.

II. CHANNELING RESULTS

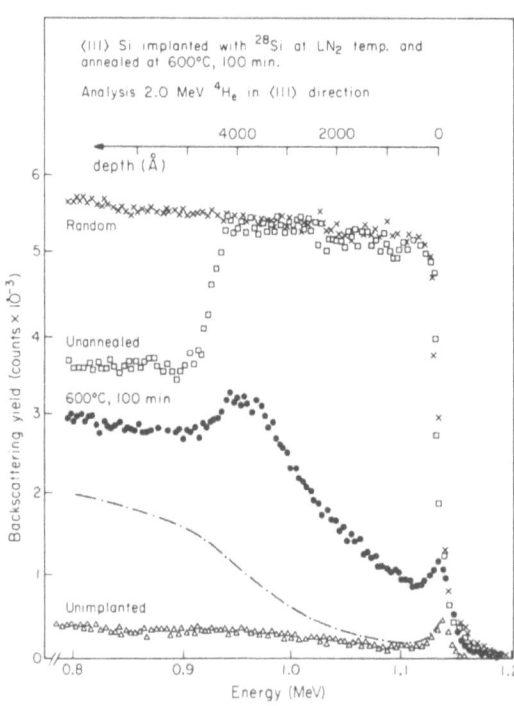

Fig. 1. *Channeling-backscattering Spectra of 2 MeV He^4*

Backscattering energy spectra for a 2.0 MeV ^{4}He beam incident along a random and channeling direction in <111> silicon, are shown in Fig. 1. The aligned spectrum, for the implanted sample before the anneal, shows that the implanted layer is amorphous to a depth of about 4500 Å. The aligned yield just behind the amorphous layer is caused by the angular spread of the incoming particles after traversing the amorphous layer.

The full circle points show the aligned spectrum after the sample has been annealed at 600°C for 100 min. The dashed line is the amount of dechanneling in the regrown layer calculated from the backscattering spectra, using plural scattering and the standard

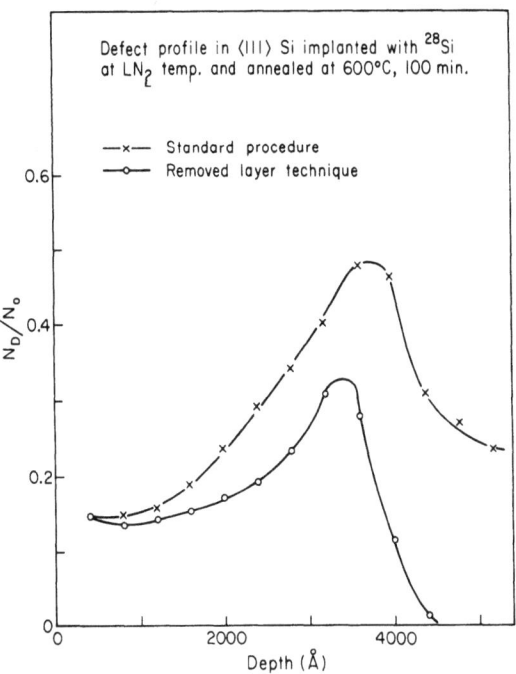

Fig. 2. *Defect Profiles for Direct Scattering Centers.*

iterative procedure.[1] The aligned yield at a depth greater than 4500 Å is clearly higher than the calculated dechanneling level. By using successive anodic oxide stripping and measuring, the minimum channeling yield behind the surface peak, one obtains direct information about the scattered fraction without making assumptions about the depth dependence of the dechanneling in the defective layer. A series of measurements as a function of the amount of material removed gives the results as seen in Fig. 2.

The defect profile, thus obtained, is given in terms of the number of direct scattering centers normalized to the density of silicon atoms N_D/N_0. The empty circle points represent the disorder distribution as measured from the surface region of the stripped samples. The cross points represent the disorder distribution calculated with the standard iterative procedure using the aligned spectrum of a nonstripped sample. Near the surface, where the assumptions for the dechanneling are not critical, the difference between the two methods is not large. However, the difference becomes larger and larger with depth where the assumptions about the dechanneling mechanism are more relevant. The tail in the calculated distribution is related to the method involved and has a nonphysical origin. To explain the high dechanneling in the regrown layer, it is necessary to know the structure of the defects in more detail.

The possible effects of the geometric arrangement of defects in the reordered layer may be observable through measurements of channeling-back-scattering yields along two different principle axis directions. It is seen in Fig. 3, however, that the yield along the <110> is the same as along <111> if the additional path length (from the 35° specimen tilt) for the <110> is taken into account. It is also clear from the figure that the anomalously high dechanneling yield behind the defect layer is present for both orientations and is not associated with a particular axis chosen for the channeling analysis. Further information on the defects can be obtained from transmission electron microscopy.

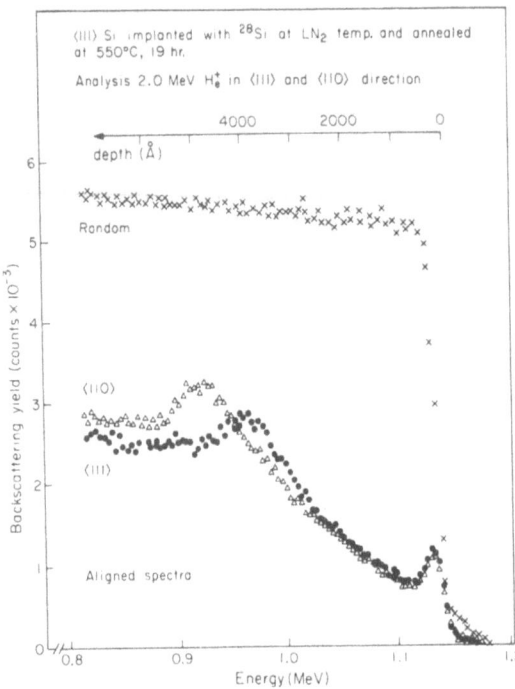

Fig. 3. *Comparison of <111> and <110> Channeling Spectra.*

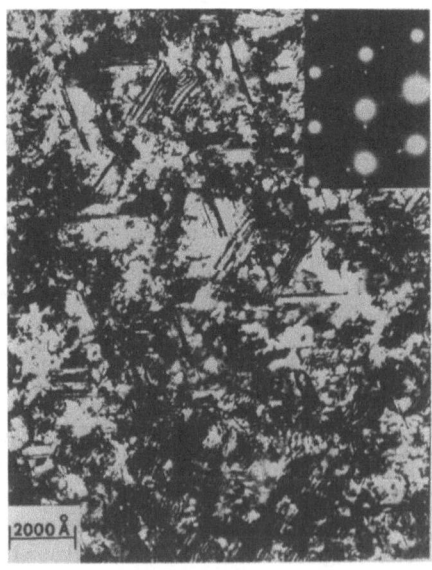

Fig. 4. *Bright Field Image of First 1500 Å of Reordered Layer.*

III. TRANSMISSION ELECTRON MICROSCOPY (TEM)

TEM specimens are prepared by jet etching with a 10% HF and 90% HNO_3 solution. Fig. 4 is a bright field image in a <110> orientation showing the basic defect structure present in the regrown layer. Fault structures are observed in this micrograph for three principal habit directions. These appear in the micrograph as being oriented along the horizontal direction and along both diagonals. The <110> zone diffraction pattern associated with this bright field micrograph exhibits clear satellite spots around the matrix spots. These satellite spots occur at a 1/3 distance between the matrix spots and are interpretable on the basis of a twinned lattice with {111} twin planes.[6] It was also determined that the angle between two of the twin habits was 110°. This is very near the 109.5° expected for the trace of the two {111} planes parallel to the <110> zone. The other twin set is inclined to the <110> zone and is evident from the extinction fringes present in the micrograph. It is concluded therefore that the faulted structures are twins. The three habit sets in the micrograph are twins lying on the three {111} planes that are independent of the

DEFECT STRUCTURES IN RECRYSTALLIZED AMORPHOUS LAYERS

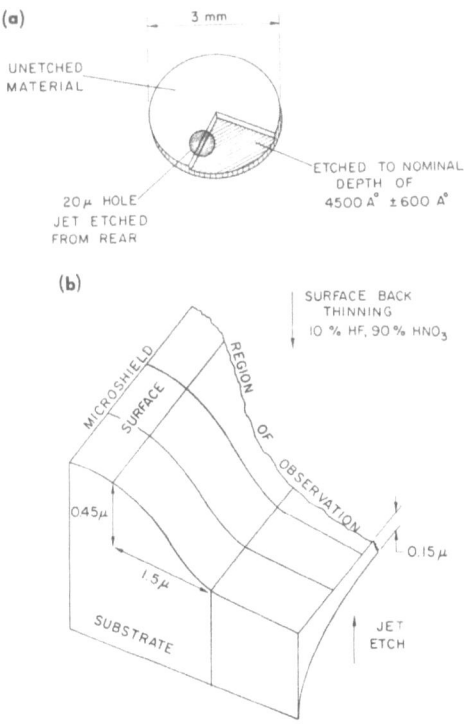

Fig. 5. *Geometry of Etched Specimens for TEM Depth Analysis.*

specific (111) plane which is normal to the regrowth direction. In another orientation, twins were observed for this (111) plane perpendicular to the regrowth direction; however, they were equiaxial in shape and were at least one order of magnitude less dense than the other twin sets. It is also clear from the micrograph in Fig. 3 that a substantial amount of localized strain is mixed with the twinned material. This black-white contrast is identified as being associated with the strain field arising from clustered point defects.

In order to correlate the channeling results with TEM, it is necessary to get microscopy data as a function of depth in the reordered layer. A simple way to do this is to combine surface etching with jet polishing from the back side of the specimen to produce a thin slice of material angling its way through the reordered layer. This procedure results in a lateral image of the defects as a function of depth. A schematic representation of the specimen geometry used in this procedure is shown in Fig. 5 (a). A disk of 3 mm diameter (the size of the TEM specimen) is cut from the original annealed piece. A section of the surface is etched to a depth corresponding to the interfacial region at 4500 Å (± 600 Å). A hole of about 20μ is then jet etched from the rear so that a puncture occurs in the near vicinity of the transition between the etched and nonetched zone. The result is a geometric contour with a shape as depicted in Fig. 5 (b). A typical result of this preparation technique is given in Fig. 6. These are two dark field images formed from two different twin spots (see upper inset of figure). The micrographs reveal a transition from large twins near the surface to very small twins near the interface for both habit sets being observed. The arrows in the micrographs define a line passing along the interface at 4500 Å. The substrate appears free of twin defects in agreement with the results obtained from the combined layer removal and channeling measurements (see Fig. 2).

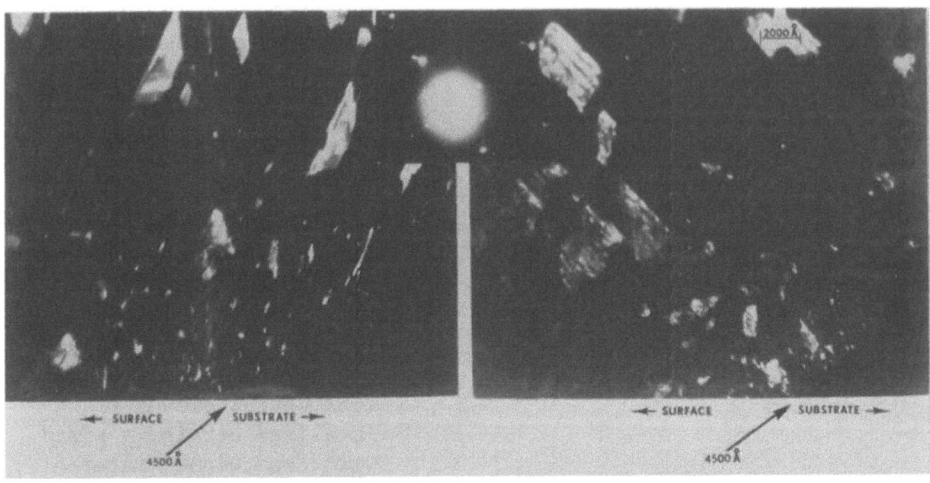

Fig. 6. Dark Field Image of Twins.

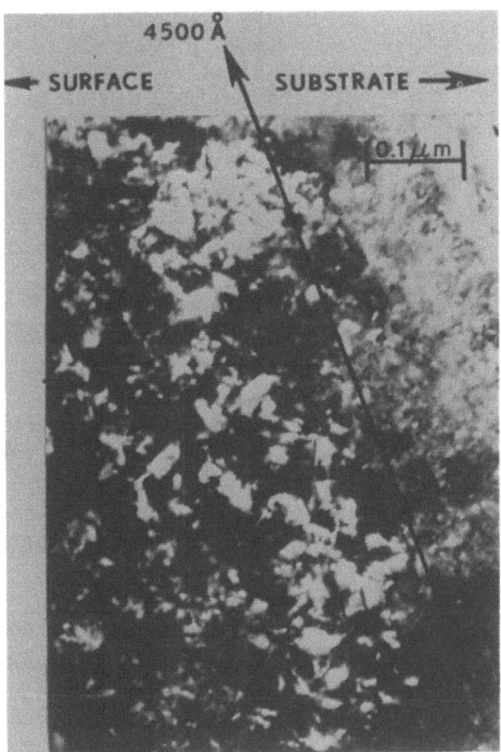

Fig. 7. Strain Contrast from Point Defect Cluster Strain Fields.

Fig. 7 is a high magnification bright field image where the strain contrast is in focus. These localized strains are produced by point defect clusters which are themselves unresolvable by TEM. The density of this strain field is observed to increase going from the surface to the substrate and appears to terminate at the interface as the twins did. In addition to the strain field, some fine polycrystallinity appears in the immediate vicinity (±500 Å) of the interface. The evidence for this appears in Fig. 8 where the inset shows a combined ring and spot pattern. Close examination of this diffraction picture reveals that the rings pass through the twin spots and not the matrix spots. This suggests that the polycrystalline material is itself

DEFECT STRUCTURES IN RECRYSTALLIZED AMORPHOUS LAYERS

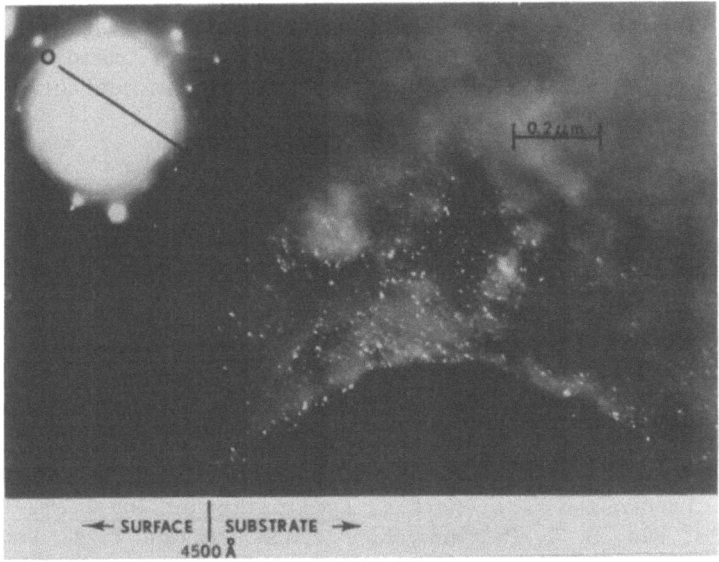

Fig. 8. Polycrystallinity Observed Near Interface.

twinned. Dark field imaging of a portion of the ring pattern shows finely dispersed polycrystalline grains in the material. The micrograph in Fig. 6 is oriented such that the interface at 4500 Å has the surface towards the left and the interfacial material to the right. The flecked region to the right of the 4500 Å mark is generally all at the interfacial base depth.

IV. CONCLUSION

The dominant defect structures in these reordered layers have been identified as twins whose dimensions change in size going from the surface to the interface. A high density of clustered defects is observed near the interface and is manifested by the presence of a heavy strain contrast in the electron micrograph image. Some fine polycrystallinity is also observed in the region of the interface between the regrown layer and the substrate. The 2 MeV He^4 channeling results indicate that, in this kind of defect arrangement, the standard analysis[1] for reducing the channeling data cannot be applied. A more direct way to examine the depth dependence of the defect distribution by channeling is to follow the change in the minimum yield as a function of layer removal. The results obtained in this way show that the number of scattering centers (N_D/N_O) is approximately constant in the first 3000 Å and increases very fast near the interface. No tail in the scattering distribution is observed to penetrate the substrate when using the stripping procedure. This agrees generally with the TEM results.

In order to account for all types of defects as are observed with TEM, it is necessary, in the channeling analysis, to recognize the contribution to dechanneling from defects that produce little or no direct scattering. These include, for example, the effects of twinning, dislocations, stacking faults, defect cluster strain fields, and mosaic structure.

REFERENCES

1. F. H. Eisen, Channeling, Ed. D. V. Morgan (J. Wiley & Sons, London, 1973), Chapter 14, p. 415.
2. P. Baeri, S.U. Campisano, G. Ciavola, G. Foti, and E. Rimini, Appl. Phys. Lett. 28, 9 (1976).
3. P. P. Pronko, Nucl. Inst. and Meth. 132, 249 (1976).
4. L. Csepregi, W. K. Chu, H. Muller, J. W. Mayer, and T. W. Sigmon, Rad. Effects 28, 227 (1976).
5. P. P. Pronko and K. L. Merkle, Application of Ion Beams to Metals, Eds. S. T. Picraux, E. P. EerNisse, and F. L. Vook. (Plenum Press, N.Y., 1974) p. 481.
6. D. W. Pashley and M. J. Stowell, Phil. Mag. 8, 1605 (1963).

ACKNOWLEDGEMENTS

The authors wish to acknowledge the assistance of T. W. Sigmon (Hewlett-Packard, Palo Alto, Calif.) for preparing the ^{28}Si implanted amorphous layers.

DEPENDENCE OF RESIDUAL DAMAGE IN "THROUGH-OXIDE" IMPLANTS ON

SUBSTRATE ORIENTATION AND ANNEAL SEQUENCE*

E.F. Kennedy,** L. Csepregi,[+] and J.W. Mayer

California Institute of Technology, Pasadena
California, 91125

and

T.W. Sigmon

Hewlett-Packard Laboratories, Palo Alto, California
94304

ABSTRACT

Channeling analysis of 200 keV ^{75}As "through-oxide" implants in crystalline silicon at a dose level of 10^{16} ions/cm^2 indicate that the orientation of the silicon, the substrate temperature during implantation, and the anneal procedure all influence the amount of residual disorder. The residual disorder is higher in <111> silicon than in <100> silicon for comparable implantations and anneal temperature sequences. Elevated substrate temperature during implantation not only leads to higher disorder after anneal but also shifts the residual disorder distribution deeper into the sample.

INTRODUCTION

High dose heavy ion implantation through oxide or nitride layers on silicon has been found to create highly disordered regions of silicon after annealing at 1000°C (1). Studies of recoil implantation yields from ^{18}O enriched SiO$_2$ layers showed that there was a high probability for forward recoil of atoms from these layers (2). Direct correlation between the high levels of residual damage and recoil oxygen implantation was shown by simulating the "through-oxide" implants with implantation of oxygen into ^{75}As implanted silicon (3). The same results were also observed for similar implants through Si_3N_4 layers (4). Transmission electron micrographs (5) showed the highly defective silicon was essentially polycrystalline in nature.

It is interesting to note that the silicon substrate material used in those previous studies was of <111> orientation and no mention was made of the substrate temperature during implant. Furthermore, the samples were reported to have been heated directly to an anneal temperature of 1000°C. It has recently been found that the anneal cycle and substrate orientation can strongly influence the amount of residual disorder found after annealing high dose implants (6,7). In this study we report the effect of the substrate orientation and temperature during implant, along with the thermal anneal history, on the disorder caused by ^{75}As implants through thin SiO_2 films.

EXPERIMENTAL

Thin films of SiO_2 (~ 500Å) suitable for MOS application were grown on <111> and <100> Czochralski grown silicon of a few ohm-cm. These wafers were then implanted with 200 keV ^{75}As ions to a dose of 10^{16} cm^{-2}. During the implantation certain of the wafers were implanted at LN_2, RT and 200°C. Care was taken to insure reliable and reproducible substrate temperatures. After implantation certain wafers were annealed in the following manner: 550°C (LTA), 1000°C (HTA), and 550°C followed by 1000°C (TSA). All anneals were performed in vacuum. Channeling and backscattering measurements were used to analyze (6) the depth distribution of disorder in these structures.

RESULTS

A. Regrowth at 550°C

In Fig. 1 we show the aligned backscattering yield for <100> silicon implanted at a substrate temperature near LN_2. The oxide layer was removed before the 2 MeV ^{4}He ion analysis. The upper portion of the figure shows the calculated ^{75}As distribution (8) and an estimate of the recoil implanted O distribution (2). After anneal at 550°C the samples contained residual disorder near the outer surface of the silicon as indicated by the aligned spectrum. The residual disorder persisted through long anneal cycles at 550°C and 600°C (2428 minutes).

For the anneal at 550°C the decrease in the disorder peak as

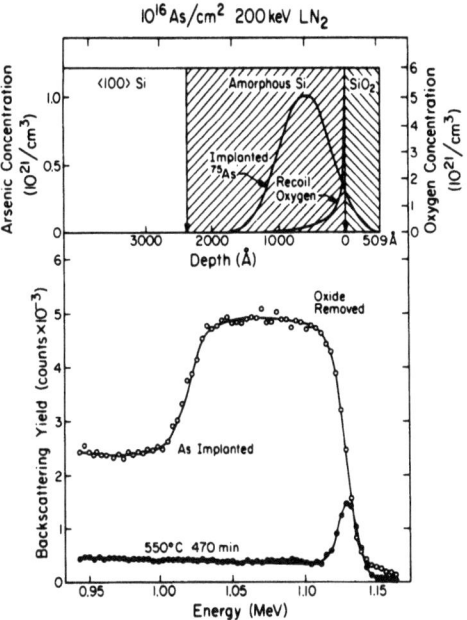

Fig. 1. Aligned RBS yields for <100> silicon implanted near LN_2 temperature. Calculated profiles of As and recoil O are also shown.

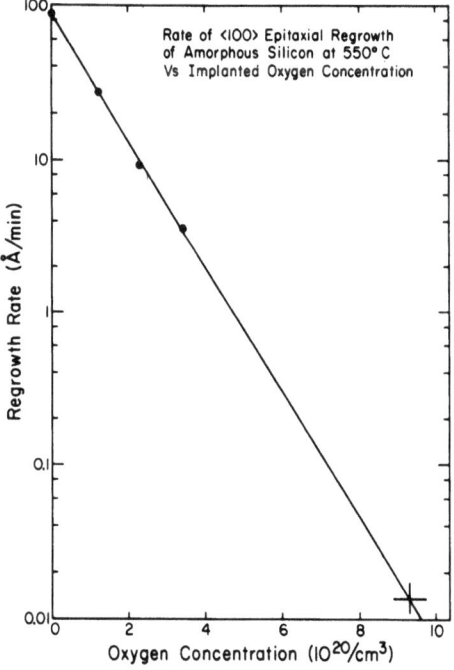

Fig. 2. Epitaxial regrowth of amorphous silicon.

a function of time between 470 and 1130 minutes corresponds to an epitaxial regrowth rate of 0.014 Å/min. We estimate the average oxygen concentration to be about 10^{21} atoms/cm^3 at the rear edge of the disordered layer after the 470 min. anneal at 550°C (Fig. 2). For this oxygen concentration, a regrowth rate of 0.014 Å/min. is consistent with the extrapolated curve for the regrowth rate vs. oxygen concentration measured previously in oxygen implanted <100> silicon (9).

B. Anneal of <100> Samples

Figure 3 shows the random and aligned spectra for silicon "through-oxide" implanted with 200 keV ^{75}As to a dose of 10^{16} ions/cm^2. The spectra for samples implanted at LN$_2$ and RT (not shown) are nearly identical and indicate the formation of an amorphous layer about 2400Å in thickness. For the sample implanted at 200°C the aligned spectra in the implanted region is reduced compared to that for the sample implanted at LN$_2$, although the dechanneling below the implanted layer is comparable for both samples. After anneal at 1000°C for 30 minutes all three samples exhibited some residual disorder; however, the spectra indicate that the disorder is appreciably higher in the sample implanted at 200°C.

Identical samples were annealed for 300 minutes at 550°C (LTA) and subsequently for 30 minutes at 1000°C (TSA). The aligned spectra in Fig. 4 indicate a high amount of residual disorder in the sample implanted at 200°C for both anneal sequences. For the RT implanted sample, the two-step anneal sequence results in a somewhat lower aligned spectra than that for the sample annealed directly to 1000°C as shown in Fig. 3.

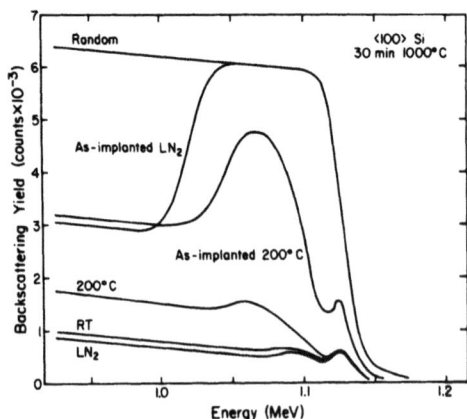

Fig. 3. "Through-Oxide" implantation of 200 keV ^{75}As. Dose 10^{16}/cm^2.

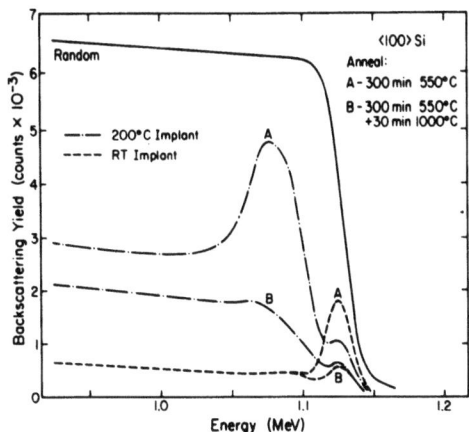

Fig. 4. Effect of low temperature and double anneal on 200 keV ^{75}As implants. Dose $10^{16}/cm^2$.

C. Anneal of <111> Samples

As observed previously (6), <111> oriented silicon samples implanted to high dose at LN_2 temperature and annealed directly to 1000°C, show high amounts of residual disorder as indicated by the aligned spectra in Fig. 5. High amounts of residual disorder are also observed in samples implanted at RT and 200°C and annealed in a two-step process at 550 and 1000°C (Fig. 6). As in the case for <100> oriented silicon, the samples implanted at 200°C have higher aligned spectra than those implanted at RT. Prolonged annealing at 600°C (1600 min.) prior to the 1000°C anneal for the LN_2 implanted sample showed little reduction in the aligned yield.

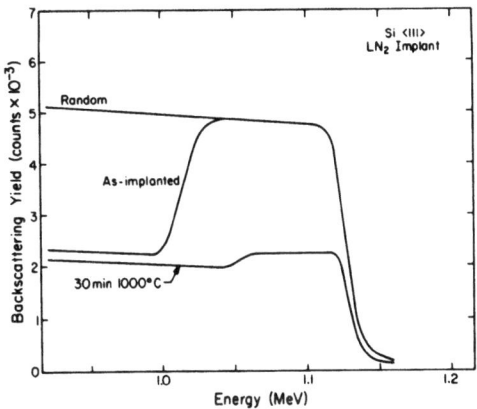

Fig. 5. Annealing of <111> Si implanted with 200 keV ^{75}As, dose $10^{16}/cm^2$.

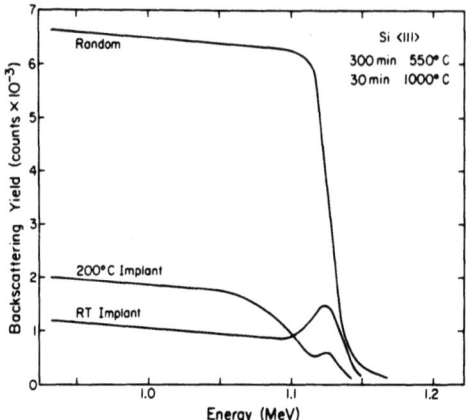

Fig. 6. Double anneal of <111> Si implanted with 200 keV ^{75}As, dose $10^{16}/cm^2$.

CONCLUSIONS

The substrate temperature during implantation and substrate orientation have a pronounced influence on the residual disorder observed after anneal of through oxide-implanted silicon. The anneal sequence also influences the residual disorder in <111> oriented samples but plays a lesser role in <100> oriented samples.

In <100> silicon implanted at RT and LN_2 the residual disorder observed after annealing at 550°C appears to be associated with the influence of oxygen on the regrowth rate. Consequently, the disorder is confined to a region within 500Å of the $Si-SiO_2$ interface. For <100> samples implanted at 200°C, the disorder after anneal at either 550 or 1000°C extends deeper into the sample. The residual disorder in <111> oriented silicon is higher than that in <100> silicon for comparable anneal sequences.

In conclusion, our results suggest that the minimum amount of disorder in "through-oxide" implants can be obtained by the use of <100> oriented silicion, by prevention of substrate heating during implantation, and use of a two-step anneal sequence. Elevated substrate temperatures during implantation not only leads to higher disorder after anneal but also shifts the stable disorder distribution deeper into the sample.

---

*Work supported in part by ONR (L. Cooper and D. Ferry), NSF (R. Hull) and Institute of Cultural Relations, Budapest, Hungary.
**Permanent Address: College of the Holy Cross, Worcester, Mass., 01610.
+Permanent Address: Central Research Institute for Physics, Budapest, Hungary.

REFERENCES

1. T.R. Cass and V.G.K. Reddi, Appl. Phys. Lett. 23, 268 (1973).
2. R.A. Moline, G.W. Reutlinger and J.C. North, in Atomic Collisions in Solids, Eds. S. Datz, B.R. Appleton and C.D. Moak (Plenum Press, New York, 1975) vol. I, p. 159.
3. W.K. Chu, H. Muller, J.W. Mayer and T.W. Sigmon, Appl. Phys. Lett. 25, 297 (1974).
4. T.W. Sigmon, W.K. Chu, H. Muller and J.W. Mayer, Proceedings of the International Conf. on Ion Implantation, Ed. S. Namba (Plenum Press, New York, 1975) p. 633.
5. R.A. Moline and A.G. Cullis, Appl. Phys. Lett. 26, 551 (1975).
6. L. Csepregi, W.K. Chu, H. Muller, J.W. Mayer and T.W. Sigmon, Rad. Effects 28, 227 (1976).
7. L. Csepregi, J.W. Mayer and T.W. Sigmon, Appl. Phys. Lett. 29, 92 (1976).
8. S. Mylorie, J.F. Gibbons and W.S. Johnson, Projected Range Statistics in Semiconductors, (Halstead Press, 1975) 2nd Ed.
9. E.F. Kennedy, L. Csepregi, J.W. Mayer and T.W. Sigmon (submitted to J. Appl. Phys.).

USE OF ION IMPLANTATION IN DEVICE FABRICATION AT HITACHI CRL

Takashi Tokuyama

Central Research Laboratory, Hitachi Ltd.
Kokubunji, Tokyo 185, Japan

ABSTRACT

In this paper, primary emphasis is placed upon finding physical and technological solutions to broaden the application of implantation to fields where implantation is not yet able to replace conventional technologies. In addition, recent results obtained from devices employing unique implantation characteristics are discussed.

INTRODUCTION

Ion implantation has been recognized for several years, as a practical processing technology in the semiconductor industry. It is already applied to most silicon device fabrication processes. However, it is not necessary to review the general status of application at this time.

In this paper, discussions are made on the technological or physical difficulties which prohibit replacing conventional technologies with implantation in two technical fields ; high dose implantation and low temperature annealing. Some recent results for devices fabricated employing unique implantation characteristics are discussed in the second place.

Implantation applications used in device fabrication are summarized in Tab. 1. In the first place, impurity doping by implantation has advantages like controllability in doping amount and profile. However, as it is well known, uniformity,

Table 1 Effects used in device fabrication
1. Impurity Doping
    Control of doping amount
    Control of profile
2. Compound Formation
    Silicon oxide, Silicon nitride
3. Disorder Introduction
    Gettering of impurities
    Control of process rate

reproducibility and in some cases the method of measureing the uniformity itself are technical problems as yet unsolved entirely. The uniformity of doping amount to $\pm 1\%$ in a wafer is required by the LSI processes of today.

There are some factors that skew the profile such as channeling, radiation enhanced diffusion and radiation retarded diffusion. Local activation of impurities which occurs in connection with the special configuration of defects results in phenomena like reverse annealing. However, these phenomena are at least controllable if the processing conditions are properly chosen.

Annealing treatment is usually carried out in connection with other process conditions and not necessarily optimized from the recovery of the damaged layer. Lots of defective structures remain after the annealing process, especially when ion dose is high or when some special impurity element like knock on oxygen atoms are introduced. These structures like dislocation lines, loops and networks are formed due to the primarily generated damages. The electrical effects of these secondary or tertiary defects are not fully understood at present. In the devices where the implanted layer becomes an active region and ion dose is rather high, for instance the base region of a bipolar transistor, implantation is still not a reliable process.

Compound formation is not widely used in silicon processing. Heavy implantation of oxygen or nitrogen to form oxides or nitrides is a candidate for use in fabricating new device structures. Introduction of disorder is intentionally applied in various parts of the silicon process. Impurity gettering is the most common application. Another application of strain introduction is control of the etching or oxidation rate. In the fabrication of complex integrated structures especially when the size of the elemental devices becomes small, local control of the processing speed becomes important.

Fig.1 shows the application field of implantations in silicon devices as a function of ion energy and dose. Low dose application like control of the threshold voltage in MOS devices or the capacitance-voltage curve of VARICAPs are nearly established. There are no problems arising from the remaining defects in this dose range after sufficient annealing. However, precise control and reproducibility of dose is essential in this range.

In the medium dose range, junction formation of the sources and drains in MOS devices, the well region of CMOS's, the channel region in junction FET's and the base region in bipolar transistors are usually carried out at 10^{14} -10^{15} /cm^2 dose. Annealing is usually done during the diffusion process to achieve specified junction depth. The limitations of implantation machines to ensure throughput is not severe in this range, and the effects of defects are rather small exept in unusual cases.

In the high dose range, machine capability for increasing the dose rate in connection with the non uniform beam heating effect, discussed by Freeman et al (1) still show some uncertainties for full use in this implantation range. However, the largest problems in this dose range are obviously the residual defects and their influence on device characteristics.

In the upper end of the high dose range, compound formation and ion beam deposition take place. Ion beam deposition is the process of depositing a thin film of a pure material with a mass separated ultra low energy (100eV) ion beam (2), however, it is not discussed in the present paper as it belongs to a different technical category.

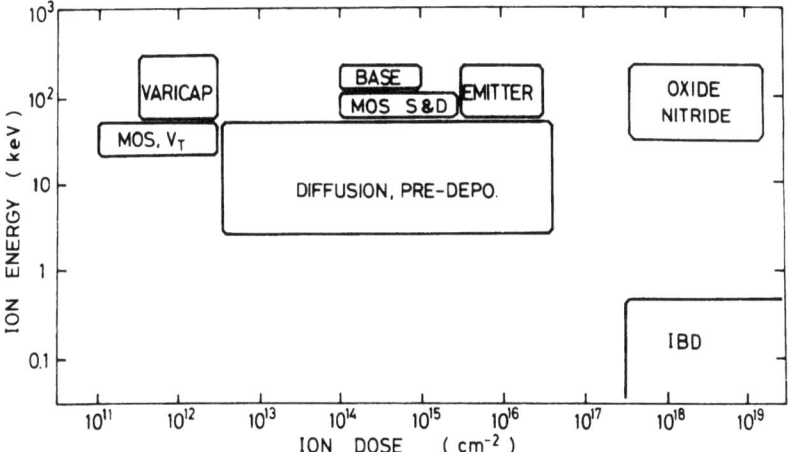

Fig.1 Application field of implantation as a function of ion dose and energy.

PROBLEMS STILL REQUIRE RESEARCH

High Dose Implantation

Fig. 2 shows the effect of the annealing atmosphere on residual defects for 50 keV phosphorus implants with a dose of 3×10^{16} /cm^2 (3). If annealing is carried out in wet oxygen for 40 min. at 1100°C, the Sirtl-etched pattern shows heavy defects remaining inside of the implanted area. More importantly, dislocations are propagated far outside of the implanted area.

On the contrary, if annealing is done in dry nitrogen, we have no defects either inside or outside the implanted area. However, when looking at the samples in more detail by transmission electron micrography, dislocation networks were observed in both samples. The concentration of network defect differs for both treatments.

The carrier distribution profile is almost the same in both samples except at the surface as shown in Fig. 3. The surface difference is due to the oxidation effect. However, the depth where networks are observed differs very much. In the oxygen treated sample, dislocation nets lie more than 1 μm from the surface and this is attributed to the climb action of dislocations in reaction with the flow of vacancies in the bulk toward the silicon-SiO$_2$ interface during annealing.

These networks are not observed in phosphorus diffused samples of equivalent phosphorus concentrations or heavily silicon implanted samples. If silicon is implanted into phosphorus diffused samples, we have both networks and outside dislocations.

The results are summarized in Fig. 4. Compared with diffusion cases, the same configuration of defects was observed in implantation cases at order of magnitude smaller phosphorus concentrations. The reason why dislocations did not propagate in nitrogen annealing is considered due to the direction of strain in the implanted area.

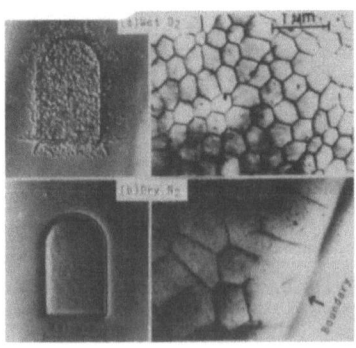

Fig.2 Defects observed by Sirtl etching and transmission electron microscope. 50 keV, 3×10^{16}/ cm^2 phosphorus implantation, 1100°C 40 min. annealing.

ION IMPLANTATION IN DEVICE FABRICATION

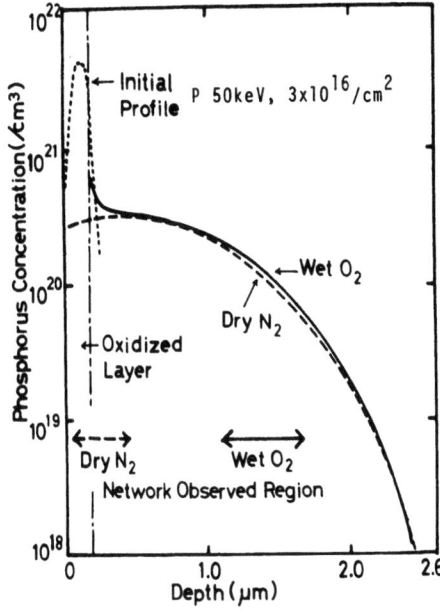

Fig. 3 Phosphorus concentration profiles and depth of defects of samples in Fig. 2.

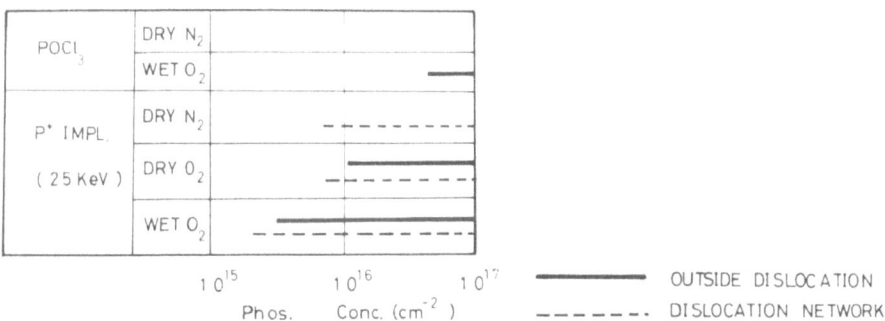

Fig. 4 Summary of defect configurations for phosphorus diffused or implanted silicon with various treatments and phosphorus concentrations. 1100°C, 40 min. treatment.

Compressive strain was detected by X-ray topographic measurement in oxygen treated samples (3). This is true in arsenic or boron implants. However, outside dislocations were observed in arsenic implants but not in boron implanted samples.

There are several approaches for minimizing both dislocation networks and outside dislocations. The first one is to reduce strains resulting from the difference in atomic radii between implanted impurity atoms and silicon. We choose germanium ions for phosphorus implantation and the results were reported elsewhere (4).

The second one is to implant into a thin surface film of poly-crystalline silicon or silicon oxide. Impurities are then diffused from the film while defects are not propagated from the film since the crystal structure is not continuous between the substrate and surface film. The results were also reported previously (5), and will not be refered to further here.

The last one is low temperature oxidation and following diffusion scheme. First, the damaged layer is oxidized at such a low temperature that defects will not propagate. The damaged layer surface is thereby turned into an oxide film. Implanted impurities do not diffuse and usually remain in the oxide. The oxidation process is carried out at 800°C in wet oxygen and followed by 1100°C diffusion in dry nitrogen atmosphere. Dislocation nets and outside dislocations were not observed after diffusion, however, the net yield of impurities was somewhat low and sheet resistivity became a little higher than expected for the ion dose.

When we apply these processes to base region fabrication of npn transistors, the following differences are observed. The results of low frequency noise measurement for various transistors are shown in Fig.5. The base regions were fabricated by implantation for all curves, and emitters by diffusion with the exception of one sample. Wet oxygen and dry nitrogen single anneal processes resulted in very large noise figures. However, low temperature oxidation and diffusion two-step treatments showed the same low noise levels as for diffused base transistors. If we implant the emitter and anneal in dry nitrogen after two-step annealing of the base, the noise figure at very low frequencies is slightly higher at present. Thus it appears that concentrated effort is needed to find a better treatment method so as to reduce or eliminate noise.

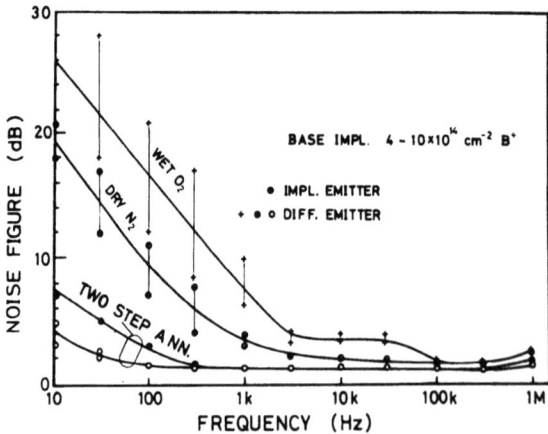

Fig. 5 Noise figure of implanted base transistors.

Low Temperature Annealing

Fig.6 shows the electrical activation characteristics and carrier mobility of medium range phosphorus implanted silicon after low temperature annealing. The doping efficiency shows the ratio of electrically active phosphorus against total implanted phosphorus. There is a peak at the $3 - 4 \times 10^{14}$ / cm^2 dose level at about 500°C.

In the lower region, range 2, the number of small amorphous islands increases proportionally with the dose and dissociation of these small amorphous islands causes phosphorus atoms to be included into the active site of the crystal lattice. However, in the higher part of the peak, range 3, such amorphous islands become so large that they can not be annealed out sufficiently at low temperatures. At the same time, the solubility curve limits the involvement of phosphorus at the peak of the phosphorus profile. Hence, doping efficiency decreases as the dose increases (6).

In the still lower dosage range, of 10^{13} / cm^2 order level, range 1, the reason why doping efficiency decreases with the increases in the dose is not clear. The possible effect of channeled and substitutionally implanted phosphorus or increase in carrier traps may be considered.

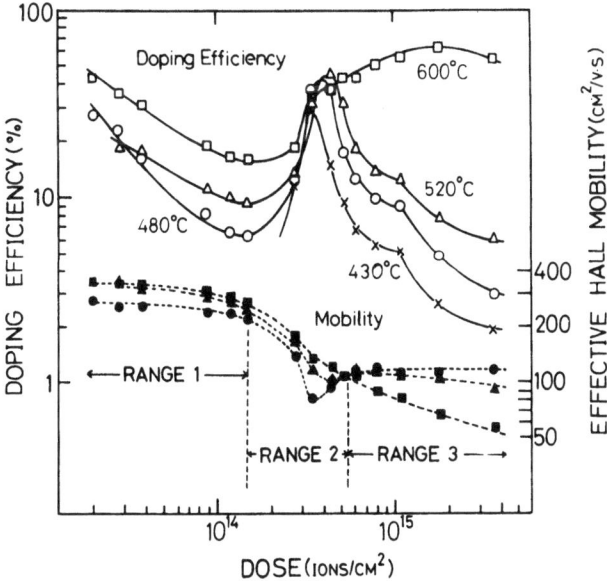

Fig. 6 Electrical activation characteristics and carrier mobility for low temperature annealed phosphorus implanted silicon. 50 keV ion energy.

If it is possible to pull up the curve in range 1, it will enable use of this low temperature process in fabricating devices. Carrier mobility is almost totally recovered in this annealing temperature range. This can be done by introducing impurities and defects independently by double implantation of phosphorus and silicon for example. There are many possible means of realizing low temperature recovery and there are also many sophisticated ways of utilizing these characteristics in device fabrication.

DEVICE APPLICATION EXAMPLES

The flow chart of a buried nitride MOS fabrication process is shown in Fig.7 (7). Nitrogen molecular ions of 10^{18} / cm^2 are implanted at 180 keV in the substrate, then annealed for 14 hrs at 1000°C. At this stage, the nitrogen implanted layer became highly resistive silicon nitride with a breakdown voltage of 160 V. The thickness of the insulating layer is about .2μm. Silicon single crystals remained on top of this. An epitaxial layer about 1 μm thick was then grown on the surface, and MOSFET's were fabricated in the epi-layer. The electrical characteristics of these FET's show a surface carrier mobility comparable to conventionally processed ones.

Two examples of through layer implantation is shown in Fig.8. (A) shows VARICAP fabrication after a diffusion and passivation process, and (B) shows base layer fabrication after emitter diffusion (8).

In the case of VARICAP, the phosphorus implanted peak should be carefully controlled from the variation of capacitance and voltage curve. For electronic tuner application of VARICAP's of

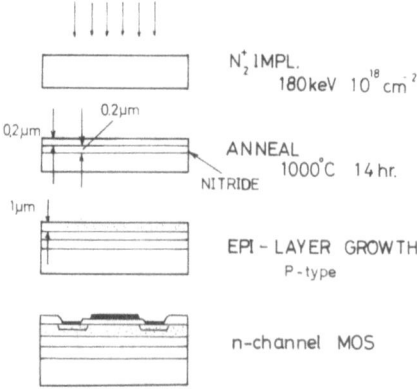

Fig. 7 Buried nitride MOS fabrication process.

UHF TV receivers, precise matching of the diodes V-C curves is required. To get such characteristic accuracy, the epitaxial and diffusion processes should of course be controlled.

Channeling of phosphorus ions is the main cause of variations on the V - C curves. We avoided channeling by not only tilting the wafer in the proper direction but also by implanting argon ions prior to phosphorus implantation. This made a thin disorderd layer on the top surface. Fig.9 is the result of capacitance variation. The capacitance of most diode chips in a wafer lie between ±1.5% at each bias voltage. C_6 and C_{12} indicate variations in the phosphorus profile, and C_{25} indicates variations in epi-layer resistivity and thickness.

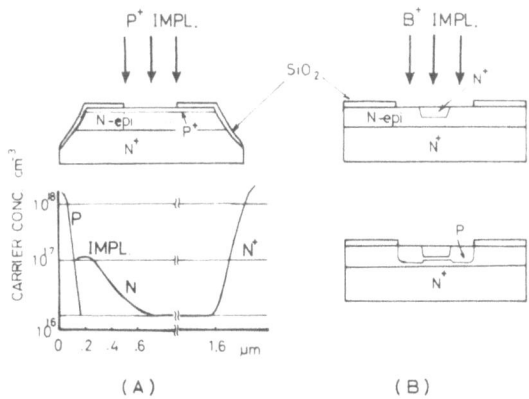

Fig. 8   Through layer implantation examples.

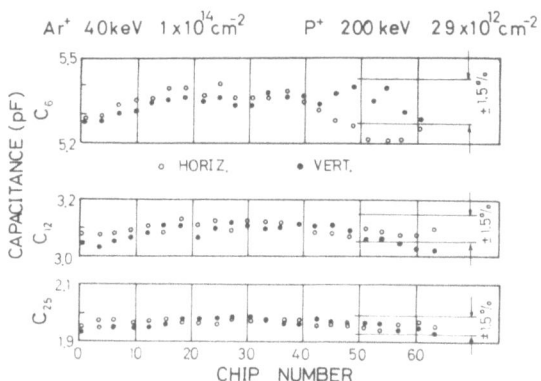

Fig. 9   Variation of capacitance values of VARICAP diodes at 6, 12 and 25 volt in a wafer with argon implanted disorderd surface layer.

In power rectifier applications, reduction of the forward voltage drop is very important to ensure minimal power loss. In Schottky diodes, the forward voltage drop is generally low compared with the p-n junction diodes because of lower barrier height. However, the reverse current is large in Schottky diodes, and large area metal-semiconductor contacts are still unreliable for power applications. There is a requirement for low loss junction type diodes.

If we realize an intermediate case between Schottky and junction diodes by implanting p-type impurities on the surface of the n-type substrate under the metal electrode, electron current flowing from the semiconductor to the metal electrode is controlled by the total p-type doping amount, provided that the electron diffusing length in the p-type layer is longer than the junction depth.

The results of our experiments are illustrated in Fig.10. In the Schottky case, the scattering of characteristics is rather large and differs for individual metal electrodes. Implantation of more than 10^{12} / cm^2 boron ions under metal electrode makes such scattering of the characteristics small. The forward voltage drop remains low compared to the p-n junction and the reverse current is also small compared to the Schottky diodes.

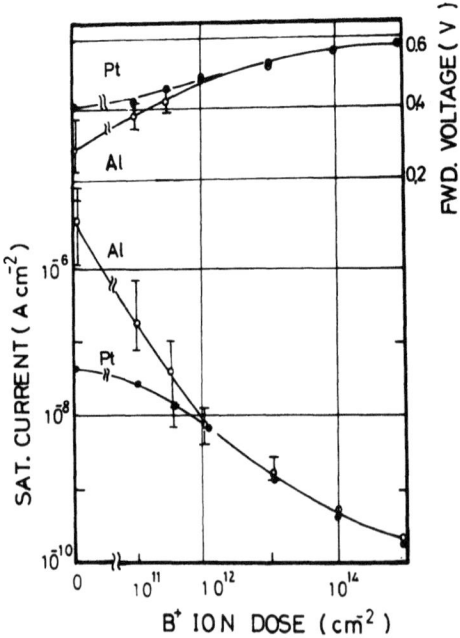

Fig. 10 Saturation current and forward voltage drop (at 0.1 mA / cm^2) of Schottky and junction diodes as a function of surface boron doping under metal electrode.

ION IMPLANTATION IN DEVICE FABRICATION

The control of the threshold voltage in MOSFET's is usually accomplished by implanting small amounts of impurities into the surface layer under the gate oxide film. In recent large scale dynamic RAM's of MOS-LSI, however, the amount of leakage current at gate voltages below the threshold voltage is rather important for obtaining appropriate memory refreshment times. This leakage current can be controlled by introducing two types of impurities into the channel region, thus changing the gate voltage versus surface induced charge curve (9).

Fig.11 is a schematic diagram of computer calculated potential curves depicting the surface layer under gate bias conditions of FET's with various impurity profiles. In the second curve, the tail of the boron profile is compensated by a phosphorus profile This limits the acceptor distribution profile only in the surface inversion layer and makes the surface induced charge versus gate voltage curve steeper.

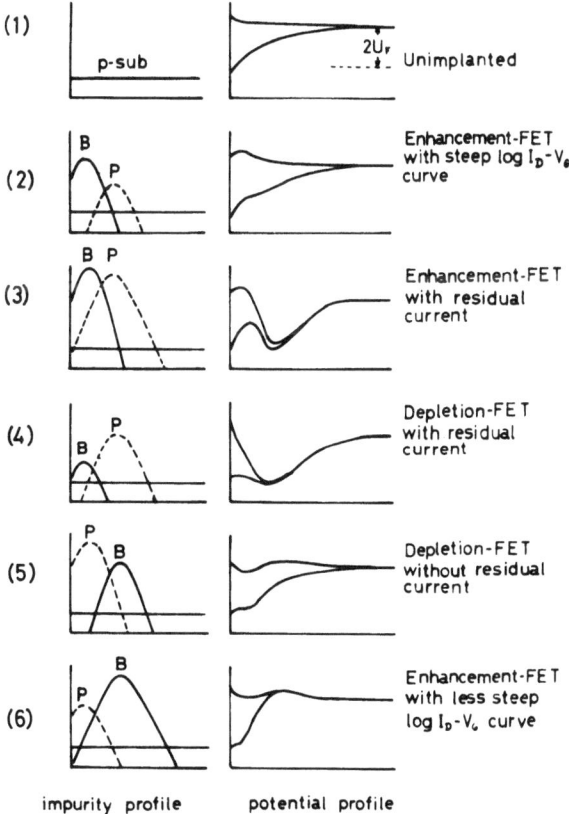

Fig. 11 Surface potential profiles that correspond to the various surface doping profiles.

Calculated examples of several surface induced charge versus gate voltage relations are shown in Fig.12. With a plus shift in the threshold voltage, double implantation of the phosphorus compensated boron results in a steep curve, indicating the reduction of the channel current below the gate threshold voltage. On the minus side, compensation of the phosphorus tail with boron makes the residual channel current low.

Many parameters are available such as ion energy, oxide thickness and substrate resistivity, so that essentially, any desired n_{SF} - V_G relation can be realized.

High voltage MOSFET's are realized by making offset gate structures to release an electric field at the gate side of the drain junction. The acceptor concentration of this offset gate area is controlled by implantation. As the concentration and structure parameters are optimized by computer calculation, we obtained a breakdown voltage of 300V for the parameters shown in Fig.13. (10)

Application of such high voltage MOSFET's in a high power MOS device utilizing a vertical drain integrated structure is shown in Fig. 14. Each mesh is 22 μm, and the source is surrounded by a channel and vertical drain (]]). A power handeling capability of 200 W was attained with a chip size of 5 x 5 mm².

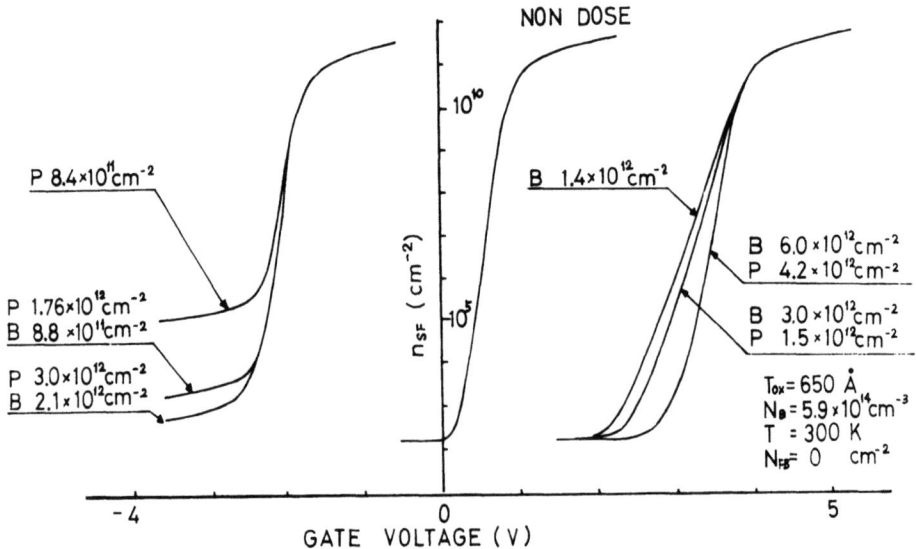

Fig. 12 Shift of $\ln(n_{SF})$ - V_G curve caused by case 2 and 5 profiles of Fig.11.

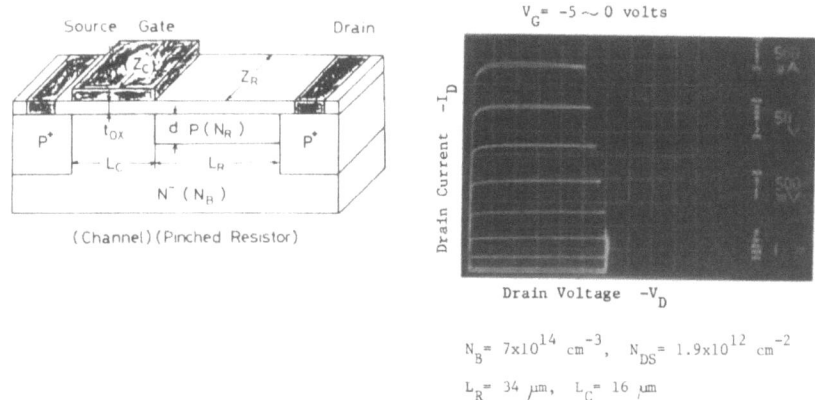

Fig. 13 Structure schematics and characteristics of high voltage MOSFET.

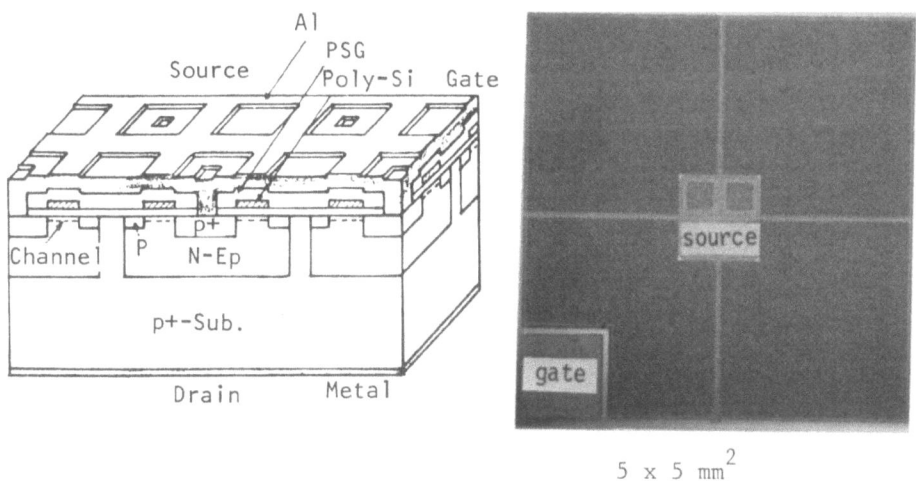

Fig. 14 Structure and surface photograph of high power MOSFET.

Lastly, shown in Fig.15 is a recent 16kbits MOS-RAM LSI. Implantation was carried out in field isolation and channel doping processes. Without implantation technology, we would not be able to attain devices of this size (12).

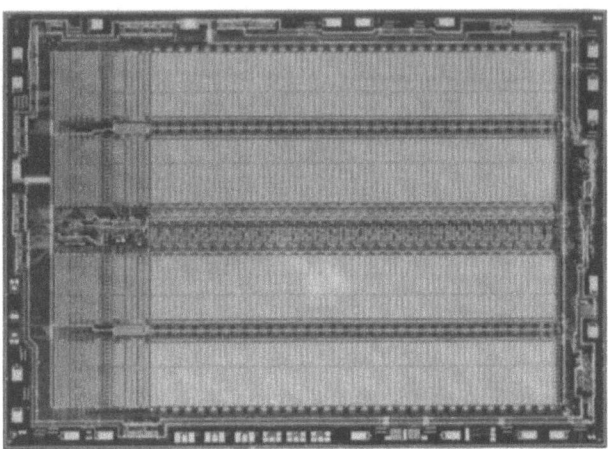

Fig. 15 16kbits silicon gate MOS-RAM LSI.

CONCLUSION

Recent silicon technologies require more and more precise control of the process paraneters and full automatic operation. Ion implantation potentially satisfies such requirements, however, compatibility or linkage between the conventional technologies is of most important and concentration of the research efforts are still required.

REFERENCES

(1) J.H.Freeman, D.J.Chivers, G.A.Gard, G.W.Hinder, B.J.Smith and J.Stephen : Proc. 4th Int'l Conf. Ion Impl. in Semi-conductors and Other Materials (Osaka 1974) pp555, Plenum Press 1975
(2) K.Yagi, S.Tamura and T.Tokuyama : Japan J. Appl. Phys. (to be published)
(3) T.Ikeda, M.Tamura, N.Yoshihiro and T.Tokuyama : Proc. 6th Conf. Solid State Devices (Tokyo 1974) Suppl. Jour. Japan Soc. Appl. Phys. $\underline{44}$ 311 (1975)

(4) N.Yoshihiro, M.Tamura and T.Tokuyama : Proc. 4th Int'l Conf. Ion Impl. in Semiconductors and Other Materials (Osaka 1974) pp571, Plenum Press 1975
(5) N.Natsuaki, M.Tamura and T.Tokuyama : Japan J. Appl. Phys. (to be published)
(6) M.Miyao, N.Natsuaki, N,Yoshihiro, M.Tamura and T.Tokuyama : Proc. 7th Conf. Solid State Devices (Tokyo 1975) Suppl. Japan J.Appl. Phys. $\underline{15}$ 57 (1976)
(7) K.Itoh and T.Tsuchimoto : "Ion Impl. in Semiconductors" Symposium held at The Inst.Phys.Chem.Res. (Wako-shi, Saitama, Japan 1975) pp43
K.Sugawara, T.Yoshimi, Y.Nakagawa and K.Itoh : J.Electrochem. Soc. $\underline{123}$ 759 (1976)
(8) T.Tokuyama, T.Ikeda and T.Tsuchimoto : Proc. 4th Congrss Microelectronics(Munchen 1970) MIKROELEKTRONIK 4 pp36, R.Oldenbourg Verlag (1971)
(9) T.Masuhara and J.Etoh : IEEE Trans. $\underline{ED-21}$ 799 (1974)
(10) I.Yoshida, T.Masuhara, M.Kubo and T.Tokuyama : Proc. 6th Conf. Solid State Devices (Tokyo 1974) Suppl. Jour Japan Soc. Appl. Phys. $\underline{44}$ 249 (1975)
(11) I.Yoshida, M.Kubo, S.Ochi and Y.Ohmura : Proc. 7th Conf. Solid State Devices (Tokyo 1975) Suppl. Japan J.Appl. Phys. $\underline{15}$ 179 (1976)
(12) K.Itoh, K.Shimohigashi, K.Chiba, K.Taniguchi and Y.Kawamoto : 1976 IEEE Int'l Solid State Circuit Conf. (Philadelphia 1976) pp140

Sb$^+$-IMPLANTED BURIED LAYER BENEATH THICK OXIDE APPLIED FOR VERTICAL FET

Y. Akasaka and K. Horie
LSI Development Laboratory, Mitsubishi Electric Corporation, Amagasaki, 661 JAPAN

G. Mitarai, Y. Hirose, K. Nomura and H. Nishiumi
Kitaitami Works, Mitsubishi Electric Corporation
Mizuhara, Itami, 664 JAPAN

ABSTRACT

Diffusion procedure of implanted Sb in Si including post-oxidation to μm-thickness is investigated in order to reduce implantation-induced damage influence and to fabricate n$^+$-buried layer beneath thick oxide. Implanted and diffused profiles of Sb are measured by ion backscattering technique and the defects are studied by Sirtl etching method. Implanted Sb atoms are accumulated at the interface of the advancing oxide front after oxidation, i.e., implanted Sb is carried inward with the oxidation speed. Because of strong segregation property, loss of implanted Sb is minimal even after the oxidation to 1 -2 μm thickness. Accumulated Sb is diffused into the substrate to form n$^+$-layer by the subsequent drive-in diffusion in the N$_2$ atmosphere. The damage effect is significantly lowered as the oxide thickness becomes larger by one or two orders magnitude than R_p of Sb ions. The n$^+$p diodes subjected to the oxidation exhibit small reverse leakages and hard breakdown. This doping method is applied to fabricating a fine mesh gate of vertical FET. The buried gate region and satisfactory triode characteristics of p-channel FET are realized.

INTRODUCTION

Sb is known to have an extremely smaller diffusion coefficient in Si and is preferable for the use of subcollectors of bipolar IC's or other fine structure devices. Sb implantation is one of the promising doping methods of Si with Sb, if it is possible to reduce the damage effect. Although oxidizing the initially damaged layer completely is an easy approach to reduce the damage introduced by ion implantation, this process is usually accompanied by a loss of implanted impurities during the oxidation. Because of its strong segregation property between Si and SiO_2, Sb is one of the best impurities which should be examined with this approach. Some previous works[1,2] pointed out that oxidation at high temperatures after implantation is not preferable since it induced the formation of expanded defects. In those works oxidation was performed at high temperatures or/and for a relatively short time to form thinner oxide layer than the thickness of initially damaged layer. This is not the present case, which is carried out at relatively low temperature (950°C-1020°C) and to form much thicker oxide layer than the initially damaged layer.

The present report gives one of the methods to obtain Sb doped n^+-layer of good quality by utilizing ion implantation and post-oxidation followed by N_2 diffusion. One of the most preferable process has been applied to fabricating the gate region of p-channel vertical power FET, and the results of device characteristics are included in this paper.

EXPERIMENTAL

The Si wafers used for this investigation were p-type, 10-20 $\Omega \cdot cm$ resistivity of (111) orientation. Sb implantation were performed at 50 keV with dose of $2 \times 10^{15}/cm^2$ in random direction at room temperature. After the implantation the sample wafers were first oxidized in the wet O_2 atmosphere at 950°C for 10, 20 and 30 minutes to various thickness which were much thicker than the projected range R_p of Sb ions. After the oxidation, Sb was driven into the substrate at 1200°C for 50 minutes in the N_2 atmosphere.

Backscattering and channeling measurements were performed after each step of the process described above, with 1.5 MeV $^4He^+$ ions. The detection angle was 150° to the incident beam direction.

For the defects observation in the implanted area with Sirtl etching, a gate definition process of vertical FET was added to the sample preparation described above in order to distinguish the implantation induced defects from possible material defects. After the photoresist process for the gate definition, the samples were exposed to 70 keV Sb implantation with doses of $2 \times 10^{14} - 1 \times 10^{15}/cm^2$. The post-oxidation and subsequent N_2 drive-in diffusion were carried out at 1020°C for 30, 60minutes and 4 hours in the wet O_2 atmosphere and at 1200°C for 3 hours, respectively. For comparison, samples receiving no oxidation but the N_2 drive-in alone were prepared. Sirtl etching was made without oxide cap after the oxidation and after the N_2 drive-in.

Electrical junction characteristics between gate and drain were also examined for various kinds of samples.

RESULTS AND DISCUSSIONS

Redistribution of Implanted Sb

Figure 1 shows backscattering spectra from Sb-implanted samples (50 keV, $2 \times 10^{15}/cm^2$) after the post-oxidation at 950°C for 10, 20 and 30 minutes. The oxide thickness measured by ellipsometry were respectively 1720, 2360 and 3300 A, which coincided with the depth of Sb peaks calculated by using the stopping power of SiO_2 experimentally determined to be 320 keV/μm. The oxidation was enhanced about twice as large as that of unimplanted samples, as we previously reported[3], and it consumed an amount of Si much greater than R_p of Sb ions (271 A)[4]. We find from the spectra that growing oxide drives Sb atoms to its front and, as a result, that a very sharp peak of Sb concentration locates at the interface of Si and SiO_2 in any sample. The concentration of Sb left in the SiO_2 was extremely low relative to that of the interface. The peak width of Sb signal was approximately 360 A and it did not change significantly with the oxide thickness. This indicates that the peak width was restricted by the resolution of the measuring system. Accordingly, the actual Sb peak at the interface should be narrower and higher than the observed spectrum. The peak height decreased gradually with an increase of the oxide thickness. This would be due to the energy straggling of the He particles.

It was also found by backscattering measurement that stripping the oxide layer before the drive-in removes almost all Sb atoms simultaneously and this suggests that Sb locates at the very interface of Si and SiO_2. This is in agreement with the result

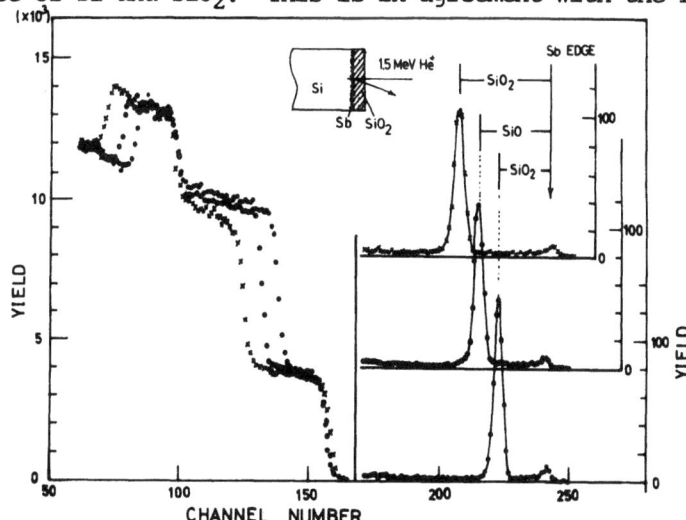

Fig. 1   Backscattering spectra from Sb implanted and post-oxidized Si.   implantation: 50 keV Sb^+, $2 \times 10^{15}/cm^2$
oxidation: 950°C, in the wet O_2, 10, 20 and 30 min.

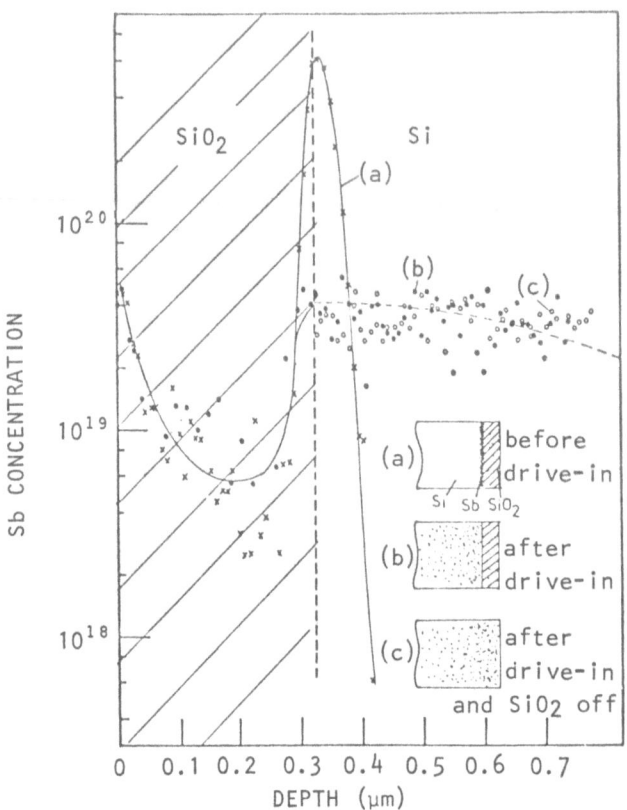

Fig.2 Depth profiles of Sb before and after the drive-in. Implantation was carried out with 50 keV Sb+ to dose of $2 \times 10^{15}/cm^2$, oxidation at 950°C in the wet O_2 ambient for 30 min., and the drive-in at 1200°C in the N_2 atmosphere for 50 min.

reported by Müller et al[5].

Figure 2 shows the depth distribution of Sb before and after the drive-in with the oxide layer on the substrate at 1200°C for 50 minutes measured with backscattering technique. Oxidation time was 30 minutes. Also shown in the figure the profile of Sb in Si after stripping the SiO_2 layer. The distribution of Sb in the SiO_2 layer did not change by the heat treatment, indicating that the diffusion coefficient of Sb in SiO_2 is very small. In contrast, Sb atoms which located at the interface before the drive-in were diffused into the substrate and the redistribution after the drive-in diffusion could be fitted to Gaussian distribution. The diffusion coefficient was calculated to be 3×10^{-13} cm^2/sec, assuming that the profile was a delta function before the drive-in. This value agreed with those reported by other workers[6,7]. The surface concentration was around 4×10^{19}/cm^3 and is very near to the solid solubility.

Channeling measurements were performed with the sample after the drive-in and stripping the oxide layer. Figure 3 shows the result of angular scan around[111] and [110] axes. The critical angles of the dips for Sb agreed with those for Si in the both axes. It was verified that Sb is highly substitutional after the drive-in, while before the drive-in Sb atoms were found to be accumulated at the interface and to locate on non-substitutional sites. It is concluded that the post-oxidation can carry Sb atoms to much deeper region than the initially damaged layer by utilizing the segregation effect.

Defect Observation by Sirtl Etching

The effect of post-oxidation on implantation-induced defects were investigated by Sirtl etching technique with the samples oxidized at 1020°C for 30, 60min, and 4 hours after the implantation. Figure 4(a) shows the Sirtl etching result in the case of 1×10^{15}/cm^2 implant with 70 keV Sb ions. The photograph in the figure shows the Sirtl-etched surfaces in the gate bonding pads, and defects observed have been confirmed to be originated from the implantation. The photograph on the left side shows a Sirtl-etched sample which was driven in the N$_2$ atmosphere alone without any oxidation. The other 6 photographs indicate the samples exposed to the oxidation to the various thickness, which were 3000, 4500 Å, and 1.2 µm, and 3 of them at the bottom side were those followed by the N$_2$ drive-in diffusion. Significantly dense shallow pits were found in the sample receiving no oxidation, and more sparse or smaller pits in the sample receiving post-oxidation to the thickness of 3000 Å before the drive-in. When the oxide becomes thick, for instance, 1.2 µm, shown on the right side, we cannot recognize any process-induced defects. Figure 4(b) shows the lower dose case which was 2×10^{14}/cm^2. It can be easily found also in this case that the growing oxide can consume the implantation-induced disorder. These results indicate that oxidation to much deeper regions than initially damaged layer is effective to eliminated the damage influence caused by shallow Sb implantation.

It was also found in electrical measurement that gate-drain junctions of the specimens subjected to the oxidation followed by the N$_2$ drive-in diffusion exhibited extremely smaller reverse leakage than that of the specimen without the oxidation.

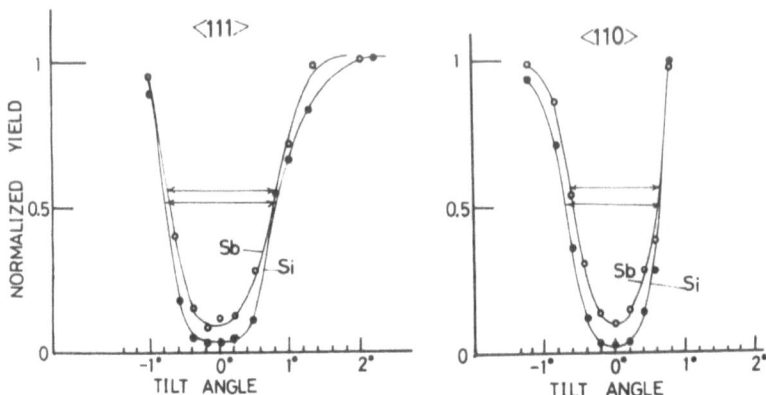

Fig. 3 Angular dependence of backscattering yields of Sb and Si after oxidation and N_2 drive-in.

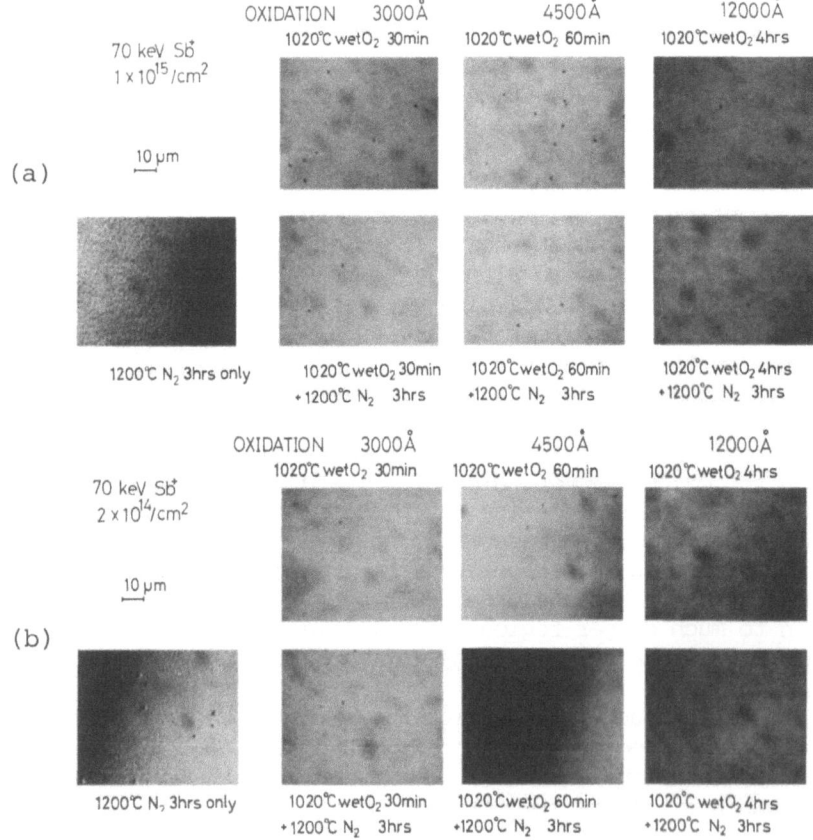

Fig. 4 Defects observation of Sb implanted layers by Sirtl etching. (a) 70 keV, $1 \times 10^{15}/cm^2$ (b) $2 \times 10^{14}/cm^2$

Application to Vertical FET

The doping method describe so far, i.e., Sb implantation followed by the oxidation and subsequent N_2 drive-in, is quite suitable in forming n-type region beneath thick oxide, and has been applied to fabricating the gate region of p-channel vertical power FET, whose structure is schematically illustrated in Fig. 5(a). A mesh type gate structure of each strip width less than few microns is required to realize satisfactory high-frequency characteristics and sufficient output power. The gate region should be buried beneath the locally oxidized layer of 1 to 2 μm thick to obtain high breakdown voltage between source and gate. The photograph in Fig. 5(b) shows the beveled section surface after lapping and Cu-staining of the gate region fabricated by the present method. A periodic black region like "teeth" is locally-oxidized layer and Sb doped n-type regions were clearly observed beneath them.

A typical drain characteristics of the p-channel vertical FET fabricated by utilizing Sb implantation is shown in Fig. 6. Satisfactory triode characteristics and output power of 30 watts were obtained. More detailed device characteristics are summarized in Table 1. We can get 37 V for BV_{SGO} and 125 V for BV_{DGO} in this

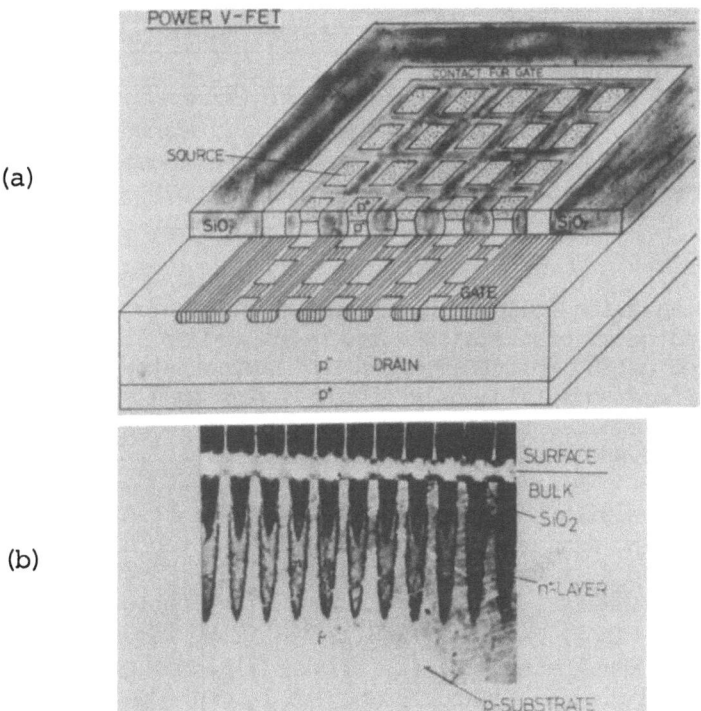

Fig. 5 (a) Schematically illustrated structure of p-ch. V-FET.
(b) Beveled surface of the gate fabricated by Sb implantation.

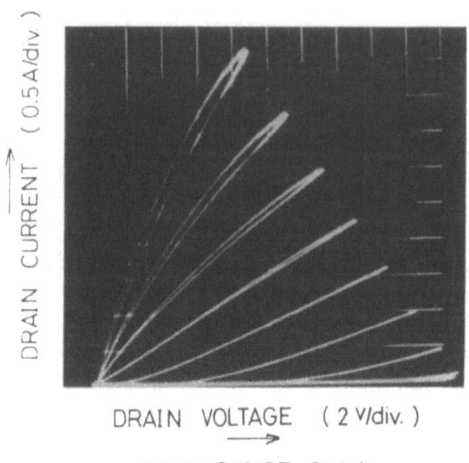

Fig.6 Drain characteristics of p-ch. V-FET fabricated by using Sb implantation.

Table 1 Device characteristics of p-ch. V-FET fabricated by Sb implantation

CHANNEL	OPERATION MODE	CHIP SIZE (mm)	V_{DGO} (V)	V_{SGO} (V)	$I_D(A)$ ($V_{DS}=-5V$, $V_{GS}=0V$)	P_d (W)	μ	r_{DS} ($V_{DS}=-5V$, $V_{GS}=0V$)
p	DEPLETION	3x3	125	37	3.1	30	3.4	1.7

structure. The results concerning the device characteristics suggests again that the oxide layer grown to much deeper region than R_p can consume the disorder caused by implantation.

SUMMARY

Sb implantation followed by oxidation and subsequent N_2 drive-in is found to be of practical use in the doping of Si with Sb. The damage induced by Sb implantation can be eliminated by oxidizing the implanted surface to sufficiently deeper region than R_p. The doping method has been applied to fabricating the n-type mesh gate of p-ch. vertical FET and successful result was obtained.

The authors wish to acknowledge the assistance of K. Tsukamoto for backscattering measurement and Sirtl etching technique. They also thank H. Sato and K. Mitsui for Sb implantation. Discussions with Dr. H. Komiya, Y. Watari and S. Kawazu were invaluable.

(1) S. Prussin; J. Appl. Phys. 45 1645 (1974)   (2) M. Tamura, H. Yoshihiro T. Ikeda; Appl. Phy. Letters 27 427 (1975)   (3) K. Nomura, Y. Hirose, Y. Akasaka, K. Horie, S. Kawazu; Proc. 4th Intern. Conf. Ion Implantation p.681 (Plenum Press, NY 1975)   (4) W.S. Johnson, J.F. Gibbons, Projected Range Statistics (Halstead Press, Penn., 1975) (5) H. Muller J. Gyulai, W.K. Chu, J.W. Mayer, T.W. Sigmon; J. Electrochem Soc. 122 1234 (1975)   (6) C.S. Fuller, J.A. Ditzenberger; J. Appl. Phys. 27 544 (1956) (7) J.J. Rohan, N.E. Pickering, J. Kennedy; J. Electrochem Soc 106 (1959) 705

ION IMPLANTED SOLAR CELLS

J.B. Neilson, T.M. Vanderwel, J. Shewchun and D.A. Thompson

Department of Engineering Physics
McMaster University
Hamilton, Ontario, L8S 4M1, Canada

Abstract

Ion implantation is investigated as a technique to fabricate solar cells on monocrystalline silicon. The effects of implanted species (As and P), implanted dopant concentration (10^{18}–10^{21} cm^{-3}), implant temperature (55° to 300°K) and annealing temperature (700° to 900°C) have been studied. Some progress has been made toward optimization of the various parameters.

Introduction

Several technologies are used in the production of solar cells. Standard diffusion techniques, MOS structures and thin film heterojunctions have all been used to make the collecting junctions. It is the aim of this paper to report on preliminary investigations into the use of ion implantation as the junction forming technology.

Much work has been done on the use of implantation doping in the fabrication of devices and integrated circuits. However, little has been done to investigate its potential in the fabrication of solar cells.[1,2] Several potential advantages of the use of ion implantation are apparent. There is greater control of the dopant distribution than with conventional diffusion techniques. In particular, very shallow junctions are possible, allowing for the formation of the "blue-shifted" cell[3], which shows an enhanced response to the short wavelength end of the spectrum. This should result in increased efficiency for terrestrial applications, and especially for space applications. Also, the geometry control available with the implantation technique facilitates the fabrication of grating cells[4], which also have an enhanced blue response and correspondingly increased efficiency.

Experimental

Cell substrates were p-type (boron doped) monocrystalline silicon, <111> oriented, with resistivities in the range 0.1 to 10 Ωcm. The implantations were carried out at McMaster using a 150 keV ion accelerator[5] fitted with a Danfysik 911A Universal Ion Source. Phosphorus and arsenic were implanted over an energy range of 20 to 80 keV with an average swept current of about 100 to 200 nanoamps per square centimetre. To achieve implant uniformity ($\lesssim$ 5%) a 2 mm. beam was swept x and y across the target. During the implant, the substrates were tilted off the aligned direction to avoid channelling. Facilities were available for implant temperatures of 40°K to 300°K.

Energies and doses for the implants were calculated using the data from Gibbons et al[6] to give a doped surface layer that was uniform with depth. A typical example is given in Figure 1. The lowest energy ion beam that was easily stabilized was found to be 20 keV. As can be seen from Figure 1, this leaves a very low dopant concentration near the surface, resulting in difficulties associated with making good ohmic contact to the device. Two possible solutions were tried: (i) following the implant a thin surface layer could be removed by growing an oxide (typically 300Å) followed by a subsequent oxide stripping using 10% HF; or (ii) a "through-the-oxide" implant could be used again followed by oxide removal resulting in a high impurity concentration at the silicon surface. By comparing stopping powers of silicon dioxide and silicon, it was found that an oxide thickness of ~130Å would be adequate. This oxide was normally removed after the anneal so that the silicon surface was protected. Both methods were tried and found equivalent, but as the oxidation rate of implanted silicon varies with different implant conditions, it was decided that the through-the-oxide implant was preferable.

Junction depth in all cases was determined from the calculated implant distribution (after Gibbons et al[6]) as shown in Figure 1. Table 1 lists calculated average concentration and junction depth for varying implant conditions.

The implanted layers were annealed in vacuum (10^{-6} torr) for 10 minutes, with anneal temperatures in the range of 700° to 900°C. Contacts were formed with evaporated aluminum in the geometry shown in Figure 2 and sintered in a nitrogen atmosphere of 400°C for 10 minutes.

TABLE 1

Species	C_B (cm^{-3})	E (keV)	ϕ (μc-cm^{-2})	E	ϕ	E	ϕ	E	ϕ	C_{AVE} (cm^{-3})	D_j (Å)
As	10^{15}	20	0.15	40	0.25	80	0.42			10^{18}	870
As	10^{15}	20	1.5	40	2.5	80	4.2			10^{19}	930
As	10^{15}	20	15	40	25	80	42			10^{20}	1000
As	2×10^{16}	20	0.15	40	0.25	80	0.42			10^{18}	775
As	2×10^{16}	20	1.5	40	2.5	80	4.2			10^{19}	850
As	2×10^{16}	20	15	40	25	80	42			10^{20}	910
As	5×10^{17}	20	0.15	40	0.25	80	0.42			10^{18}	600
As	5×10^{17}	20	1.5	40	2.5	80	4.2			10^{19}	740
As	5×10^{17}	20	15	40	25	80	42			10^{20}	825
As	10^{15}	20	15							10^{20}	340
As	10^{15}	20	15	40	25					10^{20}	570
As	10^{15}	20	15	40	25	80	42			10^{20}	1000
As	10^{15}	20	15	40	25	80	42	120	55	10^{20}	1400
P	10^{15}	20	40							10^{20}	700
P	10^{15}	20	25	40	62					10^{20}	1300
P	10^{15}	20	25	40	28	60	88			10^{20}	1800
P	10^{15}	20	25	40	48	80	115			10^{20}	2400
P	10^{15}	20	25	40	35	60	62	100	135	10^{20}	3000

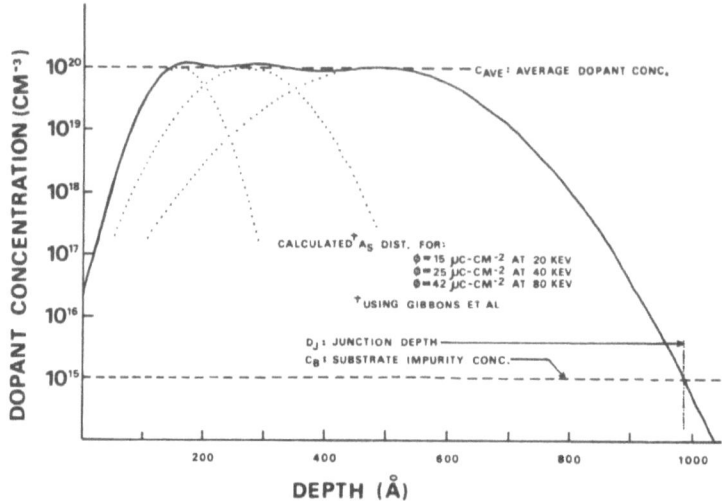

Figure 1 Calculated impurity distribution, showing several implants at different energies to achieve a uniformly doped n-type surface layer.

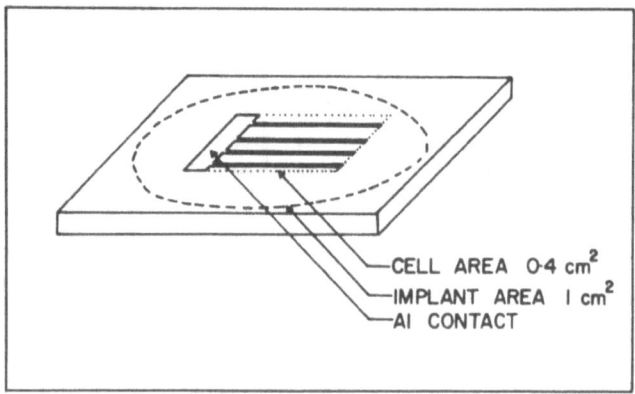

Figure 2 Experimental cell geometry.

Cell Evaluation

Cell efficiency was measured by observing the illuminated I-V characteristics. Illumination was provided by an Air Mass 2 (AM2) solar simulator consisting of four 300 watt tungsten-iodine lamps with dichroic filters. The spectrum is shown in Figure 3. Input power to the cell at AM2 is 75 mw-cm^{-2}.

To avoid fringe effects, only the area defined by the contact pad (Figure 2) was illuminated during the test. A typical I-V curve is shown in Figure 4. Significant parameters that were monitored are: the open circuit voltage, V_{oc}; the short circuit current, I_{sc}; the fill factor C_{FF}, defined as the maximum output power divided by the product of I_{sc} and V_{oc}; and the efficiency, E_{FF}, the maximum output power divided by the input power. Dark I-V characteristics were also recorded to give an indication of the quality of the p-n junction formed.

Results and Discussion

Several implantation parameters were varied, and the dependance of the cell performance was studied. Results are given here with cell efficiency as the dependent variable.

Earlier published results[7] have suggested that better electrical activity is obtained by total amorphization of the surface layer during implantation followed by epitaxial regrowth of the damaged region during the anneal stage. Hence, it was expected that at low implant temperatures, where little or no annealing occurs during the bombardment and a greater amount of damage would be allowed to accumulate, a better amorphous region would result, improving the anneal characteristics. However, this effect was not observed, as shown in Figure 5(a). All cells were fabricated on 10 Ωcm substrates and implanted with As$^+$ to a maximum energy of 80 keV, with an average concentration of 10^{20} cm^{-3} and annealed to 800°C. The dominant effect controlling efficiency in this case was the variation in I_{sc} from 14.5 ma/cm^2 at 55°K to 20 ma/cm^2 at room temperature indicating improved carrier lifetimes at the higher temperatures. Relatively constant V_{oc} and C_{FF} indicated that the barrier formation and sheet conductivity of the surface layer were not affected by the lower implant temperature.

The cells were evaluated after anneals at temperatures ranging from 700° to 900°C. As seen in Figure 5(b), from 700° to 800°C, cell performance improves with increasing anneal temperature, consistent with earlier studies[7,8] showing increased electrically active fraction of the implanted ions at higher temperatures. Over 800°C, however, V_{oc} drops off accompanied by a less rapid fall in

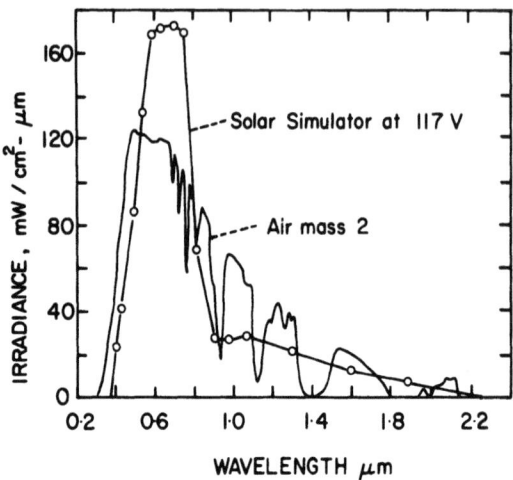

Figure 3 Spectrum of Air Mass 2 simulator used to characterize implanted solar cells

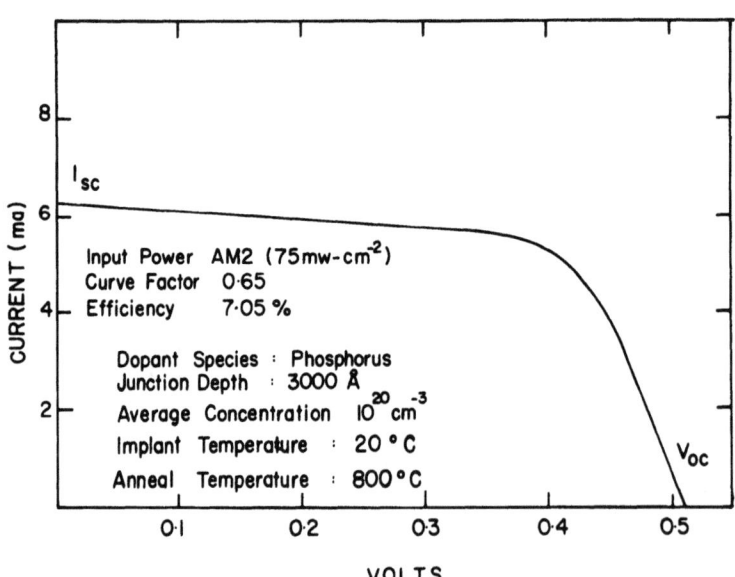

Figure 4 Illuminated current-voltage characteristic of one of the better cells. Input power source was an Air Mass 2 solar simulator.

ION IMPLANTED SOLAR CELLS

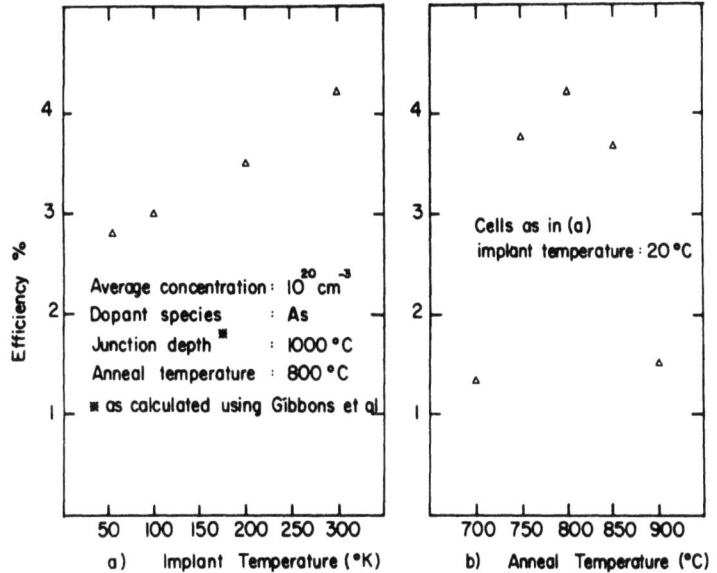

Figure 5 (a) Cell efficiency vs. implant temperature.
(b) Cell efficiency vs. anneal temperature.

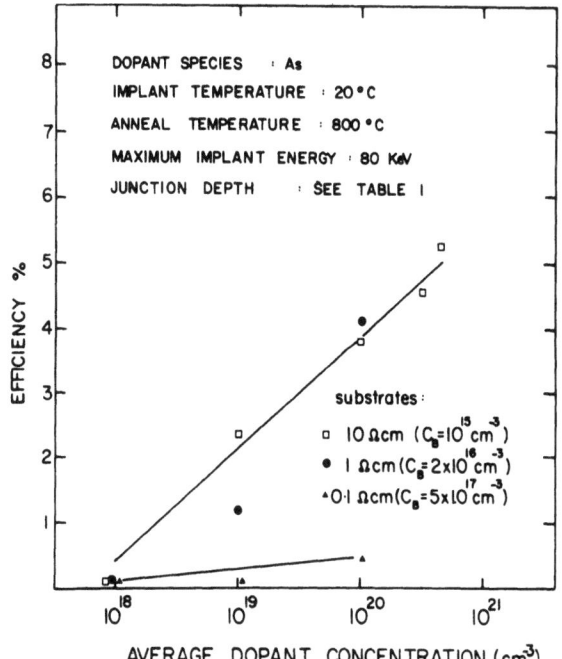

Figure 6 Cell efficiency as a function of average implanted dopant concentration and background impurity concentration. Maximum implant energy is constant, but the junction depth varies slightly as seen in Table 2.

C_{FF}. The drop in V_{oc} suggests a decrease in the junction barrier height, perhaps attributable to an enhanced diffusion effect resulting in a less abrupt junction.

The effect of both the implanted and the background impurity concentrations on cell efficiency was examined using arsenic implanted cells annealed to 800°C. The results are shown in Figure 6. It was expected that for greater dopant concentrations on both sides of the junction, the increased barrier height would result in a greater V_{oc}. However, it is clearly seen in Figure 6 that the efficiency drops off dramatically when the substrate resistivity is 0.1 Ω-cm. This is due to a reduction in carrier lifetimes, resulting in a reduced I_{sc}. In a shallow junction cell, the majority of the carriers are generated in the bulk material, and hence the bulk lifetimes and the surface resistivity are the dominating factors in cell behaviour. One would therefore expect the best results to occur for a cell with a high substrate resistivity, having long lifetimes, and a high implanted surface concentration. This is found to be true for diffused cells[9] and, as shown in Figure 6, is also consistent for implanted cells. There may be some depth dependence hidden in these results (see Table 1) but the magnitude of this effect over the range of depths here (600Å to 1000Å) should be negligible.

There are several ways in which the junction depth can affect the cell efficiency. First, surface effects are less for deeper junctions. Surface effects result from surface recombination and small local field perturbations due to inhomogeneities in junction depth and concentration. Also, the sheet resistivity of the surface layer, and hence the series collection resistivity, is lower for a thicker implanted surface layer. This would appear as an improved C_{FF}. Such effects probably account for the observations shown in Figure 7. The monotonic increase with junction depth in efficiency for the four arsenic implanted cells of Figure 7 suggests the use of phosphorus in order to obtain a deeper junction than possible with arsenic. It was originally considered that with the good match of the arsenic covalent radius with that of silicon, the strain at higher doping levels would be minimized, preserving the carrier lifetime. However, phosphorus, with its smaller radius, seems to give the better cell, even at the same calculated junction depth (40 keV P and 120 keV As). As indicated earlier, the junction depths were not measured, and the actual junction depths after annealing may differ from the calculated value, particularly for phosphorus[10,11]. It is not clearly understood whether the lifetime degradation in the implanted layer is masked by this junction depth difference, or whether it occurs at all.

Another cell characteristic dependent on junction depth is its spectral response. The photon absorption coefficient for silicon

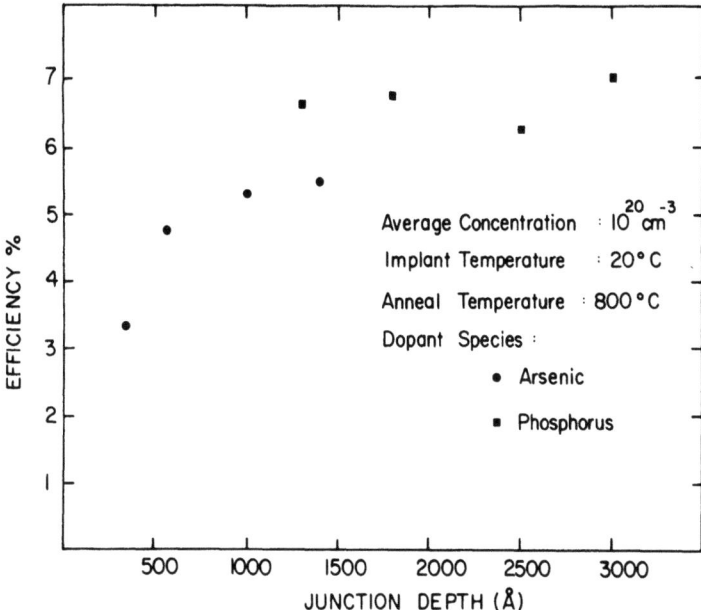

Figure 7   Cell efficiency vs. calculated junction depth. The deeper junctions were implanted with phosphorus.

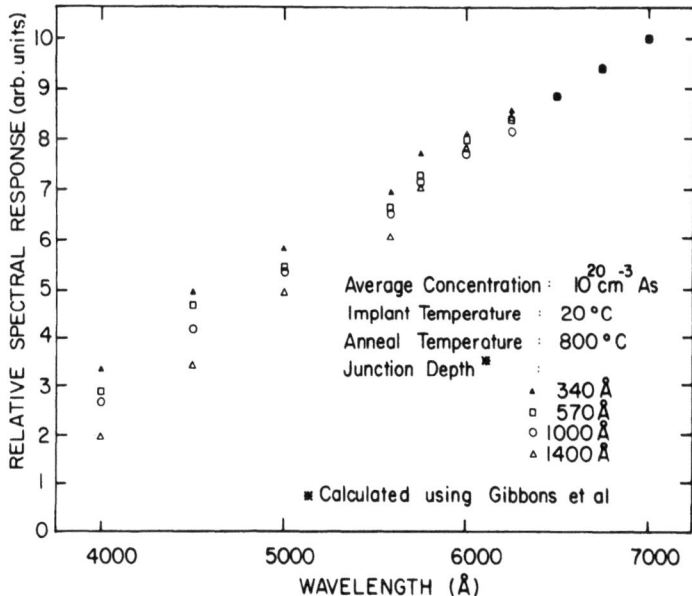

Figure 8   Relative spectral response for cells with different junction depths. Response curves are based on short circuit current, and are normalized to each other at the other red end.

increases with photon energy$^{(12)}$, leading to the absorption of most of the blue end of the solar spectrum very near the surface of the cell. The result is an enhanced blue response for the shallower junctions. Figure 8 shows the normalized spectral response (based on I_{sc}) of four cells implanted with arsenic at room temperature and annealed at 800°C. Impurity distributions are uniform to a maximum implant energy of 20, 40, 80 and 120 keV for the different cells. The four curves were normalized to coincide at 7000Å. It should be noted that although the shallower junctions showed a higher relative blue response, the overall cell efficiencies follow the behaviour of Figure 7. That is, the deeper junctions had higher efficiency.

Summary

The use of ion implantation as a solar cell fabrication technique has been investigated. A brief survey of the effects of various implantation parameters on the efficiency of the cells is presented. It was found that:
1) Low temperature implants do not improve the electrical characteristics, at least in the case of arsenic.
2) Below 800°C, cell efficiency increases with increasing anneal temperature, reflecting increased electrical activation of the impurity ions. Over 800°C, V_{OC} and C_{FF} decrease. Optimum performance was attained at 800°C.
3) The best doping combination appears to be a high implanted concentration (10^{20} cm^{-3}) and a low substrate impurity concentration (10^{15} cm^{-3}).
4) As junction depth is increased to 3000Å, cell efficiency increases, but the enhanced blue-response of the shallow junction is lost.

By applying the above, reproducable 6½ to 7% efficiency cells can be made with no antireflection coating.

The apparent difficulty expressed in point (4) has led to the proposition of the grating configuration cell$^{(4)}$. It is suggested that by implanting deep stripes and opening up the surface between, as shown in Figure 9, the advantages of the increased blue response can still be obtained without the decreased efficiency of the shallow junction. In this way, the longer bulk diffusion lengths are used to best advantage, as long as efficiency collection can still take place. This work is presently being pursued.

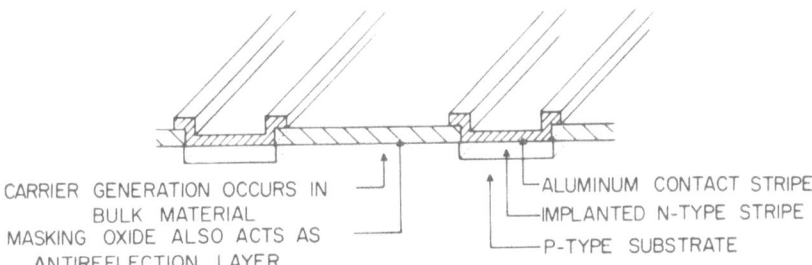

Figure 9 Proposed grating configuration cell, exploiting an enhanced blue response.

References

1. J.T. Burrill, W.J. King, S. Harrison, P. McNally, IEEE Trans. Elect. Dev. ED-14, No. 1, 10 (1967).
2. J.P. Ponpon and P. Siffert, Proc. 11th Photovoltaic Spec. Conf., 342 (1975).
3. J. Lindmayer and J. Allison, Proc. 9th Photovoltaic Spc. Conf., 83 (1973).
4. J.J. Loferski, N. Ramaganathan, E.E. Crisman and L.Y. Chen, Proc. 9th Photovoltaic Spec. Conf., 19 (1973).
5. R.S. Walker and D.A. Thompson, Nucl. Instr. & Methods 135, 489 (1976).
6. J.F. Gibbons, W.S. Johnson and S.W. Mylroie, *Projected Range Statistics*, 2nd edition, Halstead Press, 1975.
7. For review, see J.W. Mayer, L. Erikson and J.A. Davies, *Ion Implantation in Semiconductors*, Academic Press, 1970.
8. E. Baldo, F. Cappellani, and G. Restelli, Rad. Effects 19, 271 (1973).
9. M. Wolf, Energy Convers. 1112, 63-73 (1971).
10. D.A. Davies, Solid State Electron. 13, 229 (1970).
11. J.F. Gibbons, A. El-Hoshy, K.E. Manchester and F.L. Vogel, Appl. Phys. Lett. 8.2, 46 (1966).
12. J.J. Loferski, "Principles of Photovoltaic Solar Energy Conversion", 25th Annual Proc. Power Sources Conf. (1972).

THE ELECTRICAL EFFECTS OF RADIATION DAMAGE NEAR THE INTERFACE OF SCHOTTKY BARRIER CONTACTS

D V Morgan and P D Taylor

Department of Electrical and Electronic Engineering
The University of Leeds
Leeds LS2 9JT, England

ABSTRACT

The effects of radiation damage on the terminal characteristics of GaAs Schottky contacts is reported. The damage causes band-bending and carrier removal effects changing both the barrier height and the contact resistance, and increasing the minority carrier level in the barrier. Results of a defect characterisation study are summarised and compared to previously reported work. In addition the thermal stability of the damage is considered.

INTRODUCTION

Most reported work on radiation damage in semiconductors has been concerned with damage annealling, after ion implantation, so that the doping effects of the implanted ions could be observed. However, recent work has shown that this damage may be used for device processing, resulting in improved performance over more conventional techniques. Proton bombardment of GaAs, for instance, can produce insulating layers for device isolation[1]: this has been used to improve the efficiency of GaAs IMPATT diodes[2], and to increase the optical dielectric constant of GaAs in a strip region to create an optical waveguide[3]. Damage regions can enhance the diffusion of dopant ions[4], offering the potential of good control over the profile; and high field regions produced in GaAs transferred electron devices, using H or the He implantation, resulted in an improved DC-RF conversion efficiency[5]. It is, therefore, important that radiation damage effects are understood, particualry the nature and stability of the damage, and its influence on semiconductor devices.

FIG 1

Implanted ion and damage distributions in GaAs Schottky barrier for 300 KeV boron implantation, calculated from data given by Brice[17]; also shows relative position of zero bias depletion edge for barrier with
ϕ_b = 0.90 eV and
$N_D = 10^{15}$ cm^{-3}

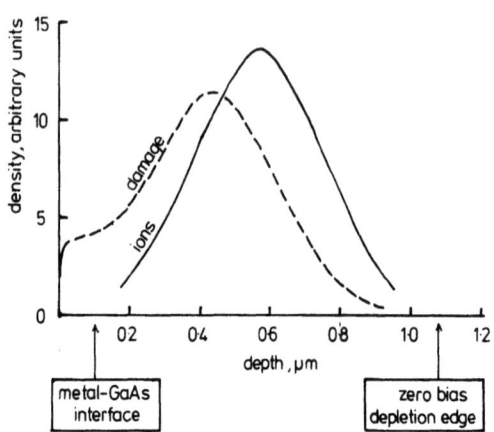

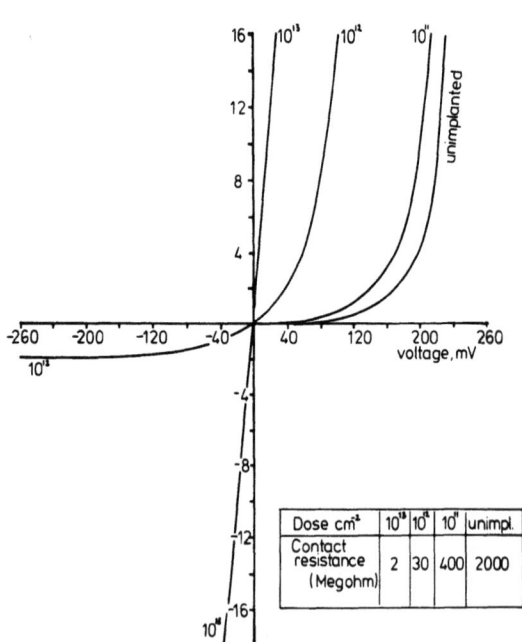

FIG 2   Current-voltage characteristics of Ni-GaAs barriers implanted with 80 KeV boron ions. The inset shows the values of contact resistance after various dose levels.

Here, we shall discuss the effects of radiation damage on both GaAs and metal-GaAs Schottky barrier contacts.

RESULTS AND DISCUSSION

The n-type GaAs ($N_d \sim 10^{15}$ cm^{-3}) barriers used here were implanted through the metal contact (1000 Å) with boron ions, which are electrically inactive in GaAs[6], to create a thin (< 1000 Å) damage layer near the interface. The damage distribution so formed is similar in shape to the implanted ion profile, both being approximately gaussian[7]. However, as with the example in Fig. 1 the damage lies closer to the surface than the ion profile, and contains a residual level of damage tailing back to the surface.

Current-voltage characteristics of Ni-GaAs barriers (Fig. 2) show a decrease in the contact resistance (dV/dI at V=0) as the implanted dose was increased: this is caused by recombination-generation effects resulting from the damage[8]. Although the damage increased the low bias conduction, the temperature dependence of the current at biases > 400 mV showed that the barrier height was in fact increased[8]: a dose of 10^{12} boron ions cm^{-2} increasing the barrier from 0.9 to 1.0 eV, for example. In previous publications[8,9], we have shown that both the barrier height change, and the increased hole concentration suggested by the observed recombination current, were the results of band-bending effects, caused by the damage, close to the interface. A further consequence of the damage is that it causes carrier removal[8].

These two effects of the damage, band-bending and carrier removal, both influence the barriers capacitance[8]. In reverse and low forward biases charge compensation causes a reduction in the capacitance; however, at high biases (> 400 mV) there is a dramatic increase in the capacitance (Fig. 3). This latter effect is a diffusion capacitance component associated with the increased minority carrier level[8].

Measurements on the terminal characteristics of Au and Al-GaAs barriers showed similar features after implantation. However, capacitance results indicated a stronger diffusion capacitance in the Au over the Al-GaAs for the same damage level: this is because the latter had the lower barrier height (0.8 eV for Al, 0.95 eV for Au) and therefore a lower minority carrier injection level.

We have performed both thermally stimulated current (TSC) and capacitance (TS CAP) measurements[9] on damaged Ni-GaAs diodes

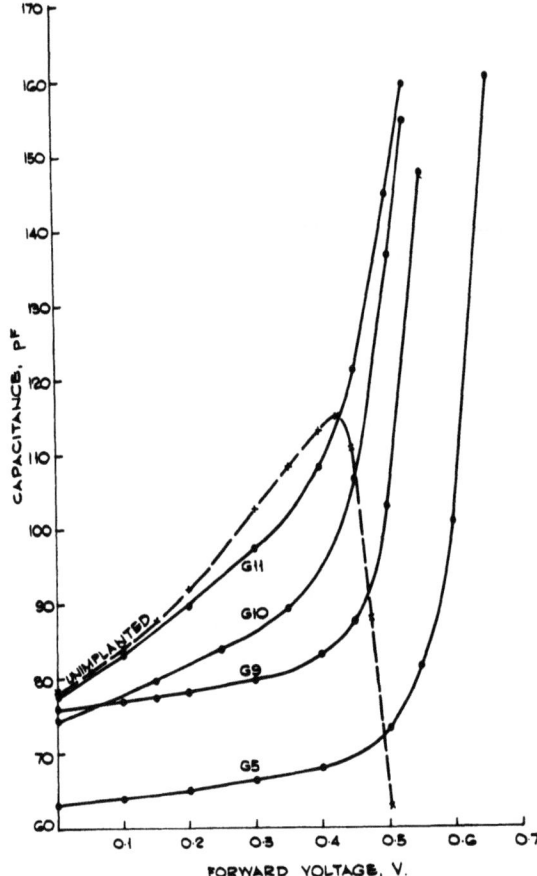

FIG 3

Forward voltage-capacitance characteristics of Ni-GaAs barriers. G11-10^{11} B ions cm^{-2} (80 KeV); G10-10^{12} B ions cm^{-2} (80 KeV); G9-10^{13} B ions cm^{-2} (80 KeV); G5-5 x 10^{11} Ge ions cm^{-2} (250 KeV)

to characterise the defect levels. The high level of damage close to the interface caused many problems in analysing TSC due to anomalous emission currents from this damage: these problems are fully discussed elsewhere[10]. Here, we summarise the results of this defect characterisation study, and compare them to other published results in Fig. 4. Apart from our results, the quoted defect levels are those resulting from electron, neutron or proton bombardment; although similar defects will result from ion implantation, this is expected to produce many more defect levels than the lighter particle irradiation.

The most frequently reported levels in Fig. 4, in the energy range E_c - 0.10 to 0.16 eV, do not feature in our results: this is a consequence of the apparatus used here, which limited the activation energies observeable to > 0.22 eV. Most of the levels reported here have been observed before, with three

ELECTRICAL EFFECTS OF RADIATION DAMAGE

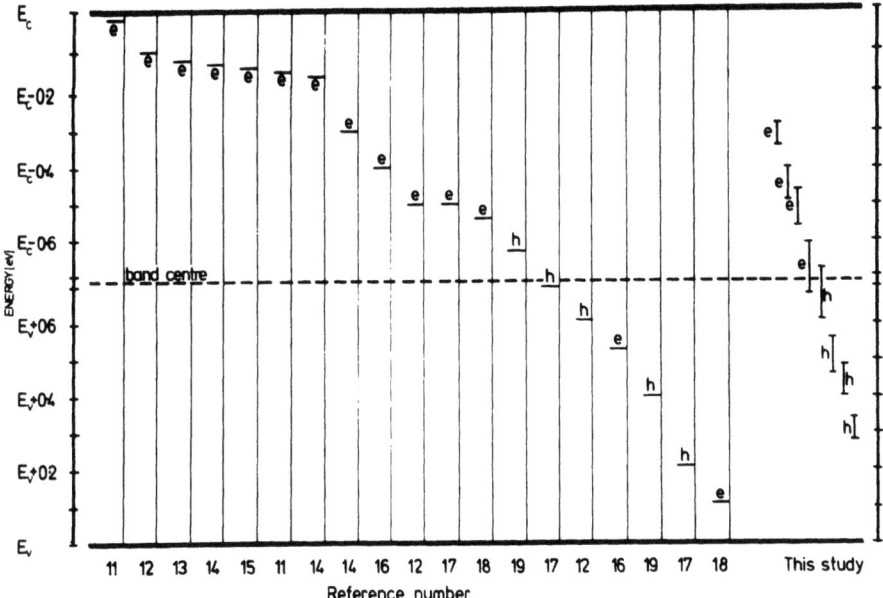

FIG. 4 Reported energy levels produced by radiation damage in n-type GaAs. Results from this study are also shown with error bars. h-hole, e-electro trap.

exceptions. The $E_c - 0.68$ eV level has not been reported previously, however, this was inherent to the GaAs used here[9]. In addition, the two traps at $E_v + 0.3$ and 0.5 eV appear only in this study. They, therefore, may be associated with the more complex defect created by ion implantation than those resulting from lighter particles.

Thermal stability measurements[20] have revealed that, except for the 0.3 eV levels, all the defect levels caused by the implantation were sufficiently removed by annealling at 400°C to be undetectable by TSC or TS CAP. We have also found[20] that, during heating, defects accumulate at the interface, although the barrier heights return to their pre-implantation values, and the diffusion capacitance is totally removed, by heating at 400°C. Current-voltage results, however, show only partial recovery after 400°C[20]: this agrees with the results of other workers[21,22] who found temperatures >700°C necessary to significantly reduce the damage.

CONCLUSIONS

In this paper we have discussed the effects of boron implantation on GaAs barriers, and have shown that the defects created by boron implantation are more complex than those observed using electron, proton or neutrons. This damage controls both the contact resistance and the barrier heights of GaAs barriers, and resulted in excess hole injection. Minority carriers can adversely affect the performance of microwave devices, resulting in a slow response time from a Schottky contact: it is therefore particularly important that such effects are minimised. We have found that less hole injection occurs after implantation in lower barrier height contacts, and that a temperature of 400°C will completely remove the diffusion capacitance which reflects the minority carrier level, although there remains a residual level of damage after heating to this temperature.

REFERENCES

1. A G Foyt, W T Lindley, C M Wolfe and J P Donnelly, Solid-St Electron, 12, 209, (1969)

2. J D Speight, P Leigh, N McIntyre, I G Groves and S O'Hara, Electron Lett, 10, 98 (1974)

3. E Garmire, H Stoll, A Yariv and R G Hunsperger, Appl Phy Lett, 21, 87 (1972)

4. G W Arnold, Proc IInd Int Conf on "Ion Implantation in Semiconductors" (eidted by I Ruge and J Graul) p 151, Springer-Verlag (1971)

5. M J Howes and D V Morgan, Proc Conf on "The Application of Ion Beams to Materials", p 91, Inst Phys, (1976)

6. D E Davies, J K Kennedy and A C Yang, Appl Phy Lett, 23, 615 (1973)

7. D K Brice, "Ion Implantation Range and Energy Deposition Distributions", Vol 1, Plenum (1975)

8. P D Taylor and D V Morgan, Solid-St Electron 19, 481 (1976)

9. P D Taylor and D V Morgan, Solid-St Electron 19, 473 (1976)

10. P D Taylor and D V Morgan, (submitted to J Phy D: Appl Phy)

11. A H Kalma and R A Berger, IEEE Trans NS-19, 209 (1972)

12. L W Aukerman, P W Davies, R D Graft and J S Shilliday, J Appl Phys, 34, 3590 (1963)

13. L W Aukerman and R D Graft, Phy Rev, 127, 1576 (1962)

14. G E Brehm and G L Pearson, J Appl Phys, 43, 568 (1972)

15. H J Stein, J Appl Phys, 40, 5300 (1969)

16. C B Pierce, H H Sander and A D Kantz, J Appl Phys 33, 3108 (1962)

17. L Borghi, P de Stefano and P Mascheretti, J Appl Phys, 43, 568 (1972)

18. J K O'Brien and J C Corelli, J Appl Phys, 44, 1921 (1973)

19. B R Pruniaux, J C North and G L Miller, Proc IInd Int Conf on "Ion Implantation in Semiconductors", (edited by I Ruge and J Graul) p 212, Springer-Verlag (1971)

20. P D Taylor and D V Morgan, Solid-St Electron (to be published)

21. G Carter, W A Grant, J D Haskell and G A Stephens, Proc IIIrd Int Conf on "Ion Implantation in Semiconductors and other Materials" (eidted by B L Crowder) p 611, Plenum (1973)

22. R G Hunsperger and E D Wolf, J Electrochem Soc, 118, 1847 (1971)

NOVEL MICROFABRICATION PROCESS WITHOUT LITHOGRAPHY USING AN ION-PROJECTION SYSTEM

R. Sacher, G. Stengl, P. Wolf, R. Kaitna

Sacher Technik Wien

A-1070 Vienna, Apollogasse 6, Austria

Abstract

An ion-projection system has been developed for a novel fabrication process without lithography for LSI-circuits based on silicon. The system is designed to meet the requirements of the new technology with sub-micrometer resolution over a chip size of 5 x 5 millimeters. A self-supporting metal-mask is illuminated by ions and imaged onto a wafer with a 10 x reduction in size. According to the thickness of the SiO_2-layer the ion energy is defined and the "exposure" dose is chosen in such a way, that the damage in the exposed oxide is sufficient for a following (dry) preferential etching technique. This etching technique is based on the effect that exposed oxide is etched faster than unexposed oxide. The difference between the etching rates is a function of the damage and depends on the type of ion and its dose.

Except for metallization the fabrication process with this ion-projection system is practicable for all semiconductor production steps where normally either photo or electron-beam lithography is used.

This report describes the ion-projection system, the fabrication of the self-supporting masks and gives a brief survey of the fabrication with the above indicated process for MOS devices.

Introduction

The structural defects created by ion implantation in a thermally grown SiO_2 film were the reason to start our examinations. These structural defects are shown by a great number of scientists to be produced by two distinct mechanisms, one dependent upon ion energy deposited in SiO_2 as atomic collision, the other dependent upon the ion energy deposited as ionization events. It has been found that from the point of view of ion implantation, any oxide which has been implanted should be annealed to recover its initial property. In unannealed SiO_2 a shift of the flat bond voltage is observed presumable due to positive charges created in the oxide by the bombardment. But the consequence of this damage of ion implantation is ultimately related to the chemical result of the dopant species. The chemical transformation of SiO_2 films with high doses of reactive ions is well established and there are extensive studies in progress. In our laboratory we investigated the possibility of creating with this chemical effect a real interesting technology for semiconductor production. This chemical effect causes in implanted SiO_2 an increased etching rate compared to the unimplanted oxide. The etching rate is dependent on the concentration level of the implanted oxide. In practice this effect can be used to produce structured oxide layers. In device technology SiO_2 layers are extensively used either as a masking layer for diffusion or as an intrinsic part of the device and according to their special purpose these layers have to be structured, what is commonly done by some lithographical technique. The same structures can be produces directly if the pattern of the layer is projected by an ion-beam onto the SiO_2.

The Ion-projection System

The main parts of the projector are the ion source, the vacuum system, the mask changing disc with the metal-masks, the accelerating tube with the two optical lens-systems and the target chamber.

Because of the demand of shock resistance two supporting frames were built, each consisting of welded iron construction damped and mounted on the floor. These two frames contain the whole ion-projection system.

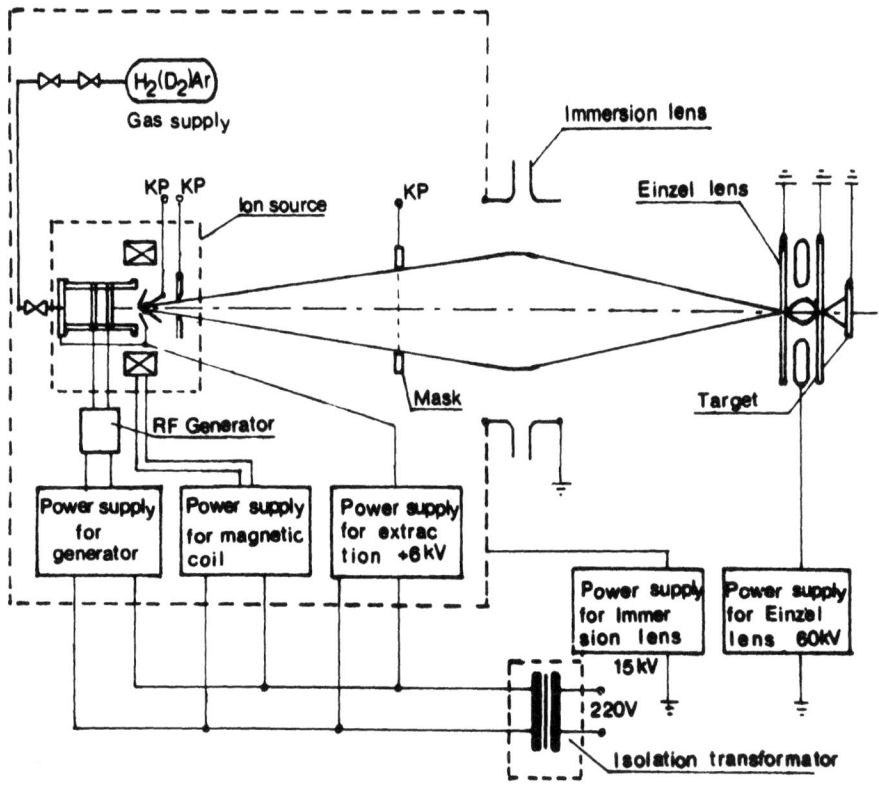

Fig.1. Schematic design of the ion-projection system

It is of utmost importance to eliminate vibration because resolution, etching acuity and picture distortion depend strongly on a vibration-free construction.

The vacuum system consist of two oil diffusion pumps with an effective pumping speed of 1000 liter/second. The target chamber and accelerating tube system is separated by a valve. The pressure in the target chamber has a typical value of 2×10^{-6} torr. This is due to Nitrogen flooding and the pumping out time is 35 minutes.

In our opinion, the generation of Hydrogen-ions by a high-frequency ion source is the best way to succeed and to disregard mass-analysis. Because of the used type of extraction in this source the traditional optical parameters allow to put a rather simple construction on the whole projecting system. The total beam current extracted from our present source is in the range from 20 µA to 200 µA. The extraction voltage can be varied up to 6 kV. The divergence of the extracted beam is dependent on the geometries and is in the range of $2°$ to $5°$. Beam diameter is 58 mm at the place of the object. Therefore a minimum of energy scattering was obtained; and the influence of colour defects of the subsequent lenses to the resolution of the total system is beyond the aperture (gap) defect. The ion beam emerging from the extractor should be relatively free of aberrations and has low divergence indicating that space charge effects should be avoided. It is virtually impossible to correct beam defects introduced by the extraction optics with the subsequent ion-optical system without encountering significant beam-current loss and additional charge defects. The ion-projection system is based on the optical principle shown in Fig. 2.

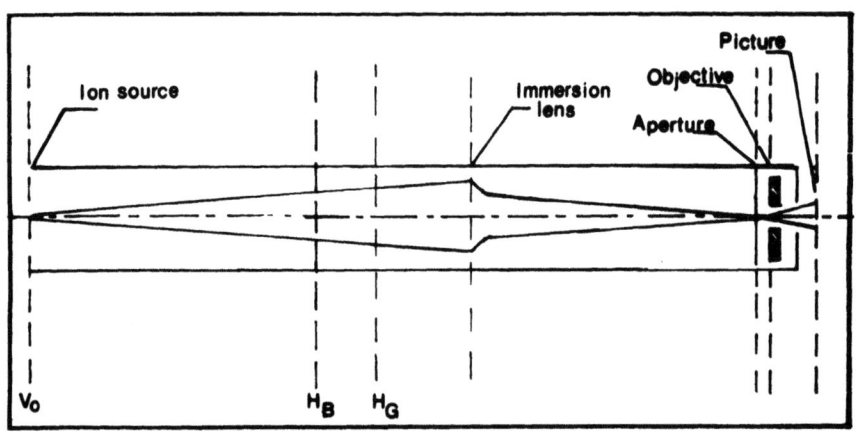

Fig.2. Principle of the ion-projection system

The extracted beam passes through a cylindrical accelerating lens (field lens) with a potential difference of 45 kV, the ion source being on high potential. The ion beam is focused using proper extraction to acceleration potential ratio. The used ratio is about 8 to 9, what means for our given ion source a beam energy between 4o keV to 52 keV. The residual types of ions are H^+, H_2^+, H_3^+, D^+, HD^+.

Field Lens

This is a symmetric immersion lens consisting of two cylinders mounted on axis each with an inner diameter of 240 mm and an outer of 300 mm. The insulation consists of PVC and holds 60 kV. In this lens the accelerating and focusing takes place. The new model has an improved lens structure and holds higher voltages. The drawback of this lens is that the principle planes in the field free area are on the lower potential side. This causes the object picture created on a definite position where there is no interaction with the field lens. Because of the double action of this lens - acceleration and focusing - there is no need for additional lenses and therefore no critical adjustment.

Einzel Lens

The demagnifying lens is an Einzel lens with the potential of the middle electrode of 52 kV in maximum. The whole system is symmetric, the outer electrodes are grounded. The supporting insulators of the lens work reliably at 55 kV. The aperture on the beam entrance is typically 1 mm, the exit diaphragm is 6 mm in diameter. The adjustment of the einzel lens in relation to the field lens is controlled by adjustment support. The HV-power supply is 60 kV, 0,5 mA and with a stability of 5×10^{-4}. The resolution of this lens is related to the diameter of the entrance aperture. The 10x reduction to the object produces a picture close to the focus. A resolution of 1 micron presupposes a diameter of aperture of 1,6 mm. To get the whole picture of 5 x 5 mm chips with full intensity of beam current, the diameter of the lens should be wider than 240 mm.

Adjustment

The singular direction is the optical axis of the field lens. The alignment of the ion source and Einzel lens occurs in relation to the field lens. For this reason the ion source can be moved within a plane. The Einzel lens support is mounted on an x-y slide and in addition it can be tilted within $\pm$ 10°. The target can

be adjusted in relation to the optical axis and this is
solved by an x-y support. The relation of wafer to
support is determined by three points, two of them on
the flat section of the wafer and the third one can be
placed depending on the size of the wafer. This adjusting
method works well. For more convenience we are working
to solve this adjustment problem with a laser-inter-
ferometer system to get exact overlap conditions between
the individual processing steps.

Mask Changing Disc and Adjustment

The mask changer consist of one turning disc, which
is guided by a three-point-bearing. Two of them are
rigid and one is spring leaded. The disc can be moved
into the right position by means of vacuum feed-through.
The turning wheel contains adjustable self supporting
projection masks. Each mask is pre-mounted in the
fashion of slides. The precise position can be adjusted
with micrometer-screws. The pre-adjustment of the masks
is done with help of an optical-bench. The pattern of
the mask is projected on a screen with a 100 fold magnification. With help of three marks it is possible to
assure the exact overlap of all masks.

Principle of Metal-mask Technology

In the ion-projection system it is necessary to
use selfsupporting masks which are produced in house by
electrolytic formation of Ni-metal on a special pre-
fabricated substrate. The substrate consists of pure
polished aluminum which is coated with a photosensitive
emulsion layer of AZ 1350 H supplied by Shipley. The
thickness of the coating is highly influential to the
result of the galvanic process. Typical values are
4,5 micron with an average exposure time of 45 seconds.
Due to the masking of the substrate the photomask is
developed and in a galvanic bath the metal-mask is
growing in those areas where there is no photoresist.
The precision of our metal-masks is $\pm$ 2 microns measured
on openings of 100 microns. The current densities are
relatively low (250 mA to 2 A). The principal fabrication steps are shown in Fig. 3. The angles of the
openings in the Ni-mask are not reproduced correctly;
Fig. 3 gives only the principle.

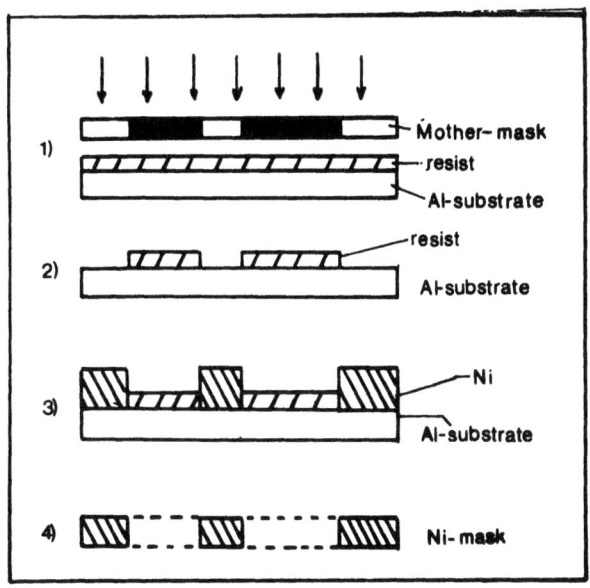

Fig.3. Fabrication steps for metal-masks. (1) Exposure of mother-mask to resist coated Al-substrate. (2) Developed photomask. (3) Electrolytic formation of the Ni-mask. (4) From Al-substrat detached Ni-mask.

All masks must be of equal thickness; because of the used fabrication process the thickness of the metal layer determines the size of the structures. There are no closed loops allowed inside a structure. Consider the relation between the openings and the bridges; the spacing between the openings is an essential question for mechanical stability. As a fairly good compromise between mechanical stability, the complexity of LSI structures, the reproducebility and the lifetime 40 microns thick Ni-masks are fabricated. A section of such a Ni-mask is shown in Fig. 4.

We use three etching systems, dry, wet and plasma. Each system has definite advantages and because of the high quality results we favor our sophisticated dry etching process. The difference in the etching rate in dependence from the ion dose shows Fig. 5 and the effective damage range is shown in Fig. 6. The diagram of Fig. 5 was taken from samples damaged with 50 keV protons with a dose range from 10^{15} to 4×10^{16} ions/cm^2 on 1 micron wet oxide on Si-wafer oriented (1,1,1).

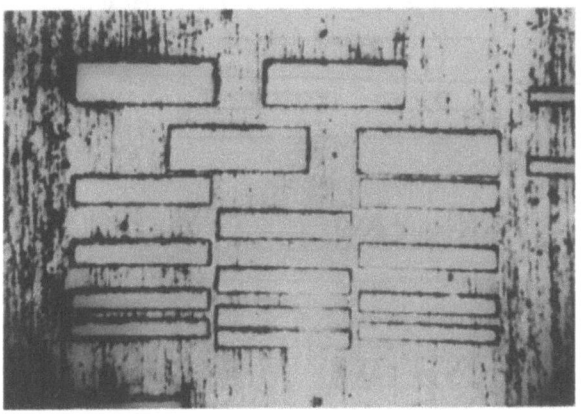

Fig.4. Section of metal-mask

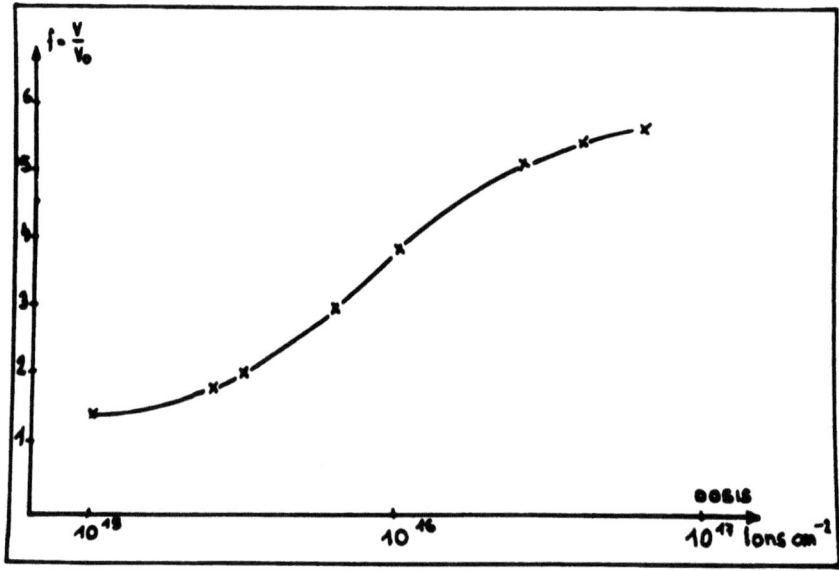

Fig.5. Etch factor dependence from implanted ion-dose in SiO$_2$.

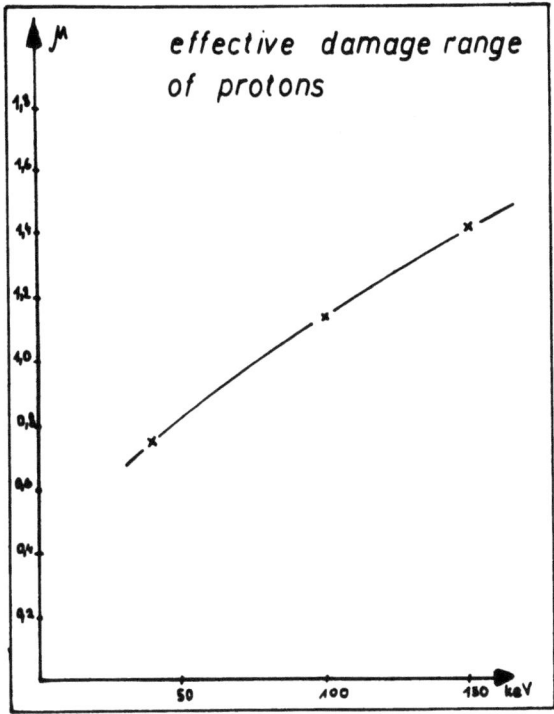

Fig.6. Effective damage range of protons in SiO_2.

The preferential etching shows a fully utilized damage range of 0,55 micron and an etching rate factor of f = 1,25 to 5. Obviously the slope of an etched structure depends from f. This dependence is shown in Fig. 7.

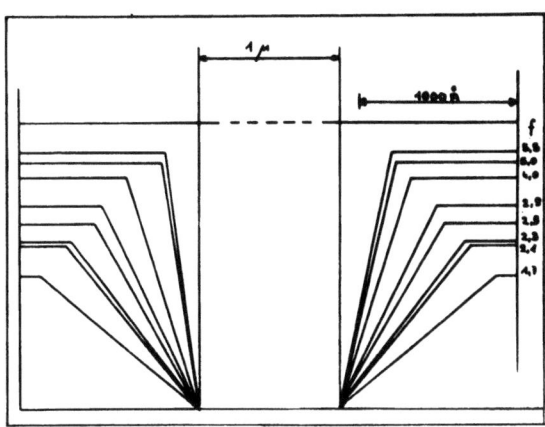

Fig.7. Dependence of slope from etch-factor.

The measurement of the etching rate was accomplished in 30 second-intervals with interferometric measurement equipment. The precision is ± 150 Å.

Experimental and Device Results

First of all it was tried to optimize the ion-projection system and to get sharp "pictures" of the projected MOS-masks (a 5 step process). The demagnification and resolution of the projection system depends on the stability of the individual system as well as on the stability of the voltages of the lenses to each other. Any change in the voltage ratio will cause the projected picture to move or the scale to be altered. To have the picture stable within the given tolerances of the used testcircuit, it should move considerable less than 1 micron, what is equivalent to a maximum 10 V change on the HV-supply. To get this stability we additionally stabilized the HV-supplies by external voltage dividers to get exact relations. Misalignment of the optics can cause even more changes of picture position, for example slight misaligning off the axis caused typically 15 microns of displacement.

The whole process showed good results when the etching caused by ion bombardment is high compared with those areas of the image without bombardment. A good deal of our effort was concerned with enhancing the

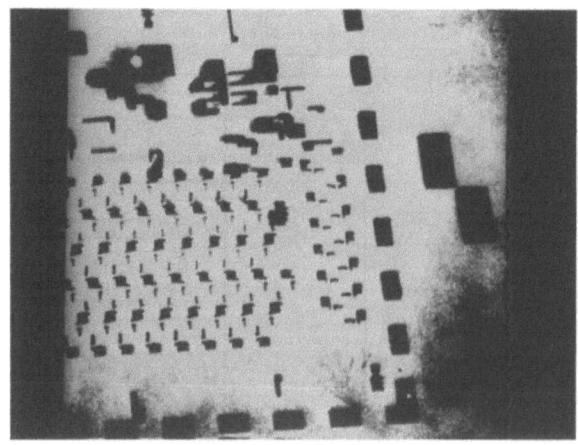

Fig.8. SEM-picture of "ion-exposed" and etched SiO_2 layer of MOS circuit.

etching rate. This is directly dependent on the ion
energy and concentration and therefore on the duration
of bombardment. From our trial etching runs with 50 keV
ion energy and a dose of 10^{12} to 10^{16} ions/cm^2 we got
several samples examined of the Univ. Vienna and Univ.
Zürich as well as in our laboratory. Fig. 8 shows some
SEM results in backscattering mode. In Fig. 9 a line-
scan through an "ion-exposed" and etched structure is
shown. The slopes are in the range of 70° to 80°.

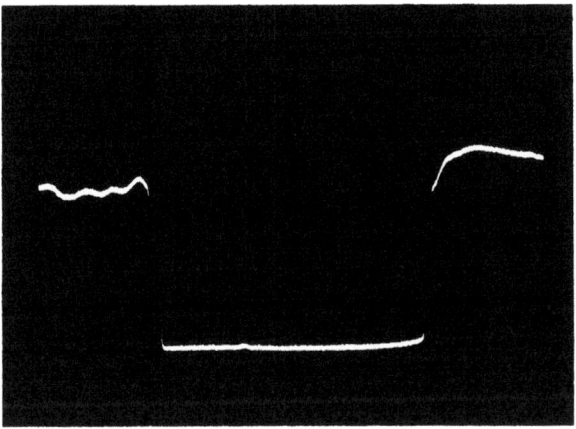

Fig.9. Line-scan through an etch-structure of sample
shown in Fig. 8.

Summary

With the above shown results and with test circuits
produced by this technology, we are convinced to have
proofed that this photolithography-free technology in
principle is feasible for high volume IC-production.
Therefore our present efforts are to optimize the ion-
projection system as well as the process-technology.
Because the ion-projection system uses shorter wavelengths, smaller structures (submicron) than those
produced with photolithography can be made. This opens
the possibility to create higher package-densities and
a greater complexity of the circuits.

FORMATION OF NEW RADIATIVE RECOMBINATION CENTER IN $Al_xGa_{1-x}As$ BY NITROGEN-ION IMPLANTATION

Y. Makita, S. Gonda, H. Tanoue and T. Tsurushima

Electrotechnical Laboratory, Tanashi Branch

5-4-1 Mukodai-machi, Tanashi-shi, Tokyo, Japan

ABSTRACT

Nitrogen (N) atoms implanted into $Al_xGa_{1-x}As$ at 350°C become new efficient radiation recombination centers, which provide 500-fold increase of the emission efficiency at 77°K compared with undoped case. This is attributed to the formation of the isoelectronic emission band situated below at the X_1 conduction band minima of $Al_xGa_{1-x}As$. A nitrogen absorption band was also found to be formed between the X_1 and Γ_1 conduction band minima. Laser emission is achieved in the N-implanted $Al_xGa_{1-x}As$ both for direct- and indirect-band-gap materials, indicating that the nitrogen isoelectronic levels are responsible for the laser oscillation.

INTRODUCTION

Ion implantation is an effective method of impurity doping, in particular, when impurity doping by other methods is not developed or difficult. Nitrogen (N) impurity in $Al_xGa_{1-x}As$ is of great interest from a viewpoint of optoelectronic application, but a doping method during crystal growth has not been developed. We have made N-doping into $Al_xGa_{1-x}As$ by ion implantation and successive annealing. As a result we have found that N in $Al_xGa_{1-x}As$ become efficient radiation recombination centers due to the formation of isoelectronic trap [1-4] and furthermore they are effective for laser oscillation in both direct [5] and indirect band-gap $Al_xGa_{1-x}As$ crystals [6]. The energy scheme of the isoelectronic bands in the N-implanted $Al_xGa_{1-x}As$ is identified by combining the results of the photoluminescence and excitation spectroscopy measurements.

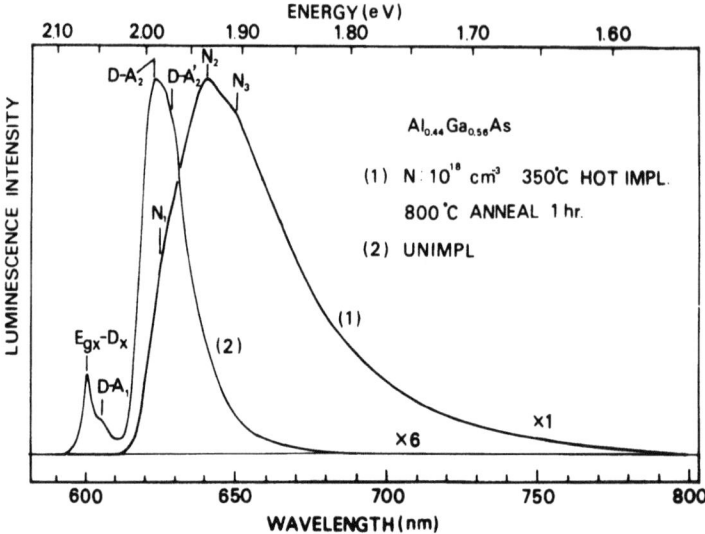

Fig.1 Photoluminescence spectra at 77°K for the N-implanted and unimplanted $Al_{0.44}Ga_{0.56}As$.

EXPERIMENTAL

The samples used in this experiment were $Al_xGa_{1-x}As$ (0<x<0.55) single crystals grown on (100) GaAs substrate by liquid-phase epitaxy (LPE), and were not intentionally doped. The composition ratio, x, was determined from photoluminescence (PL) after the empirical formula of Onton et al [7]. For the PL measurements 10-μm layers of $Al_xGa_{1-x}As$ were used. For laser platelets, two-layered $Al_xGa_{1-x}As$ was fabricated. The active region was surface layer made of 1-μm $Al_xGa_{1-x}As$ (x=0.39 or 0.46) and the underlying $Al_{0.65}Ga_{0.35}As$ (0.4μm) was used as a layer for optical confinement. Since the as-grown surface was completely wiped out, no surface treatment was made before ion implantation.

Ion implantation was accomplished at 350°C with N ions to a dose of $4 \times 10^{13} cm^{-2}$ with four different energies of 50, 80, 140 and 250keV. The theoretical calculation predicts that the nitrogen concentration of $10^{18} cm^{-3}$ is flat between 0.10μm and 0.47μm below the surface. Annealing was performed at 800°C for 1h under the flow of H_2 gas of 1 atm without dielectric coats. Under this condition the decomposition of the crystal was suppressed by H_2 gas and there was observed no trace of anomalous behavior in both emission intensity and spectra after high-temperature annealing.

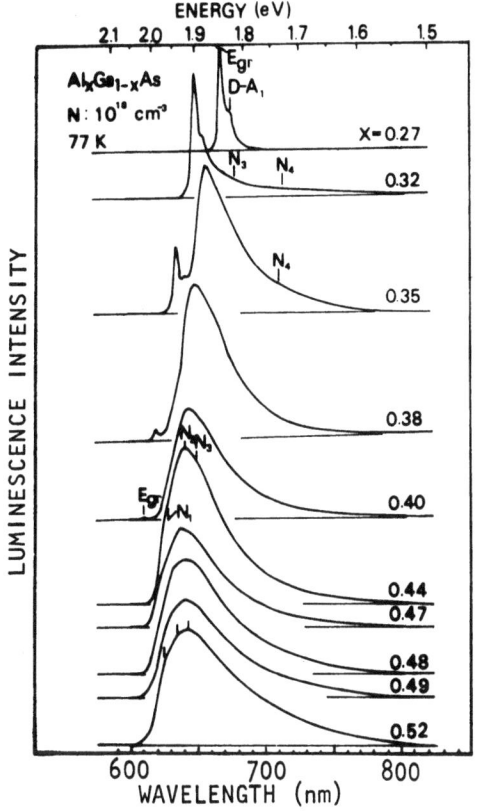

Fig.2 Photoluminescence spectra at 77°K for the N-implanted $Al_xGa_{1-x}As$ with various composition ratio x.

The PL spectra were measured at 2°K and 77°K using an Ar^+ laser as an exciting source. For excitation spectroscopy a 500-W tungsten halogen lamp and a monochromator with a spectral resolution of 0.002eV around 2.0eV were used as a light source. Concerning the measurements of laser oscillation, the N-implanted and annealed crystals were thinned down to a thickness of 100μm from the GaAs side and cleaved along (110) planes to 100-μm width to form Fabry-Perot cavity. Indium was evaporated on the GaAs side and the crystal was mounted on an indium-evaporated copper heat sink by compressing it at 300°C in Ar gas. The pump source for the laser oscillation was a pulsed dye laser (Coumarin 102) excited by a N_2 laser. The peak wavelength is 480nm and the maximum peak energy is of the order of 1kW.

PHOTOLUMINESCENCE AND THE ENERGY SCHEME OF NITROGEN IN $Al_xGa_{1-x}As$

Figure 1 shows the effect of the N-implantation in photolumi-

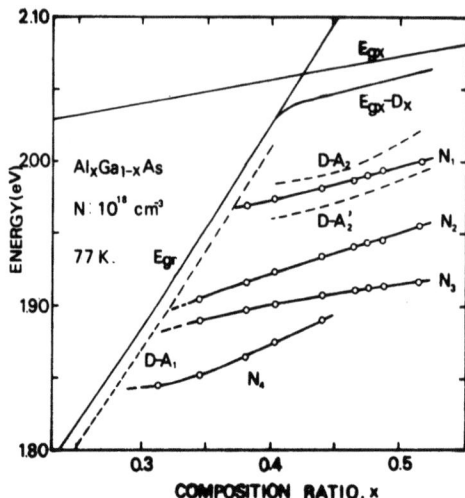

Fig.3 Energies of transitions at 77°K associated with N impurities, band edge and D-A pairs in $Al_xGa_{1-x}As$ as a function of composition ratio x.

nescence spectra for $Al_{0.44}Ga_{0.56}As$ at 77°K at the excitation level of $10W/cm^2$. Annealing at 800°C for 1h is sufficient and annealing above 800°C only produces a serious decomposition of the crystals. In the unimplanted sample, which is annealed under the same condition, the emission due to excitons bound to unidentified donors denoted by $E_{gx}-D_x$ is clearly observed together with three donor-acceptor pair emissions, $D-A_1$, $D-A_2$ and $D-A'_2$ [4]. The D-A emission was identified from the excitation-intensity dependence and the time-resolved spectroscopy [1,8]. The subscripts 1 and 2 stand for the emissions closely related with the Γ_1 and X_1 conduction-band minima, respectively [4,9]. As for the N-implanted wafer, $E_{gx}-D_x$ and D-A emissions disappear, and three new peaks denoted by N_1, N_2 and N_3 are observed with a long tail on the lower energy side. The integrated intensity (II) of the N-implanted specimen is 15 times larger than that of the unimplanted one, indicating that the emission bands formed by the implantation are efficiently radiative ones, presumably due to the formation of isoelectronic traps in the crystal. In Fig. 1, the peaks, N_1 and N_3 are rather humps. The humps, however, can be distinguished by the proper choice of excitation intensity. That is, above $10^2 W/cm^2$ N_1 peak is prominent and below $10^{-1}W/cm^2$ N_3 peak is separately observed.

Figure 2 shows PL spectra at 77°K for the N-implanted $Al_xGa_{1-x}As$ with various composition ratio x. The effect of N-implantation is appreciable for $Al_{0.32}Ga_{0.68}As$, where a long tail extending towards lower energy side is obtained with hump designated by N_4. The whole features of the N bands are more demonstrative for x=0.35. The above features are shown in Fig. 3, where

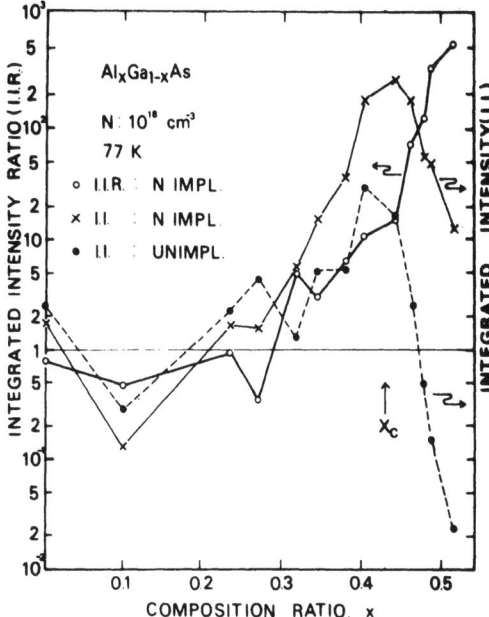

Fig.4 Integrated Intensity (I.I.) and Integrated Intensity Ratio (I.I.R.) for the N-implanted $Al_xGa_{1-x}As$ as a function of composition ratio, x.

each energy level of the N bands is plotted as a function of the composition ratio x, together with the Eg_Γ, Egx-Dx and D-A emissions. The appearence of each N band with increasing x can be understood in terms of the emergence of the N band level from the Eg_Γ band with increasing x. In Fig.3, the discontinuity of the $D-A_1$ and $D-A_2$ levels near the cross-over point (x_c=0.43) is remarkable. For the N levels no discontinuity is observed. and a good parallelism of the energy levels associated with the N bands to the X band edge (Egx) is noticed. These behaviors strongly support the following conjectures, i) the $D-A_1$ and $D-A_2$ emission bands are situated below the Γ band edge Eg_Γ and X band edge Egx, respectively, ii) the N emission bands are formed below the X_1 conduction-band minima even in the direct-band-gap minima. The second conjecture may be true, since the direct-gap-like transition should be enhanced especially in indirect-region ($x>x_c$=0.43) when the constituted emission band is situated below the X_1 conduction band minima [10].

The relative integrated emission intensity (II) of the N-free and N-implanted $Al_xGa_{1-x}As$ wafers is given in Fig.4 as a function of x. The integrated intensity ratio (IIR) (a ratio of emission intensity of the N-implanted wafer to that of the N-free wafer) is also presented. In the N-free crystal, II largely decreases for $x>x_c$ because of the indirect-gap structure. In the N-implanted

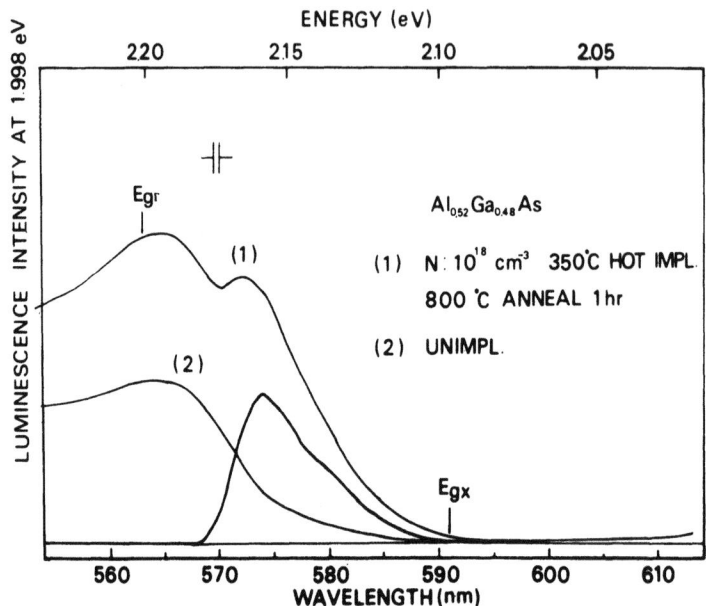

Fig.5 Excitation spectra at 2°K for the N-implanted and the unimplanted $Al_{0.52}Ga_{0.48}As$.

wafer, II slightly decreases for $x > x_c$. Thus, IIR largely increases as x is increased. A 500-fold increase of emission efficiency is attained by N implantation and successive annealing. That is, N impurities introduced by ion implantation become efficient radiation recombination centers.

At 77°K, we could not obtain a well-resolved spectrum of N_1, N_2 and N_3 bands at the moderate excitation level ($10W/cm^2$), and the energy half-width of each band is extremely broad. All of the N bands are presumably the emissions of excitons bound at N-N pairs.

EXCITATION SPECTROSCOPY

In order to investigate the optical properties of the N-implanted $Al_xGa_{1-x}As$, absorption measurements are of great importance, but no work has been presented. This was partly due to the presence of the opaque GaAs substrate and principally due to the small depth of the implanted layer, of the order of 0.5μm. The absorption coefficient, α, of $Al_xGa_{1-x}As$ with $10^{18} N/cm^3$ is of the order of $10 cm^{-1}$ [10]. To measure such a low α, a homogeneous and thick sample, of the order of 1mm, is required. When one can not prepare a thick sample, excitation spectroscopy (ES) is used as a powerful tool.

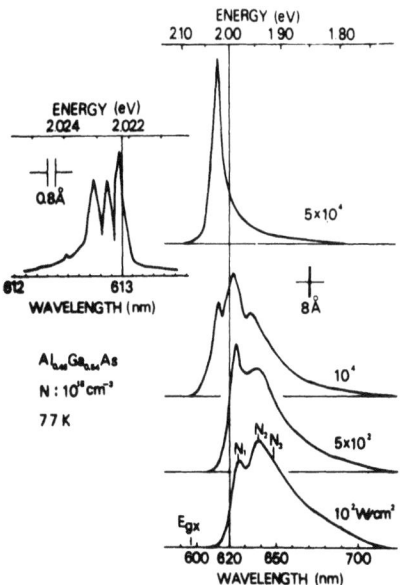

Fig.6 Laser oscillation at 77°K for the N-implanted $Al_{0.46}Ga_{0.54}As$.

In Fig.5 we present the ES spectra at 2°K for N-implanted and N-free $Al_{0.52}Ga_{0.48}As$. The emission intensity was observed at 1.9981eV. This energy corresponds to the D-A_2 emission for the N-free sample, and to the N_2 emission band for the N-implanted sample. In Fig.5, for the N-free crystal, a gradual increase of emission intensity begins at around Eg_x, revealing the indirect-band-gap feature. Maximum emission occurs at Eg_Γ. The gradual decrease on the higher energy side of Eg_Γ is due to an increase of surface recombination [11].

As for the N-implanted crystal, a distinguished shoulder was observed near 2.1634eV, together with a feeble hump at 2.0388eV. By subtracting the curve of the N-free sample from the curve of the N-implanted sample, two emission bands are obtained at 2.1585eV and at 2.1391eV. For $Al_{0.54}Ga_{0.46}As$ implanted with Ne$^+$ of $10^{18}cm^{-3}$ at 350°C, absorption bands are not observed. These absorption bands were well observed for the wafers with x≳0.44. This suggests the following important fact, i.e. absorption bands due to N impurities are formed between the Γ_1 and X_1 conduction band minima.

STIMULATED EMISSION

As a next step, we investigated the possibility of laser oscillation involving N isoelectronic traps. Excitation was performed from the N-implanted surface and the output light was detected from the cleaved side. Laser emissions were accomplished in direct-and indirect-gap, N-implanted $Al_xGa_{1-x}As$, both at 2°K and 77°K. Figure 6 shows the laser emission spectra for x=0.46 at 77°K. Laser oscillation is observed at the pumping levels over $7\times10^4 W/cm^2$, which occurs at N_1 band. The emission intensity versus pumping level curve shows a super-linear increase of emission intensity abruptly occurs at $5\times10^4 W/cm^2$. Thus, nitrogen isoelectronic traps are responsible for stimulated emission and by using the traps efficient laser oscillation can be achieved even in indirect-gap $Al_xGa_{1-x}As$.

CONCLUDING REMARKS

The optical properties of nitrogen isoelectronic impurities implanted into $Al_xGa_{1-x}As$ were investigated by photoluminescence and excitation spectroscopy. The efficient emission bands are formed below X_1 conduction band minima due to the formation of isoelectronic trap, while absorption bands are formed between Γ_1 and X_1 conduction band minima. By optical pumping method, laser oscillations were achieved in the N-implanted $Al_xGa_{1-x}As$ for both direct-and indirect-gap crystals, revealing that nitrogen isoelectronic traps are responsible for stimulated emission.

ACKNOWLEDGEMENTS

The authors wish to express their thanks to H. Ijuin and S. Mukai for crystal growth and J. Shimada for continued interest.

REFERENCES

[1] S. Gonda and Y. Makita, Appl. Phys. Lett. 27, 392 (1975).

[2] Y. Makita, S. Gonda, H. Ijuin, T. Tsurushima, H. Tanoue and S. Maekawa, Appl. Phys. Lett. 28, 103 (1976).

[3] Y. Makita, H. Ijuin, and S. Gonda, Appl. Phys. Lett. 28, 287 (1976).

[4] S. Gonda, Y. Makita, S. Mukai, T. Tsurushima and H. Tanoue, Appl. Phys. Lett. 29, 196 (1976).

[5] Y. Makita, S. Gonda and H. Ijuin, Appl. Phys. Lett. 29, 309 (1976).

[6] Y. Makita and S. Gonda, the Fifth IEEE Semiconductor Laser Conference (Mie, 1976 Sep.)

[7] A. Onton, M. R. Lorenz, and J. M. Woodall, Bull. Am. Phys. Soc. 16, 371 (1971).

[8] Y. Makita and S. Gonda, Appl. Phys. Lett. 27, 333 (1975).

[9] B. Monemar, K. K. Shih, and G. D. Petit, J. Appl. Phys. 47, 2604 (1976).

[10] P. J. Dean and R. A. Faulkner, Appl. Phys. Lett. 14, 210 (1969).

[11] P. J. Dean, Phys. Rev. 168, 889 (1976).

EFFECTS OF DUAL IMPLANTATIONS AND ANNEALING ATMOSPHERE ON

LATTICE LOCATIONS AND ATOM PROFILES OF Sn AND Sb IMPLANTED IN GaP

M. Takai, K. Gamo, T. Ishida, K. Masuda and S. Namba

Faculty of Engineering Science, Osaka University
Toyonaka, Osaka, Japan

A. Mizobuchi

Faculty of Science, Osaka University
Toyonaka, Osaka, Japan

ABSTRACT

Ion channeling, backscattering and ion-induced X-ray measurements have been done to study the effects of dual implantation and annealing atmosphere on lattice locations and atom profiles of Sn and Sb implanted in GaP.

Sn atoms implanted at 400°C are found to be ~80% on Ga substitutional sites, while Sb atoms implanted at 400°C are found to be ~100% on P substitutional sites. After annealing using an SiO_2 cap, the substitutional fraction of Sn decreases to 20%. Sn and Sb atoms implanted at room temperature are found to be on off-lattice sites even after annealing at 700°C. In case of dual implantation, the substitutional fraction of Sn after annealing is decreased by Ga implantation, while it is increased by P implantation. Diffusion of Sn implanted at 400°C is enhanced by P implantation, while it is suppressed by Ga implantation and by annealing in ampoules sealed with excess GaP.

INTRODUCTION

Recent studies on ion implantation in compound semiconductors suggest that the control of vacancy concentration has an important role in determining the implanted atom profile [1] and the doping efficiency [2]. These studies have been done by Hall effect measurements. The atom profile and the doping efficiency are determined by the lattice location of implanted atoms. Therefore lattice location measurements give more direct information.

Lattice locations of implanted impurities are probably considered to be influenced by the vacancy concentration in the implanted layer.

In the present study, the lattice-site location and the atom profile have been investigated by channeling and backscattering measurements in order to get direct information concerning the role of the vacancy in determining the lattice location of Sn and Sb implanted in GaP. Ga and P preimplantations were performed in trying to control the vacancy concentration in GaP: Ga preimplantation may produce more p-vacancies in the implanted layer, while P preimplantation may produce more Ga-vacancies. Annealing in various atmospheres was also employed to check the capability of controlling the vacancy concentration and the lattice location of Sn.

EXPERIMENTAL

S-doped GaP crystals with carrier density of $\sim 1 \times 10^{18}/cm^2$ were implanted with 70 keV Sn and Sb at 400°C and room temperature. Doses ranged from 1×10^{15} to $5 \times 10^{15}/cm^2$. For dual implantation, 50 keV Ga or 30 keV P preimplantation was performed at room temperature with the same dose of Sn. These implantation energies were chosen so that the projected range almost agrees with 70 keV Sn in GaP. Annealing was performed at 700°C for 20 minutes with an SiO_2 cap, or without cap in a P atmosphere (1.5 atm), or in a quartz ampoule sealed with excess GaP.

1.8 MeV $^4He^+$ channeling, backscattering and ion-induced X ray were used to study lattice locations of Sn and Sb. In addition to the conventional yield attenuation, angular distributions of backscattering and ion-induced X-ray yield along major axes were measured in order to distinguish two types of substitutional sites in GaP [3]. An experimental setup for ion channeling and backscattering was presented in detail elsewhere [4]. Ion-induced X-rays were measured by Si(Li) detector with an energy resolution of ~ 190 eV (FWHM).

RESULTS AND DISCUSSION

Table I shows a summary of lattice locations of Sn and Sb implanted in GaP. The lattice location is given as percentage attenuation along the <111> and the <110> axes of backscattering yield. The minimum scattering yield (χ_m) is shown as an index of lattice defects. For room temperature implantation, the percentage attenuations for Sn and Sb are zero even after annealing at 700°C. This indicates that Sn and Sb implanted at room temperature are not located on substitutional sites even after annealing. On the contrary, the percentage attenuations for hot implantation are around 70~80%. After dechanneling correction this corresponds that hot-implanted Sn and Sb are located on substitutional sites

Table I A summary of lattice locations of Sn and Sb implanted at 400°C and room temperature.

SAMPLE	%<111>	χ_m	%<110>	χ_m	$\frac{1-Y_{<111>}/Y_{<R>}}{1-\chi_m}$(%)	DOSE (cm^{-2})
Sn R.T. IMPLANT	0	0.27	0	0.23	0	2.7×10^{15}
+700°C	-	-	0	0.11	-	1.7×10^{15}
Sn 400°C H.I.	67±3	0.23	68±2	0.18	87±4	4.0×10^{15}
+700°C AN.	24±8	0.18	16±5	0.16	29±9	4.7×10^{15}
Sb R.T. IMPLANT	0	0.34	-	-	0	3.7×10^{15}
+700°C AN.	0	0.27	-	-	0	3.7×10^{15}
Sb 400°C H.I.	75±4	0.22	72±4	0.18	96^{+4}_{-9}	4.0×10^{15}
+700°C AN.	42±7	0.16	50±5	0.08	50±8	2.2×10^{15}

(1.8 MeV ^{4}He$^+$ beam)

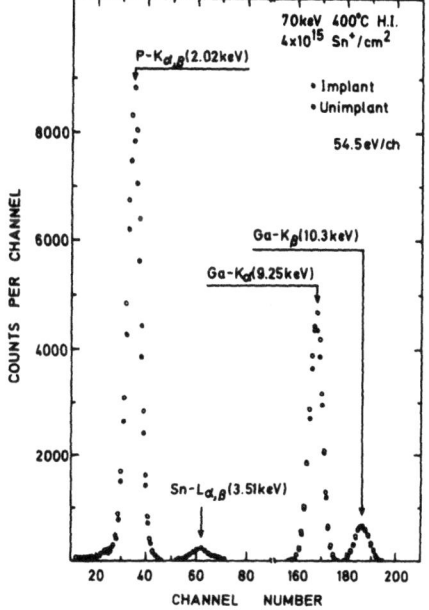

Fig. 1 Typical ion-induced X-ray spectra for unimplanted and Sn-implanted samples.

by about 80~100%. After annealing at 700°C, the substitutional fraction decreases. The concentration of Sn (~1 x 10^{21}/cm^3) exceeds the solubility (~1 x 10^{19}/cm^3), while that of Sb is below the solubility (Sb forms solid solutions of GaP$_x$Sb$_{1-x}$ with x=0~1).

Figure 1 shows typical ion-induced X-ray spectra for unimplanted and Sn-implanted GaP samples. In the X-ray spectrum of the implanted sample, Sn-L X-ray peak appears at about channel 60. Ion-induced X-ray signal from the matrix (Ga and P) was used to compare the critical half angle of channeling for the matrix with that for implanted impurities. The critical half angle of channeling ($\Psi_{1/2}$) is proportional to $\sqrt{Z_2}$, where Z_2 is the atomic number of the constituents (Ga or P). Therefore one can distinguish two substitutional sites for implanted impurities in GaP when the critical angle measurements are performed along the <110> direction, for which the

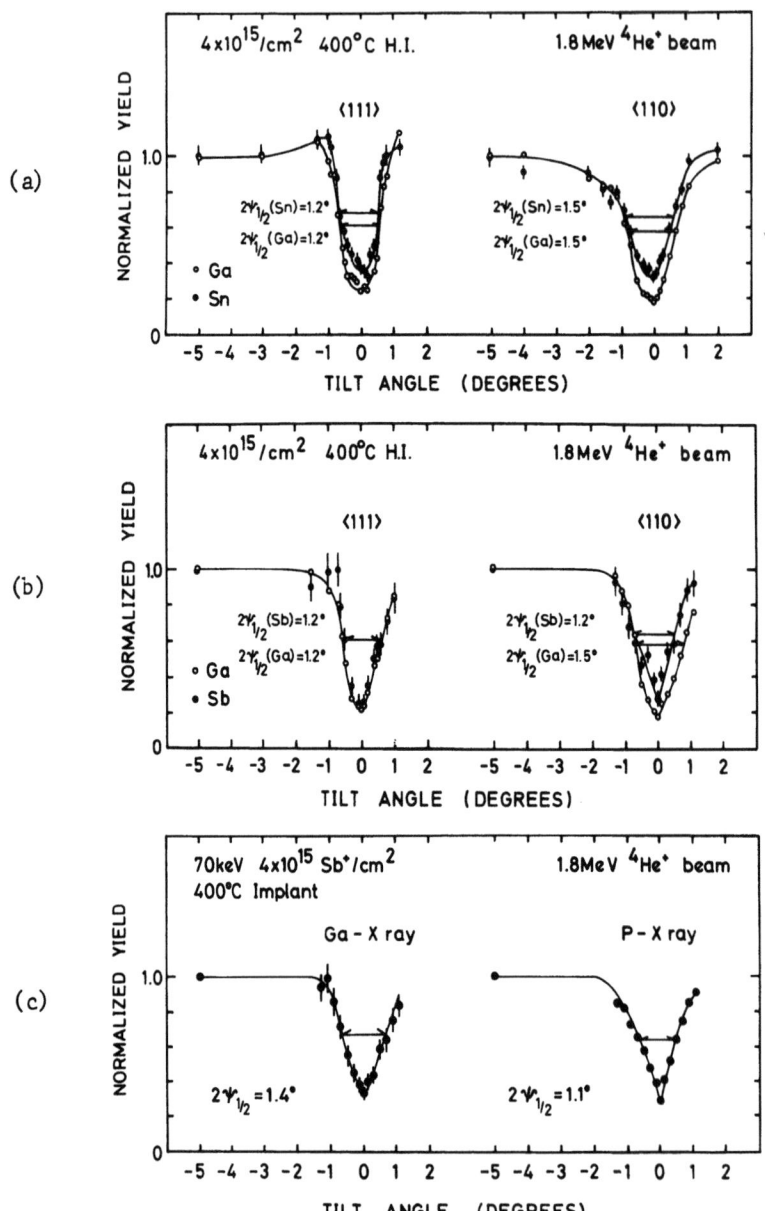

Fig. 2 Channeling angular distributions along the <111> and the <110> axes for GaP crystals implanted with Sn (a) and Sb (b), showing the orientation dependence of backscattering yield from impurities and Ga. (c) Ion-induced X-ray yields (Ga-K_α and P-$K_{\alpha,\beta}$) as a function of incident angle of 1.8 MeV $^4He^+$ along the <110> axis.

strings of atoms are made up of only one of the two types of atoms [3].

Figure 2 shows angular scans through the major axes for GaP samples implanted at 400°C with Sn and Sb at a dose of 4 x 10^{15}/cm^2. Figures 2a and 2b are the angular distribution of the back scattering yield from Ga and impurities. Fig 2c is the X-ray yield from the matrix as a function of incident angles of 4He$^+$ through the <110> axes. For the backscattering dip, the width of the Sn dip along each of the axes is the same as that of the Ga dip. The minimum scattering yield from Sn is around 0.30. These results indicate that more than 70% (nearly 80%) of Sn implanted at 400°C are located on Ga substitutional sites. For Sb, the width of the Sb dip along the <111> axis agrees with that of the Ga dip, while the width along the <110> axis is narrower than that of the Ga dip. For the X-ray dip, the width of the P dip is narrower than that of the Ga dip. A direct comparison between the Sb backscattering dip and the P X-ray dip is difficult because X-ray data include signals from greater depths as compared to backscattering data [5]. As P X-ray signals probe almost the same depth as Ga X-ray

Table II   A summary of critical half angles of channeling for Sn- and Sb-implanted GaP samples.

	<111>	<110>	
$2\Psi_{1/2}(Ga)$	1.2°±0.05°	1.5°±0.05°	R.B.S.
$2\Psi_{1/2}(Sn)$	1.2°±0.05°	1.5°±0.05°	R.B.S.
$2\Psi_{1/2}(Sb)$	1.2°±0.05°	1.2°±0.05°	R.B.S.
$2\Psi_{1/2}(Ga)$	–	1.4°±0.05°	X ray
$2\Psi_{1/2}(P)$	–	1.1°±0.05°	X ray
$\dfrac{\Psi_{1/2}(Sn)}{\Psi_{1/2}(Ga)}$	1.0±0.08	1.0±0.07	R.B.S.
$\dfrac{\Psi_{1/2}(Sb)}{\Psi_{1/2}(Ga)}$	1.0±0.08	0.80±0.06	R.B.S.
$\dfrac{\Psi_{1/2}(P)}{\Psi_{1/2}(Ga)}$	–	0.79±0.06	X ray

(1.8MeV ^{4}He$^+$ beam)

signals, the comparison can be made by taking the ratio of the X-ray angular width of P to that of Ga and the ratio of the backscattering angular width of Sb to that of Ga. The ratio of the backscattering angular width of Sb to that of Ga agrees with the ratio of the X-ray angular width of P to that of Ga. These results indicate that the Sb atoms implanted at 400°C are located on P-substitutional sites, while the Sn atoms are located on Ga-substitutional sites. Table II shows a summary of critical half angles for channeling of 1.8 MeV ^{4}He$^+$ in Sn and Sb implanted in GaP.

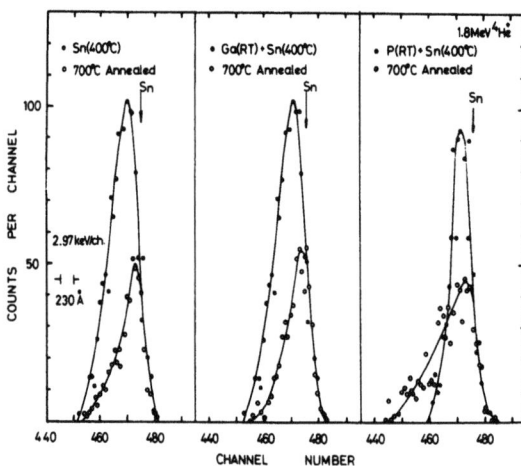

Fig. 3 Backscattered random Sn peaks for Sn implantation, Ga and Sn implantation, and P and Sn implantation before and after annealing at 700°C with SiO_2 cap.

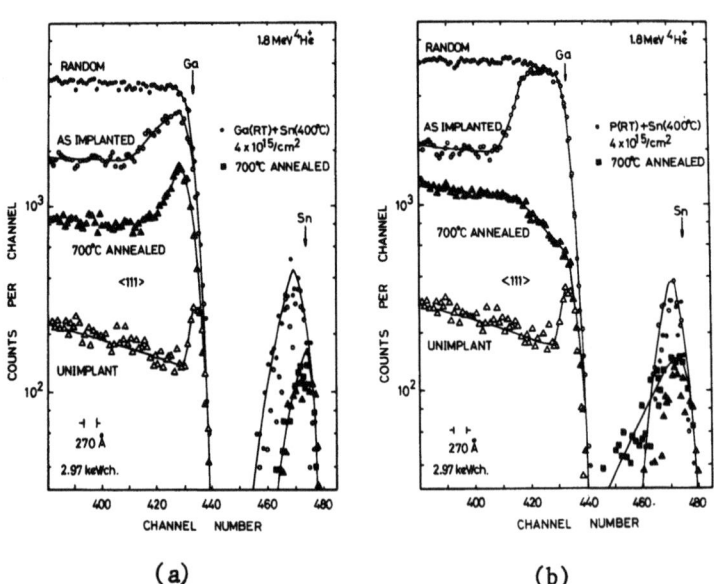

(a) (b)

Fig. 4 Backscattering spectra for GaP samples implanted with Ga and Sn (a), and with P and Sn (b) before and after annealing at 700°C.

Table III A comparison of Sn diffusion behaviors and lattice locations for various implantations and annealing.

	Anneal	Dose(/cm^2)$^{a)}$	χ_m<111>$^{b)}$	%<111>$^{c)}$	Diffusion in	Diffusion out
Ga(RT)+Sn(HI)		2.4x10^{15}	0.41	34		
	700°C	1.4x10^{15}	0.16	0	no	yes
P(RT)+Sn(HI)		2.3x10^{15}	0.32	0		
	700°C	1.9x10^{15}	0.18	27	yes	yes?
Sn(HI)		4.0x10^{15}	0.23	67		
(excess GaP)	700°C	3.8x10^{15}	0.16	19	no	yes?
	700°C	1.7x10^{15}	0.18	20	no	yes
Sn(HI)		1.2x10^{15}	0.18	54		
(P-atmosphere)	700°C	1.3x10^{15}	0.11	25	no	no

a) $\pm 0.2 \times 10^{15}$/cm^2, b) ± 0.02, c) $\pm 5\%$ (1.8 MeV ^{4}He$^+$)

Figure 3 shows the atom profiles of Sn implanted under various conditions and annealed at 700°C with SiO$_2$ cap. After annealing at 700°C, out-diffusion of Sn is observed. For samples implanted with P, diffusion into the bulk is also observed. Figure 4 shows backscattering spectra for the GaP samples implanted with Ga and Sn and with P and Sn. A large amount of lattice disorder exists in the <111> spectra for both samples before annealing, while the defect peak exists only in the spectrum for the Ga implanted sample after annealing. In the <111> spectrum for the P implanted sample, the defect peak near the surface disappears after annealing. This result suggests that more damage remains in case of the dual implantation with Ga.

The diffusion of Sn was observed for dual implantation with P, while the diffusion was not observed for dual implantation with Ga. At present it is difficult to determine the diffusion mechanism. However, one can make two possible explanations: First, the diffusion of Sn is associated with vacancy diffusion and it is enhanced by the excess Ga-vacancy produced by P implantation. Second, Sn atoms may be trapped by defects near the surface and can not migrate in the case of Ga implantation.

Table III shows a summary of diffusion behavior of Sn together with lattice location for various implantation and annealing conditions. The substitutional fraction of Sn dually implanted with Ga decreases with annealing, while that of Sn dually implanted with p increases with annealing. The diffusion does not take place during annealing when the sample is sealed with excess GaP or a P atmosphere (~1.5 atm). These annealing conditions and Ga preimplantation might be considered to suppress the Ga-vacancy formation during annealing.

CONCLUSIONS

70 keV Sn and Sb have been implanted in GaP at 400°C and room temperature with moderately high doses. Ga or P atoms were dually implanted with Sn. Annealing was performed under various conditions. Lattice locations and atom profiles were studied. From the results, the following conclusions are drawn:
1) Sn and Sb atoms implanted at 400°C are located on Ga- and P-substitutional sites respectively by 80~100%, while Sn and Sb atoms implanted at room temperature are located on off-lattice sites even after annealing at 700°C. The substitutional fraction after annealing decreases by Ga implantation, while it increases by P implantation.
2) Diffusion of Sn implanted at 400°C is enhanced by P implantation, while it is suppressed by Ga implantation and by annealing in ampoules sealed with excess GaP.

ACKNOWLEDGEMENTS

The authors are indebted to Drs. T. Tsurushima and H. Tanoue of the Electrotechnical Laboratory for implantations, and to Mr. A. Ohsawa and Mr. H. Goshi for their technical assistance in back-scattering experiments.

REFERENCES

[1] E.B. Stoneham and J.F. Gibbons, Ion Implantation in Semi-conductors, ed. by S. Namba (Plenum Press, New York, 1975) p.57
[2] T. Ambridge, R. Heckingbottom, E.C. Bell, B.J. Sealy, K.G. Stephens, and R.K. Surridge, Electronics Letters 11, 314 (1975)
[3] J.L. Mertz, L.C. Feldman, D.W. Mingay and W.M. Augustyniak, Ion Implantation in Semiconductors, ed. by I. Ruge and J. Graul (Springer-Verlag, Berlin, 1971) p.182
[4] M. Takai, K. Gamo, K. Masuda and S. Namba, Japan. J. Appl. Phys. 14, 1935 (1975)
[5] S.T. Picraux, J.A. Davies, L. Eriksson, N.G.E. Johansson and J.W. Mayer, Phys. Rev. 180, 873 (1969)

IMPLANTATION OF Be, Cd, Mg AND Zn IN GaAs AND GaAs$_{1-x}$P$_x$

R. Zülch, H. Ryssel, H. Kranz, H. Reichl, and I. Ruge

Institut für Festkörpertechnologie der FhG

Paul-Gerhardt-Allee 42, 8 München 60, Germany

ABSTRACT

Measurements of electrical activation of Be, Cd, Mg and Zn in GaAs and GaAs$_{0.6}$P$_{0.4}$ for different implantation doses of diffusion profiles of Be, Cd and Zn in GaAs are presented.

INTRODUCTION

Much work has been performed to study n-type doping of GaAs by ion implantation [1-3] since it is difficult to obtain n-doped layers by diffusion. To a much lesser extent p-type doping has been investigated [4,5] because Zn-diffusion is a convenient way to dope GaAs and other III-V compounds. This process is widely used for the production of light-emitting diodes. Nevertheless, the results obtained with Zn-diffusion are not very reproducible and therefore recently p-type doping of GaAs, GaAs$_{1-x}$P$_x$ and GaP gained more interest [6-9]. One major problem in doping these semiconductors is a protective layer to avoid decomposition of the crystal [10,11]. For p-type doping, however, this problem seems to be of less importance. The ions most suited for p-type doping of III-V semiconductors are Be, Cd, Mg and Zn. In this paper we report on results concerning the electrical activation and the diffusion behavior of these dopants in GaAs and GaAs$_{1-x}$P$_x$. As protective layers we used SiO$_2$ and Si$_3$N$_4$. The minority carrier lifetime has been measured to monitor the annealing of the crystal lattice.

EXPERIMENTAL TECHNIQUES

All implantations were performed at 150 keV in <100>-oriented crystals. n- and π-type wafers have been used. The protective layers were pyrolytic SiO_2-layers deposited at 450°C or sputtered Si_3N_4-layers. Annelaing was performed between 450°C and 850°C in a nitrogen atmosphere. For Hall-effect and sheet resistivity measurements a Van der Pauw-type pattern with pre-diffused contact pads defined by sputtered Al_2O_3-layers has been used. Differential Hall-effect and conductivity measurements have been performed using a slow etch of 1 H_2O_2 : 1 H_2SO_4 : 100 H_2O which has an etching rate of 55 nm/min. at 22°C. For $GaAs_{1-x}P_x$ ($x \approx 0.4$) only 20 parts of H_2O have been used. In this case the etching rate is 70 nm/min. at 22°C. The lifetime of the minority carriers in implanted layers was determined by means of photoluminescence. This was done by measuring the phase difference between the exciting laser light and the emitted luminescence light [11].

RESULTS

The annealing behavior of Be in π-GaAs is shown in Fig. 1 for doses between $10^{14} cm^{-2}$ and $10^{16} cm^{-2}$. Be is a very light ion (M=9) therefore little damage is caused during implantation. Up to doses of $10^{16} cm^{-2}$ nearly all Be is incorporated at electrically

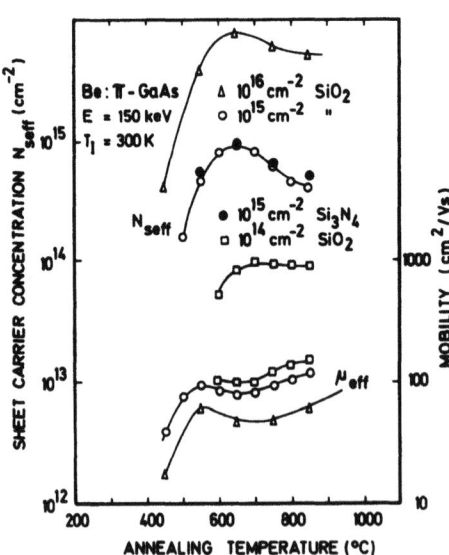

Fig. 1 Annealing behavior of Be implanted at 150 keV in π-GaAs

active lattice sites after annealing at 650°C. At higher temperatures a decrease in sheet carrier concentration can be seen from Fig. 1. The electrical activation is about the same with SiO_2 and Si_3N_4 cap layers. In Fig. 2 diffusion profiles for a 10^{15} cm^{-2} implant after annealing at 650°C - 850°C are presented. The sample annealed at 650°C shows very good correspondence to the theoretical LSS profile. R_p and ΔR_p have been calculated according to a paper of Schiott [32] to be 41.6 nm and 15.9 nm respectively at 150 keV. Also included in Fig. 2 are theoretical diffusion profiles which have been matched to the low concentration parts of the profiles. A decrease in electrical activation as already seen in Fig. 1 takes place at higher annealing temperatures. The reason for this is not yet clear but it might be due to complex formation. The mobility in these implanted layers is very good already after annealing at 650°C. The diffusion coefficient can be calculated from these measurements to be $D = 1.46 \times 10^{-4} \exp(-1.84/kT)$ cm^2 s^{-1}. In case of Si_3N_4 passivation the diffusion coeficient is somewhat larger.

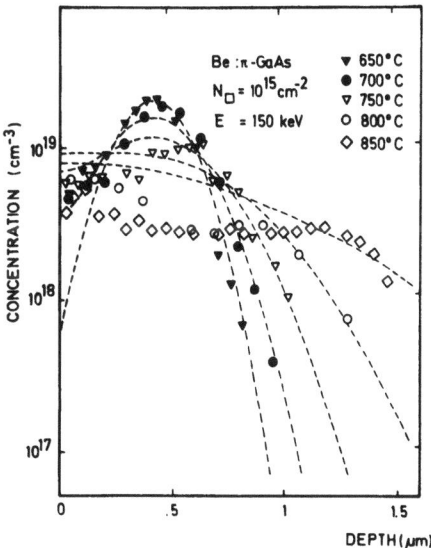

Fig. 2 Diffusion profiles of Be implanted π-GaAs.
E = 150 keV, N = 10^{15} cm^{-2}

Magnesium shows a similar annealing behavior. The total electrical activity, however, is much smaller. In Fig. 3 results for doses between 10^{14} cm^{-2} and 10^{16} cm^{-2} are depicted. As in the case of Be a reverse annealing at temperatures above 650°C

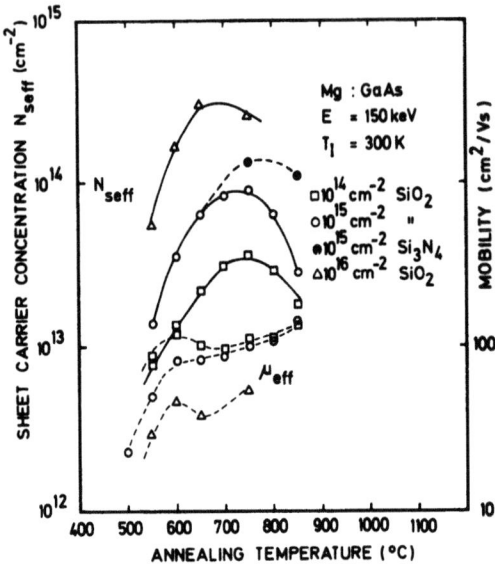

Fig. 3 Annealing behavior of Mg implanted at 150 keV in π-GaAs

to 750°C can be seen. With Si_3N_4-passivation a higher electrical activation can be obtained and the reverse annealing is smaller. From profile measurements it can be concluded that this effect is due to an outdiffusion of Mg and not due to a complex formation. The outdiffusion is smaller for Si_3N_4 passivation. The mobility of implanted layers is very poor. Only after annealing at 800°C bulk values can be obtained in the deeper parts of the profile. The diffusion coefficient of Mg was found to be D = $4.4 \times 10^{-4} \exp(-1.87/kT)$ which is very close to the value determined for Be.

Be and Mg show no difference in the annealing behavior in n- and π-doped wafers. Zn, however, shows a somewhat higher electrical activation if implanted into n-type material. Results of isochronal annealing of doses ranging from 10^{14} cm^{-2} to 10^{16} cm^{-2} are shown in Fig. 4. Again a comparison is given for Si_3N_4 as a protective layer. It results in a lower electrical activation than with a SiO_2-layer. Zn-implanted layers require a much higher annealing temperature than Be or Mg implanted layers to obtain maximum electrical activation. This is obviously due to the damage produced by these heavy ions. Zn profiles as presented in Fig. 5 show a markedly concentration-dependent diffusion behavior. The broken lines in Fig. 5 have been calculated assuming a diffusion model which takes into account a field

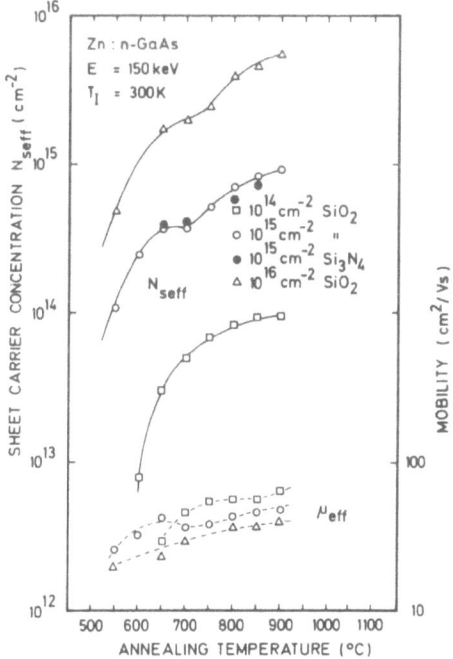

Fig. 4 Annealing behavior of Zn implanted at 150 keV in n-GaAs

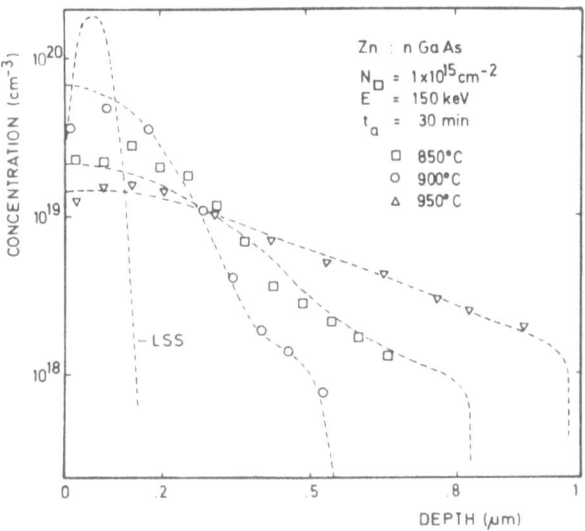

Fig. 5 Profiles of Zn implanted n-GaAs. E=150 keV, $N_\square = 10^{15}$ cm^{-2}

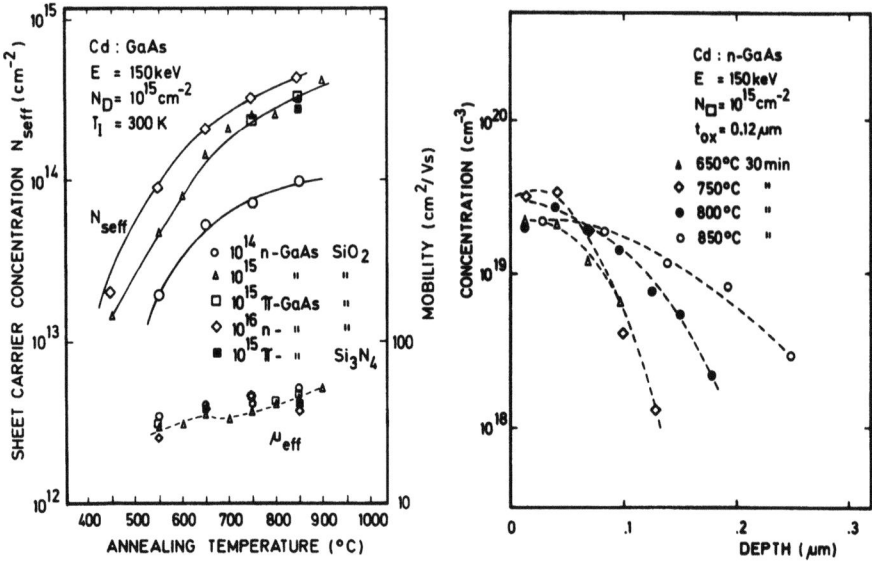

Fig. 6 Annealing behavior of Cd implanted at 150 keV in π- and n-GaAs

Fig. 7 Profiles of $10^{15} cm^{-2}$ Cd implanted at 150 keV in n-GaAs

enhancement of the diffusion process [14]. The diffusion coefficients determined from this model are four orders of magnitude smaller than found in thermal diffusion of Zn. We suppose that in case of implanted distributions with low concentration a substitutional instead of an interstitial process governs the diffusion. If an Si_3N_4-layer is used during annealing a smaller diffusion coefficient is obtained.

Cadmium has a large mass (112) and produces a lot of damage during implantation. Therefore even at high annealing temperatures the damage does not anneal out completely [15], and no saturation in electrical activation is found up to annealing temperatures of 900°C. Results for doses between $10^{14} cm^{-2}$ and $10^{16} cm^{-2}$ are shown in Fig. 6. Again with Si_3N_4-passivation a lower activation results. Only for the $10^{14} cm^{-2}$ dose a 100% activation can be obtained. This is due to the solubility limit of Cd in GaAs [16] of about $2 \times 10^{19} cm^{-3}$. The mobility in the implanted layers is very poor and never attains bulk values. In Fig. 7 diffusion profiles of a $10^{15} cm^{-2}$ implant are shown. One can clearly see the solubility and activation limit around $2 \times 10^{19} cm^{-2}$. Similar profiles have recently been published by Shin, et al [17]. The diffusion coefficient from these measurements

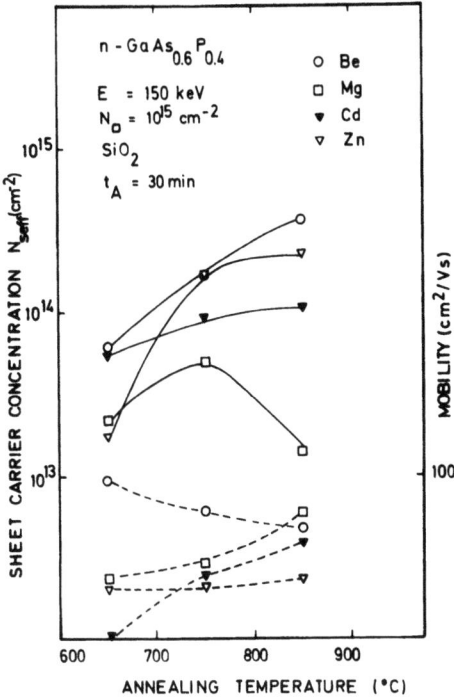

Fig. 8 Annealing behavior of Be, Cd, Mg and Zn in n-GaAs$_{0.6}$P$_{0.4}$

was D = 4.33 x 10^{-4}exp (-2.3/kT) cm^{2}s^{-1}. Again for a Si$_3$N$_4$-passivation a smaller diffusion coefficient with a preexponential term of 3.1 x 10^{-4}cm^{2}s^{-1} was found.

In GaAs$_{1-x}$P$_x$ the annealing behavior of the investigated ions and, where studied, the diffusion behavior are very similar to GaAs. In Fig. 8 the annealing behavior of 10^{15}cm^{-2} implants of Be, Cd, Mg and Zn in GaAs$_{0.6}$P$_{0.4}$ is given. Be shows no reverse annealing in contrast to the results in GaAs. The maximum activation is 35% at this dose and only 15% at a dose of 10^{16}cm^{-2}. The mobility is very high also at low annealing temperatures because of the low damage produced. Zn shows a maximum activity of 20% at 10^{15}cm^{-2} and 4% at 10^{16}cm^{-2} respectively. Cd also gets less electrically active than in GaAs, only 10% in case of 10^{15}cm^{-2}. For all ions investigated the electrical activity obtainable in GaAs$_{0.6}$P$_{0.4}$ is less than in GaAs. Up to now profiles have been measured for Zn only. One gets similar profiles as in GaAs but a much higher diffusion coefficient. Similar profiles but obviously with a still higher diffusion coefficient have been published [18].

Lifetime measurements were mainly performed on wafers which were intended for the implantation of LED's. Only one example for Zn in GaAs and GaAs$_{0.6}$P$_{0.4}$ will be given. Before implantation the lifetime in GaAs was 3 ns. The wafer was best grown 7×10^{18} cm^{-3} Si-doped from Monsanto. After implantation of 5×10^{15} cm^{-2} Zn and annealing at 600°C with a SiO$_2$ cap for more than 90 min. the lifetime in the p-type layer was measured to be 0.3 ns. In epitaxial GaAs$_{0.6}$P$_{0.4}$ the results for the same implantation and annealing parameters are 7 ns and 1 ns respectively. Results of these investigations for different ions investigated together with luminescence properties will be given elsewhere.

ACKNOWLEDGEMENTS

The authors wish to thank M. Bleier, B. Scmiedt, E. Traumüller and M. Schenk for their technical assistance.

REFERENCES

1. F. H. Eisen, in Ion Implantation in Semiconductors, S. Nambe Ed., Plenum Press, New York, p.3 (1974).
2. J. D. Sansbury and J. F. Gibbons, Rad. Eff. 6, 269(1970).
3. D. E. Davies, S. Roosild, and L. Lowe, Solid State Eelctronics 18, 733(1975).
4. R. G. Hunsperger and O. J. Marsh, J. Electrochem. Soc. 116, 488(1969).
5. R. G. Hunsperger, R. G. Wilson, and D. M. Jamba, J. Appl. Phys. 43, 1318(1972).
6. K. C. Wiemer, R. J. Dexter, and I. H. Morgan, Int. Electron. Devices Meeting, p. 8.2, Washington, Dec. 4-6, 1972.
7. Y. Ono, K. Saito and Y. Shiraki, Jap. J.Appl. Phys. 14, 1489(1975).
8. T. Itoh and Y. Oana, Appl. Phys. Lett. 24, 320(1974).
9. P. K. Chatterjee, B. C. Streetman, D. L. Kane and A. H Herzog, Int. Electron Devices Meeting, Washington, Dec. 1-3, 1975, Tech. Digest p. 187.
10. J. S. Harris. F. H. Eisen, B. Welch. R. D. Pashley, D. Sigurd, and J. W. Mayer, Appl. Phys. Lett. 21, 601(1972).
11. Y. Sato, Jap. J. Appl. Phys. 12, 242(1973).
12. G. A. Acket, W. Nijman, and H. t'Lam, J. Appl. Phys. 45, 3033(1974).
13. H. E. Schiott, Rad Eff. 6, 107(1970).
14. R. Zölch, PhD. Thesis, Techn. Univ. Munich (1976).
15. J. J. Grob, A. Ghitescu, and P. Siffert, III-V Compounds in Semiconductors, p. 611, (1972).
16. R. K. Willardson and W. P. Allred, Symposium on GaAs, paper 6, (1966).

17. B. K. Shin, D. C. Look, Y. S. Park and J. E. Ehret, J. Appl. Phys. 47, (1976).
18. H. Okabayashi, in Ion Implantation in Semiconductors, S. Namba Ed., Plenum Press, New York (1974), p. 95.

EFFECTS OF ELECTRICALLY AND OPTICALLY INACTIVE ION IMPLANTATION

IN N-GaAs$_{1-x}$P$_x$ (x ≈ 0.37) ON PHOTOLUMINESCENT PROPERTIES

H. Okabayashi

Central Research Laboratories
Nippon Electric Co., Ltd.
Takatsu-ku, Kawasaki, Japan

ABSTRACT

The effects of implantation with Ga, B, and As, which are considered to be inactive electrically and optically in GaAsP, on photoluminescent properties have been studied. From the results on the annealing behavior of near-band-gap emission, the main factors influencing the luminescence intensity recovery were concluded to be the stoichiometric effect of implanted ions for the low dose ($\lesssim 10^{13}$ cm^{-2}) implantation, and to be residual and/or secondary defects for the high dose ($\gtrsim 10^{14}$ cm^{-2}) implantation. Photoluminescence efficiency improvement was achieved by implantation with $(5\sim10)\times10^{12}$ B or Ga (group-III) ions cm^{-2} and subsequent annealing at ~750°C. The efficiency improved region was found to extend much deeper than the theoretically expected range of implanted ions. From comparison of photoluminescence spectra measured at ~80°K, it is suggested that the efficiency improvement can be attributed to reduction in the concentration of "killer" centers associating with Ga vacancies.

INTRODUCTION

In a preliminary study, it was found that implantation with 10^{13} B or Ga (group-III) ions cm^{-2} in n-GaAsP and subsequent annealing at ~750°C improved the photoluminescence intensity of as-grown crystals, but the As (group-V) ion implantation led to intensity degradation [1].
This paper describes detailed results of a study on annealing behavior of the near-band-gap emission intensity, luminescence intensity varation with depth, and spectra comparison.

EXPERIMENTAL

Experimental conditions were almost the same as those described in the previous paper [1] and summarized as follows. The $GaAs_{1-x}P_x$ (x=0.34~0.39) crystals were Te-doped, n-type and vapor-phase-epitaxial on n^+-GaAs. Implantation was carried out at room temperature and mostly at 100keV. RF-sputtered or chemical-vapor-deposited (CVD) SiO_2 films were used as a protective film on annealing. Annealing was carried out in a dry nitrogen flow. Photoluminescence measurements were made at room temperature and at ~80°K using the 514.5-nm line from an argon laser. A Spex-1702 grating monochromater and an RCA-7102 photomultiplier were used as a detector system. The relative photoluminescence intensity, I/I_0, is denoted as the ratio of the near-band-gap emission intensity for each sample, I, and its original intensity, measured before implantation, I_0.

RESULTS

Annealing Behavior of Near-Band-Gap Emission Intensity

Figure 1 shows the relative PL intensity of the near-band-gap emission for samples implanted with various Ga and B ion doses as a function of annealing temperature for 20 minute isochronal annealing. Maximum intensities, obtained after annealing at 700~750°C, of samples implanted at doses of 5×10^{12} cm^{-2} and 1×10^{13} cm^{-2} exceed one, i.e., the original value. However, the maximum intensity for the sample implanted at 1×10^{12} cm^{-2} is comparable to that of the unimplanted control sample, but is smaller than the original value. The intensity of samples, implanted at higher doses than 10^{13} cm^{-2}, decreases with increasing dose. Intensity for the sample implanted with 10^{14} Ga cm^{-2} decreases with annealing temperature above 600°C. Intensity degradation above annealing temperature of 750°C is due to SiO_2 protective film effects.

The results of the sequential isothermal annealing for samples implanted with 10^{13} Ga cm^{-2}, 10^{13} As cm^{-2}, and (10^{13} Ga+10^{13} As) cm^{-2} are shown in Fig. 2. Isothermal annealing behaviors at 500°C are similar to each other. At 600°C, the Ga, As and (Ga+As) implanted samples show different annealing behaviors, i.e., a slight increase, a slight decrease and an almost constant value in the intensity with annealing time, respectively. At 750°C, the Ga implanted sample and the As implanted sample show larger and smaller values than the original values, respectively, while the (Ga + As) implanted sample shows an intermediate value between them. These results, shown in Figs. 1 and 2, indicate that the ion species effect, i.e., stoichiometric effect, is the main factor influencing the intensity for implantation with inactive ions at low dose ($\lesssim 10^{13}$ cm^{-2}). It is noted that the unimplanted control sample shows a

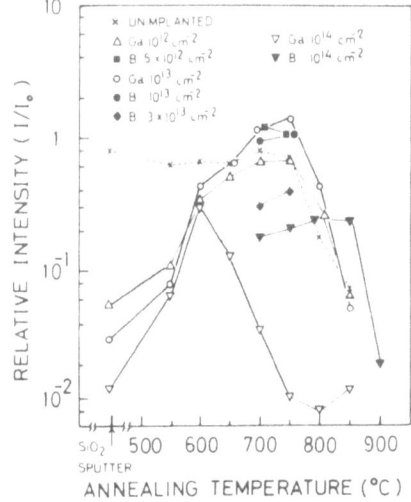
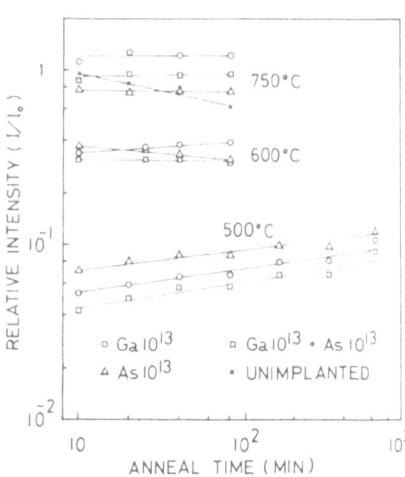

Fig. 1. Isothermal annealing (20 minutes at each temperature) of near-band-gap emission intensity.

Fig. 2. Sequential isothermal annealing of near-band-gap emission intensity.

rather distinct decrease in the intensity with annealing time.

The results of the similar isothermal annealing for samples implanted at 3×10^{13} cm^{-2} and 6×10^{13} cm^{-2} are shown in Fig. 3. Annealing at 700°C and 750°C, the (Ga+As) implanted sample shows a distinct decrease in intensity with annealing time, while the Ga implanted sample and the As implanted sample show an increase and a slight decrease in intensity at 750°C, respectively. A similar distinct decrease in intensity with annealing time was also observed in the samples implanted with 10^{14} Ga cm^{-2}, 10^{14} As cm^{-2} and (10^{14} Ga+10^{14} As) cm^{-2}, though differences in intensity, dependent on ion species, were observed. Therefore, the main factor suppressing the intensity recovery for high dose ($\gtrsim 6\times10^{13}$ cm^{-2}) implantation is not the stoichiometric effect, but residual and/or secondary defects. This is consistent with the observation in GaAs, where defects induced by implantation with heavy impurity ions at high dose ($\gtrsim 10^{14}$ cm^{-2}) are difficult to be annealed out at 700°C [2].

Near-Band-Gap Emission Intensity Variation with Depth

Near-band-gap emission intensity variation with depth was obtained by PL measurements in conjunction with step-by-step etching. The results are shown in Fig. 4. The PL intensity degraded region in the as-implanted GaAsP was also found to extend considerably

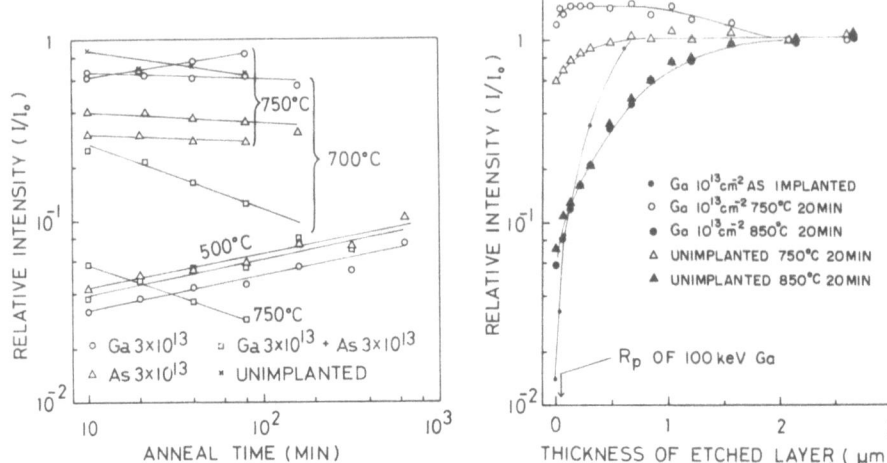

Fig. 3. Sequential isothermal annealing of near-band-gap emission intensity for samples implanted at increased doses.

Fig. 4. Near-band-gap emission intensity variation with etched layer thickness.

deeper than the theoretically expected range of implanted ions, as is known for GaAs [2] [3]. The PL intensity of the sample implanted with 10^{13} Ga cm^{-2} and annealed at 750°C for 20 minutes is higher than the original value over the region from the surface to 1.5~2 μm. It is interesting to observe that the extension of the intensity improved region is much deeper than the theoretically expected range of 100keV Ga ions in GaAsP, R_p~0.05 μm, estimated by interpolation of the calculations for GaAs and GaP [4]. The intensity degradation in the surface region in the unimplanted control sample is due to the SiO$_2$ protective film effect. After annealing at 850°C, the Ga implantation effect disappeared. Both the Ga implanted and unimplanted control samples show a similar degradation in intensity due to the SiO$_2$ protective film effect.

Photoluminescence Spectra

Figure 5 shows PL spectra measured at 80°K for samples annealed at 750°C for 80 minutes with sputtered SiO$_2$ protective films. The peak A at ~630-nm (~1.968 eV) is due to near-band-gap transition. The intensity of the peak A for the sample implanted with 10^{13} Ga cm^{-2} is considerably larger than that of the As implanted sample and that of the unimplanted control sample. The smaller intensity in the unimplanted control sample, even in comparison with that of the As implanted sample, corresponds to the results of room temperature measurements for the 750°C isothermal

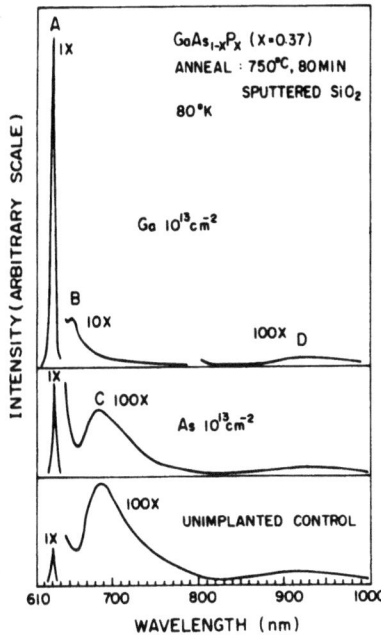

Fig. 5. 80°K photoluminescence spectra for samples annealed at 750°C for 20 minutes.

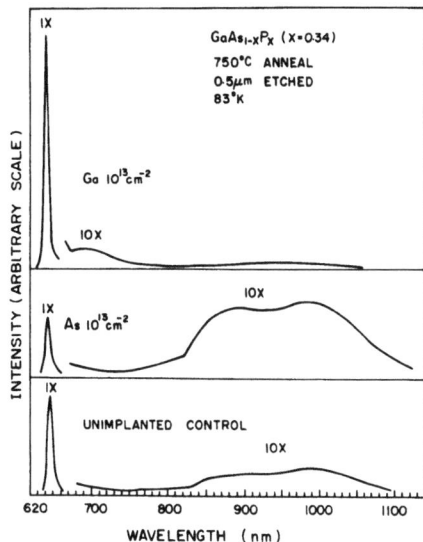

Fig. 6. 83°K photoluminescence spectra for samples etched off by ~0.5μm after annealing at 750°C for 20 minutes.

annealing shown in Fig. 2. The peak B at ~654-nm (~1.896 eV) seen in the Ga implanted sample, which has a tendency to increase with Ga dose, is weak or undetectable in the As implanted and unimplanted control samples. Therefore, peak B is considered to be associated with As and/or P vacancies or related defects. The peak C at ~686-nm (~1.808 eV) is distinct in the As implanted and unimplanted control samples, but is not separated in the Ga implanted sample. Peak C may be associated with deep impurities or Ga vacancies. The broad infrared emission peaking at 900~960-nm, which is reported to be made up of two bands [5] and associated with Ga-vacancy-donor complex defects [5] [6], is very weak in these samples. These features of the spectra, described above, are rather common in the samples implanted at 10^{13} cm^{-2} and annealed at 750°C with a SiO$_2$ protective film, though differences in the relative intensity of each peak are observed probably owing to diffences in properties of original crystals.

Some samples showed a considerable increase in the infrared emission intensity after annealing. The spectra of the samples, whose molar fraction of GaP happened to be ~0.34, are shown in Fig. 6. Although these spectra were obtained from the samples etched off by ~0.5 μm after annealing at 750°C for 20 minutes, the features seen in these spectra were also observed in the spectra obtained from the samples before etching. It is noted that the infrared emission associated with Ga-vacancy-Te complex defects, for the Ga-implanted sample, is rather weaker than that of the unimplanted control sample, while the As implanted sample shows a rather larger intensity than the unimplanted control sample. For the near-band-gap emission intensity, the Ga implanted sample shows the largest value, but the As implanted sample shows a smaller intensity than shown by the unimplanted control sample, as is expected from the order of the infrared emission intensity. Therefore, these results are understood to demonstrate that the degradation due to increase in the concentration in Ga vacancies or related defects on annealing can be suppressed by Ga implantation.

DISCUSSION

The PL intensity improvement obtained by group-III ion implantation can be mainly attributed to the internal quantum efficiency improvement of the crystal, not to reduction in the surface recombination velocity and/or in the absorption coefficient for the luminescence. This is because distinct differences in the PL spectra of the Ga implanted, As implanted, and unimplanted control samples, considered to be due to the stoichiometric effect of the implanted ions, were observed as was shown in Figs. 5 and 6. In addition to that, the PL intensity improved region was found to extend very deep by PL measurements in conjunction with step-by-step etching. Since the total emission intensity in the measured wavelength range (0.6~1.2μm), as well as the near-band-gap emission intensity, increased in the efficiency improved sample, the internal quantum efficiency improvement is considered to be due to reduction in the concentration of non-radiative "killer" centers, associated with Ga vacancies or related defects.

The reason for the very deep extension of the efficiency improved region is not clear at present, but simple explanations can be given as follows: 1. A deep in-diffusion of implanted Ga atoms to reduce "killer" centers associating with Ga vacancies or related defects. 2. Out-diffusion of "killer" centers or out-diffusion of constituent components of "killer" centers instead of in-diffusion of implanted Ga atoms.

The self-diffusion coefficient of implanted Ga atoms in GaAsP is thought not to differ extraordinarily from the value of ~10^{-21} cm^2 sec^{-1} in GaAs, obtained from extrapolation to 750°C of reported results [7]. However, the diffusion coefficient, required to explain the deep extension of the efficiency improved region, must be ~10^{-11} cm^2 sec^{-1}, assuming $\sqrt{Dt} \approx 2\mu m$. Therefore, if the in-diffusion of implanted Ga atoms is a main factor, enhancement of the Ga self-diffusion must occur. The enhancement of Ga self-diffusion may have some correlation with the deep extension of the intensity degraded region seen in as-implanted samples.

In the second explanation, implanted Ga atoms do not necessarily diffuse very deeply, because implanted and unimplanted Ga atoms can not be distinguished from each other. If it is simply assumed that out-diffusants are Ga vacancies, the Ga vacancy diffusion coefficient in GaAsP, $\sim 10^{-11}$ cm^2 sec^{-1}, required for the deep extension of the efficiency improved region, is not unprobable, since the Ga vacancy diffusion coefficient in GaAs is reported to be $\sim 10^{-8}$ cm^2 sec^{-1} [8] and $\sim 10^{-13}$ cm^2 sec^{-1} [9] at 750°C. In the Ga implanted sample, concentrations of the Ga vacancy and the As vacancy in the surface region are considered to be smaller and larger than those in the deeper region, respectively. That is, the reverse gradient in the Ga vacancy concentration is probably formed. This is favorable to the Ga vacancy out-diffusion.

To confirm these tentative explanations, however, much work must be done.

The carrier concentration profile of the efficiency improved sample showed a similar slight decrease in the surface region ($\lesssim 0.2$ μm) to that of the unimplanted control sample. This is due to the SiO$_2$ protective film effect, probably resulting from Ga out-diffusion into a SiO$_2$ film, as reported in the case of SiO$_2$ on GaAs [10]. In the deeper region ($\gtrsim 0.4$μm), the carrier concentration profile is flat and shows no difference between the efficiency improved sample and the unimplanted control sample within the experimental error of $\pm 1.5 \times 10^{16}$ cm^{-3}. If the Ga atoms, implanted at 10^{13} cm^{-2}, are assumed to be uniformly distributed over an efficiency improved region ~ 2μm thick, the average additional Ga concentration is $(2.5 \sim 5) \times 10^{16}$ cm^{-3}. This value is large enough to observe, if the number of the additional Ga atoms corresponds to the change in the carrier concentration. Therefore, the net dose effective for reduction in the "killer" center concentration may be smaller than the experimentally observed optimum dose, $(5 \sim 10) \times 10^{12}$ cm^{-2}. A deep implantation using lighter ions, such as boron, and use of a better protective film will help the quantitative understanding.

CONCLUSION

The effects on photoluminescence properties due to implantation with Ga, B, and As ions, which are considered to be inactive electrically and optically in GaAsP, have been studied.

From the results on annealing behaviors of the near-band-gap emission for samples implanted with various doses, it is concluded that the luminescence intensity of samples implanted with low doses ($\lesssim 10^{13}$ cm^{-2}) is mainly influenced by the ion species effect, i.e., stoichiometric effect, also, defects induced by high dose ($\gtrsim 10^{14}$ cm^{-2}) implantation at room temperature are difficult to anneal out. The intensity recovery for high dose is mainly suppressed by residual and/or secondary defects.

Implantation with $(5\sim 10)\times 10^{12}$ B or Ga ions cm^{-2} and subsequent annealing at $\sim 750°C$ was found to be effective in suppression of degradation on annealing and/or improvement in photoluminescence intensity. The photoluminescence intensity improvement is attributed to the internal quantum efficiency improvement, which is considered to result from reduction in the concentration of "killer" centers associated with Ga vacancies or related defects.

The efficiency improved region was found to extend much deeper than the theoretically expected range of implanted ions.

ACKNOWLEDGMENT

The author expresses his thanks to I. Hayashi, S. Asanabe, and D. Shinoda for their encouragement and helpful discussions. He is also grateful to K. Mori for his advice on photoluminescence measurements, to M. Ueki for carrying out most of the implantation, and to T. Yamagata for his help in making the measurements.

REFERENCES

1. H. Okabayashi, Appl. Phys. Lett. 28, 490 (1976).
2. J. M. Woodcock, J. M. Shannon and D. J. Clark, Solid-State Electron. 18, 267 (1975).
3. K. Aoki, K. Gamo, K. Masuda and S. Namba, Japan. J. Appl. Phys. 15, 405 (1976).
4. W. S. Johnson and J. F. Gibbons, Projected Range Statistics in Semiconductors (Stanford University Bookstore, Stanford, Cal., 1970).
5. G. Heine and M. Morgenstern, Phys. Stat. Sol. (a) 18, K139 (1973).
6. C. E. E. Stewart, J. Cryst. Growth 8, 259 (1971).
7. B. Goldstein, Phys. Rev. 121, 1305 (1961).
8. M. Toyama, Japan. J. Appl. Phys. 8, 1000 (1969).
9. S. Y. Chiang and G. L. Pearson, J. Appl. Phys. 46, 2986 (1975).
10. J. Gyulai, J. W. Mayer, I. V. Mitchell and V. Rodriguez, Appl. Phys. Lett. 17, 332 (1970).

ELECTRICAL AND PHOTOLUMINESCENCE PROPERTIES OF Be-IMPLANTED GaAs AND $GaAs_{0.62}P_{0.38}$*

P. K. Chatterjee,[‡] W. V. McLevige, B. G. Streetman and K. V. Vaidyanathan
Coordinated Science Laboratory and Department of Electrical Engineering, University of Illinois at Urbana-Champaign, Urbana, Illinois 61801

ABSTRACT

Photoluminescence and electrical measurements on Be-implanted GaAs and $GaAs_{1-x}P_x$ ($x \simeq 0.38$) indicate excellent impurity activation and lattice reordering after room temperature implantation and suitable annealing. From detailed luminescence measurements, the ionization energy of Be is estimated to be 28.4 meV in GaAs and 35 ± 3 meV in $GaAs_{0.62}P_{0.38}$. Luminescence and electrical measurements indicate that Si_3N_4 is a much better encapsulant than SiO_2 for GaAs, whereas the two encapsulants produce equivalent results for $GaAs_{1-x}P_x$.

INTRODUCTION

Ion implantation of beryllium into GaAs and $GaAs_{1-x}P_x$ has been shown to be an attractive method of acceptor doping these semiconductor compounds [1-3]. Controlled introduction of Be by standard growth and diffusion processes is very difficult [4,5]. Its small mass and relatively insignificant redistribution during anneal (for moderate concentrations) make Be a suitable acceptor for many device applications. In this paper we present the electro-optical properties of the Be acceptor in GaAs and $GaAs_{0.62}P_{0.38}$ after room temperature implantation and appropriate annealing. Low temperature (6°K) photoluminescence is used to study the optical

---

*Work supported by Joint Services Electronics Program (U.S. Army, U.S. Navy and U.S. Air Force) under Contract DAAB-07-72-C-0259, Office of Naval Research, and by Monsanto Company.

[‡]Present address: Central Research Laboratories, Texas Instruments, Inc., Dallas, Texas 75222.

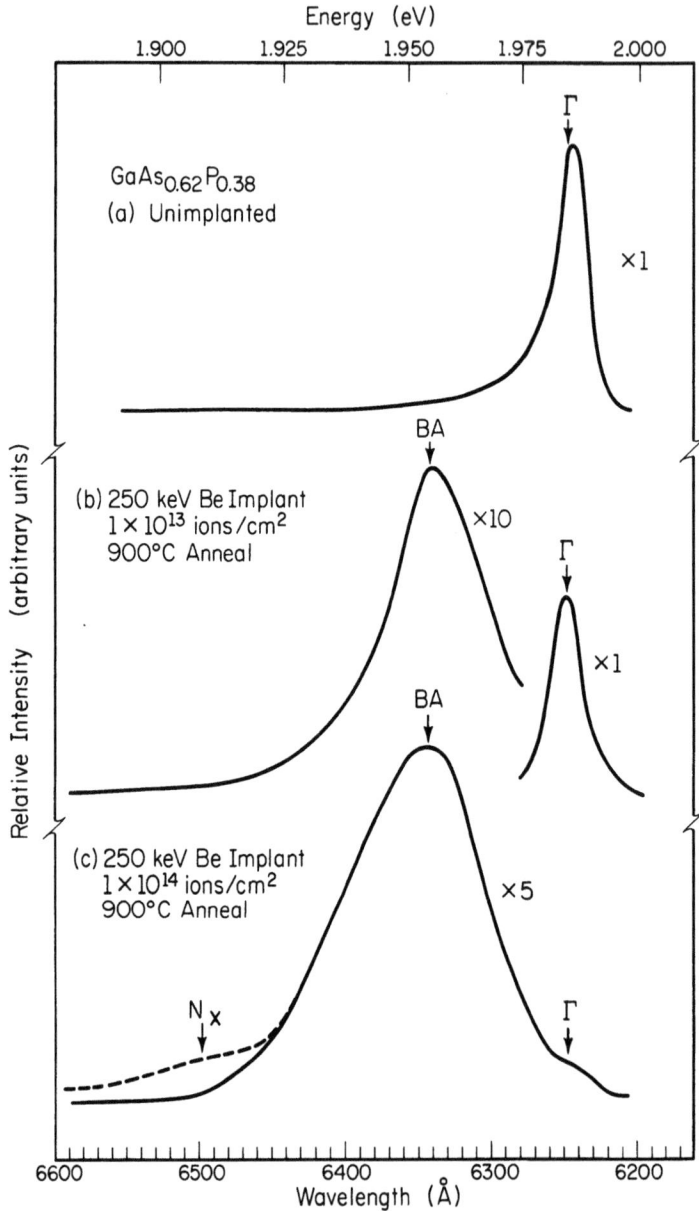

Figure 1. Typical photoluminescence spectra at 6°K: (a) unimplanted $GaAs_{0.62}P_{0.38}$; (b), (c) $GaAs_{0.62}P_{0.38}$ implanted with 250 keV Be doses shown and annealed at 900°C for 1 hr. The solid curves are obtained from samples annealed with SiO_2 encapsulation. The dashed curve in (c) shows the change in spectral distribution when Si_3N_4 encapsulation is used for the anneal.

properties of these implanted layers, and van der Pauw type double ac Hall effect and resistivity measurements are used for electrical characterization. These measurements indicate that suitably annealed Be implanted layers provide device quality material with excellent impurity activation.

PHOTOLUMINESCENCE STUDIES

The low temperature photoluminescence (PL) system used for the study of implanted layers has been described in detail elsewhere [6]. The excitation was provided by an Ar-Kr laser. The excitation wavelength was chosen to be 5145 Å for GaAs and 4880 Å for $GaAs_{0.62}P_{0.38}$ in order to provide carrier generation in the implanted region.

In Fig. 1 we compare typical PL spectra obtained from Be implanted $GaAs_{0.62}P_{0.38}$ with that obtained from an unimplanted sample from the same wafer. The peak labeled Γ has traditionally been attributed to direct recombination between the Γ_1 conduction band minimum and the valence band [7]. However, the highest emission energy PL peak (for low pumping level) in lightly doped GaAs is attributed to free exciton recombination [8]. Nelson et al. [9] have recently observed free exciton absorption in lightly doped $GaAs_{0.62}P_{0.38}$. On the basis of this observation, and consistent with the assignment in GaAs, we attribute the emission labeled Γ in Fig. 1 to free exciton recombination. In Be implanted samples [Fig. 1(b), (c) solid curves] a lower energy peak at 6340 Å (1.955 eV) grows with increasing Be dose. A similar peak at 8304 Å (1.493 eV) is observed in Be implanted GaAs (Fig. 2). We attribute this emission peak to recombination from conduction band to neutral Be acceptors (BA). Similar peaks have been observed in Zn doped GaAs [8].

From the position of the Be related peak and a detailed study of the temperature dependence of the intensity and peak energy, we estimate the binding energy of the Be acceptor to be 28.4 meV in GaAs [10] and 35 ± 3 meV in $GaAs_{0.62}P_{0.38}$ [11].

TRANSPORT MEASUREMENTS

Hall mobility and sheet carrier concentration measured for GaAs samples implanted with Be to various doses at 250 keV and annealed for 1 hour at temperatures between 600-900°C with Si_3N_4 encapsulation are shown in Fig. 3. The sheet carrier concentration increases with anneal temperature, whereas the mobility stays roughly independent in this temperature range. This is similar to the anneal data of Hunsperger et al. [2]. The maximum value of sheet carrier concentration agrees remarkably well with the implantation fluences up to $\sim 10^{14}$ ions/cm^2. At higher fluences the sheet carrier concentration is lower than the fluence, becoming about 60% of the fluence for a dose of 6×10^{14} cm^{-2}. Since the maximum impurity concentration for this implantation dose is about

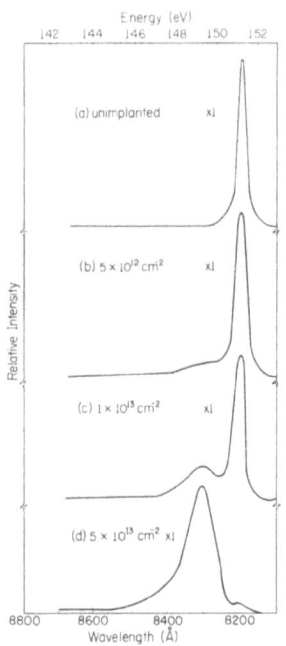

Figure 2. Typical photoluminescence spectra at 6°K: (a) unimplanted GaAs; (b-d) GaAs implanted with Be at 130 keV to doses shown and annealed at 900°C for 1 hr with Si_3N_4 encapsulation.

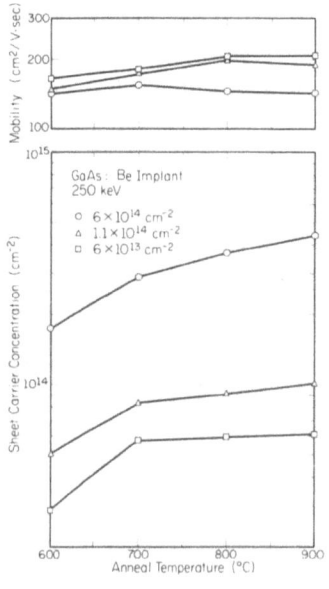

Figure 3. Hall mobility and sheet carrier concentration as a function of anneal temperature for GaAs samples implanted with 250 keV Be to the doses shown, and annealed with Si_3N_4 encapsulation.

10^{19} cm^{-3}, partial activation is not very surprising. The doping efficiency reported here is significantly higher than that obtained for donor implants in GaAs [12,13]. The measured room temperature mobilities in Be implanted samples agree very well with those measured in bulk p-type samples [6], indicating an excellent restructuring of the lattice during annealing. Similar results have been reported for GaAs$_{1-x}$P$_x$ (x ~ 0.38) samples implanted with Be [14]. As in the case of GaAs, the sheet carrier concentration increases monotonically with increasing anneal temperature and yields essentially 100% impurity activation for doses up to 1.1×10^{14} ions/cm^2. Although the Be doping efficiency in GaAs and GaAs$_{1-x}$P$_x$ (x ~ 0.38) is very similar, the mobility in GaAs$_{1-x}$P$_x$ is lower than in GaAs by almost a factor of two. This is possibly due to the alloy (As,P) disorder in ternary systems. The mobilities obtained for Be acceptors seem to be consistently higher than those reported for Zn implanted material [15].

The temperature dependences of Hall mobility in Be implanted GaAs and GaAs$_{0.62}$P$_{0.38}$ show strong impurity banding effects for Be doses as low as 1×10^{13} cm^{-2}. At lower doses, contact problems make meaningful Hall measurements difficult to obtain.

ANNEALING AND ENCAPSULATION

PL spectra from Be implanted GaAs and GaAs$_{0.62}$P$_{0.38}$ are dominated by two peaks. The exciton peak is very sensitive to lattice perfection and may be used as a monitor of the lattice reordering during the anneal. The BA peak is related to the number of optically active Be acceptor sites and serves as an excellent index of impurity activation.

Figure 4 presents the integrated intensity of the BA luminescence at 6°K (1.493 eV band) as a function of anneal temperature for GaAs samples implanted at 130 keV to a dose of 1×10^{13} ions/cm^2 and annealed in various environments. Considerable As outdiffusion takes place when GaAs samples are heated above 600°C either in air or in vacuum. The loss of As results in a defect luminescence center at 1.47 eV [16]. To prevent this loss of As, we have annealed the samples with Si$_3$N$_4$ or SiO$_2$ encapsulation or with As overpressure. The 1.47 eV defect band is not observable in these samples. As seen from Fig. 4, the integrated intensity of the Be-related band increases with anneal temperature for samples annealed with Si$_3$N$_4$ or SiO$_2$ encapsulation. In the case of samples annealed in As overpressure, the intensity of the band increases with anneal temperature between 600°C and 800°C but decreases between 800°C and 900°C, apparently due to degradation caused by pronounced Ga outdiffusion. The exciton peak recovers to the preimplanted level when the sample is annealed at 900°C with Si$_3$N$_4$ or SiO$_2$ encapsulation, but is severely degraded for samples

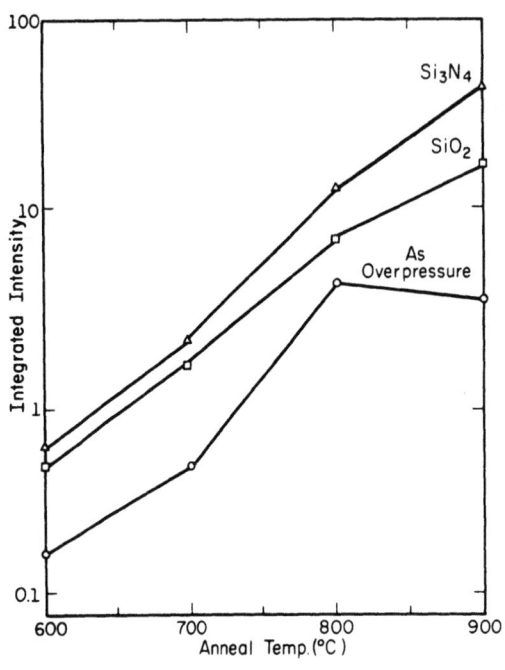

Figure 4. Integrated intensity of the 1.493 eV BA luminescence band due to Be acceptors in GaAs, as a function of isochronal anneal temperature for samples annealed with Si_3N_4 and SiO_2 encapsulation in flowing Ar, and under As overpressure.

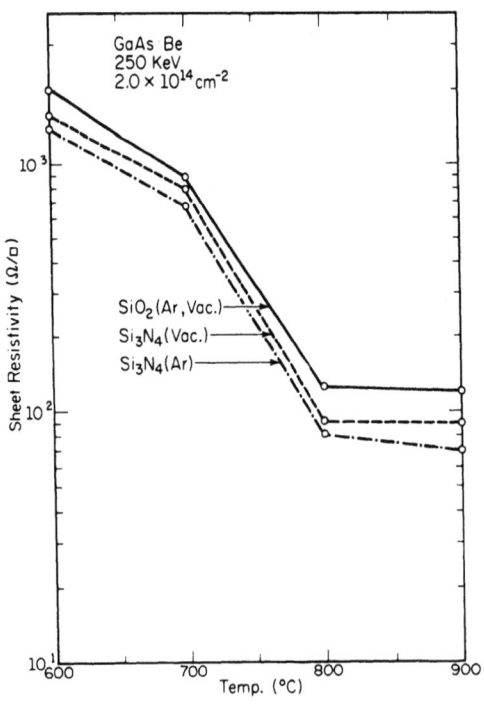

Figure 5. Sheet resistivity of GaAs samples implanted with Be as shown, as a function of isochronal anneal temperature, for anneals conducted in flowing Ar or vacuum, with SiO_2 or Si_3N_4 encapsulation.

annealed in As overpressure. The intensity of this peak in GaAs is always higher when annealing is done with Si_3N_4 encapsulation rather than with SiO_2.

The effect of various annealing environments on the electrical activation of Be-implanted GaAs is shown in Fig. 5. The sheet resistivity of the implanted layer is plotted as a function of anneal temperature for samples annealed with SiO_2 or Si_3N_4 encapsulation, under vacuum or in flowing Ar. The lowest sheet resistivity, corresponding to maximum impurity activation, is observed for the case of samples annealed with Si_3N_4 encapsulation in flowing Ar. The sheet resistivity decreases monotonically with increasing anneal temperature, indicating increasing impurity activation. Degradation of the electrical characteristics for SiO_2 encapsulation and vacuum anneals is consistent with the work of Gyulai et al. [18] and Chatterjee et al [16], who showed that significant outdiffusion of Ga into SiO_2 occurs during annealing.

Spectra presented in Fig. 1 were obtained from Be-implanted $GaAs_{0.62}P_{0.38}$ samples annealed with SiO_2 encapsulation. The intensities of both BA and Γ peaks are very similar for samples with Si_3N_4 encapsulation. However, for samples encapsulated with Si_3N_4, a lower energy peak at 6510 Å (Fig. 1(c) dashed line) is also present. This band is very similar to that reported due to recombination at nitrogen centers (N_X) with accompanying phonon-assisted transitions [17].

Electrical measurements on implanted $GaAs_{0.62}P_{0.38}$ with Si_3N_4 or SiO_2 encapsulation resulted in essentially identical sheet carrier concentration and mobility values (within $\pm$ 1%). This leads us to conclude that the encapsulating dielectric for annealing $GaAs_{1-x}P_x$ ($x \sim 0.38$) does not determine the electrical properties as critically as in the case for GaAs (Fig. 5). It is difficult to present a strong theoretical basis for this result. The presence of phosphorus in the system perhaps reduces the tendency of the crystal to form native defects by outdiffusion into the SiO_2. It may also be argued, however, that since a large density of deep level defects are present in the $GaAs_{0.62}P_{0.38}$ crystals [19], the effects of variation in the annealing encapsulation may not be significant by comparison.

SUMMARY

We have established that Be is a shallow acceptor in GaAs and $GaAs_{1-x}P_x$ on the basis of photoluminescence and electrical measurements. The implanted GaAs leads us to estimate that the binding energy of Be acceptors in GaAs is 28.4 meV. This assignment is in excellent agreement with the recently published PL data of Ashen et al. [5] on LPE grown GaAs layers. Similar experiments in Be doped $GaAs_{1-x}P_x$ ($x \sim 0.38$) provide an estimate of 35 $\pm$ 3 meV for the binding energy of Be acceptors. The error introduced in this measurement is due primarily to the uncertainty in the location of the Γ_1 conduction band minimum for the ternary alloy.

Hall effect data indicate that the hole mobility obtainable for implanted layers is as good as that reported for bulk p-GaAs samples. The mobilities obtained in p-GaAs$_{1-x}$P$_x$ ($x \sim 0.38$) are significantly lower than those in GaAs, probably because of scattering processes due to alloy (As,P) disorder. The measured hole mobilities for Be implanted layers are higher than those reported for Zn implanted GaAs$_{1-x}$P$_x$ [15]. This may reflect either the superior quality of the GaAs$_{1-x}$P$_x$ substrates used here, or improved annealing. The temperature dependence of the transport data shows effects of impurity banding which make it difficult to calculate binding energies. It is desirable to attempt these measurements on samples of low donor concentration implanted to produce uniform Be profiles with a Be concentration about 5×10^{16} cm^{-3}.

The influence of the encapsulating layer on the annealing properties is strikingly different for GaAs and GaAs$_{0.62}$P$_{0.38}$. In the case of GaAs, annealing with Si$_3$N$_4$ encapsulation leads to more efficient luminescence, better electrical activation, and a more perfect lattice than does annealing with SiO$_2$. In contrast, the electrical properties obtained from Be implanted GaAs$_{0.62}$P$_{0.38}$ is indentical for samples annealed with SiO$_2$ or Si$_3$N$_4$ encapsulation. The degradation of GaAs surfaces due to Ga out-diffusion into the SiO$_2$ layer is the apparent cause of the poorer quality of such annealed layers in Be implanted GaAs. Experimental data presented here is insufficient to explain the difference in anneal behavior of GaAs and GaAs$_{0.62}$P$_{0.38}$. Further investigations of self diffusion in the ternary would be necessary to understand this difference.

It is reasonable to conclude that Be is a very useful acceptor in GaAs and GaAs$_{1-x}$P$_x$. The acceptor binding energy is smaller than that of Zn. The p-type layers obtained by implanted Be doping have electrical and optical properties suitable for fabrication of efficient electro-optical devices.

REFERENCES

1. P. K. Chatterjee, K. V. Vaidyanathan, W. V. McLevige, and B. G. Streetman, Appl. Phys. Lett. 27, 565 (1975).

2. R. G. Hunsperger, R. G. Wilson and D. M. Jamba, J. Appl. Phys. 43, 1318 (1972).

3. P. K. Chatterjee, B. G. Streetman, D. L. Keune and A. H. Herzog, IEEE IEDM Tech. Digest, 186 (1975).

4. E. A. Potoratskii and V. M. Stukebnikov, Sov. Phys. Solid State 8, 770 (1966).

5. D. J. Ashen, P. J. Dean, D. T. J. Hurle, J. B. Mullin,
    A. M. White, and P. D. Greene, J. Phys. Chem. Solids 36,
    1041 (1975).

6. P. K. Chatterjee, Ph.D. Thesis, University of Illinois (1976).
    Available from NTIS, Springfield, VA (Report No. ADA0-25607).

7. M. G. Craford, R. W. Shaw, A. H. Herzog, and W. O. Groves,
    J. Appl. Phys. 43, 4075 (1972).

8. M. A. Gilleo, P. T. Bailey and D. E. Hill, Phys. Rev. 174,
    898 (1968).

9. R. J. Nelson, N. Holonyak, Jr., and W. O. Groves, Phys. Rev.
    B 13, 5415 (1976).

10. P. K. Chatterjee, W. V. McLevige, K. V. Vaidyanathan,
    B. G. Streetman, Appl. Phys. Lett. 28, 509 (1976).

11. P. K. Chatterjee, W. V. McLevige and B. G. Streetman,
    J. Appl. Phys. 47, 3003 (1976).

12. J. F. Gibbons and R. E. Tremain Jr., Appl. Phys. Lett. 26,
    199 (1975).

13. C. A. Stolte, IEEE IEDM Tech. Digest, p. 585 (1975).

14. P. K. Chatterjee, W. V. McLevige and B. G. Streetman,
    Solid State Electronics (To be published).

15. E. B. Stoneham, Tech. Report 4733-1, Stanford Electronics Lab.,
    (1975).

16. P. K. Chatterjee, K. V. Vaidyanathan, M. S. Durschlag and
    B. G. Streetman, Solid State Comm. 17, 1421 (1975).

17. D. J. Wolford Jr., B. G. Streetman, W. Y. Hsu, J. D. Dow,
    R. J. Nelson, and N. Holonyak, Jr., Phys. Rev. Lett. 36,
    1400 (1976).

18. J. Gyulai, J. W. Mayer, I. V. Mitchell and V. Rodriguez,
    Appl. Phys. Lett. 17, 332 (1970).

19. L. Forbes, Solid State Electronics 18, 635 (1975).

Ag-ION IMPLANTATION INTO ZnSe

D.H. Haberland, H. Nelkowski, W. Schlaak

Institut für Festkörperphysik

Technische Universität Berlin, West Germany

ABSTRACT

ZnSe crystals grown both from the melt and from the vapour phase have been implanted at 600 K with 100 keV Ag-ions. Thermoelectric probe measurements and the rectifying method confirm that these layers annealed at 800 K are p-type at room temperature. With increasing Ag-doses especially the implanted layers in low resistivity ZnSe crystals become more p-type conducting.

The spectral distribution of photoluminescence yields the expected rise of the Ag-peak (552 nm) with increasing doses. At 80 K the low resistivity samples exhibit a bright dc electroluminescence with an emission peak at 552 nm, too. Polarized in the forward direction the emission starts at 5 V, and in the reverse direction at about 20 V. The electroluminescence intensity decreases above 280 K.

The temperature dependence of photoluminescence and -conductivity as well as the TSC-curves have also been studied. The measurements will be discussed by means of a band model.

INTRODUCTION

ZnSe with a bandgap of 2.7 eV is one of the materials which can emit light in the blue-green part of the light spectrum. Until now the fabrication of a light emitting p-n-diode by diffusion seems to be very

difficult because of the self-compensation effect.
Ion implantation appears to be a promising method to
produce type conversion in II-VI compounds. By means
of this method doping with impurities does not
take place in thermodynamic equilibrium so it should
be possible to get a concentration of acceptors large
enough to have p-type conduction. The annealing temperatures for radiation damage should be low enough to
get no noticeable self compensation.

Y.S. Park, C.H. Chung and co-authors reported on
type conversion by ion implantation of ZnSe-Li [1],
ZnSe:P [2] and ZnSe:N [3] and discussed their properties. In the following we will report on implantation
of Ag-ions into ZnSe. We have chosen silver because
the acceptor level is only a few tenths of an eV above
the valence bond. Therefore, the level should be considerably populated at room temperature.

EXPERIMENTAL

For substrates we used undoped ZnSe crystals. Some
of them were grown from the melt by Eagle-Pitcher and the
others were grown in our laboratory by the halid
transport method. The as grown crystals showed a very
high resistance of about 10^{10} Ω at room temperature.
Molten Zn heat treatment for 24 h at 1220 K, a method
first proposed by Aven [4], reduced the resistivity
to the order of 1 Ω cm.

Before implanting, all samples were treated in
the same manner. They were mechanically lapped and
polished and then chemically etched in NaOH for 30 min
at 380 K. All implantations were performed with a
^{109}Ag-ion beam of about 2μA and an energy of 100 keV.

The beam, with a diameter of 1 mm, was swept over
an area of 0.4 cm^2. The misaligned ZnSe crystals were
doped with an ion dose up to 2.5 x 10^{13}/cm^2 at a substrat
temperature of 600 K. The subsequent annealing took
place for 30 min at 800 K in an atmosphere of floating
Ar-gas. The four point method was used to make the
electric measurements with electrodes in a line
consisting of indium contacts.

RESULTS

After this procedure, the implanted crystals were
investigated by thermoelectric probe measurements

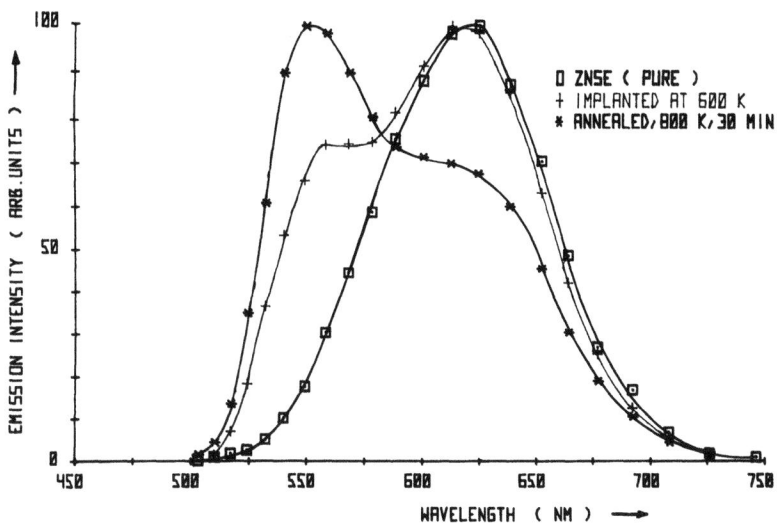

Fig. 1: Photoluminescence spectra of silver ion implantation (dose: 10^{12} cm^{-2}) in ZnSe single crystal by 450 nm excitation and at 77 K after annealing

and by the rectifying method to check the carrier type. The vapour grown crystals as well as those grown from the melt, always showed p-type conduction after Ag-implantation. With the rectifying method, the carrier type was determined at lower temperatures and under illumination as well. We observed p-type conduction on all implanted crystals.

The fig. 1 shows the spectral distribution of the photoluminescence before and after the Ag-implantation and the subsequent annealing. The intensity is normalized to the maximum of emission. The unimplanted ZnSe crystals have an emission peak at 624 nm (◻ ◻ ◻). According to Holton [5] this peak corresponds to the self-active luminescence. After the Ag-ion implantation we see an additional peak at 552 nm, which according to Halsted and Aven [6] belongs to the silver center. The second curve (+++) was recorded right after implantation; the third (✶✶✶) was registered after annealing for 30 min at 800 K in an Ar-atmosphere. After annealing, the silver centers become more effective.

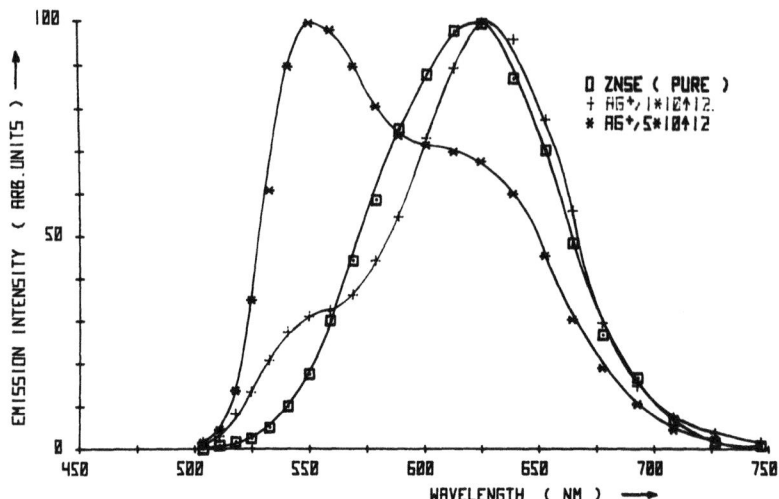

Fig. 2: Photoluminescence spectra of 100 keV silver ion implantation in ZnSe single crystal by 450 nm excitation and at 77 K.

An additional indication that the peak at 552 nm belongs to the incorporation of silver will be shown in fig. 2. These curves were recorded after annealing at 800 K. They show the spectral distribution of an unimplanted crystal (□ □ □), a second (+++) doped with 1×10^{12} cm^{-2} Ag-ions and another one (✳ ✳ ✳) doped with 5×10^{12} cm^{-2} Ag-ions. With increasing ion doses the emission peak of silver rises.

Furthermore, the thermally stimulated luminescence (TSL) and thermally stimulated current (TSC) of the crystals were measured. In fig. 3 one can see the typical change of Ag-implanted, low resistivity ZnSe crystals. The figure shows conductivity measured after excitation by 450 nm for 10 min at 77 K. The parameter is the implantation doses. The corresponding thermoluminescence measurements will not be shown because the changes are not so pronounced, but they can be interpreted in a similar manner.

Before implantation (....), we had a dark conductivity between 1 and $10 \Omega^{-1}$ cm^{-1} in the temperature range from 80 to 400 K. No difference in the conductivity could be measured before and after excitation. After Ag-implantation the dark conductivity decreased by many powers of 10 and the crystals show a clear TSC (+++).

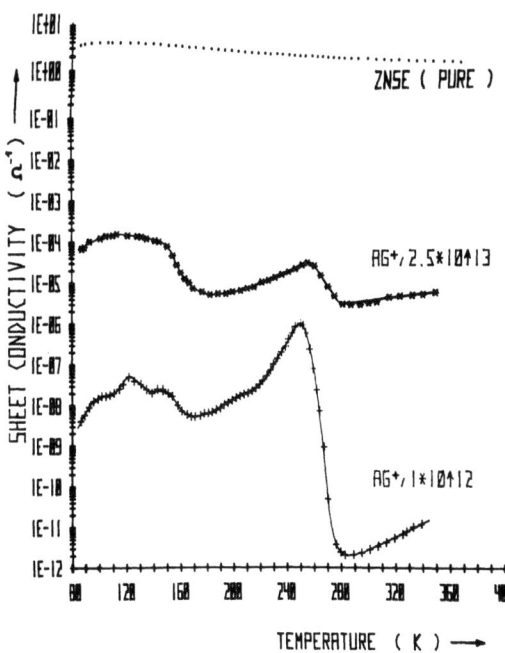

Fig. 3:
Temperature dependence of the sheet conductivity of ZnSe after excitation by 450 nm for 10 min at 77 K.

With increasing doses, the conductivity of the crystals increases again. Because of the higher dark current the peaks become less significant (✻✻✻). For all Ag-implanted ZnSe crystals one can see a pronounced peak at 250 K. This peak cannot be ovserved in the thermoluminescence measurement, because of the rapid decay of the luminescence in this temperature range.

Because of the implantation we obtain a peak at 250 K. The energetic depth E_T of an acceptor can be estimated by the well known formula $E_T = T_M/500$. In this case we get an energy level of 0.5 eV above the valence band. This agrees very well with the emission peak of Ag at 2.2 eV in the photoluminescence measurement.

As a result of the present measurement one can draw the conclusion that because of the Ag-implantation, ZnSe has become a p-type conductor. To investigate the p-n junction each side of the ZnSe diode was contacted with indium. Fig. 4 shows a typical current voltage characteristic of a diode. At 5 V in the forward biased direction, we get a greenish electroluminescence. The spectral distribution shows one peak at 552 nm. In the reverse biased direction a similar distribution of luminescence could be seen near the breakdown voltage of about 20 V.

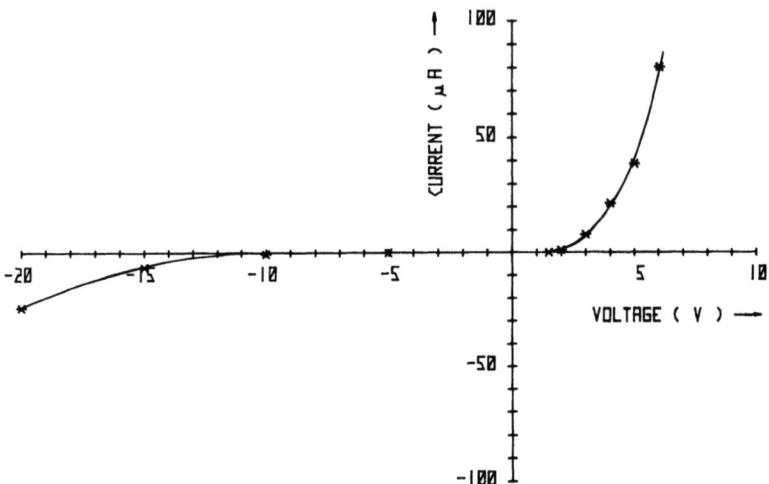

Fig. 4: Current voltage characteristic of a silver implanted ZnSe diode (dose: 2.5×10^{13} cm^{-2}) at 77 K

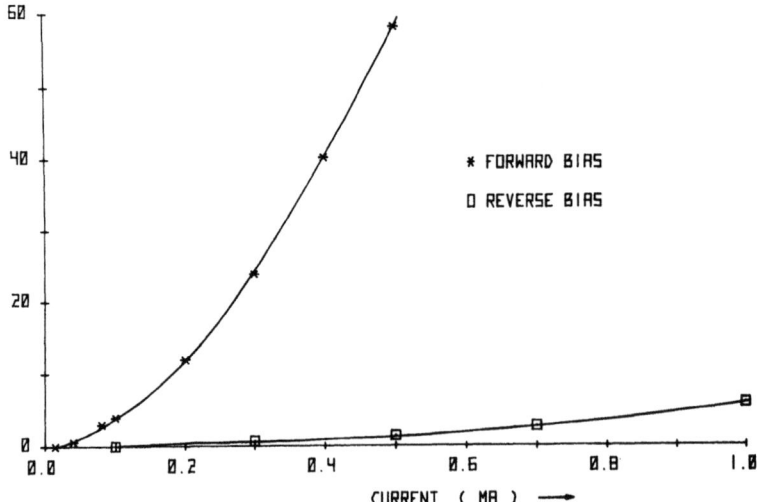

Fig. 5: Current dependence of electroluminescence intensity of a ZnSe:Ag diode (dose: 2.5×10^{13} cm^{-2})

Fig. 5 shows the dependence of the electroluminescence upon current at 77 K. In the forward direction one can see that emission rises very fast with increasing current. However, in the reverse direction there is only a weak increase. At temperatures above 280 K the electroluminescence starts to decrease.

DISCUSSION

All the results of our measurements are summarized in an energy band model of ZnSe shown in fig. 6. This figure shows the results of the measurements on the left side before and on the right after implantation.

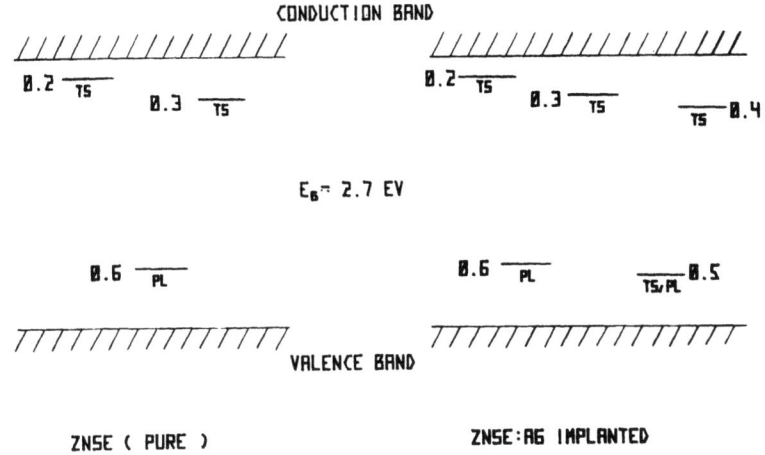

Fig. 6: Energy band model of silver implanted ZnSe (TS: thermally stimulated measurements, PL: photoluminescence results)

The initials TS and PL indicate the measuring method, TS for thermally stimulated conductivity and luminescence and PL for photoluminescence. After implantation we find two new energy levels in the forbidden energy gap. Until now the donor level at 0.4 eV could not be identified. Assuming that this level is produced by radiation damage, we will investigate how the center will be influenced by implantation of different ions. The most important result is the luminescence center or the acceptor level respectively at 0.5 eV above the valence band which were found at all Ag-implanted samples. This Ag-center was observed in the TSC as well as in the

PL measurement. It acts as an acceptor. By ion implantation, it can be incorporated in concentrations large enough to obtain good p-type conduction. This was proved by different methods to check the carrier type.

REFERENCES

1. Park, Y.S., Chung, C.H.
   Appl. Phys. Letters $\underline{16}$, 99 (1971)

2. Park, Y.S., Shin, B.K.
   J. Appl. Phys. $\underline{45}$, 1444 (1974)

3. Chung, C.H., Yoon, H.W., Kang, H.S. and Tai, C.H.
   Proc. 4th Int. Conf. on Ion Implantation, p. 253 (1974)

4. Aven, M., and Woodbury, H.H.
   Appl. Phys. Letters $\underline{1}$, 53 (1962)

5. Holton, W.C., de Wit, M., and Estle, T.L.
   Int. Symp. on Luminescence, p. 454, Munich 1965

6. Halsted, R.E., Aven, M., and Coghill, H.B.
   J. Electrochemical Soc. $\underline{112}$, 177 (1965)

RESISTANCE CONTROL OF SnO_2 FILMS BY ION IMPLANTATION

O. Tabata, S. Kimura, and Y. Sato

Government Industrial Research Institute, OSAKA

Midorigaoka-1, Ikeda, Osaka, 563, Japan

ABSTRACT

Transparent-conductive SnO_2 films have been prepared by means of ion implantation. An appropriate amount of Antimony atom is implanted into CVD polycrystalline SnO_2 films. Using this implantation doping, the sheet resistance of CVD SnO_2 films can be controlled precisely at any desirable values. A dose of about 1×10^{16} Sb ions/cm^2 brings down the sheet resistivity to such a low value as a commercial goal of 500 Ω/sq, no matter what initial sheet resistivity value is used. A tentative control equation is proposed from the experiment. The Sb atom distribution implanted into the polycrystalline SnO_2 film is also investigated by an ion microanalyzer. The peak is a typical amorphus one and the projected range is around 250 Å at 60 keV. An annealing at 650 °C does not appreciably affect the distribution of implanted atom.

INTRODUCTION

A lot of uses for transparent-conductive SnO_2 films are spreading in electronic devices, such as handy calculators, liquid-crystal watches, light-emitting diodes, solar cells, panel displays, and solid-TV as well. They all need high transparency and conductivity of the electrodes. Commercially available SnO_2 films are usually prepared by either of two methods: Physical Vapor Deposition (PVD) [1,2] or Chemical Vapor Deposition (CVD) [3,4]. The PVD film is rather weak against scratching and mechan-

Table I Comparison of SnO$_2$ Film in PVD and in CVD.

	PVD	CVD
Optical Transmittance (%)	>90	>90
Sheet Resistivity (Ω/sq) undoped doped	$10^4 - 10^6$ $\times 10^2$	$10^4 - 10^6$ $\times 10^2$
Structure	amor./poly.	crystallized
Purity	high	higher
Toughness	weak	strong

ical friction. On the contrary, the CVD film is substantially stronger. Apart from this difference in mechanical property, both films possess equivalently high transparency and good conductivity as shown in Table I, when they are doped with some proper dopants. However, there exists the difficulty of preciselycontrolling the film resistance at desired values in these techniques. This work intends to remove this difficulty by using ion implantation and to extend practical uses of ion implantation to polycrystalline film materials other than semiconductor crystals. Table I shows that the CVD film is much more suitable for this implantation experiment than the PVD film because it is excellent in its crystal structure and purity.

EXPERIMENTAL PROCEDURE AND RESULTS

The CVD pure SnO$_2$ film chosen as the sample for implantation was prepared by the following thermal decomposition reaction [5]:

$$SnCl_4 + O_2 \longrightarrow SnO_2 + 2Cl_2 \qquad (1)$$

and was grown on Pyrex plates at 700 °C. The thickness of the film is 1000 Å and the structure of the film is typically polycrystalline as shown by the X-ray diffraction pattern in Fig.2(a). The optical transparency of the film itself is about 95 % at a wave-length of 550 mμ (Curve ① in Fig.1). The sheet resistivity of such a pure film as prepared by the above CVD reaction is usually scattered in the range from 10^4 to 10^6 Ω/sq depending on reaction conditions.

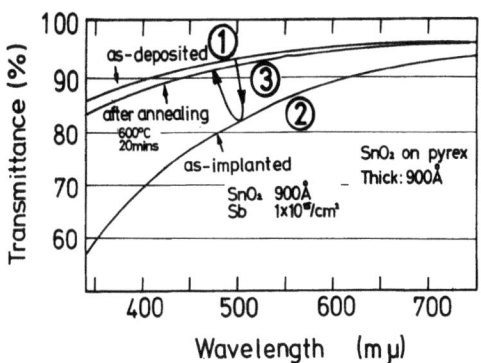

Fig. 1 Effect of Sb+ implantation followed an annealing on the optical transmittance of a CVD pure SnO_2 film.

60 keV Sb+ was implanted into the above CVD pure SnO_2 film over an appropriate range of Sb dose and subsequently an annealing process at 650 °C followed. The sheet resistivity was measured by the four probe method and the distribution profile was analyzed by an ion microanalyzer. The optical transmittance measurement and X-ray diffraction analysis also have been carried out.

Phenomena Caused by Implantation and Annealing

Once SnO_2 films on Pyrex plates are bombarded with energetic Sb ions, a huge number of crystal damages are produced in the films and the films change color to brown or yellow depending on the dose of Sb ion. Therefore, the optical transmittance over the range of visible wavelength drops considerably as shown by Curves ① and ② in Fig.1 and subsequently it almost recovers to the specific value of that prior to implantation through an annealing at a temperature higher than 500 °C.

During such optical changes as mentioned, the X-ray diffraction pattern of the film also changes. Fig.2(a) refers to the as-deposited virgin films. Some specific peaks of SnO_2 almost vanish after implantation (Fig.2(b)) and the film becomes amorphous-like. However, those peaks again grow up into the neighborhood of the original height through an annealing of 30 min at 650 °C (Fig.2(c)).

Coinciding with these optical and X-ray events, an outstanding thing happens on the sheet resistivity of the film (see Fig.3), i.e. the sheet resistivity of CVD SnO_2 film can be reduced by 3 figures or more by implantation and the following annealing, and reaches a commercial goal value of about 500 Ω/sq.

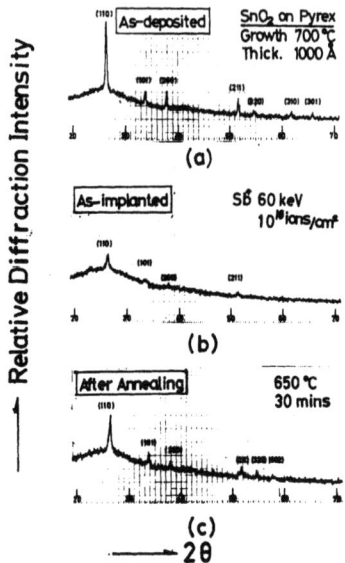

Fig.2 Variation of the X-ray diffraction pattern of a CVD SnO_2 film between as-deposited, as-implanted, and annealed.

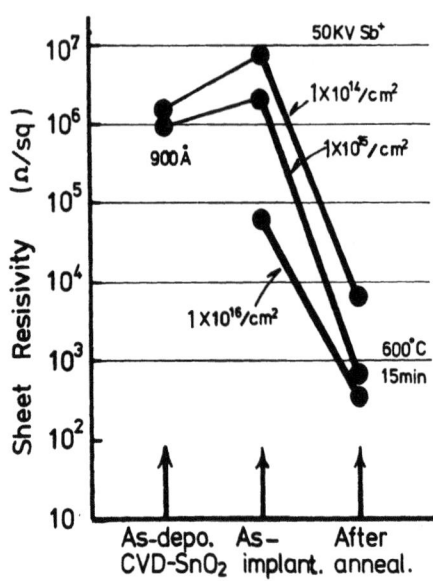

Fig.3 Doping effect of implanted Sb atom on the sheet resistivity of CVD pure SnO_2 films.

Fig.4 shows the concurrent advance of the events in sheet resistivity and in optical transmittance during an annealing with time. Curve ① shows the change of sheet resistivity with time. Curves ② and ③ show the fast ad-

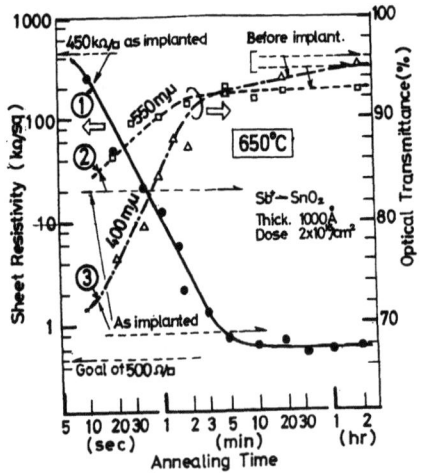

Fig.4 Annealing effects on the sheet resistivity (Curve ①) and optical transmittance (Curves ② & ③) of a CVD SnO_2 film implanted to a dose of 2×10^{16} ions/cm^2 at 60 keV.

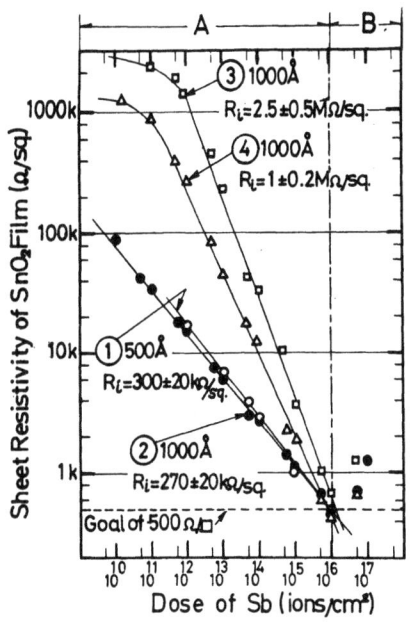

Fig. 5 Relationship between sheet resistivity and Sb dose. The implantation has been achieved with 60 keV Sb$^+$ at each total dose plotted above from 10^{10} to 10^{17} ions/cm^2. Lines ①, ②, ③, and ④ correspond to the average initial sheet resistivity of 300 kΩ/sq, 270 kΩ/sq, 1 MΩ/sq, and 2.5 MΩ/sq, respectively.

vance in recovery of the optical transmittance at wavelengths of 550 and 400 μm, respectively. The dominant variation in the resistivity and transmittance finishes within the very short period of 5 minutes.

Relationship between Sheet Resistivity and Sb Dose

Fig. 5 shows the relationship between sheet resistivity and dose of Sb atom and includes several interesting points. As for the dose region A (less than 10^{16} ions/cm^2): (1) The sheet resistivity of films with dose of Sb atom apparently changes linearly on the logarithmic scale. (2) The resistivity of all samples drops to a commercial goal value of about 500 Ω/sq with a dose of 1 x 10^{16} ions/cm^2 and settles around it, whatever the initial sheet resistivity is. (3) Lines ① and ② refer to the samples of the initial sheet resistivities of about 300 and 270 kΩ/sq, respectively. Both lines are almost in agreement because of the similarity of the initial sheet resistivity. This means that there is a good reliability of control. (4) Such shoulders as those in the upper part of Lines ③ and ④ used to appear at sufficiently higher initial sheet resistivity because of the deceleration of implantation caused by electrical charge-up on the insulating surface. In the dose region B (over 10^{16} ions/cm^2), however, the sheet resistivity again increases because of a sputtering of SnO$_2$ film which takes place during a long time implantation at a dose rate of 20 to 30 µA.

Profile of Implanted Sb Atom

The profile of implanted Sb atom was investigated with an ion microanalyzer, HITACHI IMA-2. The primary beam is O$^+$ ion of 11 keV, 3 x 10^{-7} A. The digging diameter is 800 µm and the digging rate is 70 Å as average. Fig.6 shows the concentration profile of Sb atom as implanted at energies of 30 and 60 keV and Sn atom as the major component of SnO$_2$ films. The projected range lies around 250 Å deep. The tail is somewhat like that due to channeling in a good crystal. The profile of the Sn atom is substantially flat and directly reflects the good uniformity of the CVD SnO$_2$ films. Fig.7 shows an annealing effect with time on the distribution of the Sb atom implanted into SnO$_2$ films 1000 Å thick. Fig.7(a) is as deposited. Sn atom profile is incidentally exceptional in this sample. Fig.7(b) and (c) refer to 1 and 5 min annealing, respectively. The peak moves slightly toward the surface with time at an annealing temperature of 650 °C. No appreciable diffusion is observed even after more annealing than 5 minutes, and so the doping effect of implantation is confined to a depth of 500 Å at an energy of 60 keV.

DISCUSSION

One can take such a drastic drop in sheet resistivity through the implantation-and-annealing process as shown in Fig.3 as the very evidence that the incorporation of implanted Sb atom into the structure of crystal grains composing a polycrystalline film has taken place. The fast recovery of the optical transmittance and X-ray diffraction peaks from the damage induced by ion bombardment in Figs.1 and 2 straightforwardly reflects the very rapid recovery of the crystal structure. However, some complicated damages seem to still remain, because they do not completely recover to their original crystal state.

Since the variation of sheet resistivity with the dose of Sb atoms obviously behaves in a straight line on the logarithmic coordinates, the principle of control can be represented by the following equation:

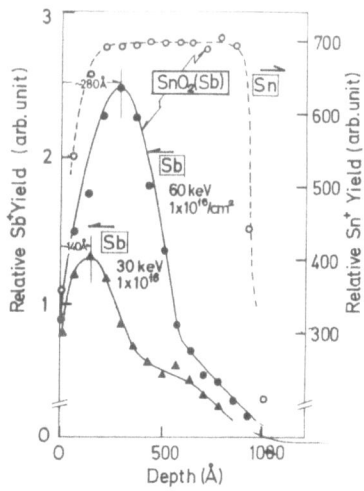

Fig.6 Depth distribution of Sb implanted at energies of 30 and 60 keV and Sn as a major component of SnO_2 film. The distribution has been obtained by ion microanalysis.

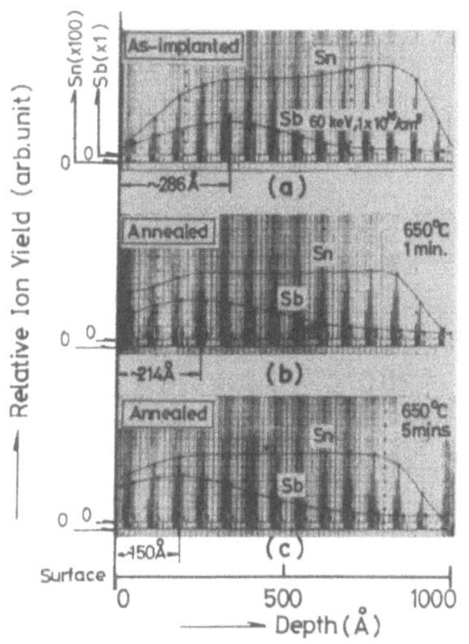

Fig. 7 Dependence of implanted Sb distribution on annealing time at 650 °C. The samples were taken from among adjacent pieces cut out of a film.

$$R = R_i \cdot D^{-a} \qquad (2)$$

where R: sheet resistivity (Ω/sq), R_i: initial sheet resistivity (Ω/sq), D: dose of Sb atom (ions/cm^2), and a: decrease-rate of sheet resistivity which is temporarily designated. The power a seems to be closely connected with the crystal structure of the film and its carrier density. Therefore, the bigger the a value is, the higher the initial resistivity is, and so the control line be-

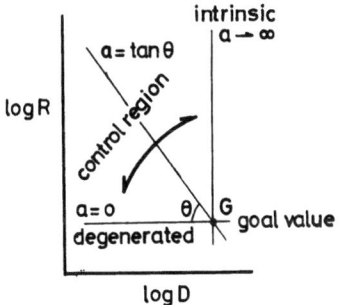

Fig.8 Control line depended on the a value.

comes steeper as the a increases. In Fig.8, Lines a = 0 and ∞ refer to a degenerated state of film which provides practically the lowest goal value of sheet resistivity and the intrinsic state, respectively.

It is evident from Fig.7 that the peak of the implanted Sb distribution does not move appreciably and the Sb atoms do not diffuse uniformly over the film thickness of 1000 Å, through an annealing for 5 min at 650°C which is close to the softening point of Pyrex plate (700°C) but not to that of SnO_2. This is true even for further long time annealing as well. In short, the implanted Sb atom distribution in CVD SnO_2 films is determined only by the acceleration energy of ion. Therefore, the implantation energy should be chosen taking account of the film thickness to obtain an effective distribution of implanted Sb atoms on the sheet resistivity.

CONCLUSIONS

The transparent-conductive SnO_2 film has been prepared by ion implantation. The ion implantation doping can achieve more perfect control of sheet resistivity than the conventional CVD doping. This proves that the ion implantation doping could be very useful even in polycrystalline materials other than semiconductor crystals.

The obtained lowest value of sheet resistivity is equivalent to a commercial goal value of around 500 Ω/sq. The profile of implanted Sb atom is a typical amorphous type and the projected range is around 250 Å deep at an energy of 60 keV in CVD polycrystalline SnO_2 films. There is no significant change in the distribution of implanted Sb atoms through an annealing for many hours at 650 °C. The incorporation of implanted Sb atom into the film structure is quite rapid within the first 5 minutes.

REFERENCES

(1) Y.Katsube et al., SHINKU, 9, 443 (1966).(in Japanese)
(2) Enrico Giani et al., J. Electrochem. Soc. 121, No.3, 394 (1974)
(3) H.A.McMasters, U.S.P. 2,429,420 Oct. 21 (1947)
(4) J.A.Aboaf et al., J. Electrochem. Soc., 120, No.5, 702 (1973)
(5) O.Tabata, Proc. 5th Int. Conf. CVD, 681 (1975)

ION-BOMBARDMENT OF AMORPHOUS SEMICONDUCTORS AND RELATED EVOLUTION OF STRUCTURAL AND ELECTRICAL PROPERTIES

M. Benmalek, J. P. Thomas, J. M. Mackowski

Institut de Physique Nucléaire de Lyon et IN2P3

43, Bd du 11 Novembre - 69621 Villeurbanne (France)

ABSTRACT

Amorphous germanium layers have been implanted with ions of different masses (Ne, Ar, Ge, Cd, Te, Xe, Au) in the energy range 20 to 300 keV and for doses between 10^{12} and 10^{16} ions.cm^{-2}

Transmission electron microscopy has revealed a structural evolution of implanted layers. At low doses $< 10^{13}$ ions.cm^{-2} white points (~30 A°) are observed and are attributed to displacement cascades. An overlapping occuring at high doses is indicated by higher transmission regions (~400 A°).

Electrical measurements of such samples show an increase of D.C. conductivity related to the evolution of the observed zones. This particularly after thermal annealing in oxygen or argon atmosphere.

Experimental parameters of the implantation (R_p, $\emptyset$, maximum depth) have been performed from backscattering of He and Li particles.

This study will show that amorphous germanium does not crystallize under irradiation by energetic heavy ions.

INTRODUCTION

If irradiation induced amorphization is now well established by several authors [1, 2] the most common hypothesis interpreting the critical dose of apparition of amorphous layers is based on the thermal spike model.

Following Naguib and Kelly [3] the main structural transformations of non-metallic solids under ion-bombardment (crystallization, amorphization, stoichiometry changes) should be governed by two criteria. In the case of amorphization the ratio of the crystallization and melting temperatures must be $T_c/T_m > 0.30$, for the first one, and the ionicity ≤ 0.47 for the second. From the first criterion amorphous Ge would be not affected by irradiation though Parsons and Balluffi [4] had reported a crystallization process induced by 70 keV Xe ions. Moreover the diameter of the crystallized zones seemed, for these authors, directly correlated with the ion range.

The aim of this work is then to present structural and electrical results obtained with amorphous Ge irradiated by several ion species. T.E.M. investigations have been performed either in bright and dark field and for doses varying between 10^{12} to 10^{16} ions.cm^{-2}. The development of zones of higher atomic order than the amorphous matrix is interpreted as well as the conductivity enhancement detected by electrical measurements. Concerning the implantation parameters, charged particles backscattering has been used allowing the implantation profile to be extracted.

EXPERIMENTAL PROCEDURE

Amorphous films were thermally evaporated from ultrapure Ge (50 Ωcm). For T.E.M. purposes substrates were either NaCl (self-supported films by floating off) or formvar films coated onto copper grids. Glass substrates have been used for electrical measurements and backscattering analysis. The thickness was determined by the vibrating quartz method as well as interferometry and backscattering.

Implantations were performed using a mass separator (Gamma Industries) of 120 keV maximum energy, multi-charged species allowing higher energy values. Samples were homogenously irradiated by means of a sweeping system, keeping for our conditions, a 80 nA/cm^2 maximum current density. Half a layer was preserved as a reference. The sample thickness was kept to 1000 A° for T.E.M. experiments but could reach 3000 A° for other measurements.

Either a Philips EM 300 Electron Microscope equipped with a heating sample holder or a JEOL 100 C has been used allowing accelerating voltages of 80 or 100 keV and magnifications up to 100,000.

The electrical resistance of the films was measured in a gap configuration using a Keithley 602 multimeter ; electrodes being evaporated Au.

In order to cover a large enough energy range of implanted ions we used 3.5 MeV α-particles and 2 MeV Li ions delivered by the 4 MeV Van de Graaff accelerator of the Institute. In the first case 300 keV Cd ions can be analyzed without interfering with the Ge front edge (5300 $A°$ analyzing depth) but with a depth resolution restricted to 330 $A°$ for our 16-20 surface barrier detector resolution. For the lower implantation energy of Cd ions (40 keV) Li ions offer the best depth resolution (170 $A°$) [5] but with restricted analyzing depth (1680 $A°$). From the experimental spectra a computer program previously described [6] allowed the concentration profile to be obtained without the distortion induced by the resolution function of the detector (HP 2116 C computer).

RESULTS and DISCUSSION

Low dose irradiated samples show bright spots as it appears on the first electron micrograph of fig. 1 (Xe - 70 keV - 6×10^{12} cm^{-2}) contrasting with the granulated aspect of the amorphous Ge layer. The evolution of the films irradiated at respectively 5×10^{13} and 10^{15} Xe/cm^2 is shown on electron micrographs 2 and 3. Higher transmission regions with diameters increasing with the dose can be noticed. No crystallization is revealed by the electron diffraction taken for the higher dose (inlet micrograph 3). This diffraction pattern is identical to those of the unimplanted film (diffuse rings).

For 70 keV Xe ions the number of zones per cm^2 and their mean diameter are given in table 1 for increasing doses. If a 10% precision is obtained for the dose we are limited to 50% for the number of zones due to the overlapping clearly appearing on the micrograph 3.

Table 1

Dose	5×10^{12}	10^{13}	5×10^{13}	10^{14}	10^{15}
N/cm^2	$\sim 10^{12}$	$\sim 10^{12}$	3×10^{11}	1.7×10^{11}	2.5×10^{10}
D_m $A°$	< 30	40 - 50	70	120 - 150	420

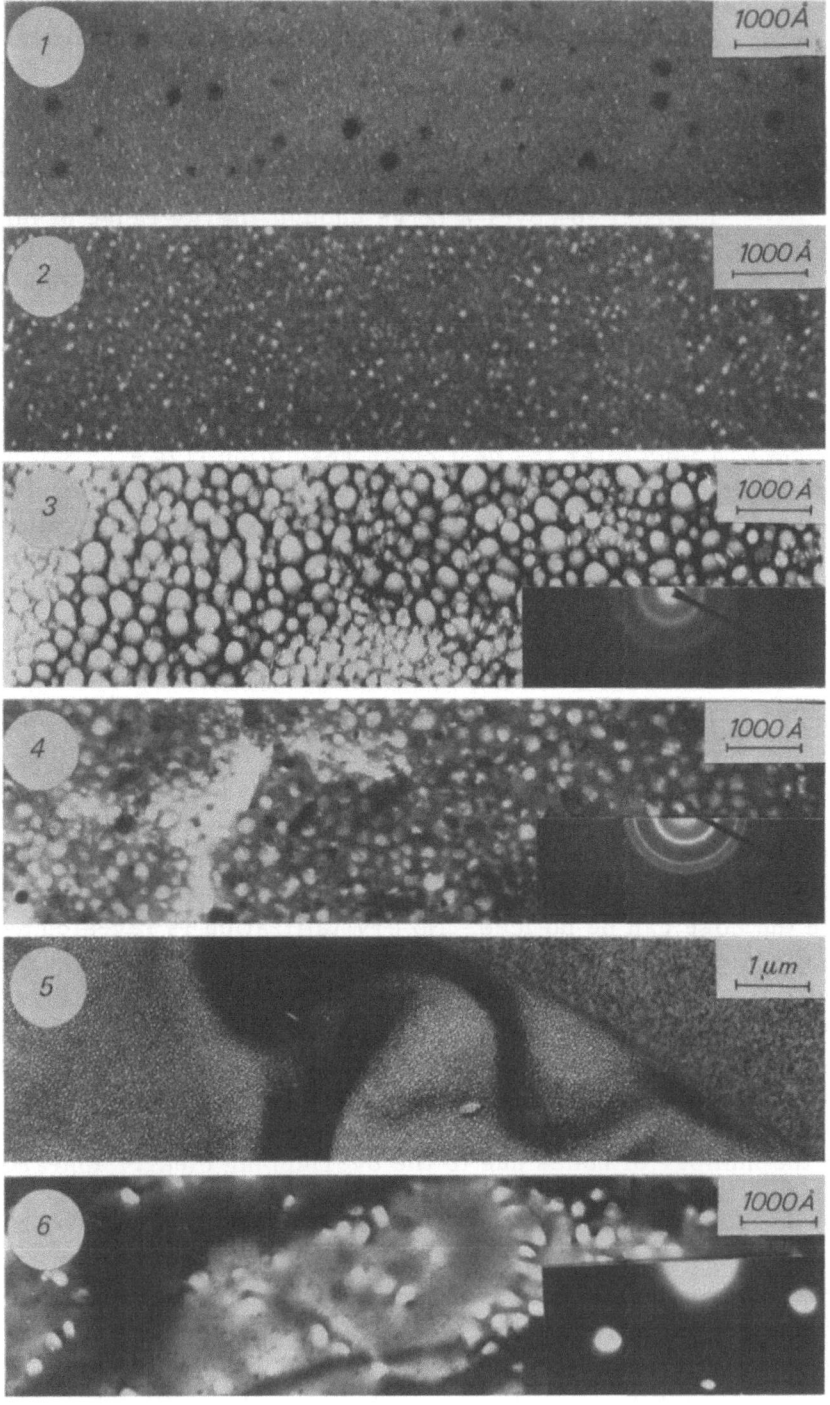

Figure 1

The mean diameter is deduced from repartition curves as represented in fig. 2 for the 10^{14} conditions and obtained by using a Zeiss TGZ 3 counter (zones are assimilated to spheres). A quasi gaussian shape can be noticed.

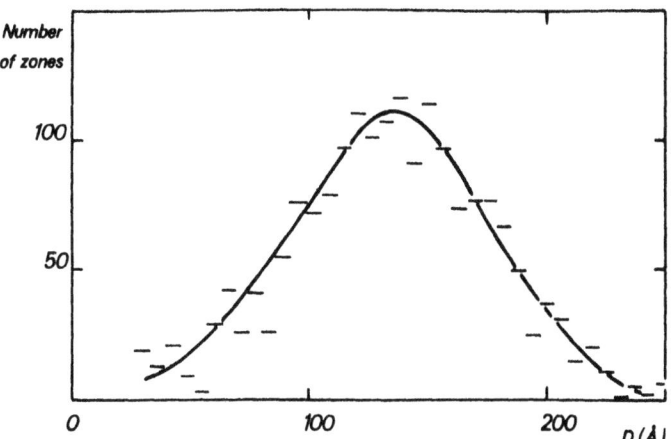

Figure 2

On the other hand it appears that only samples of thickness greater than the projected range of implanted ions exhibit a zone formation. This is pointed out irradiating layers of a given thickness with ions of energy ranging between 40 to 140 keV (420A° layer, Xe ions). Moreover at a fixed dose the zone diameter increases with the incident energy. We can then consider that zones are mainly due to the energy of damage left in the sample.

In order to investigate the ion mass effect, films have been irradiated with several ions (Ne, Ar, Ge, Cd, Te, Xe, Au) of identical energy per nucleon and for thicknesses $e > R_p$. An increase of the zone diameter with the ion mass is then clearly observed, excepted for 22 keV Ne for which no zone formation is detected even at 10^{15} cm^{-2}.

Using a modified Kinchin and Pease model proposed by Norgett et al. [7] the number of atoms displaced in a collision cascade by an ion penetrating the Ge films is given by

$$N(E) = 0.8\ E_D / 2 E_d$$

E_D is the energy of damage and E_d the displacement energy here ~ 15 eV for Ge.

For 22 keV Ne ions, $E_D \simeq 15$ keV [7] and $N(E) \simeq$ 400 atoms/ion, the number of displaced atoms appears to be insufficient to create observable zones especially if relaxation processes are important at ambiant temperature.

The dark field technique gives more informations on the structure of these zones. As a matter of fact they appear brighter and offer better diffraction than the amorphous matrix, suggesting a higher atomic order. Such behavior had been reported by Leteurtre and Soullard [8] for amorphous ZrO_2 irradiated by Kr ions.

In order to follow the evolution polycrystal→amorphous → zone polycrystalline Ge films deposited at high temperature (250°C) onto NaCl have been irradiated. At 5×10^{12} Xe/cm² for 70 keV the first step is reached and the amorphization process is revealed by haloes superimposed on the ring diffraction pattern. This critical dose for amorphization is low compared to the 10^{14} cm⁻² found for Ge single crystal [9] and the calculated values deduced from the different models [10, 11]. It is closer to the value reported by Pavlov [12] for 40 keV Xe ions in polycrystalline Ge films. Due to the role of epitaxial and channeling effects [9] in single crystals such a discrepancy is not surprising.

For polycrystalline films the zone formation is observed sooner. Micrograph 4 obtained after a 70 keV - 5×10^{13} Xe cm⁻² irradiation shows larger diameters for the zones than in micrograph 2. The inset shows the corresponding diffraction pattern. The structural transformation of the polycrystalline films is emphasized in micrograph 5 (70 keV 10^{15} Xe cm⁻²) showing large deformations in the irradiated part (left part).

In situ crystallization of these irradiated films occurs after 5 min heating at 440 ± 10°C. The mechanism of such a crystallization is identical to those described by Barna [13] and related to a surface process. For the irradiated part microcrystals are first detected in the transformed zones sustaining the hypothesis of a better organized structure. Defects remaining in this part after crystallization (micrograph 6, T = 490°C initially 70 keV - 10^{15}Xe/cm²) disappear only at higher temperature ∼ 540°C (∼ 0.6 Tm).

Since Hall effect measurements are not easy to perform on amorphous semiconductors, our investigations have beem limited ted to resistance measurements of irradiated samples. It can be pointed out from fig. 3 (70 keV - 10^{15}Xe/cm²) that in almost all the cases the relation $\ln R = A - BT^{-1/4}$ applies. Such behavior agrees with the Mott's law [14] $\sigma = \sigma_0 \exp(-T_0/T)^{1/4}$ usually found for tetrahedrally coordinated amorphous semiconductors and indicating that the conduction is due to hopping processes at low temperature.

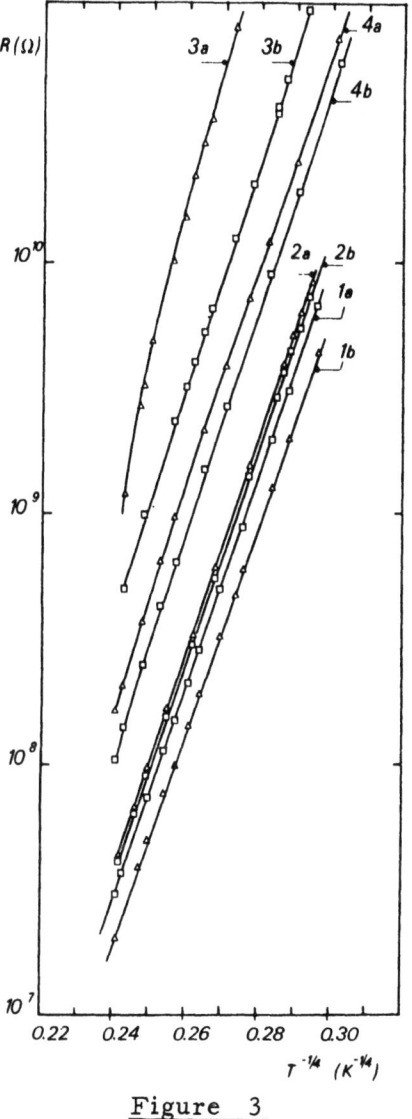

Figure 3

All samples tested after several hours of irradiation exhibit an enhancement of the conductivity as shown in Figure 3 (curves 1 with a referring to irradiated and b to non irradiated parts). This is opposite to thermal annealing behavior [15]. The irradiated sample does not recover its inital value when stored at room temperature. For example after 48 hour storage following irradiation, the resistance is still 30% lower than that of virgin material. Annealing at 60°C for 24 hours leads to almost identical curves (as shown in 2a and 2b). Annealing at 170°C for 24 hours results in a significant separation (as shown in curves 3a and 3b), the slopes remaining identical at low temperatures. Such behavior characterizes an oxygen saturation effect of the dangling bonds for amorphous semiconductors [16]. Annealing at 160°C for 55 hours in an argon atmosphere still exhibits this effect (as shown in curves 4a and 4b) but with less deviation. The major feature remains the identical value of the slopes indicating clearly the amorphous nature of the irradiated samples. The conductivity enahncement arises from irradiation induced short order defects.

Using electron microscopy low density regions have been revealed in irradiated films. This can explain an enhancement of oxygen diffusion during annealing. The discrepancy between the curves 3a and 3b would support this hypothesis. All these experimental results are in agreement with the structural modification of the irradiated layers together with the conclusion that no crystallization can be reported.

When implanted ions are heavier than the matrix, Rutherford backscattering is a very useful tool to determine precisely the dose and the concentration profile in the films used for T. E. M. and electrical test purposes. Due to the relative merits regarding analyzing depth and depth resolution [17] 3.5 MeV α-particles were used for higher implantation energies in order to avoid overlapping with the Ge front edge (like for Cd at E > 150 keV). For the lower energies Li ions at 2 MeV offer the best depth resolution achievable and allow a more precise determination of the concentration profile in such a low energy range. Figure 4 shows the backscattering spectrum of 70 keV 10^{16} Xe/cm^2 on an 1800 A° Ge layer. The xenon concentration profile is extracted from the experimental spectrum using a computer procedure [6]. As it appears on this spectrum fig. 5 the maximum implantation depth and the projected range can be deduced. For this last parameter, a comparison with theoretical values [18] shows that a fairly good agreement is found for cadmium but systematic deviations appear for xenon. An accelerated diffusion of this latter element is under investigation. All these results are given in table 2. The experimental precision for projected ranges lies between 20 and 50 A°. The determination of the maximum depth reached by the implanted ion is of particular interest to assess that particles are really almost totally stopped in the layer. This is also in support of the microscope observations leading to the displacement, cascade interpretation.

The dose has been determined using the Rutherford formula for Ge and the implant peaks as well as by charge collected during implantation. The two values agree to within 20% which corresponds to the limitation in dose uniformity of one sweeping system.

The sputtering yield which is also accessible by the backscattering technique (Ge peak before and after implantation) should be more precisely determined than by classical methods (weighing secondary ion emission, etc ..). If a good agreement with the theory [19] is found for gold (S ≃13 atoms/ion) for other ions a systematic diminution by a factor of 2 or 3 is found. Due to such small values for doses ($\leqslant 10^{15}$ cm^{-2}) and for our layers of about 700 A°, this determination must be made more carefully to increase the precision. Nevertheless we think that this method would give the best results for such low values.

ION-BOMBARDMENT OF AMORPHOUS SEMICONDUCTORS

Table 2

α 3.5 MeV (1) Li 2 MeV (2)		Cd				Xe			
E keV	40	60	80	180	250	70	100	140	300
R_p th. Å	170	230	280	540	720	240	310	410	770
R_p exp (1) (2)	150	190	260 220	610	670 680	260 220	265 250	280	650 680
Max Depth (1) (2)	~780	~900	~1300 ~1020	>1700	~2200	~960 ~900	~1160 ~920	~960	~2250

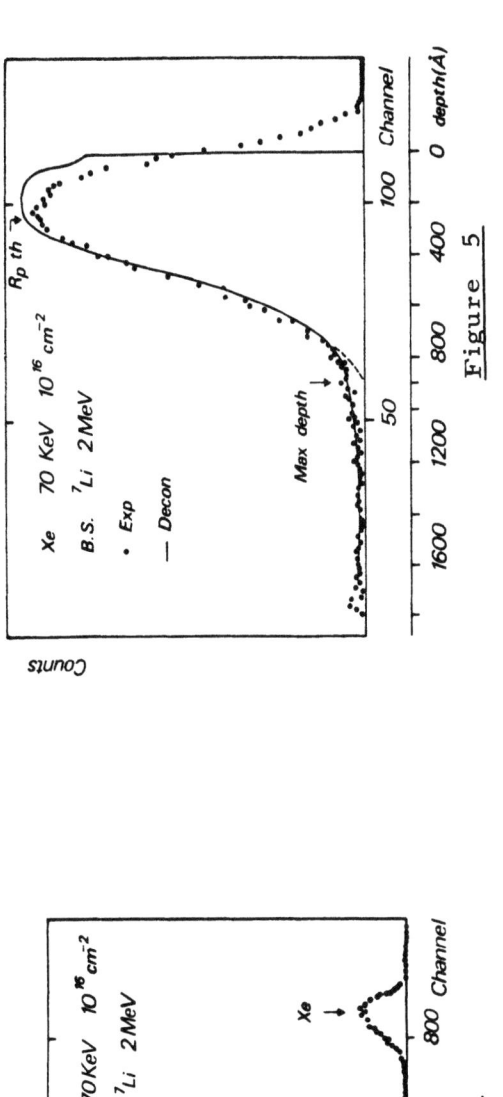

Figure 4

Figure 5

CONCLUSION

The multi-approach of the problem of amorphous Ge irradiation by heavy ions enables us to conclude that no crystallization occurs in our films. Nevertheless we can report a formation of zones with a different structure than the virgin matrix. From T.E.M. studies and using the dark field technique it appears that these zones show an atomic order higher than those of the deposited films. The enhancement of the conductivity of the irradiated samples has been correlated with the structural modifications revealed by electron microscopy. The conduction type being unchanged a complete reorganization of the implanted layer cannot be considered.

Concerning the amorphization of polycristalline Ge films a critical dose of $\sim 5\ 10^{12}$ Xe/cm^2 for 70 keV has been determined. This value is significantly lower than that is predicted by the amorphization theory.

ACKNOWLEDGMENTS

We are grateful to Professor J. Tousset for his interest in this work, to Dr. M. Maurette for helpfull discussions and to Mass Separator and Amorphous Group staffs for their technical assistance.

REFERENCES

[1] J.R. Parsons, Phil. Mag., 12, 1159, (1965)

[2] D.J. Mazey, R.S. Nelson, R.S. Barnes, Phil. Mag., 17, 1145, (1968)

[3] H.M. Naguib, R. Kelly, Rad. Effects, 25, 1, (1975)

[4] J.R. Parsons, R.W. Balluffi, J. Phys. Chem. Solids, 25, 263, (1964)

[5] J.P. Thomas, A. Cachard, M. Fallavier, J. Tardy, S. Marsaud, Ion Beam Surface Layer Analysis, Eds. Meyer, Linker, Kappeler, (Plenum Press, 1976)

[6] J.E. Engerran, Thèse Spéc. (3ème Cycle), Lyon, (1975)

[7] M.J. Norgett, M.T. Robinson, I.M. Torrens, Rapport CEA-R-4389, (1972)

[8] J. Leteurtre, J. Soullard, Rad. Effects, 20, 175, (1973)

[9] G. Holmén, P. Högberg, A. Burén, Rad. Effects, 24, 39, (1975)

[10] F. F. Morehead Jr., B. L. Crowder, Rad. Effects, 6 27, (1970)

[11] S. Furukawa, H. Ishiwara, Japan J. Appl. Phys., 11, 1062, (1972)

[12] P. V. Pavlov, Soviet Phys. Doklady, 18, 15, (1973)

[13] A. Barna, P. B. Barna, J. F. Pocza, J. Non-Cryst. Solids, 8-10, 36, (1972)

[14] N. F. Mott, Phil. Mag., 19, 835, (1969)

[15] J. A. Olley, Sol. Stat. Comm., 13, 1441, (1973)

[16] W. Beyer, J. Stuke, Phys. Stat. Sol., 30, 511, (1975)

[17] J. Tousset, J. P. Thomas, A. Cachard, Trans. of Amer. Nucl. Soc., (1975), Suppl. No 3, vol. 21

[18] W. S. Johnson, J. F. Gibbons, Projected range statistics in semiconductors (Stanford Univ. Bookstore, 1969)

[19] P. Sigmund, Phys. Rev., 184, 383, (1969)

MIS DIODE STRUCTURE IN As$^+$ IMPLANTED CdS

James A. Hutchby

NASA Langley Research Center

Hampton, Virginia

Clear evidence demonstrates that a prominent rectification behavior observed in current-voltage characteristics for As implanted and annealed, high-conductivity CdS exposed to sub-band-gap illumination is due to a metal-semiconductor junction in a platinum electrode-implanted layer interface. A typical illuminated device has a forward turnover voltage of 1.15 volts and a reverse breakdown voltage of 25-30 volts. Platinum barrier diodes fabricated in identical fashion on an unimplanted half of the substrate also display rectification behavior when exposed to sub-band-gap illumination, although the forward turnover voltage is only 0.65 volts. Placed in the dark, both diodes remain rectifying although their forward turnover voltages increase to 14 volts for the unimplanted (U) diode and 18 volts for the implanted (I) diode. Dominance of a rectifying barrier between the Pt electrodes and the implanted layer is demonstrated by over-coating a Pt contact with a liquid mixture of Ga and In, letting some of the metal contact the implanted layer surface. The Ga/In contact almost eliminates the rectifying nature of the contact, and yields a nearly linear current-voltage characteristic. However, in many samples removal of the Ga/In contact with acetone completely restores the original characteristic of the Pt electrode. The larger turnover voltage observed for the illuminated I diode may be caused by surface states modified by the As implantation and/or the presence of a thin insulating layer between the Pt pad and the implanted layer; thus the mis structure. Log I versus V measurements suggest that the dramatic decrease of the forward turnover voltages from dark to illuminated diodes is caused by optically activated hole or electron tunneling processes through the mis rectifying barrier.

INTRODUCTION

Efficient light emission processes found in some II-VI compound semiconductors have provided incentive in the past for attempting fabrication of green and blue light emitting p-n junctions from single crystal structures of CdS, ZnSe, and ZnS [1]. These materials are normally n-type, and with one possible exception [2] have not been converted to high conductivity p-type by standard doping techniques [3,4]. However, more recently type conversion efforts have been focused on ion implantation such that the literature now contains a variety of claims and denials for p-type CdS obtained by implantation of Bi [5-7], Na [8], Li [8], P [9,10], N [11]. and Sb [12] ions and p-type ZnSe by implantation of Li ions [13]. The most direct evidence of type conversion of CdS consists of positive Hall coefficients measured for 25 keV Bi implants in high-resistivity material (10^{10} Ω-cm) [5,6] and positive thermo-electric power measurements for the same Bi implants and for P implants in high-conductivity, 1 Ω-cm, material [9,10].

The initial portion of the present work including preliminary current-voltage (I-V) experiments involving all of the group VA ions (N, P, As, Sb, and Bi) indicates As implants yield the best diode rectification properties. Although contrary to previous evidence that As impurities in CdS yield only deep acceptors similar to P levels 1 eV above the valence band and may be less soluable in CdS than P [14], the preliminary evidence suggests that additional work focused on As implants could potentially yield useful p-n junction in CdS. Consequently, the present work is aimed at fabricating As implanted junctions in CdS, followed by direct determination of the majority-carrier type in the implanted layer and analysis of the diodes.

EXPERIMENTAL PROCEDURE

Samples 7 mm square by 1 mm thick were sliced from undoped boules of 5-10 Ω-cm CdS obtained from the Eagle-Picher Company (Miami, Oklahoma). Scratch-free, optical-finished surfaces, oriented perpendicular to the C-axis, were prepared by first lapping with 15 μm paper and 1 μm alumina powder followed by chemical-mechanical polishing with .05 μm Syton and 2.4 N HCl. A minimum of 175 μm was removed by the lapping procedure, and 90 μm by the chemical-mechanical polish. A final cleaning procedure involving methanol and de-ionized water rinses followed by a spin dry was used to minimize contaminants of the implantation surface.

Implantation was accomplished at room temperature using a mass analysed, electrostatically scanned and offset beam to obtain a ±1% surface uniformity. Dose rates range between 1.6×10^{-8} A/cm^2 and 1×10^{-7} A/cm^2, and the Faraday cup electron suppressor was biased to -60 volts to provide accurate fluence measurements. No effort was made to channel the ions. Half of each sample, excluding the outer edges, was As implanted at six energies ranging from 100 KeV to 10 KeV and six fluences between 6.9×10^{14} cm^{-2} and 1.0×10^{14} cm^{-2} to obtain an approximately flat As profile at a concentration of 1×10^{20} cm^{-3} over most of the implanted layer thickness. Following sputter deposition of platinum electrodes to both the implanted and unimplanted surfaces, the samples were annealed for 10 minutes in flowing nitrogen at 450°C. The total lapsed anneal time including warm-up and cool-down was 22 minutes. Finally indium electrodes were evaporated to the front and back of the unimplanted region completing a sample geometry shown in Fig. 1B.

This arrangement of the implanted - unimplanted layer boundary and the adjacent Pt and In contacts was originally chosen to simultaneously allow junction illumination and dark contacts, as shown in Fig. 1A. In this way, it was hoped that photovoltaic effects due to implanted junctions and the Pt contacts could be separated. However, the illumination mask, being fabricated from plexiglas painted black, acted as a diffuser for a small portion of the incident light thus providing a weak, but fairly uniform illumination of the entire sample.

Another structure helpful in evaluating the implanted diode is nearly identical to that shown in Fig. 1B except the Pt and In electrode materials for the unimplanted side were exchanged. This structure (EPUJ-50) has Pt contacts in symmetrical positions with respect to the implantation boundary, and thereby provides means for obtaining and comparing photovoltaic spectral response measurements for both implanted and unimplanted diodes located in similar positions.

RESULTS

Current-voltage characteristics are shown in Fig. 2 for both dark and illuminated diodes on the implanted and unimplanted halves of a CdS wafer. Concentrating first on the reverse bias

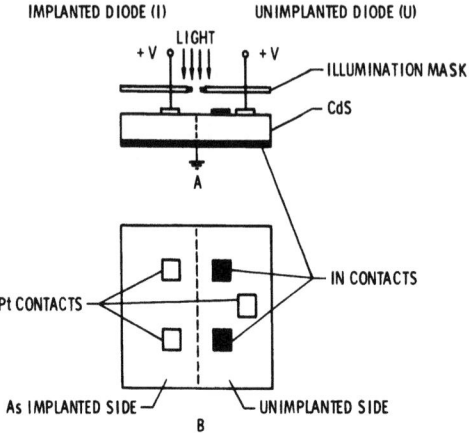

Fig. 1. Schematic diagram of CdS diode structure.

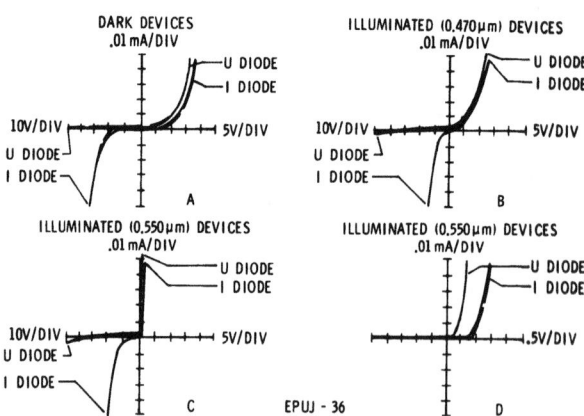

Fig. 2. I-V plots for dark and illuminated As implanted and unimplanted CdS diodes.

data, the characteristics shown in Fig. 2A for the dark diodes indicate a much lower breakdown voltage V_B for the implanted diode (I diode) biased in the reverse direction compared to the unimplanted diode (U diode). The U diode has a V_B in excess of 200V, whereas the I diode has a V_B of approximately 27V. For these voltages, the reverse characteristics appear dominated by avalanche multiplication in the depletion region in series with a fairly large, nonlinear resistance. If this is true, the large difference in V_B between the two diodes is explained by a higher concentration of ionized defects in the depletion region of the I diode.

One check on this hypothesis is to apply to the U diode an empirical universal expression for V_B derived for p-n junctions to estimate an upper limit on the concentration of ionized defects N_B^+ (cm^{-3}) in the depletion region. This defect concentration should be larger than, but comparable with, the electron concentration (n = 2.3 x 10^{15} cm^{-3}) determined from Hall coefficient measurements of the unimplanted material. As given by Sze [15], the V_B dominated by avalanche breakdown in a p-n junction is approximately

$$V_B \simeq 60(E_g/1.1)^{3/2} (N_B^+/10^{16})^{-3/4} \text{ volts} \qquad (1)$$

where E_g is the material band-gap energy in eV. For $V_B > 200V$, as determined for the U diode, $N_B^+ < 9.6 \times 10^{15}$ cm^{-3}, which is consistent with an ionized donor concentration of 2.3 x 10^{15} cm^{-3} measured for the CdS substrate.

Figures 2B and 2C show I-V characteristics of the two diodes illuminated at two different wavelengths by a Bausch and Lomb grating monochromator with a xenon source. The 0.470 μm photons, being more energetic than the 2.42 eV (0.51 μm) band-gap of CdS are strongly absorbed, whereas the opposite is true for the sub-bandgap 0.550 μm photons, which are weakly absorbed. In addition, the 0.550 μm illumination is 50-60% more intense than that at 0.470 μm. Consequently, the lower V_B for the 0.470 μm light compared to that for the 0.550 μm excitation is explained by increased photon absorption in the portion of a depletion region located near the edge of the Pt contact. Based upon the validity of Eq. 1 applied to the I diode and upon the measured breakdown voltages, the 0.470 μm radiation ionizes a minimum of 3.2 times larger average defect concentration in the depletion region than does the 0.550 μm radiation.

The forward bias I-V characteristics exhibit a somewhat contrasting behavior compared to the reverse bias results. The forward characteristics for the dark diodes are quite similar, and are somewhat alike those expected for a rectifying junction in series with a nonlinear, high-resistance or highly-compensated region. The 0.470 μm illumination has a minor, but identical affect on the forward characteristics of both the I and U diodes. However, the 0.550 μm radiatiion greatly increases the slope of both forward characteristics, and clearly distinguishes two separate turnover voltages V_T for the two diodes, as shown in Fig. 2D. The V_T for the U diode, measured as 0.65 V from extrapolation of the forward I-V slope at 0.7 mA back to the zero current voltage intercept, is most likely due to a potential barrier between the Pt contact and the CdS substrate. Determined in the same fashion, V_T for the I diode is 1.15 V, which compared to the U diode clearly indicates the implanted layer has a substantial influence on the rectification properties. This apparent V_T for the I diode increases with increasing diode current, whereas the V_T for the U diode remains relatively fixed up to the maximum diode currents examined (10 ma).

Three models possible to explain the I diode rectification property are: (a) p-n junction with the implanted layer being converted to low-conductivity p-type by the As implantation; (b) a high-low junction in which the implanted layer is a highly compensated, low conductivity, n-type region; and (c) a metal-semiconductor (m-s) structure in which the Pt-CdS interfacial-state energy distribution is modified and/or a thin, insulating, interfacial layer is introduced by the As implantation to increase the V_T of a Pt Schottky barrier [16-19].

DC Hall effect and thermo-electric power measurements do not indicate p-type conversion in the implanted layer. The Van der Pauw Hall effect measurement system, specifically designed and constructed to measure high-resistivity material, includes a 10 KGauss magnet with a sapphire-lined sample cavity having a shunt resistance of 1×10^{13} Ω connected to the external apparatus using guarded, low-noise coax leads [20]. The difference between the total signal voltage and a bucking potential obtained from a stable battery pack is plotted using a strip-chart recorder. This provides a means for separating the Hall signal from the total voltage and for removing slow variations in signal voltage due to temperature drift and other factors.

This system's capability is demonstrated by measurements on two high-resistivity samples. One sample, as received grade A CdS obtained from the Eagle-Picher Company, was measured to have a resistivity $\rho = 8.0 \pm 0.1 \times 10^8$ Ω-cm, and an electron mobility of $\mu = 211$ cm^2/volt-sec. The second sample, grade A CdS annealed at 800°C for one week in an evacuated and sealed quartz ampoule, was measured to have $\rho = 7.0 \pm 0.1\ 10^{11}$ Ω-cm and $\mu = 48$ cm^2/volt-sec. The measured sheet resistance of the second sample is $\rho = 8.6 \pm 0.1 \times 10^{12}$ Ω/□, which appears to be the upper limit of the system. The error limits of resistivity data represent fluctuations of two data points about the mean, and do not represent absolute system accuracy. However, these measurements demonstrate the system's capability for measuring high-resistivity samples. The main limitation of the system appears to be metal-semiconductor contact noise.

DC Hall measurements of high resistivity CdS (similar to the second grade A, annealed sample described above) implanted with As and annealed at 450°C were inconclusive in the sense that no Hall voltage was detected, and noisy contacts (both sputtered Pt and painted Ga/In) prevented precise resistivity measurements. However, the extremely high sample resistances between 1×10^{10} Ω and 1×10^{12} Ω did indicate a fairly insulating implanted layer, regardless of the conductivity type.

Thermo-electric power measurements using both the standard technique or an AC method [21] applied to As implanted layers on high-conductivity substrates indicate the implanted layers to be either insulating (zero thermal emf) or slightly n-type compared to strong n-type deflections for the unimplanted portion of the substrate. Since both the Hall effect and thermo-electric power measurements indicate the implanted layer is insulating, the actual conductivity type becomes inconsequential, and the p-n junction model is therefore eliminated as a possible explanation for the illuminated (0.550 µm) I-V characteristics of the I diode.

Evidence favoring the m-s rectifying contact model is shown in Figs. 3A and 3C, where I-V plots for two I diodes are shown

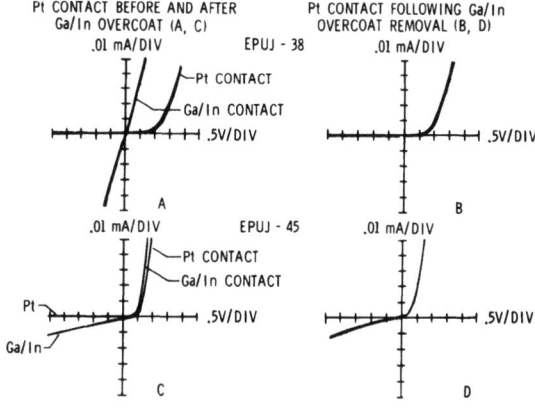

Fig. 3. I-V plots for illuminated As implanted CdS with Pt and Ga/In contacts.

for both Pt and Ga/In contacts to the implanted layer. An I-V
plot for each diode with the standard Pt electrode was first
observed (Figs. 3A and 3C, Pt contact) after which a liquid mixture of Ga and In was painted on the Pt electrode such that this
new electrode was equally divided between overlaying half the Pt
pad and directly contacting the implanted layer. The second
I-V plot labeled Ga/In contact was then obtained for the new contact. Finally, the Ga/In contact was removed with either acetone
or methanol, and the I-V characteristic for the Pt electrode
re-measured. This data is shown in Figs. 3B and 3D. Figures 3A
and 3B are fairly representative of four samples examined, whereas
Figs. 3C and 3D are representative of two samples.

Removal and restoration of the rectifying nature of the I-V
plots can result from two different models. The Ga/In may
provide an ohmic contact to the implanted layer and thereby short
circuit an electron potential barrier in a m-s diode, or the Ga/
In may contact the substrate through pin holes (or unimplanted
regions) in the implanted layer [22]. A simple analysis of the
Ga/In I-V data in Fig. 3A tends to rule out the second possiblity.
The resistance determined from the Ga/In data is 12.5 KΩ. For
measured CdS substrate resistivity and thickness of 8.1 Ω-cm
and 0.089 cm respectively, and assuming zero contact resistance,
the minimum aggregate pin-hole area required to support the Ga/In
device resistance is 5.8×10^{-5} cm^2. The aggregate linear pin-
hole dimension is therefore 7.6×10^{-3} cm, which is almost 8%
of the 0.1 cm linear dimension of the square Pt contact. Careful
cleaning and observation (under a 280X Reichert microscope with
a Nomarski attachment) of the region to be implanted clearly
avoids such large aggregate pin-hole dimensions caused by dust or
other foreign particles. In addition, the implanted regions
had no observable scratches. Therefore, the Ga/In contact clearly
appears to short-circuit a m-s rectifying barrier.

The complete model for the I diode therefore is a rectifying
Pt-CdS potential barrier, whose turnover voltage is increased by
As implantation, in series with a nonlinear, high-resistance
layer. The increase in the turnover voltage may be due to either
interface states modified by As implantation and/or to the
presence of a thin insulating region between the CdS and Pt electrode.

The precise details of the charge-conduction mechanisms
operating in the implanted and unimplanted diodes are not presently
clear, and their determination is beyond the scope of this paper.
However, additional insight to these processes for 0.550 μm
illuminated diodes is provided by log I versus V data given in
Fig. 4. The forward bias characteristics indicate linear regions
with different slopes between adjacent numbered labels on the

MIS STRUCTURE IN As⁺ IMPLANTED CdS

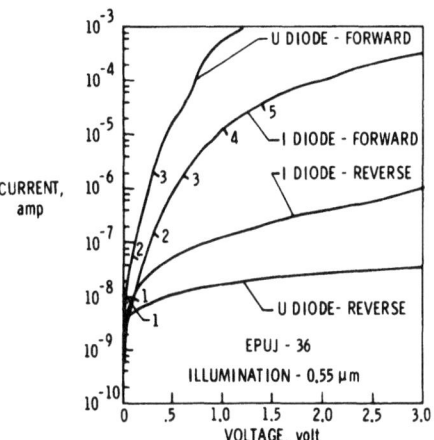

Fig. 4. Log I-V plots for As implanted and unimplanted illuminated CdS diodes.

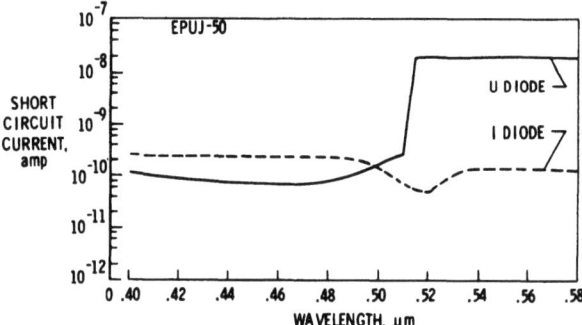

Fig. 5. Spectral Response for Adjacent As implanted and unimplanted CdS diodes.

respective curves for both diodes. Therefore, in each linear region, the diode current is proportional to exp (V/V_o) where V_o is a constant given by the appropriate log I-V slope. Values for V_o determined for the different bias regions are given in Table 1.

Two different models are normally invoked to explain an exponential I-V relationship in rectifying junctions. One is the standard thermal emission of electrons over a potential barrier giving $V_o = kT/q$, where k (Joule/°K) is Boltzmann's constant, T (degrees Kelvin) is temperature, and q (Coulomb) is the electronic charge. The second model is due to electron tunneling through a potential barrier in which V_o is a constant independent of temperature. Although considerable additional work is required, the present evidence suggests a different tunneling process for each of the linear regions involving optically activated defect centers for both the unimplanted and implanted diodes. If this is true, the light sensitivity of the non-linear, high-resistance region is linked to optically stimulated tunneling processes as opposed to any non-linear, photoconductivity or space-charge-conduction process. It is also interesting to note that the larger values of V_o are approximately integral multiples of the lowest value for each diode.

A final comparison between implanted and unimplanted diodes is given in Fig. 5 for a device fabricated with two Pt contacts on the unimplanted half adjacent to the implantation boundary. This device, EPUJ-50, is described in the EXPERIMENTAL TECHNIQUES SECTION. The short-circuit-current spectral responses were obtained using a Cary Model 14 spectrophotometer with a tungsten lamp and are normalized for constant light intensity. These results indicate a broad sub-band-gap absorption band present in both the implanted and unimplanted devices, although this response is degraded by a factor of greater than 100 in the I diode.

TABLE 1

Values of V_o (volts) determined from Log I versus V data given in Fig. 4.

REGION	U DIODE	I DIODE
1-2	.029	.068
2-3	.055	.122
3-4		.201
4-5		.371

Two sources of photo-generated carriers which can be explain this extrinsic current are holes excited from shallow defect states in the bulk semiconductor, and electrons photo-emitted from the Pt contact back into the CdS (resulting from incident photons reflected by the back CdS-In interface). The wavelength independence of the lower-energy photo-current for the U diode, however, suggests dominance of the former process, at least for this structure. The assumption that photons of these longer wavelengths probably penetrate the CdS far deeper than any m-s depletion region, and the fact that the higher energy responses (for wavelengths shorter than the 0.51 μm band gap) are similar for the diodes, suggest that the long-wavelength response degradation in the I diode is related to bulk-generated minority-carrier diffusion through a highly disordered region to reach the charge-collecting depletion region. The larger long-wavelength response of the unimplanted diode over that at shorter wavelengths may be related to deeper substrate penetration of the long-wavelength photons, thus placing the photo-generated minority carriers closer to the collecting junction (see Fig. 1A).

SUMMARY AND CONCLUSIONS

Structures fabricated by As implantation of carefully prepared, high-conductivity CdS surfaces followed by Pt deposition and $450°C$ anneal display rectifying, although substantially different I-V characteristics in both the dark and when illuminated with sub-band-gap light. Structures fabricated in identical fashion on an unimplanted portion of the substrate exhibit similar I-V characteristics, except the forward turnover voltage for an illuminated, unimplanted diode is much smaller than that for the implanted diode. Both diodes display a large decrease of the forward turnover voltage and a large decrease of an apparent series resistance between diodes measured in the dark and those exposed to illumination less energetic than the CdS energy band-gap.

Present evidence indicates the charge conduction in both structures is dominated by hole and/or electron tunneling through a metal-semiconductor potential barrier. The tunneling processes appear to be quite sensitive to sub-band-gap illumination, which cause the dramatic decreases of turnover voltages and apparent series resistences. This may be related to interaction of optically activated defect centers increasing the probability of the several tunneling transitions. This conclusion is supported by observation of substantial photo-current generated in both implanted and unimplanted diodes by sub-band-gap illumination.

The difference in the turnover voltages for the two illuminated structures is belived to be caused by interface states modified by the As implantation and/or presence of a thin insulating layer between the Pt electrode and the implanted layer, thus suggesting an mis model.

ACKNOWLEDGEMENTS

The author wishes to thank Mr. George M. Walker, Jr. for his careful preparation of the samples, Mr. John W. Burgess for precise and timely As implantations, and Mr. Tommy C. Steele for diligent measurements and helpful discussions.

REFERENCES

1. M. Aven and J. S. Prener, eds., Physics and Chemistry of II-VI Compounds, (Wiley, New York, 1967).

2. Z. K. Kun and R. J. Robinson, J. Electronic Materials 5, 23 (1976).

3. Y. S. Park, P. M. Hemenger, and C. H. Chung, Appl. Phys. Letters 8, 45 (1971).

4. J. H. Haanstra and J. Dieleman, J. Electrochem. Soc. 14, 2 (1965).

5. F. Chernow, G. Eldridge, G. Ruse, and L. Wahlin, Appl. Phys. Letters 12, 339 (1968).

6. G. Eldridge, F. Chernow, and G. Ruse, J. Appl. Phys. 44, 3858 (1973).

7. B. Tell and W. M. Gibson, J. Appl. Phys. 40, 5320 (1969).

8. B. Tell, W. M. Gibson, and J. W. Rodgers, Appl. Phys. Letters 17, 315 (1970).

9. W. W. Anderson and J. T. Mitchell, Appl. Phys. Letters 12, 334 (1968).

10. S. L. Hou, K. Beck, and J. A. Marley, Jr., Appl. Phys. Letters 14, 151 (1969).

11. Y. Shiraki, T. Shimada, and K. F. Komatsubara, J. Appl. Phys. 43, 710 (1972).

12. I. P. Akimchenko, V. S. Vavilov, V. V. Krasnopevtsev, Yu V. Milyutin, M. Harsy, and Chan Kim Loi, Sov. Phys. Semicond. 9, 19 (1975).

13. Y. S. Park and C. H. Chung, App. Phys. Letters 18, 99 (1971).

14. B. Tell, J. Appl. Phys. 41, 3789 (1970).

15. S. M. Sze, Physics of Semiconductor Devices (Wiley, New York, 1969), p. 114.

16. H. C. Card and E. H. Rhoderick, J. Phys. D. 4, 1589 (1971).

17. M. A. Green, F. D. King, and J. Shewchun, Solid-State Electron. 17, 551 (1974).

18. J. Schewchun, M. A. Green, and F. D. King, Solid-State Electron 17, 563 (1974).

19. S. J. Fonash, J. Appl. Phys. 46, 1286 (1975).

20. P. M. Hemenger, Rev. Sci. Inst. 44, 698 (1973).

21. A. G. Fischer, J. N. Carides, and J. Dresner, Solid-State Commun. 2, 157 (1964).

22. G. Eldridge, Ph.D. thesis (University of Colorado, 1970) (unpublished).

LONG RANGE MIGRATION OF DEFECTS DURING LOW TEMPERATURE BORON IMPLANTATION IN ZnTe

P. F. Engel, J. C. Pfister*, J. Marine, D. Thomas

LETI/MEP - C.E.N.G. - 85X, 38041 Grenoble CEDEX, France

I. INTRODUCTION

The electrical and optical changes induced in ZnTe by implantation of B, Al, and other ions have been shown to extend well beyond the ballistic range of these ions [1,2]. In particular, it was recently reported that implantation in ZnTe resulted in the creation of a compensated region whose width increased logarithmically with ion dose and decreased with the square root of the uncompensated acceptor concentration at the edge of this region [2]. The present paper is an attempt to understand the mechanisms leading to the extension of this region.

II. EXPERIMENTAL METHODS

The ZnTe was grown by the Bridgman technique from a Te-rich melt. After polishing and etching, each wafer was provided with a number of 3500 Å thick aluminum pads using classical photolithographic techniques. After dicing the wafer, a gold wire was bonded to each pad in order to allow both in situ capacitance measurements of the separate diodes and independent biasing during implantation. Since the capacitance decreased during implantation and the gold wire shaded the bonding area from the ion beam, a preimplantation step was included in the technology in which the active area was protected from the ion beam with a layer of photoresist and the bonding area implanted to a dose of 5×10^{13} B/cm^2 at 120 KeV. Finally the sample provided with a gold back contact was mounted in the target chamber of the ion implanter and cooled to about 90° K with LN$_2$.

The experiment consisted of a series of 140 KeV ^{11}B implantations through the Al pads followed by in situ capacitance vs. voltage measurements, all at 90° K. During implantation a positive bias was applied to some of the Al pads, thus reverse biasing the corresponding Schottky barriers. The implantation schedule is shown in Table I.

The capacitance curves were analyzed using the well-known relation [3]

$$N = 2\left[\varepsilon q \frac{d(A/C)^2}{dv}\right]^{-1}$$

where N is the concentration of uncompensated acceptors, q is the electronic charge, ε the dielectric constant, C the measured capacitance, A the active area, and v the applied voltage. Since C is also related to depletion depth x through $x = \varepsilon A/C$, a profile of net acceptors versus distance from the surface is obtained after each implantation. The value of the capacitance at zero bias was also used directly as a parameter defining the extent of the compensated region.

III. RESULTS AND DISCUSSION

Figure 1 shows the depleted depth at zero applied voltage as a function of dose for two diodes on the same sample with an initial acceptor concentration of 1.9×10^{15} cm^{-3}. One of the diodes was reverse biased with 3.4 V during implantation, the other was

Table 1. Implantation sequence and boron beam fluxes used. A capacitance vs. voltage measurement was performed after each dose shown.

Implant number	1	2	3	4	5	6	7
Total dose (cm^{-2})	10^{10}	3×10^{10}	10^{11}	3×10^{11}	10^{12}	3×10^{12}	10^{13}
Boron Flux (cm^{-2} sec^{-1})	1.7×10^9	3.3×10^9	5×10^9	5×10^9	5×10^9	5×10^{10}	5×10^{10}

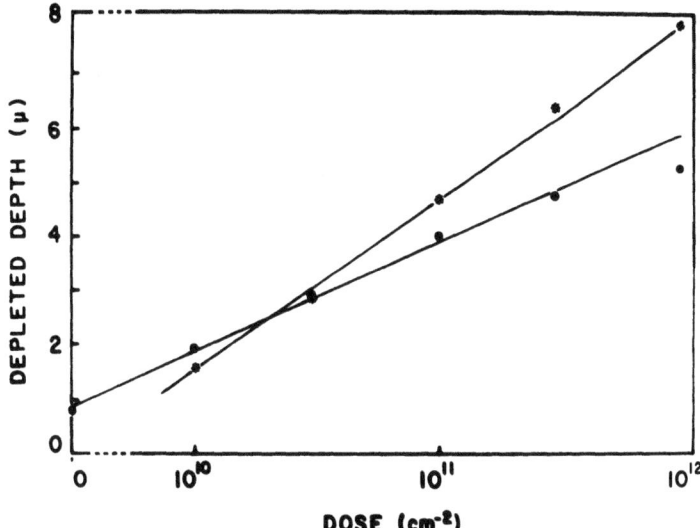

Fig. 1. Depleted depth at zero bias as a function of B dose. Dots: diode implanted without bias. Stars: diode implanted with 3.4 V reverse bias. Implantation schedule as in Table I.

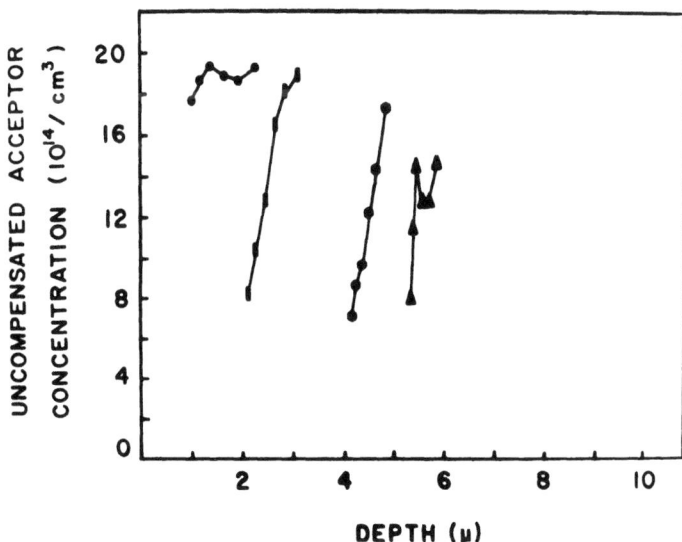

Fig. 2. Acceptor profiles for the unbiased diode of Fig. 1. Symbols are as follows:
● Before implantation   ❙ After 10^{10} B cm^{-2}
⬢ After 10^{11} B cm^{-2} ▲ After 10^{12} B cm^{-2}

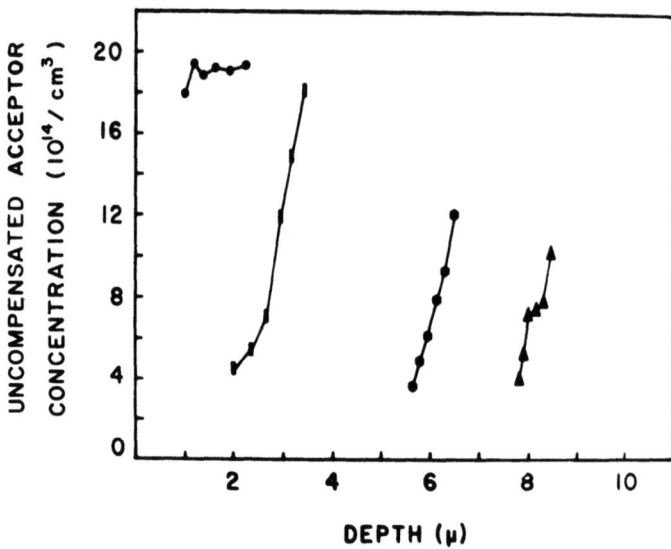

Fig. 3. Acceptor profiles for the reverse biased diode of Fig. 1. Symbols as in Fig. 2.

unbiased. The corresponding net acceptor profiles are shown in Figures 2 and 3. The boron profile obtained by using LSS parameters, is a gaussian peaking about 1200 Å inside the ZnTe with a standard deviation of 1700 Å [4,5]. Thus, the depleted zone is more than an order of magnitude wider than the region penetrated by the beam. The 10^{10} B/cm^2 curves in Figures 2 and 3 imply that if compensation is assumed to be due to B-doping, more than 10^{11} B/cm^2 would be required. These results suggest that compensation is due instead to the migration of implantation generated defects.

The effect of bias, although clearly visible from the data presented, is not as large as would be expected if the electric field were the main driving force for defect migration. In some cases, the effect is apparent only from the slopes of the acceptor profiles and not from the zero volt depletion widths. Nevertheless, in all samples the compensation due to defect migration is enhanced by an applied reverse bias. However, no significant difference has ever been observed between samples bombarded under reverse voltages between 3.4 and 8 V.

The reproducibility of the depleted depths is good to within 20% for the doses above 10^{11} cm^{-2} on samples from the same crystal. Some extra scatter appears at the lower doses, partly due to difficulties in measuring and stabilizing the lowest beam currents used. The variation of the profiles from crystal to crystal is not yet completely understood. Thus, the profiles obtained from a sample doped to 3×10^{18} P/cm^3 showed that the width of the depleted region after implantation of an unbiased diode remained of the order of the boron LSS penetration depth in ZnTe, and that fewer than 1% of the P atoms were electrically active prior to implantation. The profiles obtained from a diode on the same sample, back-biased to 3.4 V during implantation, showed a slightly larger slope, but the depleted depths at 0 V were the same in the two cases. For non-intentionally doped samples, the depleted width at zero voltage can be estimated by assuming that the quantity Nx^2 (v = 0) is constant from crystal to crystal for diodes implanted with zero applied bias [2]. The value of Nx^2 (v = 0) was found to be within a factor of 2 of 6×10^7 cm^{-2} for most samples and for B doses between 10^{11} and 10^{13} cm^{-2}.

Finally, it has been observed that the ion current density during implantation affects the measured profiles, even though this effect has not yet been systematically studied. Figure 4 shows the net acceptor profiles obtained on a diode from the same

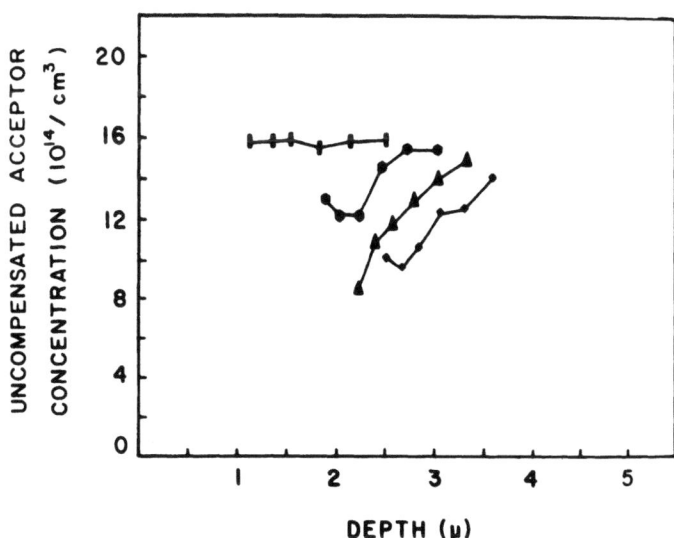

Fig. 4. Acceptor profiles for a diode implanted under 3.4 V reverse bias with 1.7×10^{10} B cm^{-2} sec^{-1}. Symbols are as follows:
I Before implantation   ● After 10^{11} B cm^{-2}
▲ After 10^{12} B cm^{-2} ◆ After 10^{13} B cm^{-2}

wafer as the samples used in Figs. 1 to 3, implanted under 3.4 V reverse bias with a B flux of 1.7×10^{10} ions/cm^2 for all doses. Comparison of Figs. 4 and 3 shows that an increase in beam current density results in a drastic decrease in the compensated depth.

IV. MODEL

The large number of parameters involved and the limited amount of data available at the present stage do not warrant quantitative comparison with a detailed model. However, some trends in the results are well enough established to justify formulating a simplified model incorporating the main physical phenomena involved while allowing a semiquantitative comparison with experiment.

The basic assumptions are as follows:

1. The implantation creates equal numbers of mobile acceptors and donors, which for the sake of definiteness will be called zinc vacancies and zinc interstitials, respectively. The generation of these defects is confined to a thin surface layer.

2. As the defects migrate away from the surface, they can interact with traps existing in the crystal prior to implantation. A trapped vacancy is an acceptor and the main effect of interstitials is to annihilate trapped vacancies, thus yielding electrically inactive centers. All traps are saturated with vacancies at the beginning of the experiment and the initial acceptor density is due to the vacancy-filled traps.

3. During implantation, steady state is reached in a region of thickness ℓ, resulting in almost complete compensation. For depths $x > \ell$, the sample is undisturbed by the implantation.

4. The progress of the above boundary is due to the flux of interstitials at $x = \ell$ annihilating the trapped vacancies. The vacancy's current flows undisturbed into the bulk since all traps are saturated.

5. The vacancies are electrically neutral so that their migration occurs by pure diffusion. No assumption is made concerning the charge of the interstitials.

The rate equation can now be written in the steady state region $(0 < x < \ell)$:

$$\frac{\partial I}{\partial t} = -\frac{\partial J_I}{\partial x} - K_1 \, IN = 0 \tag{1}$$

$$\frac{\partial V}{\partial t} = D_V \frac{\partial^2 V}{\partial x^2} - K_2 V(N_a - N) = 0 \qquad (2)$$

$$\frac{\partial N}{\partial t} = K_2 V(N_a - N) - K_1 IN = 0 \qquad (3)$$

where V and I stand for vacancies and interstitials, respectively, J_I is the interstitial current density, N_a is the original acceptor density (equal to the trap density), N is the density of trapped vacancies and K_1 and K_2 are rate constants.

The assumption of almost complete compensation means that $N \ll N_a$ and Eq. 2 can thus be uncoupled and solved to yield the vacancy concentration profile

$$V = V_0 \exp(-x/\lambda) \qquad (4)$$

where $\lambda = (D_V/K_2 N_a)^{\frac{1}{2}}$ is a diffusion length directly related to the trap density. Using Eq. 4 in Eqs. 1 and 3 yields

$$-\frac{\partial J_I}{\partial x} = K_1 IN = K_2 V(N_a - N) \simeq K_2 V N_a = \frac{J_0}{\lambda} \exp(-x/\lambda) \qquad (5)$$

where J_0 is the common value of interstitial and vacancy currents at $x = 0$. Eq. 5 implies that both currents are equal everywhere for $x < \ell$ and decay exponentially with distance. The variation of ℓ with time is calculated using assumption 4:

$$\frac{d\ell}{dt} = \frac{J_I(x = \ell)}{N_a} = \frac{J_0}{N_a} \exp(-\ell/\lambda)$$

Hence,

$$\ell = (D_V/K_2 N_a)^{\frac{1}{2}} \ln\left[\left(\frac{K_2}{N_a D_V}\right)^{\frac{1}{2}} J_0 t + c\right] \qquad (6)$$

where c is a constant of integration.

If J_0 is proportional to the beam current, then $J_0 t$ is proportional to dose. Eq. 6 thus yields the observed logarithmic dose dependence. It also shows that $N_a \ell^2$ is a physically meaningful parameter for comparison between different samples, as was previously stated [2].

The above model is clearly oversimplified. In particular, the assumption that vacancy traps are almost empty in the compensated region (implying the absence of original donors) is chiefly motivated by matehmatical convenience, and the identification of trap density with original acceptor density is very questionable, as suggested by the results on the phosphorus doped sample where the chemical doping is well above the electrical doping and, simultaneously, the defect diffusion length is much shorter than electrical doping would justify. Furthermore, the observed effect of current density can only be explained by taking into account non-linear recombination kinetics and/or non-equilibrium space charge densities due to the excess carriers or to charged defects generated by the ion beam. This comment applies also to the lack of sensitivity of the model to an applied electric field, in contrast to the experimental data. Nevertheless, this model is believed to include the basic processes which govern the creation of compensated regions during implantation in ZnTe.

V. CONCLUSIONS

The present study confirmed earlier results [1,2] on the deep penetration of implantation defects well beyond the ion range, and showed that some of these defects are charged. The logarithmic variation of compensated depth with dose and its variations with crystal perfection are accounted for in a semiquantitative way by a model involving vacancy-interstitial recombinations through traps.

ACKNOWLEDGEMENTS: Particular thanks are due to B. Schaub (CENG-LETI) who provided the ZnTe crystals and to Mrs. L. Peccoud who took care of most of the sample preparation. One of the authors (P.F.E.) wants to express his gratitude for the hospitality shown him during his stay in France.

REFERENCES

*C.E.N.G. - Département de Recherches fondamentales and Université Scientifique et Médicale de Grenoble.

1. D. Demars, M. Quillec, M. Ravetto, J. Marine and G. Guernet in Ion Implantation in Semiconductors, ed. by S. Namba (Plenum, New York, 1975), p. 235.
2. J.L. Pautrat, D. Bensahel, B. Katircioglu, J.C. Pfister and L. Revoil. To be published in Radiation Effects.
3. A.M. Goodman, J. Appl. Phys., 34, p. 329 (1962).
4. J.F. Gibbons, W.S. Johnson, S.W. Mylroie, Projected Range Statistics, 2nd edition (Halsted Press, 1975).
5. H. Ishiwara, S. Furukawa, J. Yamada and M. Kawamura in Ion Implantation in Semiconductors, ed. by S. Namba (Plenum, New York, 1975), p. 423.

STRUCTURAL REARRANGEMENT IN DIELECTRIC FILMS UNDER ION BOMBARDMENT

N. N. Gerasimenko

Institute of Semiconductors Physics
Academy of Sciences of the USSR
Siberian Branch
Novosibirsk

ABSTRACT

Structural rearrangement in silicon dioxide films on silicon under particle bombardment and annealing was studied by ESR, IR-absorption and electron-microscopy replica methods. It was found that electron bombardment primarily causes changes in the SiO_2 bond lengths and angles but heavy particle (ion) bombardment causes the production of radiation defects and the enhancement of crystallization and phase changes of the silicon dioxide films. A survey of radiation induced structural changes, radiation densification, and index of refraction modification will be presented, and the discrepancies in published data related to these topics will be discussed.

Now more and more attention is being paid to the dielectric-semiconductor interface. In particular the changes in dielectric film properties under ion bombardment are being studied because of the wide use of ion implantation for the production of discrete semiconductor devices and integrated circuits. The requirements which appeared during development of these technologies lead to emphasis on investigation of changes in electrophysical properties such as interface state density and build in charge build-up under ion bombardment. The research work carried out thus far establishes the main details of these changes and the pertinent physical models and permits optimization of the technological processes. A review of such efforts has been given by Mordkovich [1].

At the same time the nature of these changes is not yet clear and there is not enough information about structural rearrangement in dielectric films under ion bombardment and annealing.

In addition, other technologies need such an information, i.e., new directions connected with ion implantation of dielectric films; for example, ion implantation is used to: 1) increase the stability of dielectric films [2,3]; 2) increase of the radiation hardness of MOS - structures [4,5]; 3) control MNOS memory cells parameters [6]; 4) control of optical parameters of dielectric in the creation of, most importantly, optical waveguides and waveguide structures and other integrated optics elements [7,8,9,10]; 5) change the local etching rate for use in ion lithography [11], creation of windows [12], etc.

It is necessary to emphasize that in all these cases, even when the role of the implanted impurity is dominant, radiation induced structural rearrangement is very important.

We will consider structural rearrangement in silicon dioxide films which are produced by thermal oxidation of monocrystal silicon. These films are widely used in the technology of semiconductor devices and integrated circuits; the information about their properties is abundant and unambiguous. In some cases the data obtained for fused silica will be used because of no significant difference between fused silica and thermal silicon dioxide films in the property change under irradiation.

The direct indication of structural rearrangement after ion bombardment is the change in density of the SiO_2 film. There are some contradiction in the data presented in literature on this problem. On the one hand it is found that the silicon dioxide film thickness increases after irradiation with phosphorus, arsenic and argon ions bombardment [13]; on the other hand a lot of the experimental results are presented on densification of fused silica [7,8,14] thermal silicon dioxide films [15]. This densification correlated with the increase in the refraction index n. One of the possible causes of this contradiction could be the dose dependence of the change in the index of refraction (Fig. 1). Figure 2 shows the Eernisse-Norris data on stress in SiO_2 films; this stress is correlated with the index of refraction and exhibits the same sort of dose dependence. This data shows that this dependence is not monotonic, first increasing, then - decreasing in the high dose region. It is in this high dose region we expect the thickness increase.

The film microstructure rearrangement peculiarities after ion bombardment permit additional remarks about the contradiction which was discussed above. An interesting correlation has been discovered between the index of refraction change and the appearance of paramagnetic centers in the irradiated fused silica (Fig. 3). This correlation gives the opportunity to monitor amount and depth dependence of the structural changes which determine the index of refraction. Investigation of the ion irradiated silicon dioxide films found two

paramagnetic centers which have been found in irradiated quartz
(Fig. 4): E' (g_1 = 2,000; g_2 = 2,002; g_{av} = 2,001; H = 3,2 gs) –
the oxygen vacancy; and DI (g_1 = 2,0017; g_2 = 2,0065; g_3 = 2,064)
[18], which is connected with the existence of free oxygen in the
matrix. The annealing temperatures of these centers (Fig. 5) differ, but both recover in that temperature interval in which the
densification created by displacement damage disappears [14,15].

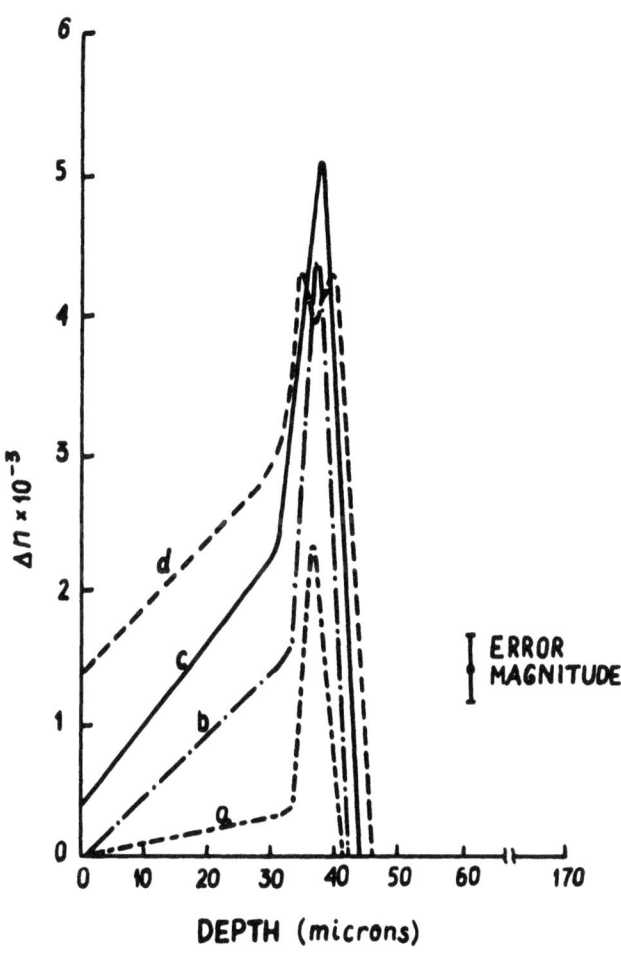

Fig. 1. Index of refraction profiles vs depth for the following
fluences in ions/cm^2: (a) 2.3x10^{16}; (b) 5.0x10^{16};
(c) 1.0x10^{17}; (d) 2.03x10^{17}. Details of the index change
within a few microns of the surface are obscured by rounding effects due to polishing [14].

Paramagnetic centers can be partially removed (∼ 80%) by ultraviolet light irradiation (mercury bulb spectra), but defects connected with E' - centers are not removed, but are transformed into a nonparamagnetic state; a subsequent fast electron irradiation (E = 3,5 Mev, D = 10^{16}cm^{-2}) recreates the E' - centers whereas fast electron irradiation without ion preirradiation can not produce E' - centers. The range distribution of the E' and DI centers coincide with the energy loss range distribution.

Useful information about structural rearrangement in irradiated films can be obtained by investigation of the infrared absorption, spectra, especially in the spectral range near 9μ, where there is the main Si-O vibration band. The shape of this band and its maximum position reflect the damage connected with the change of the Si-O bond length, the angle in the top of the silicon-oxygen tetrahedra and long range order. In particular it has been shown [11,13]

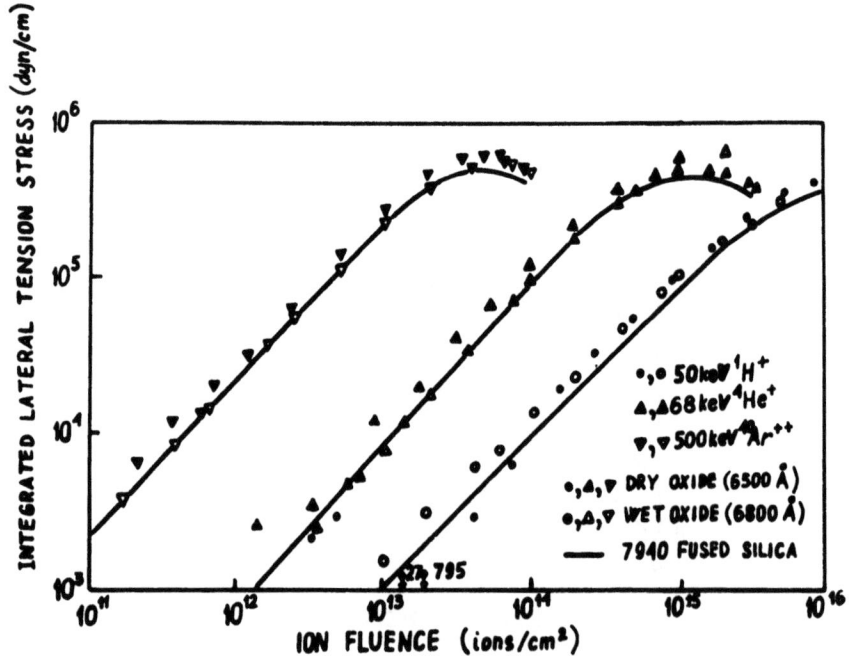

Fig. 2. Integrated lateral stress in dyn/cm vs ion fluence data for H, He, and Ar ions incident on wet and dry oxide layers. The dashed lines show comparable data for high-purity bulk fused silica [15].

that ion bombardment leads to a shift of the band to longer wavelengths which indicates the change in the tetrahedra angle [19,20] and the breakdown in the polymeric tetrahedra chains [21]. We have compared the IR-spectra of samples irradiated with electrons and heavy particles to establish the difference between structural changes produced by elastic and inelastic particles energy loss: (Fig. 6 and 7). Both kinds of energy loss lead to SiO_2 compaction [14,15].

The IR absorption spectrum of silicon oxide films consists of three main bands with maxima at 1090, 820 and 465 cm^{-1}. Most changes after irradiation take place in the main band $\nu = 1090$ cm^{-1} and this band was investigated most carefully by us, but we can comment on the other bands: the 820 cm^{-1} band intensity is almost

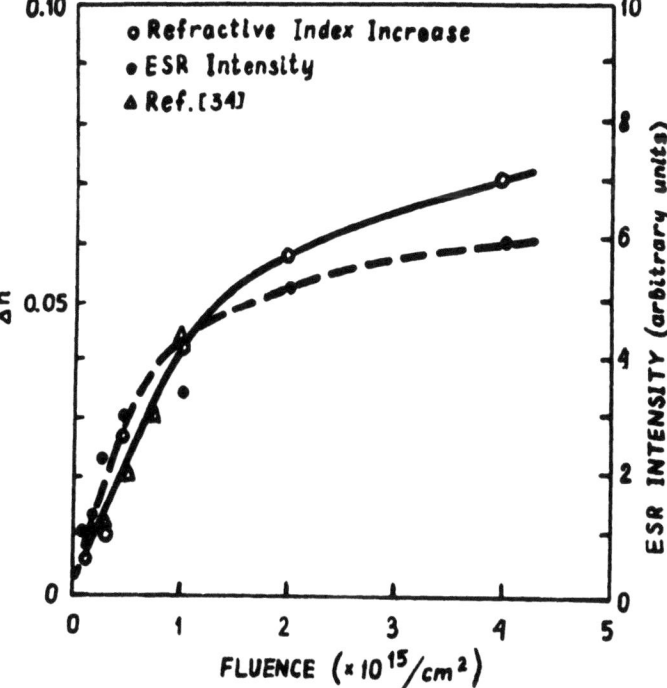

Fig. 3. Li^+ ions dose dependence of the index of refraction and paramagnetic centers [7].

unchanged after irradiation but there is a visible shift to longer wavelengths; the 465 cm^{-1} band shape changes little, but its intensity decreases.

Irradiation with B^+ and P^+ ions leads to the following changes in the 9μ band: the intensity decreases, the maximum shifts to long wavelengths and the band width increases. Comparison infrared

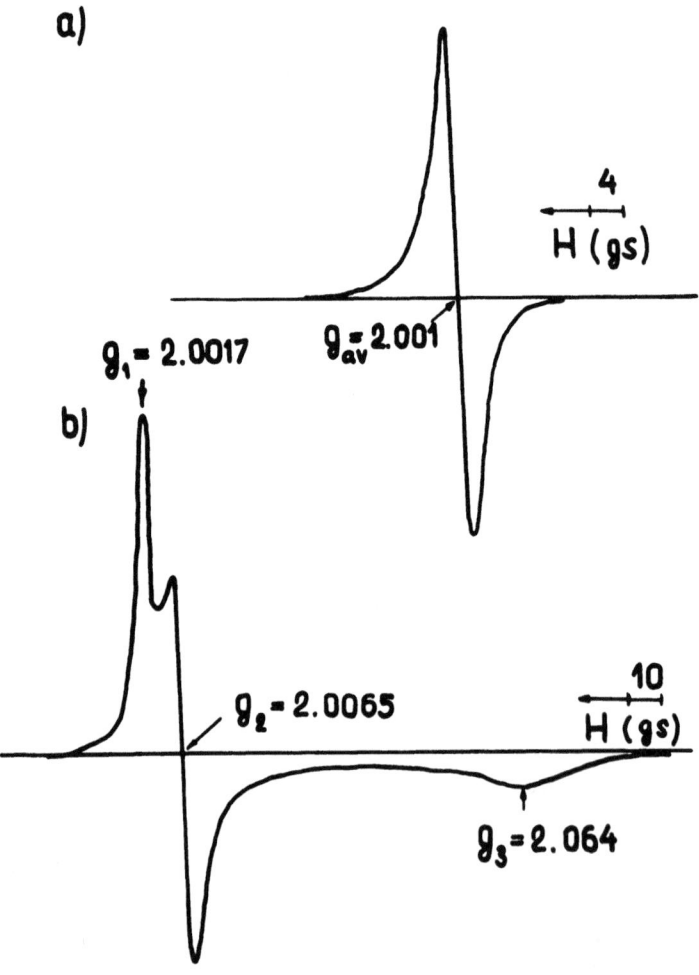

Fig. 4. ESR (Q band) spectra of the ion irradiated silicon dioxide film. a) - spectrum taken at 300°K; b) - irradiated sample after annealing at T = 600°K for 10 min. in a helium atmophere; the spectrum was taken at 77°K (DI - center) [16].

measurements, which directly measure the difference between the irradiated sample and a reference sample, show two bands (1030 and 1100 cm^{-1}), whose positions and shape are dose independent.

Low dose proton irradiation as well as electron and γ-irradiation lead to an intensity increase in the range from 1020-1130 cm^{-1}. The comparison spectrum changes with increase in proton dose; the dip in the shortwave length region appears. There is no difference between the spectra obtained from proton and ion (B$^+$,P$^+$) irradiated

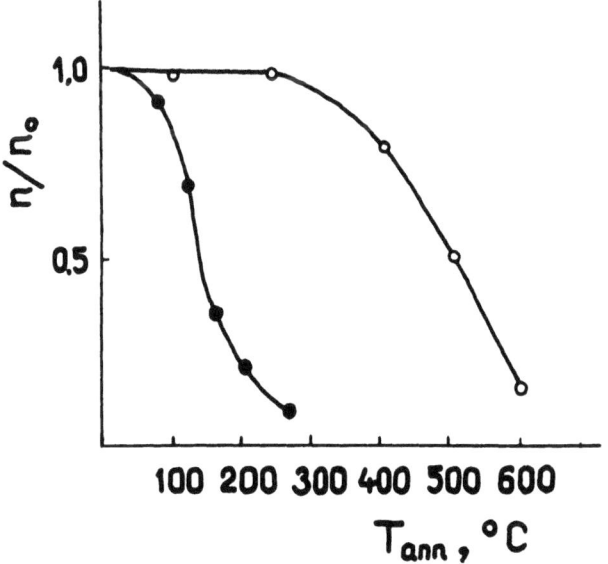

Fig. 5. Isochronal annealing (10 min) of the paramagnetic centers in a Helium atm.; closed circles - E' - centers, open circles - DI centers [16].

films after a proton dose $10^{16} cm^{-2}$. For electron and γ-irradiated films there is less absorbtion in the long wavelength region of the 9μ band and more in the 1080-1120 cm^{-1} region. There is no saturation in the dose dependence of the IR spectrum change of electron and γ-irradiated films.

The main changes in the 9μ IR spectra of irradiated SiO_2 films consist of the following: 1) The band shifts to higher frequencies

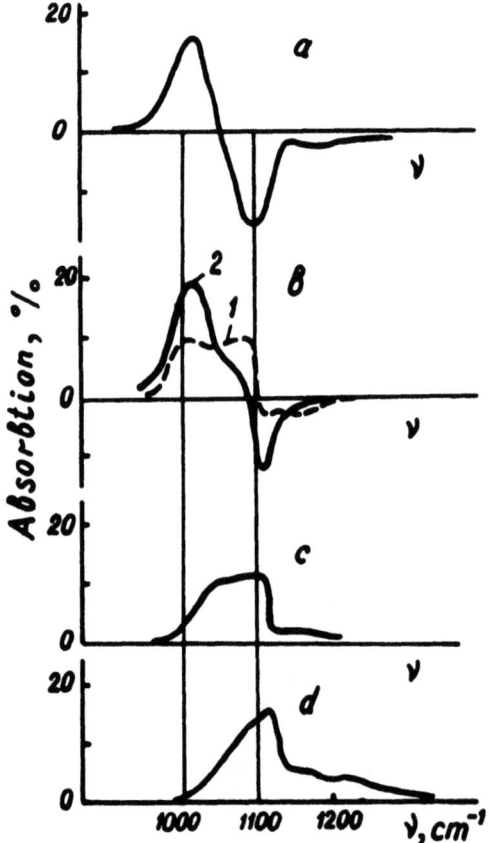

Fig. 6. Infrared comparison spectra of the thermal grown SiO_2 films (of thickness-h) irradiated with: a - boron ions (E=33 kev) h = 0,2μ; b - protons (E=40 kev), h = 0,9μ; c - electrons (E=500 kev) and d - γ- rays, h = 0,9μ.

under electron and γ-irradiation, which can be due to a length increase in the polymeric tetrahedra chains [22]. The increase in this band is due to the increase in the number of Si-O bonds. 2) The shift to lower frequency of this band under ion bombardment is due to the breakdown of the tetrahedral chains and distortion of the tetrahedra; both of these processes can lead to densification. The decrease in absorption arises from a decrease in the number of the Si-O bonds due to the dominance of the atomic collision energy

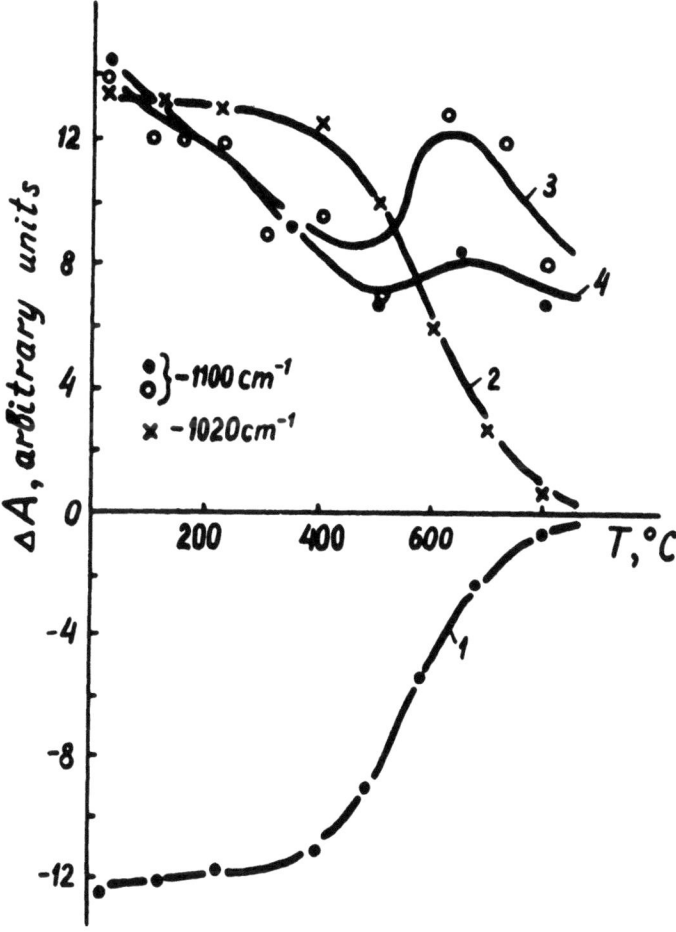

Fig. 7. Isochronal annealing of the IR spectra from thermal grown SiO_2 films irradiated with: 1 - boron, and 2 - phosphorus ions; 3 - electrons and 4 - γ-rays.

losses. 3) At small doses H^+-irradiation yields results like e^- irradiation, but at large doses, like ion-irradiation.

During isochronal annealing the changes in the IR spectra after heavy particles implantation (where the dominant defects are due to the atomic collision energy losses) anneal at the temperatures lower than do the radiation changes after light particles irradiation (where the defects are primarily due to ionization energy losses). This annealing correlates with the annealing of the change in the refractive index and the densification due to ionization and atomic collision energy losses [14,15].

The clearest results in the study of structural changes are obtained with electron diffraction and electron (EM) microscope methods. Pavlov and coworkers [23-25] discovered by electron diffraction that ion bombardment results in a decrease in the crystallization temperature of SiO_2 and Si_3N_4 films; they associated this result with the decrease in the activation energy for self-diffusion as consequence of changes in the interatomic distances. We studied the process of structural change in thermal SiO_2-films after Ar^+- and P^+-implantation and annealing using EM with gold-carbon replicas (Fig. 8,9, and 10).

A distinctive surface roughness appears as a result of chemical etching of SiO_2-films implanted with a dose $\sim 10^{15} cm^{-2}$; the roughness height increased and size decreased (Fig. 8) as the dose increased. This etching behavior persists through the implanted region and correlates with the atomic collision energy loss region. The roughness is due to a phase-separation process which is well studied in glasses. In this layer partial crystallization takes place during bombardment at 100°C creating small spheres ($\sim 10^4/cm^2$- see Fig. 9). Annealing at 600°C stimulated crystallization in this layer. Another peculiarity of implanted films appears near the Si-SiO_2 interface during annealing. Annealing at 850°C results in the creation of a network similar to that which occurs in the formation of two separated phases in glasses, and in the formation of polycrystalline silicon particles in the unimplanted oxides, but in the case of implanted oxides this annealing yields a typical network of a crystalline phase (Fig. 10), this phase having the same interplanar distances as coesite (a high pressure phase [7]). Increasing the annealing temperature to 1020°C results in practically identical structures for the implanted and unimplanted films, i.e., the same as implanted films annealed at 850°C.

It is known in films obtained by the high temperature process that nonstoichiometry exists near the Si-SiO_2 interface; one can also show that this layer is strained. Both these causes promote network formation. Crystallization in the high pressure phase (coesite) argues that large structural strains exist in this layer because 40 to 140 kbar pressures and 700 to 2000°C temperatures are necessary for the synthesis of this phase [27]. As grown stresses which arise as a result of the ordering process in the amorphous layer of pyrolitic films are as large as 4.10^3 bar [30]. This stress is smaller in thermal grown oxide and is not enough for the creation of the coesite phase. Therefore a kinetic model of stress creation seems preferable. In this model the phase lamination process proceeds with the specific volume of one phase increasing, yielding the formation of a dense phase.

In conclusion we will try to consider the general picture of the structural rearrangement in the irradiated SiO_2 films, taking into account the experimental facts mentioned above: 1) Densification under irradiation can occur due to ionization and knock-on

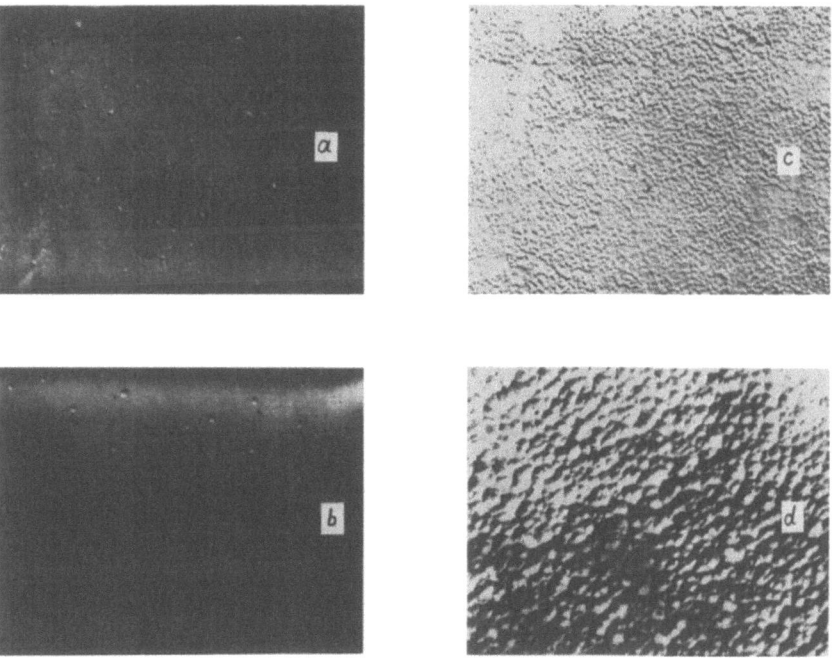

Fig. 8. Replica micrographs (x 10000) of (a) nonirradiated SiO_2 films and films irradiated with Ar^+ ions with energy 40 kev and the following doses (cm^{-2}): (b) 10^{13}, (c) 10^{15}, and (d) 5.10^{15}.

energy losses, but there are different densification mechanisms in the two cases. 2) The IR absorption data shows that ionization can lead to the improvement of film quality by the increased number of Si-O bonds and the increase in the chain length of the polymeric SiO_4 tetrahedra - in this case the densification occurs throughout the whole volume of the film and the whole system becomes more stable as shown by the delay in recovery to high temperature. 3) Elastic collisions which dominate during ion bombardment lead to different results, i.e., radiation damage, such as the breaking of bonds, the distortion of bond angles, the generation of vacancies and free atoms, plays the main role. The model proposed for the explanation of the densification in this latter case is one in which the oxygen atoms created by collisions with bombarding ions fill up voids leading to distortion of the tetrahedra [31,32]. Thermal spikes play a very important role in this case [33]; these are regions at the end of the ions tracks (with dimensions of ~ 100 A) in which a pulse of temperature to $\sim 1000°C$ and of pressure to $\sim 10^9$ dyne cm^{-2} takes place during a very short time $\sim 10^{-11}$ sec. Such a region can nucleate the new phase formation because of the enhancement of phase separation and crystallization, and point defects can also stimulate both of these processes.

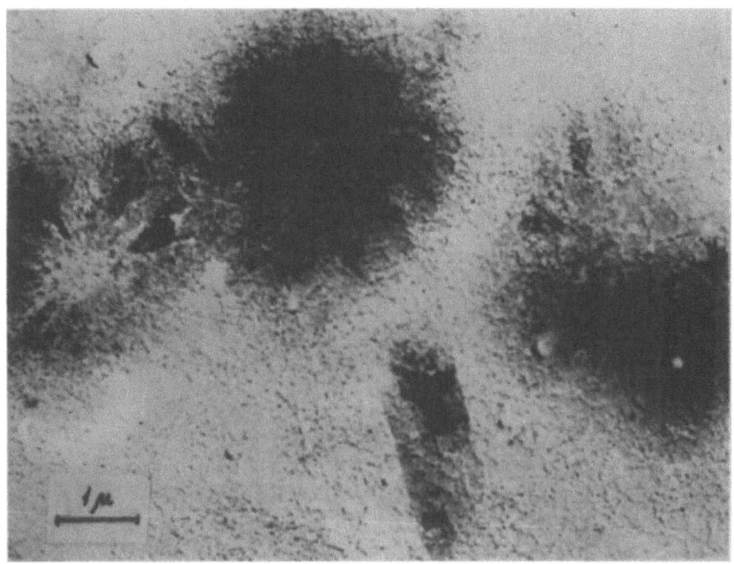

Fig. 9. A micrograph (x 10000) of crystallites created during Ar^+ ion bombardment: $E = 20$ kev, $D = 2.10^{15} cm^{-2}$, $T_{irr} = 100°C$.

Our investigation of the microstructure shows that densification is connected with a phase separation in the SiO_2 film which is stimulated by ion bombardment. The increase in the visible refractive index is connected with the generation of dense phase inclusions.

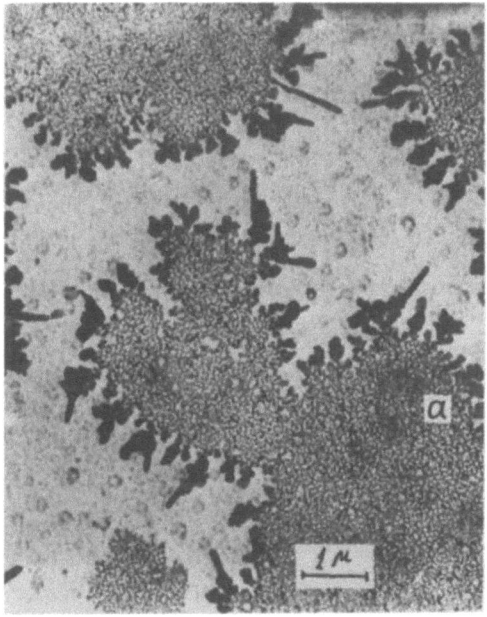

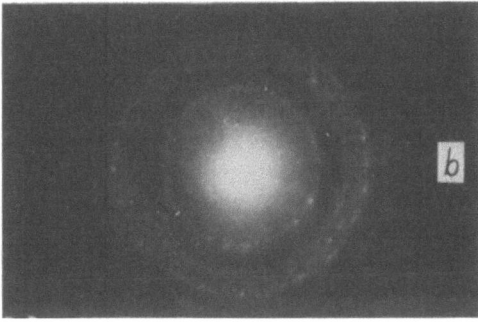

Fig. 10. A micrograph (using a gold-carbon replica) of a SiO_2 film irradiated with P^+ ions: $E = 30$ kev, $D = 3.10^{16} cm^{-2}$ and annealed at 850°C for 30 min.

The author is indebted to G. M. Tseitlin and G. P. Lebedev for fruitful discussions. The data in Figs. 6 and 7 were obtained by the author with A. A. Nesterov and L. S. Smirnov. The results in Figs. 8-10 were obtained by the author with V. A. Ivanchenko and G. P. Lebedev.

REFERENCES

1. V. N. Mordkovich, Elektronnaya tekhnika, Seriya 2, Poluprovodnikovye pribory, $\underline{8}$, 63 (1974).

2. N. V. Ashkinadze, L. K. Dumish and V. B. Dubro. Elektronnaya tekhnika, Seriya 2, Poluprovodnikovye pribory, $\underline{5}$, 3 (1975).

3. N. N. Gerasimenko, V. I. Fedchenko and V. N. Gashtol'd. Elektronnaya tekhnika, Seriya 2, Poluprovodnikovye pribory, vyp. 8 (1976).

4. E. Harari and B.S.H. Royce, IEEE Trans. Nucl. Sc. $\underline{N5-20}$, N6 (1973).

5. N. N. Gerasimenko, L. V. Lezheiko, E. V. Lyubopytova, O. N. Kuznetsov, L. S. Smirnov and F. L. Edel'man, Mikroelektronika, $\underline{3}$, 467 (1974).

6. N. N. Gerasimenko, G. P. Lebedev, G. P. Kuryshev and B. I. Usova, Mikroelektronika, $\underline{5}$, in press.

7. H. Aritomo, T. Ikegami, T. Nishimura, K. Masuda and S. Namba, Annual Rep. of Namba Lab. 1972, p. 18, Osaka Univ. Press.

8. R. D. Standley, W. M. Gibson and J. W. Rogers, Appl. Optics $\underline{11}$, 1313 (1972).

9. V. V. Vasil'ev, N. N. Gerasimenko, G. M. Tseitlin, A. A. Nesterov, V. G. Pan'kin and V. P. Panov, Materialy Vsesoyuznoi konferentsii po nelineinoi optike, Tbilisi, 1976.

10. N. N. Gerasimenko, G. M. Tseitlin, V. G. Pan'kin, K. K. Svitashev and T. I. Kovalevskaya, Zhurnal prikladnoi spektroskopii, $\underline{26}$, in press.

11. I. P. Akimchenko, V. V. Galkin, V. V. Krasnopevtsev, V. S. Krasheninnikov, Yu. V. Milyutin and A. V. Spitsyn. Mikroelektronika, $\underline{2}$, 166 (1973).

12. R. A. Moline, R. R. Buckley, S. E. Haszko and A. U. McRae, IEEE Trans. on Electron Div. $\underline{ED-21}$, 840 (1973).

13. C. R. Fritzche and W. Rothemund, J. Electrochem. Soc. 119, 1243 (1972).

14. H. M. Presby and W. L. Brown, Appl. Phys. Lett., 24, 511 (1974).

15. E. P. EerNisse and C. B. Norris, J. Appl. Phys., 45, 5196 (1974).

16. B. I. Vikhrev, N. N. Gerasimenko and G. P. Lebedev, Mikroelektronika, 5, in press.

17. R. A. Weeks, J. Appl. Phys., 27, 1376 (1956).

18. E. A. Zamotrinskaya, L. A. Torgashinova and V. F. Anufrienko, Neorganicheskie materially, 8, 1136 (1976).

19. S. Mochizuki and N. Kamai, Solid State Comm., 11, 763 (1972).

20. I. Simon, J. Amer. Ceram. Soc. 40, 150 (1957).

21. V. A. Kolesova, Optika i spektroskopiya, 6, 38 (1959).

22. W. A. Pliskin, Appl. Phys. Lett., 7, 158 (1965).

23. P. V. Pavlov, E. V. Shitova, E. I. Zorin and N. A. Galkina, Kristallografiya, 18, 609 (1973).

24. N. A. Genkin, E. V. Shitova, G. S. Khokhlova, T. M. Zotova and V. M. Genkin, Elektronnaya tekhnika, ser. Materialy, vyp. 9 (1975).

25. P. V. Pavlov, N. A. Genkina, E. V. Shitova and D. I. Tetelbaum, Phys. Stat. Sol. (a) 29, 303 (1975).

26. J. W. Cahn, J. Chem. Phys. 42, 93 (1965).

27. R. B. Sosman, *The phases of silica*, Rutgers Univ. Press, New Brunswick, 1964, pp. 275-276, 307-308.

28. T. W. Sigmon, W. K. Chu, E. Lugujjo and J. W. Mayer, Appl. Phys. Lett., 24, 105 (1974).

29. J. S. Johannesen, W. E. Spicer and Y. S. Strausser, Appl. Phys. Lett., 27, 452 (1975).

30. E. M. Trukharov, E. B. Gorokhov and S. I. Stenin, Phys. Stat. Sol. (a) 33, 435 (1976).

31. W. Primak and R. Kampwirth, J. Appl. Phys. 39, 5651 (1968).

32. W. Primak, J. Appl. Phys. 43, 2745 (1972).

33. A. E. Gorodetskii, G. A. Kachurin and L. S. Smirnov, Fiz. Tekh. Poluprov. 2, 927 (1968).

34. S. E. Miller, IEEE J. Quantum. El. QE-8, 199 (1972).

35. G. B. Kreffe and G. W. Arnold in *Ion Implantation in Semi. conductors and Other Materials*, Ed. by B. L. Crowder (Plenum, N.Y. 1973) p. 523.

ABOUT THE DETERMINATION OF LATTICE DEFECTS

IN BACKSCATTERING EXPERIMENTS

       Yves Quéré

       Centre d'études nucléaires

       Fontenay-aux-Roses 92 France

ABSTRACT

Backscattering experiments in channeling conditions are often used to determine so called "damage profiles" in implanted crystal. It is suggested that an energy analysis of the dechanneling cross section might be of some help to determine which is the predominant type of defect.

INTRODUCTION

Any kind of implantation experiment creates lattice defects in the wake of the implanted ions. The elementary defect is the Frenkel pair which consists of a vacancy and an interstitial. These point defects tend to aggregate, due to the temperature and/or the irradiation-induced lattice vibration; the result being the formation of dislocation loops, cavities, bubbles or more poorly defined clusters.

All these defects have a definite influence on the physical properties of the implanted region either directly or through their interaction with the implanted species. It may then be necessary to know which type of defect has been created. In this respect <u>electron microscopy</u> is a valuable tool (see for example [1]), but this method is both destructive and experimentally nontrivial. <u>Backscattering</u>, on the other hand, is generally easy to perform (in the implantation chamber itself) but shows essentially no sensitivity to the type of defect which has been created. At most one may expect to get an indication--more than a measurement--about the degree of lattice disorder [2].

This short paper essentially will develop the idea that a measurement of the energy dependence of the backscattering cross sections may give some more information on the type of defect present in the implanted region. We shall briefly examine the cases of dislocation loops (perfect or imperfect), cavities (or bubbles) and interstitial atoms.

DISLOCATION LOOPS

No theoretical calculation exists on the dechanneling cross section of a dislocation loop. Consider a perfect circular loop of Burgers vector b and radius R . The cross section may be written as

$$2\pi R \lambda$$

where λ is the dechanneling width of the dislocation. λ is expected to be a function of E, R, b, and of the density of dislocations ρ :

$$\lambda = \lambda(E, R, \rho, b)$$

The width λ has been evaluated for <u>straight dislocations</u> (R = ∞) either by computer simulation, including the different possible phases of oscillating channelons but only for discrete cases [3], or by an analytical calculation in the oversimplified case of nonoscillating channelons [4-5]. In both cases λ is found to increase linearly with $E^{\frac{1}{2}}$.

The simplest possible argument about the R-dependence of λ is the following. Knowing from [4-5] that λ is proportional to $b^{\frac{1}{2}}$, i.e. to the square root of the atomic displacements for a straight dislocation, and remembering that the displacements around a small loop are proportional to bR^2 , we can estimate that λ is proportional to $Rb^{\frac{1}{2}}$. Since λ is bound to tend to a constant value for $R \to \infty$, the following formula can be proposed [6]:

$$\lambda = \frac{bdaE}{\alpha Z_1 Z_2 e^2}^{\frac{1}{2}} [1-\exp{-(R/R_o)}] \qquad (1)$$

where d is the atomic density along the channel, a the Thomas Fermi radius, α a numerical constant depending on the character of the dislocation and R_o a constant length. This empirical formula essentially presents the advantage of giving the $Rb^{\frac{1}{2}}$ dependence at small values of R and the constant value obtained in [4-5] for large R's.

Very few experimental data are available to test both the numerical value and the E- and R- dependence predicted in (1). In the case of straight dislocations (R = ∞), a good agreement between experiment and formula (1) has been obtained in [7] and

[8]; whereas a strong discrepancy appears in [9] where the experimental value seems much larger than predicted by (1). For dislocation loops, only one determination has been published [6]: for 5 MeV α particles planarly channeled in aluminium, λ is found equal to 6 Å for $\overline{R}$ = 120 Å. In the same conditions but for R = ∞, a much larger value (λ = 140 Å) has been measured [8] in general agreement with formula (1) in which R_o, in this particular case, is equal to 5000 Å.

There are fewer experiments concerning the E dependence of λ. In one of them [7] a $E^{\frac{1}{2}}$ dependence is claimed in agreement with (1) but in another one [9] λ seems to decrease with increasing energy.

There exist no data on the ρ-dependence of λ but one should expect a rather strong decrease of λ when ρ increases due to dipole effects of dislocations with respect to each other.

STACKING FAULTS

In the preceding section, the dislocation loops were supposed to be perfect. In fact, small dislocation loops may often be imperfect, i.e. contain a stacking fault. In this case, the dechanneling is the sum of the two contributions of the dislocation and of the fault.

Dechanneling by stacking faults is studied in detail in [10], both theoretically and experimentally. It is mainly due to the abrupt change of potential energy at the plane of the fault. The dechanneling cross section is shown to be essentially independent of the energy.

BUBBLES AND CAVATIES

These are well known clusters which can be encountered in irradiated solids.

Dechanneling by bubbles is due essentially to the free surface at the re-entry of the particle in the solid after its flight in the bubble. The influence of the gas contained in the bubble is negligible. Dechanneling gives a very convenient measurement of the specific free surface created in the sample [11-13]. The room temperature ageing of implanted-helium bubbles (diameter 20 Å) has been observed by this method [13].

For a given experiment the dechanneling cross section of a bubble depends on the angle distribution of the channeled flux. In the oversimplified case of a harmonic potential (planar channeling) and a uniform angle distribution, the dechanneling cross section of a spherical bubble of radius R is

$$\pi R^2 (1 - \frac{\pi}{4} \frac{c}{b})$$

where c is the maximum possible oscillation amplitude and b the half width of the channel.

Again, as in the case of stacking faults, the dechanneling cross section is energy-independent.

INTERSTITIAL ATOMS

Implantation creates interstitials: either self interstitials of the matrix, generally too mobile to keep their individuality, or solute implanted atoms in an interstitial position.

The problem of the total dechanneling is complex, including the influences of both the interstitial itself and the distortion of the lattice. For the case in which the distortion may be neglected, a dechanneling cross section proportional to $E^{-\frac{1}{2}}$ is predicted theoretically [10,14]. A classical E^{-1} Rutherford dependence is expected in the observation of particles which have been backscattered specifically by an interstitial. In both cases, the cross sections decrease with increasing particle energy.

CONCLUSION

The problem of specifically analyzing the implantation lattice defects by means of a backscattering experiment remains formidable and unsolved. A possible way to go a little further than just observing the existence of "lattice disorder" might consist in analyzing the energy dependence of the backscattered yield. Of course, we should not expect to observe a simple law for the energy dependence. This is due both to the oversimplification of the above mentioned calculations and to the coexistence in any implanted sample of different types of defects. (The case of the faulted dislocation loop, which should give a E^p dependence with $0<p<\frac{1}{2}$, is characteristic of this difficulty.)

In any case, it will be necessary to test the simple ideas developed here by some experiments where the energy dependence of a <u>well defined</u> population of specific defects could be measured.

ACKNOWLEDGEMENTS

It is a pleasure to thank P. P. Pronko and S. Rothmann for very interesting discussions and their communication of unpublished results.

REFERENCES

1. M. O. Ruault, Thèse, Université d'Orsay (1975)
   I. Nashiyma, P. P. Pronko and K.L Merkle, Rad. Effects 29, 95(1976).

2. Y. Quéré, Rad. Effects 28, 253(1976).

3. D. V. Morgan and D. Van Vliet, Atom. Coll. Phenom. in Solids, p. 476, North Holland (1970).

4. Y. Quéré, Phys. Stat. Sol. 30, 713(1968).

5. Y. Quéré, Ann. Physique (Paris), 5, 105(1970).

6. G. Chalant, Note CEA-1902 (1976).

7. E. Rimini, S. V. Campisano, G. Foti, P. Baeri and S. T. Picraux, Int. Conf. on Ion Beam Surface Layer Analysis, Karlsruhe (1975).

   G. Foti, S.T. Picraux, S. V. Campisano, E. Rimini and R. A. Kant, This Conference (session III).

8. J. Leteurtre, N. Housseau and Y. Quéré, J. Physique 32, 205 (1971).

9. P. Pronko and S. Rothman, Private communication of unpublished results.

10. J. Mory, Rapport CEA-R-4745 (1976).

11. D. Ronikier-Polonsky, Note CEA N-1834(1975).

12. D. Ronikier-Polonsky, G. Désarmot, N. Housseau and Y. Quéré, Rad. Effects 27, 81(1975).

13. G. Désarmot, Rapport CEA, in press (1976).

14. J. C. Jousset, J. Mory and J. J. Quillico, J. Physique, lettres, 35, L-229(1974).

HEAVY ION RANGES IN SILICON AND ALUMINIUM

W. A. Grant, D. Dodds, J. S. Williams,
C. E. Christodoulides, R. A. Baragiola* and D. Chivers+

Department of Electrical Engineering
University of Salford, Salford M5 4WT, Lancs, U.K.

ABSTRACT

Range profiles for heavy ions implanted into Si and Al have been investigated using high resolution Rutherford backscattering. Projectile masses and energies were chosen to cover the reduced energy range $.01 < \varepsilon < 0.8$ where nuclear stopping should dominate the slowing down process. Data for more than 25 species implanted into Si at energies up to 400 keV have shown that experimentally measured Rp values exceed theoretical (LSS) predictions by about 12-15% at high ε ($\sim .5$) and by up to 80% at low ε ($\sim .01$). Differences between the ranges in the various substrates employed (single crystal Si, and evaporated and large-grain polycrystalline Al) have also been revealed. Since inner-shell energy-loss effects have been reported to give rise to oscillations in Rp for rare earth ions implanted into Si and Al targets, particular attention has been given to measuring the ranges of ions having $57 < Z_1 < 66$. No evidence for range oscillations was found in either Si or Al targets.

INTRODUCTION

In recent years the employment of Rutherford backscattering (RBS) has provided a wealth of experimental data on heavy ion

---

* Centro Atomico Bariloche, San Carlos de Bariloche, Rio Negro, Argentina.

+ A.E.R.E. Harwell, Oxfordshire, England.

ranges in light substrates both at high energies (> 100 keV)[1-3] and, when high resolution detection systems are utilised, at low energies.[4-7] The low energy region ($\epsilon < 0.5$ in reduced energy units[8]) is of particular interest, since theory[8] predicts that elastic (nuclear) stopping should dominate the slowing down process. Additionally, for $.002 < \epsilon < .1$, Schiøtt[9] has suggested that a simple power law potential ($V(r) \propto r^{-3}$) is a good approximation to the more accurate Thomas-Fermi potential employed by LSS. Recent experimental range data in Si for this energy region[4,6] provide substantial agreement with the form of the potential proposed by Schiøtt, but the experimental ranges tend to exceed predicted LSS values by 15-30%. Other workers[10] have reported on oscillatory dependence of range on atomic number Z_1 for the species $57 < Z_1 < 67$ in both Si and Al and explain this result in terms of an additional stopping term arising from inner-shell excitation effects.

In this study the aim has been to examine a few aspects of heavy ion ranges in Si and Al for the low energy region where nuclear stopping should predominate (i.e. $\epsilon < .5$). We report here on three significant features listed below:

i) A compilation of the ranges of over 25 species at various energies in Si substrates. Here, it is possible to obtain an empirical expression for experimental ranges in Si where no significant Z_1 effects are seen.

ii) Substrate (Z_2) effects have been examined by detailed comparison of heavy ion ranges in Si and Al. Additionally, ranges in two forms of Al target (evaporated thin films and bulk polycrystalline substrates) have been compared.

iii) Oscillations in ranges for 100 keV rare earth ions $57 < Z_1 < 66$ implanted into Si and Al targets have been investigated.

EXPERIMENTAL

Prior to implantation, as received* single crystal silicon wafers needed no further surface preparation. One set of Al targets were prepared for implantation by flash evaporation of Al (wrapped around a W filament) onto as received Si wafers. The pressure during evaporation was kept below 5×10^{-5} Torr and the oxygen content of the ~ 2000 Å thick films was less than 3% as measured by RBS. A few other impurities, detected by RBS (for films evaporated onto Carbon substrates), were present in concentrations of less than about 1%. Bulk Al targets were cut from

---

* Supplied by Dow Corning Co.

polycrystalline sheet and prepared for implantation by careful mechanical and chemical polishing procedures to ensure a flat surface, followed by an anneal in flowing dry Ar at 750°C for 1 hour. A second set of bulk Al targets were prepared using less rigorous surface preparation procedures so that the effect of surface topography on most probable range measurement could be gauged.

Several species were implanted into the Si and Al targets at energies between 10-400 keV on both the University of Salford and the Harwell isotope separators. Implantation conditions (dose, dose rate, etc.) were chosen as described previously[6] to facilitate range parameter measurement by RBS. Special precautions were taken during the rare earth implantations to ensure high purity (usually > 99%) of the implanted species.

Low angle RBS[11] was employed to measure range parameters and the typical geometrical configuration is shown in Fig. 1(a). Details of the experimental arrangement have been previously reported.[6] In the present measurements a liquid-nitrogen-cooled cold shield was employed to minimise carbon contamination during analysis with 2 MeV He$^+$ ions. Depth scales were calculated from the backscattered energy data using helium stopping cross sections compiled by Ziegler and Chu.[12]

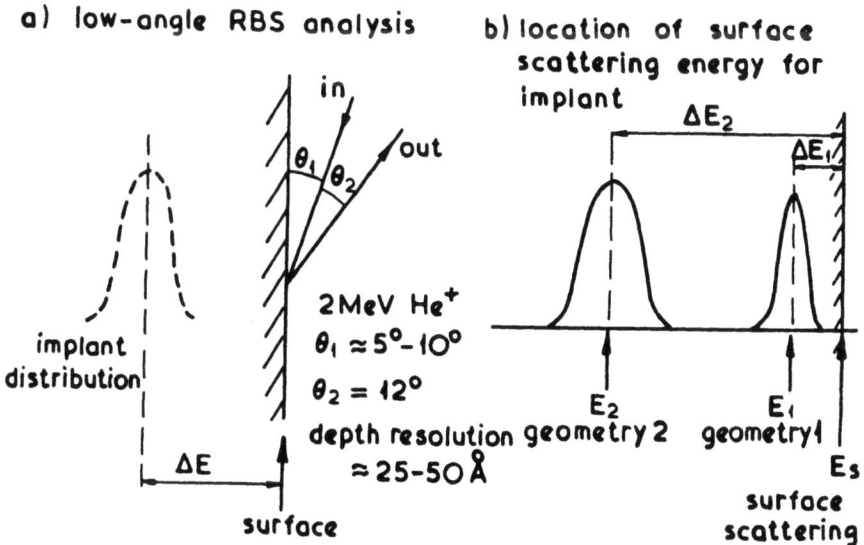

Figure 1   RBS geometry and impurity surface location for range measurements

Location of the surface backscattering energy from RBS spectra usually requires precise calibration of the system electronics with the aid of a "surface" marker, ideally a trace amount of the implanted species evaporated onto the target surface. However, in the present measurements a simple procedure was adopted which did not require accurate calibration of the backscattered energy scale. The method involves collecting spectra at two different RBS geometries and is described with the aid of Fig. 1(b). The implanted profiles obtained from the two RBS geometries are separated in backscattered energy, as shown, since the path lengths of the probe beam through the implanted surface layer to the peak of the implant distribution are different. The surface scattering energy, E_s, and the peak position in depth, R_p, are constant for both geometries and can be found from the energy-to-depth conversion relations for each geometry shown below:

$$E_1' = E_s - E_1 = S_1 R_p \quad \ldots\ldots (1)$$
$$E_2' = E_s - E_2 = S_2 R_p \quad \ldots\ldots (2)$$

where E_1 and E_2 are the respective energies corresponding to the profile peak and S_1 and S_2 are the backscattering factors for each geometry. The backscattering factor is merely a function of known stopping powers[12] and geometrical factors (θ_1 and θ_2) and is easily calculated for a particular geometry. Hence, E_s and R_p, the two unknowns in equations (1) and (2), can be found uniquely. The range, R_p, is consequently given by

$$R_p = \frac{E_2 - E_1}{S_1 - S_2}, \quad \ldots\ldots (3)$$

which does not require a prior knowledge of the surface scattering energy for the implanted species.

RESULTS AND DISCUSSION

(i) Ranges in Si

Fig. 2 summarises the range data for 27 species implanted into Si. In addition to our previously reported R_p measurements[6] (indicated by the dotted line), considerable new data is included for a wider range of species and energies. The results, as previously, are plotted in terms of the LSS dimensionless parameters ρ_p, ϵ which, according to theory, provides an almost universal curve for all Z_1, Z_2 and energy combinations. However, a small mass dependence does arise from plotting projected range parameters, ρ_p, instead of total range, ρ, and this results in the LSS envelope (based upon Thomas-Fermi interaction potentials), as shown in

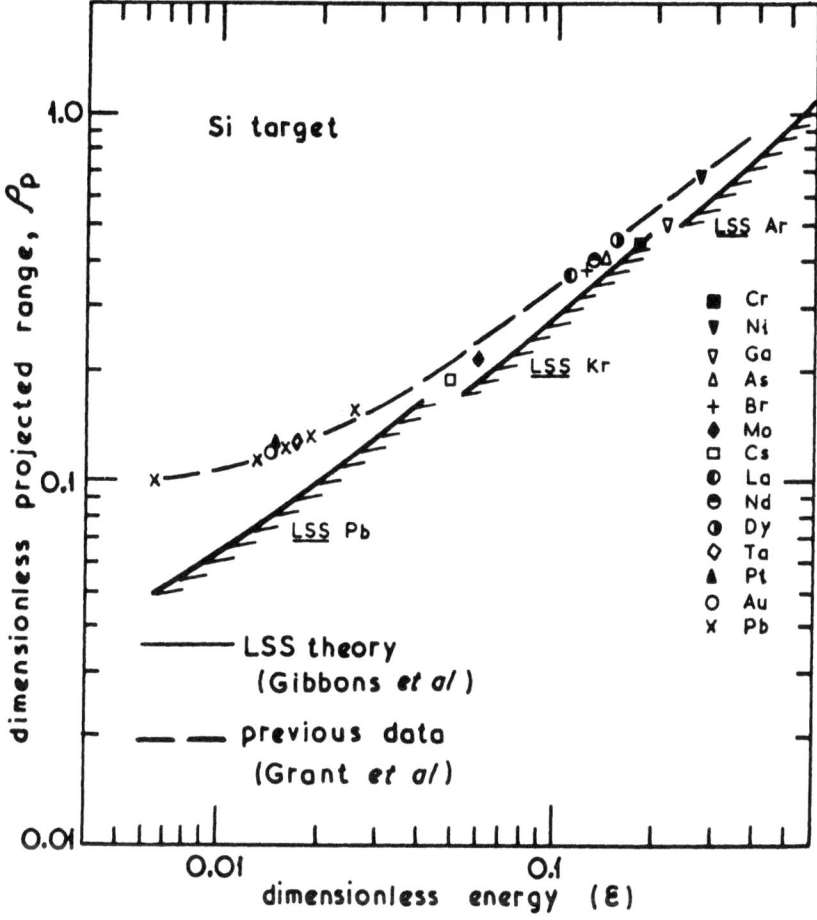

Figure 2 ρ_p/ε plot for various ions implanted into Si

Fig. 2. Theoretical range values have been taken from the tables of Gibbons et al.[13]

Our results support previous conclusions that experimentally measured R_p values exceed LSS predictions, based upon Thomas-Fermi potentials, for low energy implants into Si substrates. In the dimensionless energy range $0.02 < \varepsilon < 0.5$ an empirical curve can be drawn through our data which satisfies the relation

$$\rho_p = 1.75 \, \varepsilon^{2/3} \qquad \ldots \ldots (4)$$

The energy dependence of the above relation suggests a potential $V(r) \propto r^{-3}$ which is in agreement with the approximate form of the Thomas-Fermi potential[9] in the ε range covered by our results. However, the constant (1.75) of equation (4) exceeds that predicted using a Thomas-Fermi potential by 15-25%.

In the low ε region ($\varepsilon < .02$), the experimental ranges show a greater departure from theory (50-70%) with the energy dependence approaching $\varepsilon^{\frac{1}{2}}$ and suggesting a more screened potential than Thomas-Fermi. The present results, for the whole of the ε region investigated, are in substantial agreement with previous data.[4]

(ii) Substrate Effects

Fig. 3 illustrates on a ρ_p/ε plot the energy dependence of projected range for Si, evaporated Al, and bulk polycrystalline Al targets. Only Pb range data are shown over the energy range 20-400 keV to eliminate possible complications arising from Z_1 effects. Significant differences between the three types of substrate can be observed. Qualitatively, the Pb ranges in Si and evaporated Al exceed LSS (Thomas-Fermi) predictions whereas the bulk Al results closely follow LSS. Unfortunately, the limited Al data preclude a meaningful evaluation of the energy dependence to compare with the behaviour identified for Si targets in (i).

The magnitude of the discrepancy between experimental ranges and theory for the three types of target is better illustrated from the data presented in Fig. 4, where the ranges of 100 keV rare earth ions implanted into each target are plotted as a function of Z_1. Here, the magnitude of the departure from LSS (Thomas-Fermi) can be quantified: the ranges in Si exceed theory by an average of 15%, those in evaporated Al by ~ 8%, whereas the ranges in bulk Al are in good agreement with theory. Although there would appear to be evidence for a substrate (Z_2) effect in comparing measured ranges in Si and Al, the more surprising result is the apparent differences between the two types of Al substrate.

Careful consideration of the influence of the small but measurable oxygen and impurity content of the evaporated Al films and of the effects of surface topography on range measurements in bulk Al targets suggests that such experimental artifacts cannot account for the reproducible 8% differences in range which we observe. Hence, the explanation for our result would seem to involve structural differences between the two types of Al target. The influence of channelling during implantation has been considered and it would be expected that the large-grain bulk targets would be more amenable to channelling effects than the presumably small-grain evaporated films. This should result in some enhanced penetration in the bulk targets (which has been observed experimentally as a deeply penetrating tail to the implant profile), whereas our measured most-probable ranges show the opposite trend and are greater in the evaporated targets.

The influence of structural target differences (between Si and Al and between different Al substrates) on heavy ion penetration in the nuclear stopping region warrants investigation in light of the surprising substrate differences revealed in this study.

(iii) Z_1 oscillations in Si and Al

Fig. 4 basically summarises the range data which set out to investigate Z_1 oscillations in the ranges of 100 keV rare earth ions implanted into Si and Al. Little can be said about the clearly observed smooth variation in range with Z_1 except that the amplitude of the range oscillations reported by other workers[10] lie well outside our experimental errors and should, if present, be readily seen. We have also measured[14] the standard deviation in projected range for rare earth ion profiles in Si and again no anomalous oscillatory behaviour is observed. Recent measurements[15] of inner-shell excitation effects for rare earth ions implanted into Si and Al indicate that excitation energy-loss processes should be negligible compared with nuclear energy-loss effects, even where inner-shell level matching occurs between projectile and target atoms. On this basis no oscillatory Z_1 behaviour would be expected in projected range, consistent with our experimental results.

CONCLUSIONS

i) Measurements of the projected range of 27 species implanted into Si to cover the energy range 10-400 keV satisfy the empirical relationship (in LSS reduced parameters) $\rho_p = 1.75 \varepsilon^{2/3}$, for the reduced energy range $.02 < \varepsilon < .2$. Typically, in this energy range, measured ranges exceed those based upon LSS theory (Thomas-Fermi potential) by 15-30%.

ii) At lower energies ($\varepsilon < 0.02$) a greater departure from theory is found. This suggests that a more screened potential, such as the Lenz-Jensen, is required to describe the collision process.

iii) A substrate effect for low energy ranges has been found. Typically, for 100 keV rare earth ions, measured ranges in Si targets are ~ 15% greater than theory whereas ranges in evaporated (thin film) Al targets are ~ 8% greater. Furthermore, corresponding

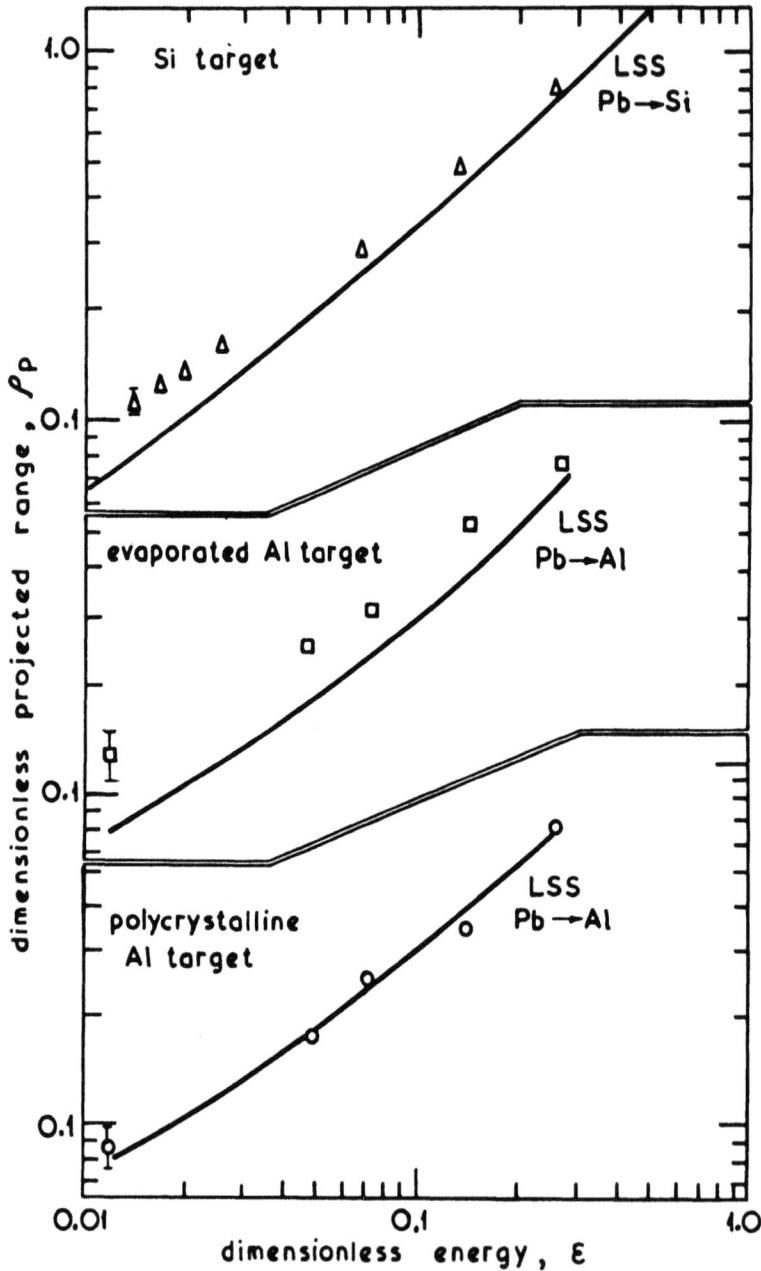

Figure 3 $\rho p/\varepsilon$ plot for various energy Pb+ implantations into Si and Al targets

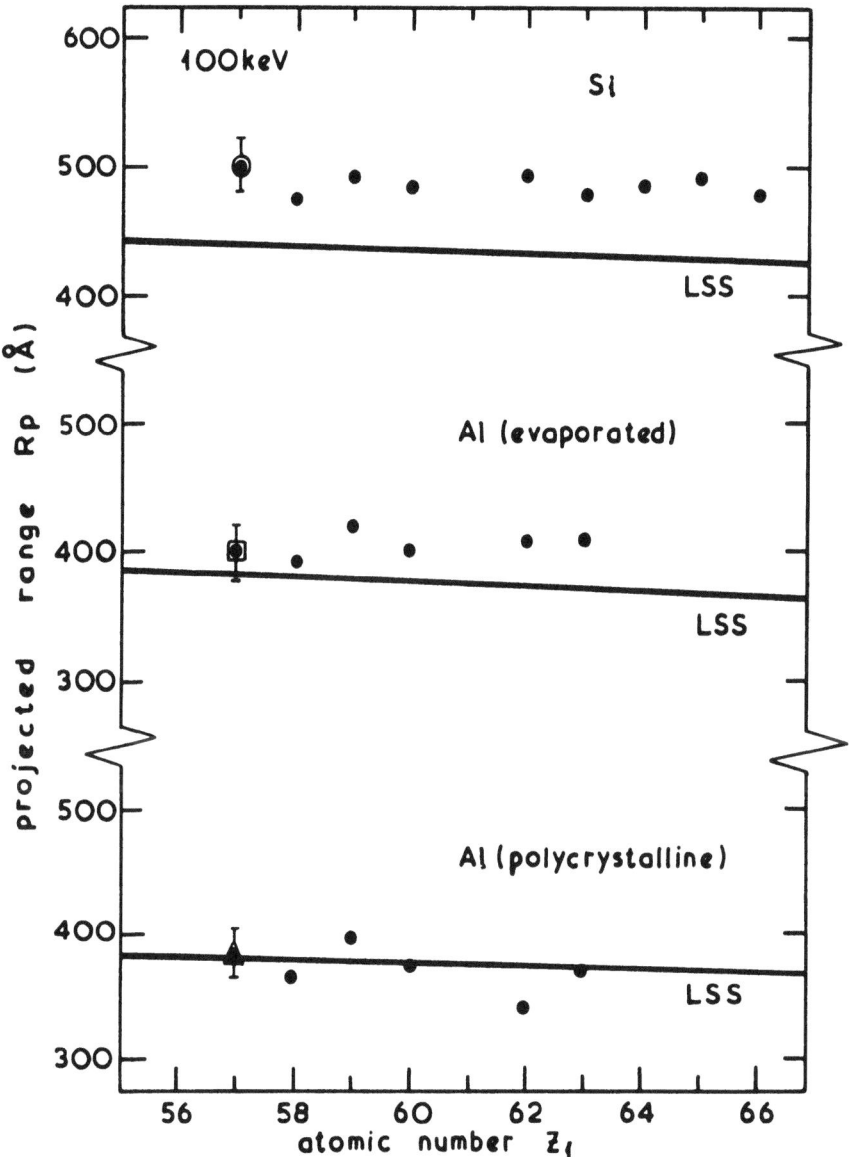

Figure 4 Projected ranges for 100 keV rare earth ions in Si and Al targets

ranges in bulk polycrystalline Al substrates are in good agreement with theory based upon a Thomas-Fermi potential.

iv) No Z_1 oscillations in range for 100 keV rare earth ions implanted into Si and Al targets have been observed.

Both SRC and NATO are acknowledged for financial support.

REFERENCES

1. W. K. Chu, B. L. Crowder, J. W. Mayer and J. F. Ziegler, Proc. Int. Conf. on Ion Implantation in Semiconductors and Other Materials, Yorktown Heights, N.Y.,(1972) Plenum Press, N.Y. (1973) p. 225
2. G. W. Nielson, B. W. Farmery and M. W. Thompson, Phys. Lett. 46A, 45 (1973)
3. D. Santry and R. D. Werner, 9th Int. Conf. on Electromag. Isotope Separators, Israel, 1976, to be published.
4. A. Feuerstein, S. Kalbitzer and H. Oetzmann, Phys. Lett. 51A, 165 (1975). Also, H. Oetzmann et al, Proc. Int. Conf. Ion Beam Surface Layer Analysis, Karlsruhe, Germany (1975). Publ. Plenum Press, N.Y. (1976), p. 245
5. J. S. Williams and W. A. Grant, Rad. Effects 25, 55 (1975)
6. W. A. Grant, J. S. Williams and D. Dodds, Proc. Int. Conf. Ion Beam Surface Layer Analysis, Karlsruhe (1975), Plenum Press, N.Y. (1976), p. 235
7. H. H. Andersen, J. Bøttiger and H. Wolder Jorgensen, Appl. Phys. Lett. 26, 678 (1975)
8. J. Lindhard, M. Scharff and H. E. Schiøtt, Kgl. Danske Videnskab Selskab, Mat-Fys. Medd. 33, 14 (1963)
9. H. E. Schiøtt, Proc. Int. Conf. Ion Implantation, Thousand Oaks (1970) Gordon & Breach, London (1971), p. 197
10. M. W. Thompson and G. W. Neilson, Phys. Lett. 49A, 151 (1974) Also, G. W. Neilson and M. W. Thompson paper presented at Int. Conf. on Atomic Collisions in Solids, Amsterdam (1975)
11. J. S. Williams, Nucl. Instrum. Meth. 126, 205 (1975)
12. J. F. Ziegler and W. K. Chu, Atomic and Nuclear Data Tables 13, 463 (1974)
13. J. F. Gibbons, W. S. Johnson and S. W. Mylroie, Projected Range Statistics, 2nd Edition, Halsted Press, 1975
14. R. A. Baragiola, D. Chivers, D. Dodds, W. A. Grant and J. S. Williams, Phys. Lett. 56A, 371 (1976)
15. I. V. Mitchell and W. N. Lennard, private communication.

RANGE DISTRIBUTIONS AND ELECTRONIC STOPPING POWERS OF ENERGETIC $^{14}N^+$ IONS

D. G. SIMONS, D. J. LAND, J. G. BRENNAN* and M. D. BROWN

Naval Surface Weapons Center

White Oak, Silver Spring, Maryland 20910

ABSTRACT

The energy dependence of the electronic stopping power, S_e, of 200 to 1600 keV ^{14}N in Fe, Ni and Zr has been inferred from the distribution profiles. The form $S_e = AE^{\frac{1}{2}}$ produces within the framework of the LSS transport theory close agreement between experimental and theoretical values for both the first and second moments of the concentration distributions. Values of A of 79.8, 69.0 and 76.1 $(keV)^{\frac{1}{2}}/(mg/cm^2)$ were determined for Fe, Ni and Zr respectively. Theoretical models to determine the Z_2 dependence of S_e on the basis of Hartree-Fock-Slater rather than Thomas-Fermi atomic wave functions reproduce to good accuracy values for one target relative to another. However, they require at least one parameter, an overall scale factor, dependent on the atomic number of the incident ion. Results for three models are presented.

INTRODUCTION

A comprehensive experimental-theoretical program to determine and predict the range distributions and stopping powers of energetic ^{14}N ions implanted in low-to-middle Z targets is in progress at the Naval Surface Weapons Center. We have previously reported our results on the Z_2 dependence of 800 keV ^{14}N ions in targets from carbon (Z = 6) through molybdenum (Z = 42)[1] and have now expanded this investigation to include targets through tellurium (Z = 52). In this paper we report on our studies of the energy dependence of the range distributions and stopping powers of ^{14}N in targets of Fe, Nr and Zr from 200 to 1600 keV, and also give a brief review of some of the results of our theoretical studies.

We use the $^{14}N(p,\gamma)^{15}O$ resonance reaction at 1061 keV to probe the implant distributions by measuring the gamma-ray yield distribution as a function of proton energy. A split-Gaussian concentration distribution vs depth is assumed whose parameters are obtained by unfolding the gamma-ray yield distribution taking into account the energy width of the incident proton beam, the Breit-Wigner gamma-ray resonance width and the energy straggling of the proton beam in the target.[2] These concentration distributions are found to be shallower and narrower than those predicted by the LSS theory. In addition, the distributions are generally skewed, falling off more rapidly at deeper depths. From the concentration distributions we infer a value of the electronic stopping power, S_e, by assuming a multiplicative correction to the LSS value of S_e; [3] that is, at this point of our investigation we assume that $S_e \propto E^{\frac{1}{2}}$. The results of this work show that S_e exhibits an oscillatory behavior as a function of Z_2 similar to those for incident protons and ^{4}He [4] and similar to the Z_1 oscillations.[5] This behavior is attributed to the shell structure of both the incident and target atoms.

EXPERIMENTAL

Since we had assumed $S_e \propto E^{\frac{1}{2}}$, we expanded our studies to investigate the velocity dependence of the range distributions and stopping powers of ^{14}N in targets of Fe, Ni and Zr. These targets were selected because they were near the middle, a minimum and a maximum in the S_e vs Z_2 curve. Representative concentration profiles for Zr are shown in Fig. 1. These profiles are normalized to the same area. We see that they become wider and more skewed as the energy is increased. Similar distributions are observed for Fe and Ni. We analyzed our data in three ways: 1) by least square fitting to the value of the first moment of the concentration distribution, R_p, to determine the constant A in $S_e = AE^{\frac{1}{2}}$; 2) similarly to determine the constant B and p in $S_e = BE^p$ and and 3) a determination of S_e from

$$S_e = [1 - \lambda_{TR}^{-1}(E)R_p(E)] \frac{dE}{dR_p} - NS_{TR}(E),$$

where

$$NS_{TR} = N \int d\sigma_n T \cos\phi \quad \text{and}$$

$$\lambda_{TR}^{-1} = N \int d\sigma_n (1 - \cos\phi)$$

are corrections arising from the nuclear stopping power. Whereas this latter approach appears to be preferable since it depends directly on the $R_p(E)$ curve, in practice $\frac{dE}{dR_p}$ had to be determined more accurately than our experimental data permitted. Thus, the

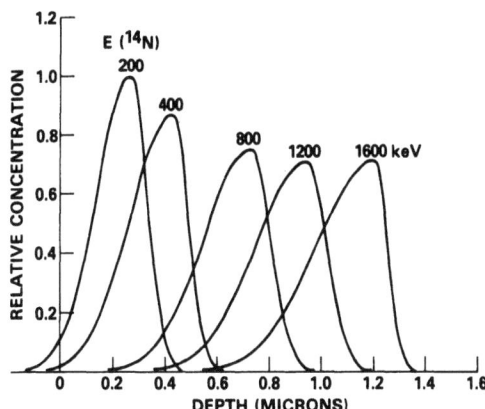

Figure 1. Concentration profiles of $^{14}N^+$ ions implanted in zirconium at selected energies. All curves are normalized to the same area.

other approaches were used to establish the functional form of $S_e(E)$. Although we were able to obtain good fits to R_p with both forms, the second moments determined from the value of S_e through the LSS transport theory provided a better fit to the data with $S_e = AE^{\frac{1}{2}}$. The difference in these fits were particularly pronounced for the Fe targets. We show the first moment (R_p) and second moment (ΔR_p) as a function of velocity for ^{14}N implanted in Zr in Fig. 2. The solid curves are the values of R_p and ΔR_p obtained from fitting the form of $S_e = AE^{\frac{1}{2}}$ using the LSS transport theory. The dashed curves are the predicted values from the LSS theory. It is particularly interesting to note how much better the fits are to the second moment when more accurate values of S_e are used. Thus, it is the value of S_e which is the major factor for the determination of both R_p and ΔR_p. The values of A, B and p obtained for S_e in keV/(mg/cm^2) are given in Table I along the corresponding LSS values.

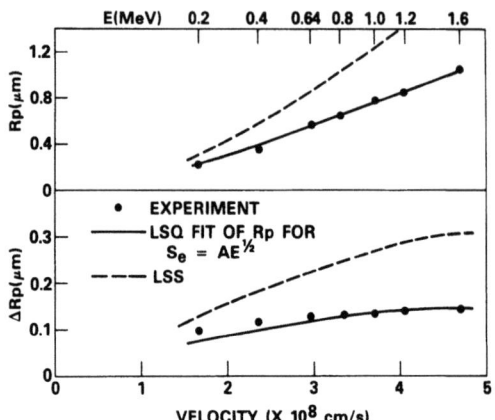

Figure 2. Velocity dependence of projected range and range straggling of $^{14}N^+$ implanted in zirconium.

Table I

Parameters for energy dependence of electronic stopping power for incident nitrogen ions. Units of S_e are in keV/(mg/cm^2)

	BE^p		$AE^{\frac{1}{2}}$	LSS
	B	p	A	
Fe	31.9	0.65	79.8	63.4
Ni	44.6	0.57	69.0	43.5
Zr	82.6	0.49	76.1	43.5

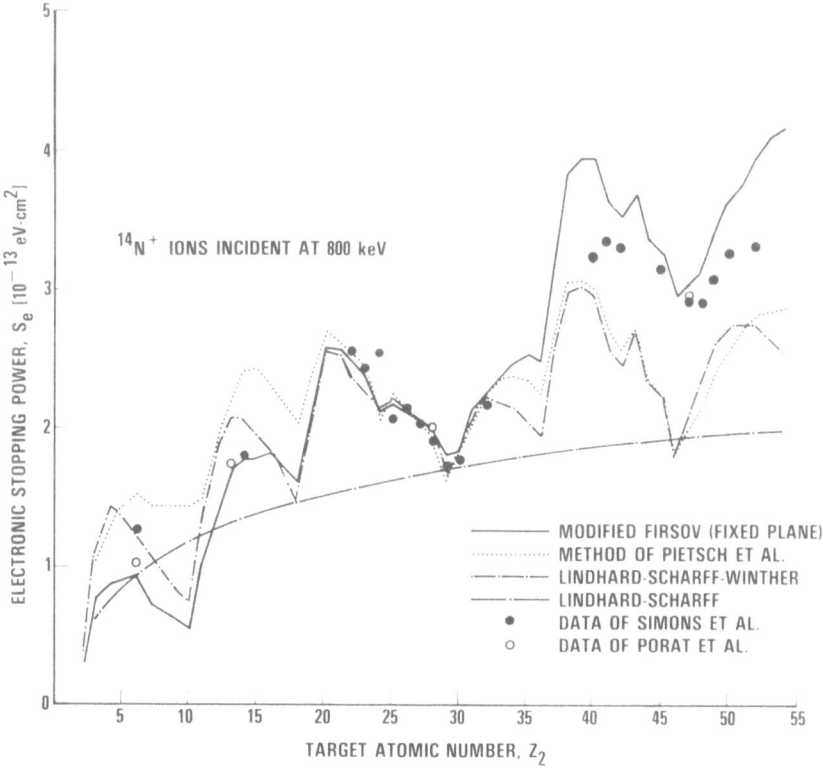

Figure 3. Comparison of experimental data of Z_2 dependence of the electronic stopping power of 800 keV ^{14}N with results of theoretical models based on the modified Firsov method, the Lindhard-Scharff-Winther method and the method of Pietsch et al.

THEORETICAL

We now give a brief review of our theoretical studies related to this problem. We have considered three models for determining the electronic stopping power as a function of Z_2: 1) the Lindhard-Scharff model modified by Pietsch et al. [6]; 2) the Lindhard-Scharff-Winther model [7], and 3) a modified Firsov model [8]. In all of these models Hartree-Foch-Slater rather than Thomas-Fermi wave functions have been used. The results of these models along with our experimental data for S_e as a function of Z_2 are shown in Fig. 3. We see that the results of all of these methods show similar Z_2 oscillations and all give relatively good fits to the experimental data. However, the Firsov model does appear to give the best overall fit. Each of the three methods requires at least one adjustable parameter.

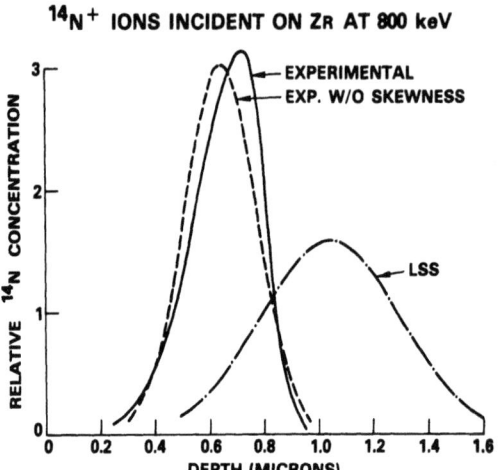

Figure 4. Concentration profiles of 800 keV $^{14}N^+$ implanted in zirconium. The experimental curve with and without skewness have the same first and second moments. All curves are normalized to the same area.

The general conclusions for our ^{14}N implantation studies are as follows:
1) S_e is the major factor for determining range distributions in the energy range from 200 to 1600 keV.
2) All the theoretical approaches require an adjustable parameter.
3) S_e proportional to $E^{\frac{1}{2}}$ is appropriate for ^{14}N implantations.
4) While the nuclear stopping power is necessary for the determination of implantation distributions, it is of secondary importance, and its magnitude and form as given by Lindhard is adequate in this energy range.

5) The LSS transport theory is a reliable method for determining R_p and ΔR_p.

6) While the distributions are skewed and indicate a necessity for calculating higher moments, such moments are of secondary importance and appear to be required only if it is necessary to predict very accurate distributions. This situation is illustrated in Fig. 4 which shows the distribution profiles for 800 keV ^{14}N in Zr. While the LSS predicted distribution is greatly different than that measured, the symmetric Gaussian distribution using R_p and ΔR_p from the adjusted value of S_e is close to the measured, skewed distribution.

REFERENCES

[*] Also Catholic University of America, Washington, D. C. 20017

[1] D. G. Simons, D. J. Land, J. G. Brennan and M. D. Brown, Phys. Rev. A 12, 2383 (1975); in Ion Beam Surface Layer Analysis edited by Meyer et al. (Plenum, New York, 1976) p863 and to be published.

[2] D. J. Land, D. G. Simons, J. G. Brennan and M. D. Brown in Ion Beam Surface Layer Analysis edited by O. Meyer et al. (Plenum, New York, 1976) p851.

[3] J. Lindhard and M. Scharff, Phys Rev. 124, 128 (1961).

[4] M. Bader, R. E. Pixley, F. S. Moyer and W. Wholung, Phys. Rev. 103, 32 (1965); D. W. Green, J. N. Cooper and J. C. Harris, Phys. Rev. 98, 466 (1955); W. K. Lin, H. G. Olson and D. Powers, Phys. Rev. B8, 1881 (1973).

[5] J. H. Ormrod, J. R. MacDonald and H. E. Duckworth, Can. J. Phys. 43, 275 (1965); P. Hvelplund and B. Fastrup, Phys. Rev. 165, 408 (1968); F. H. Eisen, Can. J. Phys. 46, 561 (1968).

[6] W. Pietsch, U. Hauser and W. Neuwirth, Nucl. Instr. and Methods 132, 79 (1976).

[7] J. Lindhard and M. Scharff, K. Dan. Vidensk. Selsk. Mod. Fys. Medd 34, No. 4 (1964).

[8] O. B. Firsov, Zh. Ehsp. Teor. Fiz. 36, 1517 (1959) [Sov. Phys.-JEPT 9, 1076 (1969)]; I. M. Cheshire, G. Dearnaley and J. M. Poate, Phys. Lett. 27A, 318 (1968); Proc. Roy. Soc. A311, 47(1969). I. M. Cheshire and J. M. Poate (p351); and C. P. Bhalla, J. N. Bradford and G. Reese (p361) in Atomic Collision Phenomena in Solids edited by D. W. Palmer et al. (North-Holland, Amsterdam, 1970).

A THEORETICAL APPROACH TO THE CALCULATION OF IMPURITY PROFILES FOR ANNEALED, ION IMPLANTED B IN Si

A. Chu and J. F. Gibbons

Stanford University

Stanford, California 94305

ABSTRACT

A three-stream diffusion model for boron in silicon is proposed. The model is capable of predicting ordinary diffusion, proton enhanced diffusion and the annealing behavior of room-temperature-implanted boron when appropriate restrictions on dose and annealing temperature are obeyed. Parameters required in the model were selected so that the model would correctly predict the impurity profiles that are obtained under conventional thermal diffusion conditions. The same parameter set is then used in the annealing calculation for implanted boron.

The computed profiles and electrical activities as a function of time compare very well with experimental data published in the literature. When the implantation dose exceeds 10^{15} ions/cm^2 and/or the annealing temperature is below 900°C, precipitation effects may occur. We outline a systematic approach to incorporate these effects into the basic three stream model.

I. INTRODUCTION

Recently Anderson and Gibbons have proposed a two stream diffusion model for the ordinary diffusion of boron [1] and proton enhanced diffusion [2] of boron in silicon. This model appears to contain the key attributes for explaining the annealing behavior of ion implanted boron in silicon. For instance, activation of boron during annealing requires at least two forms of boron, electrically active and inactive; and the damage enhanced diffusion during initial stages of annealing is qualitatively similar to the enhanced diffusion produced by protons.

It is the purpose of this paper to show that by extending the two stream diffusion model to include positively charged vacancies, we can calculate the annealed profiles and the activation of boron in silicon. The resulting three stream diffusion model will predict ordinary diffusion, proton enhanced diffusion and the annealing behavior of boron that is ion implanted into silicon at room temperature and subsequently annealed.

Naturally, a full description of boron annealing is a complicated matter and the model presented here deals with only a part of the overall picture. In particular we will concentrate on ion doses in the range below 10^{15} ions/cm^2 and annealing temperatures of 900°C-1000°C. The reasons for these limitations will be apparent from a consideration of Table 1, where we have abstracted some of the global features of the data presented by Hofker et al. [3]. These authors have measured boron carrier concentration profiles and the total boron profiles under a number of dose and annealing conditions of interest to us. As indicated in the table, their dose range extends from 10^{14} ions/cm^2 to 10^{16} ions/cm^2 and their annealing temperature range from 800°C to 1000°C. As we will show, the three stream diffusion model to be developed here is capable of predicting the annealing behavior in the lower left corner of Table 1; i.e. for doses of 10^{14} ions/cm^2 to 10^{15} ions/cm^2 and anneal temperatures of 900°C to 1000°C. The adjacent cases require extensions of the model that are straightforward but not of interest for our present purpose. For instance, in the 10^{16} ions/cm^2 dose cases, the boron peak concen-

TABLE 1. ISOCHRONAL ANNEALING RESULTS.
ANNEALING TIME = 35 MINUTES.

ANNEAL TEMP °C	Dose, ions/cm^2		
	10^{14}	10^{15}	10^{16}
800°C	Boron Near Profile ~25% Activity	Adsorption Peak	Boron <16%
900°C	3 STREAM DIFFUSION MODEL APPLIES		Precipitates when 40%
1000°C	(B$^-$, V$^+$, B$^-$V$^+$) 100% Activity		$C_B > C_{SS}$ 65%

tration exceeds solid solubility. Therefore there will be precipitation of boron. In the lower dose cases, there is a fraction of inactive boron near the peak that is similar to a precipitate, when the anneal is carried out at 800°C. However, the boron concentration is below the solid solubility limit for these cases, so this form of "precipitate" is probably boron trapped by dislocation networks that do not fully anneal at 800°C in 35 minutes. For these cases we will have to include a new species and pertinent reaction kinetics.

II. DISCUSSION OF INITIAL CONDITIONS FOR THE CALCULATIONS AND THE EXPERIMENTAL DATA TO BE PREDICTED

Our purpose in formulating a model is of course to develop a theoretical basis from which impurity profiles and electrical activities can be calculated. It is useful to begin with a brief summary of the initial conditions for the calculation and the experimental data we wish the theory to predict.

The as-implanted profile as determined by Secondary Ion Mass Spectrometry is shown in Figure 1. The vacancy concentration profile is calculated from the Frenkel-pairs produced by the 70 Kev incident boron ions as they deposit energy into atomic processes [4] along their trajectories. In what follows we will make the conserva-

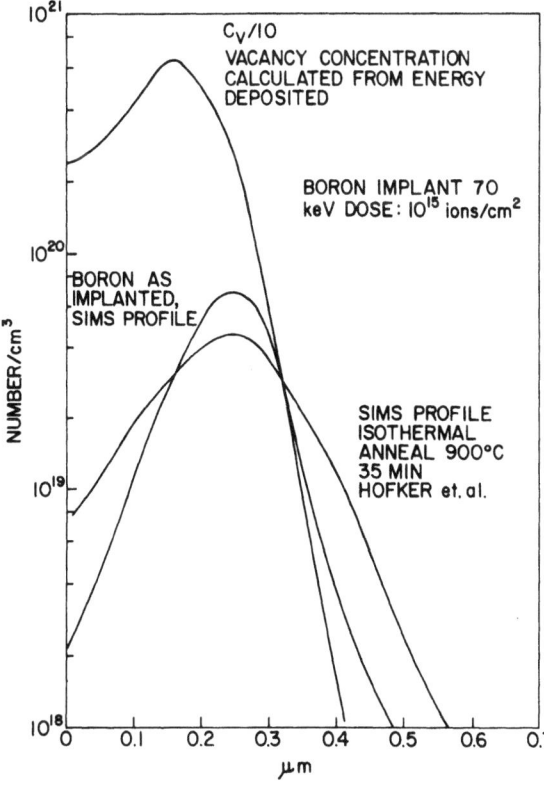

Fig. 1 As implanted Boron Profile, calculated vacancy concentration profile and the annealed profile at 900°C for 35 min.

tive assumption that as the boron atoms come to rest, they will all combine with nearby vacancies to form boron-vacancy (BV) pairs, which are electrically inactive. Hence, at the outset of the anneal the electrical activity is zero. Although a non-zero activity is more reasonable, it will also require an initial profile for the electrically active boron. At present, until we obtain short-annealing-time electrical carrier profiles, we do not wish to introduce this unknown. We further assume that the remaining high vacancy concentration will condense into clusters and dislocation loops. Room temperature annealing will occur during the implantation, in agreement with backscattering experiments [5], and what remains will anneal very quickly at temperatures of 900°C and higher, as indicated by our unpublished T.E.M. work [6]. Therefore we will use the thermal equilibrium concentration of vacancies as an initial condition in our analysis. This assumption will surely be inadequate for a liquid nitrogen implant or for the case of a heavy ion or high dose implant that produces an amorphous damage layer. For the cases to be dealt with, however, the assumption of a thermal equilibrium concentration of vacancies seems plausible.

Our model will describe the rapid diffusion and conversion of boron-vacancy pairs into electrically active substitutional boron, producing at given times profiles and electrical acitivites that we can use to compare with experimental results. Hofker et al. [3] have measured electrical carrier and total boron profiles after 35 minutes of annealing at various temperatures and implanted doses. Seidel and MacRae [7] have performed isothermal annealing experiments in which a 1.5×10^{15} ions/cm^2 dose, room temperature implant was annealed at three temperatures. They reduced all their electrical activation data to an anneal temperature of 855°C by shifting time with an exponential factor with an activation energy of 5 eV. We have scaled their 855°C data to 900°C using the same technique. The result is shown in Figure 2.

We assume that a realistic annealing model must provide not only reasonable prediction of the annealed boron impurity profile but also a reasonable prediction of the Seidel-MacRae annealing curve.

III. DEVELOPMENT OF THE MODEL

The three species in this model are substitutional (electrically active boron, boron-vacancy pairs (electrically inactive), and positively charged vacancies. Electrically-active boron is assumed to diffuse substitutionally by means of random encounters with neutral vacancies. The boron-vacancy (BV) pair is assumed to diffuse much more rapidly. The diffusion of boron then becomes a weighted diffusion with contributions from a slow, electrically active fraction (substitutional boron) and a fast, electrically inactive fraction (the boron-vacancy pair). Hence a large fraction of boron vacancy

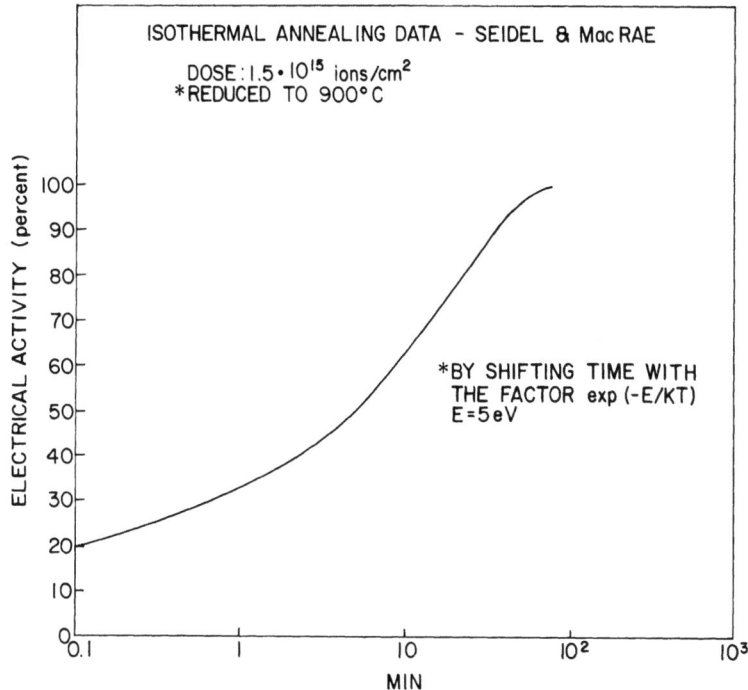

Fig. 2 Electrical activity vs time. Scaled experimental isothermal annealing data at 900°C by Seidel and MacRae.

pairs results in low electrical activity and fast diffusion, and conversely a large fraction of substitutional boron results in high electrical activity and slow diffusion. Annealing will proceed with the conversion of a large population of BV-pairs into substitutional boron.

The model consists of three diffusion equations (Eq. 1) modified by terms representing the boron-boron vacancy pair reaction; in the positive vacancy equation an additional term is introduced to represent the tendency for positive vacancies to reach the thermal equilibrium concentration in the silicon crystal. Next we will describe the features of this model and the methods used to select the various constants required in the equations.

A) <u>Diffusion Coefficients - Only Functions of Temperature.</u>
We assume that substitutional boron diffuses by the same mechanism as the self diffusion of silicon. Therefore we use the self diffusion coefficient of silicon for D_B. At 900°C this value is approximately 1.6×10^{-21} cm^2/sec. For positive vacancies we use the diffusion coefficient from Seidel and MacRae [7], which at 900°C is $D_V = 4.7 \times 10^7$ cm^2/sec. Finally, the diffusion coefficient for BV-pairs is estimated from measurements of enhanced diffusion by Hofker, et al., [3] and Anderson and Gibbons [2]. Hofker and cowork-

3-STREAM DIFFUSION MODEL

$$\frac{\partial C_B}{\partial t} = D_B \frac{\partial^2 C_B}{\partial x^2} + \frac{C_{BV} - k_o C_B C_{V+}}{\tau} \qquad \text{(a)}$$

$$\frac{\partial C_{V+}}{\partial t} = D_{V+} \frac{\partial^2 C}{\partial x^2} + \frac{C_{BV} - k_o C_B C_{V+}}{\tau} - \frac{C_{V+} - C_{V+eq}}{\tau_V} \qquad \text{(b)} \quad (1)$$

$$\frac{\partial C_{BV}}{\partial t} = D_{BV} \frac{\partial^2 C_{BV}}{\partial x^2} - \frac{C_{BV} - k_o C_B C_{V+}}{\tau} \qquad \text{(c)}$$

ers measure the enhanced diffusion coefficient after 35 minutes of annealing at 900°C to be 1.4 to 1.7×10^{-14} cm^2/sec. Since this is an average value it provides only a lower limit for the value of D_{BV}. As an alternative, Anderson and Gibbons estimated the value of D_{BV} to be 5.7×10^{-14} cm^2/sec at 750°C from proton-enhanced experiments. We use for the present work $D_{BV} = 7 \times 10^{-14}$ cm^2/sec.

B) <u>The Thermodynamic Reaction</u>. Boron and positive vacancies react to form BV-pairs according to the chemical equation 2.

$$B + V^+ \rightleftarrows BV \qquad (2)$$

At thermal equilibrium the concentrations of reactants and product are related by the thermodynamic equation 3.

$$C_{BV} = k_o C_B C_V \qquad (3)$$

The generation or disappearance of reactants or products that result from the reaction is modeled by first order kinetics with the departure from equilibrium as the driving force and a time constant.

The equilibrium constant for the reaction k_o is calculated from equation 3 as follows. Under equilibrium conditions, such as in an ordinary diffusion, the electrical activity is near 100%. Assuming 98% activity and rearranging equation 3 we obtain an expression for k_o

$$k_o = (C_{BV}/C_B)(1/C_V) \qquad (4)$$

Estimating $C_{Vo}(T)$ from Seidel and MacRae [8] to be 1.25×10^{11}/cm^3 and substituting in Eq. 4 we find $k_o = 1.63 \times 10^{-13}$ cm^3.

CALCULATION OF IMPURITY PROFILES 717

C) <u>Equilibrium Concentration of Positively Charged Vacancies.</u>
The tendency to reach equilibrium is again modelled by first order
kinetics and a time constant. Equation 5 gives the dependence of
the concentration of positive vacancies on the energy level E_{V^+},
the Fermi level E_F and the concentration of neutral vacancies $C_{Vo}(T)$
which is a function of temperature only.

$$C_{V+eq} = C_{vo}(T) \exp\left(\frac{E_{V^+} - E_F}{KT}\right) \tag{5}$$

The Fermi level is calculated using the Maxwell-Boltzmann approximation instead of the Fermi-integral of order 1/2. Hence doping
degeneracy effects are not thus far accounted for.

To obtain τ and τ_V this set of space equations is solved
by numerical analysis techniques with boundary conditions for an
ordinary thermal diffusion. Values for τ and τ_V are then chosen to
fit experimental ordinary diffusion profiles. The values of τ and
τ_V are 450 and 10^{-4} sec respectively. It is worthwhile to emphasize
that the apparent concentration-dependent diffusion coefficient of
boron is achieved in this model through the Fermi level. A high
substitutional boron concentration shifts the Fermi level closer to
the valence band edge, thus increasing the concentration of positive
vacancies ($E_V = 0.35$ eV above the valence band). This in turn increases the ratio of BV-pairs to substitutional boron, thus increasing the overall diffusion of boron.

D) <u>The Relation Between Diffusion and Electrical Activity.</u>
The two forms of boron in this model have dichotomous attributes.
Namely substitutional boron is electrically active and slow difus-
ing; on the other hand the boron-vacancy pair is electrically inactive and fast diffusing. Because of this dichotomy and because
the diffusion of the total boron is a weighted diffusion of these
two species, a relation between diffusion and electrical activity
ought to exist. We assume near 100% electrical activity in the
selection of parameters to fit thermal diffusion experiments. This
ensures that when the abnormally high concentration of boron-vacancy
pairs that is produced by the implantation has annealed, the enhanced
diffusion (associated with the initally low electrical act.) will
relax to ordinary diffusion which is simply the near-thermal-equil-
ibrium solution of the set of equations. The analytical analogue
of this discussion is carried out in Appendix A. The boron and BV
pair equations are combined and the predominance of D_{BV} over D_B is
used to arrive at a single approximated diffusion equation for total
boron. In this equation we can identify the overall diffusion coefficient D_{exp} with a product of D_{BV} and the fractional concentration
of BV pairs. Furthermore the fractional concentration can be expressed in terms of electrical activity. Hence we arrive a a diffusion coefficient (the only one accessible in an actual experiment,
hence the subindex), which is a function of the electrical activity.

$$D_{exp} = D_{BV} \frac{C_{BV}}{C_{BV}+C_B} = D_{BV}(1 - \text{elec. actv.}) \qquad (6)$$

At the outset of the anneal the assumed electrical activity is zero, the value of D_{exp} is then initially equal to D_{BV}. During anneal, as the electrical activity increases D_{exp} will decrease. Finally when the electrical activity approaches 98%, D_{exp} will approach the value of 1.4×10^{-15} cm^2/sec, the ordinary diffusion coefficient of boron at 900°C measured in thermal diffusion experiments.

IV. COMPARISON OF CALCULATED AND EXPERIMENTAL RESULTS

With the choice of parameters described in the previous section, Eq. 1 can be used to calculate the diffusive redistribution of boron under ordinary conditions of thermal diffusion. These parameters, together with the initial conditions just discussed, then permit us to solve for the annealing behavior of ion implanted boron. The results of such a calculation for 10^{14} ions/cm^2 dose implant is shown in Figure 3. The as-implanted profile-determined experimentally by Hofker et al. [3] is represented by the dotted line. Upon annealing, the BV-pairs in this profile diffuse and convert into substitutional boron. Hence a sequence of electrically active boron profiles will develop with time. Figure 3 shows calculated profiles after 1, 3, 10 and 35 minutes of annealing with corresponding electrical activities of 10, 28, 65 and 94 percent. The total boron concentration at 35 min. is represented by triangles; the fit to the solid line representing the experimental S.I.M.S. profile is excellent. Since the ordinary diffusion of boron at 900°C for 35 minutes would only modify the as-implanted profile very slightly, the fit between the calculated results and the deeper experimental profile is indicative of the existence of enhanced diffusion which progressively diminishes to the ordinary diffusion rate as the electrical activity approaches 100 percent. The calculated results are thus in good agreement with the experimental observations of Hofker and coworkers [3].

Figure 4 shows the results of a calculation performed with the same set of parameters for a 10^{15} ions/cm^2 dose, which produces an order-of-magnitude increase in the impurity and damage concentrations. The results depicted are analogous to the previous example except for some loss in the quality of the fit; the calculated total boron concentration is again represented by triangles. We attribute this difference to the doping degeneracy effects that were not correctly included in our Fermi-level calculation. The Fermi level is calculated using the Maxwell-Boltzmann approximation to the Fermi integral of order 1/2. Consequently the Fermi level was much too close to the valence band and the concentration of positive vacancies was too high. This in turn caused the relative concentration of BV pairs to be high and finally the overall effect of excessive diffusion.

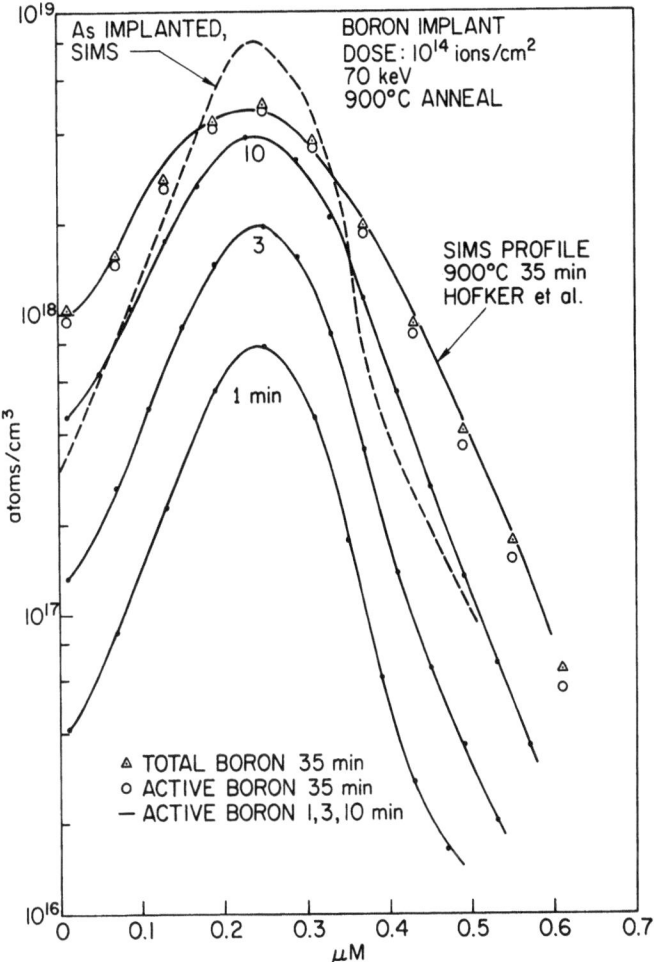

Fig. 3 10^{14} ions/cm^2 Dose Implant - Evolution of calculated active boron profiles at 1, 3, 10, 35 min. in a 900°C isothermal-anneal and comparison of the calculated total boron concentration profile with the experimental S.I.M.S. profile by Hofker at 35 min.

To verify this assumption, we have estimated the error factor in the concentration of positive vacancies and used it to reduce the diffusion coefficient of boron-vacancy pairs in a subsequent calculation. The new calculated 35 min. total boron concentration is represented by the squares in Figure 4. As expected, the agreement with the experimental profile is improved.

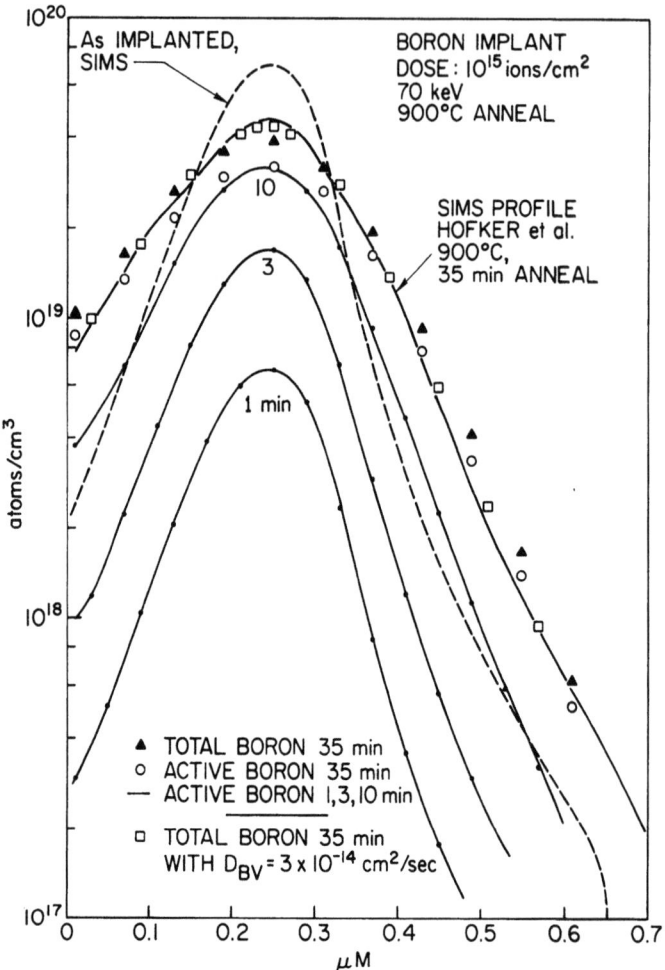

Fig. 4 10^{15} ions/cm^2 Dose Implant - Evolution of calculated active boron profiles at 1, 3, 10, 35 min. in a 900°C isothermal-anneal and comparison of the calculated total boron concentration profile with the experimental S.I.M.S. Profile by Hofker at 35 min.

The comparison of the experimental electrical activation of boron versus time by Seidel and MacRae [6] with the calculated results from the previous examples is shown in Figure 5. The agreement is good for long times. For short times we attribute the difference to our initial condition of zero electrical activity at the outset of the anneal. As discussed previously, we expect improvements with a non-zero initial condition for the electrical activity.

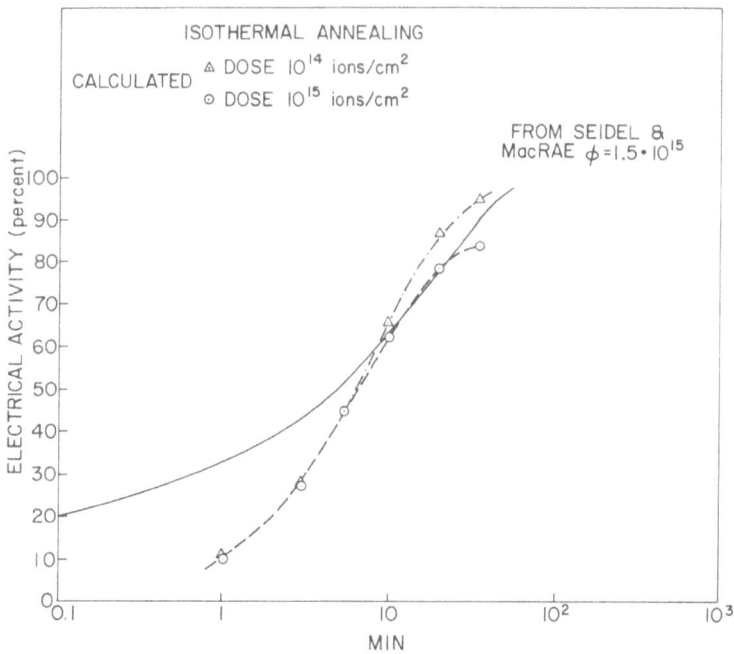

Fig. 5   Comparison of calculated electrical activity vs time with experimental isothermal results by Seidel and MacRae.

V.   DISCUSSION

The three stream diffusion model is capable of predicting boron impurity profiles under conditions of ordinary thermal diffusion, proton enhanced diffusion and radiation enhanced diffusion of the sort that occurs when boron is implanted at room temperature and subsequently annealed. In this latter case the model is valid when the dose and energy of the boron implant are such that

(a) the solid solubility limit for B in Si is not exceeded at the annealing temperature.

(b) an amorphous damage layer is not produced by the implantation.

It is also necessary for the annealing temperature to be sufficiently high so that gross damage is removed in the first few minutes of the annealing cycle.

From a practical standpoint these restrictions do not impose a serious limit on the applicability of the theory since for B in Si they will be met for implantation energies above ~ 70 keV, doses below ~ 10^{15} ions/cm^2 and annealing temperatures above ~ 900°C.

In reference to Table I, for 70 keV B in Si, when the aforementioned implantation and annealing restrictions are not observed, a new species and additional kinetic terms need to be incorporated to the model. We have preliminary results for the low temperature case that are encouraging. However we are still awaiting experimental results to further support assumptions that we make. The high dose case and also the liquid nitrogen temperature implant case require modelling of the annealing kinetics of the amorphous layer produced during implantation.

In conclusion we believe the model will be successful in predicting a number of important cases of annealing. Furthermore we believe the model can be expanded as suggested in Section I to include more complex annealing situations, and that it will then satisfactorily predict profile shapes and electrical activities in these cases as well.

REFERENCES

[1] J. R. Anderson and J. F. Gibbons, "New Boron Diffusion in Silicon," Appl. Phys. Letters 28, 184 (1976).

[2] J. R. Anderson and J. F. Gibbons, "Measurements of Diffusion Parameters in Silicon Using Proton Enhanced Diffusion," Appl. Phys. Letters, in press.

[3] W. K. Hofker, H. W. Werner, D. P. Oosthoek, H. A. M. de Grefte, "Influence of Annealing on the Concentration Prof. of Boron Implantations in Si," Applied Physics 2, 265-278, Springer-Verlag, 1973.

[3'] W. K. Hofker, H. W. Werner, D. P. Oosthoek, N. J. Koenan, "Boron Implantations in Silicon: A Comparison of Charge and Boron Conc. Profiles," Applied Physics 4, 125-133, Springer-Verlag, 1974.

[4] D. K. Brice, "Spatial Distribution of Energy Deposited Into Atomic Processes in Ion Implanted Silicon," Proc. of the First Int. Conf. on Ion Imp. in Semiconductors, Thousand Oaks, CA, Gordon and Breach Pub., New York, 1970.

[5] J. E. Westmoreland, J. W. Meyer, F. H. Eisen, B. Welch, "Production and Annealing of Lattice Disorder by 200 Kev Boron Ions," Appl. Phys. Letters 15, no. 9, Nov. 1969.

[6] Tom Magee, personal communication.

[7] T. E. Seidel and A. U. MacRae, "The isothermal Annealing of Boron Implanted Silicon," Proc. of the First Int. Conf. on Ion Imp. in Semiconductors, Thousand Oaks, CA, Gordon and Breach Pub., New York, 1970.

[8] T. E. Seidel and A. U. MacRae, "Some Properties of Boron Implanted Silicon," Trans. of the Metallurgical Society of AIME, Vol. 245, 497, March 1969.

APPENDIX A - CALCULATION OF AN EFFECTIVE DIFFUSION COEFFICIENT

Equations A1 and A2 are equations 1a and 1c in Section III rewritten in terms of fluxes, J_B and J_{BV}.

A1
$$\frac{dC_B}{dt} = \frac{dJ_B}{dx} + \left(\frac{C_{BV} - k_o C_B C_{V+}}{\tau}\right)$$

A2
$$\frac{dC_{BV}}{dt} = \frac{dJ_{BV}}{dx} - \left(\frac{C_{BV} - k_o C_B C_{V+}}{\tau}\right)$$

Addition of eq. A1 and A2 yields:

A3
$$\frac{d}{dt}(C_B + C_{BV}) = \frac{d}{dx}(J_B + J_{BV})$$

Expressing the fluxes in terms of concentration gradients and diffusion coefficients we obtain eq. A4.

A4
$$J_B + J_{BV} = D_B \frac{\partial C_B}{\partial x} + D_{BV} \frac{\partial C_{BV}}{\partial x}$$

For this equation we can show that the last term dominates as follows: D_{BV} is much greater than D_B, at 900°C; their values are 7×10^{-14} cm²/sec and 1.6×10^{-21} cm²/sec, respectively. At the outset of the anneal, $C_{BV} > C_B$ and $\frac{\partial C_{BV}}{\partial x} > \frac{\partial C_B}{\partial x}$, hence we can neglect the first term. As the annealing proceeds the diffusion of boron will approach ordinary diffusion conditions, namely,

$$\left(\frac{\partial C_{BV}}{\partial x} \bigg/ \frac{\partial C_B}{\partial x}\right) \sim \frac{C_{BV}}{C_B} \sim .02$$

The inequality between gradients has reversed; however because D_{BV} is much greater than D_B, the last term still overwhelms the first one. We can then approximate the total flux as in Eq. 5:

A5
$$J_B + J_{BV} \simeq D_{BV} \frac{\partial C_{BV}}{\partial x}$$

We now let ξ be the fractional concentration of BV pairs, which is a function of the electrical activity α:

CALCULATION OF IMPURITY PROFILES

A6
$$\xi = \frac{C_{BV}}{C_{BV}+C_B} = 1 - \alpha$$

then:

A7
$$C_{BV} = \xi (C_{BV} + C_B)$$

and the gradient becomes:

A8
$$\frac{\partial C_{BV}}{\partial x} = \frac{\partial \xi}{\partial x}(C_{BV} + C_B) + \xi \frac{\partial (C_{BV} + C_B)}{\partial x}$$

We can now identify several situations in which the first term in Eq. A8 is negligible in comparison with the last one. For instance, under equilibrium conditions we can rewrite Eq. 3 in Section III as:

$$\frac{C_{BV}}{C_B} = k_o \, C_{V+}$$

Then substitution of the above equation in the expression for $1/\xi$ yields

$$\frac{1}{\xi} = 1 + \frac{C_B}{C_{BV}} = 1 + \frac{1}{k_o {}^+C_{V+}}$$

where:

$$C_{V+} = C_{vo}(T) \, \exp \frac{E_{V+} - E_F}{kT}$$

In general the Fermi level, as a function of the acceptor concentration, will be a function of distance; however for annealing temperatures near 900°C and the boron concentration below $\sim 10^{17}$ atoms/cm^3, this dependence is very weak, hence $C_{V+} \sim C_{vo}$, $\partial \xi/\partial x \sim 0$ and the first term in eq. A8 is negligible. An analogous case arises when the annealing temperature is high, above 1100°C. In this case the Fermi level is fixed in the middle of the bandgap.

Under non-equilibrium conditions, the comparison of the terms in Eq. A8 can be performed numerically. For the particular annealing case in Section IV, the results of the calculation show that the approximation in Eq. A8 is correct under non-equilibrium and equilibrium conditions, failing only under <u>equilibrium conditions</u> when the boron concentration is near and above 10^{18} atoms/cm^3. In other words, when the boron concentration is high and the Fermi level is a function of distance, $\frac{\partial \xi}{\partial x}$ may no longer be small.

For cases in which $d\xi/dx$ may be neglected, we may then simplify

Eq. (8) to

$$A9 \qquad \frac{\partial C_{BV}}{\partial x} \simeq \xi \frac{\partial(C_{BV}+C_B)}{\partial x}$$

Substitution of eq. A9 in eq. A5 and eq. A5 in eq. A3 then yields

$$A10 \qquad \frac{d(C_B+C_{BV})}{dt} = (D_{BV} \cdot \xi) \frac{d^2(C_{BV}+C_B)}{\partial x^2}$$

where $(D_{BV} \cdot \xi)$ can be identified with the overall diffusion coefficient D_{exp}. hence:

$$A11 \qquad D_{exp} = D_{BV}\, \xi = D_{BV}\,(1 - \alpha)$$

ACKNOWLEDGEMENTS

The authors would like to acknowledge their indebtedness to the National Science Foundation and the Advanced Research Projects Agency for support of the work reported above. One of us (A. Chu) would like to acknowledge his indebtedness to Hewlett-Packard for graduate fellowship support at Stanford University.

BORON PROFILES AND DIFFUSION BEHAVIOR IN SiO_2-Si STRUCTURES

H. Ryssel and H. Kranz

Institut für Festkörpertechnologie

Paul-Gerhardt-Allee 42, 8 München 60, Germany

J. Biersack

Hahn-Meitner-Institut

Glienicker Strasse, 1 Berlin 39, Germany

K. Müller and R. A. Henkelmann

Institut für Radiochemie der TU München

8046 Garching, Germany

ABSTRACT

Boron profiles in SiO_2-Si structures have been investigated by means of the the $^{10}B(N,\alpha)^7Li$-reaction and by Hall-effect and sheet-resistivity measurements combined with anodic stripping. Implantations were performed into thermally oxidized silicon with subsequent annealing, and into bare silicon which was annealed after implantation either in an oxidizing or an inert atmosphere. At 900°C the segregation coefficient of boron at the SiO_2-Si interface is between 10 and 15. The range and range straggling of ^{10}B were found to be indistinguishable from those of ^{11}B. Therefore for measuring purposes, ^{10}B can be used instead of ^{11}B.

INTRODUCTION

There are many methods for the determination of implanted dopant distributions in silicon. The electrically active ions can be determined by means of Hall-effect and sheet resistivity measurements combined with an anodic sectioning technique [1]. At low concentrations capacitance vs voltage measurements can be applied. The total concentration of the implanted ions can be determined by backscattering, SIMS measurements, or activation analysis and some other scarcely used methods. In this paper we compare profiles of ^{10}B in silicon measured by the $^{10}B(n,\alpha)^7Li$-reaction and by the Hall-effect and sheet resistivity method. The (n,α)-reaction is especially interesting for the determination of boron distribution in layered structures, e.g., SiO_2 on silicon, or if an additional dopant species is involved where electrical methods cannot be used. Therefore we investigated SiO_2-Si structures and also the influence of implanted arsenic on the boron distribution during annealing. Results on the latter will be published elsewhere.

The $^{10}B(n,\alpha)^7Li$-reaction was first used by Ziegler [2,3] who stated that a detection sensitivity of about 3 ppm could be obtained. Due to the high neutron flux of the reactor in Grenoble used in these experiments an unprecedented detection sensitivity of 10^{11} cm^{-2} boron can be achieved [4], thus enabling studies of all boron concentrations which are relevant in semiconductor technology. The (n,α)-reaction produces monoenergetic α-particles with an energy of 1471 keV. The energy of the α-particles outside the sample is a direct measure of the boron depth as stopping powers of α-particles are well known [6].

EXPERIMENTAL TECHNIQUES

The samples used for the experiments were <111>- oriented antimony-doped silicon wafers. During implantation all wafers were tilted 7°±2° to avoid channeling. The implantation energy was varied between 30 keV and 180 keV; the dose used was 5×10^{14} cm^{-2}. Annealing was performed at 900°C either in an inert N_2 atmosphere for 30 min. or oxidizing in steam for 30 min. and 120 min. to get about 100 nm and 300 nm oxide, respectively, or dry oxidizing for 300 min. to get about 50 nm SiO_2. Electrical measurements were made on structures using a Van der Pauw geometry published earlier [5]. The (n,α)-measurements were done at the high-flux reactor of the ILL Grenoble with a thermal flux of 10^9 cm^{-2}s^{-1}. A collimator made from ^6Li enriched LiF was used. The α-particles were measured by means of a solid state detector and of the usual electronics. The resolution of the system was better than 16 keV. To convert the energy scale to a depth scale the stopping power data of Ziegler and Chu [6] have been used. Calibration of absolute

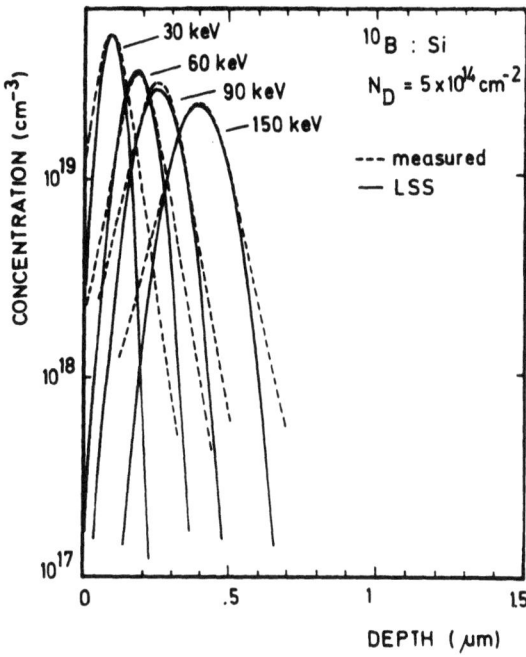

Fig. 1 ^{10}B-profiles in silicon. Energy 30 keV, 60 KeV, 90 keV, 150 keV. Dose 5×10^{14} cm^{-2}. For comparison Gaussian profiles with adjustment of R_p and ΔR_p are included.

boron concnetration was obtained through vapor deposited thin films of B and LiF.

RESULTS

Implants with energies between 30 keV and 180 keV were done into bare silicon to measure the range and the standard deviation of ^{10}B as compared to the usually used ^{11}B, and to compare electrical profiles to the total boron profiles measured by the (n,α)-mehtod. In Fig. 1 profiles are shown which have not been annealed prior to the (n,α)-measurements. It can be seen that R_p and ΔR_p for the ^{10}B implants can be described within a few percent by the ^{11}B range data. Older (n,α)-results of Crowder, et al, [3] and Ziegler, et al, [2] cannot be compared to these results because a smaller value of stopping power for α-particles has been used in their work. For the electrical measurements the samples have been annealed at 900°C for 30 min., therefore only R_p has been extracted. Gaussian profiles calculated with the measured values of R_p and ΔR_p are included in Fig. 1 for comparison. One can see that the profiles exhibit nearly exponential tails on both sides. Towards the bulk this is probably due to ions which have been scattered into channels [8]. Towards the surface this behavior has been explained by several authors [9,10, and references therein]. They found

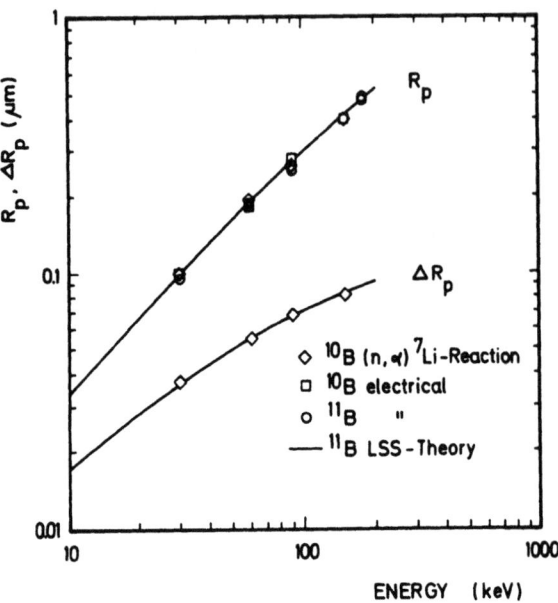

Fig. 2 Range and range straggling of ^{10}B and ^{11}B. Theoretical values from Gibbons, et al,[7].

that such skew profiles with decreased slope towards the surface are well accounted for by considering the theoretically predicted higher moments of the distribution. In Fig. 2 the values of R_p and ΔR_p for ^{10}B as measured by the (n,α)-reaction and by the electrical method are shown and compared to values of R_p for ^{11}B as measured by the electrical method. The range of ^{10}B and ^{11}B are identical within the experimental error which may be caused by the uncertainty in the stopping power data used to convert the energy scale to a depth scale in case of the (n,α)-measurements, the depth of the removed layers in case of electrical profiling, or the energy of the accelerator. Ohmura and Koike have stated in a recent paper that the range of ^{10}B is smaller than the range of ^{11}B [11]. Although our measurements of ^{11}B show more scatter than the ^{10}B measurements, we do not get this difference. Other published data on R_p or ΔR_p of ^{11}B in silicon show a large scatter of about ±10% around the LSS curve so that a comparison would be useless. Therefore it seems to be justified to use tabulated ^{11}B range data for the behavior of boron implanted layers. Moreover, a mass separation between ^{10}B and ^{11}B during the implantation is not necessary.

The (n,α)-method is very convenient to study the range distribution of boron in SiO_2 and the segregation of the boron at

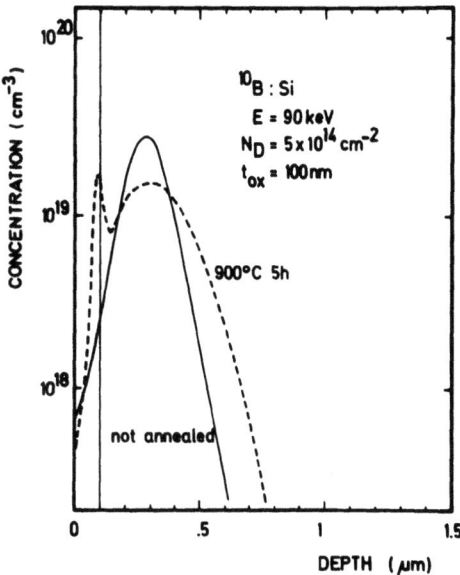

Fig. 3 Profile of ^{10}B implanted with 90 keV through 100 nm of SiO_2 before and after annealing at 900°C for 5 hours.

the SiO_2-Si interface. One limitation, however, is the limited spatial resolution of the method. Therefore only lower limits of the segregation coefficient can be estimated. In Fig. 3 a typical profile of a 90 keV implantation through 100 nm of SiO_2 and annealed at 900°C for 5 hours is shown. No abrupt step at the SiO_2-Si interface can be seen. A computer simulation would be one way to determine more precisely the segregation coefficient. Another method is to compare the (n,α)-profiles with electrical profiles which possess a better resolution at the surface of the silicon. This has been done in case of oxidizing annealing and will be discussed later. Implantations through oxide layers have also been used to determine whether there exists a step in the boron distribution at the interface or not. This step has been predicted by Furukawa, et al [12] as a result of different stopping powers in two-layered structures. From the unannealed profiles no indication of a step or a discontinuity in the slope can be detected. Monte Carlo simulations have shown that the difference in stopping power between SiO_2 and silicon is too small to result in a visible effect. Moreover, it was found that in case of larger differences in the stopping power only a discontinuity in the slope and no step should result. The reason for this is that particles, before coming to rest, slow down along a random walk path of flight. The discrepancy with the SIMS results of Combasson, et al, [13] are not yet understood.

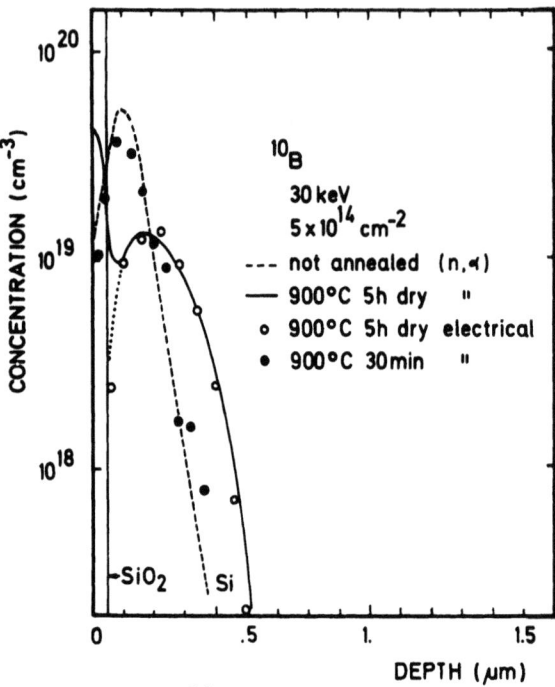

Fig. 4 Profile of ^{10}B implanted with 30 keV. --not annealed; •900°C 30 min. anneal, electrical measurement; — 900°C 5 hour anneal, (n,α)-reaction; ○ 900°C 5 hour anneal, electrical measurement.

Very often implanted layers are annealed in an oxidizing ambient. In Fig. 4 a profile obtained after dry oxidation of about 50 nm SiO_2 at 900°C for 5 hours in a N_2 atmosphere with a very low oxygen concentration is presented. To measure the segregation coefficient electrical profiles were used to determine the boron concentration at the interface. The most realistic segregation coefficient determined in this way is about 10 to 15 at 900°C. In Table 1 the estimated segregation coefficients from the implants through oxide and from the last

Table 1. Segregation coefficient measured by (n,α)-reaction and a combination of (n,α)- and electrical methods.

Energy (keV)	impl. through 100 nm SiO_2	impl. through 300 nm SiO_2	dry oxidation 50 nm SiO_2
30	10	–	10-15
60	>3	–	5
90	>2,5	4	7
120	–	2	–
150	–	–	10

experiments are given. Experimental conditions are very important to establish reliable values of the segregation coefficient. Recently three papers have been published concerning the segregation of boron in silicon [14 - 16]. In two papers indirect methods were used, i.e. profile measurements in the silicon only, and sheet resistivity measurements [14,15]. In one paper SIMS-measurements have been used to evaluate the concentrations on both sides of the interface [16]. All these results show very different results which range from about 1.7 to 17 at 900°C. This is probably due to variations in experimental conditions, for instance, concentration, drive-in times and the mathematical computation of the data. We therefore investigated additionally layers which have been annealed in a steam atmosphere at 900°C for 30 min. and 120 min. to obtain 100 nm and 300 nm SiO_2 respectively. No step at the interface can be seen. This is a clear indication that one has to consider reaction kinetics for the evaluation of the segregation coefficient. Further work is in progress at different temperatures and under different experimental conditions.

All measurements presented in this paper have been conducted either without annealing or after annealing at 900°C. Therefore the diffusion coefficient of ^{10}B could be measured at 900°C only. Its value of 4×10^{15} cm^2 s^{-1} is well within those published for ^{11}B at this temperature [14,15]. In SiO_2 a determination of the diffusion coefficient was not possible at this low temperature.

CONCLUSION

The ^{10}B (n,α) ^7Li-reaction as compared to Hall-effect measurements combined with anodic stripping has been used to investigate the behavior of implanted boron during annealing in inert or oxidizing atmosphere. The segregation coefficient at 900°C was found to be between 10 and 15 for slow oxidation. The ranges of ^{10}B and ^{11}B are indistinguishable in the energy range investigated. The (n,α)-reaction therefore can be used to study the diffusion and segregation behavior of boron in SiO_2-Si structures. The detection limit of the method is around 10^{11} cm^{-2} with a spatial resolution of 20 nm which is better than the resolution of the capacitance vs voltage method which is the only method with a comparable detection limit, since its resolution is limited to several Debeye lengths.

REFERENCES

1. J. W. Mayer, L. Eriksson and J. A. Davies, Ion Implant. in Semicond., New York, Academic Press (1970)
2. J. F. Ziegler, G. W. Cole, and J.E.E. Baglin. J. Appl. Phys. 43, 3809 (1972).

3. B.L. Crowder, J.F. Ziegler, and G.W. Cole in *Ion Implant. in Semicond. and Other Materials*, New York, Plenum Press. B.L. Crowder Ed. p. 257 (1973).
4. K, Müller, R. Henkelmann and H. Boroffka, Nucl. Instr. Meth. 128, 417(1975).
5. H. Ryssel, K. Schmid and H. Müller, J. Phys. E: Sci. Inst. 6., 492(1973).
6. J.F. Ziegler and W.K. Chu, Atom. Data and Nucl. Tables 13, 463(1974).
7. J.F. Gibbons, W.S. Johnson, S.W. Mylroie, *Projected Range Statistics*, Dowden, Hutchinson & Ross, Inc., Stroundsburg, Pa. (1975).
8. P. Blood, G. Dearnaley, and M.A. Wilkins, J. Appl. Phys. 45, 5123(1975).
9. W.K. Hofker, D.P. Oosthoek, N.J. Koeman, and H.A.M. de Grefte, Rad. Effects 24, 223(1975).
10. J. P. Biersack, D. Fink, p. 211 in *Ion Implantation in Semiconductors and Other Materials*, Plenum Press, New York, B. L. Crowder Ed. (1973).
11. Y. Ohmura and K. Koike, Appl. Phys. Lett. 26, 221(1975).
12. S. Furukawa and H. Ishiwara, J. Appl. Phys. 43, 1268(1972).
13. J.L. Combasson, J. Bernard, G. Guernet, N. Hilleret and M. Bruel, in *Ion Implantation in Semiconductors and Other Materials*, Plenum Press, New York, B.L. Crowder Ed., p. 285 (1973).
14. S.P. Murarka, Phys. Rev. B12, 2502(1975).
15. J.L. Prince and F.N. Schwettmann, J. Electrochem. Soc. 121, 705(1974).
16. J.W. Colby and L.E. Katz, J. Electrochem. Soc. 123, 409(1976).

ANOMALOUS REDISTRIBUTION OF ION-IMPLANTED DOPANTS*

H. B. Dietrich and J. Comas

Naval Research Laboratory

Washington, D. C. 20375

ABSTRACT

Based on our studies to date, the redistribution of implanted Al in Si can, in general, be placed into 3 regions which are fluence and damage dependent: (1) little diffusion at anneals as great as 900°C; (2) enhanced diffusion; and (3) structured redistributions. We have extended our studies on the anomalous diffusion of implanted dopants and have noted general trends. Recent observations of implanted Be into Si show some similarities to implanted Al in Si. Studies made on the ion implantation of Be into GaAs over an extensive range of fluences also showed the same general redistribution behavior as Al in Si. A wide range of data has been obtained for Si and GaAs on the anomalous redistribution effects and we will discuss these data in the context of generalized models. In this paper, we will present an overview of the atomic profiles of implanted dopants as a function of implant fluence, damage and anneal treatment. Atomic profiles and electrically active profiles will be compared and discussed.

INTRODUCTION

The term anomalous redistribution is intended to encompass any redistribution which deviates from that predicted by a Fickian model governed by the generally accepted diffusion coefficient. This then would include such phenomena as out diffusion during

---

*This work is supported by the Office of Naval Research and the Naval Electronic Systems Command.

elevated temperature anneals, outdiffusion during the reordering of an amorphous layer, trap-limited diffusion in the presence of residual damage and anistropic enhanced diffusion. This paper is primarily concerned with: (1) the transition from regimes of normal diffusion to regimes of marked enhanced diffusion or the onset of what shall be called a delocalization phenomena and (2) the presence, subsequent to anneal, of pronounced structure in the atomic profiles of implanted and/or background impurities.

Anomalous redistribution effects have been reported for column III dopants in ion-implanted Si. Hofker et al.[1] have observed these effects for B and we have recently published results obtained from SIMS and backscattering analysis for B, Al, Ga and In.[2] Two distinct types of atomic redistributions have been observed in our studies. When a heavy damage layer is present, a multiple peaked structure sets in subsequent to 500 to 900°C anneals of substrates which have comparatively high concentrations of the implanted dopants.

For Al at lower implant fluences (10^{13} to $10^{14} cm^{-2}$) marked enhanced diffusion effects have been observed. Recently, C-V measurements have been used by Wilson[3] and at NRL to establish that at fluences in the order of $10^{12} cm^{-2}$ implanted Al profiles are even more stable than predicted on the basis of the accepted diffusion coefficient for Al in Si. The anomalous redistribution effects observed at the higher fluences were not observed for the Al implanted at low fluences for anneal temperatures as high as 900°C.

RESULTS AND DISCUSSIONS

Redistribution Effects

In Fig. 1 Al profiles in Si measured after a 30 min, 700°C anneal are shown as a function of dose. The profiles were measured by secondary-ion mass-spectroscopy (SIMS). A background correction has been made on these profiles and they have been normalized to the dose. Hence, for each profile the relative scale has the same proportionality to the as-implanted profile. The following observations can be made: (1) at fluences on the order of $10^{13} cm^{-2}$ Al exhibits markedly enhanced diffusion with motion both toward the surface and into the bulk; (2) as the fluence is increased to 10^{14} cm^{-2} and greater, the Al begins to remain in the implanted region. However, it is at first concentrated on the bulk side of the distribution and the concentration tails off into the crystal; (3) a greater fraction of the Al is retained in the as-implanted region as the fluence is increased (by 7.5 X $10^{14} cm^{-2}$ virtually 100% of the Al is retained in the implanted region) and (4) at 7.5 X 10^{14} cm^2 a structure is beginning to appear in the atomic profile.

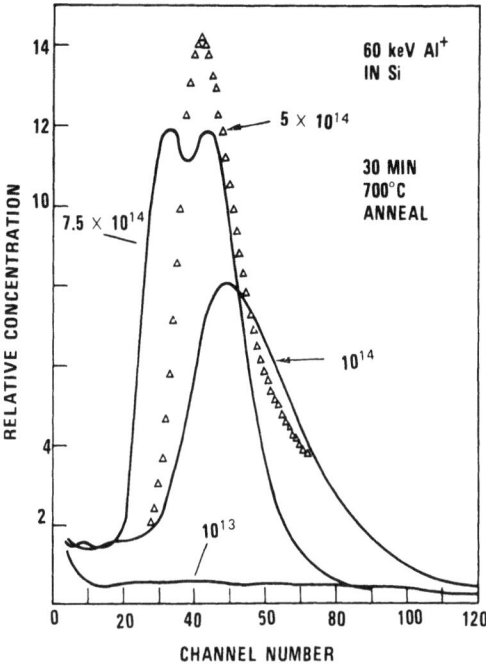

FIG. 1. Relative concentration of Al in Si as a function of implant fluence after a 700°C, 30 min anneal.

The above structure is developed fully at fluences of $10^{15} cm^{-2}$ and greater. The nature of the structure is shown in Fig. 2, which are atomic profiles as a function of anneal temperature obtained from $2 \times 10^{15} cm^{-2}$, 60 keV Al implanted samples.[2] Distinct structural changes are observed after the 480°C anneal and complete unfolding is established after an anneal of 600°C. For anneal temperature ≥ 800°C, the profiles are diffused versions of the 600°C profile.

In summary: (1) at low fluences (on the order of $10^{12} cm^{-2}$) Al has been shown to be well behaved, (2) in the 10^{13} - $5 \times 10^{13} cm^{-2}$ range Al undergoes a significant delocalization and (3) at fluences on the order of $10^{15} cm^{-2}$ the Al is again retained in the as-implanted region of the crystal and the atomic profile is structured.

A second system we have studied in some detail is Be implanted into GaAs. Again at low fluences the implanted Be is quite well behaved remaining in approximately the as-implanted Gaussian even after a 30 min, 800°C anneal. As the fluence is increased the

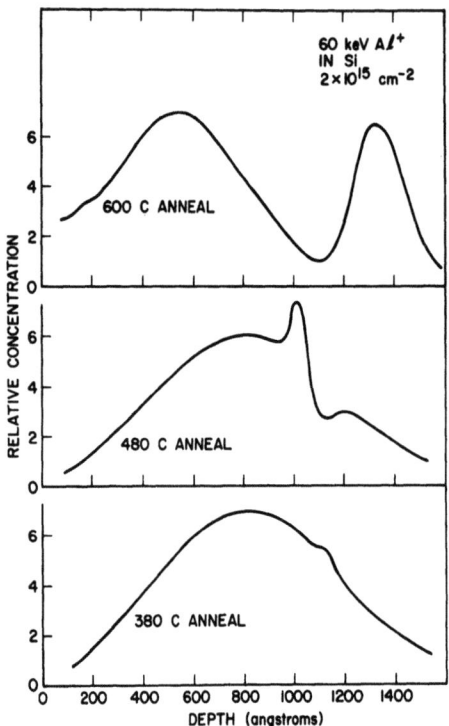

FIG. 2. Relative concentration of Al as a function of depth from samples implanted to a fluence of $2 \times 10^{15} cm^{-2}$ and annealed at 380, 480 and 600°C for 30 min. The profiles have been normalized.

FIG. 3. Relative concentration of Be in GaAs (Si-doped, n-type) as a function of implant fluence. The samples were implanted with 100 keV Be and pyrolytically grown SiO_2 layers were used as encapsulants for the 800°C, 30 min anneal treatments.

onset of a delocalization phenomena is again observed and at still higher fluences structure begins to develop in the atomic profile. Fig. 3 shows the Be profile observed in these various regimes. It is important to note that only a fraction of the Be is present in the high fluence case in which the structure is observed. In summary, then, we have found that for Be in GaAs redistribution sets in at fluences on the order of $10^{14} cm^{-2}$ at an implant energy of 100 keV, but that the precise value seems to be substrate dependent. Secondly, structure begins to appear in the Be profile at fluences on the order of $5 \times 10^{15} cm^{-2}$ but that even here only a fraction of the implanted Be is retained in the implanted region.

In the case of the Be/GaAs system, the results indicate that the redistribution occurs as the result of solubility related effects. This position seems to be borne out by the correlation of the electrically active profiles and the SIMS atomic profiles.[4] The same may be true for the Al/Si system but here the picture is less clear. The peak concentration at $10^{13} cm^{-2}$ is approximately an order of magnitude below the accepted value of the solid solubility limit for Al in Si and the damage from the Al is heavy enough to complicate the issue.

Structure Profiles

In ion-implanted Si structured profiles have been observed for implanted Be, B, Al, Ga, and In. Although the bulk of the data has been taken by secondary-ion mass-spectroscopy, the presence of the structure has been confirmed by p-γ resonance profiling for Al and by Rutherford backscattering for Ga. Hence, the structure is not related to a SIMS artifact.

In Si, implanted B, Al, Ga and In have all been found to exhibit similar structured profiles subsequent to anneal. For these ions, the structure has only been observed in the presence of a comparatively localized, high-damage layer approaching that necessary for amorphous layer production. Correlated SIMS and channeling measurements have shown that the structure in the Al profile unfolds in concert with the anneal of the damage. Other work on the Al/Si system has shown that: (1) if the damage layer extends over a region large compared to the width of the atomic profile, the structure does not appear; (2) that the Al will decorate a damage layer displaced from the as-implanted distribution in a similar structured manner; (3) that a low fluence (10^{13} - 5 X $10^{13} cm^{-2}$) implant will manifest the same behavior as a high-fluence implant if a damage region is super-imposed over the atomic profile; (4) that the structure in the latter case exists even if the damage layer is put in and annealed to 700°C before the Al implant and processing; and (5) that the structure washes out at anneal temperatures of 800-1000°C. Where comparisons have been made B, Ga and In have been found to behave similarly.

We have recently examined the Be/Si system. Our results differ from those of Hurrle and Schulz[5] in that we observe an atomic distribution of Be after annealing consisting of several well defined peaks. In Fig. 4 Be atomic profiles obtained from an unannealed and a 700°C, 30 min annealed sample are shown. The samples were implanted with 100 keV Be to a fluence of 5 X $10^{15} cm^{-2}$. At low fluences ($10^{13} cm^{-2}$, 100 keV) marked diffusion is observed, but this is not unexpected in that Be is known to be a rapid diffuser in Si. Damage associated effects for the low fluence Be implants were investigated by super-imposing an amorphous layer, produced by Ne implantation, over the Be atomic profile. Be was retained in the damaged region, whereas without the damaged layer significant out diffusion occurred.

At implant fluences on the order of 5 X 10^{14}cm^{-2}, Be is beginning to concentrate in the implanted region and the atomic profile is structured subsequent to a 700°C anneal. Here again the structure begins to washout in the 800°C region. The important observation here is that in this fluence range, the Be induced damage is quite diffused and yet structure is observed in the atomic profile. Similar results have been noted with high temperature proton irradiations of uniformly doped boron substrates. Hence, here are two cases, and Be in GaAs is a third, where structure is observed at damage levels which are far short of that required for amorphous layer production.

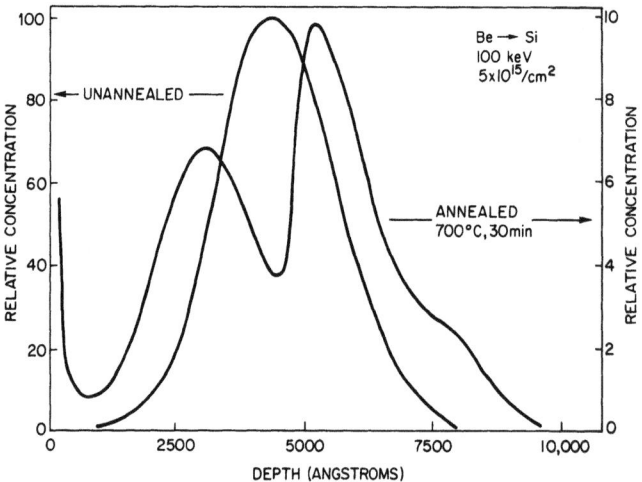

FIG. 4. Relative concentration of Be vs depth for an unannealed and a 700°C, 30 min annealed sample. The samples were implanted with 100 keV Be$^+$ to a fluence of 5 X 10^{15}cm^{-2}.

Boron in Si has also been found to deviate from the norm in a second manner. Hofker et al.[1] have observed significant redistribution effects for B co-implanted with Ar or As. We have not observed similar effects for B co-implanted with Al. In fact, in this case, the Al redistribution was curtailed in the region where the B had been implanted. Another point of interest, although it is difficult to assess its significance, is that ^{31}P is the only n-type dopant reported to show these redistribution effects and it has shown a comparatively weak effect.

We have made some correlations between the SIMS data and electrical measurements obtained from samples implanted and annealed under similar conditions. Our Al into Si data are in agreement with the enhanced diffusion effects observed electrically by Baron et al.[6] and Itoh et al.[7] Low fluence Be implants into Si (300 keV, $5 \times 10^{12} cm^{-2}$) have been recently C-V profiled by Wilson[3] and the results have been compared to the SIMS atomic profiles. The significant decrease in electrical activity observed in samples annealed to temperatures above 600°C is corroborated by the Be losses observed in the atomic profiles.

The Be electrical activity and SIMS atomic profiles in GaAs have been compared over a fluence range extending from 1×10^{13} to $5 \times 10^{15} cm^{-2}$.[4,8] The results indicate that the redistribution effects are fluence dependent. The electrical and SIMS data showed that for samples implanted at low fluences ($< 10^{14} cm^{-2}$) and annealed, there were no major Be losses or distribution changes. Samples implanted to higher fluences ($>5 \times 10^{14} cm^{-2}$) had electrically active and atomic profiles which indicated that major delocalization of Be had occurred due to the anneal treatment.

Finally, let us consider the models which have been advanced as possible explanations for the observations. It has been suggested that the structure results from: (1) a zone refining type action of the amorphous layer; (2) thermal diffusion enhanced by amorphous layer induced stress or (3) an interplay between substitutional and defect associated atoms with varying diffusion coefficients.

On the basis of our work to date, it is clear that the structure does not result from a zone refining action of the amorphous layer. The fact that one sees the structure set in even if the damaging implant is annealed at 700°C prior to the Al implant rules this out despite the fact that the structure does unfold in concert with the anneal of the damage. The most plausible explanation at this point is the combination of an enhanced diffusion process, which may be strain induced but clearly is not always such, and subsequent tie up at defect sites resulting in trap-limited diffusion in discrete portions of the crystal.

ACKNOWLEDGEMENTS

The authors gratefully acknowledge the cooperation of Larry Plew of the Naval Weapons Support Center, Crane, in obtaining the SIMS data. We wish also to thank Ms. Elizabeth Tarrents of the Air Force Materials Laboratory for supplying the GaAs used in the Be implantation study.

REFERENCES

1. W. K. Hofker, H. W. Werner, D. P. Oosthoek, and N. J. Koeman, IV International Implantation Conference, 1974, Japan.

2. H. B. Dietrich, W. H. Weisenberger, and J. Comas, Appl. Phys. Letts. $\underline{28}$, 182 (1976).
3. R. G. Wilson private communications.
4. J. Comas, L. Plew, P. K. Chatterjee, W. V. McLevige, K. V. Vaidyanathan, and B. G. Streetman, 5th Int. Impl. Conference, 1976, Boulder, Colo.
5. A. Hurrle and M. Schulz, Inst. Phys. Conf. Ser. N. $\underline{23}$, 474, (1975).
6. R. Baron, G. A. Shifrin, and O. J. Marsh, J. Appl. Phys. $\underline{40}$, 3702 (1969).
7. T. Itoh, T. Inader, and K. Kanekawa, Appl. Phys. Letts. $\underline{12}$, 244 (1968).
8. "Ion Implanation of Wide Bandgap Semiconductors", L. Anderson, Hughes Research Laboratories Report, 1975, Contract No. N00014-74-C-0158, Naval Electronic Systems Command.

INDEX

α-particles	22, 27
absorption coefficient	271, 272, 608
acoustic surface wave device	257, 259
activation energy	70, 90, 92, 384
alloy	57, 59, 63, 167, 169-171, 173, 176, 182, 191, 203, 205, 209
aluminum	11-19, 50, 95, 176, 192, 197, 199, 202, 210, 231-235, 239, 242, 245, 247, 248, 266, 267, 270, 272, 273, 320-323, 327, 329, 363, 365, 366, 369, 373, 544, 663, 693
amorphous	11, 15-18, 21, 24-26, 28, 29, 31, 33, 36, 39-46, 50, 52, 54, 132, 134, 136, 149, 167, 169, 171, 172, 213, 221, 223, 225, 258, 259, 264, 301, 303, 350
annealing	11-19, 21-29, 31, 33, 36, 38-44, 65, 67, 68, 70, 72, 74, 77, 79, 81, 82, 85-87, 89-93, 95, 96, 98, 101, 103, 105, 107-109, 111-113, 123, 125, 130, 132, 134, 138, 141, 142, 144-146, 148, 149, 159, 160, 164, 165, 168, 173-175, 215, 225-227, 240, 245, 302, 312, 379, 382-384, 386, 389, 391-393, 399, 401, 405, 407, 417, 423, 424, 427, 430, 432-434, 445, 446, 453, 456, 457, 459, 461-463, 467, 468, 472, 475, 480, 491, 493, 494, 497, 498, 501, 511, 512, 516, 520-522, 525, 543, 547, 555, 559, 575, 576, 578,

annealing (cont)	585-587, 591-593, 595, 596, 598, 599, 603-611, 615, 617, 618, 621, 623, 631, 634, 636, 643, 671, 673, 680, 711, 712, 714, 718, 727, 728, 739
anodic	203, 505, 727, 733
antimony	50, 54, 55, 192, 194, 197, 199, 321, 376, 376, 381, 382, 535-539, 541, 542, 585, 629
antistructure defects	87
areal mass	232, 233
argon	21, 39-46, 58, 61, 62, 128, 143, 151, 202, 203, 223, 285, 295, 307-309, 320, 326, 327, 348, 527, 637, 672
arsenic	50, 54, 77, 78, 81-84, 86, 87, 108, 115, 116, 122-127, 130, 149-157, 327, 334, 341, 375, 376, 381, 391, 398, 447, 467, 523, 544, 550, 552, 672
atomic diffusion	167-170, 173
auger electron spectroscopy	57-59, 61-64
backscattering	21-25, 28, 39, 40, 43, 51, 57, 74, 80, 174, 213, 227, 240, 242, 248, 370, 376, 377, 381, 462, 471, 473, 475, 477, 478, 480, 486, 487, 493, 504, 512, 535-538, 585, 586, 589-591, 637, 638, 644, 687, 688, 690, 693, 696, 714, 728, 736, 739
band-edge-emission	84
bandgap	66, 129
beryllium	131, 132, 134-138, 141-148, 173, 175, 176, 259, 611
bipolar devices	493, 520
Boltzman transport equation	333, 335
bonding	57
boron	11-13, 19, 21, 31-33, 36, 38, 50, 65-68, 70-72, 74, 77, 78, 81-87, 259, 307, 344, 392, 409-411, 414, 415, 454, 457, 523, 529, 530, 557, 560, 609, 663, 664, 666, 711-716, 718-721, 725, 727-729, 731, 733, 740
Bragg angle	410

INDEX

cadmium	92, 107-109, 113, 123-128, 130-132, 134-138, 644
capacitance-temperature	69
capacitance-voltage measurements	3, 67, 68, 82, 117, 120, 151
carrier concentration profiles	92, 96, 157
carrier mobility	525, 526
cathodoluminescent	80
channeled backscattering spectrum	171, 401, 404, 405, 407, 484, 489, 490, 505
channeled proton-induced X-ray	285, 290
channeling	51, 61, 99, 100, 143, 161, 170, 214, 216, 227, 239, 248, 252, 253, 258, 291, 376, 377, 379, 381, 462, 466, 467, 471, 483, 494, 503, 507, 509-512, 520, 536, 539, 544, 585-587, 590, 634, 642, 687, 698, 728, 739
chlorine	32, 33, 36, 38,
chromium	90, 95, 96, 98, 101, 108, 111, 116, 131, 132, 201, 202, 205, 209, 210
clustered point defect	507, 509
cobalt	202, 239, 242, 245
coimplantation	77
collision cascade	319, 322, 323, 401, 405, 407
compensation	82, 83, 86, 87, 92, 107, 113, 116, 119
copper	57, 58, 60, 61, 63, 124, 169-171, 173-176, 369
corrosion	201, 202, 209, 210
critical temperature	64
crystal lattice	409
damage	2, 31-33, 36, 38, 39, 65, 90, 108, 157, 159, 239, 244, 322, 409
damage profile	244, 245, 267, 272, 309, 491
Debye length	120, 163
dechanneling	245, 247, 250, 251, 254
defect	8, 36, 65, 66, 70, 73, 74, 77, 78, 80, 81, 83, 86, 87, 107, 123, 125-128, 131, 132, 134, 136-138, 217, 229, 398, 399, 401, 413, 414, 417, 438, 442, 457, 459, 668

defect center	13, 22, 83, 85, 86, 105, 248, 462, 463, 465, 468, 504, 510
defect cluster	
defect density	36
defect mobility	467
defect structure	131, 136, 138
degeneracy	136, 149
density of states	60
depletion region	456
depth distribution	159, 160, 166
device	38, 50, 77, 519
diamond	44, 295
diffraction pattern	40, 42, 44, 45, 192, 642
diffusion	1, 3-9, 23, 32, 36, 83, 90, 100, 105, 109, 111, 123, 125, 130, 142, 146, 484
diffusion coefficient	1, 2, 4, 6-9, 23, 89, 92, 125, 168, 174, 175, 208, 236, 538, 596, 598, 599, 717, 736
diffusion length	144, 669, 670
diffusion profile	109, 113
diffusion rates	167, 173
diode	66-69, 71, 74, 159
dislocation	36, 227, 244, 245, 247, 392, 398, 465, 467, 473
dislocation lines	465, 498, 501, 520
dislocation loops	132, 134, 137, 138, 214, 219, 248, 393, 415, 467, 468, 472, 480, 487, 495, 497, 501, 687-690, 714
dislocation networks	495, 498, 501, 522, 523
disorder	15, 27, 28, 49, 50, 221, 407
divacancy	401, 402, 407, 414, 424, 432, 433, 457-459
dual implantation	585, 591
dysprosium	171, 213-217, 220, 221
Einzel lens	567
electrical activity	33, 107, 113, 116, 350, 472
electroluminescence	621, 627
electromechanical coupling coefficient	261
electron diffraction	40, 132, 134, 171, 192, 233, 234, 258, 348, 350, 471, 497, 498, 639, 680
electron microprobe	175
electron microscopy	31, 33, 40, 213, 493, 643, 646, 671, 680, 687

INDEX

electron spin resonance	39-42, 46, 286, 671
electronegativity	50, 56, 170,
electronic stopping power	164
ellipsometry	51, 52, 305, 306, 309, 392, 494, 537
encapsulating gallium arsenide	96
encapsulation	89, 107, 108, 115, 116, 122, 142, 143, 146, 615, 617, 618
enhanced diffusion	177
epitaxial	24, 25, 27, 28, 33, 40, 44, 45, 85, 119
epitaxial layer	92, 95, 96, 115, 117, 120
epitaxial recrystallization	424, 461, 462
epr	383, 384, 386, 417, 418, 421, 424, 427-430, 432, 434, 436, 438, 441, 442
equilibrium alloy	167
equilibrium phase diagram	168
eutectic	58
Fermi energy	61
Fermi level	454, 456, 458, 459, 717, 718, 725
Fick's law	92, 208, 209, 211
field effect transistors	89, 95, 96
fluorine	32, 33, 36, 38, 77
free exciton recombination	613
Frenkel defect	266
Frenkel pair	687, 713
gallium	23, 24, 28, 49, 50, 54, 58, 86, 87, 90, 108, 112, 123, 128, 129, 143, 149-154, 156, 157
gas sputtering	235
germanium	18, 40, 45, 49, 50, 52, 53, 85, 127, 445, 447, 523, 637, 644
gold	57-63, 170, 321
grain boundaries	27
Gunn diode	97
Gunn effect	89
Gunn type digital devices	96, 97
Hall coefficient	143, 653
Hall effect	3, 11, 13, 17, 107, 109, 113, 141-144, 447, 449, 594, 613, 618, 642, 654, 655, 728
Hall mobility	90, 91, 109, 132, 134-136, 151, 613, 615

helium	40, 50, 58, 61, 168, 259, 370, 376, 381
helium neon laser	78-80
hydrogen	77, 78, 80, 85, 87, 90, 115, 116, 122, 124, 143, 483, 484, 489, 491
Impatt diode	97
impurity scattering	135
index of refraction	672
indium	21-25, 27-29, 577, 622, 651
infrared absorption	436
infrared spectroscopy	305
integrated optics	159
interference microscopy	306
interferometry	638
intermetallic alloy	167-169, 171, 177
internal friction	435-437, 441, 442
interstitial diffusion	8, 18
ion trapping	236
ionization energies	13
iridium image converter	80
iron	176, 201, 205, 210, 221
irradiation-enhanced diffusion	176
isoelectric traps	578, 582
isolation layer	159, 164
knock-on effects	236, 375, 379, 386-388, 391, 396, 398, 520
krypton	327
laser	445-450
lateral spread	163, 164, 166
lattice	16, 21, 49, 50, 55, 144, 170, 214, 235, 239, 244, 245
lattice constant	168
lattice damage	21, 167, 169, 267, 288, 290, 417, 421, 427-429, 431, 432, 434, 435, 466, 467
lattice defect	13, 267, 461, 472, 687
lattice disorder	220, 221, 298, 690
lattice distortion	227
lattice mobility	136
lattice parameter	412
lattice scattering	13, 135
lattice solution	170
lead	213-216, 219, 220, 221
low temperature annealing	519
LSS	32, 97-100, 109, 111, 136, 143, 201, 207, 211, 301, 308,

INDEX

LSS (cont.)	324, 666, 667, 693, 694, 696-698, 703-705, 709
luminescence	80, 147
magnesium	131, 132, 134-138, 595
manganese	143
metastable	167-170, 172
microblister	483, 487, 489-491
microdefects	503
micrographs	132
microstructures	137
microtwining	36, 498
microwave devices	97
mobility	5, 13, 14, 17, 32, 66, 89, 92, 93, 95, 96, 109, 112, 134, 137, 138, 141, 142, 144, 148, 149, 151, 154, 156, 170, 446, 450, 613, 617, 618
molecular ions	31, 32, 38
molybdenum	223-225, 227, 229, 352
morphology	151
MOS device	19, 493, 521
neon	227, 267
neutron radiation damage	182, 306
nickel	169, 171, 182, 201, 202, 207-210, 213-217, 219-221
nickel-aluminum alloy	181, 182
niobium	350, 359
nitrogen	8, 21, 24, 33, 51, 77, 78, 151, 223-227, 229, 231, 232, 234, 236, 248, 376, 381, 526, 575, 576, 582, 594
nonradiative centers	125-127
nonradiative defects	85
nucleation	197, 200
ohmic contact	132
optical absorption	268, 285-287, 291, 293
optical phonon	136
optical waveguide	555, 672
optically active lattice damage	273
optoelectronic devices	142
oxygen	3, 50, 51, 54, 74, 85, 229, 231, 232, 236, 244, 265, 266, 307, 320-322, 341, 342, 347, 348, 354, 358, 359, 363-366, 369-373, 386, 387, 391, 392, 396, 398, 401, 403, 417, 422-424, 435, 439-442, 459,

oxygen (cont.)	467, 468, 512, 516, 637, 643, 682, 698
oxygen recoil	327, 343
palladium	348
paramagnetic center	383, 672-674
paramagnetic defect	421, 429
paramagnetic resonance	386
passivation	49, 52-55
phase diagram	59, 167, 173, 176
phase formation kinetics	295
phase transformation	295
phonon	127
phosphorus	1, 3-9, 11-13, 15, 17, 19, 50, 99, 386, 391, 392, 396-399, 418, 454, 471, 472, 477, 480, 522, 523, 525-527, 530, 544, 617, 670, 672
photoconductivity	65, 412, 658
photoluminescence	77, 78, 83, 84, 86, 113, 123, 124, 138, 141, 142, 144, 575, 576, 582, 603, 606, 609-611, 613, 617, 621, 623, 625, 627
piezoelectric	257, 264
platinum	651
point defect	170, 214, 245, 442, 471, 472, 478, 480, 508, 682
precipitate	45, 131, 134, 136, 169, 181, 183-186, 188, 191-193, 195-200, 234
predeposition	49
profile	1-6, 78, 82, 89, 98-101, 103, 108, 115, 116, 118-120, 125, 152-156
projected range	25, 66, 81, 115, 117, 118, 120, 122, 124, 125, 136, 160, 164, 166, 202, 203, 259, 301, 320, 323, 325, 372, 391, 392, 396, 399, 536, 629, 641, 699
proton	36, 79-83, 86, 87, 159-161, 163-166
proton enhanced diffusion	711
proton excited X-ray analysis	201
radiation annealing	285
radiation damage	1, 6, 8, 9, 32, 50, 55, 116, 168, 208, 209, 236, 239, 285, 301, 320, 322, 329, 383, 471,

radiation damage (cont.)	472, 475, 483, 484, 555, 557, 622, 682
radiation defect	436, 445, 671
radiation enhanced diffusion	186, 422, 520
radiation enhanced nucleation	188
radiation hardness	672
radiative recombination center	575
radio tracer profile	2, 97
range distribution	2, 320, 364, 370, 371, 702
range oscillation	699
range profile	333, 351, 693
reactive sputtering	445
recoil dissolution	196
recoil implantation	320-322, 327, 329, 333, 334, 342, 347, 348, 352, 354-357, 359, 363-366, 369-373, 375, 376, 381, 511
recoil oxygen	397
recoil yield	319
residual disorder	511, 514-516
resistivity	13
reverse leakage current	32
rocking curve	409, 410, 415
Rutherford backscattering	50, 52, 58, 61, 62, 138, 214, 215, 220, 221, 224, 271, 285, 288, 290, 363, 364, 695
sapphire	11, 265-270, 285, 286, 291
scanning electron microscope	78-80,
scattering	136, 335, 336, 338, 356, 484
Schottky barrier	117, 143, 151, 381, 654, 664
Schottky barrier c-v measurements	104
Schottky barrier contact	375, 377
Schottky barrier diode	78, 161, 376, 528
Schottky barrier Fets	97
Schottky barrier gate Gunn effect devices	89, 95
Schottky electrode	163
secondary ion emission	644
secondary ion mass spectrometry	1, 2, 115-117, 141-144, 146-148, 231, 364, 713,

secondary ion mass
  spectrometry (cont.) 718, 728, 733, 736, 739, 741
secondary ion microprobe 82
Seebeck effect 379
selenium 98, 149-154, 156, 157
semiconductor devices 19
sheet carrier density 89, 92, 93, 109
sheet hole concentration 136
sheet resistance 11, 13, 15-19, 22, 31-36, 107, 109, 113, 132, 151, 379, 381, 394, 493-495, 497, 501, 524, 550, 594, 617, 629, 630, 631, 633-636, 727, 728, 733
silicon 1-3, 8, 11-19, 21-29, 31, 32, 34-37, 39-45, 49-66, 69-71, 74, 78, 80, 81, 83-86, 99, 115-120, 122, 124, 125, 142, 143, 145, 149, 150-152, 154-156, 295, 323, 326, 327, 329, 342, 363, 364, 366, 369, 373, 375, 376, 381, 382, 384, 386, 388, 391, 393, 394, 396, 398, 409, 413, 417, 418, 424, 427, 428, 430, 435, 436, 439-441, 445, 447, 453, 454, 457, 459, 461, 462, 467, 471, 472, 480, 510, 512, 515, 516, 519, 523, 526, 535, 671, 693, 712, 715, 728, 729, 731
silicon dioxide 1, 3, 8, 33, 78, 79, 85, 89, 90, 96, 116, 124, 132, 134, 306
silicon nitride 78, 81, 97, 98, 101, 103-105, 107, 108, 116, 143, 146, 151
silicon on sapphire 11-14, 16-19
silver 143, 352
solar cell 543, 552
solid solubility 21, 28, 111, 113, 145, 148, 167, 168, 170, 173-176, 195, 381, 539, 713, 721, 739
space-charge scattering 136, 566
spin-lattice relaxation
  times 430
sputtering 40, 62, 100, 207, 234, 347, 348, 356, 358

INDEX

stacking faults	473, 487, 489, 510
stoichiometry	156, 157, 266,
stopping power	216, 250, 267, 342, 343, 544, 703, 730
straggling	224, 342, 537, 727
structure	40
substitutional	28, 167, 169, 170
sulphur	89, 90, 92, 94-101, 103-105, 115, 116, 120, 122, 149, 354, 358
superconducting	167, 169, 172, 223
superconducting transition temperatures	224, 226
surface alloy	201-204, 209-211
surface recombination velocity	608
surface waveguides	264
surface wave resonator	264
tantalum	350
target chamber	109
tellurium	78, 85, 98, 107, 108, 111-113, 145, 149, 446
thermally stimulated conductivity	65, 66, 69-71, 73, 74, 557, 621, 624, 627
thermo-electric power	654, 655
thermoluminescence	276-277, 279, 280, 283, 624
Thomas-Fermi potentials	697, 698, 702
tin	585, 591
titanium	50, 56
transconductance	95
transferred electron device	39, 40, 44, 555
transition temperatures	172
transmission electron microscopy	31, 33, 36, 39, 40, 42, 43, 131, 132, 136, 171, 173, 176, 181, 182, 191, 192, 213, 214, 216, 217, 219-221, 231, 233, 247, 248, 251, 254, 351, 391, 392, 394, 422, 423, 461-463, 471, 473, 476, 477, 478, 480, 483, 487, 489, 490, 493, 495, 503, 507-510, 512, 522, 637, 638, 644, 646, 714
transmission infrared spectroscopy	306
trapping center	66

tungsten	170, 171
twin structure	39, 44, 46
twinned lattice	506
vacancy	17, 18, 77, 81, 83, 87, 112, 123, 124, 126, 128, 129, 149
vacancy clusters	457
vacancy loops	245
van der Pauw	13, 32, 66, 90, 132, 143, 151, 154, 594, 613, 654
vapor-phase-epitaxial	116, 117, 120
vertical power FET	536, 541
X-ray diffraction	631
X-ray double reflection	409
xenon	124, 213-215, 285, 290, 363, 364, 366, 369-373, 644
zinc	44, 123, 131, 132, 134-138, 176, 244, 247, 248, 447

If you have any concerns about our products,
you can contact us on
ProductSafety@springernature.com

In case Publisher is established outside the EU,
the EU authorized representative is:
**Springer Nature Customer Service Center GmbH
Europaplatz 3, 69115 Heidelberg, Germany**

Printed by Libri Plureos GmbH
in Hamburg, Germany